Electrical Engineering Reference Manual

for the PE Exam

Fifth Edition

Raymond B. Yarbrough, PhD, PE

Professional Publications, Inc. • Belmont, CA

ELECTRICAL ENGINEERING REFERENCE MANUAL
Fifth Edition

Printed in the United States of America

Professional Publications, Inc.
1250 Fifth Avenue, Belmont, CA 94002
(650) 593-9119
www.ppi2pass.com

Current printing of this edition: 8

Library of Congress Cataloging-in-Publication Data
Yarbrough, Raymond B.
    Electrical engineering reference manual for the PE exam / Raymond
  B. Yarbrough. -- 5th ed., rev. and reprinted.
        p.   cm.
    Rev. ed. of: Electrical engineering reference manual. 5th ed.
  1990.
    ISBN 1-888577-04-5 (hardcover)
    1. Electric engineering--Examinations, questions, etc.
  2. Electric engineering--United States--Examinations--Study guides.
  I. Yarbrough, Raymond B. Electrical engineering reference manual.
  II. Title.
  TK169.Y37   1996
  621.3'076--dc20   96-44290
                                                            CIP

TABLE OF CONTENTS

4 TIME AND FREQUENCY RESPONSE

5 POWER SYSTEMS

6 TRANSMISSION LINES

HOW TO USE THIS BOOK

1 QUICKSTART

If you are in a hurry to use this book and only want to read one paragraph, here it is:

Start anywhere. The chapters are independent. Solve every practice problem. Use the index extensively. Good luck.

2 HOW PRACTICING ENGINEERS AND STUDENTS CAN USE THIS BOOK

How you use this book depends on what you intend to use it for. If you are a practicing engineer or an engineering major and have purchased this book as a general reference handbook for your library, it will probably sit in your bookcase until you have a specific need to remember something you learned a while ago. Then, you can use the index to find material that will help you.

If you are preparing for the NCEES PE examination in electrical engineering, here are a few suggestions:

- Obtain the companion Solutions Manual. This book was meant to be used with it. (The solutions manual is separate so that this book can still be used as an examination reference in those states that prohibit collections of solved problems.

- Become familiar with the format of the PE exam. Start your review by reading chapter 17, "Engineering Licensing."

- Become intimately familiar with this book. This includes knowing the order of the chapters, the approximate locations of important figures and tables, the contents of the appendices, and so on.

- Know which subjects in this book are not covered in the PE exam. Several chapters (e.g., Mathematics, Systems of Units, and Engineering Economic Analysis) in this book are supportive and do not cover actual exam topics. These chapters provide background for the other chapters and other types of problems.

- Some subjects appear in more than one chapter. You should use the index liberally to learn all there is about a particular subject. Most subjects have secondary or tertiary indexing, which means you should be able to find entries no matter how you look them up.

- Start your review of a subject by skimming the chapter to familiarize yourself with the subjects before starting the practice problems so that you will know the location of each subject if you need a quick review.

- It isn't necessary to solve every end-of-chapter practice problem, but I suggest that you try to. The number of practice problems you solve will depend on how much time you have and how skilled you are in each area. To do all the problems requires approximately 15 to 20 hours of preparation time per chapter.

- Learn to pace yourself in an exam situation by watching the clock when you solve the "Timed" end-of-chapter problems. The timed problems are representative of the subject matter, length, and complexity of the essay-type problems you will find on the actual examination. You don't actually have to start a stopwatch when you do these problems; just be aware of your rate of progression.

- Learn what slows you down. Are you spending time learning to use your calculator? Are you hampered by not having enough reference books? Is the terminology new to you, or is language a barrier?

3 HOW INSTRUCTORS CAN USE THIS BOOK

If you are teaching a review course for the PE examination without the benefit of recent, first-hand exam experience, you can use the material in this book as a guide to prepare your lectures. You should emphasize

PROFESSIONAL PUBLICATIONS ● Belmont, CA

the subjects in each chapter and avoid most subjects omitted. You can feel confident that subjects omitted from this book are rarely, if ever, found on the PE exam.

I have always tried to overprepare my own students. For that reason, the examples and practice problems in this book are sometimes more difficult and varied than actual examination problems. Also, you will appreciate that it is more efficient in a lecture or while doing practice problems to cover several procedural steps in one practice problem than to ask numerous simple "one-liners." That is the reason there are no multiple-choice problems in this book.

There are many end-of-chapter practice problems for each major examination subject. In my courses, all of the problems are assigned. To do all the problems requires approximately 15 to 20 hours of preparation per week over the 14-week course.

"Capacity assignment" is the goal of my courses. If you assign 15 hours of practice problems as homework, and a student can only put in 10 hours of preparation that week, that student will have worked to his or her capacity. After the PE examination, your students will honestly say that they could not have prepared any more than they did in your course.

I have found that a 14-week format works well for a PE review course. Each week there is a three-hour lecture with a short intermediate break. The following table outlines a basic course format. You can, of course, add more lecture time or weeks, and that will certainly be appreciated by the students because this is a fast-paced course. However, I don't think you can cover the full breadth of material in much less time or in many fewer weeks.

Typical PE Review Course Format

meeting	subject covered
1.	Introduction, Exam Format, Mathematics
2.	Linear Circuit Analysis
3.	Waveform Measurement, Time and Frequency Response
4.	Amplifier Applications
5.	Semiconductor Circuits
6.	Waveshaping, Logic, and Data Conversion
7.	Rotating Machines
8.	Control Systems
9.	Digital Logic
10.	Power Systems
11.	Transformers
12.	Fault Calculations
13.	Transmission Lines
14.	Exam Strategy; National Electric Code

Homework assignments in my courses are not individually graded. Instead, the students obtain the accompanying Solutions Manual and have the solutions to all practice problems in advance. However, each student must turn in a completed set of problems for credit each week. I personally address all special needs or questions that are written on the assignments.

I have tried to order the course lectures in a logical, progressive manner. Lecture coverage of some examination subjects is necessarily brief. Other subjects are not covered at all. For example, engineering economic analysis has no lecture. This is consistent with the absence of the subject in the NCEES test plan, even though elements of economic decision making continue to appear in the examination.

Any skipped chapters and their end-of-chapter practice problems are assigned as floating assignments to be made up in the students' "free time."

I strongly believe in exposing my students to a realistic sample examination, but I no longer administer an in-class mock exam. Since the review course usually ends only a few days before the real PE examination, I hesitate to make students sit for several hours in the late evening to take a "final exam." Rather, I distribute and assign a take-home sample examination at the first meeting of the review course. (I use the same *Electrical Engineering Sample Examination* publication that is available to your courses from Professional Publications.)

If a practice test is to be used as an indication of preparedness, ask your students not to look at it prior to taking it. Looking at the practice examination or otherwise using it to direct their review may produce unwarranted specialization in subjects contained in the practice examination.

There are many ways to organize a PE review course depending on your available time, budget, intended audience, resources, and enthusiasm. However, all good course formats have the same result: the students struggle with the workload during the course, and then breeze through the examination after the course.

Michael R. Lindeburg, PE
Belmont, CA

PREFACE AND ACKNOWLEDGMENTS
for the Fifth Edition

This edition of the *Electrical Engineering Reference Manual* provides a single reference for the broad field of electrical engineering. Both fundamental and advanced subjects are included, and most solution techniques will be recognized by electrical engineering graduates of the past 20 years. However, the explanations are sufficiently detailed for learning solution techniques that are new.

This edition has a substantial amount of new material as well as further development of the material covered in the earlier editions. The new material includes sections on Fourier transforms—including the fast Fourier transform, commonly encountered impedances, 3-phase power measurements, Butterworth filters, VSWR measurements, antennas, data conversion, system modeling, and transport delay. Expanded coverage includes measurement circuits, frequency response, filter characteristics, and active filters. The material on amplifiers has been expanded and split into two chapters: the first emphasizes fundamentals of electronic circuits and devices, and the second deals with amplifier applications, including nonlinear and high-frequency aspects. Correspondingly, 118 new end-of-chapter problems have been added.

Working electrical engineering graduates will find this reference manual useful when faced with problems in areas they have not had to deal with recently. In addition, it will be helpful in preparing for the National Council of Examiners for Engineering and Surveying (NCEES) license examination for electrical engineers.

The contents have been extensively tested in actual classroom examination review courses. Explanations, examples, and problems are completely up-to-date and are representative of the areas covered in recent examinations. However, most examinations contain a few problems requiring very specialized knowledge that is too limited in scope to be included in a reference work such as this.

A reference book of this scope and character is a concentrated team effort, requiring a variety of skills as well as effective communication among people of diverse backgrounds. The people of Professional Publications are, as the name implies, thoroughly professional in putting together a technical work such as this. As an organization they are completely competent, and as individuals they are a joy to work with.

The preparation of this book has been largely inspired by the hundreds of engineering graduates who have reviewed for the professional engineering examination in my classes throughout the years. These working engineers have provided me with insight to the needs of engineers who have been out of school for a long time and, through their feedback, have helped me understand the pattern of examination questions.

I would like to single out a few of the many people who have made this edition possible. First and foremost is Lisa Rominger, who capably guided the project through all phases and should be given accolades for her ability to deal with a variety of personalities. She is the exemplar of tact. Mary Christensson somehow untangled the maze of revision with its pitfalls of replacements and renumbering.

For their careful and competent review and editing, I wish to thank Kenneth Nelson, Shelley Arenson of Professional Publications, and the staff at Rosenlaui Publishing Services. The technical reviewer, Kenneth Nelson, and the proofreaders vastly reduced the number of both manuscript and typesetting errors.

On the administrative side, I thank Wendy Nelson for her continuing encouragement and good humor. In the same vein, I thank Michael Lindeburg, who supplied the chapters on economics, engineering licensing, and systems of units.

Finally, I wish to dedicate this book to my wife, Judy, who drastically altered vacation plans more than once. And I thank her for her continued support in this work.

Raymond B. Yarbrough, PE

PROFESSIONAL PUBLICATIONS ● Belmont, CA

1 MATHEMATICS

1 REAL NUMBERS

Positional number systems are used to designate the value of a real number according to the positions of its digits. In the *decimal (base 10) system*, the digits are *zero* through *nine*. The position of a particular digit within a string determines the contribution of that digit to the total value of the number. For example *three* in the decimal number 347 has a place value of $3 \cdot 10^2$, *four* has a value of $4 \cdot 10^1$, and *seven* has a value of $7 \cdot 10^0$. Thus, the integer number 347 can be written as

$$(347)_{10} = (3 \cdot 10^2) + (4 \cdot 10^1) + (7 \cdot 10^0) = 300 + 40 + 7$$

This method can be applied to any positional number system. Digits of an integer are weighted by the base raised to powers determined by their positions in the string. At the right-most digit, the weighting factor is the base raised to the zero power. The digit second from the right is multiplied by the base raised to the first power. The third digit is multiplied by the base squared. Then, for the base b integer string, $N_n N_{n-1} N_{n-2} \ldots N_i \ldots N_2 N_1 N_0$, the value given to a digit n_i, is

$$\text{value of } i^{th} \text{ member } = N_i \cdot b^i \qquad 1.1$$

Fractional numbers can be similarly represented. The exponent i is negative, with a value equal to the digit's position from the right-most integer digit. For example, in the number 6.25, *five* is two places to the right of *six*, the right-most integer digit. So, the weighting for 5 is 10^{-2}. The proper name for the decimal point is the *radix point*. The radix point is used to separate the integer part of a number from its fractional part.

Example 1.1

Determine the decimal value of the *octal* (base 8) number 347.16.

$(347.16)_8$

$$= (3 \cdot 8^2) + (4 \cdot 8^1) + (7 \cdot 8^0) + (1 \cdot 8^{-1}) + (6 \cdot 8^{-2})$$

$$= (3 \cdot 64) + (4 \cdot 8) + (7 \cdot 1) + (1 \cdot 0.125)$$

$$\quad + (6 \cdot 0.015625)$$

$$= 192 + 32 + 7 + 0.125 + 0.09375$$

$$= (231.21875)_{10}$$

2 COMPLEX NUMBERS

A *complex number* has a real part and an imaginary part. The *imaginary part* is identified by a coefficient j, where $j = \sqrt{-1}$. Thus, a complex number can be written as $a + jb$. A complex number can be represented on a *complex plane* by graphing coordinates which correspond to the real and imaginary coefficients. Figure 1.1 shows a complex plane in which complex number $a + jb$ is represented as a point.

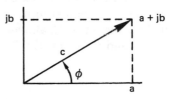

Figure 1.1 Complex Plane

A complex number can also be represented as a two-dimensional vector in the complex plane. Thus, the complex number can be written in *polar* or *phasor form* as

$$a + jb = c\angle\phi$$

$$c = \sqrt{a^2 + b^2} \qquad 1.2a$$

$$\phi = \arctan\left(\frac{b}{a}\right) \qquad 1.2b$$

The orthogonal sides of the triangle can be found from trigonometry.

$$a = c \cos \phi \qquad 1.3$$
$$b = c \sin \phi \qquad 1.4$$

Combining equations 1.2, 1.3, and 1.4, the complex number can be written as equation 1.5.

$$a + jb = c \cos \phi + jc \sin \phi$$
$$= c(\cos \phi + j \sin \phi) \qquad 1.5$$

Euler's equation $(e^{j\phi} = \cos \phi + j \sin \phi)$ can be used to write this as an exponential.

$$a + jb = ce^{j\phi} \qquad 1.6$$

The notations $\angle \phi$ and $e^{j\phi}$ can generally be used interchangeably, although the exponential form requires that ϕ be in radians for computational purposes. The $\angle$ notation is used with angles in degrees.

The *complex conjugate* of a complex number can be plotted on the complex plane as a mirror image with respect to the real axis. The conjugate of a complex number A is written as A^*. If $A = a + jb$, then $A^* = a - jb$. In the polar and exponential forms, the angle is negated to obtain the conjugate.

$$A = a + jb = c\angle\phi$$
$$= ce^{j\phi} = c(\cos \phi + j \sin \phi) \qquad 1.7a$$
$$A^* = a - jb = c\angle - \phi$$
$$= ce^{-j\phi} = c(\cos \phi - j \sin \phi) \qquad 1.7b$$

Addition and *subtraction* of complex numbers is most conveniently done in rectangular form. If the numbers are in polar form, they should be converted to rectangular form before being added.

Multiplication and *division* of complex numbers can be done in either polar or rectangular form, as long as both numbers are in the same form. Examples of each operation in rectangular and polar forms follow.

Example 1.2

If $A = 4 + j5$ and $B = 2 - j3$, what is $A \cdot B$?

$$(4 + j5) \cdot (2 - j3)$$
$$= (4)(2) + (4)(-j3) + (j5)(2) + (j5)(-j3)$$
$$= 8 - j12 + j10 - j^2 15$$

However, $j^2 = -1$, so that $-j^2 15 = +15$.

$$A \cdot B = (8 + 15) + j(-12 + 10)$$
$$= 23 - j2$$

Example 1.3

Convert A and B in example 1.2 to phasor and exponential forms before multiplying

$$A = \sqrt{4^2 + 5^2} \angle \arctan\left(\frac{5}{4}\right) = 6.4\angle 51.34°$$
$$= 6.4e^{j0.896}$$
$$B = \sqrt{2^2 + 3^2} \angle \arctan\left(\frac{-3}{2}\right) = 3.6\angle - 56.31°$$
$$= 3.6e^{-j0.983}$$

Note that the angle in the phasor form is in degrees, and in the exponential form it is in radians.

$$A \cdot B = (6.4 \cdot 3.6)\angle(51.32 - 56.31)$$
$$= 23.04\angle - 4.97$$
$$= 23.04e^{-j0.087}$$

Division of complex numbers can be considered as the multiplication of the numerator by the reciprocal of the denominator. Division of complex numbers in rectangular form is accomplished by multiplying numerator and denominator by the complex conjugate of the denominator.

Example 1.4

Find the reciprocal of $A = 4 + j5$.

$$A^{-1} = \frac{1}{4 + j5} \cdot \frac{4 - j5}{4 - j5} = \frac{4 - j5}{4^2 + 5^2}$$
$$= \frac{4 - j5}{41}$$
$$= 0.0976 - j0.1220$$

Example 1.5

Divide $B = 2 - j3$ by $A = 4 + j5$.

$$\frac{B}{A} = \frac{2 - j3}{4 + j5} \cdot \frac{4 - j5}{4 - j5} = \frac{8 - j10 - j12 + j^2 15}{41}$$
$$= \frac{-7 - j22}{41}$$
$$= -0.1707 - j0.5366$$

Division of complex numbers in the phasor and exponential forms is much simpler, as illustrated by example 1.6.

Example 1.6

Divide $B = 3.6e^{-j0.983}$ by $A = 6.4e^{j0.896}$.

$$\frac{B}{A} = \frac{3.6e^{-j0.983}}{6.4e^{j0.896}} = \frac{3.6}{6.4}e^{-j0.983}e^{-j0.896}$$

$$= 0.5625e^{-j1.879}$$

3 ALGEBRA

A. Quadratic Equations

Quadratic equations frequently arise in electrical problems. The usual form of a quadratic equation is

$$ax^2 + bx + c = 0 \qquad \text{1.8}$$

The values of x which satisfy equation 1.8 are called the *roots* or *zeros* of the equation. These roots are found from equation 1.9.

$$r_1, r_2 = \frac{-b \pm \sqrt{b^2 - 4ac}}{2a} \qquad \text{1.9}$$

Equation 1.8 can then be rewritten as equation 1.10.

$$a(x - r_1)(x - r_2) = 0 \qquad \text{1.10}$$

The variables r_1 and r_2 can be real or imaginary depending on the coefficients a, b, and c.

- If $b^2 - 4ac > 0$, then there are two different real roots.

- If $b^2 - 4ac = 0$, then there are two identical real roots equal to $-b/2a$.

- If $b^2 - 4ac < 0$, then there are two complex conjugate roots with real part $-b/2a$ and imaginary parts of $\pm\sqrt{(c/a) - (b/2a)^2}$.

B. Exponentiation

$$x^a x^b = x^{a+b} \qquad \text{1.11}$$

$$x^{-a} = \frac{1}{x^a} \qquad \text{1.12}$$

$$x^{(1/a)} = \sqrt[a]{x} \qquad \text{1.13}$$

$$(x^a)^b = x^{ab} \qquad \text{1.14}$$

$$x^{(b/a)} = \sqrt[a]{x^b} = (\sqrt[a]{x})^b \qquad \text{1.15}$$

C. Logarithms

If $b^x = y$, then $x = \log_b y$. For example, if $y = 10^x$, then $x = \log_{10} y$, $\ldots$, or if $e^x = y$, then $x = \log_e y = \ln y$. Other identities that may be of value are listed here.

$$\log_b y^a = a \log_b y \qquad \text{1.16}$$

$$a^x = b^{(x \log_b a)} \qquad \text{1.17}$$

$$\log_a y = (\log_b y)(\log_a b) \qquad \text{1.18}$$

$$\log_b xy = \log_b x + \log_b y \qquad \text{1.19}$$

$$\log(x + jy) = \log\left[\left(\sqrt{x^2 + y^2}\right)\left(e^{j\left[\arctan\left(\frac{y}{x}\right)\right]}\right)\right]$$

$$= \log\left(\sqrt{x^2 + y^2}\right) + \log\left(e^{j\left[\arctan\left(\frac{y}{x}\right)\right]}\right)$$

$$= \log\left(\sqrt{x^2 + y^2}\right) + j(\log e) \cdot \arctan\left(\frac{y}{x}\right)$$

$$\text{1.20}$$

D. Partial Fraction Expansion

A ratio of two polynomials can be rewritten as a series of terms which are the partial fractions, each fraction having terms such as $(x - r)$ in the denominator, where the constants r are the roots of the original denominator.[1] For example, $x^3 + 4x^2 + 3x$ has roots at $x = 0$, $x = -1$, and $x = -3$.

$$\frac{6x^2 + 15x + 3}{x^3 + 4x^2 + 3x} = \frac{6x^2 + 15x + 3}{x(x+1)(x+3)}$$

$$\frac{6x^2 + 15x + 3}{x(x+1)(x+3)} = \frac{A}{x} + \frac{B}{x+1} + \frac{C}{x+3}$$

The partial fraction expansion is found by the method of *undetermined coefficients*. The terms in the right-hand side are cross multiplied and are then placed over a common denominator.

$$\frac{A}{x} + \frac{B}{x+1} + \frac{C}{x+3}$$

$$= \frac{A(x^2 + 4x + 3) + B(x^2 + 3x) + C(x^2 + x)}{x(x+1)(x+3)}$$

$$= \frac{(A + B + C)x^2 + (4A + 3B + C)x + 3A}{x(x+1)(x+3)}$$

Matching the coefficients of x^n in the numerator terms results in simultaneous equations in three unknowns.

$$A + B + C = 6$$
$$4A + 3B + C = 15$$
$$3A = 3$$

The solution to these simultaneous equations is $A = 1$, $B = 3$, and $C = 2$. Therefore,

$$\frac{6x^2 + 15x + 3}{x^3 + 4x^2 + 3x} = \frac{1}{x} + \frac{3}{x+1} + \frac{2}{x+3}$$

---

[1] Before a partial fraction expansion can be carried out, the degree of the numerator must be lower than the degree of the denominator. Otherwise, it will be necessary to reduce the numerator's degree through long division.

This partial fraction expansion procedure is adequate as explained when the roots of the denominator are real and distinctly different from one another. When the roots repeat, the denominator terms will be $(x - r)^n$, $(x - r)^{n-1}$, and so on, to and including $(x - r)$. For example,

$$\frac{x^2 + 4x + 2}{x(x+1)^2} = \frac{A}{x} + \frac{B}{x+1} + \frac{C}{(x+1)^2}$$
$$= \frac{(A+B)x^2 + (2A+B+C)x + A}{x(x+1)^2}$$

(The solution to this illustration is $A = 2$, $B = -1$, and $C = 1$. Try it.) When the complex conjugate roots occur (i.e., the denominator will not factor into real terms), the form of the resulting partial fraction will be

$$\frac{Ax + B}{x^2 + ax + b} \qquad 1.21$$

Partial fraction expansions are particularly useful in the solution of differential equations by the use of Laplace transforms.

Long division is required prior to partial fraction expansion whenever the degree of the numerator polynomial is greater than or equal to the degree of the denominator polynomial. This is illustrated by example 1.7.

Example 1.7

Use long division to reduce the degree of the numerator polynomial for

$$\frac{s^2 + 3s + 2}{s^2 + 2s + 1}$$

$$\begin{array}{r} 1 \\ s^2 + 2s + 1 \overline{)s^2 + 3s + 2} \\ \underline{s^2 + 2s + 1} \\ s + 1 \end{array}$$

$s + 1$ is the remainder term. Since the degree of the remainder term is less than the degree of the denominator, the original expression can be written as

$$\frac{s^2 + 3s + 2}{s^2 + 2s + 1} = 1 + \frac{s + 1}{s^2 + 2s + 1}$$

4 TRIGONOMETRIC IDENTITIES

The common trigonometric functions can be approximated by infinite series when the angle x is expressed in radians. (Remember that $180° = \pi$ radians.)

$$\sin x = x - \frac{x^3}{3!} + \frac{x^5}{5!} - \frac{x^7}{7!} + \dots \qquad 1.22$$

$$\cos x = 1 - \frac{x^2}{2!} + \frac{x^4}{4!} - \frac{x^6}{6!} + \dots \qquad 1.23$$

$$\tan x = x + \frac{x^3}{3} + \frac{2x^5}{15} + \frac{17x^7}{315} + \dots$$
$$\text{for } |x| < \frac{\pi}{2} \qquad 1.24$$

Additional identities are presented here for convenience.

$$\tan x = \frac{\sin x}{\cos x} \qquad 1.25$$

$$\cot x = \frac{1}{\tan x} \qquad 1.26$$

$$\csc x = \frac{1}{\sin x} \qquad 1.27$$

$$\sec x = \frac{1}{\cos x} \qquad 1.28$$

$$\sin^2 x + \cos^2 x = 1 \qquad 1.29$$

$$1 + \tan^2 x = \sec^2 x \qquad 1.30$$

$$1 + \cot^2 x = \csc^2 x \qquad 1.31$$

$$\sin 2x = 2 \sin x \cos x \qquad 1.32$$

$$\tan 2x = \frac{2 \tan x}{1 - \tan^2 x} \qquad 1.33$$

$$\cos 2x = 2 \cos^2 x - 1$$
$$= \cos^2 x - \sin^2 x$$
$$= 1 - 2 \sin^2 x \qquad 1.34$$

$$\sin(x + y) = \sin x \cos y + \cos x \sin y \qquad 1.35$$

$$\sin(x - y) = \sin x \cos y - \cos x \sin y \qquad 1.36$$

$$\cos(x + y) = \cos x \cos y - \sin x \sin y \qquad 1.37$$

$$\cos(x - y) = \cos x \cos y + \sin x \sin y \qquad 1.38$$

$$2 \sin x \sin y = \cos(x - y) - \cos(x + y) \qquad 1.39$$

$$2 \cos x \cos y = \cos(x - y) + \cos(x + y) \qquad 1.40$$

$$2 \sin x \cos y = \sin(x + y) + \sin(x - y) \qquad 1.41$$

5 ANALYTIC GEOMETRY

A major use of analytic geometry in electrical engineering problems is mathematical representation of graphic information. Much of this involves straight-line approximations to device characteristics. Equations for a *straight line* in x-y space can be in several forms. Refer to figure 1.2 which shows a straight line with y-intercept b, x-intercept a, and slope m. (A negative slope is shown.)

general form:

$$Ax + By + C = 0 \qquad 1.42$$

slope-intercept form:

$$y = mx + b \qquad 1.43$$

two-point form:

$$\frac{y - y_1}{x - x_1} = \frac{y_2 - y_1}{x_2 - x_1} = m \qquad 1.44$$

intercept form:

$$\frac{x}{a} + \frac{y}{b} = 1 \qquad 1.45$$

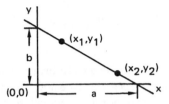

Figure 1.2   A Straight Line in x-y Space

6   DIFFERENTIAL CALCULUS

Some of the frequently encountered derivatives are listed here. The terms u and v are functions of x. The terms a and b are real constants.

$$\frac{d}{dx}a = 0 \qquad 1.46$$

$$\frac{d}{dx}x^a = a \cdot x^{a-1} \qquad 1.47$$

$$\frac{d}{dx}a \cdot u = a \cdot \frac{du}{dx} \qquad 1.48$$

$$\frac{d}{dx}u^a = a \cdot u^{a-1} \cdot \frac{du}{dx} \qquad 1.49$$

$$\frac{d}{dx}uv = u \cdot \frac{dv}{dx} + v \cdot \frac{du}{dx} \qquad 1.50$$

$$\frac{d}{dx}\frac{u}{v} = \frac{1}{v} \cdot \frac{du}{dx} - \frac{u}{v^2} \cdot \frac{dv}{dx}$$

$$= \left(v \cdot \frac{du}{dx} - u \cdot \frac{dv}{dx}\right) \Big/ v^2 \qquad 1.51$$

$$\frac{d}{dx}\sin u = \frac{du}{dx}\cos u \qquad 1.52$$

$$\frac{d}{dx}\cos u = -\frac{du}{dx} \cdot \sin u \qquad 1.53$$

$$\frac{d}{dx}\log_b u = \log_b(e) \cdot \frac{1}{u} \cdot \frac{du}{dx} \qquad 1.54$$

$$\frac{d}{dx}\ln u = \frac{1}{u} \cdot \frac{du}{dx} \qquad 1.55$$

$$\frac{d}{dx}b^u = b^u \cdot \ln b \cdot \frac{du}{dx} \qquad 1.56$$

$$\frac{d}{dx}e^u = e^u \cdot \frac{du}{dx} \qquad 1.57$$

Differentiation is used to find *maxima* and *minima* of functions. At a maximum or minimum, a function will have a zero slope. Therefore, the first derivative of the function must be zero at a maximum or minimum. Another point which will have zero slope is a *point of inflection*. Here, the slope is the same sign on either side but zero at the point of inflection.

To distinguish between maxima, minima, and points of inflection, it is necessary to evaluate the second derivative of the function at the point where the first derivative is zero. If the second derivative is negative, the point is a *maximum*. If the second derivative is positive, the point is a *minimum*. If the second derivative is zero, the point is a point of inflection.

Example 1.8

Find a maximum or minimum of $u = x^2 + 2x + 3$.

$$\frac{du}{dx} = \frac{d}{dx}(x^2 + 2x + 3) = 2x + 2$$

The point sought is where

$$\frac{du}{dx} = 0$$

or at $x = -1$.

To determine whether this point is a maximum or a minimum, the second derivative is taken.

$$\frac{d^2u}{dx^2} = \frac{d}{dx}(2x + 2) = 2$$

This is positive at $x = -1$, so u is minimum at $x = -1$.

7   INTEGRAL CALCULUS

Some integrals which occur frequently in electrical engineering problems are listed here. The letters u and v are functions of x, while the letters $a, b,$ and c are constants. n is an integer constant.

$$\int x^a dx = \frac{x^{a+1}}{a + 1} + c, \text{for } a \neq -1 \qquad 1.58$$

$$\int \frac{1}{x}dx = \ln x + c \qquad 1.59$$

$$\int u\,dv = uv - \int v\,du \qquad 1.60$$

$$\int e^{ax}dx = \frac{1}{a}e^{ax} + c \qquad 1.61$$

$$\int \sin(bx)dx = -\frac{1}{b}\cos(bx) + c \qquad 1.62$$

$$\int \cos(bx)dx = \frac{1}{b}\sin(bx) + c \qquad 1.63$$

$$\int e^{ax}\cos(bx)dx = \frac{e^{ax}}{a^2 + b^2}[a\cos(bx)$$
$$+ b\sin(bx)] + c \qquad 1.64$$

$$\int e^{ax}\sin(bx)dx = \frac{e^{ax}}{a^2 + b^2}[a\sin(bx)$$
$$- b\cos(bx)] + c \qquad 1.65$$

A common application of integration is to find the *average value* of a function. For a function u, the average value from $x = a$ to $x = b$ is

$$\bar{u} = \frac{1}{b-a}\int_a^b u\,dx \qquad 1.66$$

Example 1.9

Find the average value of $u = 2x + 5\sin^2 x$ over the interval $x = -1$ to $x = 2$. Use the trigonometric identity $\cos 2x = 1 - 2\sin^2 x$ and solve for $\sin^2 x$. (x must be expressed in radians.)

$$\sin^2 x = \frac{1}{2}(1 - \cos 2x)$$

$$u = 2x + \frac{5}{2}(1 - \cos 2x)$$

$$= 2x + 2.5 - 2.5\cos 2x$$

From equation 1.66 with $a = -1$ and $b = 2$,

$$\bar{u} = \frac{1}{2-(-1)}\left[2\left(\frac{x^2}{2}\right) + 2.5x - 2.5\left(\frac{1}{2}\sin 2x\right)\right]_{x=-1}^{x=2}$$

$$= \frac{1}{3}\left[(2^2 - (-1)^2) + 2.5(2 - (-1))\right.$$
$$\left. - 1.25[\sin 4 - \sin(-2)]\right]$$

$$= \frac{1}{3}\left[3 + 7.5 - 1.25[-0.757 - (-0.909)]\right]$$

$$= \frac{1}{3}[10.5 - 1.25(0.152)] = 3.436$$

8 DIFFERENTIAL EQUATIONS

First- and second-order linear differential equations can be solved by classical methods, or with the aid of Laplace transforms. Classical solutions to linear differential equations of the first and second order are presented here. Such differential equations often occur in DC switching problems. In these problems, the equations will have the form of equation 1.67, where a, b, c, and d are constants.

$$a\frac{d^2 x}{dt^2} + b\frac{dx}{dt} + cx = d \qquad 1.67$$

A. First-Order Differential Equations

The first-order differential equation contains only the first derivative of the independent variable (i.e., a in equation 1.67 is zero). The solution is

$$x = \frac{d}{c} + Ae^{-ct/b} \qquad 1.68$$

A is an unknown coefficient. In order to determine the value of A, it is necessary to know the value of x, or one of its derivatives, for a known value of t. The initial value of x is frequently known. In this case, $x = x_0$ at $t = 0$.

$$x_0 = \frac{d}{c} + Ae^0$$
$$= \frac{d}{c} + A \qquad 1.69$$

So, the coefficient is

$$A = x_0 - \frac{d}{c} \qquad 1.70$$

The solution to the differential equation is

$$x = \frac{d}{c} + (x_0 - \frac{d}{c})e^{-ct/b} \qquad 1.71$$

Example 1.10

Given a current described by $2\,di/dt + 5i = 10$, determine i for $t > 0$, if $i = -1$ at $t = 0$.

Here $b = 2$, $c = 5$ and $d = 10$.

Then $c/b = 2.5$, $d/c = 2$, and $i_0 - (d/c) = -3$. From equation 1.71,
$$i = 2 - 3e^{-2.5t}$$

B. Second-Order Differential Equations

There are three distinct cases to be considered when the coefficient a in equation 1.67 is non-zero.

case 1: $b^2 - 4ac$ is positive (the overdamped case)

case 2: $b^2 - 4ac$ is zero (the critically damped case)

case 3: $b^2 - 4ac$ is negative (the underdamped case)

• Case 1: Overdamped: $b^2 - 4ac > 0$

If $b^2 - 4ac > 0$, then the solution to equation 1.67 is

$$x = \frac{d}{c} + Ae^{-t/\tau_1} + Be^{-t/\tau_2} \qquad 1.72$$

$$\frac{1}{\tau_1} = \frac{b + \sqrt{b^2 - 4ac}}{2a} \qquad 1.73$$

$$\frac{1}{\tau_2} = \frac{b - \sqrt{b^2 - 4ac}}{2a} \qquad 1.74$$

A and B are undetermined coefficients which must be evaluated from the initial conditions. The usual situation is to know x and its derivative at $t = 0$. From this information it is possible to obtain two equations in two unknowns and solve for the undetermined coefficients A and B. The derivative of equation 1.72 is

$$\frac{dx}{dt} = \frac{-A}{\tau_1}e^{-t/\tau_1} - \frac{B}{\tau_2}e^{-t/\tau_2} \qquad 1.75$$

Example 1.11

A current is described by

$$\frac{d^2i}{dt^2} + 5\frac{di}{dt} + 4i = 2$$

at $t = 0$, $i = 0$ and $di/dt = 0.8$. Determine the current for $t > 0$.

Referring to equation 1.67, $a = 1, b = 5, c = 4$, and $d = 2$. $b^2 - 4ac = 25 - 16 = 9$, so this is the overdamped case. From equations 1.73 and 1.74,

$$\frac{1}{\tau_1} = \frac{5 + 3}{2} = 4$$

$$\frac{1}{\tau_2} = \frac{5 - 3}{2} = 1$$

$$\text{and } \frac{d}{c} = 0.5$$

i and di/dt can be found from equations 1.72 and 1.75 respectively.

$$i = 0.5 + Ae^{-4t} + Be^{-t}$$

$$\frac{di}{dt} = -4Ae^{-4t} - Be^{-t}$$

Substituting the initial conditions,

$$0 = 0.5 + Ae^0 + Be^0 = 0.5 + A + B$$

$$0.8 = -4Ae^0 - Be^0 = -4A - B$$

The simultaneous solution to these two equations is $A = -0.1$ and $B = -0.4$. Therefore, the current as a function of time is

$$i = 0.5 - 0.1e^{-4t} - 0.4e^{-t}$$

• Case 2: Critically damped: $b^2 - 4ac = 0$

In the critically damped case, the solution to equation 1.67 is

$$x = \frac{d}{c} + Ae^{-bt/2a} + Bte^{-bt/2a} \qquad 1.76$$

A and B are undetermined coefficients which must be evaluated from the initial conditions. The derivative of equation 1.76 is

$$\frac{dx}{dt} = -\frac{bA}{2a}e^{-bt/2a} + Be^{-bt/2a}$$

$$\qquad - \frac{bB}{2a}te^{-bt/2a} \qquad 1.77$$

If the conditions are given at $t = 0$, then

$$x_0 = \frac{d}{c} + A \qquad 1.78$$

$$\left.\frac{dx}{dt}\right|_{t=0} = -\frac{bA}{2a} + B \qquad 1.79$$

Example 1.12

A current is described by the differential equation

$$\frac{d^2i}{dt^2} + 2\frac{di}{dt} + i = 0.25$$

Determine i as a function of time if $i = 0$ and $di/dt = 0.8$ at $t = 0$.

This is the critically damped case since $a = 1$, $b = 2$, $c = 1$ and $d = 0.25$, so that $b^2 - 4ac = 0$. From equation 1.76, $b/2a = 1$, and $d/c = 0.25$, so that the skeleton solution is

$$i = 0.25 + Ae^{-t} + Bte^{-t}$$

At $t = 0, i = 0$. From equation 1.78,

$$i_0 = 0 = 0.25 + A$$

$$A = -0.25$$

At $t = 0$, $di/dt = 0.8$. From equation 1.79,

$$\left.\frac{di}{dt}\right|_{t=0} = 0.8 = -A + B$$

$$B = 0.55$$

The current as a function of time is

$$i = 0.25 - 0.25e^{-t} + 0.55te^{-t}$$

- Case 3: Underdamped: $b^2 - 4ac < 0$

In this case the solution of equation 1.67 is

$$x = \frac{d}{c} + Ae^{-\alpha t}\cos\beta t + Be^{-\alpha t}\sin\beta t \qquad 1.80$$

$$\alpha = \frac{b}{2a} \qquad 1.81$$

$$\beta = \frac{\sqrt{4ac - b^2}}{2a} \qquad 1.82$$

The derivative of x is

$$\frac{dx}{dt} = (-\alpha A + \beta B)e^{-\alpha t}\cos\beta t$$
$$- (\beta A + \alpha B)e^{-\alpha t}\sin\beta t \qquad 1.83$$

At $t = 0$, equations 1.80 and 1.83 become

$$x_0 = \frac{d}{c} + A \qquad 1.84$$

$$\left.\frac{dx}{dt}\right|_{t=0} = -\alpha A + \beta B \qquad 1.85$$

Example 1.13

A voltage is described by $d^2v/dt^2 + dv/dt + 2.5v = 50$. $v = 30$ and $dv/dt = 2.5$ at $t = 0$. Determine v for $t > 0$.

Referring to equation 1.67, $a = 1, b = 1, c = 2.5$ and $d = 50$. Then $b/2a = 0.5$, $b^2 - 4ac = -9$ (the underdamped case), and $d/c = 20$. From equation 1.80,

$$v = 20 + Ae^{-0.5t}\cos 1.5t + Be^{-0.5t}\sin 1.5t$$

Substituting the initial conditions in equations 1.84 and 1.85 respectively,

$$30 = 20 + A$$
$$2.5 = -0.5 \times 10 + 1.5B$$

Therefore, $A = 10$ and $B = 5$. The voltage as a function of time is

$$v = 20 + 10e^{-0.5t}\cos 1.5t$$
$$+ 5e^{-0.5t}\sin 1.5t$$

9 FOURIER ANALYSIS

Periodic functions are conveniently analyzed by means of *Fourier series. Phasor analysis* is performed for each

component, and superposition is used to combine the components.

A periodic function with a period T can be represented by the Fourier series given in equation 1.86.

$$f(t) = \frac{a_0}{2} + \sum_{n=1}^{\infty} a_n \cos\frac{2\pi n}{T}t$$
$$+ \sum_{n=1}^{\infty} b_n \sin\frac{2\pi n}{T}t \qquad 1.86$$

The Fourier coefficients $a_0, a_n,$ and b_n can be found from equations 1.87, 1.88, and 1.89.

$$a_0 = \frac{2}{T}\int_{t_1}^{t_1+T} f(t)dt \qquad 1.87$$

$$a_n = \frac{2}{T}\int_{t_1}^{t_1+T} f(t)\cos\frac{2\pi n}{T}t\ dt \qquad 1.88$$

$$b_n = \frac{2}{T}\int_{t_1}^{t_1+T} f(t)\sin\frac{2\pi n}{T}t\ dt \qquad 1.89$$

Example 1.14

A rectangular wave has voltage levels of 0 and 5 volts, equal times at both levels, and a period of 10 seconds (i.e., 5 seconds at 5 volts and 5 seconds at 0 volts). The $t = 0$ line bisects the 5 volt level. Obtain a Fourier series for this wave. One cycle of the wave can be represented by the function

$$f(t) = 0, -5 < t < -2.5$$
$$= 5, -2.5 < t < 2.5$$
$$= 0, 2.5 < t < 5$$

Substituting $t_1 = -5$ and $T = 10$ into equations 1.87, 1.88, and 1.89 gives the coefficients $a_0, a_n,$ and b_n.

$$a_0 = \frac{2}{10}\int_{-5}^{5} f(t)dt$$

$$= \frac{1}{5}\int_{-2.5}^{2.5} 5dt = 5$$

$$a_n = \frac{1}{5}\int_{-2.5}^{2.5} 5\cos\frac{2\pi n}{10}t\ dt$$

$$= \frac{10}{\pi n}\sin\frac{\pi n}{2}$$

$$b_n = \frac{1}{5}\int_{-2.5}^{2.5} 5\sin\frac{2\pi n}{10}t\ dt$$

$$= \frac{-5}{\pi n}\left(\cos\frac{\pi n}{2} - \cos\left(\frac{-\pi n}{2}\right)\right) = 0$$

Evaluating the first few a coefficients,

$$a_1 = \frac{10}{1\pi}\sin\frac{\pi}{2} = \frac{10}{\pi}$$

$$a_2 = \frac{10}{2\pi}\sin\pi = 0$$

$$a_3 = \frac{10}{3\pi}\sin\frac{3\pi}{2} = -\frac{10}{3\pi}$$

$$a_4 = \frac{10}{4\pi}\sin 2\pi = 0$$

In general, the a coefficients are

$$a_{2m} = 0$$

$$a_{2m-1} = \frac{10(-1)^{m-1}}{(2m-1)\pi}$$

$$m = 1, 2, 3, 4 \ldots .$$

The infinite Fourier series which approximates the original square wave is

$$f(t) = \frac{5}{2} + \sum_{m=1}^{\infty} \frac{10(-1)^{m-1}}{(2m-1)\pi}\cos\frac{(2m-1)2\pi}{10}t$$

10   TRANSIENT ANALYSIS BY LAPLACE TRANSFORMS

Laplace transform analysis is similar to operational calculus if the operator s is interpreted as the derivative operator d/dt or $\mathbf{D}$. The advantage of the Laplace method is that the initial conditions are automatically included in the analysis. The Laplace transform of a function $f(t)$ is written $F(s)$.

$$F(s) = \mathcal{L}\{f(t)\} = \int_0^{\infty} f(t)e^{-st}dt \qquad 1.90$$

The inverse Laplace transform of $F(s)$ is $f(t)$.

$$f(t) = \mathcal{L}^{-1}\{F(s)\} = \frac{1}{2\pi j}\int_{c-j\infty}^{c+j\infty} F(s)e^{st}ds \qquad 1.91$$

Equations 1.92, 1.93, and 1.94 are useful in writing network relations.

$$\mathcal{L}\left\{\frac{d}{dt}f(t)\right\} = sF(s) - f(0^+) \qquad 1.92$$

$$\mathcal{L}\left\{\int_0^t f(t)dt\right\} = \frac{1}{s}F(s) \qquad 1.93$$

The Laplace transform of the nth derivative is

$$\mathcal{L}\left\{\frac{d^n}{dt^n}f(t)\right\} = s^n F(s) - s^{n-1}f(0^+)$$

$$- s^{n-2}\frac{df(0^+)}{dt} - s^{n-3}\frac{d^2 f(0^+)}{dt^2} - \cdots$$

$$- s\frac{d^{n-2}f(0^+)}{dt^{n-2}} - \frac{d^{n-1}f(0^+)}{dt^{n-1}} \qquad 1.94$$

Table 1.1 provides a useful set of transform pairs.

Example 1.15

Obtain the Laplace transform of the differential equation

$$10 = (10^{-5})\frac{d^2 V}{dt^2} + (5\cdot 10^{-3})\frac{dV}{dt} + V$$

The initial conditions are $v = 5$, $\frac{dv}{dt} = 10^3$ at $t = 0$.

The terms are transformed using table 1.1 to obtain

$$\frac{10}{s} = 10^{-5}[s^2 V(s) - 5s - 1000]$$

$$+ 5\cdot 10^{-3}[sV(s) - 5] + V(s)$$

Once a differential equation has been transformed, it is necessary to solve for the transformed variable in terms of s.

Example 1.16

From example 1.15 obtain $V(s)$ as a function s.

First, multiply the answer of example 1.15 by 10^5.

$$\frac{10^6}{s} = s^2 V(s) - 5s - 1000 + 500sV(s) - 2500 + 10^5 V(s)$$

Next, gather terms and solve for $V(s)$.

$$\frac{10^6}{s} + 5s + 3500 = (s^2 + 500s + 10^5)V(s)$$

$$V(s) = \frac{5s^2 + 3500s + 10^6}{s(s^2 + 500s + 10^5)}$$

The next step is to rearrange the function of s into terms that are shown in the right-hand column of table 1.1. This will typically require that the function of s be factored into first and second order terms using the *partial-fraction expansion* method. Partial fractions have been covered earlier in this chapter.

Table 1.1
Laplace Transform Pairs

$f(t)$	$F(s)$
$df(t)/dt$	$sF(s) - f(0^+)$
$d^2 f(t)/dt^2$	$s^2 F(s) - sf(0^+) - \frac{df}{dt}(0^+)$
$\int_0^t f(t)dt$	$(1/s)F(s)$
1 or $u(t)$(unit step)	$1/s$
$u(t - \tau)$	$e^{-s\tau}/s$
$\frac{d}{dt}u(t) = \delta(t)$	1
$\delta(t - \tau)$	$e^{-s\tau}$
t	$1/s^2$
e^{-at}	$1/(s + a)$
te^{-at}	$1/(s + a)^2$
$\sin bt$	$b/(s^2 + b^2)$
$\cos bt$	$s/(s^2 + b^2)$
$e^{-at} \sin bt$	$b/[(s + a)^2 + b^2]$
$e^{-at} \cos bt$	$(s + a)/[(s + a)^2 + b^2]$

The function $u(t)$ is zero before $t = 0$ and one thereafter; the function $u(t - \tau)$ is defined as zero before $t = \tau$ and one thereafter. The function $\delta(t)$ is the time derivative of $u(t)$ and is infinite at $t = 0$, but has an area of one.

Example 1.17

Separate the function $V(s)$ from example 1.16 into its partial fractions. From table 1.1, the useful forms are

$$\frac{1}{s}; \quad \frac{s + a}{(s + a)^2 + b^2}; \quad \text{and} \quad \frac{b}{(s + a)^2 + b^2}$$

The function can be written as

$$V(s) = \frac{5s^2 + 3500s + 10^6}{s[(s + 250)^2 + (10^5 - 250^2)]}$$

Therefore, $a = 250$ and $b^2 = 10^5 - 250^2$, or $b = 194$. The partial fraction expansion is

$$V(s) = \frac{A}{s} + \frac{B(s + 250) + C(194)}{(s + 250)^2 + 194^2}$$

The two terms are cross multiplied to obtain the original denominator.

$$V(s) = \frac{As^2 + 500As + 10^5 A + Bs^2 + 250Bs + 194Cs}{s(s^2 + 500as + 10^5)}$$

Next, equate the coefficients of the various powers of s in the numerator of this last expression with the coefficients from the original expressions.

$$s^2 : A + B = 5$$
$$s^1 : 500A + 250B + 194C = 3500$$
$$s^0 : 10^5 A = 10^6$$

A is easily found to be 10. Since $A + B = 5$, then $B = -5$. C is subsequently found to equal -1.29. Finally $V(s)$ can be written as

$$V(s) = \frac{10}{s} - 5\left[\frac{s + 250}{(s + 250)^2 + 194^2}\right]$$
$$- 1.29\left[\frac{194}{(s + 250)^2 + 194^2}\right]$$

The final step is to obtain the inverse transform from the transform pairs in table 1.1.

Example 1.18

Obtain $V(t)$ from $V(s)$ in example 1.17.

$$V(t) = 10 - 5e^{-250t}\cos 194t - 1.29e^{-250t}\sin 194t$$

A. Initial Value Theorem

When a Laplace transform $F(s)$ is known, a corresponding time function can be evaluated at $t = 0$ by the *initial value theorem*, equation 1.95.

$$f(0^+) = \lim_{s\to\infty} sF(s) \qquad\qquad 1.95$$

Example 1.19

Evaluate the function

$$V(s) = \frac{5s^2 + 3500s + 10^6}{s(s^2 + 500s + 10^5)}$$

of example 1.16 at $t = 0^+$. Use the initial value theorem to obtain $v(t)$ at $t = 0^+$.

$$v(0^+) = \lim_{s\to\infty} \frac{5s^3 + 3500s^2 + 10^6 s}{s^3 + 500s^2 + 10^5 s}$$

Dividing both numerator and denominator by s^3, the following is obtained.

$$v(0^+) = \lim_{s\to\infty} \left[\frac{5 + \dfrac{3500}{s} + \dfrac{10^6}{s^2}}{1 + \dfrac{500}{s} + \dfrac{10^5}{s^2}} \right] = 5$$

B. Final Value Theorem

When a Laplace transform $F(s)$ is known, the corresponding time function can be evaluated at $t = \infty$ by the final *value theorem*, equation 1.96.

$$f(\infty) = \lim_{s\to 0} sF(s) \qquad\qquad 1.96$$

Example 1.20

Determine $v(\infty)$ for the function $V(s)$ from example 1.16.

$$v(\infty) = \lim_{s\to 0} \frac{5s^3 + 3500s^2 + 10^6 s}{s^3 + 500s^2 + 10^5 s}$$

$$= \lim_{s\to 0} \frac{5s^2 + 3500s + 10^6}{s^2 + 500s + 10^5}$$

$$= \frac{10^6}{10^5} = 10$$

11 FINITE FOURIER SERIES

The Fourier series for a periodic function of time as given in equation 1.86 can also be written in a polar form as

$$f(t) = \sum_{k=-\infty}^{\infty} c_k e^{j(2\pi/T)kt} \qquad\qquad 1.97$$

In equation 1.97, k has been substituted for n of equation 1.87. The coefficients c_k of equation 1.97 are complex numbers and can be found in the same way the sine and cosine coefficients of equation 1.86 were found using the following formula:

$$c_k = \frac{1}{T} \int_{t=t_1}^{t_1+T} f(t)e^{-j(2\pi/T)kt}dt \qquad\qquad 1.98$$

Signal processing and analysis by computers has become a major activity, and computers are limited to discrete time samples of continuous time signals. Thus, a computer deals with a sequence of values sampled from a signal waveform. The samples are taken at a constant rate, so the samples are separated by a fixed time interval.

A computer calculation of a signal's frequency spectrum is done by taking a fixed number of samples and performing numerical integrations, or their equivalent, to find the Fourier coefficients c_k following the ideas of equation 1.98.

The number of samples used in the calculation is an even number, and for computational reasons, that number is an integer power of 2, i.e., 2^m, where m is an integer. The interval during which the samples are taken is again T. Thus, the interval of time between samples is $T/2N$. This replaces dt in equation 1.98. The $2N$ samples are numbered from 0 to $(2N-1)$, using an index p. Thus, starting the time interval at the time of the first sample, time t in equation 1.98 is replaced by $(T/2N)p$, with p incrementing with each sample from 0 to $2N-1$. Finally, we replace the integration of equation 1.98 with a summation over the index p, to give

$$C_k = \frac{1}{T} \sum_{p=0}^{2N-1} f(pT/2N)e^{-j(2\pi/T)k(pT/2N)}(T/2N)$$

or

$$C_k = \frac{1}{2N} \sum_{p=0}^{2N-1} f(pT/2N)e^{-j\pi kp/N} \qquad\qquad 1.99$$

Once the coefficients are obtained, the sampled function can be represented as in equation 1.97.

$$f(pT/2N) = \sum_{k=0}^{2N-1} C_k e^{j\pi kp/N} \qquad 1.100$$

The coefficients for the finite Fourier coefficients have been changed to capital letters C_k to distinguish them from the continuous time periodic coefficients of equation 1.98. In a computer analysis of a signal, it is assumed that the function is periodic, while in practice it is not. However, with sufficient data points the signal can be characterized by these coefficients if it is statistically stationary in time.

Example 1.21

Find the coefficients of a signal represented by the following eight samples:

p:	0	1	2	3	4	5	6	7
$f(pT/2N)$:	0	0	8	0	0	0	0	0

which is an impulse in the third sampling interval. Using equation 1.99,

$$C_k = \frac{1}{8} \sum_{p=0}^{8-1} f(pT/8) e^{-j\pi kp/4}$$

$$= \frac{1}{8}[0+0+8e^{-jk\pi/2}+0+0+0+0+0] = e^{-jk\pi/2}$$

Then,

$$C_0 = 1, \quad C_1 = e^{-j\pi/2}, \quad C_2 = e^{-j\pi}, \quad C_3 = e^{-j3\pi/2}$$
$$C_4 = e^{-j2\pi}, \quad C_5 = e^{-j5\pi/2}, \quad C_6 = e^{-j3\pi}$$
$$C_7 = e^{-j7\pi/2}$$

To reconstruct the original samples, apply equation 1.100.

$$f(pT/2N) = \sum_{k=0}^{7} C_k e^{jkp\pi/4}$$

$$f(pT/2N) = 1 + e^{-j\pi/2}e^{jp\pi/4} + e^{-j\pi}e^{jp\pi/2}$$
$$+ e^{-j3\pi/2}e^{j3p\pi/4} + e^{-j2\pi}e^{jp\pi}$$
$$+ e^{-j5\pi/2}e^{j5p\pi/4} + e^{-j3\pi}e^{j3p\pi/2}$$
$$+ e^{-j7\pi/2}e^{j7p\pi/4}$$

It is left as an exercise for the reader to show that this result matches the original eight samples. By substituting a value of p from zero to seven, it can be shown that the corresponding function values are obtained.

A notable feature of the foregoing example is the number of multiplications necessary. There were eight complex multiplications for each coefficient, and eight coefficients were found, although the first one was simply the average value, which did not require multiplications. For a straightforward computer algorithm (brute force), a total of 64 complex multiplications would be required. In general, the direct method requires $4N^2$ multiplications, which can require extensive computer time when the sample size is in the tens to hundreds of thousands.

In this example, only eight coefficients ($2N$), including the average value were found. The reason for this is simply that with only eight data points it was possible to solve for only eight unknowns. If the example were extended to find the next eight coefficients, duplications of values would occur. For example, in solving for C_{k+2N} the following expression is obtained:

$$C_{k+2N} = \frac{1}{2N} \sum_{p=0}^{2N-1} f(pT/2N) e^{-j\pi(k+2N)p/N}$$

$$= \frac{1}{2N} \sum_{p=0}^{2N-1} f(pT/2N) e^{-j\pi kp/N} e^{-j2\pi p}$$

But as $e^{-j2\pi p} = 1$, in general, the result is

$$C_{k\pm2Nm} = C_k \qquad 1.101$$

m is an integer. This fact has ramifications when signals contain frequencies beyond the capabilities of the finite Fourier series. With $2N$ samples taken over a period T, the sampling frequency is $2N/T$. The Nyquist sampling criterion states that the highest frequency component that can be resolved is half the sampling frequency, or N/T. If there are components of frequencies higher than N/T, they will cause errors in the coefficients obtained, a phenomenon called *aliasing*.

A hidden clue to this can be observed in the fact that the higher coefficients C_k are the complex conjugates of the lower coefficients. While this is not obvious from example 1.21, it can be seen in the following.

Examine the relationship between the first half of the coefficients and the last half, beginning with the last and going to the first:

$$C_{2N-k} = \frac{1}{2N} \sum_{p=0}^{2N-1} f(pT/2N) e^{-j\pi(2N-k)p/N}$$

$$C_{2N-k} = \frac{1}{2N} \sum_{p=0}^{2N-1} f(pT/2N) e^{-j2\pi p} e^{+j\pi kp/N}$$
$$\qquad 1.102$$

Some careful consideration of equation 1.102 will reveal that the complex conjugate of C_k is found by

$$C_{2N-k} = \frac{1}{2N} \sum_{p=0}^{2N-1} f(pT/2N)e^{+j\pi kp/N} = C_k^* \quad 1.103$$

Thus, only N of the coefficients are needed, as the others can be obtained from their complex conjugates.

In example 1.21, it appears that coefficients have been found for frequency components higher than permitted by the Nyquist criterion. However, taking the results of equations 1.101 and 1.102 together, a suitable way of viewing those results can be obtained.

Example 1.22

Since the method used in example 1.21 correctly reconstructs the signal, but violates the Nyquist criterion, re-cast the function of equation 1.100 to limit it to frequencies of the Nyquist frequency and less.

Taking a straightforward approach, since C_4 is the coefficient for the Nyquist frequency, first consider C_5, the next higher frequency. From equation 1.101 it can be seen that

$$C_5 = C_{5-8} = C_{-3}$$

and from equation 1.102 or 1.103:

$$C_5 = C_3^*$$

Thus,

$$C_{-3} = C_3^* = C_5$$

Following the same logic, $C_{-2} = C_2^* = C_6$, and $C_{-1} = C_1^* = C_7$. To place the finite Fourier series into the proper range, a negatively subscripted coefficient is used with each appropriate (negative) frequency.

With a term-by-term substitution into the result of example 1.21, the result is

$$
\begin{aligned}
f(pT/2N) =& 1 + e^{-j\pi/2}e^{jp\pi/4} + e^{-j\pi}e^{jp\pi/2} \\
&+ e^{-j3\pi/2}e^{j3p\pi/4} + e^{-j2\pi}e^{jp\pi} \\
&+ e^{j3\pi/2}e^{-j3p\pi/4} + e^{j\pi}e^{-jp\pi/2} \\
&+ e^{j\pi/2}e^{-jp\pi/4}
\end{aligned}
$$

The results of example 1.22 can be generalized by obtaining coefficients with subscripts from $-(N-1)$ to N, and recasting equation 1.100 to

$$f(pT/2N) = \sum_{k=1-N}^{N} C_k e^{j\pi kp/N} \quad 1.104$$

This assures that the Nyquist criterion is satisfied, and allows the construction of sine and cosine series using Euler's equation.

12 FAST FOURIER TRANSFORMS

The *fast Fourier transform (FFT)* is simply an efficient way to obtain the finite Fourier series, the latter sometimes being called the discrete Fourier transform. Additional development of the FFT can be found in textbooks devoted to digital signal processing or digital filters. The motivation for using the FFT is that it reduces the order of multiplications from N^2 to $N\log_2 N$, which can be a factor of 100 or more for a large sample size. This means the calculation using the FFT takes the order of only 1% of the time required for the method used in example 1.21. As implementation of the FFT is largely a programming exercise, only a limited development has been presented here.

The finite Fourier series as given in equation 1.99 is repeated for discussion.

$$C_k = \frac{1}{2N} \sum_{p=0}^{2N-1} f(pT/2N)e^{-j\pi kp/N}$$

In carrying out the indicated steps for each coefficient C_k, many operations are repeated, leading to the order of $(2N)^2$ multiplications. The FFT must duplicate the results, including every term of equation 1.99. The duplication of effort is avoided by a regrouping of the operations as follows.

For a particular k, the series of equation 1.99 is written out in order of the index p, giving $2N$ terms. This series is then divided into L consecutive segments of D terms, which requires that $LD = 2N$. Then new summations are created by combining the corresponding terms of each segment into sub-summations, using a new index m_p. Each sub-summation will contain L terms, so the index m_p will range from zero to $L-1$. The elements of these sub-summations are separated by D in the series of equation 1.99. There will be D of these sub-summations for each C_k, and the p_oth sub-summation can be written as

$$S_{k,p_o} = \sum_{m_p=0}^{L-1} f[(p_o + m_p D)T/2N]e^{-j\pi k(p_o+m_p D)/N}$$

$$1.105$$

Then equation 1.99 can be duplicated in this new form by summing these D summations over the index p_o, which ranges from zero to $D-1$:

$$C_k = \frac{1}{2N} \sum_{p_o=0}^{D-1} S_{k,p_o} \quad 1.106$$

Next comes the key step in achieving the reduction of computation time. This was independently rediscovered by Cooley and Tukey. It is seen in equation 1.105 that p has become $p_o + m_p D$.

Similarly, a replacement of k can be made $k_o + m_k L$, where k_o has the range from zero to $L-1$, and m_k has the range zero to $D-1$ to cover all necessary values of k. Substituting into the exponential of equation 1.105, and by expanding the function, the following expression is obtained.

$$e^{-j\pi k_o p_o/N} e^{-j\pi k_o m_p D/N} e^{-j\pi m_k p_o L/N} e^{-j\pi m_k m_p LD/N}$$

$$1.107$$

As $LD = 2N$ and m_k and m_p are integers, the last term of expression 1.107 is equal to one, and thus disappears. The first and third terms of expression 1.107 are independent of m_p and can be brought out from the summation of equation 1.105. Also note that $D/N = 2/L$ and $L/N = 2/0 \rightarrow D$, so the exponents of expression 1.107 can be simplified. Thus, equation 1.105 can be rewritten as follows:

$$S_{k_o+m_k L, p_o} = e^{-j\pi k_o p_o/N} e^{-j2\pi m_k p_o/D}$$

$$\sum_{m_p=0}^{L-1} f[(p_o T/2N) + (m_p T/L)] e^{-j2\pi k_o m_p/L}$$

$$1.108$$

Extracting from equation 1.108 a term $G(k_o, p_o)$

$$G(k_o, p_o) = e^{-j\pi k_o p_o/N} \sum_{m_p=0}^{L-1} f[(p_o T/2N)$$

$$+ (m_p T/L)] e^{-j2\pi k_o m_p/L} \qquad 1.109$$

Then equation 1.106 can be rewritten as

$$C_{k_o+m_k L} = \frac{1}{2N} \sum_{p_o=0}^{D-1} G(k_o, p_o) e^{-j2\pi m_k p_o/D} \qquad 1.110$$

The savings in computation can now be examined. Computations are made in two steps: first the G terms of equation 1.109 are calculated, which requires L multiplications for each combination of k_o and p_o. Regardless of how D and L are chosen, there are $2N$ combinations of k_o and p_o, so that a total of $2NL$ multiplications are required for the first phase.

The second phase calculates the C_k values of equation 1.110 using the G values from the first phase. There are D multiplications for each C, and there are $2N$ C's to be obtained, for a total of $2ND$ multiplications. Thus, the total number of multiplications involved here are $2N(L + D)$, as compared with $(2N)^2$ for the finite Fourier series method of equation 1.99. As an example,

with $2N = 10^4$, making $D = L = 10^2$, $(2N)^2 = 10^8$, while $2N(L + D) = 2 \times 10^6$. So, the FFT method in its rudimentary form depicted here requires only 2% of the time required to calculate the coefficients for 10^4 sample points.

With additional development, further reduction can be obtained by repeating the process just outlined and carefully selecting the starting values of D and L.

The interested reader is referred to: *The Fast Fourier Transform*, E. O. Brigham, Prentice Hall, 1974.

13 THE FOURIER TRANSFORM

The Fourier transform is used in much the same way as the Laplace transform, and may be developed from the finite Fourier series defined in equations 1.99 and 1.104, with T and the number of samples extending from negative to positive infinity. Without going into the details, the transformation from the time domain to k-space, which is also called the *frequency domain* with ω replacing k, corresponding to equation 1.99 is

$$F(\omega) = \int_{t=-\infty}^{\infty} f(t) e^{-j\omega t} \, dt \qquad 1.111$$

The transformation from the frequency domain to the time domain corresponding to equation 1.104 is

$$f(t) = \frac{1}{2\pi} \int_{\omega=-\infty}^{\infty} F(\omega) e^{jt\omega} \, d\omega \qquad 1.112$$

Note the term 2π must appear in the pair of transformations, but while its exact placement is not consistent from one table of transforms to the next, it is important to note its placement when using a particular table of transform pairs. Here it has been placed in the definition of the transform of $F(\omega)$ to $f(t)$, usually referred to as the *inverse transform*.

The derivations and proofs of certain properties of Fourier transform pairs are beyond the scope of this book. Table 1.2 provides some of those properties as a set of fundamental transformation identities.

A set of common transformations is provided in Table 1.3. These are usually called Fourier transform pairs. The first two, $T1$ and $T2$, are readily obtained from equation 1.111. The third, $T3$, is obtained from equation 1.112. $T4$ and $T5$ are obtained from $T3$ and Euler's equation. The remainder are obtainable from the residue theorem of complex variable theory.

Table 1.2
Fundamental Fourier Transformation Identities

$I1.$ $\qquad f(t) \longleftrightarrow F(\omega)$

$I2.$ $\qquad A f(t) \longleftrightarrow A F(\omega)$

$I3.$ $\qquad f_1(t) + f_2(t) \longleftrightarrow F_1(\omega) + F_2(\omega)$

$I4.$ $\qquad f(t+c) \longleftrightarrow e^{jc\omega} F(\omega)$

$I5.$ $\qquad \dfrac{df(t)}{dt} \longleftrightarrow j\omega F(\omega)$

$I6.$ $\qquad \displaystyle\int_{y=-\infty}^{t} f(y)\,dy \longleftrightarrow \dfrac{1}{j\omega} F(\omega)$

$I7.$
$$\int_{g=-\infty}^{t} \int_{h=-\infty}^{g} f(h)\,dh\,dg \longleftrightarrow \frac{1}{(j\omega)^2} F(\omega)$$

$I8.$ $\qquad t^n f(t) \longleftrightarrow j^n \dfrac{\partial^n}{\partial \omega^n} F(\omega)$

$I9.$ $\qquad e^{jat} f(t) \longleftrightarrow F(\omega - a)$

$I10.$ $\qquad f(t)\cos at \longleftrightarrow \dfrac{1}{2}[F(\omega - a) + F(\omega + a)]$

$I11.$
$$\int_{t=-\infty}^{\infty} f_1(t)\,f_2(y-t)\,dt \longleftrightarrow F_1(\omega)F_2(\omega)$$

a and c are constants, y is an independent variable, which can be a constant. g and h are dummy variables, and n is an integer.

Fourier transforms provide a useful tool in the analysis of linear continuous time systems. Such systems have output signals which are uniquely related to the input signals (including noise). Such a system can be characterized by its output time response to an input of a unit impulse. This response is symbolized by a time function $h(t)$, as depicted in figure 1.3(a). Of course an impulse applied at another time will have a similar response, but translated in time an equal amount. Thus, for an input of $\delta(t-\tau)$, the output will be $h(t-\tau)$.

In general, in a linear system, the output response at any time τ will be proportional to the amplitude of the input $x(\tau)$. This is indicated in figure 1.3(b).

An input signal, $x(t)$ can be broken into an infinite number of impulses, summed in an integral in the form

$$x(t) = \int_{\tau=-\infty}^{\infty} x(\tau)\delta(t-\tau)\,d\tau \qquad 1.113$$

So that the corresponding output signal becomes

$$y(t) = \int_{\tau=-\infty}^{\infty} x(\tau)h(t-\tau)\,d\tau \qquad 1.114$$

The form of equation 1.114 is that of a convolution integral, and this leads to the primary application of the Fourier transform. In table 1.2, identity I11 has the convolution integral in the $f(t)$ column. Thus, a convolution integral in the time domain becomes a multiplication in the frequency domain. For the transformation of $x(t)$ resulting in $X(\omega)$, and the transformation of $h(t)$ resulting in $H(\omega)$, the convolution of $x(t)$ and $h(t)$ is transformed to $X(\omega)H(\omega)$.

Table 1.3
Some Useful Fourier Transform Pairs

$I1.$ $\qquad f(t) \longleftrightarrow F(\omega)$

$T1.$ $\qquad \delta(t+a) \longleftrightarrow e^{ja\omega}$

$T2.$
$$u(t+a) - u(t+b) \longleftrightarrow \frac{e^{ja\omega}}{j\omega} - \frac{e^{jb\omega}}{j\omega}$$

$T3.$
$$u(t+a) - u(t-a) \longleftrightarrow \frac{\sin a\omega}{\omega}$$

$T4.$ $\qquad e^{-at}u(t) \longleftrightarrow \dfrac{1}{j\omega + a}$

$T5.$ $\qquad e^{jat} \longleftrightarrow 2\pi\delta(\omega - a)$

$T6.$ $\qquad \sin at \longleftrightarrow j\pi[\delta(\omega + a) - \delta(\omega - a)]$

$T7.$ $\qquad \cos at \longleftrightarrow \pi[\delta(\omega + a) + \delta(\omega - a)]$

$T8.$ $\qquad \dfrac{t}{t^2 + y^2} \longleftrightarrow j\pi e^{y\omega}$

$T9.$ $\qquad \dfrac{y}{t^2 + y^2} \longleftrightarrow \pi e^{y\omega}$

$T10.$ $\qquad \tan^{-1}\dfrac{t}{y} \longleftrightarrow \dfrac{\pi}{j\omega} e^{y\omega}$

$T11.$ $\qquad \dfrac{1}{2}\ln(t^2 + y^2) \longleftrightarrow \dfrac{\pi}{\omega} e^{y\omega}$

$T12.$
$$\frac{1}{2}\ln[1 + (t/y)^2] \longleftrightarrow \frac{\pi}{\omega} e^{y\omega}$$

a and b are constants, y is an independent variable and may be a constant.

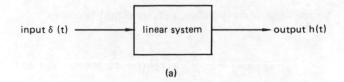

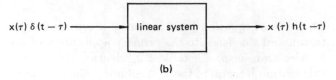

Figure 1.3 (a) Linear System With Unit Impulse Response $h(t)$, (b) Response for Impulse at Time τ of Amplitude $x(\tau)$

Example 1.23

A linear system has a unit step response $h(t) = te^{-t}u(t)$ and it receives an input pulse signal of $u(t) - u(t-1)$. Obtain the output time response $y(t)$, for this input.

The Fourier transform of the input signal is found from $T2$ of table 1.3:

$$x(t) = u(t) - u(t-1) : X(\omega) = \frac{1 - e^{-j\omega}}{j\omega}$$

There is no transform given for $h(t)$, however it can be generated using the identity $I8$ of table 1.2 with $n = 1$, and the transformation $T4$ of table 1.3:

$$H(\omega) = j\frac{\partial}{\partial\omega}(j\omega + 1)^{-1} = (j\omega + 1)^{-2}$$

Then the transformation of the time output is

$$Y(\omega) = X(\omega)H(\omega) = \frac{1 - e^{-j\omega}}{j\omega(j\omega + 1)^2}$$

Following the partial fraction expansion method for Laplace transforms,

$$Y(\omega) = \frac{1 - e^{-j\omega}}{j\omega} - \frac{1 - e^{-j\omega}}{j\omega + 1} - \frac{1 - e^{-j\omega}}{(j\omega + 1)^2}$$

or

$$Y(\omega) = \frac{1 - e^{-j\omega}}{j\omega} - \frac{1}{j\omega + 1} + \frac{e^{-j\omega}}{j\omega + 1}$$
$$- \frac{1}{(j\omega + 1)^2} + \frac{e^{-j\omega}}{(j\omega + 1)^2}$$

The first term on the right of this equation is recognizable as the transform of the input signal.

The second term, from transform $T4$ of table 1.3, is the transform of $-e^{-t}u(t)$. The third term is found from $T4$ and identity $I4$ of table 1.2 to be $+e^{-(t-1)}u(t-1)$.

The fourth term is recognized from the transform of $h(t)$ to be $-te^{-t}u(t)$. Applying this result with identity $I4$, the last term is found to be $+(t-1)e^{-(t-1)}u(t-1)$.

Then the inverse transform of $Y(\omega)$ is

$$y(t) = [1 - e^{-t} - te^{-t}]u(t) - [1 - e^{-(t-1)}$$
$$- (t-1)e^{-(t-1)}]u(t-1)$$

14 DETERMINANTS

Matrices are rectangular arrays of numbers which are associated with simultaneous linear equations. (Symbols and subscripted letters are used to indicate the actual numbers.) The arrays are two dimensional, laid out as rows and columns. The subscripts identify the position of a particular member of the array. The first subscript gives the row position and the second gives the column position. Thus a member with the subscript 23 will belong in the second row and the third column.

Each position in the array is associated with a sign. Those positions whose subscripts add to an even number are associated with a positive sign, and those whose subscripts add to an odd number are associated with a negative sign. The arrays for which *determinants* are calculated must have an equal number of rows and columns.

As an example, consider a square 4×4 matrix called $\mathbf{A}$.

$$\mathbf{A} = \begin{bmatrix} a_{11} & a_{12} & a_{13} & a_{14} \\ a_{21} & a_{22} & a_{23} & a_{24} \\ a_{31} & a_{32} & a_{33} & a_{34} \\ a_{41} & a_{42} & a_{43} & a_{44} \end{bmatrix}$$

The signs associated with the positions of $\mathbf{A}$ are

$$\begin{bmatrix} + & - & + & - \\ - & + & - & + \\ + & - & + & - \\ - & + & - & + \end{bmatrix}$$

The fundamental determinant operation can be carried out on a 2×2 matrix such as the matrix $\mathbf{B}$. The determinant of $\mathbf{B}$ is indicated by writing the original matrix between straight vertical lines.

$$\mathbf{B} = \begin{bmatrix} b_{11} & b_{12} \\ b_{21} & b_{22} \end{bmatrix}$$

$$\det \mathbf{B} = \begin{vmatrix} b_{11} & b_{12} \\ b_{21} & b_{22} \end{vmatrix}$$

$$= b_{11}b_{22} - b_{21}b_{12}$$

For a 3×3 determinant, an algorithm can be used to simplify the determinant calculation. (This algorithm is valid only for 3×3 determinants. It will produce erroneous results for any other size matrix.)

For a 3×3 matrix, first copy columns 1 and 2 and place them in the positions of columns 4 and 5, respectively. Then, the determinant can be taken by adding all of the diagonals downward to the right and subtracting the diagonals upward to the right.

$$\mathbf{C} = \begin{bmatrix} c_{11} & c_{12} & c_{13} \\ c_{21} & c_{22} & c_{23} \\ c_{31} & c_{32} & c_{33} \end{bmatrix}$$

$$\det \mathbf{C} = \begin{vmatrix} c_{11} & c_{12} & c_{13} & c_{11} & c_{12} \\ c_{21} & c_{22} & c_{23} & c_{21} & c_{22} \\ c_{31} & c_{32} & c_{33} & c_{31} & c_{32} \end{vmatrix}$$

$$= c_{11}c_{22}c_{33} + c_{12}c_{23}c_{31} + c_{13}c_{21}c_{32}$$

$$- c_{31}c_{22}c_{13} - c_{32}c_{23}c_{11} - c_{33}c_{21}c_{12}$$

In general, higher order determinants are found by the use of *minors*. Each position in a matrix has a minor. The minor of a position is the determinant of one less dimension created by eliminating the row and column of the position. Thus, the minor of position 11 in the matrix $\mathbf{C}$ above is obtained by eliminating row 1 and column 1 from $\mathbf{C}$. This 2×2 minor is indicated by the symbol m with the subscripts of the position. For example, m_{32} is the minor of position row-3 and column-2.

$$m_{11} = \begin{vmatrix} c_{22} & c_{23} \\ c_{32} & c_{33} \end{vmatrix}$$

$$m_{32} = + \begin{vmatrix} c_{11} & c_{13} \\ c_{21} & c_{23} \end{vmatrix}$$

$$\text{cofactor}_{ij} = (-1)^{i+j} \text{minor}_{ij}$$

The rigorous method of calculating a determinant involves expanding the matrix along any row or column. The determinant is found by summing the products of elements in any row or column and each element's cofactor, thus, assigning the sign according to the position. Det $\mathbf{C}$ can be obtained by either of the following:

$$\det \mathbf{C} = c_{11}m_{11} - c_{12}m_{12} + c_{13}m_{13}$$

$$= -c_{12}m_{12} + c_{22}m_{22} - c_{32}m_{32}$$

Any other expansion along other rows or columns will also work. The usual method is to expand along the first row or first column. The sign associated with each expansion term comes from the position—positive when the subscripts add to an even number, and negative when they add to an odd number.

Determinants of larger matrixes require multiple levels of expansion until the minors are reduced to two or three dimensions. Then, the methods given for calculating the determinants can be used. For example, the determinant of matrix $\mathbf{A}$ given previously can be found by expanding along the first column.

$$\det \mathbf{A} = a_{11}m_{11} - a_{21}m_{21}$$

$$+ a_{31}m_{31} - a_{41}m_{41}$$

$$\det \mathbf{A} = +a_{11} \begin{vmatrix} a_{22} & a_{23} & a_{24} \\ a_{32} & a_{33} & a_{34} \\ a_{42} & a_{43} & a_{44} \end{vmatrix} - a_{21} \begin{vmatrix} a_{12} & a_{13} & a_{14} \\ a_{32} & a_{33} & a_{34} \\ a_{42} & a_{43} & a_{44} \end{vmatrix}$$

$$+ a_{31} \begin{vmatrix} a_{12} & a_{13} & a_{14} \\ a_{22} & a_{23} & a_{24} \\ a_{42} & a_{43} & a_{44} \end{vmatrix} - a_{41} \begin{vmatrix} a_{12} & a_{13} & a_{14} \\ a_{22} & a_{23} & a_{24} \\ a_{32} & a_{33} & a_{34} \end{vmatrix}$$

The resulting 3×3 determinants can be obtained by the methods given above. For larger matrixes, the process becomes sufficiently involved as to warrant the use of a computer.

15 LINEAR SIMULTANEOUS EQUATIONS

System analysis involving linear or piecewise linear elements only will result in simultaneous linear equations with constant coefficients. For example, a set of four equations in four unknowns might be

$$f_1 = a_{11}x_1 + a_{12}x_2 + a_{13}x_3 + a_{14}x_4$$

$$f_2 = a_{21}x_1 + a_{22}x_2 + a_{23}x_3 + a_{24}x_4$$

$$f_3 = a_{31}x_1 + a_{32}x_2 + a_{33}x_3 + a_{34}x_4$$

$$f_4 = a_{41}x_1 + a_{42}x_2 + a_{43}x_3 + a_{44}x_4$$

The f's are known functions, and the x's are variables. This is set into a matrix equation as follows: The f's are formed into a column matrix on the left side of the equation. The a's are extracted into a 4×4 matrix which multiplies a column matrix of the x's.

$$\begin{bmatrix} f_1 \\ f_2 \\ f_3 \\ f_4 \end{bmatrix} = \begin{bmatrix} a_{11} & a_{12} & a_{13} & a_{14} \\ a_{21} & a_{22} & a_{23} & a_{24} \\ a_{31} & a_{32} & a_{33} & a_{34} \\ a_{41} & a_{42} & a_{43} & a_{44} \end{bmatrix} \begin{bmatrix} x_1 \\ x_2 \\ x_3 \\ x_4 \end{bmatrix}$$

The solution for any of the variable (x's) is obtained by *Cramer's rule*. This rule uses determinants as follows:

$$x_j = \frac{f_1 m_{1j} - f_2 m_{2j} + f_3 m_{3j} - f_4 m_{4j}}{\det \mathbf{A}}$$

The determinants of the minors take on the positional signs, and $\mathbf{A}$ is the 4×4 matrix of the a terms. This is illustrated with a numerical example involving a 3×3 system of simultaneous equations.

Example 1.24

A system is described by the set of equations given. Determine the values of the variables x, y, and z which simultaneously satisfy the three equations.

$$10 = 3x + y + 2z$$
$$5t = x - 2y + z$$
$$\sin t = -x + y - z$$

These are placed in matrix form

$$\begin{bmatrix} 10 \\ 5t \\ \sin t \end{bmatrix} = \begin{bmatrix} 3 & 1 & 2 \\ 1 & -2 & 1 \\ -1 & 1 & -1 \end{bmatrix} \begin{bmatrix} x \\ y \\ z \end{bmatrix}$$

Cramer's rule is used to find the simultaneous solution values.

$$x = \frac{+10\begin{vmatrix} -2 & 1 \\ 1 & -1 \end{vmatrix} - 5t\begin{vmatrix} 1 & 2 \\ 1 & -1 \end{vmatrix} + \sin t\begin{vmatrix} 1 & 2 \\ -2 & 1 \end{vmatrix}}{+3\begin{vmatrix} -2 & 1 \\ 1 & -1 \end{vmatrix} - 1\begin{vmatrix} 1 & 2 \\ 1 & -1 \end{vmatrix} + (-1)\begin{vmatrix} 1 & 2 \\ -2 & 1 \end{vmatrix}}$$

$$= \frac{10(2-1) - 5t(-1-2) + \sin t(1+4)}{3(2-1) - 1(-1-2) + (-1)(1+4)}$$

$$= \frac{10 + 15t + 5\sin t}{3 + 3 - 5} = 10 + 15t + 5\sin t$$

$$y = \frac{-10\begin{vmatrix} 1 & 1 \\ -1 & -1 \end{vmatrix} + 5t\begin{vmatrix} 3 & 2 \\ -1 & -1 \end{vmatrix} - \sin t\begin{vmatrix} 3 & 2 \\ 1 & 1 \end{vmatrix}}{\det \mathbf{A}}$$

$$= \frac{-10(-1+1) + 5t(-3+2) - \sin t(3-2)}{1}$$

$$= 0 - 5t - \sin t$$

$$z = \frac{+10\begin{vmatrix} 1 & -2 \\ -1 & 1 \end{vmatrix} - 5t\begin{vmatrix} 3 & 1 \\ -1 & 1 \end{vmatrix} + \sin t\begin{vmatrix} 3 & 1 \\ 1 & -2 \end{vmatrix}}{1}$$

$$= \frac{+10(1-2) - 5t(3+1) + \sin t(-6-1)}{1}$$

$$= -10 - 20t - 7\sin t$$

APPENDIX A

SI Prefixes

prefix		symbol
exa	10^{18}	E
peta	10^{15}	P
tera	10^{12}	T
giga	10^{9}	G
mega	10^{6}	M
kilo	10^{3}	k
hecto	10^{2}	h
deka	10^{1}	da
deci	10^{-1}	d
centi	10^{-2}	c
milli	10^{-3}	m
micro	10^{-6}	μ
nano	10^{-9}	n
pico	10^{-12}	p
femto	10^{-15}	f
atto	10^{-18}	a

PROFESSIONAL PUBLICATIONS ● Belmont, CA

PRACTICE PROBLEMS

1. Determine the decimal value of the binary number 1101.011.

2. Given $A = 4 + j5$ and $B = 2 - j3$, obtain the product $A \cdot B^*$.

3. Given $A = 4 + j5$, obtain the product $A \cdot A^*$ using (a) polar and (b) rectangular operations.

4. Given $B = 2 - j3$, find the reciprocal B^{-1}.

5. Given $A = 4 + j5$ and $B = 2 - j3$, find the ratio $\frac{A}{B}$.

6. Given $A = 6.4e^{j0.896}$ and $B = 3.6e^{-j0.983}$ find the ratio A/B and compare your answer to problem 5.

7. Obtain the partial fraction expansion of

$$\frac{1}{x(x+1)}$$

8. Obtain the partial fraction expansion of

$$\frac{4x^2 + 5x + 2}{x^2(x+1)}$$

9. Obtain the partial fraction expansion of

$$\frac{5x^2 + 11x + 20}{x(x^2 + 2x + 5)}$$

10. Use long division to reduce the degree of the numerator order for

$$\frac{s^3 + 4s^2 + 6s + 2}{(s+1)^2}$$

11. Use trigonometric identities to obtain expressions for (a) $\cos \theta/2$, and (b) $\sin \theta/2$.

12. Three curves are shown. Obtain straight-line approximations for them over the specified intervals.

- Curve (a) over the interval of $1 < I < 10$.

- Curve (b) over the intervals $0 < I < 4.5$ and $5.5 < I < 10$.

- Curve (c) over the intervals $0 < I < 2$ and $3 < I < 4.5$.

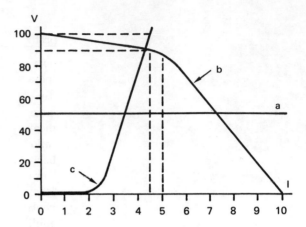

13. Find the local maximum or minimum of

$$u = 2 \sin x + 3 \cos x$$

in the region of $-90° < x < 90°$. Determine whether it is a minimum or a maximum.

14. Determine the average value of the function

$$e^{-2x} \cos \pi x$$

over the interval $x = 0$ to $x = 1$.

15. A voltage is described by $0.01 \, dv/dt + 0.0025v = 0$. $v = 100$ at $t = 0$. Determine v for $t > 0$.

16. Given a current described by

$$\frac{d^2 i}{dt^2} + 5\frac{di}{dt} + 4i = 2$$

with $i = 1.7$ and $di/dt = 0$ at $t = 0$, determine the current for $t > 0$.

17. Given

$$\frac{d^2 i}{dt^2} + 2\frac{di}{dt} + i = 0.25$$

and conditions at $t = 0$ of $i = 0.8$ and $di/dt = 0$, determine i for $t > 0$.

18. Given

$$\frac{d^2 v}{dt^2} + 3\frac{dv}{dt} + 6.25v + 12.5 = 0$$

and conditions at $t = 0$ of $v = 10$ and $dv/dt = 0$, determine v for $t > 0$.

19. A rectangular wave has voltage levels of 0 and 5 volts, equal times at both levels, and a period of 10 seconds (i.e., 5 seconds at 5 volts and 5 seconds at 0 volts). Obtain a Fourier series for this wave, taking $t = 0$ as the instant that the voltage rises from the zero level to the 5 volt level.

20. Obtain the Laplace transformed equation for the differential equation

$$0.02 = 10^{-6}\frac{d^2i}{dt^2} + 2 \times 10^{-4}\frac{di}{dt} + i$$

$$i(0) = 0.5$$

$$\frac{di}{dt}(0) = 50$$

21. From the answer to problem 20, obtain $I(s)$ as a function of s.

22. Using partial fraction expansion, separate the function $I(s)$ of problem 21 into its component parts.

23. Using table 1.1, obtain $i(t)$ from your solution of problem 22.

24. Referring to example 1.21, verify that the finite Fourier series of the example matches the sample values by substituting values of p corresponding to the samples, from 0 through 7.

25. A voltage waveform is sampled in equal time increments over a period of 10 seconds. The resulting samples are: 5,5,5,5,0,0,0,0. Obtain the discrete Fourier series for this voltage and compare the results with those of problem 19. Note, with eight samples $N = 4$ and the sampling frequency is $8/T$. Thus, the Nyquist frequency is N/T or 0.4, so the guidelines of example 1.22 should be observed. Hint: Use Euler's equation to compare these coefficients with the first few of problem 19.

26. A particular finite Fourier transform will employ 4096 sampled values. Compare the computation time in terms of the number of multiplications necessary using a "brute force" method which utilizes the complex conjugates versus a fast Fourier transform algorithm such as that illustrated in section 12.

27. A particular signal waveform (a ramp) is zero before $t = 0$, rises linearly to a value of 5 at $t = 10$ seconds, and remains at 5 for the foreseeable future. (a) Using direct integration determine its Laplace transform. (b) Using direct integration determine its Fourier transform.

28. Refer to the waveform of problem 27 and recognize that this waveform is the integral of a pulse. Using identities, theorems, and transform pairs, obtain (a) its Laplace transform, and (b) its Fourier transform.

29. The signal of problem 27 is applied to a system which has an impulse response $u(t)e^{-0.2t}\sin 0.2t$. Obtain the output response $y(t)$ using Laplace transform methods.

30. The signal of problem 27 is applied to a system which has an impulse response $u(t)e^{-0.2t}\sin 0.2t$. Obtain the output response $y(t)$ using Fourier transform methods.

2 LINEAR CIRCUIT ANALYSIS

Nomenclature

A	area or chain parameter	meters^2,–
B	chain parameter	–
B	magnetic flux density	teslas
C	capacitance or chain parameter	farads,–
C_{ab}	Seebeck coefficient	–
D	chain parameter	–
E	electric field intensity	volts/meter
F	force	newtons
g	hybrid parameter	–
h	hybrid parameter	–
i	instantaneous current	amps
I	electric current magnitude	amps
k	magnetic coupling coefficient, or constant multiplier	–
l	length	meters
L	inductance	henrys
M	mutual inductance	henrys
n	charge carrier density	meters^{-3}
N	number of turns	–
q	charge	coulombs
r	small signal resistance parameter	ohms
R	resistance	ohms
s	Laplace transform variable	–
t	time	seconds
T	temperature	degrees
v	small signal voltage	volts
v	velocity	meters/second
V	voltage magnitude	volts
w	width	meters
W	energy	joules
y	small signal admittance parameter	mhos
Y	admittance	mhos
z	small signal impedance parameter	ohms
Z	impedance	ohms

Symbols

α	common base transistor current gain	–
α_{20}	temperature coefficient of resistivity	1/°C
β	common emitter transistor current gain	–
ϵ	permittivity	farads/meter
μ	permeability	henrys/meter
ω	radian frequency	radians/second
ρ	resistivity	ohm-meters
ϕ	flux	webers

Subscripts

b	transistor base or common base
c	transistor collector, common collector, or cold
C	capacitance
e	transistor emitter or common emitter
f	forward
h	hot
i	input
L	inductance, load
m	maximum
o	output
r	reverse
R	resistance
s	source
Th	Thevenin
x	x-component
y	y-component
z	z-component

PROFESSIONAL PUBLICATIONS ● Belmont, CA

1　REVIEW OF FUNDAMENTALS

Circuit analysis is the fundamental method of electrical engineering, both for understanding existing electrical systems and for designing new ones. A competence in these methods is an implicit prerequisite for all the following material in this book, with the exception of the chapter on Engineering Economics. Circuit analysis is based on Kirchhoff's two circuit laws and the voltage-current relationships of the various circuit elements. Only ideal resistors and ideal sources for circuit elements are considered here, in order to stress the fundamental notions.

A.　Ideal Resistors

Resistance is modeled by an ideal linear element which is called an *ideal resistor*. The voltage across an ideal resistor (hereafter simply called resistor) is a constant times the current flowing through the resistor.

$$v = Ri \qquad 2.1$$

v is the voltage in units of volts, i is the current in units of amperes, and R is the resistance of the resistor in units of ohms.

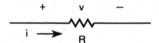

Figure 2.1　Circuit Symbol for a Resistor, Showing Sink Notation Polarity to Correspond with Equation 2.1

Equation 2.1 is valid only when the polarity of the voltage drop is assigned in the direction of the current flow. This is illustrated in figure 2.1. Assigning the voltage drop polarity to correspond with the direction of current flow is sometimes called *sink* notation, as it corresponds to a positive power flow into the element. In general, all passive circuit elements are to have their voltage polarity assigned in the direction of assigned current flow, i.e., in sink notation.

B.　Ideal Independent Voltage Sources

An (ideal) independent voltage source maintains its voltage independent of the current flowing through it. This is one of the source models to be used. In general, the independent voltage is a function of time, although it may be constant. The current-voltage relationship is

$$v_s = v_s(t) \qquad 2.2$$

The subscript s is used generically to indicate an independent source.

The assigned current direction through an independent voltage source is arbitrary, but it is usually assigned in the direction that corresponds to a positive power flow out of the source, which is called source notation for polarity. This is illustrated in figure 2.2.

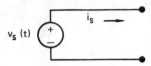

Figure 2.2　Circuit Symbol for an Ideal Voltage Source Showing Source Notation for Voltage Polarity and Current Direction

C.　Independent Current Sources

An (ideal) independent current source maintains the current flow independent of the voltage across it. In general, the independent current is a function of time, although it may be constant. The current-voltage relationship is

$$i_s = i_s(t) \qquad 2.3$$

The assigned voltage polarity across an independent current source is usually assigned in source notation as illustrated in figure 2.3.

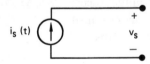

Figure 2.3　Circuit Symbol for an Ideal Current Source with Voltage Polarity Assigned in Source Notation

D.　Controlled or Dependent Voltage Sources

A controlled voltage source symbol is as shown in figure 2.4. Its controlling variable may be a voltage, current, or other physical quantity such as light intensity, temperature, velocity, etc. Controlled sources are often called *dependent sources*. The controlled voltage source of 2.4(a) is controlled by a voltage, v_a, elsewhere, and μ is a constant. The voltage source of 2.4(b) is controlled by a current, i_a, elsewhere, and r_m is a constant.

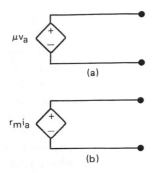

Figure 2.4 Circuit Symbol for Controlled Voltage Sources: (a) Voltage Controlled Voltage Source, (b) Current Controlled Voltage Source

E. Controlled or Dependent Current Sources

A controlled current source symbol is as shown in figure 2.5. Its controlling variable may be a voltage, current, or other physical quantity such as light intensity, temperature, velocity, etc. The controlled current source of 2.5(a) is controlled by a voltage, v_a, elsewhere, and g_m is a constant. The current source of 2.5(b) is controlled by a current, i_a, elsewhere, and β is a constant.

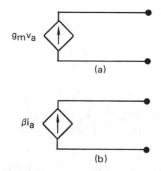

Figure 2.5 Circuit Symbol for Controlled Current Sources: (a) Voltage Controlled Current Source, (b) Current Controlled Current Source

2   LINEAR CIRCUIT ELEMENTS

The basic circuit elements are resistance, capacitance, inductance, mutual inductance, and ideal transformers.

A. Resistance

Resistance is the opposition of current flow in a conductor. It is the ratio of voltage to current for constant voltage and current. Resistance of a conductor depends on the *resistivity*, ρ, of the conductor material. However, the resistivity is a function of temperature and has

been found to vary approximately linearly with temperature. Resistivity is usually referenced to 20°C and is written as ρ_{20}. ρ_{20} has values of 1.8×10^{-8} ohm-meters for copper and 1.08×10^{-6} ohm-meters for nichrome. The *temperature coefficient of resistivity* is also measured at 20°C and is given the name α_{20}. It has values of 3.9×10^{-3} per degree for copper and 17×10^{-5} per degree for nichrome. The expression for the resistivity at a temperature T other than 20°C is

$$\rho_T = \rho_{20}[1 + \alpha_{20}(T - 20)] \qquad 2.4$$

The resistance of a conductor of length l and cross-sectional area A is

$$R = \frac{\rho_T \times l}{A} \qquad 2.5$$

Ohm's law relates the voltage and current in a circuit to the circuit resistance.

$$V = IR \qquad 2.6$$

The *power* dissipated by a resistor is

$$P = I^2 R = \frac{V^2}{R} \qquad 2.7$$

Resistances in series add linearly, and resistances in parallel add inversely. This second property gives rise to the use of the parameter *conductance* which is the reciprocal of resistance. Conductances in parallel add linearly.

$$R_{\text{series}} = R_1 + R_2 + R_3 + \cdots \qquad 2.8$$

$$\frac{1}{R_{\text{parallel}}} = \frac{1}{R_1} + \frac{1}{R_2} + \frac{1}{R_3} + \cdots \qquad 2.9$$

$$G = \frac{1}{R} \qquad 2.10$$

$$G_{\text{parallel}} = G_1 + G_2 + G_3 + \cdots \qquad 2.11$$

Example 2.1

A resistor constructed from AWG (American Wire Gage) #30 copper wire is to have a resistance of one ohm at 100°C. Determine the length of wire necessary to obtain this resistance. What will be the resistance of the wire at 20°?

From Appendix A, in chapter 6, AWG #30 wire has a cross-sectional area of 100 circular mils. *Circular measure* is the diameter squared, so the diameter is 10 mils

(0.01 inches). In SI units, the diameter is 2.54×10^{-4} meters and the cross-sectional area is $5.07 \times 10^{-8} \text{m}^2$.

$$\rho_{100} = 1.8 \times 10^{-8}[1 + 3.9 \times 10^{-3}(100 - 20)]$$

$$= 2.36 \times 10^{-8} \text{ohm-m}$$

$$l = \frac{R \times A}{\rho_{100}} = \frac{1.0 \times 5.07 \times 10^{-8}}{2.36 \times 10^{-8}}$$

$$= 2.15 \text{ m}$$

$$R_{20} = R_{100} \times \frac{\rho_{20}}{\rho_{100}} = 1 \times \frac{1.8}{2.36}$$

$$= 0.763 \text{ ohms}$$

B. Capacitance

Capacitance is the ability to store an electrical charge. It is calculated for parallel plates as the *permittivity* (ϵ) multiplied by the surface area of the plates and divided by the plate separation. The permittivity is the product of the *permittivity of free space* (ϵ_0) and the dielectric constant (ϵ_r) of the material separating the parallel plates. The permittivity of free space has a value of $1/(36\pi \times 10^9)$. The dielectric constant for linear materials varies between 1 and 10.

$$\epsilon = \epsilon_0 \epsilon_r \qquad 2.12$$

The current into the positive terminal of a capacitor is the time rate of increase in stored charge. If the current direction is chosen as positive when flowing into the positive terminal of a capacitor as shown in figure 2.6, the voltage-current relationship for a capacitor is

$$i = C\frac{dv_C}{dt} \qquad 2.13$$

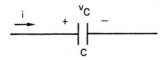

Figure 2.6 Voltage Polarity and Current Directions for Capacitors

The energy stored in a capacitor is

$$W = \frac{1}{2}Cv_C^2 \qquad 2.14$$

As energy cannot change instantaneously, the voltage cannot change instantaneously either. This fact is of major importance in determining the transient behavior of circuits containing capacitors.

Example 2.2

A voltage source is connected to a parallel combination of a resistance and a capacitance as shown. Assume that the switch has been open for a long time so that the capacitor has no charge on it prior to $t = 0$ when the switch is closed. Determine (a) the current through the capacitor at the instant the switch is closed, (b) the time derivative of the capacitor voltage at the instant the switch is closed, and (c) the voltage that will exist across the capacitor after a long time.

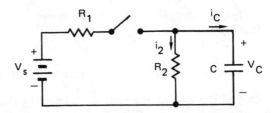

As the capacitor is uncharged at the instant before the switch closes ($t = 0^-$), it must also be uncharged at the instant after the switch closes ($t = 0^+$). Thus, the voltage across the capacitance remains zero. This causes the capacitor to act as a short circuit across R_2. All of the current flows initially through the capacitor instantaneously. This current is

$$i_C(0^+) = \frac{V_s}{R_1}$$

Since $i = C\frac{dv}{dt}$,

$$\frac{dv(0^+)}{dt} = \frac{V_s}{R_1 C}$$

The capacitance will become fully charged after a long time, and the voltage across it will no longer be changing. Thus, dv/dt and i_C are zero. All the current flows through R_2, and the capacitor acts as an open circuit to the current. The current through R_2 is

$$\frac{V_s}{R_1 + R_2}$$

The voltage across the capacitor is the same as across R_2.

$$V_C(\infty) = i_2 R_2 = V_s \frac{R_2}{R_1 + R_2}$$

C. Inductance

Inductance is a property which resists changes in current. Inductance is measured in units of *henrys*. For a closed magnetic circuit such as the toroid shown in figure 2.7, the inductance is

$$L = \mu \frac{N^2 A}{l} \qquad 2.15$$

μ is the *permeability* of the medium, N is the number of turns on the winding, A is the core cross-section area, and l is the average flux path length.

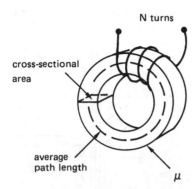

Figure 2.7   Inductance of a Toroid with N Turns

The general voltage-current relationship is

$$V = L \frac{di}{dt} \qquad 2.16$$

The energy stored in an inductance is

$$W = \frac{Li^2}{2} \qquad 2.17$$

Energy is unable to change instantaneously, consequently, current is also unable to change instantaneously. This concept is used to determine the transient behavior of circuits containing inductors.

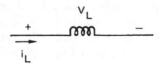

Figure 2.8   Defined Voltage and Current Directions for an Inductance

Example 2.3

A voltage source is connected to a parallel combination of resistance and inductance by a switch which closes at $t = 0$. Assume that the switch has been open for a long time so no energy is stored in the inductance just prior to the closing of the switch. Determine (a) the inductance current, (b) the inductance voltage, and (c) the inductance time rate of change of current immediately after the switch closing. Determine (d) the current in the inductance and (e) the voltage across the inductance after a long period of time.

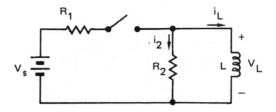

As there is no stored energy just prior to the switch closing, there is no current flowing at that time, nor is any current flowing at the instant the switch is closed. Therefore,

$$i_L(0^+) = 0$$

The inductance acts as an open circuit at the instant the switch closes, so the current is all shunted through R_2. The current is then

$$\frac{V_s}{R_1 + R_2}$$

The voltage across R_2 is the current times its resistance. The same voltage appears across the inductance.

$$v_L(0^+) = V_s \frac{R_2}{R_1 + R_2}$$

As $v_L = L(di/dt)$, the rate of current change in the inductor is

$$\frac{di_L(0^+)}{dt} = \left(\frac{V_s}{L} \right) \left(\frac{R_2}{R_1 + R_2} \right)$$

After a long time the current will settle down to a steady value, so di_L/dt is zero. There will be no voltage drop across the inductance, and the inductance will behave as a short circuit across R_2. All the current will then flow through the inductance.

$$i_L(\infty) = \frac{V_s}{R_1}$$
$$v_L(\infty) = 0$$

Mutual inductance occurs when two or more coils share a magnetic flux. This is the case when two or more wires are wound on the same magnetic core. The reference direction of the flux is arbitrarily chosen, and each winding is *phased* by determining the current direction which will cause flux in the reference direction. The reference direction of the flux is related to a coil current by the right-hand rule: Extend the right-hand thumb in the flux reference direction, then curl the fingers which point in the direction of current flow causing flux in the reference direction. A symbol (typically a dot) is placed at the end of the coil where the current causing flux in the reference direction enters. This is done on each coil.

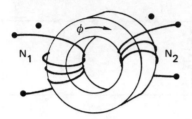

Figure 2.9 Mutually Coupled Coils Having Mutual Inductance

If the device is shown in a schematic drawing as in figure 2.9, the current and voltage directions are given references according to the dots: positive currents go into the dotted end of each coil and positive voltages appear at the dotted terminal. Following the defined directions strictly, the voltage-current relationships are:

$$v_1 = L_1 \frac{di_1}{dt} + M \frac{di_2}{dt} \qquad 2.18$$

$$v_2 = M \frac{di_1}{dt} + L_2 \frac{di_2}{dt} \qquad 2.19$$

In a system involving mutual inductance, M, the total stored energy is

$$W = \frac{1}{2}L_1 i_1^2 + \frac{1}{2}L_2 i_2^2 + M i_1 i_2 \qquad 2.20$$

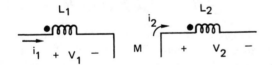

Figure 2.10 Circuit Symbols and Directions for Mutually Coupled Coils

The energy cannot change instantaneously, but it is permissable for both currents to change abruptly. The mutual inductance is always less than the geometric mean of the two self-inductances, L_1 and L_2, because the mutual flux is always less than the total flux. The ratio of the mutual inductance to the geometric mean is called the *coupling coefficient*, k.

$$k = \frac{M}{\sqrt{L_1 L_2}} \qquad 2.21$$

An *ideal transformer* is a simplified model of coupled coils, assuming $k = 1$ and a high enough permeability

to make the inductances insignificant in the circuit calculations. All flux from one coil is assumed to link all the turns of the other coil so that the voltage per turn of each coil is identical at each instant in time.

$$\frac{v_1}{N_1} = \frac{v_2}{N_2} = \frac{v_3}{N_3} = \cdots \qquad 2.22$$

Also, the algebraic sum of the *magnetomotive force* terms is zero. (Positive currents are defined by the dots.)

$$N_1 i_1 + N_2 i_2 + N_3 i_3 + \cdots = 0 \qquad 2.23$$

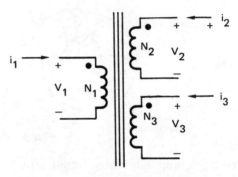

Figure 2.11 Circuit Symbols and Directions for an Ideal Transformer

An ideal transformer is lossless and stores no energy. Its functions are to change voltage levels and match impedances in circuits, as well as to provide equal and opposite waveforms in some applications. The ideal transformer model adequately describes the behavior of a transformer in many circuits. This is not true in power circuit problems where efficiency, regulation, or power factor are being determined.

Example 2.4

A transformer consists of a primary winding with 500 turns and two secondary windings of 125 turns and 36 turns. The 125 turn secondary winding has 60 ohms connected to its terminals, and the 36 turn secondary winding has 3 ohms connected to its terminals. If the primary winding is connected to a 120 volt, 60 Hz source, determine the current rating for each winding.

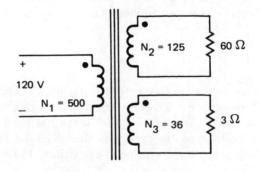

From equation 2.22, the secondary voltages are:

$$v_2 = \frac{N_2 v_1}{N_1} = \frac{125 \times 120}{500}$$
$$= 30 \text{ V}$$
$$v_3 = \frac{N_3 v_1}{N_1} = \frac{36 \times 120}{500}$$
$$= 8.64 \text{ V}$$

The currents in the secondaries can then be found from the resistances.

$$i_2 = \frac{-v_2}{R_2} = \frac{-30}{60}$$
$$= -0.5 \text{ A}$$
$$i_3 = \frac{-v_3}{R_3} = \frac{-8.64}{3}$$
$$= -2.88 \text{ A}$$

The negative signs on the currents are necessary because of the defined current directions (i.e., into the dotted ends of the coils).

From equation 2.23, the primary current is

$$N_1 i_1 + N_2 i_2 + N_3 i_3 = (500 \times i_1)$$
$$- (125 \times 0.5) - (36 \times 2.88)$$
$$= 0$$
$$i_1 = 0.332 \text{ A}$$

The minimum current ratings are:

332 mA for winding #1

500 mA for winding #2

2.88 A for winding #3

3 LINEAR SOURCE MODELS

Electrical power sources are typically nonlinear, but for purposes of circuit analysis they are modeled as linear devices. The model includes a constant or ideal voltage or current source, in combination with a linear impedance. Such sources are classified as *independent sources* to distinguish them from the *dependent sources* of active devices which are not actually sources of power.

A third source type is the *transducer* which produces a voltage or current signal dependent on non-electric quantities. *Thermocouples* are transducers that produce voltages as functions of temperature; *photodiodes* are transducers that provide currents according to the level of illumination; *microphones* are transducers that provide voltages as functions of sound intensity.

Independent sources may be modeled by a linear approximation to measured values. For example, a voltage-current characteristic curve for a direct current source (e.g., a battery or generator) is shown in example 2.5. A straight line approximation to this characteristic curve can be obtained by analytic geometry.

Example 2.5

Obtain an expression for the characteristic curve shown over the range from 8 to 12 amps. Obtain linear circuit models from the derived expressions.

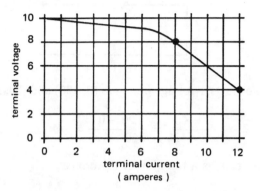

The two-point method is used by taking the pairs (8,8) and (4,12).

$$\frac{i-8}{v-8} = \frac{12-8}{4-8} = -1$$

Either voltage or current can be chosen as the dependent variable

$$v = -i + 16 \qquad (A)$$
$$i = -v + 16 \qquad (B)$$

Equation A is a loop voltage equation where the terminal voltage v is the sum of two terms, one independent of the current and the other dependent on the current. The dependent term is modeled as a 1 ohm resistance in series with the 16 volt source. Equation B is a node current equation where the terminal current, i, is the difference in two terms. One term is constant at 16 amps and is modeled as a 16 amp source. The other is dependent on the terminal voltage, v, and is modeled as a shunt resistance of 1 ohm across the terminals.

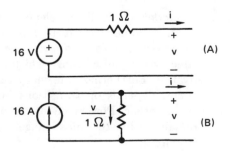

Source characteristics are sometimes given in terms of *open-circuit* or *no-load* voltage, and the current with some known load resistance or impedance. These two parameters can be used to establish a linear model.

Example 2.6

A DC generator has a no-load voltage of 125 volts and a full-load voltage of 120 volts at its rated current of 20 amps. Obtain a linear circuit model for the generator.

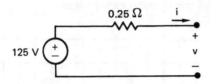

The two current-voltage pairs are (0, 125) and (20, 120).

$$\frac{i-0}{v-125} = \frac{20-0}{120-125} = -4$$

Solving for v as a function of i produces

$$v = 125 - 0.25i$$

The generator can be modeled as a voltage source of 125 volts in series with a resistance of 0.25 ohms.

The percentage change of voltage from the no-load to full-load condition is called the *voltage regulation*. This is a generator parameter often given together with full-load voltage and current.

$$\text{voltage regulation} = \frac{v_{\text{no-load}} - v_{\text{full-load}}}{v_{\text{full-load}}} \qquad 2.24$$

Example 2.7

Find the percentage regulation for the source of example 2.6.
$$\text{regulation} = \frac{125-120}{120} \times 100 = 4.17\%$$

Piecewise linear models of nonlinear elements can be obtained by a method similar to that used in obtaining linear models of power sources. However, the results are only applicable over a limited range which may not be known in advance. Furthermore, in many cases the working of a circuit depends on transitions from one nonlinear region to another. Power sources, on the other hand, usually remain within the linear range described by the equivalent circuit.

Example 2.8

A typical silicon diode is approximated by a *forward voltage drop* of 0.6 volts at negligible current and 1.5 to 2 volts at the rated current. Obtain an equivalent circuit for a silicon diode rated 10 watts at 5 amps.

Power is $V_{\text{rated}} \times I_{\text{rated}}$, so the voltage at rated current is $V_{\text{rated}} = 10/5 = 2$ volts. Then the two (v, i) pairs are $(0.6, 0)$ and $(2, 5)$. The two-point equation is

$$\frac{v-0.6}{i-0} = \frac{2-0.6}{5-0} = 0.28$$

Solving for v yields

$$v = 0.6 + 0.28i$$

The equivalent circuit is shown. This model is valid only when the current is positive and the voltage exceeds 0.6 volts.

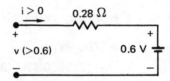

A *zener diode* is typically a silicon diode operated in the reverse direction. It essentially operates as an open circuit for low reverse voltages (defined as the positive direction) and has a voltage known as the *zener* or *avalanche voltage* at which the current abruptly increases when further voltage is applied. Zener diode characteristics may be specified graphically, but are more frequently specified by the breakdown (zener) voltage, and by the related current and power.

Example 2.9

Obtain an equivalent circuit for a 6-volt zener diode with a rated current of 0.75 amps and rated power of 5 watts.

Taking the *reverse voltage* and *reverse current* as positive, the voltage at rated power is

$$V_{\text{rated}} = \frac{5}{0.75} = 6.67 \text{ volts}$$

The two (v, i) pairs are then (6, 0) and (6.67, 0.75). The two-point method then yields

$$\frac{v-6}{i-0} = \frac{6.67-6}{0.75-0} = 0.893$$

The v-i equation is

$$v = 6 + 0.893i$$

The equivalent circuit is shown.

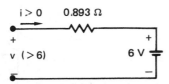

It should be noted that the equivalent circuits for non-linear circuit elements are different from circuits for power sources in the assigned current directions and are, therefore, drawn in a different manner.

Dependent or *controlled sources* which depend on circuit quantities (e.g., voltage or current) are necessary in the analysis of linear *active circuits* such as transistor amplifiers. Controlled sources typically have three or more terminals. An example is the two-winding transformer modeled by equations 2.18 and 2.19. These equations can be treated as two loop equations, which then produce an equivalent circuit as shown in figure 2.12. The voltage generated in loop 1 is due to the current in loop 2, and the voltage generated in loop 2 is due to the current in loop 1. Such a circuit is said to be *bi-lateral*.

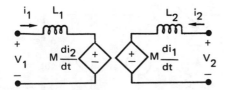

Figure 2.12 Equivalent Circuit Model for a Mutually Coupled Circuit Using Controlled Sources

It is unusual to represent circuits with controlled sources as shown in figure 2.12. It is simpler to represent them as coupled circuits with the coupling included as common circuit elements. The most common instance is of *unilateral devices* (e.g., transistors) in which the collector circuit only mildly affects the base circuit, while the behavior of the collector circuit is principally controlled by the base circuit. In these cases, the circuit can be described by a linear relationship in matrix form. The matrix terms can be impedance, admittance, or mixed parameters.

A *three-terminal device* can be described by the matrix equation 2.25. The voltages are measured with respect to the third terminal.

$$\begin{bmatrix} v_1 \\ v_2 \end{bmatrix} = \begin{bmatrix} z_i & z_r \\ z_f & z_o \end{bmatrix} \times \begin{bmatrix} i_1 \\ i_2 \end{bmatrix} \qquad 2.25$$

The subscripts on the *z-parameters* stand for *input, reverse, forward,* and *output*. The convention is that terminal 1 is the input and terminal 2 is the output. Equation 2.25 is a voltage relationship. It can be implemented in two series circuits with an impedance and a current controlled voltage source as shown in figure 2.13.

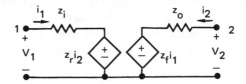

Figure 2.13 Equivalent Circuit Voltage of Controlled Sources

The *z-parameters* result in current controlled voltage sources in series with impedances. Equation 2.25 can also be written in terms of currents, which then results in voltage controlled current sources in parallel with admittances. This is the *y-parameter* formulation.

Example 2.10

A three-terminal device has the following *y-parameters*: $y_i = 0.02$, $y_r = -0.0002$, $y_f = 0.95$, $y_0 = 0.0001$. Obtain an equivalent circuit for this device.

The *y-parameter* equations can be written as equation 2.25. Voltages and currents are interchanged, and the *y-parameters* are substituted for the *z-parameters* with the same subscripts.

$$i_1 = 0.02v_1 - 0.0002v_2$$
$$i_2 = 0.95v_1 + 0.0001v_2$$

These are node current equations. Each equation defines the connections to one node, and the voltages are measured with respect to the reference node. The equivalent circuit is shown. Admittances are in *mhos*.

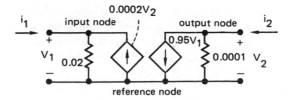

Junction transistors are usually specified by a mixed set of small-signal parameters. These are called *hybrid parameters* because the equivalent circuit consists of a series input circuit and a parallel output circuit. Since the hybrid parameters are usually associated with a common emitter configuration, they carry an additional subscript, e. The parameters are h_{ie}, h_{re}, h_{fe}, and h_{oe}. h_{re} is typically negligible and is ignored. The useful equations are:

$$v_{be} = h_{ie}i_b \qquad 2.26$$

$$i_c = h_{fe}i_b + h_{oe}v_{ce} \qquad 2.27$$

The equivalent circuit corresponding to equations 2.26 and 2.27 is shown in figure 2.14. This circuit is the most useful of all transistor equivalent circuits. It will be used extensively in the chapter covering linear electronics.

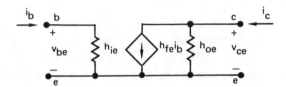

Figure 2.14 Hybrid Equivalent Circuit for a Junction Transistor in the Common Emitter Configuration

Transducers can usually be treated as controlled sources, with the controlling variable being the non-electric quantity converted to an electric signal. Such devices have an output circuit similar to those of three-terminal devices, but a circuit is not usually included in the input.

A silicon *photodiode* is constructed so that the PN junction is exposed to visible light. Typical characteristic curves of the photoelectric effect are shown in figure 2.15. The curves show that with the diode reverse biased, the current will be roughly proportional to the illumination. The current is approximately constant for any voltage above a certain level. Applying a high enough voltage, therefore, will cause the photodiode to behave as an essentially linear controlled current source.

In cases where the curves are not available, it is usually adequate to use a single point (I_{light}, i) where I_{light} is the illumination in footcandles (fc) and i is the current. The *dark current* (the current for no illumination), i_0, is assumed to be zero. In cases where this is not a reasonable assumption, the dark current must be known. In cases of high illumination levels, it is possible to ignore the constant current generator, i_0.

Example 2.11

For the curves shown in figure 2.15, construct the equivalent circuit.

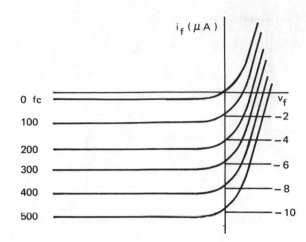

Figure 2.15 Typical Photodiode Characteristics

The dark current is about 0.7 μA, and the current is about 10.3 μA at 500 fc. The two-point method is used to find an equation of current versus illumination.

$$\frac{i - 0.7}{I_{light} - 0} = \frac{10.3 - 0.7}{500 - 0} = 0.0192$$

$$i = 0.7 + 0.0192\,I_{light}$$

The equivalent circuit is shown. The biasing is also shown, but it is not necessary in the equivalent circuit as the voltage is in series with a current source.

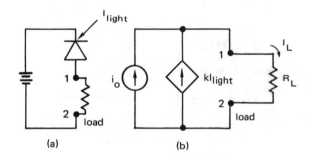

(a) biasing circuit for photodiode of figure 2.15
(b) equivalent circuit.

A *phototransistor* has characteristics similar to the photodiode, but it has a much larger gain (k) because of the current gain (β) of the transistor. Thus, the values on the current axis would be approximately β times those on figure 2.15.

Diode characteristics are dependent on temperature, as indicated in figure 2.16. For a temperature-dependent device, behavior can be modeled by a circuit with a temperature-dependent voltage source. To model such a device, it is necessary to obtain the voltage-temperature pairs to generate an equation. Then, the voltage is modeled as two sources, one dependent and the other independent of the temperature.

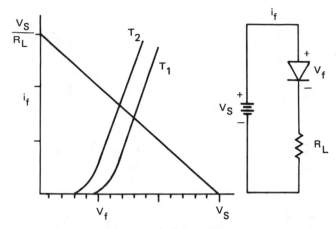

Figure 2.16 Temperature Characteristics for a Diode

Example 2.12

The characteristics shown in figure 2.16 have voltage divisions of 0.2 volts and current divisions of 1 amp. Obtain an equivalent circuit for a diode that accounts for the temperature dependence with $T_1 = 20°C$ and $T_2 = 80°C$.

At T_1 the (v_f, i_f) points are $(0.95, 0)$ and $(1.4, 1.5)$. The resulting equation is

$$v_{f1} = 0.95 + 0.3i_f$$

As the intercept of the straight line approximation at T_2 is $v_f = 0.6$, and the slope is essentially the same, the voltage is

$$v_{f2} = 0.6 + 0.3i_f$$

The only difference in the two voltages is the intercept which can be paired with the temperatures as $(0.95, 20°)$ and $(0.6, 80°)$. Then the two-point method gives

$$\frac{V_0 - 0.95}{T - 20} = \frac{0.6 - 0.95}{80 - 20} = -0.0058$$
$$v_f = 1.067 - 0.0058T + 0.3i_f$$

The equivalent circuit is shown.

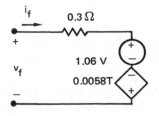

Thermocouples generate a voltage dependent on the difference in temperatures between two junctions of dissimilar metals. This is known as the *Seebeck effect* and is a nonlinear function of the form

$$V_{ab} = C_{ab}(T_h - T_c) \qquad 2.28$$

V_{ab} is the voltage generated by the temperature difference between the hot and cold junctions, C_{ab} is the *Seebeck coefficient*, T_h is the hotter junction's temperature, and T_c is the colder junction's temperature. Equation 2.28 is linear, but the coefficient C_{ab} is nonlinear as indicated in figure 2.17.

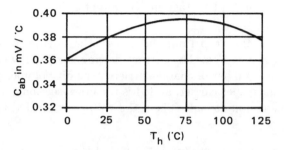

Figure 2.17 Seebeck Coefficient versus Temperature

Such devices can be considered linear only over the range of temperatures where the coefficient is flat (i.e., from 50° to 100°C). Over that range, the voltage is accurately predicted by equation 2.28. The equivalent circuit is shown in figure 2.18. The resistance of the wires and junctions is assumed negligible.

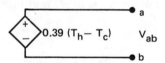

Figure 2.18 Equivalent Circuit for Thermocouples Operating with Hot Temperature from 50° to 100°C

The *Hall effect* is a consequence of the *Lorentz force equation*:

$$\mathbf{F} = q(\mathbf{E} + \mathbf{v} \times \mathbf{B}) \qquad 2.29$$

$\mathbf{F}$ is the force on a charge q moving at velocity $\mathbf{v}$ in an electric field $\mathbf{E}$ and magnetic field with flux density $\mathbf{B}$ ($\mathbf{F}, \mathbf{e}, \mathbf{v}$, and $\mathbf{B}$ are vectors). In a conductor with the current flow in the x-direction and with the magnetic flux density in the z-direction as shown in figure 2.19, the force is zero. The electric field must be in the direction shown.

$$-E_y = v_x B_z \qquad 2.30$$

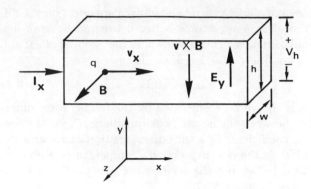

Figure 2.19   Geometry for the Calculation of the Hall Voltage in a Conductor

The voltage drop from the top to the bottom of the conductor is $E_y h$. The current flowing is $nqv_x A$, where n is the number of charge carriers per unit volume, q is the charge per carrier, and A is the cross-sectional area of the conductor wh. Then, the voltage generated is

$$V_{\text{Hall}} = \left(\frac{I}{w}\right)\left(\frac{1}{nq}\right) B_z \qquad 2.31$$

Example 2.13

Copper has an electron density of 10^{28} electrons per cubic meter. The charge on each electron is 1.6×10^{-19} coulombs. For a copper conductor with width 5 mm, height 5 mm, and length 10 cm, determine the Hall voltage for a current of 1 amp when the magnetic field intensity is 1 amp/m. $B = \mu_0 H$, and $\mu_0 = 4\pi \times 10^{-7}$.

With $H = 1$ amp/m, $B = 4\pi \times 10^{-7}$.

$$V_{\text{Hall}} = \frac{1 \times 4\pi \times 10^{-7}}{(5 \times 10^{-3})(10^{28})(1.6 \times 10^{-19})}$$
$$= 1.57 \times 10^{-13} \text{ volts}$$

4   CIRCUIT REDUCTION TECHNIQUES

A. Source Conversion

An electrical source can be modeled either by a voltage source in series with an impedance or by a current source in parallel with an impedance. These models are equivalent to one another in that they have the same terminal voltage-current relationships under the condition that

$$v_{\text{Th}} = i_N Z_{\text{Th}} \qquad 2.32$$
$$i_N = \frac{v_{\text{Th}}}{Z_{\text{Th}}} \qquad 2.33$$

With this equivalency, it is possible to convert a source from one form to the other. This is an essential operation in circuit reduction.

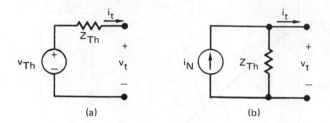

Figure 2.20   (a) Equivalent Voltage Source, (b) Equivalent Current Source

Example 2.14

A source is modeled as a voltage of $v = 10\cos(20t)$ with an impedance consisting of a 4 ohm resistance and a 0.15 henry inductance in series. Obtain the equivalent current source model.

First, convert the voltage source to phasor form, taking $\cos 20t$ as the reference.

$$v_{\text{Th}} = 10\angle 0°$$

Since $\omega = 20$, the inductive reactance is

$$20 \times 0.15 = j3 \text{ ohms}$$

The impedance is $4 + j3$, which should be converted to polar form.

$$Z_{\text{Th}} = 4 + j3 = 5\angle 36.87°$$

The value of the current source is found from the equation 2.33.

$$I_N = \frac{10\angle 0°}{5\angle 36.87°} = 2\angle -36.87°$$

This can be converted to the time form

$$i_N = 2\cos(20t - 36.86°)$$

The equivalent source is this current source in parallel with the original impedance as in figure 2.20(b).

There is confusion as to the relative *directions* of the sources in equivalent circuits. A rule is helpful:

- **Rule 2.1** Both the original and the equivalent must have the same *open-circuit voltage polarity*. Also, both the original model and the equivalent must cause a short-circuit current to flow in the same direction in the short circuited condition.

B. Series and Parallel Combination Rules

The models discussed in section A consist of pure voltage or current sources, and impedances. The rules for the combinations of these elements are listed here.

- **Rule 2.2** A pure voltage source model cannot be placed in parallel with another pure voltage source model, as the result would be indeterminant. (This rule also applies to pure current source models in series.)

- **Rule 2.3** A pure voltage source in parallel with a pure current source makes the current source redundant. The voltage is set by the pure voltage source, and the pure voltage source will absorb all the current from the current source.

- **Rule 2.4** A pure current source in series with a voltage source makes the voltage source redundant. The current is set by the current source, and the current source can absorb all the voltage from the voltage source.

- **Rule 2.5** Pure current source models in parallel can be combined algebraically with Kirchhoff's current law.

- **Rule 2.6** Pure voltage sources in series can be combined algebraically with Kirchhoff's voltage law.

- **Rule 2.7** Impedances in series add directly. Impedances in parallel add reciprocally.

C. Partial Reductions

It is sometimes convenient to reduce a network with a number of sources to a simpler circuit by making combinations described in the previous section. This is done to check analytic results, as each step of the reduction can be easily checked. It is particularly useful when the impedances involved are purely resistive.

- Parallel Voltage Sources

Combining two voltage sources in parallel is best accomplished by first converting both to equivalent current sources. Then, the resulting parallel current sources can be added algebraically, and the parallel impedances can be added reciprocally. It will then be possible to convert the resulting current source to its equivalent voltage source.

Example 2.15

Two voltage sources as shown are to be combined into a single source model.

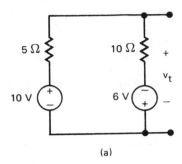

(a)

First, each voltage source is converted to an equivalent current source.

$$i_{N1} = \frac{10}{5} = 2 \,(\text{upward})$$

$$i_{N2} = \frac{6}{10} = 0.6 \,(\text{downward})$$

$$= -0.6 \,(\text{upward})$$

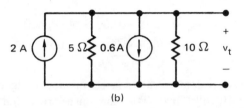

(b)

These two current sources are combined in parallel, with the upward direction taken as positive. The result is 1.4 amps.

The two parallel resistances are combined reciprocally to give

$$\frac{1}{R} = \frac{1}{5} + \frac{1}{10} = \frac{3}{10}$$

The combined impedance is 10/3 ohms.

The reduced source is shown. It could be converted to a voltage source.

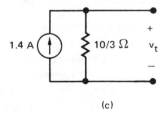

(c)

• Series Current Sources

Current source models in series first should be converted to equivalent voltage sources and then combined by adding the equivalent voltage source values algebraically. The impedances should be added directly.

Example 2.16

Two current source models are shown. These are to be combined into a single source model.

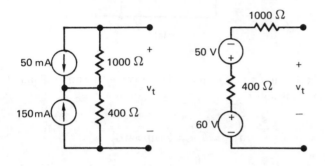

First, each current source is converted to an equivalent voltage source.

$$v_{Th1} = 0.05 \times 1000 = 50 \,(\text{down})$$
$$v_{Th2} = 0.15 \times 400 = 60 \,(\text{up})$$

These equivalent voltage sources are added algebraically to give 10 volts (up). The resistances are added directly to give a total of 1400 ohms.

The resulting reduced source is then as shown.

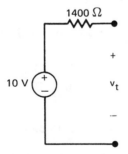

• Iterative Method of Circuit Reduction

The techniques of reducing parallel voltage source models and series current source models can be extended to an iterative method of circuit reduction where an entire network is reduced to a single source and impedance. This is useful with phasor analysis or where capacitance and/or inductance is present in a complex network and

the transient voltage or current is of interest. It is also useful in the analysis of circuits containing nonlinear or active elements.

Example 2.17

The network shown has capacitance and inductance which are neither in series nor in parallel. The voltage source, v_s, is a signal source model, and it is a variable.

Simplify the circuit as much as possible and find the voltage across the capacitance.

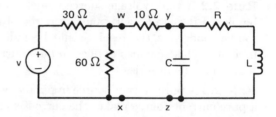

Note the nodes marked w, x, y, and z. These will be used for intermediate steps in the iterative procedure.

The portion of the network to the left of nodes w-x is first converted to an equivalent current source model. The 30 and 60 ohm resistances are then combined in parallel. The circuit to the left of w-x is shown.

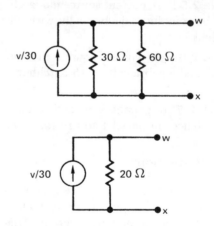

The current source model to the left of w-x is next converted to an equivalent voltage source. The 20 ohm resistance can be combined in series with the 10 ohm resistance between w and y to further reduce the circuit.

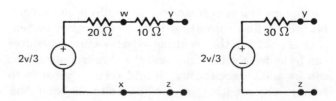

Finally, the voltage source model is converted to an equivalent current source model. This circuit form is used to establish the circuit equations.

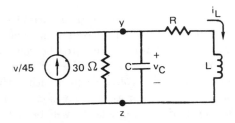

The circuit equations are written in terms of the capacitance voltage and inductance current—the states of the two energy storage elements. The current summation at the node above the capacitance, C, is

$$\frac{v_s}{45} = \frac{v_C}{30} + C\left(\frac{dv_C}{dt}\right) + i_L$$

The voltage equation around the right-hand loop is

$$v_C = Ri_L + L\left(\frac{di_L}{dt}\right)$$

These two equations are solved most easily by methods of Laplace transformations as outlined in chapter 1. If a sinusoidal analysis was required, the impedance of the inductance and resistance in the right-hand branch would be combined in parallel with the 30 ohm resistance to further simplify the circuit.

D. Wye-Delta and Delta-Wye Conversion

It is sometimes convenient (particularly in power networks) to make a more complex conversion involving three terminals. The basic conversions are called *wye-delta* and *delta-wye* transformations. They are alternatively knows as the *tee-pi* and *pi-tee* transformations, respectively.

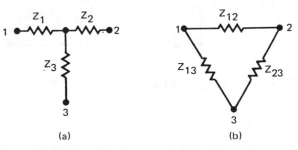

Figure 2.21 (a) Wye (Tee) Network, (b) Delta (Pi) Network

Two forms of a circuit are shown in figure 2.21. Note that for the wye form, the impedances are single subscripted according to the terminal to which they

are connected. The delta impedances are double subscripted according to the two terminals to which they are connected.

These networks are equivalent as long as equations 2.34, 2.35, and 2.36 are valid.

$$Z_{12} = \frac{Z_1Z_2 + Z_1Z_3 + Z_2Z_3}{Z_3} \qquad 2.34$$

$$Z_{13} = \frac{Z_1Z_2 + Z_1Z_3 + Z_2Z_3}{Z_2} \qquad 2.35$$

$$Z_{23} = \frac{Z_1Z_2 + Z_1Z_3 + Z_2Z_3}{Z_1} \qquad 2.36$$

Equations 2.34, 2.35, and 2.36 can be used to convert a wye (tee) circuit to a delta (pi) circuit. In order to convert in the other direction, equations 2.37, 2.38, and 2.39 are needed.

$$Z_1 = \frac{Z_{12}Z_{13}}{Z_{12} + Z_{13} + Z_{23}} \qquad 2.37$$

$$Z_2 = \frac{Z_{12}Z_{23}}{Z_{12} + Z_{13} + Z_{23}} \qquad 2.38$$

$$Z_3 = \frac{Z_{13}Z_{23}}{Z_{12} + Z_{13} + Z_{23}} \qquad 2.39$$

Example 2.18

In example 2.17, a wye configuration, repeated here for convenience, consisted of 30, 60, and 10 ohm resistances. (Note the terminal designations.) Convert this configuration to a delta structure.

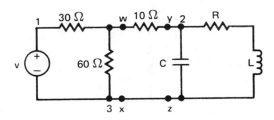

$$Z_{12} = \frac{(30 \times 10) + (30 \times 60) + (10 \times 60)}{60}$$
$$= \frac{2700}{60} = 45\,\text{ohms}$$

$$Z_{13} = \frac{(30 \times 10) + (30 \times 60) + (10 \times 60)}{10}$$
$$= \frac{2700}{10} = 270\,\text{ohms}$$

$$Z_{23} = \frac{(30 \times 10) + (30 \times 60) + (10 \times 60)}{30}$$

$$= \frac{2700}{30} = 90 \text{ ohms}$$

The circuit is next shown after the delta structure is substituted into the circuit to the left of terminals y-z.

The 270 ohm resistance in parallel with the source, v_s, is redundant. It can be removed.

Then, the source model consisting of v_s and the 45 ohm resistance is converted to the equivalent current model as shown. The parallel resistances of 45 and 90 ohms are combined into 30 ohms. The resultant circuit is identical to the solution of example 2.17.

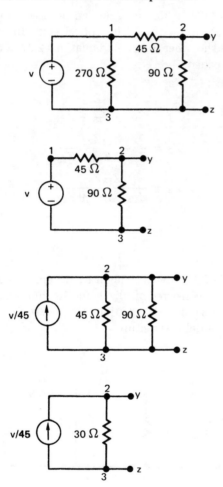

E. Redundant Impedances

It was shown in example 2.18 that a resistance placed across an ideal voltage source is redundant and may be removed. As far as the remainder of the circuit is concerned, an impedance across an ideal voltage source (which theoretically can supply an unlimited amount of current) will not affect any other voltage or current in the circuit.

This is true except for the current through the source, which will be in error if calculated using the voltage divided by the redundant impedance. This source current usually is of no concern in circuit analysis problems.

An impedance in series with an ideal current source is likewise redundant and can be removed. The current source provides the same current to the remainder of the circuit regardless of whether the series impedance is present or not. The voltage across either the ideal current source or the series combination of the ideal current source and the impedance will be determined by the remainder of the circuit.

F. Voltage Dividers

A *voltage divider* circuit is one in which a source is in series with two impedances, and the voltage of interest is across one of the two impedances. The voltage divides between the two impedances proportionally to their values. This is an important technique in finding the Thevenin equivalent of a partial circuit. It shortens the number of steps in circuit reduction problems.

Figure 2.22 shows a basic voltage divider circuit. A simple analysis will show that the voltage across Z_2 is

$$v_{\text{Th}} = v_1 \left(\frac{Z_2}{Z_1 + Z_2} \right) \qquad 2.40$$

This is also the Thevenin equivalent voltage looking into the terminals across Z_2, (i.e., into terminals a-b as seen from the right). The Thevenin equivalent impedance looking into those same terminals is the parallel combination of Z_1 and Z_2.

$$Z_{\text{Th}} = \frac{Z_1 \times Z_2}{Z_1 + Z_2} \qquad 2.41$$

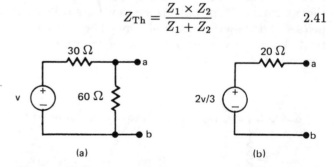

(a)   (b)

Figure 2.22   (a) Circuit, (b) Thevenin Equivalent

Example 2.19

The circuit shown in figure 2.22(a) is part of the circuit analyzed in examples 2.17 and 2.18 with $Z_1 = 30\Omega$ and $Z_2 = 60\Omega$. The Thevenin equivalent voltage is found from equation 2.40.

$$v_{\text{Th}} = \frac{60}{60 + 30} v_s = \frac{2v_s}{3}$$

The Thevenin equivalent impedance is found from equation 2.41.

$$Z_{\text{Th}} = \frac{30 \times 60}{30 + 60} = 20 \text{ ohms}$$

This coincides with the values found in examples 2.17 and 2.18.

5 NETWORK THEOREMS

Many system design problems include fixed portions of networks which are to be incorporated, but not changed. Each fixed portion can usually be described by the characteristics seen from its terminals. The fixed portions of the networks can be simplified to Thevenin or Norton equivalents. This simplification is always possible when the system is linear or piecewise linear.

A. Thevenin's Theorem

Thevenin's theorem states that any circuit composed of linear impedances, power sources, and linear controlled sources can be replaced with a single voltage source and a series impedance. The resulting voltage source is called the *Thevenin equivalent voltage source*. The resulting impedance is called the *Thevenin equivalent impedance*.

The Thevenin equivalent circuit is found in the following manner.

> step 1: With the terminals of the circuit open circuited, determine the voltage across the terminals.
>
> step 2: With the terminal short circuited, determine the short-circuit current.

The open-circuit voltage is the Thevenin equivalent voltage source, and the ratio of the open-circuit voltage to the short-circuit current is the Thevenin equivalent impedance.

A shortcut to the Thevenin equivalent impedance is to find the impedance looking into the terminals with all power sources altered: voltage sources are short circuited and all current sources are open circuited. In this process, all controlled sources are retained as is.

B. Norton's Theorem

Norton's theorem is the dual of Thevenin's theorem. Under the same linear constraints, the circuit can be replaced by a Norton equivalent current source and an equivalent impedance in parallel with the current source. The *Norton equivalent current* is the short-circuit current. The impedance is the same as found in the Thevenin equivalent circuit.

Example 2.20

A simplified small-signal amplifier circuit is shown. The voltage source, v_{in}, is a variable signal source and can be considered to be an ideal voltage source. The transistor has parameters $h_{\text{ie}} = 500$, $h_{\text{re}} = 0$, $h_{\text{fe}} = 50$, and $h_{\text{oe}} = 0.0001$. Determine the Thevenin equivalent circuit (as seen from terminals a and b) in terms of the unknown turns N_1 and N_2 on the right-hand transformer.

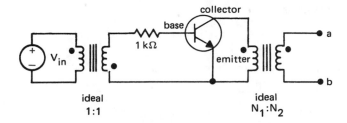

Method 1: The straightforward method involves finding the open-circuit voltage and the short-circuit current. The ideal transformer relations are:

$$\frac{V_1}{N_1} = \frac{V_2}{N_2}$$
$$N_1 i_1 + N_2 i_2 = 0$$

All quantities are referred to the dots as previously discussed. For the input transformer, the voltage at the dot on the secondary is the same as at the dot on the primary, with a phase reversal. The current is driven into the dot on the primary side, so the current is out of the dot on the secondary side.

The equivalent transistor model contains a resistance of 500 ohms for h_{ie}, a controlled current source $50i_b$ for h_{fe}, and a parallel resistance of 10 kΩ for h_{oe}.

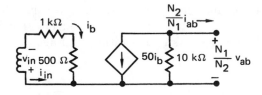

First, determine the open-circuit voltage across the output terminals. $i_{ab} = 0$, and all currents from the current source go through the 10k resistor.

$$\frac{N_1 v_{ab}}{N_2} = -50i_b \times 10^4$$

From the input loop, it is apparent that

$$i_b = \frac{-v_{\text{in}}}{1500}$$

Then, by substituting this into the previous expression containing i_b, the open-circuit voltage is

$$v_{ab}|_{\text{o.c.}} = \left(\frac{N_2}{N_1}\right) 333.3 v_{\text{in}}$$

In the short-circuit condition, v_{ab} is zero. The 10K collector resistor is shorted out causing all of the current source current to flow through the short circuit.

The short-circuit current in the primary output transformer is

$$\left(\frac{N_2}{N_1}\right) i_{ab}|_{\text{s.c.}} = -50 i_b$$

Since i_b has been determined to be $-v_{\text{in}}/1500$,

$$i_{ab}|_{\text{s.c.}} = \left(\frac{N_1}{N_2}\right) \frac{v_{\text{in}}}{30}$$

The Thevenin equivalent impedance is found by taking the ratio of the open-circuit voltage to the short-circuit current.

$$Z_{\text{Thevenin}} = 10^4 \left(\frac{N_2}{N_1}\right)^2$$

The Thevenin equivalent voltage source is $v_{ab}|_{\text{o.c.}}$

$$v_{\text{Thevenin}} = 333.3 \left(\frac{N_2}{N_1}\right) v_{\text{in}}$$

Method 2: With this method, the open-circuit voltage is used as the voltage source, but the Thevenin equivalent impedance is found by looking into the output terminals with all power and signal sources short circuited and all controlled sources operating. (If there were current sources, they would be open circuited.) The impedance is found by assuming a voltage on the output terminals, v_{ab}, finding the current that would result, and then taking the ratio of voltage to current.

With the signal source short circuited, the voltage, v_{in}, is zero, and so the current from the controlled current source will also be zero since i_b is zero. The current $N_2 i_{ab}/N_1$ is

$$\frac{N_1 v_{ab}}{N_2 \times 10^4}$$

The Thevenin equivalent impedance is

$$Z_{\text{Thevenin}} = \frac{v_{ab}}{-i_{ab}} = \frac{v_{ab} \times 10^4}{\left(\frac{N_1}{N_2}\right)^2 v_{ab}}$$

$$= \left(\frac{N_2}{N_1}\right)^2 \times 10^4$$

C. Maximum Power Transfer Theorem

The *maximum power transfer theorem* says that maximum power will be transferred from a fixed network to an impedance that is the complex conjugate of the Thevenin equivalent impedance of the fixed network. In example 2.20, the impedance is a pure resistance with no imaginary part, so the complex conjugate is the same as the Thevenin equivalent impedance.

Example 2.21

A 60 Hz alternator generates a single-phase voltage of 120 volts (rms) and has a synchronous reactance of 0.25 ohms. The alternator feeds a line which is essentially resistive with an impedance of 0.1 ohm. The line is terminated with a 2:1 step-up transformer. What should the secondary impedance of the transformer be in order to have maximum power transfer to the load?

Maximum power will be transferred when the Thevenin source impedance seen by the load is the complex conjugate of the load. With the source fixed in this example, the maximum power transfer theorem describes how to obtain maximum power transfer. The real part of the load impedance must be the same as the real part of the source, and an imaginary impedance which is the negative of the imaginary part of the source impedance as seen from the load, must be added.

Since there is a transformer, the source impedance can be transferred to the load side and matched, or the load impedance can be transferred to the source side and matched. The impedance attached to the secondary is transformed to the primary side by multiplying by the square of the turns ratio. This is the ratio of turns on the winding looking into the turns on the winding that the impedance is attached to. Thus, if you look into N_{in} turns, and an impedance Z is attached to N_z turns, the impedance seen is

$$\left(\frac{N_{\text{in}}}{N_z}\right)^2 Z$$

In this example, the primary impedance is $0.1 + j0.25$. The synchronous reactance is an inductive reactance appearing in series with the alternator. Since the turns ratio is 2:1 as seen from the secondary side (step-up means the secondary has more turns than the primary), the impedance seen from the load (secondary) side is

$$2^2 \times (0.1 + j0.25) = 0.4 + j1$$

The load impedance for maximum power transfer is the complex conjugate.

$$Z_{\text{load}|\text{max power transfer}} = 0.4 - j1 \text{ ohms}$$

D. The Superposition Theorem

The *superposition theorem* states that the response to several signs or sources can be calculated separately (setting all other independent sources to zero), and then added algebraically to obtain the total response. This theorem is valid only for linear circuits. The superposition theorem is of most value when several signal sources of different frequencies are applied to a circuit.

Example 2.22

The circuit shown has $Z_1 = 1$ μF and $Z_2 = 200$ mH. The voltage source is represented by the partial Fourier series

$$v_s = 20 \cos 1000t - 5 \cos 4000t$$

Determine the voltage across the 100 ohm resistor.

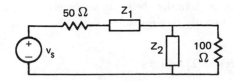

Working with the 1000 rad/sec source first, the impedances are $Z_1 = -j10^3$, and $Z_2 = j200$ ohms at this frequency. The Thevenin equivalent circuit seen from the 100 ohm resistor is then a voltage source.

$$V_{\text{Th}} = v_s \frac{Z_2}{50 + Z_1 + Z_2} = \frac{j200}{50 - j800} v_s$$
$$= -0.250 v_s \angle - 3.58°$$

The Thevenin equivalent impedance seen from the 100 ohm resistance is Z_2 in parallel with $(50 + Z_1)$.

$$Z_{\text{Th}} = \frac{(50 + Z_1) \times Z_2}{50 + Z_1 + Z_2}$$
$$= \frac{(50 - j1000) \times j200}{50 - j800}$$
$$= 3.1 + j249.8$$

The voltage across the 100 ohm resistance can be obtained for the 1000 rad/sec component of v_s.

$$v_{100}(1000) = \frac{100}{100 + Z_{\text{Th}}} V_{\text{Th}}$$
$$= \frac{100 \times (-0.250 v_s) \angle - 3.58°}{103.1 + j249.8}$$
$$= -0.0925 v_s \angle - 71.15°$$

Expressed trigonometrically, this voltage is

$$1.85 \cos(1000t + 108.85°)$$

For the component of v_s at 4000 rad/sec the impedances are $Z_1 = -j250$ and $Z_2 = j800$. For variety, the two loop equations are written in matrix form.

$$\begin{bmatrix} v_s \\ 0 \end{bmatrix} = \begin{bmatrix} (50 + j550) & -j800 \\ -j800 & (100 + j800) \end{bmatrix} \begin{bmatrix} i_1 \\ i_2 \end{bmatrix}$$

i_1 is the clockwise left-hand loop current, and i_2 is the clockwise right-hand loop current. This is solved for i_2 with the determinant method explained in chapter 1, as $100i_2$ is the voltage desired.

$$i_2 = 0.00354 v_s \angle 65.1°$$

The voltage across the 100 ohm resistor is

$$v_{100}(4000) = 0.354 v_s \angle 65.1°$$

This can be expressed trigonometrically as

$$-1.77 \cos(4000t + 65.1°)$$

Then applying the superposition theorem,

$$v_{100} = v_{100}(1000) + v_{100}(4000)$$
$$= 1.85 \cos(1000t + 108.85°)$$
$$-1.77 \cos(4000t + 65.1°)$$

E. Miller's Theorem

Miller's theorem states that when an admittance such as Y_{ab} is connected between two nodes, the voltage at the second node (i.e., node b) is linearly dependent on that of the first node (i.e., node a), as follows:

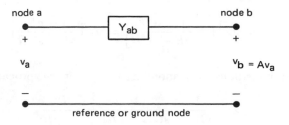

The linking admittance, Y_{ab}, can be replaced by two admittances, one from each node to the reference (ground) node (as are Y_a from node a to ground and Y_b from node b to ground) as indicated in the following:

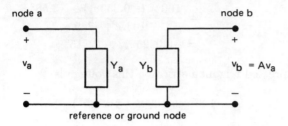

The relationships are as follows:

$$Y_a = Y_{ab}(1 - A)$$

$$Y_b = Y_{ab}\left(1 - \frac{1}{A}\right)$$

This is illustrated with an example useful in *unilateralizing* amplifier circuits.

Example 2.23

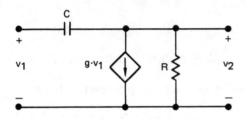

Determine A via Kirchhoff's current law applied at the output node:

$$sC(v_1 - v_2) = g \cdot v_1 + \frac{1}{R}v_2$$

From which

$$\frac{v_2}{v_1} = -\frac{gR - sCR}{1 + sCR} = A$$

As $Y_a = Y_{ab}(1 - A)$,

$$Y_a = sC\left[1 + \frac{gR - sCR}{1 + sCR}\right] = sC\left[\frac{1 + gR}{1 + sCR}\right]$$

And, as $Y_b = Y_{ab}(1 - \frac{1}{A})$,

$$Y_b = sC\left[1 + \frac{1 + sCR}{gR - sCR}\right] = sC\left[\frac{1 + gR}{gR - sCR}\right]$$

The impedance corresponding to Y_a is its reciprocal:

$$Z_a = \frac{1}{Y_a} = \frac{1}{sC(1 + gR)} + \frac{R}{1 + gR}$$

which corresponds to a capacitor of $C(1 + gR)$ capacitance in series with a resistance $R/(1 + gR)$.

The impedance corresponding to Y_b is its reciprocal:

$$Z_b = \frac{1}{Y_b} = \frac{gR}{sC(1 + gR)} - \frac{R}{1 + gR}$$

This corresponds to a capacitor of $C[1 + (1/gR)]$ capacitances in series with a negative resistance $-R/(1 + gR)$.

The unilateralized circuit is one with no feedback, and is shown below:

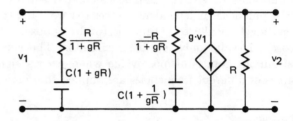

In many cases, the effect of the linking admittance is ignored in determining the factor A. This can be a source of serious error, as an analysis would reveal that the positive and negative resistances $R/(1 + gR)$ would not then appear in the resultant model.

For a large value of gR such as $gR \gg 1$, the resistances become positive and negative $1/g$, which for a junction transistor is virtually the same as the emitter resistance, which is usually negligible.

F. Current Dividers

Parallel resistors divide the current among themselves in proportion to the inverse of their resistances. Every resistor in a parallel combination shares the same voltage, say V, so that one of the resistors with resistance R has a current flow of V/R. The total current is the sum of such terms to include all resistors in the parallel combination. By using the conductances, i.e., the reciprocals of the resistances, the fraction of the current flowing through one resistance in a parallel combination is the ratio of its conductance to the sum of the conductances of the combination.

Example 2.24

A set of four parallel resistors will have a total current of

$$i_{\text{total}} = G_1v + G_2v + G_3v + G_4v = (G_1 + G_2 + G_3 + G_4)v$$

The various currents are:

$$i_1 = G_1 v, \ i_2 = G_2 v, \ i_3 = G_3 v, \ i_4 = G_4 v$$

Then, the ratios of the current to the total current are:

$$\frac{i_1}{i_{\text{total}}} = \frac{G_1}{G_1 + G_2 + G_3 + G_4}$$
$$\frac{i_2}{i_{\text{total}}} = \frac{G_2}{G_1 + G_2 + G_3 + G_4}$$
$$\frac{i_3}{i_{\text{total}}} = \frac{G_3}{G_1 + G_2 + G_3 + G_4}$$
$$\frac{i_4}{i_{\text{total}}} = \frac{G_4}{G_1 + G_2 + G_3 + G_4}$$

The fraction of the total current that a resistor in parallel with others takes is the ratio of its conductance to the sum of all the conductances it is in parallel with.

6 TWO-PORT NETWORKS

Two-port networks such as transformers and transistors may be either three-terminal or four-terminal devices. They must be linear, or piecewise linear to be treated under the theory presented in this section. A representation of such a network is shown in figure 2.23 with the current directions and voltage polarities. There are four variables: v_1, i_1, v_2, and i_2. The port with the subscript 1 is called the *input port* and the corresponding terminals are the *input terminals*. The terminals with subscript 2 are called *output terminals* and make up the *output port*.

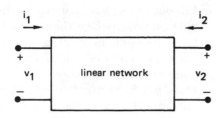

Figure 2.23 Two-Port Network Showing Defined Directions

Two-port theory is particularly important in circuits which employ feedback, such as amplifiers. The four port variables can be arranged into dependent and independent pairs. For a linear network, the distinction is arbitrary. The result is a set of parameters, such as the z-parameters used to describe transistors. In example 2.22, the independent variables are the currents, and the dependent variables are the voltages.

By choosing voltage as the independent variable, *y-parameters* are obtained. By choosing the input current

and the output voltage as independent variables, the *h-parameters* are obtained. Generally, the parameters are identified by double subscripts in standard matrix notation with the first subscript corresponding to the row and the second subscript to the column. Thus, the z-parameters are written

$$\begin{bmatrix} v_1 \\ v_2 \end{bmatrix} = \begin{bmatrix} z_{11} & z_{12} \\ z_{21} & z_{22} \end{bmatrix} \begin{bmatrix} i_1 \\ i_2 \end{bmatrix} \qquad 2.42$$

The z-parameters are particularly useful when the parameters of two networks are known, and when the input and output currents are the same for the two networks. This is illustrated in figure 2.24 for a common emitter transistor with an emitter resistor. The two sets of z-parameters are combined by addition, since the input and output voltages are the sum of the dependent variables. The upper network is indicated by unprimed variables and parameters; the lower network is indicated by primed variables. Equation 2.42 applies to the upper network. By priming all quantities, it also applies to the lower network.

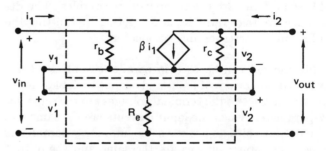

Figure 2.24 Combination of Two-Port Networks with Common Currents

Example 2.25

Determine the z-parameters for the upper and the lower networks shown in figure 2.24, and find the z-parameters for the entire circuit.

For the upper network, the equations are:

$$v_1 = r_b i_1$$
$$i_2 = \beta i_1 + \frac{v_2}{r_c}$$

The first equation is already in the z-parameter form. The second equation must be manipulated to get v_2 as the dependent variable.

$$v_2 = -\beta r_c i_1 + r_c i_2$$

Thus the z matrix for the upper network is

$$\mathbf{Z} = \begin{bmatrix} r_b & 0 \\ -\beta r_c & r_c \end{bmatrix}$$

For the lower network,

$$v_1' = (i_1 + i_2)R_e$$
$$v_2' = (i_1 + i_2)R_e$$

Note that $i_1' = i_1$, and $i_2' = i_2$. Thus the z' matrix for the lower network is

$$\mathbf{Z}' = \begin{bmatrix} R_e & R_e \\ R_e & R_e \end{bmatrix}$$

Since $v_{\text{in}} = v_1 + v_1'$ and $v_{\text{out}} = v_2 + v_2'$, the z matrixes can be added to give the z-parameters for the entire circuit.

$$\begin{bmatrix} v_{\text{in}} \\ v_{\text{out}} \end{bmatrix} = \begin{bmatrix} r_b + R_e & R_e \\ -\beta r_c + R_e & r_c + R_e \end{bmatrix} \begin{bmatrix} i_1 \\ i_2 \end{bmatrix}$$

With the z-parameters, the input currents are the same for both circuits. The output currents are also the same. These currents are the independent variables. The dependent voltage variables add together, permitting the addition of corresponding parameters.

This method can be used in parallel circuits for three other combinations of input and output variables: for y-parameters when the terminal voltages are the same, for h-parameters when the input currents are the same and the output voltages are the same, and for g-parameters when the input voltages are the same and the output currents are the same.

$$\begin{bmatrix} i_1 \\ v_2 \end{bmatrix} = \begin{bmatrix} g_{11} & g_{12} \\ g_{21} & g_{22} \end{bmatrix} \begin{bmatrix} v_1 \\ i_2 \end{bmatrix}$$

The corresponding g-parameter circuit model is shown in figure 2.25.

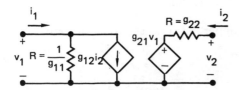

Figure 2.25 g-Parameter Circuit Model

It is also possible to choose independent variables which permit *cascading* of networks. The parameters are then called *chain parameters*. The dependent variables are input voltage and current. The independent variables are the output voltage and the negative of the output current. The output current is negated to cause it to be the same as the following network's input current.

$$\begin{bmatrix} v_1 \\ i_1 \end{bmatrix} = \begin{bmatrix} A & B \\ C & D \end{bmatrix} \begin{bmatrix} v_2 \\ -i_2 \end{bmatrix} \qquad 2.43$$

7   NETWORK ANALYSIS

This section shows how linear network equations can be generated by inspection, bypassing the tedious process of writing Kirchhoff law relationships.

Node voltage equations are generated by taking all voltages with respect to a reference node, and writing the Kirchhoff current law equations in terms of those node voltages. The process can be quickly accomplished by the following steps.

step 1: Use Norton's theorem to convert all voltage sources with series impedances to current sources with shunt impedances. Voltage sources without series impedances between two nodes make one of the nodes a *dependent node*. The voltage at a dependent node is the voltage at the other node plus the value of the voltage source. Voltage at dependent nodes is known.

step 2: Assign voltage variables to all nodes except the reference node, including the dependent nodes.

step 3: Set up an empty square *admittance matrix* with the same number of rows and columns as the number of independent and dependent nodes, but not including the reference node.

step 4: Fill out the square matrix. At each diagonal position (where the row and column have the same number) insert the sum of the admittances connected directly to the node with that number. At other positions, place the negative of the sum of admittances that directly connect the nodes having the same subscripts as the row and column of the position.

step 5: To the right of the square matrix construct the column matrix for the node voltages. Enter them in order. Include the dependent node voltages.

step 6: To the left of the square matrix, construct the column matrix for the sources. Enter the sum of the current sources feeding directly into the node with the same numeric subscript as the row. Current sources feeding directly out of a node are negative. Include controlled (dependent) sources as well as independent sources.

The above procedure is adequate whenever there are no dependent sources and no dependent node voltages.

Example 2.26

For the network shown, obtain the network matrix equation in terms of the node voltages.

Number the nodes from left to right. The reference node is the line across the bottom of the diagram.

step 1: Convert all voltage sources with series impedances. The 5 volt source in series with the 10 ohm resistance converts to a 0.5 amp current source connected between the same two nodes.

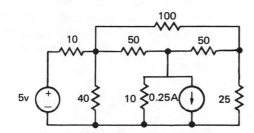

step 2: Assign the voltage subscripts at the nodes. The results of step 1 and step 2 are shown. Resistance has been converted to conductance (admittance) in the figure shown.

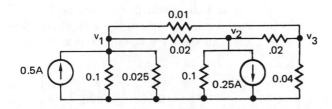

step 3: This step calls for a 3 × 3 admittance matrix.

columns

		1	2	3
	1	[		]
rows	2			
	3	[		]

step 4: Fill out the admittance matrix. Work first along the main diagonal with the node self-admittance y_{11}, (Take the connected admittances counterclockwise.)

$$y_{11} = 0.1 + 0.025 + 0.02 + 0.01$$
$$= 0.155 \qquad (node\ 1)$$
$$y_{22} = 0.02 + 0.1 + 0.02 = 0.14$$
$$(node\ 2)$$
$$y_{33} = 0.01 + 0.02 + 0.04 = 0.07$$
$$(node\ 3)$$

The negatives of the interconnecting admittances are placed in the corresponding rows and columns.

$$y_{12} = y_{21} = 0.02$$
$$y_{13} = y_{31} = 0.01$$
$$y_{23} = y_{32} = 0.02$$

The result is then

$$\begin{bmatrix} 0.155 & -0.02 & -0.01 \\ -0.02 & 0.14 & -0.02 \\ -0.01 & -0.02 & 0.07 \end{bmatrix}$$

step 5: The single column node voltage matrix is

$$\begin{bmatrix} v_1 \\ v_2 \\ v_3 \end{bmatrix}$$

step 6: Step 6 calls for a one column, three row source matrix.

row 1: 0.5A enters node 1.

row 2: −0.25A enters node 2. (0.25A leaves node 2)

row 3: there are no sources connected to node 3.

$$\begin{bmatrix} 0.5 \\ -0.25 \\ 0 \end{bmatrix}$$

The final result is

$$\begin{bmatrix} 0.5 \\ -0.25 \\ 0 \end{bmatrix} = \begin{bmatrix} 0.155 & -0.02 & -0.01 \\ -0.02 & 0.14 & -0.02 \\ -0.01 & -0.02 & 0.07 \end{bmatrix} \begin{bmatrix} v_1 \\ v_2 \\ v_3 \end{bmatrix}$$

If there are controlled sources in the network, an additional step is required.

step 7: When *controlled sources* are present, they are placed in the column source matrix as if they were independent sources. It is then necessary to obtain the controlling variable in terms of the independent node voltages.

For example, suppose the 0.25A independent source had been a controlled source whose value was one quarter of the current flowing from left to right in the left-hand 50 ohm resistance. It would then be necessary to express that current in terms of the node voltages $(v_1 - v_2)/50$. The current source would then be given a value of

$$0.25 \times \frac{(v_1 - v_2)}{50} = 0.005(v_1 - v_2)$$

The source matrix would be

$$\begin{bmatrix} 0.5 \\ -0.005(v_1 - v_2) \\ 0 \end{bmatrix}$$

The row in the source matrix that contains the independent variables must be modified. For any independent variable appearing in the column source matrix, subtract the coefficient of that variable from the admittance term in the same row and in the column having the same subscript as that variable.

The controlled source of $-0.005(v_1 - v_2)$, for example, appears in the source matrix in the second row. The coefficient of v_1 is -0.005, and that quantity must be subtracted from the term in the first column of the second row (i.e., -0.02). The coefficient of v_2 is $+0.005$, so that quantity must be subtracted from the term in the second column and the second row (i.e., 0.14). This also removes the controlled source from the source matrix. The resulting matrix equation for the example is

$$\begin{bmatrix} 0.5 \\ 0 \\ 0 \end{bmatrix} = \begin{bmatrix} 0.155 & -0.02 & -0.01 \\ -0.015 & 0.135 & -0.02 \\ -0.01 & -0.02 & 0.07 \end{bmatrix} \begin{bmatrix} v_1 \\ v_2 \\ v_3 \end{bmatrix}$$

The presence of *dependent nodes* requires an additional step. Dependent nodes can only be approximated in practice, as there is some impedance to any real source. However, some highly regulated power supplies and flat compounded DC generators approach being impedanceless sources, which produces dependent nodes.

The method of treating such nodes is to include them within the original admittance matrix equations. Rather than being given a distinct name, however, they are carried in the independent variable (node voltage) column matrix as an adjacent node voltage plus the intervening voltage. The treatment involves a step called *node merging*.

    step 8: When dependent nodes are present, the admittances of those node voltages must be (1) added to the admittance of the node voltages that appear in their place, and (2) the product of the internode voltage

sources and the admittances must be subtracted from the source terms for those rows.

As an example, suppose there were a dependent node in example 2.26 from a 5 volt source connected from node 1(+) to node 2(−). A term would then appear in the second row of the matrix equation.

$$\begin{bmatrix} 0.5 \\ -0.25 \\ 0 \end{bmatrix} = \begin{bmatrix} 0.155 & -0.02 & -0.01 \\ -0.02 & 0.14 & -0.02 \\ -0.01 & -0.02 & 0.07 \end{bmatrix} \begin{bmatrix} v_1 \\ v_1 - 5 \\ v_3 \end{bmatrix}$$

The $v_1 - 5$ term results from the 5 volt impedanceless source between nodes 1 and 2, with v_2 being 5 volts less than v_1. The terms in the second column of the admittance matrix are the coefficients of the voltage at the node 2. The equation of the first row is

$$0.5 = 0.155v_1 - 0.02(v_1 - 5) - 0.01v_3$$

The two coefficients of v_1 are 0.155 and -0.02, which must be added in order to obtain the correct coefficient. Then, as v_1's coefficients are in column 1 and v_2's coefficients are in column 2, the column 2 values must be added to the column 1 values and the result put in column 1. This step removes v_1 from the second row position in the node voltage column matrix.

$$\begin{bmatrix} 0.5 \\ -0.25 \\ 0 \end{bmatrix} = \begin{bmatrix} 0.135 & -0.02 & -0.01 \\ 0.12 & 0.14 & -0.02 \\ -0.03 & -0.02 & 0.07 \end{bmatrix} \begin{bmatrix} v_1 \\ -5 \\ v_3 \end{bmatrix}$$

The second column is composed of the coefficients of -5, which belong on the source side of the matrix. For the row 1 equation, this is done by multiplying the known voltage (in this case -5) by its coefficients and subtracting the result from the source term in the same row.

$$\begin{bmatrix} 0.5 - (-5 \times (-0.02)) \\ -0.25 - (-5 \times 0.14) \\ 0 - (-5 \times (-0.02)) \end{bmatrix} = \begin{bmatrix} 0.135 & -0.01 \\ 0.12 & -0.02 \\ -0.03 & 0.07 \end{bmatrix} \begin{bmatrix} v_1 \\ v_3 \end{bmatrix}$$

To complete the node merging, it is necessary to add the rows of the two nodes that have been merged. In this case, nodes 1 and 2 were merged, so rows 1 and 2 must be added.

$$\begin{bmatrix} 0.4 + 0.45 \\ -0.1 \end{bmatrix} = \begin{bmatrix} 0.135 + 0.12 & -0.01 + (-0.02) \\ -0.03 & 0.07 \end{bmatrix} \begin{bmatrix} v_1 \\ v_3 \end{bmatrix}$$

$$\begin{bmatrix} 0.85 \\ -0.1 \end{bmatrix} = \begin{bmatrix} 0.255 & -0.03 \\ -0.03 & 0.07 \end{bmatrix} \begin{bmatrix} v_1 \\ v_3 \end{bmatrix}$$

If you began with a symmetrical matrix, as you should have, you will finish with a symmetrical matrix.

Loop current analysis often can also be carried out by inspection. Loop current analysis, however, is restricted to planar networks, i.e., those which have no crossing wires. The method consists of the following steps:

step 1: Convert all current sources with parallel impedances to voltage sources with series impedances using Thevenin's theorem. If there are current sources without parallel impedances, node voltage analysis is recommended.

step 2: Assign loop currents to all meshes (loops) of the network. Assume the directions of all currents to be clockwise. Identify each loop current by a numeric subscript corresponding to the mesh.

step 3: From left to right, set up a column source matrix with the number of rows equal to the number of loops, a square impedance matrix with the number of rows and columns equal to the number of loops, and a column loop current matrix with the loop currents entered in numerical order.

step 4: In each row of the source matrix, place the sum of the *source voltage rises* going around the loops. The number of rows of the source and current matrixes are the same. Voltage rises in the current direction are positive. Include controlled voltage sources in the source matrix.

step 5: In the diagonal positions of the impedance matrix, place the sum of all the impedances in the loops having the respective numeric subscripts.

step 6: In non-diagonal positions, place the negative of the impedances common to the two loops with the different subscripts.

step 7: If there are controlled sources, their voltages will have been assigned to the column source matrix in step 4. It is necessary to obtain the controlling variable in terms of the loop currents. When this is done, the coefficients of the controlling loop currents are subtracted from the term with the column subscript of that current and the row subscript of the impedance matrix.

Example 2.27

Perform a loop current analysis of the circuit shown.

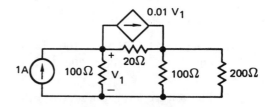

step 1: The first step is to convert the current sources to voltage sources. The 1 amp independent source on the left has a parallel resistance of 100 ohms. The Thevenin equivalent is a 100 volt source in series with a 100 ohm resistance. The Thevenin equivalent of the controlled current source with a 20 ohm shunting resistance is a controlled voltage source of $20 \times 0.01 v_1$ in series with the 20 ohm resistance.

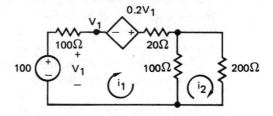

step 2: The loop currents are chosen in the clockwise direction and are numbered as shown.

step 3: This step is illustrated here.

$$\begin{bmatrix} \ \\ \ \end{bmatrix} = \begin{bmatrix} \ \\ \ \end{bmatrix} \begin{bmatrix} i_1 \\ i_2 \end{bmatrix}$$

step 4: Since their polarities are in the current direction, the sources in loop 1 add to $100 + 0.2 v_1$. This is entered in row 1 of the source matrix. Loop 2 has no sources, so enter a zero in row 2.

$$\begin{bmatrix} 100 + 0.2 v_1 \\ 0 \end{bmatrix} = \begin{bmatrix} \ \\ \ \end{bmatrix} \begin{bmatrix} i_1 \\ i_2 \end{bmatrix}$$

step 5: Loop 1 has a total of 220 ohms resistance, so enter 220 in the top diagonal position of the impedance matrix. In loop 2, there is a total of 300 ohms, so enter 300 in the lower diagonal position.

step 6: The impedance common to loops 1 and 2 is 100 ohms, so enter -100 in the non-diagonal positions of the impedance matrix.

The results of steps 5 and 6 are

$$\begin{bmatrix} 100 + 0.2v_1 \\ 0 \end{bmatrix} = \begin{bmatrix} 220 & -100 \\ -100 & 300 \end{bmatrix} \begin{bmatrix} i_1 \\ i_2 \end{bmatrix}$$

step 7: The voltage, v_1, is the source voltage (100) minus the voltage drop in the 100 ohm resistance in series.

$$v_1 = 100 - 100i_1$$

v_1 is multiplied by 0.2 to obtain

$$0.2v_1 = 20 - 20i_1$$

The constant 20 is added to the 100 in row 1 of the column source matrix. $-20i_1$ must be transferred across the equals sign. This is done by subtracting -20 from the proper coefficient of i_1 in the impedance matrix. The row is 1, and the column of i_1 coefficients is 1 in the impedance matrix. Therefore, the 220 in the impedance matrix has -20 subtracted from it.

$$z_{11} = 220 - (-20) = 240$$
$$\begin{bmatrix} 120 \\ 0 \end{bmatrix} = \begin{bmatrix} 240 & -100 \\ -100 & 300 \end{bmatrix} \begin{bmatrix} i_1 \\ i_2 \end{bmatrix}$$

8   IMPEDANCE ANALYSIS

This section will review the method of evaluating impedances of circuit elements under steady-state (DC and AC) and transient conditions.

DC steady-state impedance analysis is based on the fact that the voltages and currents in a circuit are constant. This means that the time derivatives of all voltages and currents are zero.

$$\frac{dv}{dt} = 0 \qquad\qquad 2.44$$

$$\frac{di}{dt} = 0 \qquad\qquad 2.45$$

The DC voltage-current relationships are developed here.

- Resistance
$$Z_R|_{DC} = R \qquad\qquad 2.46$$

- Inductance

$$v = L\frac{di}{dt} \qquad\qquad 2.47$$

$$\frac{di}{dt} = 0 \qquad\qquad 2.48$$

$$v = 0 \qquad\qquad 2.49$$

$$Z = \frac{v}{i} \qquad\qquad 2.50$$

$$Z_L|_{DC} = 0 \,(\text{a short circuit}) \qquad 2.51$$

- Capacitance

$$i = C\frac{dv}{dt} \qquad\qquad 2.52$$

$$\frac{dv}{dt} = 0 \qquad\qquad 2.53$$

$$Z_C|_{DC} = \infty \,(\text{an open circuit}) \qquad 2.54$$

Thus, in steady-state analysis of DC circuits, inductances are treated as short circuits and capacitances are treated as open circuits.

AC steady-state impedances are most easily evaluated by using the *phasor* representation of voltages and currents. Phasor conventions are derived from the definition of a complex number in exponential form.

$$e^{j\omega t} = \cos\omega t + j\sin\omega t \qquad 2.55$$

$$j = \sqrt{-1} \qquad\qquad 2.56$$

The real part of $e^{j\omega t}$ is $\cos\omega t$, and the imaginary part is $\sin\omega t$. $e^{j\omega t}$ can be used to represent a general sinusoid rather than specific sine or cosine waveforms. The time derivative of $e^{j\omega t}$ is

$$\frac{d}{dt}e^{j\omega t} = j\omega e^{j\omega t} \qquad\qquad 2.57$$

If a sinusoidal current is represented in the exponential form, then

$$i = I_m e^{j\omega t} \qquad\qquad 2.58$$

$$\frac{di}{dt} = j\omega I_m e^{j\omega t} = j\omega i \qquad 2.59$$

For a general sinusoidal voltage,

$$v = V_m e^{j\omega t} \qquad\qquad 2.60$$

$$\frac{dv}{dt} = j\omega V_m e^{j\omega t} = j\omega v \qquad 2.61$$

Sinusoidal impedances can also be defined using exponential forms.

- Resistance
$$Z_R|_{AC} = R \qquad 2.62$$

- Inductance
$$v = L\frac{di}{dt} = j\omega L i \qquad 2.63$$

$$Z_L|_{AC} = \frac{v}{i} = j\omega L \qquad 2.64$$

- Capacitance
$$i = C\frac{dv}{dt} = j\omega C v \qquad 2.65$$

$$Z_C|_{AC} = \frac{v}{i} = \frac{1}{j\omega C} \qquad 2.66$$

For *transient analysis* of linear circuits, a more general impedance form is useful. Currents and voltages are given the form e^{st}. s is a complex variable of the form $\sigma + j\omega$; it is the same as the Laplace transform variable. The voltage and current forms are $V_m e^{st}$ and $I_m e^{st}$ respectively, so the time derivatives are:

$$\frac{di}{dt} = sI_m e^{st} = si \qquad 2.67$$

$$\frac{dv}{dt} = sV_m e^{st} = sv \qquad 2.68$$

The general forms of the impedances are found from the voltage-current relationships. These impedance forms are important in transient analysis and in system analysis involving stability.

- Resistance
$$Z_R = R \qquad 2.69$$

- Inductance
$$v = L\frac{di}{dt} = sLi \qquad 2.70$$

$$Z_L = \frac{v}{i} = sL \qquad 2.71$$

- Capacitance
$$i = C\frac{dv}{dt} = sCv \qquad 2.72$$

$$Z_C = \frac{v}{i} = \frac{1}{sC} \qquad 2.73$$

9 COMMONLY ENCOUNTERED IMPEDANCES

A number of simple combinations of resistance, capacitance and inductance have been tabulated here for convenience. These are expressed in Laplace s-notation.

- Parallel R-C
$$Y = \frac{1 + sCR}{R}$$

$$Z = \frac{R}{1 + sCR}$$

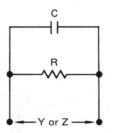

- Series R-C
$$Z = \frac{1 + sCR}{sC}$$

$$Y = \frac{sC}{1 + sCR}$$

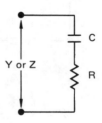

- Series R-L
$$Z = R\left(1 + \frac{sL}{R}\right)$$

$$Y = \frac{1}{R\left(1 + \dfrac{sL}{R}\right)}$$

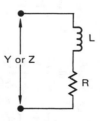

- Parallel R-L

$$Y = \frac{1 + \dfrac{sL}{R}}{sL}$$

$$Z = \frac{sL}{1 + \dfrac{sL}{R}}$$

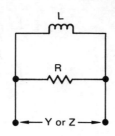

- Parallel L-C

$$Y = \frac{1 + s^2LC}{sL}$$

$$Z = \frac{sL}{1 + s^2LC}$$

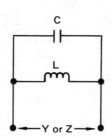

- Series L-C

$$Z = \frac{1 + s^2LC}{sC}$$

$$Y = \frac{sC}{1 + s^2LC}$$

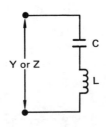

- Series R-L-C

$$Z = \frac{1 + sCR + s^2LC}{sC}$$

$$Y = \frac{sC}{1 + sCR + s^2LC}$$

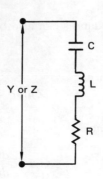

- Parallel R-L-C

$$Y = \frac{1 + \dfrac{sL}{R} + s^2LC}{sL}$$

$$Z = \frac{sL}{1 + \dfrac{sL}{R} + s^2LC}$$

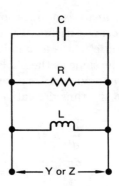

- Resistive Inductor in Parallel With C

$$Y = \frac{1 + sCR + s^2LC}{R\left(1 + \dfrac{sL}{R}\right)}$$

$$Z = \frac{R\left(1 + \dfrac{sL}{R}\right)}{1 + sCR + s^2LC}$$

- Resistive Inductor in Parallel With a Lossy Capacitor

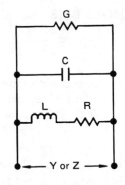

$$Y = \frac{1+GR}{R} \; \frac{1 + \dfrac{s(CR+GL)}{1+GR} + \dfrac{s^2 LC}{1+GR}}{1 + \dfrac{sL}{R}}$$

$$Z = \frac{R}{1+GR} \; \frac{1 + \dfrac{sL}{R}}{1 + \dfrac{s(CR+GL)}{1+GR} + \dfrac{s^2 LC}{1+GR}}$$

PRACTICE PROBLEMS

Warmups

1. Using the same diameter nichrome wire as in example 2.1, determine the length needed to obtain a resistance of 1 ohm at 100°C. Determine the resistance at 20°C.

2. For the circuit in example 2.2, assume the switch has been closed for a long time and is opened at $t = 0$. Determine (a) the current through the capacitor at the instant the switch is opened, (b) the time derivative of the capacitor voltage at the instant the switch is opened, and (c) the voltage that will exist across the capacitor after a long time.

3. For the circuit in example 2.3, assume that the switch has been closed for a long time (the final conditions of example 2.3 apply), and the switch is opened at $t = 0$. Determine at the instant the switch opens (a) the current in the inductance, (b) the time rate of change of the current in the inductance, and (c) the voltage across the inductance.

4. Repeat example 2.4 with the 120 volt source connected to winding 2 and the 60 ohm load attached to winding 1.

5. Obtain a linear model of voltage versus current for the characteristic curve shown in example 2.5 over the current range of 0 to 6 amps. Obtain the current source and voltage source equivalent circuits.

6. A 5 volt, 25 amp DC source has voltage regulation of 1%. Obtain an equivalent circuit for this source.

7.

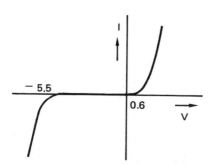

All silicon diodes have a reverse breakdown voltage essentially the same as the avalanche voltage of a zener diode. This results in three distinct regions of operation: one for the normal forward region, one for the avalanche region, and an intermediate region where the device behaves essentially as an open circuit. Obtain the linear models in these three regions for a 1 amp, 2 watt silicon diode with a reverse breakdown voltage of 5.5 volts and a maximum reverse voltage of 5.6 volts.

8. A three-terminal device is described by the following z-parameter equations:

$$v_{in} = 250i_{in} + 5i_{out}$$
$$v_{out} = -100i_{in} + 25i_{out}$$

Obtain an equivalent circuit for this device.

9. Common base transistor configurations are often described in terms of *common base y-parameters*: y_{ib}, y_{rb}, y_{fb}, and y_{ob}. (The b's in the subscripts indicate that the parameters were obtained from a common base configuration with the emitter as the input terminal and the collector the output terminal.) A common circuit model for the transistor used in this configuration is shown. Obtain the y-parameters for this common base circuit.

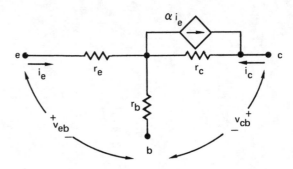

10. For a photodiode with characteristics shown in figure 2.15 and used in the circuit for example 2.11 with a 5 volt source and 10,000 ohm load, determine the load voltage as a function of illumination level. Over what range of illumination levels will this result be valid? Assume the curves extend to −5 volts on the third quadrant of the characteristic curves.

11. Repeat example 2.12 for $T_1 = 0°C$ and $T_2 = 100°C$.

12. Use the dimensions given in example 2.13, but assume the conductor is an n-type semiconductor with 10^{19} electrons per cubic meter. The current is 0.1 amp. Obtain an equivalent circuit for a transducer using the Hall voltage to measure magnetic field intensity. Assume that the voltage pickups are capacitive so that there is no connection of the output to the current circuit.

13. For the circuit of example 2.20, obtain the Norton equivalent circuit.

14. Consider the circuit of example 2.20 with an 8 ohm speaker attached to terminals a-b. Show that the necessary turns ratio (N_1/N_2) to provide maximum power transfer is 35.36.

15. The elements Z_1 and Z_2 are interchanged in example 2.22, and the voltage, v_s, takes on the value of

$$20 + 5 \cos 2236t \text{ volts}$$

Find the voltage across the 100 ohm resistor.

16. The circuit of example 2.25 has $r_b = 200$ ohms, $\beta = 50$, $r_c = 2500$ ohms, and $R_e = 10$ ohms. Find the voltage gain v_{out}/v_{in} when a 1000 ohm load is placed across the v_{out} terminals. Note that the output current i_2 is $-v_{out}/1000$ for this load.

17. Using y-parameters, obtain the total circuit y-parameters for the circuits indicated by the dashed lines. Hint: first find the y-parameters of the two indicated two-port networks, then combine them to obtain the total network y-parameters.

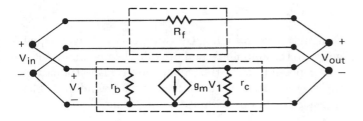

Concentrates

1. In the circuit shown in example 2.22, the voltage source is operating at a frequency of 2000 rad/sec. Specify the impedances Z_1 and Z_2 so that maximum power is transferred to the 100 ohm load resistance. (Hint: Z_1 and Z_2 must be reactive or else they will absorb some of the power.) In specifying Z_1 and Z_2, give their values in microfarads and/or millihenries. There are two possible sets of answers.

2.

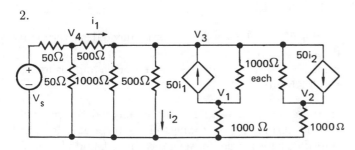

For the circuit shown, obtain the node voltage equations in matrix form. Solve these equations for v_1 and v_2. (The actual output is $v_1 - v_2$.) Obtain the voltage gain v_{out}/v_s for this emitter-coupled amplifier.

3. For the circuit shown, solve for the current in the 200 ohm resistance connected to the secondary of the ideal transformer.

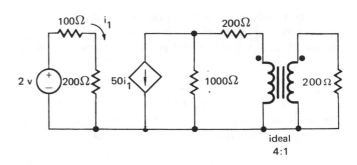

4. A 60 Hz source has its voltage measured under various loads, with the results shown below:

load	voltmeter reading
open circuit	120 volts rms
60 Ω resistance	72 volts rms
60 Ω pure capacitance	360 volts rms
60 Ω pure inductance	51.43 volts rms

Determine the voltage for a load of 120 ohms pure capacitance.

5. A bridge network is used to measure the temperature of a chemical bath.

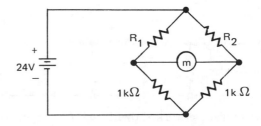

R_1 and R_2 are identical thermistors with negative temperature coefficients of resistance of $-4\%/°C$. At $25°C$ they are both 1500 ohms. The meter (m) has an internal resistance of 800 ohms. Resistance R_2 is held at $25°C$. (a) Determine the meter current when R_1 is at $20°C$. (b) Determine the meter current when R_1 is at $30°C$.

6. An oscillator circuit is shown. Determine the minimum value of K such that $e_0 = E$. Also, find the frequency at which $e_0 = E$.

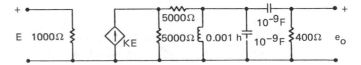

7. A transformer with 100 primary turns, which has a winding resistance of 10 ohms, is operating with an applied primary voltage of 9 volts (rms) at a current of 3 mA (also rms).

Suddenly the current increases to about 30 mA while the terminal voltage remains about the same. Assuming the increase in current is due to a short circuit in the primary winding, how many turns have been shorted?

8. An infinite two-dimensional array of one ohm resistors are laid out as indicated. Determine the resistance between any two adjacent nodes, such as node a and node b as indicated.

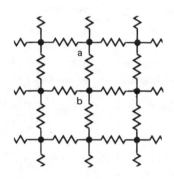

9.

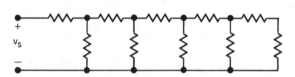

For the ladder network shown, each of the resistances is 1 ohm. Determine the input resistance as seen from v_s, and the output resistance seen by the right-hand resistance.

10. For the circuit shown, R_L is the load resistance.

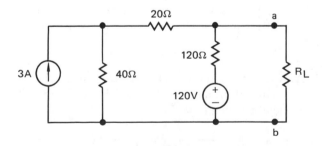

(a) Find the Thevenin equivalent circuit as seen by the load resistance (i.e., to the left of terminals a and b).

(b) Find the value of load resistance which will permit the maximum power to be received by the load resistance. Determine the power R_L would receive in this case.

(c) Determine the power delivered to an R_L which is half the value found in part b.

(d) Determine the power delivered to an R_L which is twice the value found in part (b).

Timed

1. A weak signal source with an output impedance of 3500 ohms must feed two amplifiers; one with an input impedance of 50 ohms, and the other with an input impedance of 70 ohms. All of these impedances are pure real numbers.

It has been decided to use a transformer to couple the signals to the amplifiers, with the intention of maximizing the power transmitted to them. It has also been decided that the two amplifiers are to receive equal input power.

Design a three-winding transformer to deliver equal power to the amplifiers, and provide the maximum total power possible from the signal source. Determine the turns ratios needed for the transformer windings. The transformer can be assumed to be ideal.

2. Three windings of a linear transformer are evaluated with an impedance measuring device. Winding #1 is found to have an inductance $L_1 = 2$ mH, and its terminals are tagged or marked a and b. Winding #2 is measured to have 2.5 mH of inductance, and its terminals are marked or tagged c and d. Winding #3 has a measured inductance of 3 mH, and its terminals are marked e and f.

In the fourth test, terminal b is connected to terminal c, and terminal d is connected to terminal e. The impedance measured between terminals a and f is found to be 1.5 mH.

In the fifth test, terminal b is connected to terminal d, and terminal c is connected to terminal e. The impedance measured between terminals a and f is found to be 3.5 mH.

In the sixth test, terminal a is connected to terminal d, and terminal c is connected to terminal e. The impedance measured between terminals b and f is found to be 21.5 mH.

Determine the mutual inductances between the three pairs of windings.

3. In the circuit shown, the input signal source has a voltage of 150 mV and frequency of 10^6 radians per second.

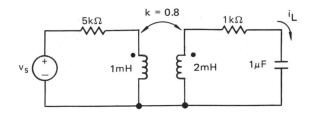

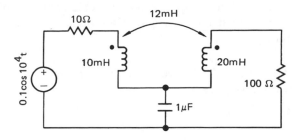

(a) Determine the output impedance seen from the capacitor.

(b) Determine the value of the current i_L.

(c) Change the capacitance value to obtain the largest possible amplitude of i_L.

4. A 100 mile long communication line uses 100 volt power supplies, one at each end, to provide a signal to be observed at the other end. The receiver at each end has an input impedance of 600 ohms resistive. The composite wire and ground resistance varies between 15 ohms per mile and 20 ohms per mile. The line has a maximum shunt conductance of 25 μmhos for one mile of the conductor, which of course increases with length.

(a) Draw a T-model for the circuit, with the conductance lumped together in the middle of the T, and half the line plus ground resistance on either side of the conductance.

(b) Calculate the current drawn from a 100 volt battery driving the line from one end.

(c) Calculate the current reaching the 600 ohm receiver at the other end.

5. A particular power source can be considered to be composed of several DC power supplies and resistances, so that it is a linear system within certain limits. For our purposes we can presume those limits are observed. A pair of terminals is available on this system. The voltage at these terminals is 28 volts when nothing else is connected to it except the voltmeter, which has an internal resistance of 1000 ohms. When a resistance of 100 ohms is connected across these same terminals, the voltmeter measures 25 volts when it is connected to make the voltage measurement across the 100 ohm resistance.

(a) Devise the simplest possible equivalent circuit to represent the two terminals of the power source above.

(b) Assuming your model is valid under short circuit conditions, what short-circuit current could these terminals provide?

6. Determine the current through the 100 ohm resistor on the right.

7. A certain oscilloscope has an input resistance of 1 Megohm and an input capacitance of approximately 10 picofarads. A ten-to-one attenuator is obtained by connecting a 10 Megohm resistance in series with the oscilloscope terminals. Under these conditions it is found that the oscilloscope is unable to follow signals at one megahertz in frequency.

(a) For a step input from an ideal voltage source, determine the time until the oscilloscope input voltage reaches 70% of its final value.

(b) Design a compensator consisting of a shunt capacitance across the 10 megohm resistance so that the oscilloscope can follow input signals well above one megahertz.

8.

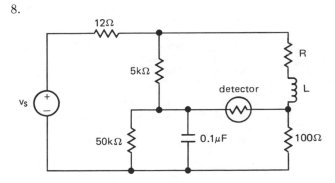

The Bridge circuit shown is in balance, in that the voltage across the detector is zero. Determine the values of R and L. V_S is a sinusoid.

9.

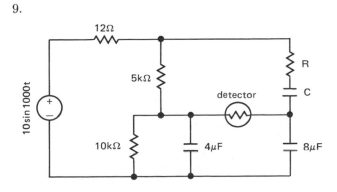

PROFESSIONAL PUBLICATIONS ● Belmont, CA

The Bridge circuit shown is in balance, in that the voltage across the detector is zero. Determine the values of R and C.

10. Determine the Thevenin equivalent circuit as seen from the terminals x-y.

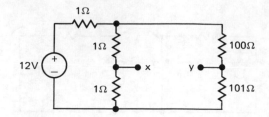

3 WAVEFORMS, POWER, AND MEASUREMENTS

PART 1: PERIODIC SIGNALS

Nomenclature

a_0	Fourier coefficient	–
a_n	Fourier coefficient	–
b_n	Fourier coefficient	–
B	susceptance	mhos
$\mathbf{f}$	a function	–
f	frequency, or function value	Hz,–
$\mathbf{F}$	a phasor function	–
G	conductance	mhos
I	phasor current	amps
I^*	complex conjugate of I	amps
j	imaginary operator, $\sqrt{-1}$	–
n	an integer	–
p	instantaneous power	watts
PF	power factor	–
Q	reactive power	volt-amperes reactive
R	resistance	ohms
RMS	root-mean-square value	–
S	apparent power	volt-amperes
t	time	seconds
T	period of periodic function	seconds
v	instantaneous voltage	volts
V	phasor voltage	volts
X	reactance	ohms
Y	admittance	mhos
Z	impedance	ohms

Symbols

ϕ	an angle	degrees or radians
θ	an angle	degrees or radians
ω	radian frequency	radians/ second

Subscripts

C	capacitance or capacitive
f	fundamental
fs	full-scale
L	inductance or inductive
p	peak value
s	source
thr	threshold

A periodic signal is one which repeats continuously. Examples are constant currents and voltages, sinusoids, square waves, pulse trains, and triangular and sawtooth waveforms. Periodic signals can be investigated with Fourier series analysis.

Several different parameters are useful in describing periodic signals, depending on the particular waveform and the detail desired. Parameters that describe periodic signals include the average value, root-mean-square (RMS) value, period or time between repetitions, frequency content or spectrum, and peak value (or some other significant measure of the *size* of the waveform).[1]

1 AVERAGE VALUES

The *average value* of a periodic waveform is the first (zero-frequency) term of the Fourier series described in chapter 1. If $\mathbf{f}(t)$ is a periodic function which repeats itself each T seconds, then the average value of $\mathbf{f}(t)$, f_{avg}, is given by equation 3.1. (t_1 is some convenient time for the evaluation of the integral.)

---

[1] The RMS value is also known as the *effective* and *heating value*.

$$f_{\text{avg}} = \frac{1}{T} \int_{t_1}^{t_1+T} \mathbf{f}(t)\, dt \qquad 3.1$$

Since the act of integration can be interpreted as finding the area under a curve, equation 3.1 divides the net area of the waveform by the period T. This is illustrated in figure 3.1. For the function shown, the left-most shaded area is a positive area, and the shaded area on the right is a negative area. The negative area is subtracted from the positive area, and the average value is the net remaining area divided by the period T.

$$f_{\text{avg}} = \frac{\text{positive area} - \text{negative area}}{T} \qquad 3.2$$

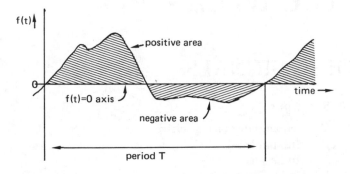

Figure 3.1   Areas Used to Find the Average Value

Example 3.1

A *constant function* (i.e., $\mathbf{f}(t) = \text{constant}$) has an arbitrary period. It can be thought of as having any convenient period. In this example, the function is a constant voltage of 3 volts. A period of T is arbitrarily marked on the time axis. From equation 3.1, the average voltage is

$$v_{\text{avg}} = \frac{1}{T} \int_{t=0}^{T} 3\, dt = \frac{3t}{T}\Big|_{t=0}^{t=T}$$

$$= \frac{3(T-0)}{T} = 3 \text{ volts}$$

Using equation 3.2, it is seen that the positive area (shaded) is $3 \times T = 3T$. Dividing by T, the value of 3 is obtained. The obvious result of this example is that the average value of a constant function is that constant.

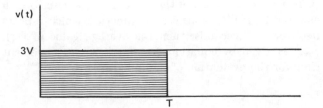

The average value of a *pure sinusoid* is zero since it has equal positive and negative areas, as shown in figure 3.2.

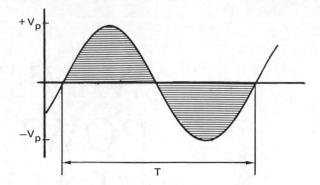

Figure 3.2   A Pure Sinusoid

A *uniform pulse train* with a pulse width of ΔT seconds and a period of T seconds is shown in figure 3.3. The maximum voltage is V_{max} and the minimum is V_{min}. The net area is

$$V_{\text{min}}T + (V_{\text{max}} - V_{\text{min}})\Delta T$$

This, when divided by the period T (equation 3.2), gives an average value of

$$V_{\text{min}} + (V_{\text{max}} - V_{\text{min}})\frac{\Delta T}{T}$$

Using integration and equation 3.1,

$$v_{\text{avg}} = \frac{1}{T} \int_{t=0}^{\Delta T} V_{\text{max}}\, dt + \frac{1}{T} \int_{t=\Delta T}^{T} V_{\text{min}}\, dt$$

$$= V_{\text{max}}\frac{\Delta T}{T} + V_{\text{min}}\frac{T - \Delta T}{T} \qquad 3.3$$

This is identical to the value found by the area method. If $V_{\text{min}} = 0$, then

$$v_{\text{avg}} = V_{\text{max}}\frac{\Delta T}{T} \qquad 3.4$$

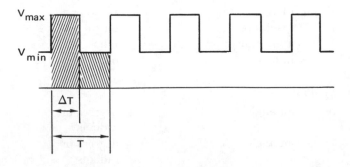

Figure 3.3   A Uniform Pulse Train

If $V_{\min} = -V_{\max}$, and $\Delta T = T/2$ (a square wave), $v_{\text{avg}} = 0$.

A *triangular waveform* is shown in figure 3.4 with $t = 0$ taken at the minimum and $t = T$ at the following minimum. The average value can be found by the area method (equation 3.2).

$$v_{\text{avg}} = \left(\frac{1}{2T}\right)(V_{\max} - V_{\min})T + \left(\frac{1}{T}\right)V_{\min}T$$

$$= \frac{1}{2}(V_{\max} + V_{\min}) \qquad 3.5$$

Equation 3.5 applies to all forms of triangular waves, regardless of whether $V_{\min}$ is positive or negative.

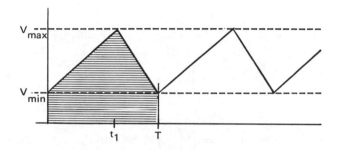

Figure 3.4 A Triangular Wave

For a *rectified sinusoid*, one half of the wave has been eliminated.[2] For purposes of analysis, it is usually assumed that the negative part has been eliminated. An ideally rectified sinusoid is shown in figure 3.5. During the interval $0 < t < T$, the voltage can be represented as

$$v(t) = \begin{cases} V_p \sin \omega t & (t < \frac{T}{2}) \\ 0 & (\frac{T}{2} < t < T) \end{cases} \qquad 3.6$$

$$\omega = \frac{2\pi}{T} \qquad 3.7$$

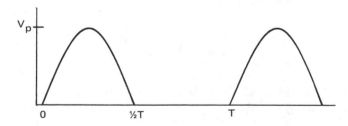

Figure 3.5 A Rectified Sinusoid

Equation 3.1 can be used to find the average value. The following calculation recognizes that $\omega T = 2\pi$.

---

[2] Rectified sinusoids are also known as *half-wave sinusoids*.

$$v_{\text{avg}} = \frac{1}{T}\int_{t=0}^{T/2} V_p \sin \omega t\, dt + \frac{1}{T}\int_{t=T/2}^{T} 0\, dt$$

$$= \frac{V_p}{\omega T}(-\cos \omega t)\Big|_{t=0}^{T/2}$$

$$= \frac{V_p}{\pi} \qquad 3.8$$

For a full-wave rectified sinusoid, the period is half as long. The average is $2V_p/\pi$.

Real *rectifier diodes* will not pass current until a threshold voltage is reached. The rectified voltage is the difference between the sinusoid and the threshold value. This is illustrated in figure 3.6 for full-wave rectification. The first graph shows the ideal rectified waveform and the threshold level; the second graph shows the net voltage.

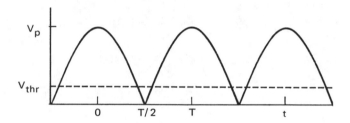

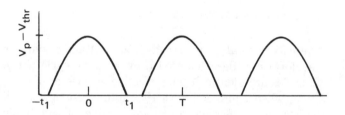

Figure 3.6 Determination of the Average Value of a Sinusoid After Full-Wave Rectification with Non-Ideal Diodes

Because of symmetry, half the average value can be found from equation 3.1 by integrating from $t = 0$ to $t = T/2$. The waveform is represented by equation 3.9.

$$v(t) = \begin{cases} V_p \cos \frac{\pi}{T}t - V_{\text{thr}} & \text{for } 0 < t < t_1 \\ 0 & \text{for } t_1 < t < \frac{T}{2} \end{cases} \qquad 3.9$$

At $t = t_1$,

$$V_p \cos \frac{\pi}{T}t_1 = V_{\text{thr}} \qquad 3.10$$

$$\frac{1}{2}v_{\text{avg}} = \frac{1}{T}\int_{t=0}^{t_1}\left(V_p \cos \frac{\pi}{T}t - V_{\text{thr}}\right)dt$$

$$= \frac{1}{T}\left(\frac{T}{\pi}\right)V_p \sin \frac{\pi}{T}t_1 - V_{\text{thr}}\left(\frac{t_1}{T}\right) \qquad 3.11$$

From equation 3.10 and $\sin^2 x + \cos^2 x = 1$,

$$\sin\left(\frac{\pi}{T}\right) t_1 = \sqrt{1 - \left(\frac{V_{\text{thr}}}{V_p}\right)^2} \qquad 3.12$$

$$\frac{t_1}{T} = \frac{1}{\pi} \arccos\left(\frac{V_{\text{thr}}}{V_p}\right) \qquad 3.13$$

$$v_{\text{avg}} = \frac{2}{\pi} V_p \sqrt{1 - \left(\frac{V_{\text{thr}}}{V_p}\right)^2}$$

$$- \frac{2}{\pi} V_{\text{thr}} \arccos\left(\frac{V_{\text{thr}}}{V_p}\right) \qquad 3.14$$

The arccos in equation 3.14 must be expressed in radians. If it is assumed that V_{thr} is negligible (i.e., diodes are ideal), the following errors are generated.

$$1\% \text{ error for} \frac{V_{\text{thr}}}{V_p} = 0.0064$$

$$2\% \text{ error for} \frac{V_{\text{thr}}}{V_p} = 0.0128$$

$$5\% \text{ error for} \frac{V_{\text{thr}}}{V_p} = 0.0321$$

$$10\% \text{ error for} \frac{V_{\text{thr}}}{V_p} = 0.0650$$

2   RMS VALUES

The *effective value* of a steady-state voltage or current waveform can be calculated from a single cycle, since all cycles are identical. The effective voltage or current value, when combined with the circuit resistance, will determine the average power for a cycle. For a single complete cycle, the effective value is calculated from the average of the square of the signal. The effective value then, is the square Root of the Mean of the Square of the signal, hence the name RMS.

$$f_{\text{RMS}}^2 = \frac{1}{T} \int_{t_1}^{t_1+T} f^2(t)\, dt \qquad 3.15$$

T in equation 3.15 is the period of the repeating function $f(t)$. t_1 is a convenient time to begin the evaluation of the integral.

Example 3.2

Calculate the RMS of a constant 3 volt signal.

Using equation 3.15, the RMS value is

$$v_{\text{RMS}}^2 = \frac{1}{T} \int_0^T 3^2\, dt = \left(\frac{1}{T}\right) 9(T - 0)$$

$$v_{\text{RMS}} = \sqrt{9} = 3 \text{ volts}$$

For a pure *sinusoidal voltage*,

$$v(t) = V_p \cos(\omega t + \phi) \qquad 3.16$$

V_p is the peak value or amplitude, $\omega = 2\pi/T$ is the radian frequency, and ϕ is the arbitrary phase angle. From equation 3.15,

$$v_{\text{RMS}}^2 = \frac{1}{T} \int_0^T V_p^2 \cos^2(\omega t + \phi)\, dt \qquad 3.17$$

$$\cos^2 x = \frac{1}{2}(1 + \cos 2x) \qquad 3.18$$

$$v_{\text{RMS}}^2 = \frac{1}{T} \int_0^T \frac{1}{2} V_p^2\, dt + \frac{1}{T} \int_0^T \frac{1}{2} V_p^2 \cos(2\omega t + 2\phi)\, dt$$

$$= \frac{1}{2} V_p^2 + 0 \qquad 3.19$$

$$v_{\text{RMS}} = \sqrt{\left(\frac{1}{2}\right)} V_p = 0.707 V_p \qquad 3.20$$

The RMS value of a *uniform pulse train* as shown in figure 3.3 is

$$v_{\text{RMS}}^2 = \frac{1}{T} \int_0^{\Delta T} V_{\text{max}}^2\, dt + \frac{1}{T} \int_{\Delta T}^T V_{\text{min}}^2\, dt$$

$$= V_{\text{max}}^2 \frac{\Delta T}{T} + V_{\text{min}}^2 \frac{T - \Delta T}{T}$$

$$= \left(V_{\text{max}}^2 - V_{\text{min}}^2\right) \frac{\Delta T}{T} + V_{\text{min}}^2 \qquad 3.21$$

If $V_{\text{min}} = 0$, then

$$v_{\text{RMS}} = V_{\text{max}} \sqrt{\frac{\Delta T}{T}} \qquad 3.22$$

If the pulse train is a square wave with $V_{\text{min}} = -V_{\text{max}}$, then $v_{\text{RMS}} = V_{\text{max}}$.

For simplification, the *triangular* wave shown in figure 3.4 is redefined in figure 3.7.

$$v(t) = \begin{cases} \dfrac{-(V_{\text{max}} - V_{\text{min}})}{t_a} t + V_{\text{min}} & (-t_a < t < 0) \\[2mm] \dfrac{V_{\text{max}} - V_{\text{min}}}{t_b} t + V_{\text{min}} & (0 < t < t_b) \end{cases}$$

$$3.23$$

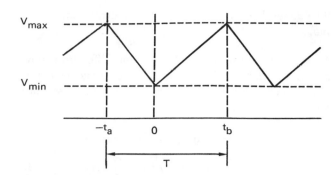

Figure 3.7 Triangular Wave Redefined

The application of equation 3.15 to figure 3.7 is

$$V_{\text{RMS}}^2 = \frac{1}{T}\int_{-t_a}^{0}\left(-\frac{V_{\max}-V_{\min}}{t_a}t+V_{\min}\right)^2 dt$$

$$+\frac{1}{T}\int_{0}^{t_b}\left(\frac{V_{\max}-V_{\min}}{t_b}t+V_{\min}\right)^2 dt$$

$$=\frac{1}{3}(V_{\max}^2+V_{\min}^2+V_{\max}V_{\min}) \qquad 3.24$$

If $V_{\min}=0$ or if $V_{\min}=-V_{\max}$, then

$$v_{\text{RMS}}=\frac{V_{\max}}{\sqrt{3}} \qquad 3.25$$

For the *half-wave rectified sinusoid* shown in figure 3.5, equation 3.15 becomes

$$v_{\text{RMS}}^2 = \frac{1}{T}\int_{0}^{T/2} V_p^2 \sin^2\omega t\, dt$$

$$=\frac{V_p^2}{T}\int_{0}^{T/2}\frac{1}{2}(1-\cos 2\omega t)\,dt$$

$$=\frac{1}{4}V_p^2 \qquad 3.26$$

Thus, $v_{\text{RMS}}=(1/2)V_p$.

A full wave rectified sinusoid has the same RMS value as a pure sinusoid. The RMS value of a sinusoid after full-wave rectification using diodes with a threshold can also be calculated. From figure 3.6, the application of equation 3.15 is

$$\frac{1}{2}v_{\text{RMS}}^2 = \frac{1}{T}\int_{0}^{t_1}(V_p\cos\omega t-V_{\text{thr}})^2\,dt$$

$$=\frac{1}{T}\int_{0}^{t_1}\left(\frac{1}{2}V_p^2(1+\cos 2\omega t)\right.$$

$$\left.-2V_pV_{\text{thr}}\cos\omega t+V_{\text{thr}}^2\right)dt$$

$$=\frac{1}{T}\left(\frac{1}{2}V_p^2 t_1+\frac{V_p^2}{4\omega}\sin 2\omega t_1\right.$$

$$\left.-2\frac{V_pV_{\text{thr}}}{\omega}\sin\omega t_1+V_{\text{thr}}^2 t_1\right) \qquad 3.27$$

However,

$$\omega T = \pi \qquad 3.28$$

$$\sin 2\omega t_1 = 2\sin\omega t_1\cos\omega t_1 \qquad 3.29$$

Therefore,

$$\frac{1}{2}v_{\text{RMS}}^2 = \left(\frac{1}{2}V_p^2+V_{\text{thr}}^2\right)\frac{t_1}{T}$$

$$+\frac{1}{\pi}\sin\omega t_1\left(\frac{1}{2}V_p^2\cos\omega t_1-2V_pV_{\text{thr}}\right) \qquad 3.30$$

Substituting equations 3.10, 3.12, and 3.13,

$$v_{\text{RMS}}^2 = \frac{1}{2}V_p^2\left(\left(1+\frac{2V_{\text{thr}}^2}{V_p^2}\right)\left(\frac{2}{\pi}\right)\arccos\frac{V_{\text{thr}}}{V_p}\right.$$

$$\left.-\frac{6V_{\text{thr}}}{\pi V_p}\sqrt{1-\left(\frac{V_{\text{thr}}}{V_p}\right)^2}\right) \qquad 3.31$$

If

$$\frac{V_{\text{thr}}}{V_p}=0$$

then

$$\arccos 0 = \frac{\pi}{2}$$

Then

$$v_{\text{RMS}}^2 = \frac{1}{2}V_p^2$$

This is the same as for a complete sinusoid.

3   PARAMETERS OF FOURIER SERIES

All periodic functions can be represented by a *Fourier series*. This series, as discussed in chapter 1, can be written in the form

$$f(t)=\frac{1}{2}a_o+\sum_{n=1}^{\infty}a_n\cos\frac{2\pi n}{T}t+\sum_{n=1}^{\infty}b_n\sin\frac{2\pi n}{T}t \qquad 3.32$$

The *combined form* can also be used.

$$f(t)=\frac{1}{2}a_0+\sum_{n=1}^{\infty}(a_n^2+b_n^2)^{1/2}\cos\left(\frac{2\pi n}{T}t-\phi_n\right) \qquad 3.33$$

$$\phi_n=\arctan\left(\frac{b_n}{a_n}\right) \qquad 3.34$$

The average value of the function represented by the Fourier series is

$$f_{\text{avg}}=\frac{1}{2}a_0$$

The RMS value is

$$f_{\text{RMS}} = \sqrt{\left(\frac{1}{2}a_0\right)^2 + \frac{1}{2}\sum_{n=1}^{\infty}(a_n^2 + b_n^2)} \qquad 3.35$$

Example 3.3

A voltage waveform is approximated by the truncated Fourier series $v(t)$. Determine (a) the average value, (b) the fundamental frequency, and (c) the RMS value.

$$v(t) = 5 + 3\cos 100t + 0.5\cos 300t + 0.1\cos 600t$$
$$+ 4\sin 100t + 0.2\sin 200t$$

(a) The value of $(1/2)a_0$ is 5, therefore, $v_{\text{avg}} = 5$.

(b) The lowest frequency, other than zero, is 100 rad/sec, so the fundamental frequency is

$$\frac{100}{2\pi} = 15.9 \text{ Hz}$$

(c) From equation 3.35,

$$v_{\text{RMS}}^2 = 5^2 + \frac{1}{2}(3^2 + 0.5^2 + 0.1^2 + 4^2 + 0.2^2)$$
$$= 37.64$$
$$v_{\text{RMS}} = 6.13 \text{ volts}$$

Frequency spectra are generally defined in terms of energy content per unit frequency. For a Fourier series with discrete frequencies, it is more convenient to define energy per second or average power as a function of frequency. Since power is proportional to the square of the RMS signal value and the Fourier series is a series of pure sinusoids, the power spectra can be constructed from the RMS values of each of the sinusoids. If ω_f is the fundamental radian frequency, then

$$P(0) = \left(\frac{1}{2}a_0\right)^2 \qquad 3.36$$

$$P(\omega_f) = \frac{1}{2}(a_1^2 + b_1^2) \qquad 3.37$$

$$P(2\omega_f) = \frac{1}{2}(a_2^2 + b_2^2) \qquad 3.38$$

$$P(n\omega_f) = \frac{1}{2}(a_n^2 + b_n^2) \qquad 3.39$$

$$\omega_f = \frac{2\pi}{T} \qquad 3.40$$

$P(0)$ is the spectral value at $\omega = 0$; $P(\omega_f)$ is the spectral value at $\omega = \omega_f$; and $P(2\omega_f)$ is the spectral value at $\omega = 2\omega_f$.

The general term is $P(n\omega_f)$, the spectral value at $\omega = n\omega_f$.

Signal spectra for voltage or current differ from power spectra in that the RMS value of each frequency component is used. For example, the RMS values associated with a voltage are:

$$v(0) = \sqrt{P(0)} \qquad 3.41$$

$$v(\omega_f) = \sqrt{P(\omega_f)} \qquad 3.42$$

$$v(2\omega_f) = \sqrt{P(2\omega_f)} \qquad 3.43$$

$$v(n\omega_f) = \sqrt{P(n\omega_f)} \qquad 3.44$$

Example 3.4

A voltage signal is approximated by the truncated Fourier series

$$v(t) = 2 + 1.5\cos 1000t + 0.3\sin 2000t$$
$$+ 0.5\cos 3000t + 0.1\sin 4000t$$

Plot the power and voltage spectra for this signal.

The power spectrum is found from equations 3.36 through 3.39.

$$P(0) = \left(\frac{1}{2}a_0\right)^2 = 4$$

$$P(1000) = \frac{1}{2}(1.5^2 + 0^2) = 1.125$$

$$P(2000) = \frac{1}{2}(0^2 + 0.3^2) = 0.045$$

$$P(3000) = \frac{1}{2}(0.5^2 + 0^2) = 0.125$$

$$P(4000) = \frac{1}{2}(0^2 + 0.1^2) = 0.005$$

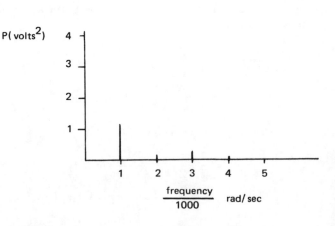

The voltage spectrum is found from equations 3.41 through 3.44.

$$v(0) = \sqrt{4} = 2$$
$$v(1000) = \sqrt{1.125} = 1.061$$
$$v(2000) = \sqrt{0.045} = 0.212$$
$$v(3000) = \sqrt{0.125} = 0.354$$
$$v(4000) = \sqrt{0.005} = 0.071$$

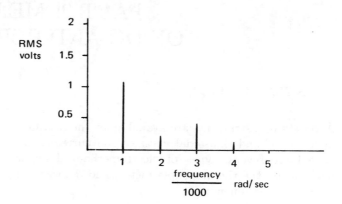

PART 2: MEASUREMENT OF DC AND PERIODIC SIGNALS

4 INTRODUCTION

A variety of instruments are available for the measurement of DC and sinusoidal voltages and currents. Included are several types of electromechanical meter movements, the thermocouple meter, and a variety of electronic instruments.

Electromechanical instruments respond to a voltage or current through the mechanical rotation of a rotor with an attached pointer. These instruments are calibrated so that specific angles of rotation correspond to specific voltage or current levels, and a scale is so marked.

Because of their mechanical inertia, these instruments are normally inadequate in the 5–20 Hz frequency range. Significant inductance limits usefulness to a maximum frequency between 100 and 2500 Hz, depending on the movement type. Such instruments are primarily used in DC or power systems.

Electronic instruments can also employ an electromechanical movement, or they may use a digital readout system. Due to the inclusion of special signal conditioning circuits, electronic meters can be used at much higher frequencies than electromechanical movements.

5 PERMANENT MAGNET MOVING COIL INSTRUMENTS

The *permanent magnet moving coil meter movement*, known as the *d'Arsonval movement*, produces an electromagnetic torque proportional to the current flowing in its rotating coil. The pointer is attached to the coil, as is a restraining spring. Coil torque is proportional to the current; the mechanical spring torque is proportional to the angular displacement, but opposite in direction. The pointer will come to rest at the position where the torques are equal and opposite.

The d'Arsonval movement is used primarily to measure average or DC currents and voltages. A similar arrangement is used in data recording systems, where the pointer is replaced with a pen or thermal stylus. Such *pen-recorders* are capable of responding to waveforms with frequencies up to about 15 Hz, although special designs employing compensation can extend the useful range to as much as 100 Hz.

A specific d'Arsonval meter movement is designed to show a full-scale pointer deflection for some specific current, I_{fs}. The usual range of full-scale coil currents is 50 microamps to 10 milliamps. Usually the lower full-scale current requires more turns of wire on the coil, producing a correspondingly higher coil resistance.

Another parameter in ammeter applications is the DC voltage necessary to cause I_{fs} to flow. A standard value of 50 to 100 millivolts (usually 50) is used. This typically requires an additional resistance, called a *swamping resistor*, in series with the coil.

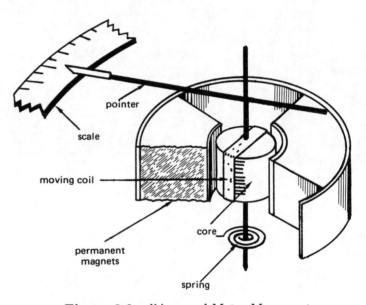

Figure 3.8 d'Arsonval Meter Movement

A. DC Voltmeters

A d'Arsonval meter movement can be incorporated into a *DC voltmeter* (with a full-scale voltage, V_{fs}) by connecting a series resistance to limit coil current to I_{fs} when V_{fs} is applied.

$$\frac{R_{coil} + R_{ext}}{V_{fs}} = \frac{1}{I_{fs}} \qquad 3.45$$

Equation 3.45 shows that a full-scale swing with a voltage greater than $R_{coil} I_{fs}$ can be obtained by proper selection of R_{ext}. The quantity $1/I_{fs}$ is fixed for a particular meter movement. It is called the *sensitivity*, with units of ohms/volt.

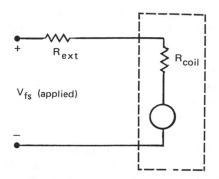

Figure 3.9 DC Voltmeter

Common DC voltmeters ($I_{fs} = 1$ mA) have a sensitivity of 1000 ohms per volt, while more expensive models have sensitivities of 5000, 10,000, and 20,000 ohms/volt.

Accuracy is another design consideration. Movements are available with standard accuracies of 0.5, 1, 2, and 5%. Accuracy is defined as the maximum error relative to the full-scale value, regardless of the current at which that error occurs. In precision instruments, a calibration curve is provided to convert a reading from the voltage measured.

Example 3.5

A 1000 ohm/volt, 2% meter movement has a coil resistance of 200 ohms. Specify the external resistance necessary to provide a full-scale reading of 5 volts with an overall accuracy of at least 3%.

Use equation 3.45 with $V_{fs} = 5$ and $R_{coil} = 200$.

$$200 + R_{ext} = (1000)(5)$$
$$R_{ext} = 4800 \text{ ohms (nominal)}$$

To account for the accuracy required, the maximum error introduced by an error in resistance will occur at full-scale voltage. Since the voltage at which maximum error occurs in the meter is unknown, it is assumed to occur at the same value as the resistance error, i.e., at full-scale voltage. Therefore, the maximum error that can be introduced by R_{ext} is 1%. As R_{ext} is much greater than R_{coil}, a 1% tolerance on R_{ext} is the maximum that can be allowed. Therefore, R_{ext} must be a 4800 ohm, 1% resistor.

B. DC Ammeters

The d'Arsonval movement, when used as an ammeter, is typically specified in terms of the voltage needed to give full-scale pointer deflection. The standard value for full-scale deflection is 50 millivolts, which is more than the coil usually needs to cause I_{fs} to flow. Therefore,

an external resistance, called a *swamping resistor*, is added in series with the coil as shown in figure 3.10. The value of the swamping resistance can be found from equation 3.45 with I_{fs} and V_{fs} known.

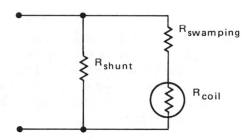

Figure 3.10 Design of a DC Ammeter Using a Swamping Resistance and a Shunt Resistance for a 50 mV System

The standard ammeter is designed so that the shunt will have 50 mV across it, and the coil will have I_{fs} flowing when the sum of the coil and shunt currents are equal to the desired full-scale current.

$$I_{\text{design full-scale}} = \frac{V_{fs}}{R_{shunt}} + I_{fs(\text{movement})} \qquad 3.46$$

Example 3.6

A 1% d'Arsonval movement has coil and swamping resistances adding to 200 ohms. The full-scale voltage is 50 mV. Determine (a) the full-scale current when no shunt resistance is connected, and (b) the shunt resistance and its tolerance to produce a 1 amp full-scale current meter with an overall accuracy of at least 2%.

(a) $$I_{fs} = \frac{50 \times 10^{-3} V}{200 \text{ ohms}} = 2.5 \times 10^{-4} \text{amps}$$

From equation 3.46,

(b) $$R_{shunt} = \frac{50 \times 10^{-3}}{1 - (2.5 \times 10^{-4})} = 0.05001 \text{ ohms}$$

R_{shunt} must be 0.05 ohms with a tolerance of 1%.

6 THE ELECTRODYNAMOMETER MOVEMENT

Electrodynamometer movements differ from the d'Arsonval movement in that the permanent magnet of the d'Arsonval movement is replaced by a fixed field coil. It has a moving coil, pointer, and scale similar to the d'Arsonval, but the scale is non-linear (square-law) in voltmeter or ammeter applications.

The field and moving coils produce magnetic fields which interact when both are carrying currents. The magnetic fields of the two coils try to align with one another, producing an instantaneous torque on the moving coil that is proportional to the product of the two currents.

The mechanical portion of the system, including inertia and springs, causes the pointer to move into an equilibrium position which is proportional to the average of the product of the two coil currents. Thus, the electrodynamometer movement is an RMS meter when the same current is flowing in both coils.

Electrodynamometers can be used to measure RMS values of voltage and current waveforms in the range of 25 Hz to 2500 Hz. The lower limit is due to the movement's attempt to follow instantaneous values; the upper limit is due to the inductive reactance becoming significant.

The electrodynamometer can be used to measure average values of DC voltages and currents since the average and RMS values are identical. They can also be configured as *wattmeters* over the same frequency range in which AC voltmeters and ammeters are used. They can be calibrated using precisely known DC values, after which they will retain the calibration for AC signals over their entire frequency range. Electrodynamometers, however, generally require more power for operation than comparable d'Arsonval meters.

7　MOVING-IRON INSTRUMENTS

Instruments which make use of the attraction or repulsion of magnetic materials in the presence of a magnetic field are called *moving-iron instruments*. These instruments can be used as voltmeters or ammeters similar to the d'Arsonval movement. They are useful in the measurement of AC signals in the range of 25 to 125 Hz. They are essentially RMS reading (i.e., they use the square-law scale).

They are less useful for DC measurements because of a tendency to acquire significant remnant magnetization. They are typically used to measure voltages and currents in power systems. Sensitivity and power consumption are intermediate to d'Arsonval and electrodynamometer movements.

8　THERMOCOUPLE INSTRUMENTS

When two dissimilar metal wires are joined in a complete circuit, and a temperature differential between the two joints is maintained, a current will flow. The current will be proportional to the temperature differential

between the two junctions. True RMS measurements of electric currents can be made using this thermocouple effect.

One junction (the *hot junction*) is placed in physical, but not electrical, contact with a wire carrying an unknown electric current. The *cold junction* is connected to a heat sink which remains at the ambient temperature. The current heats the wire and the hot junction. The temperature rise is proportional to the square of the unknown current. The current generated by the thermocouple is measured by a d'Arsonval movement properly calibrated with a square-law scale. The result is a true RMS ammeter useful for frequencies up to 100 MHz.

9　RECTIFIERS, D'ARSONVAL METERS, AND OPAMPS

The linearity and reliability of d'Arsonval meter movements make them desirable for use in inexpensive AC meter applications. As the d'Arsonval movement responds only to average values, the AC signal must be modified in some way to allow such a measurement. Diodes have been extremely useful in such applications, but they have a *threshold* voltage level which results in a nonlinear relation of the meter current to the rectified voltage.

A simple rectifier meter circuit and its voltage current relation is shown in figure 3.11.

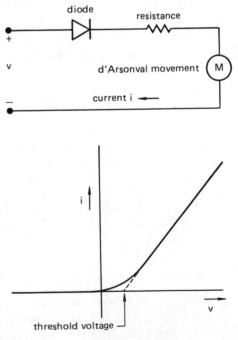

Figure 3.11　Half-Wave Rectifier Meter Circuit and *v-i* Characteristic

Operational amplifiers are widely used in instrumentation systems to overcome such nonlinearities, as well as to avoid circuit loading by the measuring system. The use of such amplifiers is usually under conditions which allow them to be approximated by *ideal operational amplifiers*, which are called ideal opamps. A more complete treatment of operational amplifiers is given in chapter 9.

The essential characteristics of the ideal opamps are that both the gain and the input impedance are extremely large, while the output impedance is negligibly small. Thus, a model of the ideal opamp is as follows:

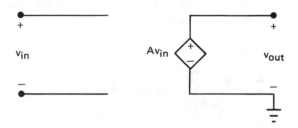

The symbol of the opamp is as follows:

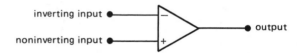

The opamp uses two DC power supply voltages, for example, +16 and −16 volts. The output voltage can approach these power supply voltages, but cannot quite reach them, so that for +16 and −16 volt DC supply voltages, the output voltage range would be from about −15 to +15 volts. The linear range of the output voltage is typically two volts less than the difference in the DC supply voltages, i.e., $16 - (-16)$ volts in this example. Thus, for the opamp with a typical gain, A, of about 10^5 to be acting in its linear mode, the maximum range of the input voltage, v_{in}, is a few microvolts. With an input impedance on the order of 10^7 ohms, the amplifier in linear operation will have a maximum input current on the order of a few picoamperes, which is negligible in all cases.

Operational amplifiers are used with negative feedback, which requires the output be connected to the inverting input terminal through some impedance. A half-wave rectifying circuit similar to that of figure 3.11, but incorporating an opamp, is shown in 3.12 with its v-i characteristic.

The notable difference between the characteristics of the half-wave rectifier of fig 3.11 and the *precision half-wave rectifier* of figure 3.12 is the nonlinearity identified by the threshold level in fig 3.11. When the input is

postive, the circuit of figure 3.12 has an output equal to its input which is the reason it is called a precision rectifier.

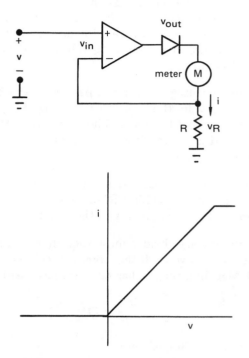

Figure 3.12 Precision Half-Wave Rectifier Meter Using Opamp

The analysis of the circuit of figure 3.12 is as follows:

(a) As the input resistance is essentially infinite, there is no current flowing into either of the opamp input terminals. Thus, the meter current, i, flows through the resistance, R.

(b) With a positive input voltage, v, the voltage v_{out} is positive, and current flows through the diode. As the output voltage, v_{out}, is finite and the gain is infinite, the opamp input voltage, v_{in}, is essentially zero. Thus, the voltage, v_R, must be the same as v.

(c) Therefore, the voltage $v_R = v = iR$, and $i = v/R$.

(d) With a negative input voltage, v, the output voltage will go negative, but the diode blocks current flow, making the voltage v_R zero. This causes the voltage v_{in} to be the same as v, producing a saturated negative output. This, however, does not cause a current flow.

Example 3.7

Determine the average current flowing through the meter as a function of the RMS value of a pure sinusoid voltage, and thereby relate the scale factors to the full-scale current of the meter movement.

Specifications: the meter current is 0.1 mA, the full-scale for the voltmeter is to be 150 v(RMS), and the DC power supply voltages are +16 and −16 volts. The input impedance of the resulting voltmeter is to be at least 1 Megohm.

As the input voltage is to reach peak values of $150\sqrt{2}$ or about 212 volts (both positive and negative), while the opamp will not be able to follow more than 15 volts (positive and negative), it is necessary to use a voltage divider at the input.

Arbitrarily choose to make an input peak voltage v of 14 volts correspond to the 150 VRMS input, and to provide an average current of 0.1 mA to the meter movement.

The input voltage divider must then divide 212 volts peak down to 14 volts. If the larger resistance is chosen to be 1 Megohm, the smaller resistor must satisfy

$$\frac{212r}{r + 10^6} = 14$$

from which $r = 70.7$ Kilohms.

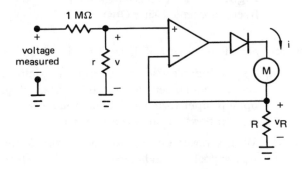

Since, the average value of a half-cycle of a pure sinusoid is the peak divided by π, for an average current of 0.1 mA the peak should be $\pi \times 0.1$ or 0.314 mA. This peak current must flow through R to develop 14 volts to match v. Therefore $0.314 \times 10^{-3}R$ must be 14, or $R = 44.6$ Kilohms.

Full-wave rectification without the threshold problem can be obtained using a bridge enclosing the meter movement in the feedback circuit. This is indicated in figure 3.13.

Example 3.8

The circuit of figure 3.13 is designed to provide full-scale meter current when the input signal is a pure sinusoid with an RMS value of 10 volts. A pulse train such as that shown in figure 3.3 with parameters $V_{max} = 4.3$

volts, $V_{min} = 0.7$ volts, and $\Delta T/T = 0.5$, is applied at the input of the circuit of figure 3.13. Determine the reading the meter will give, and compare that to the actual RMS value of the signal.

First, the average value of the signal is found from equation 3.3 to be 2.5 volts. The meter is calibrated to read the RMS value of a sinusoid that has an RMS value that is $\pi/2\sqrt{2}$ or 1.11 times the average of the full-wave rectified sinusoid. Thus, the meter will read $1.11 \times 2.5 = 2.78$ volts.

Secondly, the RMS value of the pulse train is found from equation 3.21 to be 3.08 volts. Therefore, using the circuit of figure 3.13 to measure the waveform of figure 3.3 produced an error of 9.74%.

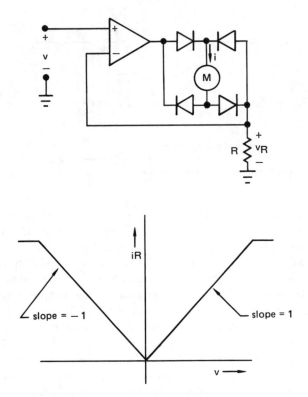

Figure 3.13 Full-Wave Rectifier Meter Using Opamp

10 PEAK DETECTION USING OPAMPS AND D'ARSONVAL METERS

Figure 3.14 shows a fundamental *peak detector* circuit, consisting of a precision diode, as the opamp and diode combination is called, and a capacitance.

The analysis of the circuit is the same as that previously discussed and shows that the output voltage, the capacitance voltage, v_C in this case, will follow the input voltage, v, whenever v is more positive than v_C.

When v is less than v_C, the output of the opamp will be negative so that the diode will be reverse biased, and no current can flow.

In practice, it is necessary to provide a resistance across the capacitance so that it can discharge at some pre-selected rate which is slow enough as to not affect the accuracy of the measurement.

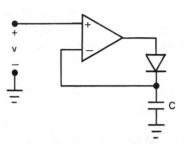

Figure 3.14 Peak Detector

Example 3.9

A peak detector can be combined with the half-wave rectifier meter to create a peak detecting voltmeter as indicated in figure 3.15. Design the voltmeter for a peak input voltage, v, of 14 volts to cause a full-scale meter deflection for a full-scale current of 0.1 mA. R_1 and C are to be chosen so that the capacitor voltage will drop 1% during one cycle of a 60 Hz input signal. Thus, the design of R_2 is 44.6 Kilohms, the same as in example 3.7.

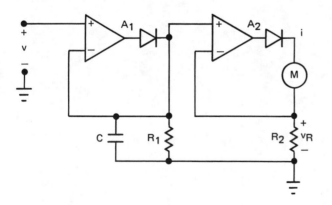

Figure 3.15 Peak Detecting Voltmeter

The design of C and R_1 is to achieve a 1% drop in capacitance voltage during the 1/120 of a second that no diode current is flowing from the amplifier, A_1. The voltage across the capacitor is controlled by the equation.

$$v_C = V_0 e^{-t/RC} \text{ where } R = R_1$$

Setting $t = 1/60$ and $v_C = 0.99V_0$, RC is found to be 1.658. Using a 1.0 microfarad capacitor, R_1 must be 1.658 Megohms.

The circuit shown can be calibrated to any sinusoidal voltage that has a peak value of 14 volts or less. For voltages greater than 14 volts peak, a voltage divider circuit can be incorporated. The RMS value of the voltage will, of course, be the peak divided by $\sqrt{2}$.

Example 3.10

The pulse train of example 3.8 is applied to the voltmeter of example 3.9. What will the meter interpret the RMS value of this signal to be? Compare the result to that of example 3.8.

The peak value of the pulses is 4.3 volts, so the meter will interpret the voltage to be $4.3/\sqrt{2}$ or 3.04 VRMS.

The meter of example 3.8 gave a reading of 2.78 VRMS for a 9.74% error, while the meter of example 3.9 results in an error of 1.28% from the actual RMS value of 3.08 volts.

11 DIGITAL METERS

Ordinary *digital meters* use circuitry similar to electronic analog meters. Either peak or peak-to-peak values are obtained. One of several schemes is used to display the reading on a numerical display. Accuracy of such meters is typically one-half of the least significant binary digit. Digital meters have the same problems with non-sinusoidal signals as the electronic meters calibrated to read RMS values of sinusoids.

Digital meters incorporating microprocessors can be programmed to respond to any desired parameter of a waveform. In particular, they can be used when a waveform is not a pure sinusoid.

One popular form of digital meter uses what is called the *dual slope integrator* technique which involves the use of an opamp, a *comparator*, two electronic switches, and a counter. The circuit is shown in figure 3.16.

The comparator is essentially an opamp with its output switched between 0 volts and 5 volts depending on whether the input is positive or negative.

The counter has four inputs: enable A which is the comparator output, enable B which must be high to permit the counter to count, a clear signal, and the clock.

The counter has been designed to count clock pulses when both enable terminals are high. Under that condition the least significant digit is incremented with each clock pulse. If the counter is reset at $t = 0$ and is permitted to continue counting for a period T, it will reach its full count.

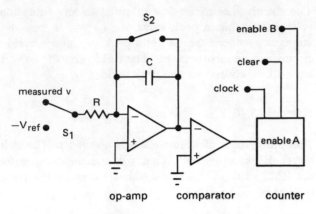

Figure 3.16 Dual-Slope A-D Converter

The analysis begins with the switch S_2 closed and switch S_1 connected to the measured voltage input position. When switch S_2 is opened, the opamp input voltage v_{in} will remain zero so that the current flowing through R is v/R, which is also the capacitance current. The capacitance voltage drops as a function of time and has the value

$$v_C = \frac{1}{C} \int \frac{v}{R}\, dt = \frac{1}{RC} \int v\, dt$$

Note that the output voltage of the opamp is $-v_C$ since the left terminal of the capacitor is at virtual ground and the capacitor voltage drops to the right, making the output voltage negative. This condition continues for a time period T, where the output voltage is the negative integral of the input voltage.

As long as the negative input to the comparator remains negative, the comparator output will be +5 volts. However, during this first integration time, enable B is low, keeping the counter from incrementing.

At the end of the integration time T, the counter is reset to zero and enable B is changed to +5 volts, allowing the counter to begin counting from zero. At the same instant S_1 is switched to the position of the negative voltage, $-V_{\text{ref}}$. The opamp now begins to integrate $-V_{\text{ref}}$, but with a starting value set by the integral of the measured voltage. The capacitor is then discharged at a constant rate, causing the opamp output voltage to approach zero. When the output just passes zero, the comparator output voltage will switch to zero which will prevent the counter from continuing to count clock pulses.

V_{ref} represents the full-scale voltage. If v is equal to V_{ref}, the counter will count throughout the entire period T to its full count value. When the input voltage is less than V_{ref}, the discharging by $-V_{\text{ref}}$ will reach zero before the full count period T, and the count will be proportional to the actual value of v. In fact, the count will be v/V_{ref} times the full-scale value.

PART 3: POWER WITH SINUSOIDAL CIRCUITS

12   REVIEW OF PHASORS

Phasor representation of steady-state sinusoids and the accompanying impedances permits the analysis and design of complicated systems without having to use differential equations. The phasor representation was previously introduced in chapter 2. The *complex plane* is shown in figure 3.17. Sinusoidal voltages, currents, and impedances can be represented by points on the complex plane. The horizontal axis represents real values, and the vertical axis represents imaginary values. Since *complex numbers* contain real and imaginary parts, the plane of points with real and imaginary coordinates is called a complex plane.

Two forms of complex quantities are used: *rectangular* and *polar forms*. A complex value $\mathbf{F}$ with a real part A and an imaginary part B is written in the rectangular form as

$$\mathbf{F} = A + jB \qquad 3.47$$

$$j = \sqrt{-1} \qquad 3.48$$

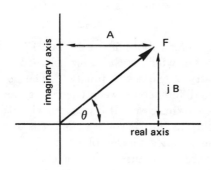

Figure 3.17   The Complex Plane

$\mathbf{F}$ can be written in the polar form as

$$\mathbf{F} = Ce^{j\theta} \qquad 3.49$$

$\mathbf{F}$ is a point in the complex plane, but it is typically represented in a vector-like fashion. The vector is directed to the actual point from the origin.

$$C = \sqrt{A^2 + B^2} \qquad 3.50$$

$$A = C\cos\theta \qquad 3.51$$

$$B = C\sin\theta \qquad 3.52$$

$$\tan\theta = \frac{B}{A} \qquad 3.53$$

Phasor representation of sinusoidal voltages and currents is based on the polar form with the angle θ being a function of time.

$$v = V_{\max}\cos(\omega t + \phi) \qquad 3.54$$

Substituting θ for $\omega t + \phi$, v becomes the *real part* of $V_{\max}e^{j(\omega t + \phi)}$. If the function *sine* is used rather than *cosine*, the *imaginary part* of $V_{\max}e^{j(\omega t + \phi)}$ is obtained. Thus, the exponential form of a sinusoid can be used to simplify calculations. The real or imaginary part of the calculated voltage or current is obtained, depending on whether the *cosine* or the *sine* function, respectively, is used. Since

$$e^{j(\omega t + \phi)} = e^{j\omega t}e^{j\phi}$$

the *fixed part* can be separated from the *time variable part* in this notation. These are shown in figure 3.18 for several values of ωt. Notice that the magnitude of $e^{j\omega t}$ remains equal to 1, but the angle increases (rotating counterclockwise) linearly with time. All functions having a component $e^{j\omega t}$ are assumed to rotate counterclockwise with an angular velocity of ω on the complex plane.

Multiplying $e^{j\omega t}$ and $e^{j\phi}$ results in the addition of the exponents: $e^{j(\omega t + \phi)}$. Thus, the product gives the points on figure 3.18(b) for times corresponding to those of figure 3.18(a).

Voltage can be represented as

$$v = V_{\max}e^{j\phi}e^{j\omega t} \qquad 3.55$$

In electric circuits involving sinusoids it is more convenient to deal with RMS values of voltage and current rather than with peak values, and with angles in degrees rather than in radians. (Angles in the exponential form must be in radians for mathematical calculations.) Thus, an alternative representation, which is called the *effective value phasor notation* has evolved.

$$v = v_{\text{RMS}}\angle\theta \qquad 3.56$$

$$v_{\text{RMS}} = \frac{V_{\max}}{\sqrt{2}} \qquad 3.57$$

The *reference phasor* is one whose value is known. All other voltages and currents have their positions on the complex plane measured from the reference phasor at the time that the reference phasor is on the positive real axis. The relative angles between phasor voltages and currents will remain constant for sinusoidal steady-state operation.

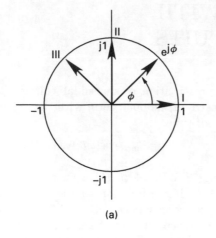

(a)

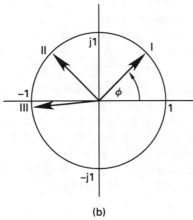

(b)

Figure 3.18 (a) $e^{j\phi}$ for all time and $e^{j\omega t}$ for (I) $\omega t = 0$, (II) $\omega t = \pi/2$, and (III) $\omega t = 3\pi/4$; and (b) showing $e^{j(\omega t + \phi)}$ for (I) $\omega t = 0$, (II) $\omega t = \pi/2$, and (III) $\omega t = 3\pi/4$

13 PHASOR POWER

Power into a pair of terminals is the instantaneous product of the voltage and current. With the voltage at an angle θ and the current at an angle ϕ, the power is

$$p(t) = [V_{\max} \cos(\omega t + \theta)] \times [I_{\max} \cos(\omega t + \phi)] \quad 3.58$$

Using trigonometric identities, equation 3.58 becomes

$$p(t) = \frac{V_{\max} I_{\max}}{2} [\cos(\theta - \phi) + \cos(2\omega t + \theta + \phi)] \quad 3.59$$

The first cosine term in the brackets is a constant. The second term is a sinusoidal time function with an average of zero. Thus, the average power is

$$P = V_{\text{RMS}} I_{\text{RMS}} \cos(\theta - \phi) \quad 3.60$$

Care must be taken when calculating the average power using phasor notation.

$$v = V_{\text{RMS}} \angle \theta \quad 3.61$$
$$i = I_{\text{RMS}} \angle \phi \quad 3.62$$
$$P \neq iv = V_{\text{RMS}} I_{\text{RMS}} \angle (\theta + \phi) \quad 3.63$$

Equation 3.63 does not give the power, since the power angle should be the difference between the voltage and the current angles as in equation 3.60. This is handled by equation 3.64.

$$P = \mathbf{Re}\{VI^*\} \quad 3.64$$

Re means the real part of the bracketed quantity. I^* is the complex conjugate of the phasor I. The complex conjugate is obtained by negating the angle in polar form. In rectangular form it is obtained by negating the imaginary part.

$$VI^* = (V_{\text{RMS}} \angle \theta) \times (I_{\text{RMS}} \angle - \phi)$$
$$= V_{\text{RMS}} I_{\text{RMS}} \angle (\theta - \phi) \quad 3.65$$

The real part of equation 3.65 is

$$P = V_{\text{RMS}} I_{\text{RMS}} \cos(\theta - \phi) \quad 3.66$$

The *power factor* (abbreviated p.f. or PF) is the quantity $\cos(\theta - \phi)$, which is the same as $\cos(\phi - \theta)$ due to the symmetry of the *cosine* function. The *power factor angle* is the difference between the voltage and current angles ($\theta - \phi$ in this case). It is positive when the voltage *leads* the current (i.e., is counterclockwise from the current on the complex phasor plane) and negative when the voltage *lags* the current (i.e., is clockwise from the current on the complex phasor plane) as indicated in figure 3.19.

The *power triangle* consists of the product VI^* and its real and imaginary components. The components are added on the complex plane, but the plane is not a phasor plane because average rather than instantaneous values are used. In most power problems the voltage is the reference quantity. Then, the power triangle is in the first quadrant when the current lags the voltage (i.e., an inductive circuit) and is in the fourth quadrant when the voltage lags the current (i.e., a capacitive circuit).

Figure 3.20(a) shows the phasor and power diagrams for an inductive circuit, and figure 3.20(b) shows the diagrams for a capacitive circuit. The phasors are shown stopped at the instant the reference voltage reaches the positive real axis.

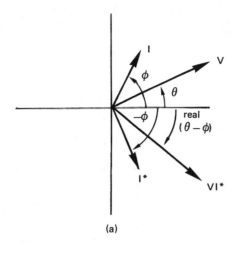

(a)

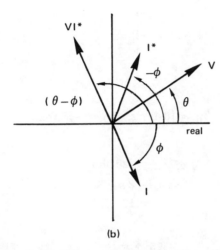

(b)

Figure 3.19 Phasor Representations of $V\angle\theta$, $I\angle\phi$, $I^*\angle-\phi$, and the Angle $(\theta-\phi)$ for (a) Current Leading Voltage (Capacitive) Circuit and (b) Voltage Leading Current (Inductive) Circuit

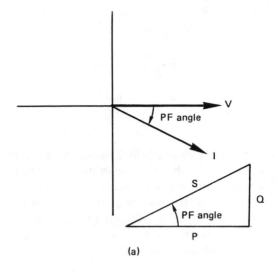

(a)

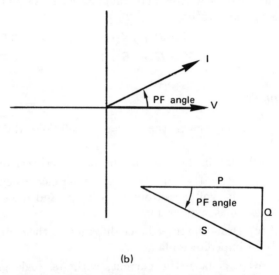

(b)

Figure 3.20 (a) Inductive Circuit with V Leading I, (b) Capacitive Circuit with I Leading V

The power triangle is constructed from the *average power*, P, in watts (W) measured along the real axis; the average *apparent power*, S, in volt-amperes (VA) as the hypotenuse of the triangle; and the *reactive power*, Q, in volt-amperes reactive (VAR) in the imaginary direction.

$$S(\text{volt-amps}) = P(\text{watts}) + jQ(\text{volt-amps-reactive})$$
$$= VI^* \qquad 3.67$$
$$|S|^2 = |P|^2 + |Q|^2 = I_{\text{RMS}}^2 V_{\text{RMS}}^2 \qquad 3.68$$
$$\text{PF} = \frac{|P|}{|S|} \qquad 3.69$$

Reactance is the ratio of voltage to current for an inductor or capacitor in a circuit operating with a sinusoidal

voltage. The *impedance* is j times the reactance, symbolized by jX. The negative reciprocal of reactance is *susceptance*.

$$B = \frac{-1}{X} \qquad 3.70$$

These terms are related. For an inductor,

$$Z_L = jX_L = j\omega L$$
$$= j2\pi fL \qquad 3.71$$
$$Y_L = \frac{1}{Z_L} = \frac{1}{j\omega L}$$
$$= \frac{-j}{\omega L} = \frac{-j}{2\pi fL}$$
$$= -jB_L \qquad 3.72$$

For a capacitor,

$$Z_C = \frac{1}{j\omega C} = \frac{-j}{\omega C}$$

$$= \frac{-j}{2\pi f C} = -jX_C \qquad 3.73$$

$$Y_C = \frac{1}{Z_C} = j\omega C$$

$$= jB_C \qquad 3.74$$

In the sinusoidal steady state, impedances are considered as a real part plus j times an imaginary part. The real part of the impedance is the *resistive part*, and the imaginary part is the *reactive part*. The same is generally true for admittances, although the real part should be called the *conductive part*, and the imaginary part should be called the *susceptive part*.

Z and Y can be written in rectangular form as equations 3.75(a) and 3.75(b).

$$Z = R + jX \text{(ohms)} \qquad 3.75(a)$$

$$Y = G + jB \text{(mhos)} \qquad 3.75(b)$$

Example 3.11

In the circuit shown, the voltage is 230 volts RMS at 60 Hz.

  (a) Obtain the reactance of each reactive element.

  (b) Using the current, I, as a reference, calculate the voltages around the loop, and draw the phasor diagram.

  (c) Draw the impedance diagram on the complex impedance plane.

  (d) Calculate the current, with its angle given with respect to the voltage, V_s.

  (e) Determine the power factor.

  (f) Determine the power triangle.

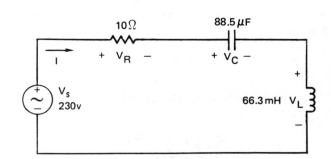

(a)
$$X_C = \frac{-1}{377C} = -29.97$$

$$\approx -30 \text{ ohms}$$

$$X_L = 377L = 24.995$$

$$\approx 25 \text{ ohms}$$

(b)
$$V_R = 10I, \quad V_C = -j30I,$$

and

$$V_L = j25I$$
$$V_s = V_R + V_C + V_L$$
$$= (10 - j5)I$$

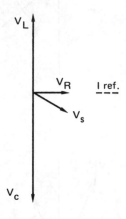

(c)
$$Z_R = 10 \text{ ohms}, \quad Z_C = -j30 \text{ ohms},$$

and

$$Z_L = j25 \text{ ohms}$$
$$Z_{\text{loop}} = Z_R + Z_C + Z_L$$
$$= 10 - j5$$

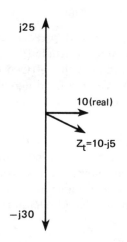

(d)
$$V_s = Z_{\text{total}}I$$
$$230\angle 0° = (10 - j5)I = (11.18\angle - 26.57°)I$$
$$I = \frac{230\angle 0°}{11.18\angle - 26.57°} = 20.57\angle 26.57°$$

(e)
$$\text{PF} = \cos(\angle V - \angle I) = \cos(0 - 26.57°)$$
$$= 0.894$$

(f) $|S| = |V| \times |I| = 230 \times 20.57$
        $\qquad = 4731\,\text{VA}$

$P = |S|\cos(-26.57°) = 4230\,\text{W}$

$Q = |S|\sin(-26.57°) = -2112\,\text{VAR}$

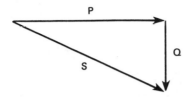

14   POWER FACTOR CORRECTION

Correction in power circuits is generally accomplished by placing *shunt capacitance* across the load terminals, (i.e., in parallel with the load). A capacitance is usually required to correct the power factor because most power circuits are inductive. The power factor may be specified as *leading* (capacitive) or as *lagging* (inductive).

The power angle is positive for inductive loads involving lagging current and negative for capacitive loads where the current is leading. However, care must be taken when the power factor is known, since the cosines of both positive and negative angles less than 90° are positive. In the absence of specific information about the leading or lagging nature, it is assumed that the load is inductive.

The power triangle for inductive currents (current lagging voltage) is in the first quadrant, and for capacitive currents (current leading the voltage) the power triangle is in the fourth quadrant.

The fundamental method used in power factor correction is to draw a new line current which cancels all or part of the reactive component already flowing to the original load. The *in-phase component* of the current accounts for the power, P, delivered to the load.

$$P = V \times (\text{component of current in phase with } V)$$
$$= VI\cos\,\theta = VI \times \text{PF} \qquad 3.76$$

The component of the current that is *out-of-phase* with V constitutes the reactive component.

$$Q = VI\sin\theta = VI\sqrt{1-\text{PF}^2} \qquad 3.77$$

The component of current out-of-phase with the voltage is

$$I_{\text{reactive}} = \frac{Q}{V} \qquad 3.78$$

The current needed to produce a desired power factor can be determined from the in-phase component of the current, P/V, as

$$I_{\text{corrected}} = \left(\frac{P}{V}\right)\frac{\sqrt{1-\text{PF}^2}}{\text{PF}} \qquad 3.79$$

PF is the power factor specified for the corrected system.

The correcting current provides the difference between the new reactive component of the current and the pre-existing reactive component of the load current Q/V. The *corrected current* is the pre-existing reactive component of the current minus the correcting current.

$$I_{\text{correcting}} = \left(\frac{P}{V}\right)\frac{\sqrt{1-\text{PF}^2}}{\text{PF}} - \frac{Q}{V} \qquad 3.80$$

The reactance necessary to do the correction is

$$X_{\text{correcting}} = \frac{V}{I_{\text{correcting}}}$$
$$= \frac{V^2}{\dfrac{P\sqrt{1-\text{PF}^2}}{\text{PF}} - Q}$$

Q is the initial reactive power in VARS, P is the power (unchanged after the correction takes place), and PF is the desired power factor.

Example 3.12

A load operates at 60 Hz, 440 V, 15 kVA, and 10 kW. Determine the capacitance necessary to correct the power factor to 0.95. Determine the VA rating of the capacitor.

The in-phase component of the load current is

$$\frac{P}{V} = \frac{10,000}{440} = 22.7\,\text{amps}$$

The amplitude of the current is the volt-amperes divided by the voltage.

$$\frac{S}{V} = \frac{15,000}{440} = 34.1\,\text{amps}$$

The reactive component can be determined by trigonometry.

$$I^2_{\text{reactive}} = (34.1)^2 - (22.7)^2$$
$$= 647.5 = (25.4)^2$$

The new lagging current is

$$I_{\text{corrected}} = \frac{22.7}{0.95}\sqrt{1-(0.95)^2} = 7.5$$

The correcting current is

$$I_{correcting} = 7.5 - 25.4 = -17.9 \text{ amps (leading)}$$

The reactance necessary to produce such a current is

$$\frac{-440}{17.9} = -24.6 \text{ ohms (reactive)}$$

The capacitive reactance is $1/2\pi fC$. At 60 Hz, $2\pi f = 377$,

$$C = \frac{1}{377 \times 24.6} = 1.08 \times 10^{-4} \text{ F}$$
$$= 108 \,\mu\text{F}$$

Since the current through the correcting capacitance is 18 amps, the volt-ampere rating is

$$(440)(17.9) = 7876 \text{ VA (say 8 kVA)}$$

15 THREE-PHASE POWER SYSTEM DEFINITIONS

Three-phase voltage generators provide three sinusoidal voltages 120 degrees apart. These are designated with one as a reference, typically called phase A. The others are phase B and phase C. There are two arrangements of the phase sequence, one of which is called ABC, which has definitions:

$$v_a(t) = V_p \cos \omega t$$
$$v_b(t) = V_p \cos(\omega t - 120°)$$
$$v_c(t) = V_p \cos(\omega t - 240°) = V_p \cos(\omega t + 120°)$$

The phasor diagram is shown in figure 3.21(a).

The other possible phase sequence is called ACB, and has definitions:

$$v_a(t) = V_p \cos \omega t$$
$$v_b(t) = V_p \cos(\omega t - 240°) = V_p \cos(\omega t + 120°)$$
$$v_c(t) = V_p \cos(\omega t - 120°)$$

The phasor diagram is shown in figure 3.21(b).

The phase voltages of the generator are connected to three lines either in a *wye* or a *delta* connection. The lines are designated A, B, and C according to the positive polarities of the phase voltages connected to them. The voltage of three-phase systems is specified by the line-to-line voltages, with the positive polarity given in the first subscript and the negative polarity in the second.

In the wye connection, as shown in figure 3.22(a), the generator voltages are connected to a point called the *neutral*, marked with "n."

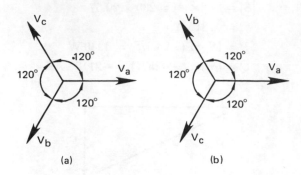

Figure 3.21   Phasor Diagrams for Three-phase Voltages: (a) Phase Sequence ABC, (b) Phase Sequence ACB

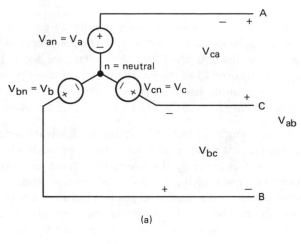

(a)

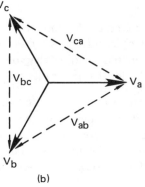

(b)

Figure 3.22   (a) Wye Connected Generators, (b) Phasor Diagram for an ABC Phase Sequence

16 TWO-WATTMETER METHOD OF THREE-PHASE POWER MEASUREMENT

For a three-phase, three-wire system it is necessary to use only two wattmeters to determine the total power being delivered. In addition, when the system is balanced, the voltage-amperes-reactive can be found from the two-wattmeter method.

Using RMS phasors, the apparent power supplied to a load is $S = P + jQ$, where P is the average power and Q is the *volt-amperes-reactive* (VARS). The apparent power is obtained from the voltage and current is

$$S = P + jQ = VI^*   \qquad 3.82$$

V is the phasor voltage drop, and I^* is the complex conjugate of the phasor current.

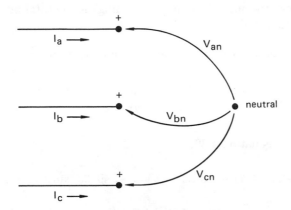

Figure 3.23 Three-Phase, Three-Wire Lines with Artificial Neutral

Figure 3.23 shows a portion of a three-wire system with an *artificial neutral point* defined to isolate the power of each line. The apparent power or volt-amperes for the three lines can be added to obtain

$$S = V_{an}I_a^* + V_{bn}I_b^* + V_{cn}I_c^*   \qquad 3.83$$

As the neutral is artificial, the three line currents must add to zero: $I_a + I_b + I_c = 0$. Since the currents are complex numbers, their real parts must add to zero, while their imaginary parts must add to zero independently. Thus,

$$I_a^* + I_b^* + I_c^* = 0$$

or

$$I_c^* = -I_a^* - I_b^*   \qquad 3.84$$

Equation 3.83 can then be rewritten to eliminate I_c^*:

$$S = (V_{an} - V_{cn})I_a^* + (V_{bn} - V_{cn})I_b^*   \qquad 3.85$$

The voltage drop from line a to line c is V_{ac} which is recognized via Kirchhoff's voltage law to be

$$V_{ac} = V_{an} - V_{cn}$$

Similarly, the voltage drop from line b to line c is

$$V_{bc} = V_{bn} - V_{cn}$$

Thus, equation 3.85 becomes

$$S = S_a + S_b = V_{ac}I_a^* + V_{bc}I_b^*   \qquad 3.86$$

for $S_a = V_{ac}I_a^*$ and $S_b = V_{bc}I_b^*$.

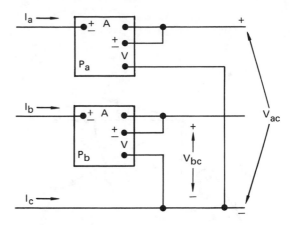

Figure 3.24 The Connections for the Two-Wattmeter Method of Measuring Power in a Three-Phase, Three-Line Power System

The total power is the real parts of S_a and S_b, which can be measured with a wattmeter arrangement as indicated in figure 3.24. The $\pm$ terminal of each wattmeter is connected so the positive direction of current flow is into the $\pm$ current terminal (indicated by A on each wattmeter), and with the $\pm$ voltage terminal (indicated by V on each wattmeter) connected to the line carrying that current. The total power delivered by the three-phase line is then the sum of the powers read on the two wattmeters.

$$P = P_a + P_b   \qquad 3.87$$

The result given in equation 3.87 is valid regardless of whether the loads are balanced or not. However, for the case where the loads are balanced, even more information can be obtained from the wattmeter readings. This will be demonstrated in the next section.

17   TWO-WATTMETER METHOD WITH BALANCED LOADS

An impedance having a phasor voltage drop, V, and phasor current, I, in the direction of the voltage drop, with the current leading the voltage by a phase angle θ, has an average power of $|V|\,|I|\cos\theta$. For a balanced three-phase load, the power associated with each line current and line-to-neutral voltage is $|V_{\text{line-neutral}}|\,|I_{\text{line}}|\cos\theta$, where this θ is the phase lead of the line current with respect to the associated line-neutral voltage. For a balanced load, all currents have the same relative angle to their associated line-neutral voltages, with the phasor relations shown in figure 3.25 for an ABC sequence, and a leading power angle θ.

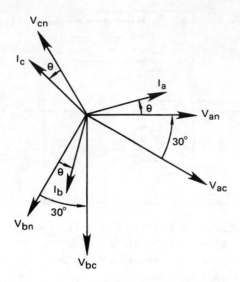

Figure 3.25 Phasor Diagram of a Balanced Three-Phase Load with Leading Power Angle θ

As can be seen in figure 3.25, the power measured by the upper wattmeter in figure 3.24 will be

$$P_a = |V_{ac}| \, |I_a| \cos(30° + \theta) \qquad 3.88$$

and by the lower wattmeter

$$P_b = |V_{bc}| \, |I_b| \cos(30° - \theta) \qquad 3.89$$

Of course, $|V_{ac}| = |V_{bc}| = V_{\text{line-line}} = \sqrt{3}V_{\text{line-neutral}}$, and $|I_a| = |I_b| = I_{\text{line}}$. With the balanced load, the total power is $3|V_{\text{line-neutral}}| \, |I_{\text{line}}| \cos \theta$. With a leading current, the Volt-Amperes Reactive (VARS) is $-3|V_{\text{line-neutral}}||I_{\text{line}}| \sin \theta$.

Adding equations 3.88 and 3.89, and making the substitutions indicated,

$$P_a + P_b = \sqrt{3}|V_{\text{line-neutral}}| \, |I_{\text{line}}|[\cos(30° + \theta) + \cos(30° - \theta)]$$

From equation 1.40,

$$\cos(30° + \theta) + \cos(30° - \theta) = 2\cos 30° \cos \theta = \sqrt{3}\cos \theta$$

And

$$P_a + P_b = \sqrt{3}|V_{\text{line-neutral}}| \, |I_{\text{line}}|\sqrt{3}\cos \theta$$
$$P_a + P_b = 3|V_{\text{line-neutral}}| \, |I_{\text{line}}| \cos \theta \qquad 3.90$$

Thus, the sum of the power readings is the total power as expected.

If equation 3.89 is subtracted from equation 3.88,

$$P_a - P_b = \sqrt{3}|V_{\text{line-neutral}}| \, |I_{\text{line}}|[\cos(30° + \theta) \\ - \cos(30° - \theta)]$$

From equation 1.39,

$$\cos(30° + \theta) - \cos(30° - \theta) = -2\sin 30° \sin \theta = -\sin \theta$$

Thus, by multiplying the difference in power readings by $\sqrt{3}$,

$$\sqrt{3}(P_a - P_b) = -3|V_{\text{line-neutral}}| \, |I_{\text{line}}| \sin \theta \qquad 3.91$$

The difference in the power readings multiplied by the square root of three yields the total VARS for a balanced system. It is important that the phase sequence is known, and that the meters be connected as indicated in figure 3.24 for phase sequence ABC. For phase sequence ACB, the negative is taken.

$$\text{phase sequence ABC: total VARS} = \sqrt{3}(P_a - P_b) \\ 3.92$$
$$\text{phase sequence ACB: total VARS} = \sqrt{3}(P_b - P_a) \\ 3.93$$

PRACTICE PROBLEMS

Warmups

1. Using integration, show that the average value of a pure sinusoidal is zero.

2. A triangular wave has

$$V_{min} = -V_{max} = -10 \text{ V}$$

After half-wave rectification, find its (a) average value, and (b) RMS value.

3. An irregular pulse train repeats itself after 50 ms. The pulses are 1 ms in duration, 2 volts in amplitude, and a total of 10 occur during any 50 ms cycle. Determine the average and RMS values of this waveform.

4. A peak-to-peak detecting AC digital voltmeter shows four significant decimal figures. Determine the resolution (precision) as a percentage of full-scale.

5. A sinusoidal has its peak value at $t = 1$ ms, and a period of 10 ms. Express this signal as cosine and sine functions.

6. The voltage $10\cos(100t + 25°)$ is applied to a resistance of 25 ohms and an inductance of 0.5 henrys in parallel. Find the RMS value and the phase angle of the sum of the currents.

7. Construct the power triangle for problem 6.

8. Specify a capacitance in farads to completely correct the power factor of the circuit of problem 6 (i.e., to make $Q = 0$).

9. A voltmeter is designed using a 1.00 mA d'Arsonval movement having 435 ohms of wire resistance and a series resistor of 9565 ohms. Considering all significant figures given, determine the full-scale voltage of this meter and its accuracy at half of full-scale.

10. A DC voltmeter with 20,000 ohms internal resistance measures the voltage across a resistance as 10.0 volts. The current through the parallel combination of the resistor and voltmeter is measured as 1.00 mA by a milliammeter with 100 ohms of internal resistance. The resistor is fed from a source with a Thevenin resistance of 600 ohms. Determine the resistance of the resistor and the current that would flow in it if the meters were removed from the circuit.

Concentrates

1. A voltmeter is being designed to measure voltages in the full-scale ranges of 3, 10, 30, and 100 volts DC. The meter movement to be used has an internal resistance of 50 ohms and a full-scale current of 1 mA. Using a four contact multiposition switch, design the voltmeter.

2. An AC voltmeter consists of a d'Arsonval meter with a full-scale current of 0.125 mA and a series resistance of 10^6 ohms. A full-wave bridge of silicon diodes ($V_{thr} = 0.6$ volts) is used to rectify the AC voltage. The full-scale needle deflection is 50 degrees. Give the scale increments in degrees from 0 volts to 100 volts in 10 volt increments for the RMS value of a pure sinusoid.

3. A train of rectangular pulses is applied to an ideal low-pass filter circuit. The pulse height is 5 volts, and the duration of each pulse is 2 ms. The repetition period is 10 ms. The low-pass filter has a cut-off frequency of 500 Hz. What percentage of the signal power is available at the output of the filter?

4. A full-wave rectifier-type VTVM (vacuum tube voltmeter) is set to an RMS AC scale with a range of 50 volts. The meter is connected to a symmetrical (zero-average) triangular waveform of 100 volts peak-to-peak. What does the meter read?

5. An impedance receives a line current which lags the voltage by 30 degrees. When the voltage across the impedance is 100 volts (RMS), the impedance dissipates 200 watts. Specify the reactance of a capacitance to be placed in parallel with the impedance which would make the line current be in phase with the voltage.

6. A parallel combination of a resistance (10 ohms), capacitance (88.5 microfarads), and inductance (66.3 millihenrys) has 60 Hz, 230 volts (RMS) applied. Obtain the (a) reactances of C and L, (b) admittance of each circuit element, (c) phasor diagram for the currents, using the applied voltage as the reference, (d) admittance diagram for the circuit, including the total admittance, (e) input current as a sine function, taking the applied voltage as reference (Is the circuit inductive or capacitive?), (f) power factor, and (g) power triangle.

7. A certain voltage is known to have a triangular waveform. It is measured by a DC voltmeter as 2.0 volts. It is also measured by a full-wave rectifier meter calibrated for RMS values of pure sinusoids as 6.85 VRMS. Determine the maximum and minimum values of the voltage waveform.

8. A pulse train has two voltage levels; one is zero and the other is positive. The train is measured with a DC meter as 0.5 volt, and by a peak detecting meter calibrated to read RMS voltage of pure sinusoids as 6.36 volts. Determine the duty cycle of the pulse train. (The duty cycle is the percentage of the time the signal is at its high value.)

9. For the three-phase circuit, show that the three meters connected as shown below will measure the total power as $P_a + P_b + P_c$.

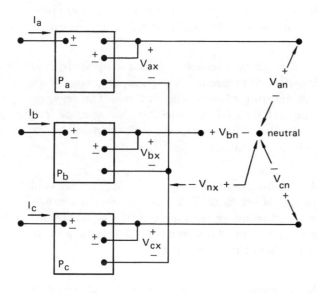

10. The measurement system below is used for a balanced load of 1 kW with a lagging power factor of 0.8. Determine the wattmeter readings.

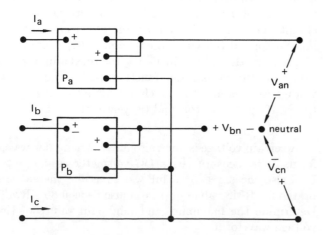

Timed

1. An average-responding meter with an RMS calibrated scale is placed in a full-wave ideal diode bridge. A voltage $v(t) = 6 + 10\cos 1000t$ is measured. Determine the (a) RMS value, (b) waveform of the rectified wave, and (c) reading of the meter.

2. A triangular saw-tooth wave has a maximum of 12 volts and a minimum of -4 volts. Determine the meter readings for (a) a d'Arsonval voltmeter, (b) a meter consisting of a d'Arsonval meter and an ideal diode with the meter scale calibrated to read RMS voltage, (c) a peak-to-peak electronic voltmeter calibrated to read the RMS voltage of a sinusoid, and (d) an electrodynamometer voltmeter.

3. An ammeter is being designed to measure currents over the five ranges indicated in the accompanying illustration. The indicating meter is a 1.00 milliampere movement with an internal resistance of 50 ohms. The total resistance $(R_1 + R_2 + R_3 + R_4 + R_5)$ is to be 1000 ohms. Specify the resistances R_a and R_1 through R_5.

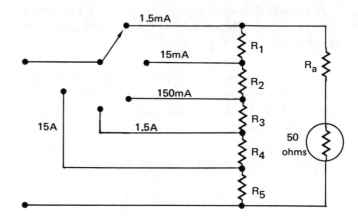

4. A triangular voltage waveform measures 5 volts with a d'Arsonval meter movement. A peak-to-peak reading, RMS calibrated, electronic voltmeter measures the same waveform as 7 volts. What are the maximum and minimum values of this voltage waveform?

5. A load is connected to a voltage of 1320 volts at 60 Hz. The load dissipates 100 kW with a 0.867 lagging power factor. Specify the capacitance needed to correct the power factor to (a) 0.895 lagging, and (b) 0.95 leading.

(Give the voltage and volt-ampere-reactive ratings at 60 Hz for the capacitors.)

6. For the measurement system below, the phase sequence is ABC. Determine the readings of P_a and P_c in terms of the total P and Q for a balanced system.

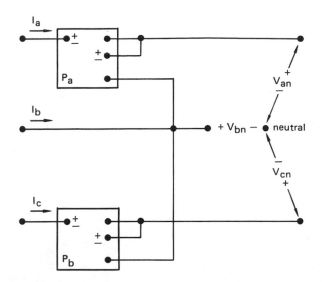

7. A particular true-RMS voltmeter having both a voltage scale and a power scale (calibrated in dBm) is used for low power measurements of RF (radio frequency) signals. Power measurement in dBm is relative to 1mW:

$$P_{\mathrm{dBm}} = 10 \log \frac{P}{0.001} = 10 \log 1000 P$$

The voltmeter is correctly calibrated to read both the voltage and the power in dBm in the situation where the voltage is measured across a 600 ohm resistance. For this reason, the scale points of zero dBm and 0.7746 VRMS coincide.

Thus, the power is calculated from the voltage using the formula

$$P_{\mathrm{dBm}} = 10 \log 1000 \frac{V^2}{600} = 2.2185 + 20 \log V$$

When the meter is used to read voltages across other resistors, the power value is incorrect by a certain constant value depending on the value of the resistor. Determine the correction as a function of the resistance.

8. The signal shown below is measured with the following voltmeters: (a) DC voltmeter, (b) an RMS reading AC voltmeter using a d'Arsonval meter in a full-wave bridge in the feedback circuit of an opamp, (c) an RMS reading AC voltmeter using a d'Arsonval meter in series with a diode in the feedback circuit of an opamp,

(d) a true RMS voltmeter such as an electrodynamometer, (e) an RMS reading AC voltmeter using a peak detector, and (f) an RMS reading AC voltmeter using a peak-to-peak detector. Determine the reading on each meter.

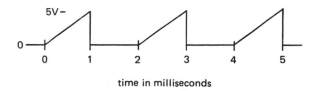

time in milliseconds

9. A periodic voltage with equally spaced samples shown in the table is measured by a number of different voltmeters such as those listed in problem 8. Determine which of the meters would give the following readings (rounded to two significant digits). (a) 1.4, (b) 0.29, (c) 1.9, (d) 1.3 (Note: It is possible that more than one meter would give the same reading.)

sample:	1	2	3	4	5	6	7
voltage:	1.0	2.0	2.0	1.0	−1.0	−2.0	−1.0

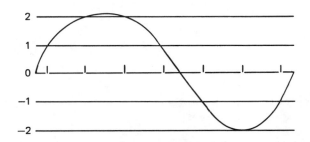

10. A digital meter uses the number system base b, and has n digits. The noise is one half the place value of the least significant digit. Determine the *signal to noise* ratio for the case where the reading is half of the full-scale value. Obtain the signal to noise ratios for the following cases:

n	b	S/N (db)
3	10	20 log (S/N) =
4	8	
8	2	
3	16	

4 TIME AND FREQUENCY RESPONSE

Nomenclature

C	capacitance	farads
e	base of the natural logarithm	2.718
f	a function of t	–
F	a function of s	–
i	instantaneous current	amps
l	constant current	amps
$I(s)$	Laplace transformed current (usually subscripted)	amp-sec
j	$\sqrt{-1}$	–
L	inductance	henrys
n	an integer	–
q	a function of t	–
Q	quality factor	–
R	resistance	ohms
s	the Laplace transform variable, or a root for natural response	–, 1/sec
t	time	seconds
$u(t)$	the unit step function	–
v	instantaneous voltage	volts
V	(unsubscripted) voltage	volts
$V(s)$	Laplace transformed voltage (usually subscripted)	volt-sec
z	impedance	ohms

Symbols

α	exponent	1/sec
β	natural frequency	rad/sec
ζ	damping coefficient	–
τ	time constant	sec
ϕ	an angle	radians
ω	angular frequency	rad/sec
ω_1	lower half power frequency	rad/sec
ω_2	upper half power frequency	rad/sec

Subscripts

C	capacitance
h	high frequency
l	low frequency
L	inductance
m	mid-frequency
n	natural
N	Norton
o	initial value, or resonant
s	source, or signal
ss	steady state
t	terminal
Th	Thevenin

1 INTRODUCTION

This chapter reviews circuit analysis techniques based on the previous chapter. Some basic circuit reduction methods are reviewed by means of examples. The classical solution of differential equations for transient analysis is covered as a review of the physical interpretation of circuit conditions. The application of Laplace transforms expands on the previous material and relates to the classical solutions. Finally, the Laplace transform approach to sinusoidal steady-state conditions and a discussion of transfer functions broadens the treatment begun in chapter 3.

2 FIRST-ORDER TRANSIENT ANALYSIS

A. Introduction

There are three classes of transient problems based on the nature of the energy sources: DC switching transients, AC switching transients, and pulse transients. When switches (or other devices performing switching functions) operate, they change the circuit topology, either connecting or disconnecting different parts of the networks, hence the name *switching transients*. *Pulse transients* do not involve topology changes in the circuit, as only the voltage or current waveforms of independent sources are changed.

The energy storage elements in electric circuits are inductors and capacitors. A network which contains only one energy storage element can be described by a first-order differential equation. The use of Thevenin or Norton equivalent circuits will result in a standard form which can be readily solved.

A *capacitive first-order circuit* is described by a differential equation using either the capacitor current or the voltage as the dependent variable. The voltage is most conveniently used because it results in a true differential equation. (The capacitor current equation is an integral equation.) The capacitor voltage is also preferred because the energy stored is a function of the capacitor voltage.

$$W = \frac{1}{2}Cv_C^2 \qquad 4.1$$

After the determination of the Thevenin equivalent voltage source and resistance, the capacitance voltage is described by the equation

$$v_{Th} = i_{Th}R_{Th} + v_C \qquad 4.2$$

As i_{Th} is the capacitance current, it can be determined from the capacitor's voltage-current relationship.

$$i_{Th} = C\frac{dv_C}{dt} \qquad 4.3$$

Substituted into the previous equation, this produces

$$v_{Th} = R_{Th}C\frac{dv_C}{dt} + v_C \qquad 4.4$$

An *inductive first-order circuit* is best described in terms of the current for the same reasons that the capacitor voltage was chosen. The inductance current is the state variable, and the describing equation in terms of the current is

$$v_{Th} = R_{Th}i_L + L\frac{di_L}{dt} \qquad 4.5$$

The energy stored in an inductor is

$$W = \frac{1}{2}Li_L^2 \qquad 4.6$$

B. First-Order Solutions

The solution of first-order differential equations contains a *natural response* term in exponential form and a *forced response* term with the same form as the forcing function. The forcing function is v_{Th} in equations 4.4 and 4.5. The general form for the first-order differential equation is

$$f(t) = b\frac{dq}{dt} + cq \qquad 4.7$$

$f(t)$ is the forcing function and is a general function of time, b and c are constants, and q is the quantity which is to be determined. The solution is

$$q(t) = Ae^{-t/\tau} + \text{a function of } f(t) \qquad 4.8$$

τ is a time constant and is equal to

$$\tau = \frac{b}{c} \qquad 4.9$$

For a first-order R-C circuit, the time constant is $R_{Th}C$.

For a first-order R-L circuit, the time constant is L/R_{Th}.

The classical solution of a differential equation is to first determine the *steady-state value* and then to adjust the natural response coefficient (i.e., A in equation 4.8) to match known conditions at some time—typically at the onset of the transient.

C. First-Order Switching Transients

First-order switching transients result from the closing or opening of a switch. A switching, whether by actual switches, relay contacts, or nonlinear device operation (such as a saturating or cutoff transistor), results in a change of the circuit topology. The simplest case is where a DC power supply is switched into or out of a network. Figure 4.1 is an illustration.

Example 4.1

The network shown in figure 4.1 has $V = 10$ volts, $R_1 = 10$ ohms, $R_2 = 10$ ohms, and an inductance of 0.15 henrys. Determine the steady-state inductance current with the switch (a) closed, and (b) open.

With the switch closed, figure 4.1(b) applies. $v_{Th} = 5$ volts, and $R_{Th} = 5$ ohms. The circuit equation is

$$5 = 5i + 0.15\frac{di}{dt}$$

For a DC circuit, the steady-state current is constant. Therefore,

$$\frac{di}{dt} = 0$$
$$5 = 5i + 0$$
$$i = 1 \text{ amp}$$

With the switch open, figure 4.1(c) applies.

$$v_{Th} = 0$$
$$R_{Th} = 10$$

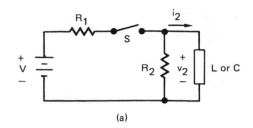

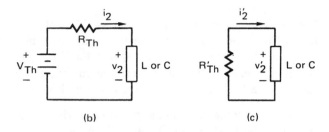

Figure 4.1 (a) First-Order Network with One Energy Storage Element, (b) Thevenin Equivalent when Switch is Closed, (c) Thevenin Equivalent Circuit when Switch is Open

The system equation is given by equation 4.5.

$$0 = 10i + 0.15\frac{di}{dt}$$

The steady-state DC current is constant, so $di/dt = 0$.

$$0 = 10i + 0$$
$$i = 0$$

Example 4.2

The network shown in figure 4.1 has $V = 10$ volts, $R_1 = 10$ ohms, $R_2 = 10$ ohms, and a capacitance of $1\,\mu F$. Determine the steady-state capacitance voltage with the switch (a) closed, and (b) open.

The Thevenin values are the same as in example 4.1. From equation 4.4,

$$5 = 5 \times 10^{-6}\frac{dv_C}{dt} + v_C$$

For DC steady-state, all voltage must be constant.

$$\frac{dv_C}{dt} = 0$$
$$5 = 0 + v_C$$
$$v_C = 5 \text{ volts}$$

With the switch open, figure 4.1(c) applies. Using $v_{Th} = 0$, equation 4.4 yields

$$0 = 10^{-5}\frac{dv_C}{dt} + v_C$$

Again,

$$\frac{dv_C}{dt} = 0$$
$$0 = 0 + v_C$$
$$v_C = 0$$

The *sinusoidal steady-state* case is approached similarly. However, the forcing function is a sinusoid.

$$v_{Th} = V_m \cos(\omega t + \theta) \qquad 4.10$$

Thus, equation 4.7 becomes

$$V_m \cos(\omega t + \theta) = b\frac{dq}{dt} + cq \qquad 4.11$$

Example 4.3

Figure 4.1 has $v = 10\cos(1000t + 60°)$ volts, $R_1 = 10$ ohms, $R_2 = 10$ ohms, and an inductance of 0.15 henrys. Determine the steady-state current with the switch (a) closed, and (b) open.

With the switch closed, the Thevenin resistance is 5 ohms.

$$X_L = 1000 \times 0.15 = 150$$

The phasor impedance is $5 + j150$. The Thevenin equivalent voltage in phasor form is $V_{Th} = 5\angle 60°$. The phasor impedance is $Z = 150\angle 88.1°$.

$$i = \frac{V_{Th}}{Z} = 0.033\angle{-28.1°}$$
$$i = 0.033\cos(1000t - 28.1°)$$

When the switch is opened, there will be no steady-state current.

Example 4.4

Figure 4.1 has $v = 10\cos(1000t + 60°)$ volts, $R_1 = 10$ ohms, $R_2 = 10$ ohms, and a capacitance of $1\,\mu F$. Determine the steady-state capacitance voltage with the switch (a) closed, and (b) open.

Equation 4.4 can be put into phasor form.

$$v_{Th} = (j\omega R_{Th}C + 1)v_C$$
$$= 5\angle 60°$$
$$\omega R_{Th}C = 10^3 \times 5 \times 10^{-6} = 5 \times 10^{-3}$$

So,

$$1 + j\omega R_{\mathrm{Th}}C = 1\angle 0.3°$$

and,

$$v_C = 5\angle 59.7°$$

This result can be transformed to the time form.

$$v_C = 5\cos(1000t + 59.7°)$$

The steady-state voltage for the switch open case is again zero.

Once the steady-state solution for the switching circuit is known, the *total solution* is obtained by matching conditions at the switching time for the two circuits. The conditions that must match are the inductance current (for inductive circuits) and the capacitance voltage (for capacitive circuits).

Stored energy cannot change instantaneously, so an inductance current must remain the same during switching. Similarly, a capacitance voltage must remain the same during switching.

The usual procedure is to determine inductance current or capacitance voltage just prior to switching and to set that value just after switching. This information is used in equations 4.12 and 4.13.

$$i_L(t) = Ae^{-t/\tau} + I_{L,\text{steady state}} \qquad 4.12$$
$$v_C(t) = Ae^{-t/\tau} + V_{C,\text{steady state}} \qquad 4.13$$

When the time of switching (t_s) and the steady-state values before and after switching are known, the transient coefficient A of equations 4.12 and 4.13 can be found. This is illustrated in examples 4.5 through 4.8.

Example 4.5

The network in figure 4.1 has $V = 10$ volts, $R_1 = 10$ ohms, $R_2 = 10$ ohms, and an inductance of 0.15 henrys as in example 4.1. This network has had the switch open for a long time, and the switch is closed at $t = 0$. Find i_L after the switch closes.

Having the switch open for a long but unspecified time implies that steady-state conditions have been reached. Thus, $i_L = 0$. This also will be the initial value, $i_L(0^+)$.

The steady-state current with a closed switch was found to be 1 amp in example 4.1. The time constant is

$$\tau = \frac{L}{R} = \frac{0.15}{5}$$
$$= 0.03$$

Equation 4.12 can be written as

$$i_L(t) = Ae^{-t/0.03} + 1$$

At $t = 0^+$, $i_L(0^+) = 0$ and $e^0 = 1$, so

$$0 = A + 1$$
$$A = -1$$

Thus for $t > 0$,

$$i_L(t) = 1 - e^{-t/0.03}$$

Example 4.6

The network in figure 4.1 has values $V = 10$ volts, $R_1 = 10$ ohms, $R_2 = 10$ ohms, and capacitance of 1000 μF. The switch has been closed for a long time and it is opened at $t = 0$. Determine the capacitance voltage as a function of time after the switch opens.

The initial voltage is 5 volts at the instant the switch is opened. The final voltage is zero. From equation 4.13 at $t = 0$ with $v_C(0) = 5$, A is found to be 5.

The time constant is

$$\tau = R_{\mathrm{Th}}C = 10 \times 1000 \times 10^{-6}$$
$$= 10^{-2}$$
$$v_C(t) = 5e^{-100t}$$

For sinusoidal forcing functions, the procedure is similar. However, the precise instant of switching must be given in order to derive an exact solution.

Example 4.7

As in example 4.3, the network of figure 4.1 has values $V = 10\cos(1000t + 60°)$, $R_1 = 10$ ohms, $R_2 = 10$ ohms, and an inductance of 0.15 henrys. The switch has been closed for a long time and is opened at $t = 0$. Determine the inductance current after the switch opens.

The steady-state current was found in example 4.3 to be $0.033\cos(1000t - 28.1°)$ for the switch closed. As the switch opens at $t = 0$, the initial inductance current will be

$$i(0^+) = 0.033\cos(0 - 28°) = 0.029$$

The final current will be zero. From equation 4.12, at $t = 0$,

$$e^{-t/\tau} = 1$$

So $A = i(0^+)$.

$$i(t) = 0.029e^{-t/0.015}$$

Example 4.8

As in example 4.3, the network of figure 4.1 has values $V = 10\cos(1000t + 60°)$, $R_1 = 10$ ohms, $R_2 = 10$ ohms, and an inductance of 0.15 henrys. The switch has been open for a long time and is closed at $t = 0$. Determine the inductance current after the switch closes.

For example 4.3, the initial current in the inductance is 0. The steady-state current is $0.033\cos(1000t - 28.1°)$. From equation 4.12,

$$i(t) = Ae^{-t/0.03} + 0.033\cos(1000t - 28.1°)$$

At $t = 0^+$, $i(0^+) = 0$. Since $e^0 = 1$,

$$0 = A + 0.033\cos(-28.1°)$$

or $A = -0.029$.

$$i(t) = -0.029e^{-t/0.03} + 0.033\cos(1000t - 28.1°)$$

D. First-Order Pulse Transients

First order pulse transients are not caused by switching changes in network topology. Rather, the transients are caused by the pulses from the sources. Pulses are represented by *unit steps* superimposed on one another as shown in figure 4.2.

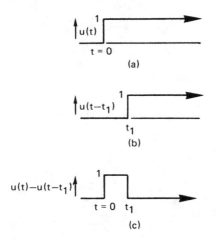

Figure 4.2 (a) Unit Step $u(t)$, (b) Unit Step $u(t - t_1)$, (c) Pulse $u(t) - u(t - t_1)$

The unit step $u(t - t_1)$ is defined by equation 4.14.

$$\begin{aligned} u(t - t_1) &= 0 \quad t < t_1 \\ &= 1 \quad t > t_1 \end{aligned}$$

$$4.14$$

Any pulse train can be constructed from a series of unit steps. A single pulse beginning at $t = 0$ and ending at $t = t_1$ is shown in figure 4.2(c). Because the networks are linear, it is convenient to determine the response to each unit step and apply superposition by adding the solutions. The response to a unit step is called the *step response*. Since the network is linear, the step responses are all the same, but are only translated in time by the appropriate amount.

The response of an energy storage network to a unit step at time t_1 is determined in terms of the state variable of the energy storage device—inductance current or capacitance voltage. If other quantities are required, they can be obtained by use of circuit relations.

The unit step is zero for all time up to t_1, so there is no initial inductance current or capacitor voltage. Then, the response to the unit step at $t = t_1$ is

$$v_C(t) = u(t - t_1)\left(1 - e^{-(t-t_1)/\tau}\right) \qquad 4.15$$

$$i_L(t) = \frac{1}{R_{\text{Th}}}u(t - t_1)\left(1 - e^{-(t-t_1)/\tau}\right)$$

$$4.16$$

τ is $R_{\text{Th}}C$ for the capacitance circuit and L/R_{Th} for the inductance circuit.

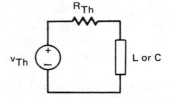

Figure 4.3 Single Energy Storage Network

Example 4.9

A capacitance of $100\,\mu\text{F}$ sees a Thevenin resistance of 1000 ohms and a step voltage which changes from 0 to 10 volts at $t = 1$. Find the capacitance voltage.

The voltage required is not limited to $t > 1$.

$$RC = \tau = 10^3 \times 100 \times 10^{-6}$$
$$= 10^{-1}$$

From equation 4.15,

$$v_C(t) = 10u(t - 1)\left(1 - e^{-10(t-1)}\right)$$

The factor of 10 is necessary to scale the unit step to step of height 10.

Example 4.10

The pulse source in example 4.9 puts out two 10 volt pulses 10 ms in length and separated by 20 ms. Determine the capacitance voltage.

The voltage source can be represented as

$$v = 10u(t) - 10u(t - 0.01) + 10u(t - 0.03) - 10u(t - 0.04)$$

The response to a general step of 10 volts is 10 times equation 4.15. Using superposition, the response is

$$v_C(t) = 10u(t)\left(1 - e^{-10t}\right)$$
$$- 10u(t - 0.01)\left(1 - e^{-10(t-0.01)}\right)$$
$$+ 10u(t - 0.03)\left(1 - e^{-10(t-0.03)}\right)$$
$$- 10u(t - 0.04)\left(1 - e^{-10(t-0.04)}\right)$$

3 SECOND-ORDER TRANSIENT ANALYSIS

A. Introduction

Networks containing two independent energy storage elements require a second-order differential equation to describe them. Such a differential equation can be written in terms of any circuit variable. However, inductance current and capacitance voltage are preferred because they relate directly to the initial conditions.

The general form of a second-order differential equation is

$$f(t) = a\frac{d^2q}{dt^2} + b\frac{dq}{dt} + cq \qquad 4.17$$

There are three possible cases, depending on the magnitudes of a, b, and c.

- Case 1: Overdamped $b^2 > 4ac$

Overdamped transients die out without oscillation. The solution will have two transient exponential parts and a steady-state part. The form of the solution is

$$q(t) = Ae^{s_1 t} + Be^{s_2 t} + q_{ss} \qquad 4.18$$
$$s_1 = \frac{-b + \sqrt{b^2 - 4ac}}{2a} \qquad 4.19$$
$$s_2 = \frac{-b - \sqrt{b^2 - 4ac}}{2a} \qquad 4.20$$

q_{ss} is the steady-state value of the variable q.

The coefficients A and B in equation 4.18 are found from the initial and steady-state values of q and dq/dt. Assuming the transient is initiated at $t = 0^+$, the initial values $q(0^+)$ and $q'(0^+)$ as well as the values of q_{ss} and q'_{ss} must be determined.

$$q(0^+) = A + B + q_{ss} \qquad 4.21$$
$$q'(0^+) = s_1 A + s_2 B + q'_{ss} \qquad 4.22$$

The solution of equations 4.21 and 4.22 determines A and B.

$$A = \frac{1}{2}\left(1 + \frac{b}{\sqrt{b^2 - 4ac}}\right)[q(0^+) - q_{ss}]$$
$$+ \left(\frac{a}{\sqrt{b^2 - 4ac}}\right)[q'(0^+) - q'_{ss}] \qquad 4.23$$

$$B = \frac{1}{2}\left(1 - \frac{b}{\sqrt{b^2 - 4ac}}\right)[q(0^+) - q_{ss}]$$
$$- \left(\frac{a}{\sqrt{b^2 - 4ac}}\right)[q'(0^+) - q'_{ss}] \qquad 4.24$$

Example 4.11

A system is described by the differential equation

$$12 = \frac{d^2v}{dt^2} + 5\frac{dv}{dt} + 4v$$

The initial conditions are $v = -5$ and $dv/dt = 2$. Determine the voltage v for $t > 0$.

In steady state, dv/dt and d^2v/dt^2 are both zero, so the forcing function is a constant 12. The steady-state value is

$$v_{ss} = \frac{12}{4} = 3$$
$$v'_{ss} = 0$$

From equations 4.19 and 4.20,

$$s_1 = \frac{-5 + \sqrt{25 - 16}}{2}$$
$$= \frac{-5 + 3}{2} = -1$$
$$s_2 = \frac{-5 - \sqrt{25 - 16}}{2}$$
$$= \frac{-5 - 3}{2} = -4$$

From equations 4.23 and 4.24,

$$A = \frac{1}{2}\left(1 + \frac{5}{3}\right)(-5 - 3) + \frac{1}{3}\left(2 - 0\right) = -10$$

$$B = \frac{1}{2}\left(1 - \frac{5}{3}\right)(-5 - 3) - \frac{1}{3}\left(2 - 0\right) = 2$$

Then, from equation 4.18,

$$v(t) = -10e^{-t} + 2e^{-4t} + 3$$

Example 4.12

A system is described by

$$12\cos 2t = \frac{d^2v}{dt^2} + 5\frac{dv}{dt} + 4v$$

Initial conditions are $v = -5$ and $dv/dt = 2$ at $t = 0$. Determine v for $t > 0$.

Assume a steady-state solution of $v_{ss} = A\cos(2t + \phi)$. After successive differentiating,

$$12\cos 2t = -4A\cos(2t + \phi) - 10A\sin(2t + \phi)$$
$$+ 4A\cos(2t + \phi)$$

From this,

$$A = 1.2$$
$$\phi = \frac{-\pi}{2}$$
$$v_{ss} = 1.2\sin 2t$$

For the natural response, assume

$$v_n = Be^{s_1 t} + Ce^{s_2 t}$$

Substitute e^{st} into the describing equation (while ignoring $12\cos 2t$, as it is taken care of by the steady-state response).

$$0 = (s^2 + 5s + 4)e^{st}$$

Then, $s_1 = -1$ and $s_2 = -4$.

$$v(t) = 1.2\sin 2t + Be^{-t} + Ce^{-4t}$$

$$\frac{dv}{dt} = 2.4\cos 2t - Be^{-t} - 4Ce^{-4t}$$

These equations are evaluated at $t = 0$ and matched with initial conditions.

$$-5 = 0 + B + C$$
$$2 = 2.4 - B - 4C$$

Or, $B = 6.8$ and $C = 1.8$.

$$v(t) = 1.2\sin 2t - 6.8e^{-t} + 1.8e^{-4t}$$

- Case 2: Critically Damped $b^2 = 4ac$

A second-order system needs to have two transient components. With only one time constant, the solution is

$$q(t) = Ae^{st} + Bte^{st} + q_{ss} \qquad 4.25$$

$$s = \frac{-b}{2a} \qquad 4.26$$

The initial values at $t = 0^+$ are

$$q(0^+) = A + 0 + q_{ss} \qquad 4.27$$
$$q'(0^+) = sA + B + q'_{ss} \qquad 4.28$$

Equations 4.27 and 4.28 can be solved to yield A and B.

$$A = q(0^+) - q_{ss} \qquad 4.29$$

$$B = q'(0^+) - q'_{ss} + \frac{b}{2a}A \qquad 4.30$$

Example 4.13

A system is described by the differential equation

$$20 = \frac{d^2 i}{dt^2} + 4\frac{di}{dt} + 4i$$

Initial conditions are $i = 0$ and $di/dt = 4$ at $t = 0$. Determine i for $t > 0$. The steady-state current is

$$i_{ss} = \frac{20}{4} = 5$$

The initial current was given as $i_0 = 0$. From equation 4.26,

$$s = \frac{-4}{2} = -2$$

From equations 4.29 and 4.30,

$$A = 0 - 5 = -5$$
$$B = 4 - 0 + 2(-5) = -6$$

Then,

$$i(t) = 5(1 - e^{-2t}) - 6te^{-2t}$$

Example 4.14

Repeat example 4.13 with $f(t) = 20\sin 4t$ as the forcing function.

Assume $i_{ss} = A\sin(4t + \phi)$ and substitute into the defining equation.

$$20\sin 4t = A[-16\sin(4t + \phi)$$
$$+ 16\cos(4t + \phi) + 4\sin(4t + \phi)]$$

This implies $A = -1$ and $\phi = 53.13°$.

This system is critically damped with $s = -2$ (see equation 4.25).

$$i(t) = -\sin(4t + 53.13°) + Be^{-2t} + Cte^{-2t}$$

At $t = 0$, $i = 0$, so

$$0 = -\sin 53.13° + B$$

From this, B is found to be 0.8.

$$\frac{di}{dt} = -4\cos(4t + 53.13°) - 2Be^{-2t} + Ce^{-2t} - 2Cte^{-2t}$$

At $t = 0$, $di/dt = 4$, so

$$4 = -4\cos 53.13° - 2B + C$$

This determines $C = 8$. Therefore,

$$i(t) = -\sin(4t + 53.13°) + 0.8e^{-2t} + 8te^{-2t}$$

- Case 3: Underdamped $b^2 < 4ac$

In this case, there is an oscillatory transient conveniently represented by

$$q = Ae^{-\alpha t}\cos\beta t + Be^{-\alpha t}\sin\beta t + q_{ss} \qquad 4.31$$

$$\alpha = \frac{b}{2a} \qquad 4.32$$

$$\beta = \frac{\sqrt{4ac - b^2}}{2a} \qquad 4.33$$

And, for a constant $f(t)$,

$$q_{ss} = \frac{f(t)}{c} \qquad 4.34$$

The initial conditions result in

$$q(0^+) = A + q_{ss} \qquad 4.35$$
$$q'(0^+) = -\alpha A + \beta B + q'_{ss} \qquad 4.36$$

Solving for A and B,

$$A = q(0^+) - q_{ss} \qquad 4.37$$

$$B = \frac{\alpha}{\beta}A + \frac{1}{\beta}[q'(0^+) - q'_{ss}] \qquad 4.38$$

An alternative solution has the form

$$q = Ce^{-\alpha t}\cos(\beta t - \phi) + q_{ss} \qquad 4.39$$

$$C = \sqrt{A^2 + B^2} \qquad 4.40$$

$$\tan\phi = \frac{B}{A} \qquad 4.41$$

Example 4.15

A system is described by

$$100 = 5\frac{d^2v}{dt^2} + 6\frac{dv}{dt} + 5v$$

At the time of switching, the derivative of the voltage is 3, and the voltage is 25. Determine the voltage after the switching at $t = 0$.

In this case, a is 5, b is 6, and c is 5. Then,

$$b^2 - 4ac = -64$$

So the system is underdamped.

From equations 4.32 and 4.33,

$$\alpha = \frac{6}{2 \times 5} = 0.6$$

$$\beta = \frac{\sqrt{64}}{2 \times 5} = 0.8$$

The form of $v(t)$ is

$$v(t) = Ae^{-0.6t}\cos 0.8t + Be^{-0.6t}\sin 0.8t + v_{ss}$$

v_{ss} is found from the original equation.

$$v_{ss} = \frac{100}{5} = 20$$

A is found from equation 4.37.

$$A = 25 - 20 = 5$$

B is found from equation 4.38.

$$B = 0.75(5) + 1.25(3 - 0) = 7.5$$

Then, $v(t)$ is

$$v(t) = 5e^{-0.6t}\cos 0.8t + 7.5e^{-0.6t}\sin 0.8t + 20$$

In the alternative form of equation 4.39, $v(t)$ is

$$v(t) = 20 + 9.01e^{-0.6t}\cos(0.8t - 56.31°)$$

Example 4.16

The forcing function in example 4.15 is replaced by $120\cos 2t$. Determine $v(t)$ for $t > 0$.

A different approach is used to find the steady-state response. Assume

$$v_{ss} = M\cos 2t + N\sin 2t$$

This is then differentiated and used in the describing equation.

$$120\cos 2t = (-20M + 12N + 5M)\cos 2t$$
$$+ (-20N - 12M + 5N)\sin 2t$$

By equating coefficients of the sine and cosine terms,

$$\text{from cos:} \quad 120 = (-15M + 12N)$$
$$\text{from sin:} \quad 0 = (-15N - 12M)$$

Solving for M and N,

$$v_{ss} = -4.9\cos 2t + 3.9\sin 2t$$

The derivative is

$$\frac{dv_{ss}}{dt} = 7.8\cos 2t + 9.8\sin 2t$$

The values at $t = 0$ are:

$$v_{ss}(0^+) = -4.9$$

$$\frac{dv_{ss}(0^+)}{dt} = 7.8$$

From equations 4.37 and 4.38,

$$A = 25 - (-4.9) = 29.9$$

$$B = \frac{0.6}{0.8}(29.9) + \frac{1}{0.8}(3 - 7.8) = 16.4$$

Then, $v(t)$ for $t > 0$ is

$$v(t) = 29.9e^{-0.6t}\cos 0.8t + 16.4e^{-0.6t}\sin 0.8t$$
$$- 4.9\cos 2t + 3.9\sin 2t$$

This can be simplified by combining terms of the same frequencies.

$$v(t) = 34.1e^{-0.6t}\cos(0.8t - 28.7°)$$
$$- 6.26\cos(2t - 38.5°)$$

B. Step Response of Second-Order Systems

The *step response of second-order systems* for zero initial conditions is useful in the analysis of pulse transients. The final condition with a step input is a constant.

$$\frac{f(t)}{c} = q_{ss} \qquad 4.42$$

For a unit step, the final condition is $1/c$. Thus, for zero initial conditions the unit step response is found by setting $f(t) = 1$ in equation 4.17.

- Case 1: Overdamped $b^2 > 4ac$

$$s_1 = \frac{-b + \sqrt{b^2 - 4ac}}{2a} \qquad 4.43$$

$$s_2 = \frac{-b - \sqrt{b^2 - 4ac}}{2a} \qquad 4.44$$

$$q_{ss} = \frac{1}{c} \qquad 4.45$$

Then, from equations 4.23 and 4.24,

$$q(t) = -\frac{1}{2c}\left(1 + \frac{b}{\sqrt{b^2 - 4ac}}\right)e^{s_1 t}$$
$$- \frac{1}{2c}\left(1 - \frac{b}{\sqrt{b^2 - 4ac}}\right)e^{s_2 t} + \frac{1}{c}$$
$$\qquad 4.46$$

- Case 2: Critically Damped $b^2 = 4ac$

From equations 4.29 and 4.30,

$$q(t) = -\frac{1}{c}e^{-st} - \frac{b}{2ac}te^{-st} + \frac{1}{c} \qquad 4.47$$

$$s = \frac{-b}{2a} \qquad 4.48$$

- Case 3: Underdamped $b^2 < 4ac$

$$\alpha = \frac{b}{2a} \qquad 4.49$$

$$\beta = \frac{\sqrt{4ac - b^2}}{2a} \qquad 4.50$$

From equations 4.37 and 4.38,

$$q(t) = \frac{1}{c} - \frac{1}{c}e^{-\alpha t}\cos\beta t - \frac{\alpha}{\beta c}e^{-\alpha t}\sin\beta t \qquad 4.51$$

Of course, the angles must be in radians.

Example 4.17

A network has an output current described by

$$0.2V_s = 20\frac{d^2 i}{dt^2} + 15\frac{di}{dt} + 6i$$

V_s is a 10 volt pulse lasting 0.1 second. Determine the current 0.2 second after the pulse has ended.

$b^2 - 4ac = 225 - 480$, so the system is underdamped.

$$\alpha = \frac{15}{2 \times 20} = 0.375$$

$$\beta = \frac{\sqrt{4ac - b^2}}{2a} = \frac{\sqrt{255}}{40}$$
$$= 0.4$$

The unit step response is found from equation 4.51.

$$i(t) = \frac{1}{6}\left(1 - e^{-0.375t}\cos 0.4t - 0.939e^{-0.375t}\sin 0.4t\right)u(t)$$

The pulse is represented by

$$V_s = 10u(t) - 10u(t - 0.1)$$
$$0.2V_s = 2[u(t) - u(t - 0.1)]$$

The response to the first step is:

$$i_1(t) = \frac{2}{6}\Bigg(1 - e^{-0.375t}\cos 0.4t$$
$$- 0.939e^{-0.375t}\sin 0.4t\Bigg)u(t)$$

The response to the second step is

$$i_2(t) = -\frac{2}{6}\Bigg[1 - e^{-0.375(t-0.1)}\cos 0.4(t-0.1)$$
$$- 0.939e^{-0.375(t-0.1)}\sin 0.4(t-0.1)\Bigg]u(t-0.1)$$

The total current is $i_1 + i_2$. At $t = 0.3$ (0.2 after the pulse), the current is

$$i(0.3) = \frac{1}{3}\Bigg[1 - e^{-0.375\times0.3}\cos(0.4\times0.3)$$
$$- 0.939e^{-0.375\times0.3}\sin(0.4\times0.3)\Bigg]$$
$$- \frac{1}{3}\Bigg[1 - e^{-0.375\times0.2}\cos(0.4\times0.2)$$
$$- 0.939e^{-0.375\times0.2}\sin(0.4\times0.2)\Bigg]$$
$$= 0.0023$$

C. Initial Conditions in Second-Order Systems

Initial conditions in second-order systems must be found from the conditions just prior to switching. A method of finding them is to isolate the portion of the network containing the two energy storage elements. Then, the Thevenin equivalent of the remaining portion is found.

The energy stored in a capacitance or inductance cannot change during the switching. When a capacitance voltage and/or inductance current is known prior to switching, the initial conditions are then known for those elements.

Series LC circuits, as indicated in figure 4.4, will have zero steady-state current. The capacitance blocks the DC current and eventually absorbs the entire Thevenin equivalent voltage. With initial current I_0 and initial voltage V_0 across the capacitance, the circuit differential equation is

$$V_{\text{Th}} = R_{\text{Th}}i + L\frac{di}{dt} + v_C \qquad 4.52$$

$$i = C\frac{dv_C}{dt} \qquad 4.53$$

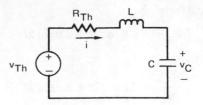

Figure 4.4 Series LC and Thevenin Equivalent Circuit

The differential equation in terms of v_C is

$$v_{\text{Th}} = LC\frac{d^2v_C}{dt^2} + R_{\text{Th}}C\frac{dv_C}{dt} + v_C \qquad 4.54$$

The initial conditions result in $v_C(0^+) = V_0$ and

$$C\frac{dv_C(0^+)}{dt} = I_0 \qquad 4.55$$

$$\frac{dv_C(0^+)}{dt} = \frac{I_0}{C} \qquad 4.56$$

Example 4.18

A series combination of 1 henry and 10 μF is connected to a Thevenin equivalent circuit having a 10 volt source in series with a 500 ohm resistance. The initial current is 0.01 amp, and the drop across the capacitance is initially 5 volts. Determine v_C for $t > 0$.

Calculate the products for use with equation 4.54.

$$LC = 1 \times 10^{-5} = 10^{-5}$$
$$R_{\text{Th}}C = 500 \times 10^{-5} = 5 \times 10^{-3}$$

The differential equation is

$$10 = 10^{-5}\frac{d^2v_C}{dt^2} + 5 \times 10^{-3}\frac{dv_C}{dt} + v_C$$

Using $a = 10^{-5}, b = 5 \times 10^{-3}$, and $c = 1$,

$$b^2 - 4ac = -1.5 \times 10^{-5}$$

This is an underdamped circuit. From equations 4.49 and 4.50,

$$\alpha = 250$$
$$\beta = 194$$

Therefore, the form of $v_C(t)$ is

$$v_C(t) = Ae^{-250t}\cos 194t + Be^{-250t}\sin 194t + 10$$

From equation 4.56,

$$\frac{dv_C(0^+)}{dt} = \frac{0.01}{10^{-5}} = 10^3$$

Next, simultaneously solve for A and B.

$$v_C(0^+) = 5 = A + 10$$

$$\frac{dv_C(0^+)}{dt} = 10^3 = -250A + 194B$$

So, $A = -5$, and $B = -1.29$. The capacitance voltage is

$$v_C(t) = 10 - 5e^{-250t} \cos 194t - 1.29e^{-250t} \sin 194t$$

A *parallel LC circuit* will have a DC steady-state response of zero volts because of the inductance. The network shown in figure 4.5 is described by the differential equation

$$\frac{v_{Th}}{R_{Th}} = \frac{v_C}{R_{Th}} + C\frac{dv_C}{dt} + i_L \qquad 4.57$$

$$v_C = v_L = L\frac{di_L}{dt} \qquad 4.58$$

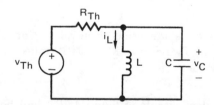

Figure 4.5 Parallel LC and Thevenin Equivalent Circuit

By use of equation 4.58, the differential equation can also be written in terms of the inductance current.

$$\frac{v_{Th}}{R_{Th}} = \frac{L}{R_{Th}}\frac{di_L}{dt} + LC\frac{d^2i_L}{dt^2} + i_L \qquad 4.59$$

$$\frac{di_L(0^+)}{dt} = \frac{v_C(0^+)}{L} \qquad 4.60$$

Example 4.19

The parallel LC circuit shown in figure 4.5 has an inductance current $i_L(0^+) = 0.5$ amp and capacitance voltage $v_C(0^+) = 5$ volts. $v_{Th} = 10$ volts, $R_{Th} = 500$ ohms, $L = 0.1$ henry, and $C = 10$ μF. Determine the inductance current after $t = 0$.

The differential equation is given by equation 4.59.

$$\frac{10}{500} = \frac{0.1}{500}\frac{di_L}{dt} + 10^{-6}\frac{d^2i_L}{dt^2} + i_L$$

or

$$0.02 = 10^{-6}\frac{d^2i_L}{dt^2} + 2 \times 10^{-4}\frac{di_L}{dt} + i_L$$

$$b^2 - 4ac = 4 \times 10^{-8} - 4 \times 10^{-6} = -3.96 \times 10^{-6}$$

Thus, the circuit is underdamped, and

$$\alpha = \frac{b}{2a} = \frac{2 \times 10^{-4}}{2(10^{-6})} = 100$$

$$\beta = \sqrt{\frac{c}{a} - \alpha^2} = \sqrt{10^6 - 10^4}$$

$$= 995$$

The solution for i_L is

$$i_L(t) = Ae^{-100t}\cos 995t + Be^{-100t}\sin 995t + 0.02$$
$$i_L(0^+) = 0.5 = A + 0.02$$

So, $A = 0.48$.

Differentiating the expression for $i_L(t)$,

$$\frac{di_L(t)}{dt} = -100(Ae^{-100t}\cos 995t + Be^{-100t}\sin 995t)$$
$$+ 995(-Ae^{-100t}\sin 995t + Be^{-100t}\cos 995t)$$

At $t = 0$,

$$\frac{di_L(0^+)}{dt} = -100A + 995B$$

$$= \frac{5}{0.1} = 50$$

This can be solved for B since A is known.

$$B = \frac{50 + 100A}{995} = 0.098$$

Thus, the inductance current is

$$i_L(t) = 0.02 + 0.48e^{-100t}\cos 995t$$
$$+ 0.098e^{-100t}\sin 995t$$

4 LAPLACE TRANSFORM ANALYSIS

A. Introduction

Laplace transforms are a convenient general form for system analysis. Laplace transforms can be applied either by taking the transform of a differential equation which has been obtained in the classical way, or by analyzing a transformed circuit.

The first step in the classical approach is to obtain the system differential equations. The next step is to obtain the Laplace transform of the differential equations and of all voltage and current sources. The transformed equations are then solved algebraically.

With the circuit transformation method, circuit analysis is performed on transformed impedances with initial condition sources for each energy storage element. This results in the same transformed equations obtained from the classical approach.

Regardless of which approach is taken, the transformed equations are broken into their partial fractions to be inversely transformed. This is particularly true for transient analysis of second- and higher-order systems. For steady-state sinusoidal analysis, the equations reduce to phasor form, and graphical representations of system characteristics can be used.

The transformation of classical differential equations is the most direct method in problems where the desired quantity is an energy storage device variable. In these cases, the sources and resistances can be reduced simply and are either in a Thevenin form (with a voltage source and series resistance), or in a Norton form (with an equivalent current source and parallel resistance). The Laplace transforms can be applied directly to each variable.

$$\mathcal{L}\{f(t)\} = F(s) \qquad 4.61$$

$$\mathcal{L}\left\{\frac{df(t)}{dt}\right\} = sF(s) - f(0^+) \qquad 4.62$$

$f(t)$ can be any function of time, and $f(0^+)$ is the function value at $t = 0^+$. State variables are usually capacitance voltage and inductance current. These are energy storage variables which cannot change value instantaneously. When the functions are state variables, values cannot change instantaneously. In that case,

$$f(0) = f(0^+) \qquad 4.63$$

B. Capacitance

The voltage across a capacitance is a state variable. It is preferred as one of the variables in system equations. The capacitance voltage does not change instantaneously, so $v_C(0^+) = v_C(0)$. The capacitance voltage $v_C(t)$ and the capacitive current $C(dv_C/dt)$ are transformed according to equations 4.64 and 4.65.

$$\mathcal{L}\{v_C(t)\} = V_C(s) \qquad 4.64$$

$$\mathcal{L}\{i_C(t)\} = \mathcal{L}\{C\frac{dv_C(t)}{dt}\}$$

$$= sCV_C(s) - Cv_C(0) \qquad 4.65$$

Equation 4.65 is a key to the transformed circuit. In a transformed circuit, the variables are the Laplace transforms of the voltage and current. The elements are impedances and transformed sources. The terminals of a transformed capacitance must behave according to equation 4.65.

Such a circuit element could be modeled with two current sources as indicated in figure 4.6. However, recognizing that the current through the left-hand source is a function of the terminal voltage, it can be replaced by an admittance sC or by an impedance $1/sC$. This equivalent transformed circuit element accounts for the initial conditions by the parallel current source $Cv_C(0)$.

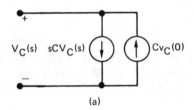

(a)

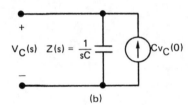

(b)

Figure 4.6   Transformed Equivalent Capacitance Circuit: (a) Two-Source Model, (b) Impedance and Source Model

The equivalent circuit of figure 4.6(b) will be used in general Laplace analysis in this chapter.

C. Inductance

Inductance current is similarly a state variable. It is preferred as one of the variables when an inductance is a part of a circuit. The inductance current $i_L(t)$ and the inductance voltage $L(di_L/dt)$ are transformed by equations 4.66 and 4.67.

$$\mathcal{L}\{i_L(t)\} = I_L(s) \qquad 4.66$$

$$\mathcal{L}\{v_L(t)\} = \mathcal{L}\left\{L\frac{di_L(t)}{dt}\right\}$$

$$= sLI_L(s) - Li_L(0) \qquad 4.67$$

The terminals of a transformed inductance must behave according to equation 4.67. Such a circuit element could be modeled with two voltage sources as indicated in figure 4.7(a).

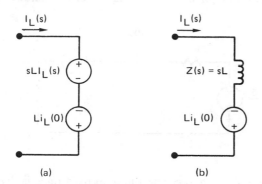

(a)   (b)

Figure 4.7 Transformed Equivalent Inductance Circuit: (a) Two-Source Model, (b) Impedance and Source Model

The voltage across the upper source is a function of the terminal current. It can be replaced with an impedance sL as in figure 4.7(b). This equivalent transformed circuit element accounts for the initial conditions with the series current source $Li_L(0)$.

The equivalent circuit of figure 4.7(b) will be used in general Laplace analysis in this chapter.

D. First-Order Systems

Consider the Laplace transformation of a differential equation such as would be obtained from a circuit such as in figure 4.3. The first step in the analysis of any circuit is to obtain the system differential equations by using Kirchhoff's voltage and current laws. For first-order systems where the state variable is the required quantity, the circuit seen from the terminals of the energy storage device is reduced to a Thevenin or Norton equivalent. The circuit equation is easily written, as illustrated in the following examples.

For a capacitance, the Thevenin form is preferred, since the final value of the capacitance voltage must be the Thevenin equivalent voltage source's final value. This provides a quick check on the resulting expression for the capacitance voltage.

Example 4.20

The reduced circuit shown represents the circuit after the switching or source changes. The circuit equation is Kirchhoff's voltage law around the single loop. The current flowing is

$$i(t) = C\frac{dv_C(t)}{dt}$$

The differential equation is

$$v_{Th}(t) = RC\frac{dv_C(t)}{dt} + v_C(t)$$

Equation 4.65 is used to transform the system equation term by term.

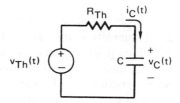

$$V_{Th}(s) = RC[sV_C(s) - v_C(0)] + V_C(s)$$

This is solved for $V_C(s)$.

$$V_C(s) = \frac{V_{Th}(s) + RCv_C(0)}{sCR + 1}$$

This resulting equation for $V_C(s)$ is solved for $V_C(t)$ by the methods of chapter 1, after obtaining the transform for the source voltage.

For an inductance current in a first-order system, the Norton equivalent source is preferred, since the final value of the inductance current will be identical to the final value of the Norton current source.

Example 4.21

A reduced circuit is shown, and the inductance current is desired.

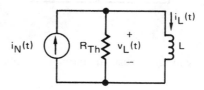

The circuit equation is Kirchhoff's current law taken at the upper node. The current through the resistance is the inductance voltage divided by the resistance.

$$i_N(t) = \frac{L}{R}\frac{di_L(t)}{dt} + i_L(t)$$

This is transformed with equations 4.66 and 4.67.

$$I_N(s) = \frac{L}{R}\left[sI_L(s) - i_L(0)\right] + I_L(s) \qquad 4.68$$

This can be solved for $I_L(s)$.

$$I_L(s) = \frac{RI_N(s) + Li_L(0)}{sL + R}$$

$I_L(t)$ can be found from $I_L(s)$ by using the inverse transformation method introduced in chapter 1.

The last two examples resulted in identical forms of the Laplace transformed expression for the state variable. In general, any first-order system will have a similar form. However, where an initial condition term is not a state variable, it will have to be evaluated at $t = 0^+$ as illustrated in section 3.

To use the convenience of state variables in circuits where the desired quantities are not state variables, circuit transformations prior to circuit analysis are required. Then, any ordinary method of circuit analysis, including network reduction, can be used. For network reduction, the impedances can be treated as resistances and combined. Initial condition sources are treated as any other independent source.

Example 4.22

The circuit shown has the switch closed at $t = 0$ after being open for a long time. Obtain the Laplace transform expression for the voltage across the 100 ohm resistance.

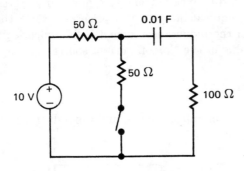

The initial voltage across the capacitance is $v_C(0) = 10$ volts.

It is necessary to convert the voltage divider consisting of the 10 volt source and the two 50 ohm resistances to the Thevenin equivalent and its transformation. These two steps result in the transformed circuit shown.

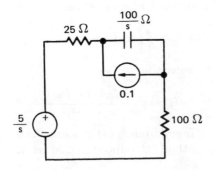

Next, the parallel impedance $(100/s)$ and the initial condition current source of 0.1 amp are converted to a Thevenin equivalent, yielding the circuit shown.

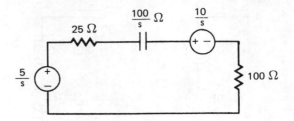

As the voltage across the 100 ohm resistance is desired, solve for its current and multiply by 100. Using Kirchhoff's voltage law around the loop yields

$$\frac{5}{s} - \frac{10}{s} = 125I(s) + \frac{100}{s}I(s)$$

This is solved for $I(s)$.

$$I(s) = \frac{\dfrac{-5}{s}}{\dfrac{100}{s} + 125} = \frac{-0.04}{s + 0.8}$$

The expression for $I(s)$ is transformed back to the time domain by using Laplace transform pairs.

$$i(t) = -0.04e^{-0.8t}$$
$$v_{100}(t) = 100i(t) = -4e^{-0.8t}$$

E. Second-Order Systems

A second-order system contains two energy storage devices. When the desired quantity is one of the state variables, the method of transforming the system differential equations (written in terms of the state variables) is appropriate. This is demonstrated in example 4.23.

Example 4.23

Begin with the reduced circuit shown.

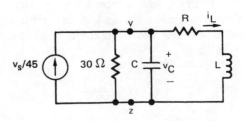

The circuit equations are specified in terms of the capacitance voltage and inductance current, the variables most easily handled by the Laplace transformation.

$$\frac{v_s}{45} = \frac{v_C}{30} + C\frac{dv_C}{dt} + i_L$$

$$v_C = Ri_L + L\frac{di_L}{dt}$$

These equations are next transformed term by term using

$$\mathcal{L}\{v_s(t)\} = V_s(s)$$

The transformed circuit equations are:

$$\frac{V_s(s)}{45} = \frac{V_C(s)}{30} + sCV_C(s) - Cv_C(0) + I_L(s)$$

$$V_C(s) = RI_L(s) + sLI_L(s) - Li_L(0)$$

The capacitance voltage is needed. Solve for the inductance current in terms of the capacitance voltage in the first equation. Substituting into the second equation,

$$V_C(s) = \frac{(R+sL)\left(\dfrac{V_s(s)}{45} + Cv_C(0)\right) - Li_L(0)}{s^2LC + s\left(\dfrac{L}{30} + RC\right) + 1 + \dfrac{R}{30}}$$

This will allow the methods of chapter 1 to be used.

When the desired variables include more than the state variables, circuit transformation is recommended.

Example 4.24

Consider the circuit previously used in example 4.23. The initial values of the state variables are 8 volts (dropping across the capacitor) and 40 mA of current downward in the inductance. Determine the response of the inductance current to a 5 volt step voltage at $t = 0$.

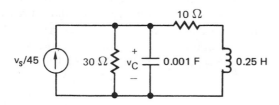

First, transform the circuit according to the models of sub-sections B and C above. The current source for the capacitance's transformed circuit is

$$0.001 \text{ farad} \times 8 \text{ volts} = 0.008 \text{ amp-sec}$$

The voltage source for the inductance's transformed circuit is

$$0.25 \text{ henry} \times 0.04 \text{ amp} = 0.01 \text{ volt-sec}$$

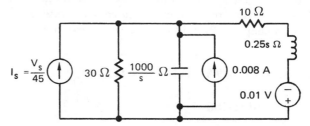

The parallel elements to the left of the 10 ohm resistance are combined into a Norton equivalent. The parallel current sources add directly, giving a total of $I_s(s) + 0.008$. The 30 ohm resistance and the parallel capacitance impedance of $1000/s$ combine to $1000/(s + 33.3)$.

Next, the Norton source is converted to a Thevenin equivalent voltage source by multiplying by the impedance of the parallel combination of 30 ohms and $1000/s$ ohms, and by substituting $V_s/45$ for I_s.

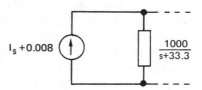

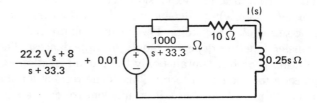

The Kirchhoff voltage law equation is written in terms of the inductance current.

$$\frac{22.2V_s + 8}{s + 33.3} + 0.01 = \left(\frac{1000}{s + 33.3} + 10 + 0.25s\right)I(s)$$

This is solved for $I(s)$.

$$I(s) = \frac{88.9V_s + 33.3 + 0.04s}{s^2 + 73.3s + 5333}$$

The function $v_s(t)$ is a 5 volt step. Therefore, $V_s = 5/s$. The denominator polynomial factors to $(s + 36.7)^2 + 63.2^2$ (to put it into the proper form for transformation). Thus, the transformed current is

$$I(s) = \frac{\dfrac{444}{s} + 33.3 + 0.04s}{(s + 36.7)^2 + 63.2^2}$$

F. Higher-Order Systems

Higher-order systems are best analyzed by transforming the circuit (as was done in the last example). Higher-order systems are encountered in control systems (covered in chapter 12), and in sinusoidally excited circuits.

5   SINUSOIDAL ANALYSIS

A. Introduction

The phasor analysis of sinusoids in chapter 3 deals with fixed frequency sinusoids. However, a Laplace analysis can be used to study the effects of *variable frequencies*, including *spectral inputs* which have multiple frequencies.

In a sinusoidal steady-state analysis, neither initial conditions nor natural transient response terms are significant. They will die out, leaving only the response to the sinusoidal signal. For sinusoidal analysis, several steps used in the previous sections can be skipped.

In particular, the Laplace transform method leads to a partial fraction expansion involving all the factors of the denominator polynomial. Terms appear for the natural response as well as for the signal. The natural response terms will include the effects of initial conditions, but the initial conditions will not affect the signal terms.

The partial fraction expansion terms for the signal (*forced response*) are dependent only on the characteristic terms of the system. These can be expressed as a transfer function dependent only on impedances. The solution procedure obtains only the partial fraction expansion for the signal term. This is the residue for the signal term. For this case, it is sufficient to replace s by $j\omega$.

B. Transfer Functions

A *transfer function* is the Laplace transform of a ratio of variables. The variables can be impedance, admittance, current ratio, or voltage ratio. Transfer functions assume that the initial conditions are zero or do not matter. For example, in example 4.23 the expression obtained was

$$V_C(s) = \frac{(R + sL)\left(\dfrac{V_s(s)}{45} + Cv_C(0)\right) - Li_L(0)}{s^2 LC + s\left(\dfrac{L}{30} + RC\right) + 1 + \dfrac{R}{30}}$$

The transfer function V_C/V_s can be obtained by eliminating the term $Cv_C(0) - Li_L(0)$ and dividing both sides by $V_s(s)$.

$$G(s) = \frac{V_C(s)}{V_s(s)}$$

$$= \frac{\dfrac{R + sL}{45}}{s^2 LC + s\left(\dfrac{L}{30} + RC\right) + 1 + \dfrac{R}{30}}$$

This is equivalent to removing the initial condition sources from the transformed circuit, solving for a variable as a function of the signal, and taking the ratio of the response to the signal. Transfer functions are used for transient and sinusoidal steady-state analysis and design. The transient application has been covered in the previous sections, so the frequency response application is now considered.

In *frequency response analysis*, the transfer function is evaluated in the *frequency domain*, rather than in the time domain. In essence, the operation requires the evaluation of the residue due to the signal in the partial fraction expansion. This is done by replacing s by $j\omega$ in the transfer function. The result is a complex number, which is normally expressed in polar form with magnitude and angle. If the time response to a sinusoidal signal is desired, the magnitude of the transfer function can be multiplied by the signal magnitude, and the angle of the transfer function added to the signal angle.

C. First-Order Response

The capacitance and inductance transformations for sinusoidal steady-state analysis use only the impedances. Initial condition sources are omitted. The capacitance initial condition current source is replaced by an open circuit, and the inductance initial condition voltage source is replaced by a short circuit.

Example 4.25

Consider the circuit of example 4.22 with the 10 volt source replaced by an arbitrary sinusoidal source $V_s(s)$. The complete circuit for sinusoidal analysis is shown. This circuit omits the initial condition source and has included the voltage divider effect of the source and resistance.

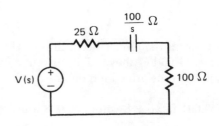

The Kirchhoff voltage law equation is

$$V_s(s) = \left(125 + \frac{100}{s}\right) I(s)$$

Solving for $I(s)$,

$$I(s) = \frac{sV_s(s)}{125s + 100}$$

The transfer function is

$$G(s) = \frac{I(s)}{V_s(s)}$$

$$= \frac{s}{125s + 100}$$

The final step is to substitute $j\omega$ for s and obtain the magnitude and angle of the transfer function as a function of the source frequency ω.

$$G(j\omega) = \frac{j\omega}{125j\omega + 100}$$

$$|G(j\omega)| = \frac{|j\omega|}{25(|5j\omega + 4|)}$$

$$|G(j\omega)| = \frac{\omega}{25\sqrt{25\omega^2 + 16}}$$

The $j\omega$ term in the numerator of $G(j\omega)$ gives a constant 90 degree angle, so

$$\angle G(j\omega) = 90° - \arctan\frac{125\omega}{100}$$

$$= 90° - \arctan(1.25\omega)$$

The result obtained here is identical to that which would have been obtained by phasor methods. Therefore, it is not necessary to use Laplace transforms to obtain general sinusoidal results.

• Bode Diagrams

Information on the frequency response of a system is typically given by plots of magnitude and angle (called *phase*) as a function of frequency. Most common is the Bode[1] plot, a plot of magnitude in decibels and angle in degrees versus the logarithm of frequency. The advantage of using decibels to represent magnitude is that decibels are a logarithmic measure. The logarithm of a product of two terms is the sum of the logarithms of the separate terms, while the logarithm of a ratio of terms is the logarithm of the numerator term minus the logarithm of the denominator term.

$$\log(xy) = \log(x) + \log(y) \qquad 4.69$$

$$\log\left(\frac{x}{y}\right) = \log(x) - \log(y) \qquad 4.70$$

The *decibel* is defined as 10 times a power ratio, but it has been loosely adopted for transfer functions in general. It can be used for ratios of voltage or current by taking the square of the ratio and ignoring the fact that the two quantities are not acting on the same impedance.

[1] Pronounced "bow'-dee".

Thus, the decibel measure of current gain, G_I, or voltage gain, G_V, can be written as in equations 4.71 and 4.72, respectively.

$$G_{I|\text{dB}} = 10\log\left|\left(\frac{I_2}{I_1}\right)\right|^2$$

$$= 20\log\left|\left(\frac{I_2}{I_1}\right)\right| \qquad 4.71$$

$$G_{V|\text{dB}} = 10\log\left|\left(\frac{V_2}{V_1}\right)\right|^2$$

$$= 20\log\left|\left(\frac{V_2}{V_1}\right)\right| \qquad 4.72$$

It is generally assumed that the decibel measure is 20 times the logarithm of the transfer function.

$$G_{|\text{dB}} = 20\log(|G(j\omega)|) \qquad 4.73$$

In general, $G(j\omega)$ consists of a numerator polynomial which can be factored into a product of terms of the form $(j\omega + z_i)$ multiplied by a constant, and a denominator polynomial which can be factored into a product of terms of the form $(j\omega + p_j)$, where z_i and p_j are, in general, complex numbers.

$$G(j\omega)$$

$$= \frac{A(j\omega + z_1)(j\omega + z_2)(j\omega + z_3)\cdots(j\omega + z_k)}{(j\omega)^n(j\omega + p_1)(j\omega + p_2)(j\omega + p_3)\cdots(j\omega + p_m)}$$

$$4.74$$

except when $z_i = 0$, $(j\omega + z_i) = z_i(1 + j\omega/z_i)$
except when $p_j = 0$, $(j\omega + p_j) = p_j(1 + j\omega/p_j)$

Thus, when the transfer function is written in decibels, it is in the form of

$$G(j\omega)_{\text{dB}} = 20\log K + \sum_{i=1}^{k} 20\log\left|1 + \frac{j\omega}{z_i}\right|$$

$$- 20n\log\omega - \sum_{j=1}^{m} 20\log\left|1 + \frac{j\omega}{p_j}\right| \qquad 4.75$$

Note,

$$\sum_{i=1}^{4} f_i = f_1 + f_2 + f_3 + f_4 \qquad 4.76$$

The constant K of equation 4.75 is related to the constant A of equation 4.74 by

$$K = A\frac{\displaystyle\prod_{i=1}^{k}(z_i)}{\displaystyle\prod_{j=1}^{m}(p_j)} \qquad 4.77$$

Note,

$$\prod_{i=1}^{4} f_i = (f_1)(f_2)(f_3)(f_4) \qquad 4.78$$

The angle of G is

$$\angle G(j\omega) = -90n + \sum_{i=1}^{k} \arctan\left(\frac{\omega}{z_i}\right) - \sum_{j=1}^{m} \arctan\left(\frac{\omega}{p_j}\right) \qquad 4.79$$

All angles are in degrees.

Most first-order systems have transfer functions in one of the following three forms:

$$G_1(j\omega) = \frac{K\left(1 + \dfrac{j\omega}{z}\right)}{1 + \dfrac{j\omega}{p}} \qquad 4.80$$

$$G_2(j\omega) = \frac{K(j\omega)}{1 + \dfrac{j\omega}{p}} \qquad 4.81$$

$$G_3(j\omega) = \frac{K}{1 + \dfrac{j\omega}{p}} \qquad 4.82$$

In decibel form,

$$G_1(j\omega)|_{dB} = 20\log|K| + 20\log\left|1 + \frac{j\omega}{z}\right|$$

$$- 20\log\left|1 + \frac{j\omega}{p}\right| \qquad 4.83$$

$$G_2(j\omega)|_{dB} = 20\log|K| + 20\log\omega$$

$$- 20\log\left|1 + \frac{j\omega}{p}\right| \qquad 4.84$$

$$G_3(j\omega)|_{dB} = 20\log|K| - 20\log\left|1 + \frac{j\omega}{p}\right| \qquad 4.85$$

The Bode diagram has magnitude in dB and phase angle in degrees plotted against the base 10 logarithm of frequency. A unit of log frequency is a decade: from $\log a$ to $\log 10a$. For example,

$$\log 10a = \log 10 + \log a$$
$$= 1 + \log a \qquad 4.86$$

Thus, a decade of frequency corresponds to one unit of $\log\omega$.

There are four types of functions in the decibel forms of the first-order transfer function depicted above. These are discussed below and shown in figures 4.8 through 4.13.

(1) $20\log|K|$ is a constant and independent of frequency, as indicated in figure 4.8.

(2) $20\log(\omega)$ is proportional to $\log(\omega)$ or $\log(f)$.

$$\log(\omega) = \log(2\pi f) = \log(2\pi) + \log(f)$$
$$= 0.798 + \log(f)$$

$$20\log(\omega) = -40 \text{ dB at } \omega = 0.01 \text{ and } \log\omega = -2$$
$$20\log(\omega) = -20 \text{ dB at } \omega = 0.1 \text{ and } \log\omega = -1$$
$$20\log(\omega) = \quad 0 \text{ dB at } \omega = 1 \quad \text{ and } \log\omega = 0$$
$$20\log(\omega) = \quad 20 \text{ dB at } \omega = 10 \text{ and } \log\omega = 1$$
$$20\log(\omega) = \quad 40 \text{ dB at } \omega = 100 \text{ and } \log\omega = 2$$

Thus, it can be seen as in figure 4.8, $20\log\omega$ has a slope of 20 dB/decade of frequency. (A *decade* of frequency is a factor of ten change.)

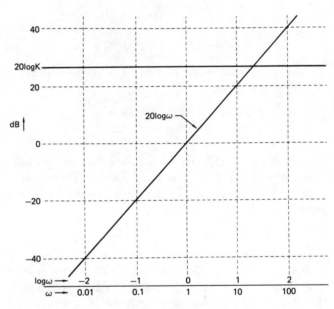

Figure 4.8   Decibel Plots of $20\log K$ and $20\log\omega$

(3) $20\log|1 + j\omega/z|$ approaches 0 dB for $\omega/z \ll 1$ (low-frequency asymptote).

$20\log|1 + j\omega/z| = 3$ dB for $w/z = 1$.

$20\log|1 + j\omega/z|$ approaches $20\log(\omega/z)$ for $\omega/z \gg 1$ (high-frequency asymptote).

$$20\log(\omega/z) = 20\log(\omega) - 20\log(z)$$

$20\log(\omega/z)$ passes through 0 dB at $\omega = z$, and has a slope of +20 dB/decade. This is indicated in figure 4.9.

(4) $-20\log|1 + j\omega/p|$ approaches 0 dB for $\omega/p \ll 1$ (low-frequency asymptote).

$-20\log|1 + j\omega/p| = -3$ dB for $\omega/p = 1$.

$-20\log|1 + j\omega/p|$ approaches $-20\log(\omega/p)$ for $\omega/p \gg 1$ (high-frequency asymptote).

$$-20\log(\omega/p) = -20\log(\omega) + 20\log(p)$$

$-20 \log(\omega/p)$ passes through 0 dB at $\omega = p$, and has a slope of -20 dB/decade. This is indicated in figure 4.10.

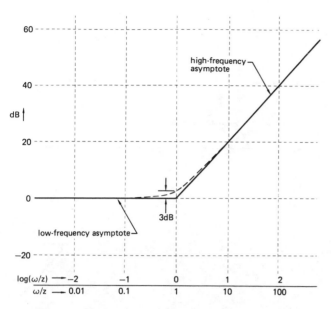

Figure 4.9 Decibel Plots of $20 \log |1 + j\omega/z|$ with Asymptotes Shown as Solid Lines and Actual Values Dashed

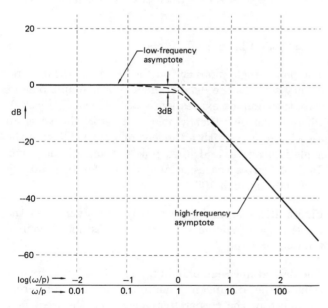

Figure 4.10 Decibel Plots of $-20 \log |1+j\omega/p|$ with Asymptotes Shown as Solid Lines and Actual Values Dashed

The function $G_1(j\omega)$ includes the gain K, a zero, and a pole.

$$G_1(j\omega) = \frac{K\left(1 + \dfrac{j\omega}{z}\right)}{1 + \dfrac{j\omega}{p}}$$

$$G_1(j\omega)|_{dB} = 20 \log |K| + 20 \log \left|1 + \frac{j\omega}{z}\right|$$

$$-20 \log \left|1 + \frac{j\omega}{p}\right|$$

In figure 4.11, the amplitude plot is shown for the case where the zero is smaller than the pole. For low frequencies, only the gain term is active. When the frequency approaches the value of the zero, the zero's effect begins to be a factor. Note in figure 4.9, the zero contributes 0 dB at low frequencies. When the frequency approaches that of the pole, the pole influence begins, which at yet higher frequencies cancels out the effects of the zero in as much as the high-frequency behavior becomes a constant at a level of $20 \log(Kp/z)$. For this situation, $K = 0.37$, $z = 0.071$, $p = 4.6$, and $Kp/z = 24.0$.

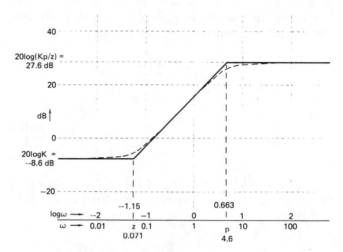

Figure 4.11 Decibel Plot of $G_1(j\omega)$ for the Case Where the Zero is Smaller than the Pole

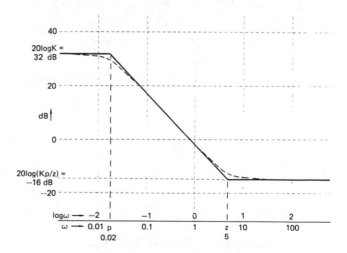

Figure 4.12 Decibel Plots of $G_1(j\omega)$ for the Case Where the Zero is Larger than the Pole

In figure 4.12, the amplitude plot of $G_1(j\omega)$ is shown for the case where the zero is larger than the pole. For low frequencies, only the gain term is active. When the frequency approaches the value of the pole, the pole's effect begins to be a factor. Note in figure 4.10, the pole contributes 0 dB at low frequencies. When the frequency approaches that of the zero, the zero influence begins, which at yet higher frequencies cancels out the effects of the pole in that the high frequency behavior becomes a constant at a level of $20\log(Kp/z)$. In this case, $K = 40$, $z = 5.0$, $p = 0.02$, and $Kp/z = 0.16$.

The function $G_2(j\omega)$ has a zero at zero frequency, but the frequency zero does not appear on the log plot since the logarithm of zero is minus infinity. The only part of the function that can be easily determined is the high-frequency asymptote, where the effects of the pole and zero have cancelled to give a value of Kp. At low frequencies where $\omega << p$, the function behaves as the linear combination of the gain and a zero at zero frequency. This is shown in figure 4.8.

$$G_2(j) = \frac{K(j\omega)}{1 + \dfrac{j\omega}{p}}$$

$$G_2(j\omega)|_{\mathrm{dB}} = 20\log|K| + 20\log\omega - 20\log\left|1 + \frac{j\omega}{p}\right|$$

The decibel plot for a value of $Kp = 4.0$ is shown in figure 4.13 with the frequency normalized to the pole frequency.

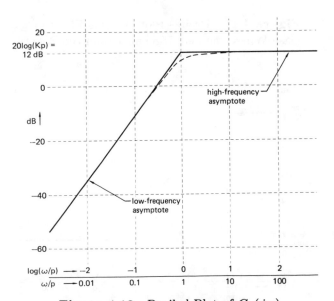

Figure 4.13   Decibel Plot of $G_2(j\omega)$

The function $G_3(j\omega)$ is very similar to that of figure 4.10, except that the low-frequency asymptote is now at $20\log K$. In figure 4.14, $K = 100$. The frequency is again scaled to the pole frequency.

$$G_3(j\omega) = \frac{K}{1 + \dfrac{j\omega}{p}}$$

$$G_3(j\omega)|_{\mathrm{dB}} = 20\log|K| - 20\log\left|1 + \frac{j\omega}{p}\right|$$

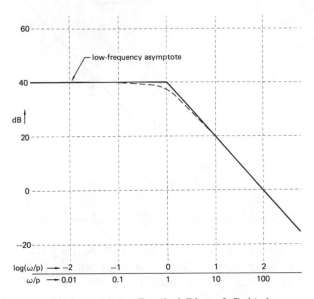

Figure 4.14   Decibel Plot of $G_3(j\omega)$

- Bode Phase Angle Plots

The phase angle plots are on the same logarithmic frequency scale as the decibel plots. For a simple pole or zero, the complex expression is of the form $(1 + j\omega/q)$, where q can be either a pole or a zero, which has an associated angle with a tangent of ω/q. For a zero the angle is positive, and for a pole the angle is negative. Table 4.1 gives the associated angles for ω/q ranging from 0.01 through 100.

Figure 4.15 shows the phase angle of a zero and the phase angle of a pole when the scale is normalized to the pole or zero.

The dashed line segments in figure 4.15 are approximations to the arctangent functions over the range from one tenth of the corner frequency (i.e., the frequency of the pole or zero is referred to as a corner frequency because the decibel plot asymptotes meet at those frequencies) to ten times the corner frequency. These straight line segments begin at an angle of zero degrees one decade before the zero or pole at zero degrees phase angle, and extend to one decade beyond the pole or zero at +90 degrees for the zero, and −90 degrees for the pole. These are also called asymptotes, and are accompanied by a low-frequency asymptote of zero degrees for frequencies less than one tenth of the corner frequency,

Table 4.1

ω/q	0.01	0.02	0.04	0.08	0.10	0.20	0.40	0.80	1.0
$\log(\omega/q)$	-2	-1.7	-1.4	-1.1	-1.0	-0.70	-0.40	-0.10	0
$\theta(\deg)$	0.6	1.1	2.3	4.6	5.7	11.3	21.8	38.7	45

ω/q	1.25	2.5	5	10	12.5	25	50	100
$\log(\omega/q)$	0.10	0.40	0.70	1.0	1.1	1.4	1.7	2
$\theta(\deg)$	51.3	68.2	78.7	84.3	85.4	87.7	88.9	89.4

and a high frequency asymptote of ±90 degrees (plus for a zero and minus for a pole) beginning at a frequency of ten times the corner frequency. The useful aspect of these approximations is they are no worse than about 6 degrees in error at any point. For this reason they are useful in approximating the phase angle contributions to the Bode diagram.

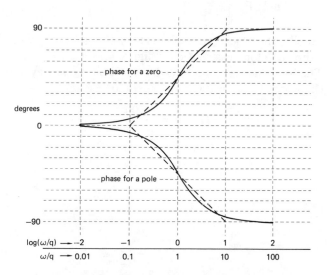

Figure 4.15 Phase Angle Plot for Pole or Zero with Normalization of Frequency

In addition, a zero at zero frequency (a factor of s in the numerator) contributes a phase angle of $+90$ degrees, while a pole at zero frequency (a factor of s in the denominator) contributes a phase angle of -90 degrees. The gain constant K can be either positive or negative. If positive, it contributes no phase angle, but if negative it contributes an angle of ±180 degrees (whichever is worse for the stability of the system).

Returning to the three common first-order transfer functions of the three following forms:

$$G_1(j\omega) = \frac{K\left(1 + \dfrac{j\omega}{z}\right)}{1 + \dfrac{j\omega}{p}}$$

$$G_2(j\omega) = \frac{K(j\omega)}{1 + \dfrac{j\omega}{p}}$$

$$G_3(j\omega) = \frac{K}{1 + \dfrac{j\omega}{p}}$$

The angle, as the angle of the zero minus the angle of the pole can be obtained by

$$\angle G_1 = \arctan\left(\frac{\omega}{z}\right) - \arctan\left(\frac{\omega}{p}\right) \qquad 4.87$$

$$\angle G_2 = 90° - \arctan\left(\frac{\omega}{p}\right) \qquad 4.88$$

$$\angle G_3 = -\arctan\left(\frac{\omega}{p}\right) \qquad 4.89$$

First consider $G_1(j\omega)$ for the case where the zero is smaller than the pole. In this situation $K = 0.37$, the zero is at 0.071, and the pole is at 4.6. This corresponds to the situation of figure 4.11, and so the asymptotic decibel plot is repeated here in figure 4.16 with the angles in degrees on the right-hand scale.

$$G_1(j\omega) = \frac{0.37\left(1 + \dfrac{j\omega}{0.071}\right)}{1 + \dfrac{j\omega}{4.6}}$$

$$\angle G_1 = \arctan\left(\frac{\omega}{0.071}\right) - \arctan\left(\frac{\omega}{4.6}\right)$$

Next consider $G_1(j\omega)$ for the case where the zero is larger than the pole. In this case, the zero is at 5.0 and the pole is at 0.02. This corresponds to the situation of figure 4.12, and so the asymptotic decibel plot is repeated here in figure 4.17 with the phase angle in degrees on the right-hand scale.

$$G_1(j\omega) = \frac{40\left(1 + \dfrac{j\omega}{5}\right)}{1 + \dfrac{j\omega}{0.02}}$$

$$\angle G_1 = \arctan\left(\frac{\omega}{5.0}\right) - \arctan\left(\frac{\omega}{0.02}\right)$$

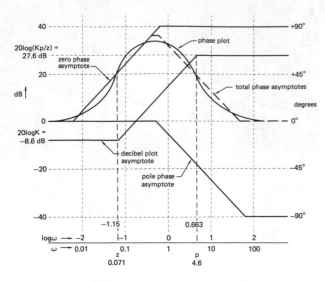

Figure 4.16 Asymptotic Decibel Plot of $G_1(j)$ with the Asymptotic and Actual Phase for the Case where the Zero is Smaller than the Pole

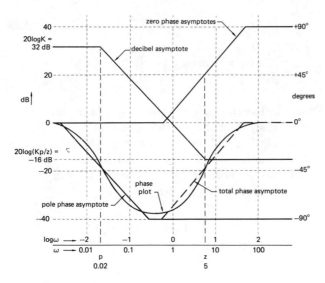

Figure 4.17 Asymptotic Decibel Plot of $G_1(j)$ Together with the Phase Plot for the Case Where the Zero is Larger than the Pole

Next, consider $G_2(j\omega)$ with $Kp = 4.0$ and the frequency scaled to the pole. This corresponds to the situation of figure 4.13, and so the asymptotic decibel plot is repeated here in figure 4.18 with phase angle in degrees on the right-hand scale.

$$G_2(j\omega) = \frac{K(j\omega)}{1 + \dfrac{j\omega}{p}}$$

$$\angle G_2 = 90° - \arctan\left(\frac{\omega}{p}\right)$$

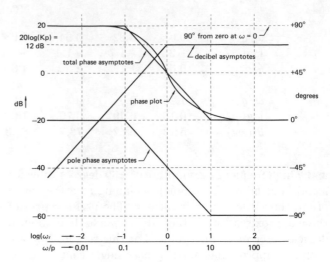

Figure 4.18 Asymptotic Decibel Plot of Function $G_2(j\omega)$ Together with the Phase Angle Plot

Finally, the phase angle of function $G_3(j\omega)$ is shown in figure 4.19 for the situation in figure 4.14 with $K = 100$. The frequency is again scaled to the pole frequency and the asymptotic decibel plot repeated.

$$G_3(j\omega) = \frac{100}{1 + \dfrac{j\omega}{p}}$$

$$\angle G_3 = -\arctan\left(\frac{\omega}{p}\right)$$

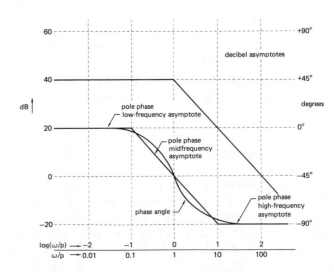

Figure 4.19 Asymptotic Decibel Plot of $G_3(j)$ Together with the Phase Angle Plot

- First-Order Filters

First-order filters are those filters limited to single-pole circuits, such as low-pass and high-pass single-time-constant filters. A low-pass filter is one which allows signals of low frequency to pass through while attenuating higher frequency signals. A high-pass filter does the opposite, attenuating lower frequency signals and passing the higher frequency signals.

The decibel plots depicted in the previous section represent the frequency response of some typical single-pole filters.

The function $G_2(s)$ is an example of a high-pass filter, and $G_3(s)$ is an example of a low-pass filter. The function $G_1(s)$ is a low-pass when the zero is larger than the pole and high-pass when the zero is smaller than the pole, although it is less desireable than $G_2(s)$ for a high-pass filter, and less desireable than $G_3(s)$ for a low-pass filter, as $G_1(s)$ fails to attenuate the undesired frequency range as well as the others.

Example 4.26

A single-time-constant low-pass filter has the same form of transfer function as $G_3(s)$ of the previous section. This can be realized in a simple form as shown here.

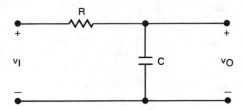

The voltage transfer function of this circuit can be found to be

$$\frac{V_o(s)}{V_i(s)} = T_v(s) = \frac{1}{1 + sCR}$$

In comparing the transfer function with that of function G_3 of the previous section, it can be seen that the pole is at $\omega = 1/RC$. The Bode decibel and phase angle asymptotes for this example are shown in figure 4.20.

Example 4.27

The circuit of example 4.25 is an example of a high-pass filter when the output voltage is taken across the 100 ohm resistor. A simpler high-pass circuit is shown here involving only a capacitor and a resistor.

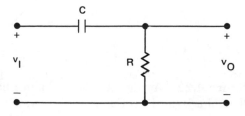

The voltage transfer function of this circuit can be found to be

$$\frac{V_o(s)}{V_i(s)} = T_v(s) = \frac{sCR}{1 + sCR}$$

In comparing the transfer function with that of function G_2 of the previous section, it can be seen that the pole is at $\omega = 1/RC$. The Bode decibel and phase angle asymptotes for this example are shown in figure 4.21.

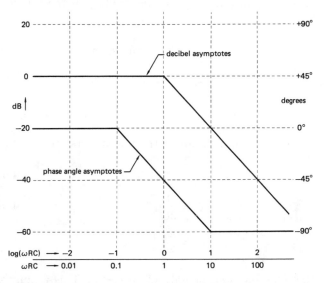

Figure 4.20 Asymptotic Decibel and Phase Angle Plots for the Low-Pass Circuit of Example 4.26

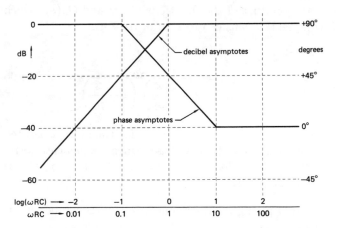

Figure 4.21 Asymptotic Bode Decibel and Phase Angle Plots for the High-Pass Circuit of Example 4.27

D. Second-Order Response

The method for solving second-order systems is similar to first-order methods. The major difference is that critically damped and underdamped responses can occur. For first-order systems, the denominator polynomial of a transfer function is limited to *simple poles*.

This means that there can be only a single factor of the form $s + a$.

In second-order systems, the denominator must be factored into its *poles*. These can be two simple pole factors such as $(s+a)(s+b)$, repeated simple poles $(s+a)^2$, or a pair of complex conjugate poles.

Denominators of transfer functions can be expressed in several convenient forms depending on the problem type.

$$s^2 + 2\zeta\omega_n s + \omega_n^2 \quad \text{(control systems)} \qquad 4.90$$

$$s^2 + \frac{\omega_o}{Q}s + \omega_o^2 \quad \text{(resonance analysis)} \qquad 4.91$$

$$(s + \alpha)^2 + \beta^2 \quad \text{(transient analysis)} \qquad 4.92$$

The relationships between the parameters are:

$$\omega_o^2 = \omega_n^2 = \alpha^2 + \beta^2 \qquad 4.93$$

$$\zeta = \frac{1}{2Q} \qquad 4.94$$

$$\alpha = \frac{\omega_o}{2Q} \qquad 4.95$$

Complex poles occur only when $\zeta < 1$. This is equivalent to $Q > 0.5$ and β being a real number.

Consideration of second-order frequency response can begin with a system that has a transfer function that is the product of the functions $G_2(s)$ and $G_3(s)$, which will yield a transfer function of the form

$$H(s) = \frac{\dfrac{s}{\omega_o}}{\left(1 + \dfrac{s}{p_1}\right)\left(1 + \dfrac{s}{p_2}\right)} \qquad 4.96$$

This is a band-pass filter which, for the poles separated by at least one decade, has corner frequencies corresponding very closely to p_1 and p_2. For $\omega_o = p_1$ this is as in figure 4.22.

By multiplying the denominator out, this can be rewritten as

$$H(s) = \frac{\dfrac{s}{\omega_o}}{1 + s\left(\dfrac{1}{p_1} + \dfrac{1}{p_2}\right) + \dfrac{s^2}{p_1 p_2}} \qquad 4.97$$

Notice there are two real denominator roots or poles: $-p_1$ and $-p_2$. As long as these poles are far apart, the corner frequencies (i.e., where the amplitude is 3 dB down from the maximum value) are at the poles. However, as the poles approach each other, this is no longer true. In order to deal with that situation, the

transfer function is generalized to allow cases of complex roots of the form

$$H(s) = \frac{\dfrac{s}{\omega_o}}{1 + \dfrac{s}{Q\omega_o} + \dfrac{s^2}{\omega_o^2}} \qquad 4.98$$

This is further simplified by multiplying the numerator and denominator by ω_o/s to obtain

$$H(s) = \frac{1}{\dfrac{\omega_o}{s} + \dfrac{1}{Q} + \dfrac{s}{\omega_o}} \qquad 4.99$$

For a frequency response solution, consider H with s replaced by $j\omega$.

$$H(j\omega) = \frac{1}{\dfrac{1}{Q} + j\left(\dfrac{\omega}{\omega_o} - \dfrac{\omega_o}{\omega}\right)} \qquad 4.100$$

It can be seen that H has its maximum magnitude of Q at $\omega = \omega_o$, and has a phase angle of zero there. It has a phase angle of $+45$ degrees and a magnitude of $Q/\sqrt{2}$ at a frequency ω_1 where $\omega_1/\omega_o - \omega_o/\omega_1 = -1/Q$. It has a phase angle of -45 degrees at the frequency ω_2 where $\omega_2/\omega_o - \omega_o/\omega_2 = 1/Q$, where again its magnitude is $Q/\sqrt{2}$. These two frequencies are called the half-power frequencies, and can be solved to be

$$\frac{\omega_1}{\omega_o} = \frac{1}{2Q}\left(\sqrt{1 + 4Q^2} - 1\right) \qquad 4.101$$

$$\frac{\omega_2}{\omega_o} = \frac{1}{2Q}\left(\sqrt{1 + 4Q^2} + 1\right) \qquad 4.102$$

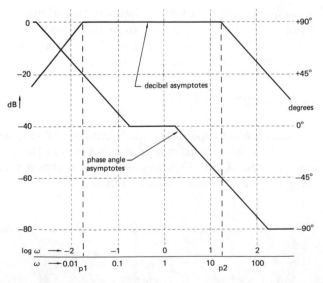

Figure 4.22 Asymptotic Decibel and Phase Plots of a Band-Pass Filter

Notice that the product of equations 4.101 and 4.102 is 1, so that the frequency ω_o is the geometric mean of ω_1 and ω_2.

$$\omega_o^2 = \omega_1 \omega_2$$

and

$$\frac{\omega_2}{\omega_o} = \frac{\omega_o}{\omega_1} \quad \text{or} \quad \frac{\omega_1}{\omega_o} = \frac{\omega_o}{\omega_2} \qquad 4.103$$

Also notice that the difference between these two equations is $1/Q$, so that the bandwidth, $\omega_2 - \omega_1 = \omega_o/Q$. The bandwidth is defined as the difference between the half-power frequencies. When the poles are real (i.e., $Q \le 0.5$), they can be found from equation 4.98 to be at

$$\frac{s}{\omega_o} = \frac{1}{2Q}\left(-1 \pm \sqrt{1 - 4Q^2}\right) \qquad 4.104$$

It can be seen that the magnitudes of the poles and the 3 dB frequencies given in equations 4.101 and 4.102 are very nearly the same for $Q \ll 0.5$. As a result, $4Q^2$ is negligible when compared to "one" in the radicals.

Example 4.28

Determine the value of Q, the pole locations, and the half-power frequencies for the cases: (a) $\omega_2 = 10\omega_1$, and (b) $\omega_2 = 100\omega_1$.

(a) From equation 4.103, $\omega_1/\omega_o = 1/\sqrt{10} = 0.316$ and $\omega_2/\omega_o = \sqrt{10} = 3.16$. From equations 4.101 and 4.102,

$$10\frac{\omega_1}{\omega_o} = 10\frac{1}{2Q}\left(\sqrt{1 + 4Q^2} - 1\right) = \frac{\omega_2}{\omega_o}$$
$$= \frac{1}{2Q}\left(\sqrt{1 + 4Q^2} + 1\right)$$

which is solved to find $Q = \sqrt{10}/9 = 0.351$. From equation 4.104 the poles are found to be at

$$\frac{s}{\omega_o} = -0.411 \text{ and } -2.435$$

(b) From equation 4.103, $\omega_1/\omega_o = 1/\sqrt{100} = 0.1$ and $\omega_2/\omega_o = \sqrt{100} = 10.0$. From equations 4.101 and 4.103,

$$100\frac{\omega_1}{\omega_o} = 100\frac{1}{2Q}\left(\sqrt{1 + 4Q^2} - 1\right) = \frac{\omega_2}{\omega_o}$$
$$= \frac{1}{2Q}\left(\sqrt{1 + 4Q^2} + 1\right)$$

which is solved to find $Q = 1/9.9 = 0.101$. From equation 4.104 the poles are found to be at

$$\frac{s}{\omega_o} = -0.102 \text{ and } -9.798$$

This demonstrates that the poles and corner frequencies are very nearly the same for widely separated poles, but

as they approach one another the corner frequencies (at -3 dB from the maximum) are further apart than the poles.

The gain plot in decibels is shown for representative values for Q in figure 4.23.

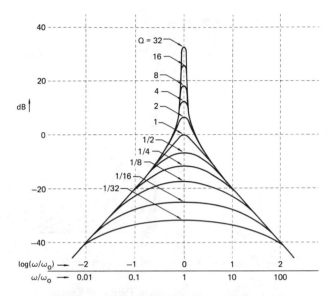

Figure 4.23 Decibel Plots of Second-Order Band-Pass Filters

Figure 4.23 shows that these functions are all asymptotic to the same low- and high-frequency asymptotes.

$$\text{low-frequency asymptote: } 20\log\left(\frac{\omega}{\omega_o}\right)$$

$$\text{high-frequency asymptote: } 20\log\left(\frac{\omega_o}{\omega}\right)$$

$$= -20\log\left(\frac{\omega}{\omega_o}\right)$$

It can be seen that these asymptotes meet at $\omega = \omega_o$ for this particular formula (equations 4.98 through 4.100).

The phase angle plots are also of interest. They become more abrupt as the poles come closer to each other. As Q increases, the poles become complex. This is illustrated in figure 4.24.

- Second-Order Filters

Low Pass: The simplest two-pole low-pass filter has the transfer function

$$G_{\text{LP}}(s) = \frac{1}{1 + \dfrac{s}{Q\omega_o} + \dfrac{s^2}{\omega_o^2}} \qquad 4.105$$

This has amplitude Bode diagrams as in figure 4.25. It has a phase plot as in figure 4.24, except that the low-frequency phase asymptote is zero degrees, and the high-frequency asymptote is −180 degrees so the phase of $G_{LP}(j\omega)$ differs from the plot of figure 4.24 only by the fixed constant of −90 degrees.

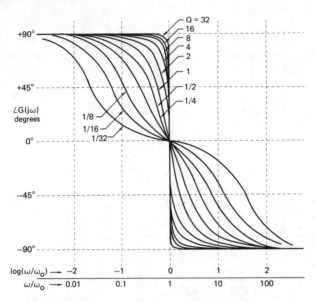

Figure 4.24   Phase Angle Plots of Second-Order Band-Pass Filters

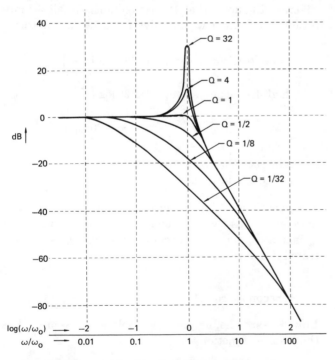

Figure 4.25   Decibel Plots of Second-Order Low-Pass Filters

High Pass: The simplest two-pole high-pass filter has the transfer function

$$G_{HP}(s) = \frac{\dfrac{s^2}{\omega_o^2}}{1 + \dfrac{s}{Q\omega_o} + \dfrac{s^2}{\omega_o^2}} \qquad 4.106$$

The Bode amplitude diagrams in figure 4.26 can be seen to be mirror images of the low-pass characteristics of figure 4.25. The phase is again as in figure 4.24, except here the low-frequency phase asymptote is at +180 degrees, and the high-frequency asymptote is 0 degrees, so the phase of $G_{HP}(j\omega)$ differs from the plot of figure 4.24 only by the fixed constant of +90 degrees.

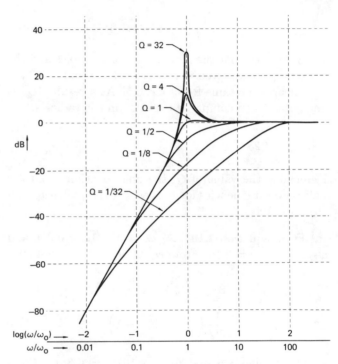

Figure 4.26   Decibel Plots of Second-Order High-Pass Filters

Band Reject: A *band-rejection filter* is one which greatly attenuates a band of frequencies, but will pass signals with frequencies lower and higher than the band. The band-reject filter will theoretically completely eliminate one frequency, the resonant frequency ω_o.

The simplest band-reject filter has the transfer function

$$H_{BR}(s) = \frac{\dfrac{s^2}{\omega_o^2} + 1}{1 + \dfrac{s}{Q\omega_o} + \dfrac{s^2}{\omega_o^2}} \qquad 4.107$$

With $s = j\omega$, this becomes

$$H_{BR}(j\omega) = \frac{1 - \dfrac{\omega^2}{\omega_o^2}}{1 - \dfrac{\omega^2}{\omega_o^2} + \dfrac{j\omega}{Q\omega_o}} \qquad 4.108$$

H_{BR} can be seen to have both low- and high-frequency asymptotes of 0 dB and 0 degrees. At $\omega/\omega_o = 1$, H_{BR} is zero, which is negative infinity in decibels.

At lower frequencies, when $\omega/\omega_o < 1$, the phase angle is negative and approaches -90 degrees as ω/ω_o approaches 1.

For higher frequencies, with $\omega/\omega_o > 1$, the phase angle is positive and approaches $+90$ degrees as ω/ω_o approaches 1.

It can be seen that the -3 dB frequencies will occur where

$$\left| \frac{\omega}{\omega_o} - \frac{\omega_o}{\omega} \right| = \frac{1}{Q} \qquad 4.109$$

At the lower -3 dB frequency,

$$\frac{\omega_1}{\omega_o} = \frac{1}{2Q} \left(\sqrt{4Q^2 + 1} - 1 \right) \qquad 4.110$$

and the phase angle is -45 degrees.

At the higher -3 dB frequency,

$$\frac{\omega_2}{\omega_o} = \frac{1}{2Q} \left(\sqrt{4Q^2 + 1} + 1 \right) \qquad 4.111$$

and the phase angle is $+45$ degrees.

The Bode amplitude diagrams for a number of values of Q are shown in figure 4.27, with the corresponding phase plots given in figure 4.28.

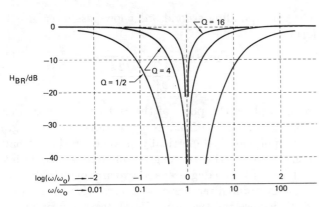

Figure 4.27 Decibel Plots for Simple Band-Reject Filters

Example 4.29

Obtain the transfer function for a band-rejection filter with a center frequency of 120 Hz and a rejection band (between -3 dB frequencies) of 30 Hz.

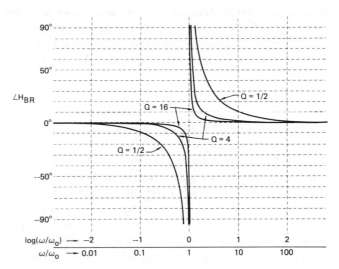

Figure 4.28 Phase Plots for Simple Band-Reject Filters

From equations 4.111 and 4.110,

$$(\omega_2 - \omega_1) = \frac{\omega_o}{Q}$$

and

$$(f_2 - f_1) = \frac{f_o}{Q}$$

and given $f_o = 120\,\mathrm{Hz}$ and $f_2 - f_1 = 30\,\mathrm{Hz}$,

$$Q = \frac{120}{30} = 4$$

$$\omega_o = 2\pi f_o = 240\pi = 754$$

The transfer function from equation 4.107 is

$$H_{BR}(s) = \frac{s^2 + 754^2}{s^2 + 0.25(754)s + 754^2}$$

or

$$H_{BR}(s) = \frac{s^2 + 568{,}500}{s^2 + 188.5s + 568{,}500}$$

E. Higher-Order Response

Higher-order systems can have combinations of simple poles and complex poles according to the number of energy storage devices in the system. There is one pole for each energy storage device (i.e., one pole for each state variable). The major problem is determining the poles, a matter of factoring a denominator polynomial of order three or higher.

The analysis requires a circuit transformation to Laplace impedances as described in the preceding sections. Once a transfer function is obtained, the task is to factor the polynomials of the numerator and denominator for purposes of obtaining graphical plots or other information. A method to obtain the roots for the third-order case, is given here.

In a *third-order system*, there must be at least one real root. Upon finding one, that value can be factored out to leave a second-degree equation whose roots can be found by the usual methods. Consider the function of equation 4.112.

$$F = s^3 + as^2 + bs + c \qquad 4.112$$

If a known value of s will make this function zero, one of the roots will be known. This can be done in a random fashion, converging on the root by recognizing that when two trials result in values of F with opposite signs, the root lies between the corresponding values of s. This process can be simplified by making good starting guesses. The function itself gives some clues to the range of possible values for the roots. The good starting guesses are found from the adjacent terms in the function.

For functions of the form of equation 4.112, high-, mid-, and low-frequency approximations can be found.

$$s_{high} = -a \qquad 4.113$$

$$s_{mid} = -\frac{b}{a} \qquad 4.114$$

$$s_{low} = -\frac{c}{b} \qquad 4.115$$

The procedure is to first evaluate F for these three values. At least one of these will result in a negative F and at least one in a positive F. The two closest values of s which give opposite signs for F are used to determine the next guess, either by intuition, simple averaging, or linear interpolation.[2]

Example 4.30

The denominator of a transfer function is

$$F = s^3 + 2s^2 + 6s + 1$$

Determine a root.

First evaluate the starting value of s.

$s_{high} = -2$
$$F = (-2)^3 + 2(-2)^2 + 6(-2) + 1 = -11$$

$s_{mid} = \frac{-6}{2} = -3$
$$F = (-3)^3 + 2(-3)^2 + 6(-3) + 1 = -26$$

$s_{low} = \frac{-1}{6}$
$$F = \left(\frac{-1}{6}\right)^3 + 2\left(\frac{-1}{6}\right)^2 + 6\left(\frac{-1}{6}\right) + 1 = +0.0509$$

---

[2] This method is an extension of the bisection method of root determination.

It can be seen that a root must exist between $s = -1/6$ and $s = -2$. As F is near 0 at $s = -1/6$, intuition is used to choose the next guess near this value, say at $-1/5 = -0.2$.

$$F = (-0.2)^3 + 2(-0.2)^2 + 6(-0.2) + 1 = -0.128$$

This indicates the root is between $s = -1/6$ and $s = -1/5$. The next guess is $s = -0.18$, which results in $F = -0.021$. $s = -0.17$ results in $F = +0.033$, indicating the root lies between -0.17 and -0.18. At $s = -0.175$, $F = +0.006$. This process is continued to the desired accuracy. The root is reasonably approximated by -0.176. After long division by the factor $(s + 0.176)$, the function is

$$F = (s + 0.176)(s^2 + 0.182s + 5.679)$$

The remaining second-order roots are complex. They can be found by factoring the quadratic term.

- Butterworth Filters

A *Butterworth filter* has a maximally flat response over the pass band, and is most easily understood by viewing its poles in the complex s-plane. These filters have the same bandwidth (3 dB frequency) regardless of the order. The first-order low-pass filter has the transfer function

$$T_{1B}(s) = \frac{1}{\dfrac{s}{\omega_o} + 1} \qquad 4.116$$

The corresponding second-order low-pass filter has the form

$$T_{2B}(s) = \frac{1}{\dfrac{s^2}{\omega_o^2} + \dfrac{\sqrt{2}\,s}{\omega_o} + 1} \qquad 4.117$$

In the s-plane the poles of the Butterworth filters are located on a circle of radius ω_o centered on the origin. The n poles are symmetrically arranged in the left half-plane to divide the 180 degrees of that semicircle into $n-1$ arcs of $180/n$ degrees, and two arcs of $90/n$ degrees. The first-order filter has $n = 1$ arcs with zero arcs of $180/1$ degrees and two arcs of $90/1$ degrees, as indicated in figure 4.29(a). For the second-order filter, there are $2 - 1$ or 1 arc of $180/2 = 90$ degrees, and 2 arcs of $90/2 = 45$ degrees, as indicated in figure 4.29(b).

Each pair of complex poles for Butterworth filters can be realized by a second-order stage with a value of Q related to the angle that a radius drawn from the origin to the complex pole makes with the imaginary axis. Figure 4.30 shows the relationship between the angle

(θ) and the real part of the coordinates of the complex poles.

$$\text{real part of complex pole location} = \frac{\omega_o}{2Q}$$

$$4.118$$

$$\sin\theta = \frac{1}{2Q}, \quad \text{or} \quad Q = \frac{1}{2\sin\theta} = 0.5\csc\theta$$

$$4.119$$

Higher-order Butterworth filters will have angles of

$$\theta_1 = \frac{90}{n},$$

$$Q_1 = 0.5\csc\left(\frac{90}{n}\right), \quad n > 1$$

$$\theta_2 = \frac{90}{n} + \frac{180}{n} = \frac{270}{n},$$

$$Q_2 = 0.5\csc\left(\frac{270}{n}\right), \quad n \geq 3$$

$$\theta_3 = \frac{90}{n} + 2\left(\frac{180}{n}\right) = \frac{450}{n},$$

$$Q_3 = 0.5\csc\left(\frac{450}{n}\right), \quad n \geq 5$$

$$\theta_4 = \frac{90}{n} + 3\left(\frac{180}{n}\right) = \frac{630}{n},$$

$$Q_4 = 0.5\csc\left(\frac{630}{n}\right), \quad n \geq 7$$

$$\text{etc.}$$

$$\theta_P = \frac{90}{n} + (P-1)\left(\frac{180}{n}\right) = \frac{[90 + (P-1)180]}{n}$$

Note, odd n's have one real pole at $s = -\omega_o$. The poles for various n's can be generated from the above formulas to obtain a set of values as in table 4.2.

- Selection of n for Butterworth Low-Pass Filters

The amplitude of the nth-order Butterworth low-pass filter is given by

$$|T_{n\text{B}}(j\omega)| = \frac{1}{\sqrt{1 + \left(\dfrac{\omega}{\omega_o}\right)^{2n}}}$$

$$4.120$$

At $\omega = \omega_o$, the amplitude is $1/\sqrt{2}$, so the gain is -3 dB. By specifying the gain (or loss) at another frequency ω_r, the order n can be calculated using equation 4.120. Let $-A$ be the specified gain in dB (or $+A$ the specified loss in dB) at $\omega = \omega_r$. Then it is a straightforward calculation to obtain n.

$$n = \frac{\log(10^{0.1A} - 1)}{2\log\left(\dfrac{\omega_r}{\omega_o}\right)}$$

$$4.121$$

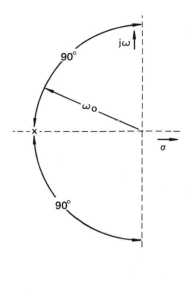

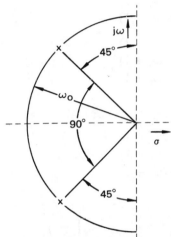

Figure 4.29 Poles for First- and Second-Order Butterworth Low-Pass Filters

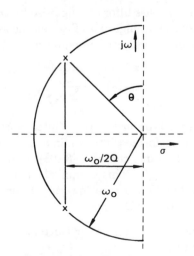

Figure 4.30 Relationship Between the Angle from the Imaginary Axis to the Pole in the Upper Half-Plane and the Quality Factor Q for a Complex Pole Pair

Table 4.2
Pole Expressions for Butterworth Filters of Order 1 Through 8

order	angles	pole expressions
$n = 1$	$\theta = 90°$	$(s/\omega_o + 1)$
$n = 2$	$\theta = 45°$	$(s^2/\omega_o^2 + 1.41s/\omega_o + 1)$
$n = 3$	$\theta_1 = 30°$	$(s^2/\omega_o^2 + s/\omega_o + 1)$
	$\theta_2 = 90°$	$(s/\omega_o + 1)$
$n = 4$	$\theta_1 = 22.5°$	$(s^2/\omega_o^2 + 0.765s/\omega_o + 1)$
	$\theta_2 = 67.5°$	$(s^2/\omega_o^2 + 1.848s/\omega_o + 1)$
$n = 5$	$\theta_1 = 18°$	$(s^2/\omega_o^2 + 0.618s/\omega_o + 1)$
	$\theta_2 = 54°$	$(s^2/\omega_o^2 + 1.618s/\omega_o + 1)$
	$\theta_3 = 90°$	$(s/\omega_o + 1)$
$n = 6$	$\theta_1 = 15°$	$(s^2/\omega_o^2 + 0.518s/\omega_o + 1)$
	$\theta_2 = 45°$	$(s^2/\omega_o^2 + 1.414s/\omega_o + 1)$
	$\theta_3 = 75°$	$(s^2/\omega_o^2 + 1.932s/\omega_o + 1)$
$n = 7$	$\theta_1 = 12.87°$	$(s^2/\omega_o^2 + 0.445s/\omega_o + 1)$
	$\theta_2 = 38.57°$	$(s^2/\omega_o^2 + 1.247s/\omega_o + 1)$
	$\theta_3 = 64.29°$	$(s^2/\omega_o^2 + 1.802s/\omega_o + 1)$
	$\theta_4 = 90°$	$(s/\omega_o + 1)$
$n = 8$	$\theta_1 = 11.25°$	$(s^2/\omega_o^2 + 0.390s/\omega_o + 1)$
	$\theta_2 = 33.75°$	$(s^2/\omega_o^2 + 1.111s/\omega_o + 1)$
	$\theta_3 = 56.25°$	$(s^2/\omega_o^2 + 1.663s/\omega_o + 1)$
	$\theta_4 = 78.75°$	$(s^2/\omega_o^2 + 1.962s/\omega_o + 1)$

Example 4.31

Determine the order of a Butterworth low-pass filter which is to have a loss of 40 dB at a frequency of $2\omega_o$.

From equation 4.121,

$$n = \frac{\log(10^4 - 1)}{2\log(2)} = 6.644$$

To meet the specifications, it is necessary to use the next higher order, or $n = 7$. Hence, a seventh-order filter is required.

• Butterworth High-Pass Filters

As seen in the previous section on second-order response, low-pass and high-pass filters are mirror images of each other. Therefore, a second-order low-pass filter with a −3 dB frequency of ω_o can be transformed into a high-pass filter with the same "corner" frequency via multiplying by $(s/\omega_o)^2$. This is indicated for a Butterworth second-order high-pass filter in equation 4.122.

$$T_{2\text{Bh}}(s) = \frac{\left(\dfrac{s}{\omega_o}\right)^2}{\dfrac{s^2}{\omega_o^2} + \dfrac{\sqrt{2}\,s}{\omega_o} + 1} \qquad 4.122$$

A first-order low-pass filter can likewise be transformed into a corresponding high-pass filter at the same corner frequency. This is indicated in equation 4.123 by

$$T_{1\text{Bhp}}(s) = \frac{\dfrac{s}{\omega_o}}{\dfrac{s}{\omega_o} + 1} \qquad 4.123$$

In general, an nth-order low-pass Butterworth filter design can be transformed into a mirror image high-pass filter having the same corner (−3 dB) frequency by replacing s/ω_o by ω_o/s. This is equivalent to multiplying the transfer function of the low-pass filter (as formulated here) by $(s/\omega_o)^n$.

The amplitude of the nth-order Butterworth high-pass filter is given by

$$|T_{n\text{Bhp}}(j)| = \frac{1}{\sqrt{1 + \left(\dfrac{\omega_o}{\omega}\right)^{2n}}} \qquad 4.124$$

Again, the formula for the required value of n to obtain a particular attenuation of $-A$ dB at a frequency ω_r (which in the high-pass case is less than ω_o) can be found to be

$$n = \frac{\log(10^{A/10} - 1)}{2 \log \left(\dfrac{\omega_o}{\omega_r} \right)} \qquad 4.125$$

• **Butterworth Band-Pass Filters**

The band-pass filter is derived from the low-pass via a transformation by substituting the expression

$$Q \left(\frac{s}{\omega_o} + \frac{\omega_o}{s} \right) \qquad 4.126$$

$$\text{for} \quad \left(\frac{s}{\omega_o} \right) \qquad 4.127$$

Low-pass terms for Butterworth filters are of the forms:

$$\frac{1}{\dfrac{s}{\omega_o} + 1} \qquad 4.128$$

and

$$\frac{1}{\left(\dfrac{s}{\omega_o} \right)^2 + \dfrac{As}{\omega_o} + 1} \qquad 4.129$$

Under this transformation they become, respectively,

$$\frac{1}{Q \left(\dfrac{s}{\omega_o} + \dfrac{\omega_o}{s} \right) + 1} \qquad 4.130$$

and

$$\frac{1}{Q^2 \left(\dfrac{s}{\omega_o} + \dfrac{\omega_o}{s} \right)^2 + AQ \left(\dfrac{s}{\omega_o} + \dfrac{\omega_o}{s} \right) + 1} \qquad 4.131$$

When $s = j\omega$, $Q(s/\omega_o + \omega_o/s) = jQ(\omega/\omega_o - \omega_o/\omega)$, and, when $s = j\omega_o$, $Q(s/\omega_o + \omega_o/s) = 0$. At $s = j\omega_o$ these terms reach their maximum values of 1. To determine the corner frequencies, equation 4.120 can be used together with the transformation to obtain

$$|T_{nBP}| = \frac{1}{\sqrt{1 + Q^{2n} \left(\dfrac{\omega}{\omega_o} - \dfrac{\omega_o}{\omega} \right)^{2n}}} \qquad 4.132$$

The magnitude of equation 4.132 becomes $1/\sqrt{2}$ or -3 dB when

$$Q^2 \left(\frac{\omega}{\omega_o} - \frac{\omega_o}{\omega} \right)^2 = 1 \qquad 4.133$$

which is when

$$\frac{\omega}{\omega_o} = \frac{1}{2Q} \left(\sqrt{4Q^2 + 1} \pm 1 \right) \qquad 4.134$$

The lower -3 dB frequency is

$$\omega_1 = \omega_o \frac{1}{2Q} \left(\sqrt{4Q^2 + 1} - 1 \right) \qquad 4.135$$

The upper -3 dB frequency is

$$\omega_2 = \omega_o \frac{1}{2Q} \left(\sqrt{4Q^2 + 1} + 1 \right) \qquad 4.136$$

The bandwidth is seen to be set entirely by Q and ω_o.

Again the formula for the required value of n to obtain a particular attenuation of $-A$ dB at a frequency ω_r which is higher than ω_o is

$$n = \frac{\log(10^{A/10} - 1)}{2 \log Q \left(\dfrac{\omega_r}{\omega_o} - \dfrac{\omega_o}{\omega_r} \right)} \qquad 4.137$$

To obtain that same attenuation of $-A$ dB at a frequency ω_r which is lower than ω_o, n can be found to be

$$n = \frac{\log(10^{A/10} - 1)}{2 \log Q \left(\dfrac{\omega_o}{\omega_r} - \dfrac{\omega_r}{\omega_o} \right)} \qquad 4.138$$

Example 4.32

Determine the expression for a third-order Butterworth band-pass filter with a Q of 2, centered at 1000 rad/sec. Also determine the half-power frequencies.

From table 4.2,

$$n = 3 \quad \theta_1 = 30° \qquad \left(\frac{s^2}{\omega_o^2} + \frac{s}{\omega_o} + 1 \right)$$

$$\theta_2 = 90° \qquad \left(\frac{s}{\omega_o} + 1 \right)$$

Thus, the low-pass filter expression is

$$T_{3Bl}(s) = \frac{1}{\left(\dfrac{s}{\omega_o} + 1 \right) \left(\dfrac{s^2}{\omega_o^2} + \dfrac{s}{\omega_o} + 1 \right)}$$

Here, $\omega_o = 1000$. Substitute $S = s/1000$, and to obtain the band-pass transfer function, for s/ω_o substitute $2(S + 1/S) = 2S + 2/S$:

$$T_{3Bbp}(S)$$

$$= \frac{1}{\left(2S + \dfrac{2}{S} + 1 \right) \left(4S^2 + 8 + \dfrac{4}{S^2} + 2S + \dfrac{2}{S} + 1 \right)}$$

$$= \frac{0.125 S^3}{(2S + 0.5S + 1)(S^4 + 0.5S^3 + 2.25S^2 + 0.5S + 1)}$$

From equation 4.106, $\omega_1\omega_2 = \omega_o^2$, and

$$\omega_2 - \omega_1 = \frac{\omega_o}{Q},$$

from which $\quad \omega_1 = \frac{\omega_o}{2Q}(\sqrt{4Q^2+1}-1)$

and $\quad \omega_2 = \frac{\omega_o}{2Q}(\sqrt{4Q^2+1}+1)$

Then,

$$\omega_1 = \frac{1000}{4}(\sqrt{4^2+1}-1) = 781 \text{ rad/sec}$$

and

$$\omega_2 = \frac{1000}{4}(\sqrt{4^2+1}+1) = 1281 \text{ rad/sec}$$

PRACTICE PROBLEMS

Warmups

1. For the circuit shown, determine the steady-state inductance current i_L for (a) the switch closed, and (b) the switch open.

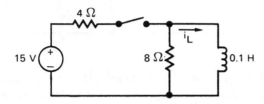

2. For the circuit shown, determine the steady-state capacitance voltage v_C for (a) the switch closed, and (b) the switch open.

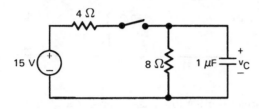

3. For the circuit shown, determine the steady-state inductance current i_L for (a) the switch closed, and (b) the switch open.

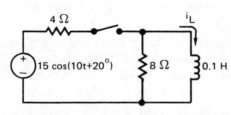

4. The circuit in problem 2 has had the switch closed for a long time. At $t = 0$, the switch opens. Determine the capacitance voltage, $v_C(t)$, for $t > 0$.

5. The circuit in problem 2 has had the switch open for a long time. At $t = 0$, the switch closes. Determine the capacitance voltage $v_C(t)$ for $t > 0$.

6. A system is described by the equation

$$12 \sin 2t = \frac{d^2v}{dt^2} + 5\frac{dv}{dt} + 4v$$

The initial conditions on the voltage are:

$$v(0) = -5$$
$$\frac{dv(0)}{dt} = 2$$

Determine $v(t)$ for $t > 0$.

7. For the circuit shown, the current source is

$$i_s(t) = 0.01 \cos 500t$$

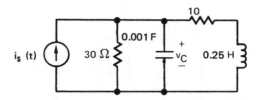

Find the transfer function

$$G(s) = \frac{V_C(s)}{I_s(s)}$$

8. For a series RLC circuit of 10 ohms, 0.5 henrys, and 100 μF, determine the resonant frequency, the quality factor, and the bandwidth.

9. For a parallel RLC circuit of 10 ohms, 0.5 henrys, and 100 μF, determine the resonant frequency, the quality factor, and the bandwidth.

10. The switch of the circuit shown has been open for a long time and is closed at $t = 0$.

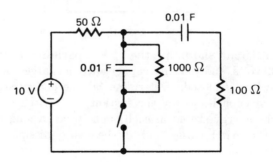

Determine the currents flowing in the two capacitances at the instant the switch closes. Give the magnitudes and directions.

Concentrates

1. A system is described by the differential equation

$$20 \sin 4t = \frac{d^2i}{dt^2} + 4\frac{di}{dt} + 4i$$

The initial conditions are:

$$i(0) = 0$$
$$\frac{di(0)}{dt} = 4$$

Determine $i(t)$ for $t > 0$.

2. The circuit shown has had the switch open for a long time. It closes at $t = 0$. Determine the values of (a) the capacitance current at $t = 0^+$, and (b) the inductance voltage at $t = 0^+$.

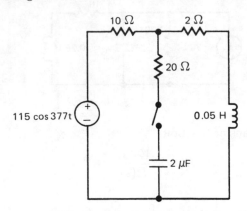

3. The circuit has an initial capacitance voltage of 20 volts and initial inductance current of 1 amp at $t = 0$. Determine the capacitance voltage for $t > 0$.

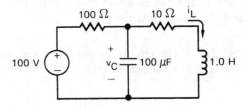

4. The circuit shown has the 1 μF capacitance charged to 100 volts, and the 2 μF capacitance uncharged before the switch is closed. Determine (a) the energy stored in each capacitance at the instant the switch closes, (b) the energy stored in each capacitance a long time after the switch closes, and (c) the energy dissipated in the resistance.

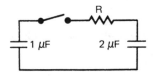

5. For the circuit shown, obtain the voltage transfer function, frequency response, and unit step voltage response.

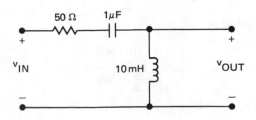

6. For the circuit shown, obtain the voltage transfer function, frequency response, and unit step voltage response.

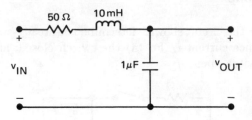

7. For the circuit shown, obtain the voltage transfer function, frequency response, and unit step voltage response.

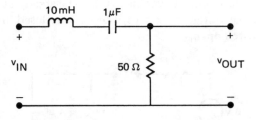

8. A second-order low-pass filter has a Q greater than 0.5, and a natural frequency of 1 radian per second (note $\omega_o = 1$). Determine the maximum gain and the maximum step voltage response for the following values of Q: 0.707, 1.0, 1.414, 2.0, 4.0, and 8.0. Determine the best correlation between the two numbers for each Q.

9. A third-order Butterworth low-pass filter has a natural frequency (ω_o) of 1000 Hz. Determine its voltage response to a unit step if its DC gain is 20 dB.

10. A particular band-reject filter has the voltage transfer function given as

$$G = \frac{s^2 + 100}{s^2 + 5s + 100}$$

Determine the Q of the filter, and the response of this filter to a unit step voltage input.

Timed

1. The circuit shown is a phase-lead compensator. Together with the amplifier, it is to provide a voltage ratio (gain) of 1 and a phase lead of ϕ_d at a frequency of ω_d. The amplifier has a voltage gain of K. Determine the values of R_1, R_2, and C in terms of K, ϕ_d, and ω_d. Discuss the limitations on the angle and the desirable ranges of values.

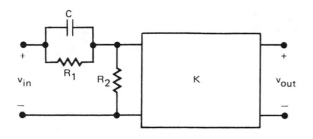

2. For the circuit shown, find (a) the system transfer function, (b) the response to a unit step at the input, and (c) the response to a unit pulse with a duration of 1 millisecond.

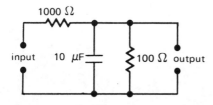

3. The circuit shown has had the switch closed for a long time. It is opened at $t = 0$. For the instant just after the switch opens ($t = 0^+$), determine the current i_1, the current i_2, the voltage v, and the time rate of change of the current i_3.

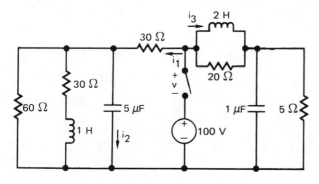

4. Specify R_2 and capacitances C_1 and C_2 to cause the following circuit to be a Butterworth low-pass filter with a -3 dB (corner) frequency of 150 Hz. The amplifier is to be considered an ideal unity gain voltage amplifier (infinite input resistance and zero output resistance).

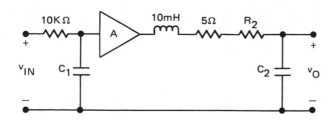

5. Given the circuit shown where the amplifiers are ideal unity-gain voltage amplifiers (infinite input resistance and zero output resistance), amplifier A3 is a summing amplifier, so its output is the sum of its two inputs. Determine the transfer function of the resulting filter, identify it and its principal parameters.

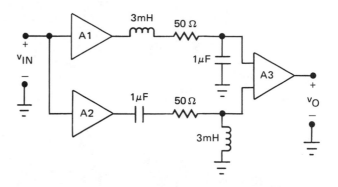

6. A Butterworth filter has the transfer function given below. Determine its corner frequency, and its response to a 1 volt step input.

$$G(s) = \frac{100,000}{s^4 + 26.13s^3 + 341.4s^2 + 2613s + 10,000}$$

7. A high-pass filter is to be designed for a -3 dB frequency of 150 Hz, and is to have an attenuation of 39 dB at 60 Hz. The filter is to have a high-frequency gain of 0 dB. Specify the transfer function. (Use a Butterworth design.)

8. A band-reject filter is to reject 120 Hz and have 150 Hz as a -3 dB frequency. Obtain the transfer function for a second-order filter to accomplish this task.

9. Given a third-order Butterworth low-pass filter with a frequency of 100 rad/sec as the -3 dB frequency, and a DC gain of unity, obtain the filter response when the input is a unit step function.

10. A set of individual band-pass filters are to be designed to have unity gain, octave bandwidths, and have at least 39 dB loss (attenuation) one decade below the lower corner frequency. These are to form a set covering the audio range from a low corner frequency of 200 Hz through six bands to 12,800 Hz. Obtain the transfer functions for each of these filters. (Note, Butterworth filters are easiest to calculate.)

5 POWER SYSTEMS

Nomenclature

a	transformer turns ratio	–
a	$0.5(-1 + j\sqrt{3})$	–
A	an ABCD parameter	volts/ volt
AWG#	American Wire Gage number	–
B	(subscripted) susceptance	mhos
B	(unsubscripted) ABCD parameter	ohms
C	an ABCD parameter	mhos
D	an ABCD parameter	amps/ amp
e	an equivalent voltage source	per unit
E_g	back EMF	volts
EMF	electromotive force	volts
G	conductance	mhos
I	current	amps RMS
j	$\sqrt{-1}$	–
L	inductance	henrys
MMF	magnetomotive force	amps/ meter
N	number of turns	–
P	power	watts
PF	power factor	–
P.U.	per unit	–
Q	reactive power	vars
R	resistance	ohms
S	apparent power	volt- amps
t	time	sec
v	instantaneous voltage	volts
V	voltage	volts
X	reactance	ohms
Y	admittance	mhos
Z	impedance	ohms

Symbols

η	efficiency	–
Δ	delta connection	–
δ	angle	degrees or radians
θ	angle	degrees or radians
ω	frequency	rad/sec

Subscripts

a	phase a
b	phase b
c	phase c, core
d	direct axis
g	generator
l	line
L	load
m	motor
n	neutral
oc	open circuit
p	primary
ps	primary to secondary
pt	primary to tertiary
q	quadrature axis
s	secondary
sc	short circuit
t	tertiary
Th	Thevenin
Y	wye connection

1 THREE-PHASE CIRCUITS

Three-phase power is widely used in power systems today. In addition to economic considerations, a three-phase generator operating under a balanced load condition produces a constant (rather than pulsating) torque. This causes less bearing wear than a single-phase generator.

Three-phase power is delivered by three-wire or four-wire systems. A mixture of the two can also be used with intervening transformers making the change from three- to four-wire arrangements. Three-wire systems use only the three power lines, while a four-wire system is made up of three power lines and a neutral. Voltages and currents in the three-phase systems are essentially sinusoidal. In a balanced system, there are 120 electrical degrees between the three phases, regardless of whether the voltages are measured line-to-line or line-to-neutral.

As sinusoids are negative for half of each cycle, it is necessary to establish reference conventions for voltage polarities and current directions. In single-phase circuits, the voltage peak can be taken as a reference, and all other voltages then can be referred to that reference.

In a three-phase system, there are three references possessing fixed relationships among them. These references are the peaks of either the generator or bus voltages, and they are referred to by the letters a, b, and c. In general, voltages are given in terms of node pair voltage drops (e.g., the *voltage drop* from node a to node b is designated as V_{ab}). This voltage is instantaneously positive whenever there is a voltage drop from node a to node b, and instantaneously negative whenever the voltage drop is from node b to node a.

Voltage reference polarities are used in phasor diagrams, in the fixed relationship to other voltages established by the generator or bus voltages. *Current directions* are given in double subscript notation. A current flowing from node a through a circuit element to node b would be designated as I_{ab}. This current would be instantaneously positive when the current was flowing from node a to node b, and would be instantaneously negative whenever the current flow was from node b toward node a. The *neutral node*, when present, is given the letter subscript n.

A current designated $I_{aa'}$ is from node a to node a'. The current $I_{a'a}$ is from node a' to node a. Thus, $I_{a'a} = -I_{aa'}$.

Three-phase voltage generators produce three voltages 120 degrees apart in phase. The three sources can be connected with a wye (Y) connection or in a delta (Δ) configuration, as shown in figure 5.1.

The voltage at node a (dropping to the neutral in the wye connection, or dropping to node b in the delta connection) is usually taken as the phasor reference. It is parallel to the positive-real axis in phasor diagrams such as figure 5.2. For the *wye connection*, there are two possible phase sequences of the voltages V_{an}, V_{bn}, and V_{cn}. These two arrangements are called *phase sequence abc* and *phase sequence acb*. All other combinations are identical to one of these, although perhaps displaced in time. Such time displacement is of no consequence when steady-state operation is considered.

The instantaneous voltages for the two sequences are defined by equations 5.1 through 5.6.

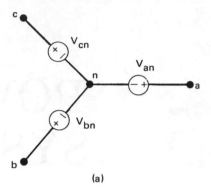

(a)

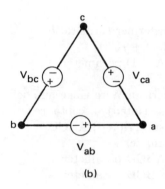

(b)

Figure 5.1 Wye and Delta Three-Phase Generator Connections

- Phase Sequence abc (wye)

$$V_{an} = V_m \cos \omega t \qquad\qquad 5.1$$
$$V_{bn} = V_m \cos(\omega t - 120°) \qquad 5.2$$
$$V_{cn} = V_m \cos(\omega t - 240°)$$
$$\quad = V_m \cos(\omega t + 120°) \qquad 5.3$$

- Phase Sequence acb (wye)

$$V_{an} = V_m \cos \omega t \qquad\qquad 5.4$$
$$V_{cn} = V_m \cos(\omega t - 120°) \qquad 5.5$$
$$V_{bn} = V_m \cos(\omega t - 240°)$$
$$\quad = V_m \cos(\omega t + 120°) \qquad 5.6$$

The phasor diagrams for the two phase sequences are shown in figure 5.2.

Note that the nodes are indicated in figure 5.2. It is also possible to add the phasor vectors in string diagrams as in figure 5.3. In loop analysis of voltages, however, it is more convenient for the nodes and phasor diagrams to have the same nodal relations. An example of this is shown in figure 5.4 where a loop from node a to node b and then to node n and back to node a is indicated. This phasor diagram shows the node and voltage relations

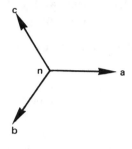

(a).

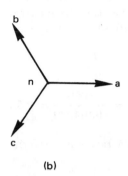

(b)

Figure 5.2 Phasor Diagrams for *abc* and *acb* Wye-Connected Generators

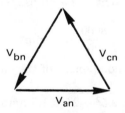

Figure 5.3 Phasor Sum of Line-to-Neutral Voltages for a Phase Sequence *abc*

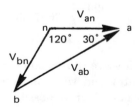

Figure 5.4 Line-to-Line Voltage from Node *a* to Node *b* for a Phase Sequence *abc*

simultaneously. The vector operation can be written in equation form, keeping in mind that the quantities are vectors and the subtraction is vector subtraction.

$$V_{ab} = V_{an} - V_{bn} \qquad 5.7$$

The relationship between the *line-to-line voltage* and the *line-to-neutral voltage* is

$$V_{ll} = \sqrt{3}V_{ln} \qquad 5.8$$

• Phase Sequence *abc*

$$V_{ab} = \sqrt{3}V_{an} \angle 30° \qquad 5.9$$
$$V_{bc} = \sqrt{3}V_{bn} \angle 30° \qquad 5.10$$
$$V_{ca} = \sqrt{3}V_{cn} \angle 30° \qquad 5.11$$

• Phase Sequence *acb*

$$V_{ab} = \sqrt{3}V_{an} \angle -30° \qquad 5.12$$
$$V_{bc} = \sqrt{3}V_{bn} \angle -30° \qquad 5.13$$
$$V_{ca} = \sqrt{3}V_{cn} \angle -30° \qquad 5.14$$

The line-to-neutral voltages are usually referred to as the *phase voltages*. If the voltages V_{ab}, V_{bc}, and V_{ca} are added together as phasors, the sum is zero. This permits the connection of generators in the delta connection, provided that the phasing is correct. To determine if the phasing is correct before closing the circuit, a voltage measurement can be made between the last two terminals being connected. If the voltage is zero, the phasing is correct and the circuit can be closed. If there is a substantial voltage (twice the voltage for one winding or phase), then the last coil has its terminals reversed.

There also are two ways to connect the three generator windings in a *delta configuration*. One connection will result in the phase sequence *abc*, while the other will result in the phase sequence *acb*. These are indicated in figure 5.5. Taking the voltage referenced to node *a* as the reference waveform, equations 5.15 through 5.20 can be derived. V_{ll} is an effective (RMS) value.

• Phase Sequence *abc* (delta)

$$V_{ab} = \sqrt{2}V_{ll} \cos \omega t \qquad 5.15$$
$$V_{bc} = \sqrt{2}V_{ll} \cos(\omega t - 120°) \qquad 5.16$$
$$V_{ca} = \sqrt{2}V_{ll} \cos(\omega t - 240°) \qquad 5.17$$

• Phase Sequence *acb* (delta)

$$V_{ab} = \sqrt{2}V_{ll} \cos \omega t \qquad 5.18$$
$$V_{bc} = \sqrt{2}V_{ll} \cos(\omega t - 240°) \qquad 5.19$$
$$V_{ca} = \sqrt{2}V_{ll} \cos(\omega t - 120°) \qquad 5.20$$

A *balanced wye-wye connection* consists of wye-connected generators and wye-connected loads consisting of three identical impedances as shown in figure 5.6.

The neutral wire is indicated by a dashed line. In a balanced condition, the instantaneous sum of the currents at nodes n and n' are each zero, so no neutral current will flow. However, the neutral should be retained to permit small amounts of unbalance.

With a balanced condition, it is possible to carry out an analysis on only one of the circuits. Such analysis is called *single-line analysis*. Having obtained the results of an analysis for one line, the real and apparent powers of the three-phase system are simply three times the single-line powers. Working with the a node, the phase currents are all given by equation 5.21.

$$I_{aa'} = I_{a'n'} = \frac{V_{an}}{Z} \qquad 5.21$$

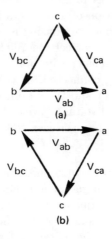

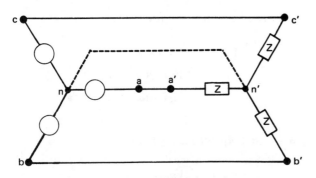

Figure 5.5 Phasor Diagrams for *abc* and *acb* Delta-Connected Generators

Figure 5.6 Wye-Wye Balanced Circuit

Example 5.1

A three-phase, 6920 volt (line-to-line) generator is connected to a balanced 24 kW wye-connected load with a power factor of 0.8. Determine (a) the line current,

(b) the generator kVA rating, and (c) the compensating reactance to bring the PF to unity.

The voltage must be converted to line-to-neutral with equation 5.8.

$$V_{ln} = \frac{6920}{\sqrt{3}} = 4000 \text{ volts}$$

The real power per phase is one-third of the generator power or 8 kW. The apparent power per phase is the real power per phase divided by the power factor. The line current (same as the phase current) is the apparent power per phase divided by the phase voltage.

$$
\begin{aligned}
I_{\text{line}} &= \frac{P_{\text{phase}}}{\text{PF} \times V_{ln}} \\
&= \frac{8000}{(0.8)4000} = 2.5 \text{ amps}
\end{aligned}
$$

The generator kVA rating is three times the phase kVA.

$$S_{\text{phase}} = \frac{8}{0.8} = 10 \text{ kVA}$$

$$\text{kVA}_{\text{rated}} = 3 \times 10 = 30 \text{ kVA}$$

The total line current is 2.5 amps. The in-phase component is

$$\frac{8000}{4000} = 2 \text{ amps}$$

The out-of-phase component of the current is

$$\sqrt{2.5^2 - 2^2} = 1.5 \text{ amps}$$

The compensating reactance must take an equal and opposite reactive current, so the reactance is

$$X_{\text{compensating}} = \frac{4000}{1.5} = 2667 \text{ ohms (reactive)}$$

The *balanced delta-wye connection* consists of a delta-connected generator and wye-connected loads. Since the system is balanced, it is possible to treat it as a wye-wye circuit after obtaining the line-to-neutral phase voltages by dividing the line-to-line voltage by $\sqrt{3}$, as was done in example 5.1.

The *balanced delta-delta connection* consists of a delta-connected source and delta-connected equal impedances as shown in figure 5.7. The current in each branch of the load is the line-to-line voltage divided by the impedance. It is, therefore, more useful to deal with admittances. The *phase currents* are the currents through the impedances, so that the line current is $\sqrt{3}$ times the phase current in magnitude. The phase currents are:

$$I_{a'b'} = V_{ab}Y \qquad 5.22$$
$$I_{b'c'} = V_{bc}Y \qquad 5.23$$
$$I_{c'a'} = V_{ca}Y \qquad 5.24$$

For an *abc* phase sequence, the line currents can be calculated as follows.

$$I_{aa'} = I_{a'b'} - I_{c'a'} = (V_{ab} - V_{ca})Y \qquad 5.25$$
$$V_{ab} - V_{ca} = \sqrt{3}V_{ab}\angle -30° \qquad 5.26$$
$$I_{aa'} = \sqrt{3}I_{a'b'}\angle -30° \qquad 5.27$$

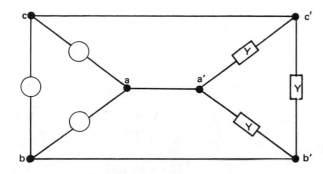

Figure 5.7 Delta-Delta Balanced Circuit

The *power* is the sum of the powers in the load admittances.

$$P = 3|V_{ab}^2| \times \text{real}\{Y\} \qquad 5.28$$
$$S = 3|V_{ab}^2| \times |Y| \qquad 5.29$$
$$Q = 3|V_{ab}^2| \times \text{imaginary}\{Y\} \qquad 5.30$$

Example 5.2

A 138 kV (RMS) line-to-line three-phase system delivers 70.7 MVA to a balanced delta load with a power factor of 0.707. Determine (a) the line current, and (b) the reactance necessary to achieve unity power factor.

Since the voltage is RMS, the apparent power to each phase is

$$S_{\text{phase}} = \frac{70.7}{3} = 23.567 \text{ MVA}$$

The phase and line currents are:

$$I_{\text{phase}} = \frac{S_{\text{phase}}}{V_{\text{line}}} = 170.8 \text{ amps}$$
$$I_{\text{line}} = \sqrt{3}I_{\text{phase}} = 295.8 \text{ amps}$$

The reactive component of the phase current is

$$I_{\text{phase, reactive}} = I_{\text{phase}}\sqrt{1 - (\text{PF})^2}$$
$$= 170.8\sqrt{1 - 0.5} = 120.8 \text{ amps}$$

The compensating reactance is

$$|X| = \frac{V_{\text{phase}}}{I_{\text{phase, reactive}}}$$
$$= \frac{138 \times 10^3}{120.8} = 1142 \text{ ohms (reactive)}$$

The *balanced wye-delta connection* consists of a wye-connected generator and a delta-connected load. The neutral from the generator cannot be connected to the load so the generator should be converted to an equivalent delta connection. This is accomplished by multiplying the line-to-neutral voltage by $\sqrt{3}$ and treating the problem as delta-delta.

2 POWER TRANSFORMERS

Power transformers are employed to step up voltage levels from generators. Power can then be transmitted at a low current (and therefore with less loss) using small conductors. At the end of the transmission line, a stepdown transformer reduces the voltage to a usable level.

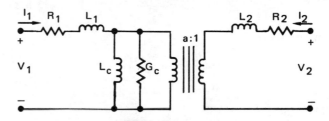

Figure 5.8 Circuit Model
for a Power Transformer

A two-winding transformer model is shown in figure 5.8. The common model parameters are:

R_1 primary winding resistance
L_1 primary winding leakage inductance
L_c core inductance
G_c core conductance
a_{ps} primary to secondary turns ratio N_p/N_s
L_2 secondary winding leakage inductance
R_2 secondary winding resistance

Transformer parameter measurements require open-circuit and short-circuit tests. These can be performed on the primary or secondary winding. The results provide approximate values for the parameters shown in figure 5.8.

A. Open-Circuit Tests

Open-circuit tests are always performed at the rated voltage of the winding to which the voltage is applied. This is usually the primary winding.

For transformers with a high voltage primary and low voltage secondary, measurements usually are made on the secondary side. The core impedances are placed on the primary side of the equivalent circuit as shown in figure 5.8.

With the secondary winding open circuited (except for a high impedance voltmeter), the primary winding impedance is negligible. A measurement of the primary voltage, current, and power will determine the parameters G_c and L_c or X_c. With the voltage drop across R_1 and L_1 negligible, the admittance is

$$Y_{\text{open circuit}} = G_c - jB_c \qquad 5.31$$

The susceptance is

$$B_c = \frac{1}{X_c} = \frac{1}{\omega L_c} \qquad 5.32$$

$$S_{\text{oc}} = V_{1,\text{oc}}^2 Y_c = V_{1,\text{oc}}^2 G_c - jV_{1,\text{oc}}^2 B_c \qquad 5.33$$

The open circuit power is

$$P_{\text{oc}} = V_{\text{oc}}^2 G_c \qquad 5.34$$

Therefore,

$$G_c = \frac{P_{\text{oc}}}{V_{1,\text{oc}}^2} \qquad 5.35$$

Furthermore,

$$Q_{\text{oc}}^2 = S_{\text{oc}}^2 - P_{\text{oc}}^2 \qquad 5.36$$

$$Q_{\text{oc}} = V_{1,\text{oc}}^2 B_c \qquad 5.37$$

Therefore,

$$B_c = \frac{\sqrt{S_{\text{oc}}^2 - P_{\text{oc}}^2}}{V_{1,\text{oc}}^2} \qquad 5.38$$

The ratio of the primary to secondary voltages with no secondary current is approximately the turns ratio.

$$a_{\text{ps}} = \frac{V_{1,\text{oc}}}{V_{2,\text{oc}}}\Big|_{I_2=0} \qquad 5.39$$

Example 5.3

An open-circuit test is performed on a 440 volt transformer winding. The results are $V_1 = 440$ volts, $P = 100$ watts, $I = 1$ amp, and $V_2 = 1100$ volts. Find the transformer parameters.

$$S = (440)(1) = 440 \text{ VA}$$

$$Q^2 = 440^2 - 100^2 = 428^2$$

$$G_c = \frac{100}{440^2} = 0.00052 \text{ mhos}$$

$$B_c = \frac{-428}{440^2} = -0.00221 \text{ mhos}$$

$$a_{\text{ps}} = \frac{440}{1100} = 0.4$$

B. Short-Circuit Tests

Short-circuit tests are used to determine the winding impedances. Short-circuit tests are always conducted at the rated current of the winding to which the voltage is applied. The secondary winding is short circuited, causing the secondary winding impedance to essentially short out the core admittance. The effective circuit is the primary winding impedance in series with the transformed secondary winding impedance.

$$Z_{\text{sc}} = R_1 + jX_1 + a_{\text{ps}}^2(R_2 + jX_2) \qquad 5.40$$

The primary voltage, V_{sc}, current, I_{sc}, and power, P_{sc} are next measured.

$$
\begin{aligned}
V_{\text{sc}}I_{\text{sc}} &= I_{\text{sc}}^2 Z \\
&= I_{\text{sc}}^2(R_1 + a_{\text{ps}}^2 R_2) \\
&\quad + jI_{\text{sc}}^2(X_1 + a_{\text{ps}}^2 X_2) \qquad 5.41
\end{aligned}
$$

However,

$$P_{\text{sc}} = I_{\text{sc}}^2(R_1 + a_{\text{ps}}^2 R_2) \qquad 5.42$$

Therefore,

$$R_1 + a_{\text{ps}}^2 R_2 = \frac{P_{\text{sc}}}{I_{\text{sc}}^2} \qquad 5.43$$

Furthermore,

$$Q_{\text{sc}}^2 = S_{\text{sc}}^2 - P_{\text{sc}}^2 \qquad 5.44$$

$$Q_{\text{sc}} = I_{\text{sc}}^2(X_1 + a_{\text{ps}}^2 X_2) \qquad 5.45$$

Therefore,

$$X_1 + a_{\text{ps}}^2 X_2 = \frac{\sqrt{V_{\text{sc}}^2 I_{\text{sc}}^2 - P_{\text{sc}}^2}}{I_{\text{sc}}^2} \qquad 5.46$$

For reasons of efficiency, power transformers are typically designed so that $R_1 = a_{\text{ps}}^2 R_2$ and $X_1 = a_{\text{ps}}^2 X_2$. If this is not the case, then further tests are needed to obtain the equivalent circuit parameters. When these relationships are valid, then

$$R_1 = a_{\text{ps}}^2 R_2 = \frac{P_{\text{sc}}}{2I_{\text{sc}}^2} \qquad 5.47$$

$$X_1 = a_{\text{ps}}^2 X_2 = \frac{Q_{\text{sc}}}{2I_{\text{sc}}^2} \qquad 5.48$$

It is possible to determine the turns ratio by measuring the short-circuit currents in the primary and secondary windings.

$$a_{\text{ps}} = \frac{I_{2,\text{sc}}}{I_{1,\text{sc}}} \qquad 5.49$$

Example 5.4

A transformer rated at 15 kVA and 1320 primary volts is subjected to a short-circuit test. The secondary current is measured as 56.8 amps, the primary voltage is 100 volts, and the power is 750 watts. Determine the equivalent circuit parameters. The rated current is the volt-amp rating divided by the voltage rating.

$$I_{\text{rated}} = \frac{15}{1.32} = 11.36 \text{ amps}$$

The apparent power and reactive powers are:

$$S = 11.36 \times 100 = 1136 \text{ VA}$$

$$Q = \sqrt{(1136)^2 - (750)^2} = 853 \text{ VAR}$$

From equation 5.47,

$$R_1 = a_{\text{ps}}^2 R_2$$

$$= \frac{750}{2 \times (11.36)^2} = 2.9 \text{ ohms}$$

From equation 5.48,

$$X_1 = a_{\text{ps}}^2 X_2$$

$$= \frac{853}{(2 \times 11.36^2)} = 3.3 \text{ ohms}$$

and from equation 5.49,

$$a_{\text{ps}} = \frac{56.8}{11.36} = 5$$

C. ABCD Parameters

ABCD parameters are convenient problem solving tools used in power transmission and distribution problems. The ABCD or *chain parameters* are of the form

$$V_{\text{in}} = AV_{\text{out}} - BI_{\text{out}} \qquad 5.50$$
$$I_{\text{in}} = CV_{\text{out}} - DI_{\text{out}} \qquad 5.51$$

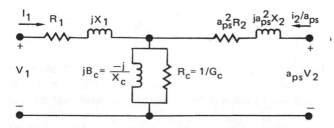

Figure 5.9 Transformer Circuit with All Quantities Referred to the Primary Winding

Circuit analysis of the transformer with all quantities referred to the primary winding, as in figure 5.9, yields equations 5.52 and 5.53.

$$V_1 = (1 + Z_1 Y_c) a_{\text{ps}} V_2$$
$$\qquad - \frac{[Z_1 + a_{\text{ps}}^2 Z_2 (1 + Z_1 Y_c)] I_2}{a_{\text{ps}}} \qquad 5.52$$

$$I_1 = (Y_c) a_{\text{ps}} V_2 - \frac{(1 + a_{\text{ps}}^2 Z_2 Y_c) I_2}{a_{\text{ps}}} \qquad 5.53$$

$$Z_1 = R_1 + jX_1 \qquad 5.54$$

$$Z_2 = R_2 + jX_2 \qquad 5.55$$

$$Y_c = G_c - jB_c \qquad 5.56$$

For the usual case of $a_{\text{ps}}^2 Z_2 = Z_1$,

$$A = 1 + Z_1 Y_c \qquad 5.57$$

$$B = Z_1 (2 + Z_1 Y_c) \qquad 5.58$$

$$C = Y_c \qquad 5.59$$

$$D = A \qquad 5.60$$

Example 5.5

A transformer has $Z_1 = 0.01 + j0.02$ and $Y_c = 0.0002 - j0.001$. Determine the ABCD parameters referred to the primary. Assume $a_{\text{ps}}^2 Z_2 = Z_1$. Then,

$$Z_1 Y_c = 2.2 \times 10^{-5} - j6 \times 10^{-6}$$

For all practical purposes this is zero.

$$A = 1$$
$$B = 2Z_1 = 0.02 + j0.04$$
$$C = Y_c = 0.0002 - j0.001$$
$$D = A = 1$$

A transformer with a load impedance attached to the secondary behaves according to $V_2 = -I_2 Z_L$. If ABCD parameters are used, then

$$V_1 = \frac{-(Aa_{\text{ps}}^2 Z_L + B) I_2}{a_{\text{ps}}} \qquad 5.61$$

$$I_1 = \frac{-(Ca_{\text{ps}}^2 Z_L + D) I_2}{a_{\text{ps}}} \qquad 5.62$$

Solving for I_1 and I_2 in terms of V_1,

$$I_2 = \frac{-a_{\text{ps}} V_1}{Aa_{\text{ps}}^2 Z_L + B} \qquad 5.63$$

$$I_1 = \frac{V_1 (Ca_{\text{ps}}^2 Z_L + D)}{Aa_{\text{ps}}^2 Z_L + B} \qquad 5.64$$

Example 5.6

The transformer of example 5.5 has a load of 0.5 ohm referred to the primary. Determine the primary input impedance.

Substitute the values from example 5.5 into equation 5.64.

$$\frac{I_1}{V_1} = \frac{(2 \times 10^{-4} - j10^{-3}) \times 0.5 + 1}{0.5 + (0.02 + j0.04)}$$

$$= \frac{1.0001 - j0.0005}{0.52 + j0.04}$$

$$Z_{\text{in}} = \frac{0.52 + j0.04}{1.0001 + j0.0005} = 0.52 + j0.043$$

ABCD parameters are particularly useful for chaining two-port networks together. Chaining is applicable to cascaded amplifiers as well as power distribution systems. For two cascaded networks, the resulting ABCD parameters are:

$$A = A_1 A_2 + B_1 C_2 \qquad 5.65$$
$$B = A_1 B_2 + B_1 D_2 \qquad 5.66$$
$$C = A_2 C_1 + C_2 D_1 \qquad 5.67$$
$$D = B_2 C_1 + D_1 D_2 \qquad 5.68$$

ABCD constants for common network forms are given in figure 5.10.

3   THE PER-UNIT SYSTEM

In power systems, values of voltage, current, VA rating, and impedance are usually expressed in a *per-unit system*, which is percent expressed as a decimal. The usual *bases* of the system are the kVA and kV ratings. The base current and impedance can then be calculated. For single-phase systems and per-phase values, the per-phase kilovoltamperes and line-to-neutral kilovolts can be used as the bases.

$$\text{base current} = \frac{\text{base kVA}_{1\phi}}{\text{base kV}_{\text{ln}}} \qquad 5.69$$

$$\text{base impedance} = \frac{\text{base } V_{\text{ln}}}{\text{base current}} \qquad 5.70$$

$$\text{base kW}_{1\phi} = \text{base kVA}_{1\phi} \qquad 5.71$$

$$\text{base kVAR}_{1\phi} = \text{base kVA}_{1\phi} \qquad 5.72$$

The quantities expressed in per-unit notation are:

$$\text{current (P.U.)} = \frac{\text{actual current}}{\text{base current}} \qquad 5.73$$

$$\text{voltage (P.U.)} = \frac{\text{actual voltage}}{\text{base voltage}} \qquad 5.74$$

$$\text{impedance (P.U.)} = \frac{\text{actual impedance}}{\text{base impedance}} \qquad 5.75$$

$$\text{power (P.U.)} = \frac{\text{actual power}}{\text{base power}} \qquad 5.76$$

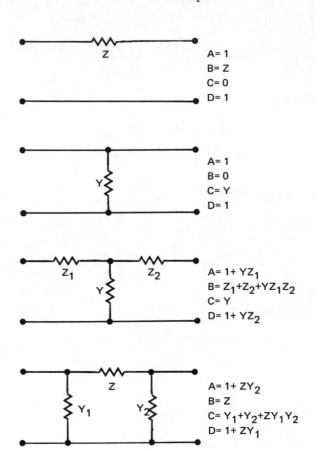

Figure 5.10   ABCD Parameters for Other Two-Port Networks: Series Impedance, Shunt Admittance, Unbalanced TEE Network, and Unbalanced PI Network

As three-phase power systems are usually specified in terms of total kVA and line-to-line voltages,

$$\text{kVA}_{1\phi} = \frac{\text{kVA}}{3} \qquad 5.77$$

$$V_{\text{ln}} = \frac{V}{\sqrt{3}} \qquad 5.78$$

Within any part of a system, all impedances must be expressed in the same per-unit system. Thus, it may be necessary to convert from another per-unit system. For instance, if the resistance and reactance of a device are given as per-unit quantities, the base is the rated kVA and kV of the device.

$$\text{base } Z = \frac{\text{base } V_{\text{ln}}}{\text{base } I} = \frac{(\text{base } V_{\text{ln}})^2}{\text{base VA}_{1\phi}} \qquad 5.79$$

The actual impedance is

$$Z = Z(\text{P.U.}) \times \frac{(\text{base } V_{\text{ln}})^2}{\text{base VA}_{1\phi}} \qquad 5.80$$

Equation 5.81 can be used to convert to a new per-unit system.

$$Z \text{ (new P.U.)} = Z(\text{P.U.}) \times \left(\frac{\text{base } V_{\text{ln}}}{\text{new base } V_{\text{ln}}}\right)^2$$
$$\times \frac{\text{new base VA}_{1\phi}}{\text{base VA}_{1\phi}} \qquad 5.81$$

Example 5.7

A 3000 kVA, 150 kV machine has a per-unit impedance of $0.1 + j0.6$. Determine the actual impedance of the machine. Express the per-unit impedance in a system which has 15 MVA and 150 kV bases.

On the basis of 3000 kVA and 150 kV, the per-phase bases are 1000 kVA and $150/\sqrt{3}$ kV $= 86.6$ kV. Then, the base current is $1000/86.6 = 11.55$ A, and the base impedance is

$$Z_{\text{base}} = \frac{86{,}600}{11.55} = 7500 \text{ ohms}$$

Then,

$$Z = 7500(0.1 + j0.6) = 750 + j4500 \text{ ohms}$$

In the new system, base kVA $= 5000$ and base kV $= 150/\sqrt{3} = 86.6$, so that equation 5.69 gives

$$\text{base } I = \frac{5000}{86.6} = 57.74$$

The base impedance for the new per-unit system is

$$\text{new base } Z = \frac{86{,}600}{57.74} = 1500 \text{ ohms}$$

Therefore,

$$\text{new } Z(\text{P.U.}) = \frac{750 + j4500}{1500} = 0.5 + j3 \text{ P.U.}$$

The per-unit system has special advantages in evaluating performance of transformers, particularly two-winding transformers, as the per-unit impedances remain the same for bases taken from the transformer ratings. Percentage impedances are per-unit impedances × 100.

Example 5.8

A 13.8 kV: 440 V, 50 kVA single-phase transformer has a leakage reactance of 300 ohms referred to the 13.8 kV side. Determine the P.U. value of the leakage reactance (a) for the high-voltage base and (b) for the low-voltage base.

The high-voltage has a kVA base of 50 kVA and a voltage base of 13.8 kV. Thus, the current base is 50 kVA/13.8 kV $= 3.62$ A, with an impedance base of 13.8 kV/3.62 $= 3810$ ohms. Then,

$$X(\text{P.U.}) = \frac{300}{3810} = 0.079$$

For the low-voltage side, the leakage reactance is adjusted by the square of the turns ratio.

$$X_l(\text{secondary}) = \left(\frac{N_2}{N_1}\right)^2 X_l \text{ (referred to primary)}$$
$$= \left(\frac{0.440}{13.8}\right)^2 \times 300 = 0.305 \text{ ohm}$$

The bases for the low-voltage side are 50 kVA and 0.44 kV. From equation 5.69, the current base is

$$\text{base } I = \frac{50{,}000}{440} = 114 \text{ amps}$$

From equation 5.70, the impedance base is

$$\text{base } Z = \frac{440}{114} = 3.86$$

Then

$$X_l(\text{P.U.}) = \frac{0.305}{3.86} = 0.079$$

Three-winding transformers present a special problem. The transformer kVA rating applies to the primary winding, while the secondary and tertiary windings have ratings which add up to the primary kVA ratings. The tests required are three partial short-circuit tests.

> *test 1*: Test 1 is performed from the primary side with shorted secondary and open tertiary windings.

$$Z_{\text{sc}_1} = R_1 + jX_1 = R_p + jX_p + a_{\text{ps}}^2(R_s + jX_s) \quad 5.82$$

The subscript s indicates quantities in the secondary circuit, and the subscript p indicates quantities in the primary circuit.

test 2: Test 2 is performed from the primary side with open secondary and shorted tertiary windings.

$$Z_{sc_2} = R_2 + jX_2 = R_p + jX_p + a_{pt}^2(R_t + jX_t) \quad 5.83$$

The subscript t indicates quantities in the tertiary circuit.

test 3: Test 3 is performed from the secondary side with open primary and shorted tertiary windings.

$$Z_{sc_3} = R_3 + jX_3 = R_s + jX_s + \frac{a_{pt}^2}{a_{ps}^2}(R_t + jX_t) \quad 5.84$$

Multiplying through by the square of the primary to secondary turns ratio (a_{ps}^2), equation 5.84 becomes

$$a_{ps}^2 Z_{sc_3} = a_{ps}^2 R_3 + ja_{ps}^2 X_3$$
$$= a_{ps}^2(R_s jX_s) + a_{pt}^2(R_t + jX_t) \quad 5.85$$

Equations 5.86 through 5.91 can be obtained by separating real and imaginary parts in equations 5.82, 5.83, and 5.84.

$$R_p = \frac{R_1 + R_2 - a_{ps}^2 R_3}{2} \quad\quad 5.86$$

$$X_p = \frac{X_1 + X_2 - a_{ps}^2 X_3}{2} \quad\quad 5.87$$

$$a_p^2 R_s = \frac{R_1 + a_p^2 R_3 - R_2}{2} \quad\quad 5.88$$

$$a_{ps}^2 X_s = \frac{X_1 + a_{ps}^2 X_3 - X_2}{2} \quad\quad 5.89$$

$$a_{pt}^2 R_t = \frac{R_2 + a_{ps}^2 R_3 - R_1}{2} \quad\quad 5.90$$

$$a_{pt}^2 X_t = \frac{X_2 + a_{ps}^2 X_3 - X_1}{2} \quad\quad 5.91$$

The primary base is used as base for problems involving feeding through a three-winding transformer.

Example 5.9

A three-winding, single-phase transformer is rated 6.8 kV, 50 kVA; 440 V, 30 kVA; and 1380 V, 20 kVA. Short-circuit tests are performed as follows:

test 1: On the 6.8 kV winding, 440 V winding short circuited and 1380 V winding open circuited:

$$Z_I = 100 + j180$$

test 2: On the 6.8 kV winding, 440 V winding open and 1380 V winding shorted:

$$Z_{II} = 120 + j200$$

test 3: On the 1380 V winding, 6.8 kV winding open and 440 V winding shorted:

$$Z_{III} = 5 + j6.6$$

Find the per-unit impedances in the primary base.

Notice that the third test is reversed from that described in the procedure (if the 440 V winding is the secondary). To overcome this problem, designate the 1380 volt winding as the secondary, so the two first tests must be reversed to use equations 5.82 through 5.91.

$$Z_1 = R_1 + jX_1 = 120 + j200$$
$$Z_2 = R_2 + jX_2 = 100 + j180$$

The turns ratio from primary to secondary is

$$a_{ps} = \frac{6.8}{1.38}$$
$$a_{ps}^2 = 24.28$$
$$a_{ps}^2 Z_3 = 121 + j160$$

Thus, from equations 5.86 through 5.91,

$$R_p = \frac{120 + 100 - 121}{2} = 49.5 \text{ ohms}$$

$$X_p = \frac{200 + 180 - 160}{2} = 110 \text{ ohms}$$

$$a_{ps}^2 R_s = \frac{120 - 100 + 121}{2} = 70.5 \text{ ohms}$$

$$a_{ps}^2 X_s = \frac{200 - 180 + 160}{2} = 90 \text{ ohms}$$

$$a_{pt}^2 R_t = \frac{-120 + 100 + 121}{2} = 50.5 \text{ ohms}$$

$$a_{pt}^2 X_t = \frac{-200 + 180 + 160}{2} = 70 \text{ ohms}$$

The base impedance is

$$\text{base } Z = \frac{(6800)^2}{50,000} = 925 \text{ ohms}$$

Then,

$$Z_p = 0.054 + j0.119 \text{ P.U. (6.8 kV winding)}$$
$$a_{ps}^2 Z_s = 0.076 + j0.097 \text{ P.U. (1.38 kV winding)}$$
$$a_{pt}^2 Z_t = 0.055 + j0.076 \text{ P.U. (440 V winding)}$$

4 ONE-LINE DIAGRAMS

The normal operation of three-phase power systems is balanced. Since balanced three-phase systems are analyzed on a per-phase basis, it is redundant to draw more than one phase. The *one-line diagram* is a simplification which omits even the neutral, leaving one line for each three-phase circuit. Certain symbols are in standard use to describe elements of power systems. Some of these are shown in figure 5.11.

Locations of circuit breakers and fuses are not shown in *load studies*. In *stability studies* of transients due to faults, the locations and characteristics of fuses and circuit breakers (as well as control relays) are essential, and they must be included on the one-line diagram.

Figure 5.12 shows a one-line diagram of a portion of a hypothetical power system. This diagram begins at the top with the *generation level*. Two synchronous generators are shown connected to a *transmission line*. The generation voltage is 6.9 kV, which is in the usual range of power generation. The generators are connected with intervening *circuit breakers* to the low-voltage sides of wye-delta transformers.

The transmission voltage level is indicated by the usual ranges. The generators are shown with their neutrals connected through surge current limiting impedances. This limits the maximum current in case of a ground fault in the generator circuit. One or more additional generators can be connected to each power bus.

There may also be local loads from the generation stations. The transmission lines interconnect the generators and lead to subtransmission distribution buses. *Subtransmission systems* are typically in a *grid form*, with interconnections between input buses, and with many paths to any distribution substation. There are usually at least two switchable inputs to a substation input bus to allow for isolation of faults while continuing service.

Each distribution substation feeds a number of distribution transformers, either in a radial, tree, or loop pattern, or via a grid system. The *distribution transformer* serves secondary mains from which individual three-phase and single-phase feeders bring the service to consumers.

Very large users may have their own substation. Intermediate size users may have a separate distribution transformer. Small users typically share a distribution transformer.

The frequency of the power is rigidly controlled by varying the prime-mover speeds. Power transferred from one generator bus to another is controlled by varying the relative phase angle between the two voltages via the rotor current magnitude. The voltage levels are largely controlled by the use of taps on the substation transformers. These taps allow minor changes in the primary to secondary turns ratios on the transformers.

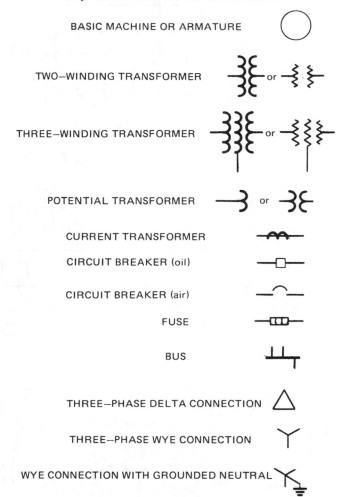

Figure 5.11 Symbols for One-Line Diagrams

5 DISTRIBUTION TRANSFORMERS

Distribution transformers are able to serve either balanced three-phase loads or unbalanced single-phase loads. However, the overall distribution system remains balanced through a statistical process of combining single-phase and other unbalances. Nevertheless, the load at an individual distribution transformer may be unbalanced, requiring more complex analysis than the single-line method.

For most line regulation calculations, power factor and primary voltage loading of the distribution transformer can be considered to be constant. This situation implies an *infinite bus*, because the change in loading from an individual distribution transformer does not change the

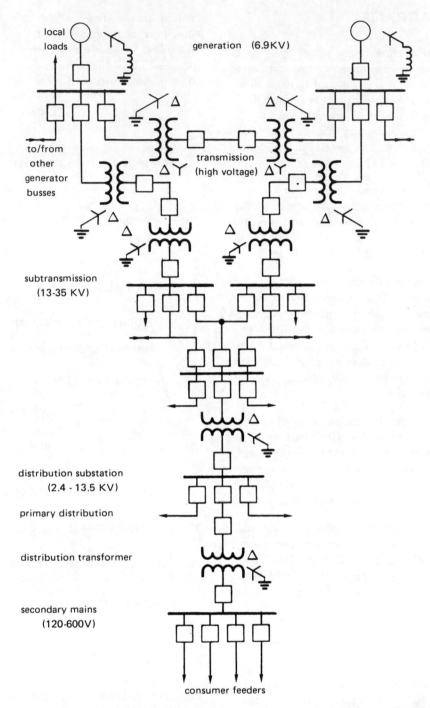

Figure 5.12   Partial One-Line Diagram of a Hypothetical Power System

primary voltage. In such calculations, it is necessary to include only the series portion of the transformer equivalent impedance (i.e., the leakage reactance and the winding resistance of the combined primary and secondary) to account for the transformer.

A. Typical Per-Unit Values

Typical per-unit values for *distribution transformer impedances* can be estimated from the following formulas

where such information is not known. The estimates will be within 25% of actual values for practical machines using the rated kVA.

- *single-phase distribution transformers:*

$$Z = [0.0252 - 0.0029 \ln(\text{kVA})]$$
$$+ j[0.010 + 0.0042 \ln(\text{kVA})] \text{ P.U.} \qquad 5.92$$

- *three-phase transformers (per-phase):*

$$Z = [0.0364 - 0.0045 \ln(kVA)]$$
$$+ j[0.026 + 0.0018 \ln(kVA)] \text{ P.U.} \qquad 5.93$$

These formulas illustrate the essentially inductive nature of transformer impedances. Most single-phase transformers are 2400 volts to 120/240 volts with ratings up to 100 kVA. Three-phase distribution transformers are typically 2400/4160 volts to 240/480 volts with ratings up to 400 kVA.

B. Voltage Regulation

Voltage regulation is conveniently calculated using per-unit quantities because the primary voltage of a transformer (referred to the secondary winding) is considered constant (i.e., an infinite bus). The difference between the primary voltage and the output of the transformer in per-unit values is the voltage regulation of the transformer when the output voltage is 1 P.U.

Example 5.10

A 50 kVA, single-phase transformer has a 440 volt secondary voltage at rated kVA and 80% lagging power factor. Determine the half-load output voltage at the same power factor. Calculate the voltage regulations going from no-load to full-load, and going from half-load to full-load.

The primary voltage of the transformer is considered constant. Then, from equation 5.92 the winding impedance is

$$Z_{\text{winding}} = 0.014 + j0.026 \text{ P.U.} = 0.030\angle 61.7° \text{ P.U.}$$

Since $\arccos(0.8) = 36.9°$, the current under full-load is $1\angle -36.9°$. Taking the transformer output voltage as 1 P.U., the primary voltage referred to the secondary is

$$V_p' = |1 + 0.030\angle 61.7° \times 1\angle -36.9°| = 1.027 \text{ P.U.}$$

To find the voltage of the transformer at half-load, assume that it is 1 P.U. and calculate what the primary voltage would be. Then, adjust proportionally from the given information.

$$V_p'' = |1 + 0.030\angle 61.7° \times 0.5\angle -36.9°| = 1.014 \text{ P.U.}$$

$$V_{\text{out}} = \frac{1.027}{1.014} \times 1 = 1.013 \text{ for 50\% output}$$

The voltage regulation from half- to full-load at PF = 0.8 is $1.013 - 1 = 0.013$ (1.3%). And from the first result, the regulation from no-load to full-load is $1.027 - 1 = 0.027$ (2.7%).

C. Power Factor Correction

Power factor correction is sometimes used to improve the power-carrying capacity of a transformer or a feeder line from a secondary main. The current capacity of a line is determined by its size and acceptable temperature rise. The power carried by a line under small power factors is less than at high power factors because the current amplitude (not the phase) determines the capacity of feeder lines and transformers.

Example 5.11

A feeder from a distribution transformer is 500 feet long. It consists of three AWG # 4/0 wires having a resistance of 0.052 ohms per 1000 feet and current capacity of 225 amps. The three-phase load is balanced and draws 120 kW at 440 volts with a 60% lagging power factor. Determine the kVA rating of capacitors capable of correcting the power factor enough to allow rated current to flow. The uncorrected current is

$$|I| = \frac{\dfrac{120,000}{3}}{0.6 \times \dfrac{440}{\sqrt{3}}} = 262.4 \text{ amps per phase}$$

Referenced to the phase voltage, the current is

$$I = 157.4 - j209.9 \text{ amps}$$

This must be corrected to a magnitude of 225 while retaining the real part of 157.5. The corrected current is

$$I_{\text{corrected}} = 157.5 - j160.7 \text{ amps}$$

The leading current required through the capacitors is

$$209.9 - 160.7 = 49.2 \text{ amps}$$

Thus, the required per-phase rating of the capacitors is

$$\text{rating} = \frac{254 \times 49.3}{1000} = 12.5 \text{ kVAR}$$

When the power factor of a transformer output is unity (or just slightly leading), the voltage regulation is minimized. This is because the voltage drop across the winding impedance is essentially in *quadrature* (90° out of time phase) with the primary voltage.

Example 5.12

A transformer rated 150 kVA, three-phase, supplies 100 kVA at a 50% lagging power factor. The transformer's winding impedance is 2% resistance and 5% reactance, based on 150 kVA. Assume 100% voltage. If the power factor is corrected to 90%, what improvement in voltage regulation will be achieved?

At 2/3 of the rated kVA and 100% of the voltage, the current will be 2/3 P.U.

$$I = 0.667\angle-60° = 0.333 - j0.577$$

Correction to 90% PF with the same in-phase component of current will result in a current of

$$I_{\text{corrected}} = 0.333 - j0.333\tan(\arccos 0.9)$$
$$I_{\text{corrected}} = 0.333 - j0.161 = 0.37\angle-25.8°$$

The uncorrected voltage drop across the winding impedance is

$$V = (0.02 + j.0.05) \times 0.667\angle-60°$$
$$= 0.054\angle68.2° \times 0.667\angle-60°$$
$$= 0.0356 + j0.0051$$

This is virtually in phase with the terminal voltage, so the primary voltage is $1.0357 + j0.005$ with a magnitude of 1.0357 P.U. This corresponds to a voltage regulation of 3.57%.

With a 90% power factor, the voltage across the transformer's winding impedance is

$$0.054\angle68.2° \times 0.37\angle-25.8° = 0.0147 + j0.0134$$

Thus, the primary voltage referred to the secondary is $1.0148 + j0.0135$. This has a magnitude of 1.0149, corresponding to voltage regulation of 1.49%. The improvement in voltage regulation is

$$3.57\% - 1.49\% = 2.08\%$$

D. Induction Motor Starting

Induction motor starting across secondary mains can cause undesirable *flicker* in lighting circuits and momentary undervoltage in the feeders. The usual starting current for induction motors is the motor power factor multiplied by six times the rated current. Upon starting, the motor is essentially inductive, as the winding reactances are on the order of five times the resistances. For larger distribution transformers, the reactance of the windings is so much greater than the resistance that only the reactance of the transformer winding needs to be included.

On starting, the transformer-induction motor combination appears as two inductances in series forming a voltage divider. The drop across the winding impedance is essentially in phase with the primary voltage, causing the worst case of voltage regulation. The drop in terminal voltage is estimated algebraically (rather than vectorially) from the drop across the winding impedance.

Example 5.13

A 250 kVA, three-phase distribution transformer has a 440 volt secondary rating and a 4% inductive reactance based on its own ratings. What size induction motor can be started directly on the line if the dip in secondary main voltage is limited to 2%? Assume a machine efficiency of 90% and a PF of 0.9 running. State your answer in horsepower.

Use the transformer base of 250 kVA and 254 volts line-to-neutral. The starting current is approximated as

$$I_{\text{starting}} = 6 \times 0.9 \times I_{\text{rated}} = 5.4 I_{\text{rated}}$$

This current can cause a 0.02 P.U. drop in voltage across the 0.04 P.U. reactance of the transformer winding.

$$5.4 I_{\text{rated}} \times 0.04 = 0.02 \text{ P.U.}$$
$$I_{\text{rated}} = 0.0926 \text{ P.U.}$$

To obtain the current in amps, use the ratings of the transformer.

$$I_{\text{rated}} = 0.0926 I_{\text{base}}$$
$$= 0.0926 \times \frac{\left(\dfrac{250 \text{ kVA}}{3}\right)}{0.254 \text{ kV}} = 30.4 \text{ amps}$$

The kVA rating of the motor is

$$30.4 \text{ A} \times 254 \text{ V} \times 3 \text{ phases} = 23.16 \text{ kVA}$$

With a 90% efficiency and a PF of 0.9, the rated horsepower of the motor is

$$\text{rated horsepower} = 23.16 \text{ kVA} \times \frac{1 \text{ hp}}{0.746 \text{ kW}}$$
$$\times \frac{0.9 \text{ kW}}{\text{kVA}} \times 0.9 = 25.15 \text{ hp}$$

6 THREE-PHASE TRANSFORMER CONNECTIONS

Voltage ratios for three-phase transformers specify the line-to-line voltages regardless of the transformer configuration. However, when turns ratios are given, the primary and secondary winding configurations must be known in order to determine the line-to-line voltages. For either a wye primary with wye secondary configuration or a delta primary with delta secondary (i.e., wye-wye or delta-delta), the turns and voltage ratios are the same. For a wye primary with delta secondary (wye-delta), the secondary voltage is

$$V_s = \frac{N_s}{\sqrt{3}N_p}V_p \qquad 5.94$$

For a delta primary with wye secondary (delta-wye),

$$V_s = \frac{\sqrt{3}N_s}{N_p}V_p \qquad 5.95$$

Distribution transformers with three-phase secondaries are usually wye-connected with a grounded neutral. This has the effect of maintaining the primary voltage referred to the secondary (V_p') when unbalanced loads are applied.

The nonlinear nature of transformers designed for minimum weight and cost causes a *third-harmonic* voltage to be generated. A delta-connected winding causes the currents from third-harmonic voltages to circulate around the delta, being essentially shorted out, and thereby not interfering with the line currents or voltages. Thus, it is essential that all transformers have at least one delta winding. Where wye-wye is required, a tertiary winding should be connected in delta and left open (no load) to suppress the third-harmonic effects.

An *open delta* or *vee transformer* is sometimes used with only two phases being connected in an incomplete delta. This is typically accomplished with two single-phase transformers (not necessarily equally rated). When an accurate analysis is required, a complete circuit diagram including the winding impedances of the two transformers is required. If less accuracy is required, the winding impedances can be ignored and two single-line analyses performed.

Example 5.14

An open-delta, three-phase transformer consisting of two single-phase transformers is operating with a balanced three-phase load of 50 kVA, 440 volts, and a lagging power factor of 0.8. Determine the required ratings of the two transformers.

Take the line-neutral voltage of the a phase as reference. Assuming an abc phase sequence with the c-a leg of the transformer missing, the voltages are:

$$V_{an} = 254\angle 0°$$
$$V_{ab} = 440\angle 30°$$
$$V_{cn} = 254\angle 120°$$
$$V_{cb} = 440\angle 90°$$

Neglecting the winding impedances, the currents are:

$$I_a = \frac{\left(\dfrac{50,000}{3}\right)}{254}\angle -\arccos 0.8$$
$$= 65.6\angle -36.9° \text{ amps}$$
$$I_c = 65.6\angle(-36.9° + 120°)$$
$$= 65.6\angle 83.1° \text{ amps}$$

With the c-a leg of the transformer missing, I_a is the current in the a-b leg, and I_c is the current in the c-b leg. If asterisks are used to represent complex conjugates, then the ratings of the two legs are:

$$S_{a\text{-}b} = V_{ab} \times I_a^* = 0.44\angle 30° \times (65.6\angle -36.9°)^*$$
$$= 0.44\angle 30° \times 65.6\angle 36.9°$$
$$= 28.87\angle 66.9° \text{ kVA}$$
$$S_{c\text{-}b} = V_{cb} \times I_c^* = 0.44\angle 90° \times 65.6\angle -83.1°$$
$$= 28.87\angle 6.9° \text{ kVA}$$

Unbalanced loads can add to the complexity of problems with open delta transformers.

Example 5.15

The transformer in example 5.14 has an additional single-phase load of 10 kW resistive connected across the c-a leg. What should be the required ratings for the transformers?

A component of current in phase with V_{ca} is added to the currents previously found. On the c line this current will be in phase with V_{ca}, while on the a line it will be 180° out of phase with V_{ca} (i.e., in phase with V_{ac}).

$$|I_{1\phi}| = \frac{10,000}{440} = 22.73 \text{ amps}$$

This current is at the angle of V_{ca}, which leads V_{ab} by 120°. It is, therefore, at 150° with respect to V_{an}, the reference taken in example 5.14. Adding this current vectorially to the currents found in example 5.14,

$$I_c = 65.6\angle 83.1° + 22.73\angle 150° = 77.40\angle 98.8° \text{ amps}$$
$$I_a = 65.6\angle -36.9° + 22.73\angle -30° = 88.21\angle -35.1°$$

The kVA for each leg is

$$S_{c\text{-}b} = V_{cb}I_c^* = 0.440\angle 90° \times 77.40\angle -98.8°$$
$$= 34.06\angle -8.8° \text{ kVA}$$
$$S_{a\text{-}b} = 0.440\angle 30° \times 88.21\angle 35.1°$$
$$= 38.8\angle 65.1° \text{ kVA}$$

7   VOLTAGE CONTROL

Voltage level is largely controlled at the distribution substation by tap-changes made under load. The taps can be on the primary or secondary sides or on both. The switching under load is done by an auto-transformer. The auto-transformer is initially connected between the present and previous tap, with its secondary adjusted to the present tap end. The other end of the auto-transformer is switched to the next tap on the transformer. Then the output (secondary) is moved by small steps to the next tap end. This establishes a new turns-ratio without interrupting the current flow.

It is sometimes necessary to bank two or more transformers. Placing two transformers in parallel will increase the current carrying capability and reliability of a system. When tap changes are made on banked transformers, very large currents in both the primary and secondary circuits can circulate. Only the leakage reactance of the parallel transformers is present to limit this circulating current.

Example 5.16

Two 1000 kVA transformers are connected in parallel. Taps are available in 2% increments above and below the nominal rated voltage. The reactances of the transformers are 10% (0.1 P.U.). If the taps are changed consecutively rather than concurrently, what will the circulating current be during the tap changes?

The per-unit voltage between taps is 0.02. When one transformer has completed the tap change and the other has not yet begun, there is a 0.02 P.U. voltage around the loop consisting of the two secondaries. This voltage across the two 0.1 P.U. reactances in series will cause a circulating current of 0.02/0.2 = 0.1 P.U. The current is essentially in quadrature with the load current (assuming that the load PF is high) so that the increase in current caused by a 10% component added in quadrature to the rated current is only about 0.5% even though the component is 10%.

That is,

$$1 + j0.1 = 1.005 \angle 5.7°$$

Banked transformers are often not identical, since systems develop over time and new capacity is added. This can lead to uneven load distribution if proper design is not observed.

Example 5.17

Two parallel transformers have 8% reactance each. One is rated 25,000 kVA and the other at 15,000 kVA. For a 30,000 kVA load with a lagging PF of 0.8, what fraction of the load will each transformer carry?

First put both transformers on the same base. The 25,000 kVA base is chosen (a combined base of 40,000 kVA would also be a reasonable choice). The reactance of the smaller transformer is then converted to the 25,000 kVA base.

$$X_{25,000 \text{ kVA}} = \frac{25,000}{15,000} \times 0.08 = 0.133 \text{ P.U.}$$

The circuit can now be modeled by two voltage sources with identical open-circuit voltages of 1 P.U. The series reactances are 0.08 and 0.133, respectively. The outputs are tied together. The result is a current division independent of the load.

$$I_{25,000 \text{ kVA}} = \frac{0.133}{0.133 + 0.08} \times I_{\text{load}} = 0.625 I_{\text{load}}$$

$$I_{15,000 \text{ kVA}} = \frac{0.08}{0.133 + 0.08} \times I_{\text{load}} = 0.375 I_{\text{load}}$$

As the output voltages are identical, each transformer carries the same percentage of load power as it does the load current.

$$S_{25,000 \text{ kVA}} = 0.625 \times 30,000 = 18,750 \text{ kVA}$$

$$S_{15,000 \text{ kVA}} = 0.375 \times 30,000 = 11,250 \text{ kVA}$$

Note that the fraction carried by each transformer is proportional to its kVA rating. This is a result of the per-unit impedance being the same for both.

8   REPRESENTATIVE VALUES OF POWER DEVICE PARAMETERS

The per-unit values of impedances and other parameters of power devices lie in a reasonably restricted range. This restricted range is dependent on the base quantities of kVA and kV ratings and on power factors. Distribution transformer winding impedances were given previously. Power transformers are designed so that per-unit resistance decreases similarly to the distribution transformers. However, the resistance is so low at high kVA levels as to be negligible in all but efficiency studies.

When the resistance is needed, an estimated linear interpolation is reasonable. With a 100 kVA (three-phase) or a 500 kVA (single-phase) transformer, the per-unit resistance is 1%; for a 10 MVA (three-phase) or a 5 MVA (single-phase) transformer, the per unit value is about 0.5%.

The leakage reactance of power transformers is approximately proportional to the high-voltage winding rating in kV. Reasonable estimates are given by equations 5.96 and 5.97.

- forced air-cooled transformers

$$X = 0.05 + 10^{-4} \text{ kV P.U.} \qquad 5.96$$

- forced oil-cooled transformers

$$X = 0.08 + 2 \times 10^{-4} \text{ kV P.U.} \qquad 5.97$$

The *magnetizing current* is approximately 1% of the rated current.

Synchronous generators have synchronous reactances of approximately 10% (0.1 P.U.) and winding resistances on the order of 1% (0.01 P.U.). Synchronous capacitors have synchronous reactances of approximately 14% (0.14 P.U.) and resistances of 1% or less.

Synchronous motors present a problem in terms of the kVA rating because motor name-plate information is in horsepower. If the efficiency and running power factor are given, then the reactance and resistance values for the synchronous generators are used after determining the rated kVA from equation 5.98.

$$\text{kVA}_{\text{rated}} = \frac{(\text{horsepower})(0.746)}{\eta(\text{PF})} \qquad 5.98$$

If neither efficiency nor power factor is given, the kVA rating is estimated to be the same as the horsepower. If unity power factor is given, the efficiency is assumed to be such that the kVA rating is 0.85 times the horsepower rating. If only the power factor is given as 80%, the kVA rating is 1.1 times the horsepower rating.

The starting current for induction motors is estimated as

$$I_{\text{starting}} = 6 \times \text{PF} \times I_{\text{rated}} \qquad 5.99$$

This is equivalent to a winding reactance of $1/(6 \times \text{PF})$ P.U. based on the machine's kVA. However, the kVA rating should be taken as the horsepower rating if no other information is given. (This implies approximately 75% efficiency.) The *running power factor* (typically around 85%) is the power factor when running at the rated current, voltage, and horsepower. It can be taken as essentially the rotor resistance divided by the slip. If the slip is not given, it can be assumed to be about 3%.

Example 5.18

A 50 hp, three-phase induction motor has a running power factor of 85%. Estimate its per-unit equivalent circuit parameters $X_e, r_s,$ and a_r^2.

The starting current will be

$$6 \times \text{PF} = 6 \times 0.85 = 5.1 \text{ P.U.}$$

The equivalent reactance is dominant in starting, so its value is estimated as

$$\frac{1}{5.1} = 0.196 \text{ P.U.}$$

The PF angle is $\cos^{-1} 0.85 = 31.8°$, and its tangent is 0.620. This is the ratio of X_e to $(r_s + (a^2 r_r)/\text{slip})$.

$$r_s + \frac{a^2 r_r}{\text{slip}} = \frac{X_e}{0.620} = 0.316 \text{ P.U.}$$

r_s is usually 1.5 to 2 times $a^2 r_r$. Since the slip was not given, it is assumed to be 0.03. Then

$$\left(2 + \frac{1}{0.03}\right) a^2 r_r = 0.316$$

$$a^2 r_r = 0.009 \text{ P.U.}$$

$$r_s = 0.018 \text{ P.U.}$$

$$X_e = 0.196 \text{ P.U.}$$

9   FAULT CALCULATIONS

In power systems, faults are expected. System reliability and preservation require that protection of equipment be a part of the initial design.

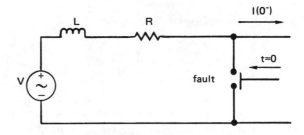

Figure 5.13   Model for an R-L Circuit with a Fault

Figure 5.13 shows a model for a system which experiences a fault. If the voltage is sinusoidal, the current is given by equation 5.101.

$$v = V_{\text{max}} \sin(\omega t + \delta) \qquad 5.100$$

$$i = \frac{V_{\text{max}}}{|Z|} \left[\sin(\omega t + \delta - \theta) - e^{-Rt/L} \sin(\delta - \theta) \right]$$

$$+ I(0^-) e^{-Rt/L} \qquad 5.101$$

$$|Z| = \sqrt{R^2 + (\omega L)^2} \qquad 5.102$$

$$\tan \theta = \frac{\omega L}{R} \qquad 5.103$$

δ is an arbitrary phase angle and $I(0^-)$ is the current flowing just prior to the fault occurrence.

The time at which the fault occurs is crucial in determining the *transient component* (i.e., the coefficient of $e^{-Rt/L}$). The worse case of the unidirectional component of fault current occurs when the pre-fault current is at its maximum value and $\sin(\delta - \theta) = 1$. Then, the two components add to equation 5.104.

$$|I_{\text{unidirectional, max}}| = \frac{V_{\text{max}}}{|Z|} + \sqrt{2}I_{\text{RMS (pre-fault)}} \quad 5.104$$

In most power systems, the fault current is much higher than the pre-fault current, so the pre-fault current can be ignored.

$e^{-Rt/L}$ can be written as $e^{-R(\omega t)/X}$ in order to study the behavior of the unidirectional component (called the DC *component*). For most power systems, the short-circuit PF is between 0.1 and 0.25, which also corresponds to a R/X ratio of 0.1 to 0.25. As one cycle corresponds to $\omega t = 2\pi$, the DC component diminishes to $1/e$ of its initial value in $X/2\pi R$ cycles. For $X/R = 4$ it is less than 1 cycle. Even for $X/R = 50$, the time constant is eight cycles. For most purposes, the unidirectional component can be ignored. Thus, the fault current after a few cycles can be expressed by equation 5.105.

$$I \approx \frac{V_{\text{max}}}{|Z|} \sin(\omega t + \delta - \theta) \quad 5.105$$

The RMS value of equation 5.105 is

$$I_{\text{RMS}} = \frac{V_{\text{RMS}}}{|Z|} \quad 5.106$$

Synchronous generators have synchronous reactances dependent on the armature current. In addition, the phase angle of the armature current has an effect on the synchronous reactance. This is due to the non-uniformity of the rotor (most pronounced in salient pole machines) and the non-uniformity of the stator structure.

At high (unity) power factors, the armature current is essentially in phase with the terminal voltage, and the armature reaction voltage is leading the generator voltage by about 45°. Within the air gap, the reluctance is somewhat higher for the armature reaction MMF than for the rotor MMF. Therefore, a lower synchronous reactance is experienced. With small lagging power factors, the armature sees approximately the same (lower) reluctances as the rotor, and thus has higher synchronous reactances.

This can be taken into account by a *two-reactance theory*. The two reactances are called the *direct* and the *quadrature reactances*. The current is divided into two components—one in phase with the general voltage E_g, and the other lagging E_g by 90°. The 90° lagging current component flows into the same reluctance as the rotor current (direct component). The current in phase with E_g acts into the reluctance that is 90° (electrically) from the rotor current (quadrature component). The quadrature reactance is always somewhat less than the direct reactance, even in non-salient machines. Thus, there are two reactances, X_d and X_q, and the armature reactance voltage jI_aX_s consists of two components.

$$jI_aX_s = jI_{a,\text{ direct}}X_d + jI_{a,\text{quadrature}}X_q \quad 5.107$$

$I_{a,\text{ direct}}$ is the component of I_a which lags E_g by 90°. $I_{a,\text{ quadrature}}$ is the component in phase with E_g.

For fault conditions, the power factor is low. The reactance to be used is X_d. Typical reactance values are given in table 5.1.

The reactance increases with current because the reluctance increases due to saturation. Under transient conditions a lower reactance is produced. The transient reactance is given by the primed quantities X_d' and X_q'. Table 5.1 gives values of X_d' useful in fault calculations.

During the initial part of the fault transient, a few cycles are required to establish the transient reactance. This period is called the *subtransient time interval*. During this interval, the effective reactance is changing from a considerably lower value to the transient reactances X_d' and X_q'. The initial reactances are called *subtransient reactances* and are identifiable by double primes X_d'' and X_q''. Table 5.1 gives values for X_d'' applicable to the pre-transient fault currents.

The generator voltage, E_g, just prior to a fault can be calculated from the synchronous reactance. However, this value will not be correct for the calculation of fault currents because of the requirement for continuity of energy. The sudden change of current in the armature winding results in a lower armature reactance, but the current through the leakage reactance must remain the same at the instant of the fault. (See equation 5.101 for $t = 0$.) Thus, for transient analysis, a change in E_g to $E_{g'}$ and to $E_{g''}$ is necessary.

These voltages are calculated from the load condition just prior to the occurrence of the fault. However, the transient reactance X_d' or subtransient reactance X_d'' are used as appropriate substitutes for the synchronous reactance X_s.

When a synchronous motor is in a system, it must be considered to be a generator for fault calculations.

Table 5.1
Typical Constants of Three-Phase Synchronous Machines
(typical values are given above the bars, and the ranges given below the bars)

	X_d	X_q	X_d'	X_d''	X_2	X_0
	(unsat.)	rated current	rated voltage	rated voltage	rated current	rated current (note 1)
two-pole turbine generators	$\dfrac{1.20}{0.95-1.45}$	$\dfrac{1.16}{0.92-1.42}$	$\dfrac{0.15}{0.12-0.21}$	$\dfrac{0.09}{0.07-0.14}$	$= X_d''$	$\dfrac{0.03}{0.01-0.08}$
four-pole turbine generators	$\dfrac{1.20}{1.00-1.45}$	$\dfrac{1.16}{0.92-1.42}$	$\dfrac{0.23}{0.20-0.28}$	$\dfrac{0.14}{0.12-0.17}$	$= X_d''$	$\dfrac{0.08}{0.015-0.14}$
salient-pole generators and motors (with dampers)	$\dfrac{1.25}{0.60-1.50}$	$\dfrac{0.70}{0.40-0.80}$	$\dfrac{0.30}{0.20-0.50}$ **note 2**	$\dfrac{0.20}{0.13-0.32}$ **note 2**	$\dfrac{0.20}{0.13-0.32}$ **note 2**	$\dfrac{0.18}{0.03-0.23}$
salient-pole generators (without dampers)	$\dfrac{1.25}{0.60-1.50}$	$\dfrac{0.70}{0.40-0.80}$	$\dfrac{0.30}{0.20-0.50}$ **note 2**	$\dfrac{0.30}{0.20-0.50}$ **note 2**	$\dfrac{0.48}{0.35-0.50}$	$\dfrac{0.19}{0.03-0.24}$
capacitors (air-cooled)	$\dfrac{1.85}{1.25-2.20}$	$\dfrac{1.15}{0.95-1.30}$	$\dfrac{0.40}{0.36-0.50}$	$\dfrac{0.27}{0.19-0.30}$	$\dfrac{0.26}{0.18-0.40}$	$\dfrac{0.12}{0.025-0.15}$
capacitors (hydrogen-cooled at $\frac{1}{2}$ psi)	$\dfrac{2.20}{1.50-2.65}$	$\dfrac{1.35}{1.10-1.55}$	$\dfrac{0.48}{0.36-0.60}$	$\dfrac{0.32}{0.23-0.36}$	$\dfrac{0.31}{0.22-0.48}$	$\dfrac{0.14}{0.03-0.18}$

Used with permission from Electrical Transmission and Distribution
Reference Book, by permission of the Westinghouse Electric Corporation.

note 1 X_0 varies so critically with armature winding pitch that an average value can hardly be given. Variations are from 0.1 to 0.7 of X_d''. Low limit is for two-thirds pitch windings.

note 2 High-speed units tend to have low reactance, and low-speed units have high reactance.

A value for the generated voltage is calculated on the basis of the transient reactance or the subtransient reactance.

Circuit breakers for specific applications are rated according to the current they can interrupt and the voltage they can withstand after interruption. The usual rating specification is in kVA or MVA. This rating is the product of the interrupt current capacity and the line-to-line kV rating.

Example 5.19

A synchronous generator and a synchronous motor are both rated at 15 MVA, 13.9 kV with 25% subtransient reactances. They are connected through a line having 0.3 + j0.4 P.U. impedance on the 15 MVA, 13.9 kV base. The motor is running at 12 MVA, 13 kV, 0.8 PF lagging. (a) Determine the equivalent generated voltages for the generator ($E_{g''g}$) and the motor ($E_{g''m}$) under the expectation of a fault. (b) Determine the subtransient fault current of the generator and the motor for a three-phase fault at the terminals of the motor. (c) Determine the MVA rating of circuit breakers to protect the machines under these conditions. The fault is on the bus side of the circuit breakers.

The synchronous motor will act as a generator into the fault while it gradually loses speed. The subtransient time interval is too short for the motor to have lost significant speed, so it can be treated as a synchronous generator for this interval.

The terminal voltage is 13/13.9 or 0.935 per unit. This is taken as the reference (at 0°). The current is

$$|I_L| = \frac{\left(\dfrac{12 \text{ MVA}}{13 \text{ kV}}\right)}{\left(\dfrac{15 \text{ MVA}}{13.9 \text{ kV}}\right)} = 0.855 \text{ P.U.}$$

$$I_L = 0.855\angle{-36.9°} = 0.684 - j0.513 \text{ P.U.}$$

With a 25% subtransient reactance, the necessary E_{gm} is

$$E_{g''m} = V_t - I_L(jx_d'') = 0.935 - (0.685 - j0.513)$$
$$\times (j0.25)$$
$$= 0.807 - j0.171 = 0.825\angle{-12°} \text{ P.U.}$$

The impedance from the motor terminals is

$$Z = Z_{\text{line}} + j0.25 = 0.3 + j0.65 = 0.716\angle{65.2°}$$

Then, the required value of $E_{g''g}$ is

$$E_{g''g} = V_t + I_L Z = 0.935 + (0.716\angle{65.2°})$$
$$\times (0.855\angle{-36.9°})$$
$$= 1.47 + j0.29 = 1.5\angle{11.1°} \text{ P.U.}$$

When the fault occurs at the motor terminals, the motor current will be $E_{g''m}/jx_d''$.

$$I_m = \frac{0.825\angle{-12°}}{0.25\angle{90°}} = 3.3\angle{-102°} \text{ P.U.}$$

The generator current will be $E_{g''g}/Z$.

$$I_g = \frac{1.5\angle{11.1°}}{0.716\angle{65.2°}} = 2.09\angle{-54°} \text{ P.U.}$$

The total fault current is

$$3.3\angle{-102°} + 2.09\angle{-54.1°} = 4.95\angle{-83.7°}$$

The three-phase breaker rating can be found from P.U. current, the base kVA, and base kV.

kVA rating of breaker (three-phase) $= I$ (P.U.)
$\times$ base kVA

The rating of the motor circuit breaker is

3.3×15 MVA $= 49.5$ MVA
at 13.9 kV three-phase rating

The rating for the generator is

2.09×15 MVA $= 31.35$ MVA
at 13.9 kV three-phase rating

Asymmetrical faults occur when there is a single line-to-ground (or line-to-neutral) fault, a line-to-line fault, or a two-line-to-ground (or neutral) fault. The three-phase fault previously discussed permits the use of per-phase calculations because of the symmetry.

When an asymmetrical fault occurs, the analysis can be performed laboriously on an unbalanced three-phase basis, or by use of *symmetrical components*. The use of symmetrical components is convenient because fault currents can be determined in terms of the pre-fault conditions and the associated *sequence impedances*. A partial development of symmetrical components is included here.

An unbalanced three-phase problem can be solved by using three sets of balanced voltage and current components. These sets consist of a positive sequence (normally *abc*), a negative sequence (normally *acb*), and a zero sequence. With an unbalanced condition, each phase current can be represented by three components: positive sequence, negative sequence, and zero sequence. These are illustrated in figure 5.14.

The *positive sequence* has been used previously. It includes balanced three-phase generators, transformers and motors.

The *negative sequence* can be considered either as a mirror image of the positive sequence with phasor rotation in the opposite direction, or as an order reversal in phase sequence (e.g., *acb* if the positive sequence is *abc*, or *abc* if the positive sequence is *acb*). The former method is preferred.

The *zero sequence* is a set of equal voltages or currents which have the same time phase in each of the three legs. The zero sequence can occur only in four-wire systems (e.g., wye connections with a neutral wire or a grounded neutral).

It is possible to calculate the three sequence components from the phasor diagram if the magnitudes and angles are known. However, the primary interest is in determining fault currents when only the generator voltages are known. For this procedure, the equivalent circuits for each phase sequence are needed.

Symmetrical components are mathematical concepts used to account for the imbalance of the circuit. They are useful in determining all currents in a system when a fault occurs. It is possible, however, to use some of the theory to determine the fault currents alone, thus

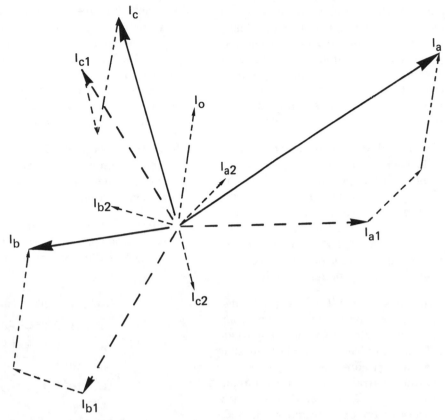

Figure 5.14 Phasor Diagram Showing Symmetrical Components
for Unbalanced Currents in a Three-Phase System

specifying breaker requirements. In order to do this, it is necessary to determine the equivalent circuits for the three sequences, so as to determine the sequence impedances.

Positive-sequence impedances are those which have been used in balanced three-phase analyses up to this point. The generators are positive-sequence by definition, so they appear only in positive-sequence equivalent circuits. The positive-sequence circuit is established with the neutral as the reference bus. The impedance of a generator can be the subtransient reactance, the transient reactance, or the synchronous reactance depending on the analysis required. For fault analysis, the subtransient reactance is used. The equivalent generator voltage as calculated in example 5.19 is appropriate.

For rotating machines, the *negative-sequence imped-ance's* contribution to the armature reaction is a field rotating in the opposite direction from the rotor field. Thus, the reactance is different from that of the positive sequence. The *negative-sequence reactance* is the average of the subtransient direct and the quadrature reactances.

$$X = \frac{1}{2}(X_d'' + X_q'') \qquad 5.108$$

Zero-sequence impedances are quite different from positive-sequence impedances. In machines, such as synchronous motors, the three components of the zero-sequence currents do not form a rotating field. The only part of the machine impedance they see is the leakage reactance, which is given in table 5.1 as X_0.

On a transmission line, the zero-sequence currents are equal, in phase with one another, and travel parallel in the three lines. The result is a single line over the ground plane or a single-phase line if there is a ground wire present. The series reactance is increased by a factor of 2 to 3.5 over the positive-sequence reactance.

The resistance of a zero-sequence line is not changed from the positive-sequence value, as it is a function of frequency and not of spacing. The only capacitance seen by zero-sequence currents is the capacitance to ground. This produces an increase in capacitive reactance by about the same factor as the increase in inductive reactance.

A zero-sequence current flowing in one line of a three-phase network must, by definition, also be flowing in the other two lines. This has drastic effects on the impedance of three-phase loads. The symmetry of the zero-sequence circuit requires that the zero-sequence voltage from any line to the reference be the same as from any other line to the reference at that point. Thus, the zero-sequence voltages across any delta-connected load must be zero, so that zero-sequence currents cannot flow in a delta-connected load. Similarly, a wye-connected load with a floating neutral cannot support zero-sequence currents, as current must sum to zero at the floating neutral.

Only a wye-connected load with a grounded neutral, either directly or through an impedance Z_n, permits zero-sequence currents. The impedance seen from the line side of such a load to ground is the load impedance, Z_L, plus three times the neutral to reference impedance. The factor 3 is required because the three zero-sequence components must flow through Z_n. Z_n will also include *ground impedance* from the actual grounding point and the zero-sequence reference point or *reference bus*.

Transformers are the most difficult for which to determine zero-sequence impedances. The key point for transformers is that primary current can flow only when secondary current flows, and vice versa. When one side of a two-winding transformer contains a wye-connected winding with a floating neutral, it is an open circuit to zero-sequence currents. Only transformers with grounded wye connections on both primary and secondary windings permit a connection from primary to secondary. In this situation, the impedance is the same as for the positive-sequence circuit.

Only a delta-connected transformer secondary winding can act as a current path for zero-sequence currents. In such a case, the same current flows around the closed path of the delta, satisfying the requirement of equal zero-sequence currents in each phase. However, the equal phase to reference bus on a delta-connected winding requires that there be zero voltage from line to line in the delta-connected secondary. This will appear as a short circuit to a grounded wye-connected primary. Conversely, a delta-connected primary will appear as a short circuit to a grounded wye-connected secondary.

Figure 5.15 shows zero-sequence impedances for various circuit configurations.

Sequence networks are obtained by the correct use of the elements outlined previously. To illustrate the process, consider the simple distribution system shown in figure 5.16(a). The positive-sequence network, as previously determined for balanced networks, is shown in figure 5.16(b).

The negative-sequence network is similar to the positive-sequence network, but it omits all positive-sequence generators and replaces X_d'' by $\frac{1}{2}(X_d'' + X_q'')$. This is shown in figure 5.16(c).

The zero-sequence network is more difficult to obtain. Care must be taken in observing ground paths. The generator has a ground path which carries $3I_0$, so for any individual phase it appears as $3Z_n$, as shown in figure 5.16(d). The wye-connected primary of transformer T_1 is ungrounded and results in an open circuit at point A. The secondary of T_1 and primary of T_2 are both delta-connected, and they isolate the transmission line from the zero-sequence network.

The delta connection of T_2 and the grounded secondary wye connection effectively ties the transformer impedances to ground. The grounded wye connections of the motor and load put them in parallel as shown.

Symmetrical components are most conveniently represented with the phasor operator **a**, which is a unit vector at an angle of 120° (similar to j, which is a unit vector at 90°).

$$\mathbf{a} = 1\angle 120° = -0.5 + j0.866 \qquad 5.109$$

$$\mathbf{a}^2 = 1\angle 240° = -0.5 - j0.866 \qquad 5.110$$

$$\mathbf{a}^3 = 1\angle 360° = 1 \qquad 5.111$$

$$\mathbf{a}^4 = a \qquad 5.112$$

$$\mathbf{a}^5 = a^2 \qquad 5.113$$

$$\mathbf{a}^6 = a^3 \qquad 5.114$$

Analysis of the symmetrical components in terms of actual circuit values makes use of equations 5.115 through 5.118.

$$I_a = I_{a0} + I_{a1} + I_{a2} \qquad 5.115$$

$$I_b = I_{b0} + I_{b1} + I_{b2} \qquad 5.116$$

$$I_c = I_{c0} + I_{c1} + I_{c2} \qquad 5.117$$

$$I_{a0} = I_{b0} = I_{c0}$$

$$= \frac{1}{3}(I_a + I_b + I_c) \qquad 5.118$$

For an *abc* sequence, the positive-sequence components are:

$$I_{b1} = \mathbf{a}^2 I_{a1} \qquad 5.119$$

$$I_{c1} = \mathbf{a} I_{a1} \qquad 5.120$$

The negative-sequence components are:

$$I_{b2} = \mathbf{a} I_{a2} \qquad 5.121$$

$$I_{c2} = \mathbf{a}^2 I_{a2} \qquad 5.122$$

Rearranging equations 5.115 through 5.122 results in equations 5.123 through 5.126.

$$I_{a0} = \frac{1}{3}(I_a + I_b + I_c) \qquad 5.123$$

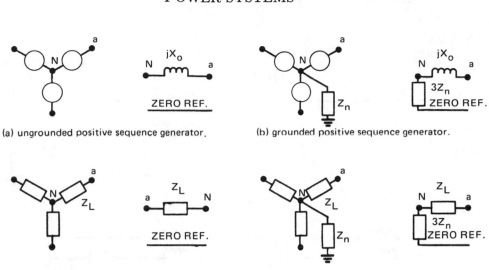

(a) ungrounded positive sequence generator.

(b) grounded positive sequence generator.

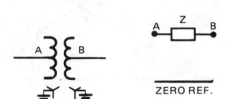

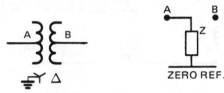

(c) ungrounded (Y or Δ) load.

(d) grounded wye load.

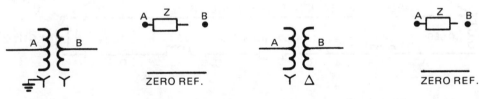

(e) Y-Y, both neutrals grounded.

(f) Y-Δ, grounded Y neutral.

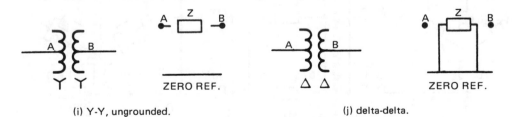

(g) Y-Y, one neutral grounded.

(h) Y-Δ, ungrounded.

(i) Y-Y, ungrounded.

(j) delta-delta.

Figure 5.15 Zero-Sequence Impedances

$$I_{a1} = \frac{1}{3}(I_a + \mathbf{a}I_b + \mathbf{a}^2 I_c) \qquad 5.124$$

$$I_{a2} = \frac{1}{3}(I_a + \mathbf{a}^2 I_b + \mathbf{a}I_c) \qquad 5.125$$

A similar development for sequence to phase voltage conversion results in equations 5.126 through 5.128.

The zero-sequence voltage reference is ground. The positive- and negative-sequence voltages are referenced to neutral.

$$V_{a0} = \frac{1}{3}(V_{an} + V_{bn} + V_{cn}) \qquad 5.126$$

$$V_{a1} = \frac{1}{3}(V_{an} + \mathbf{a}V_{bn} + \mathbf{a}^2 V_{cn}) \qquad 5.127$$

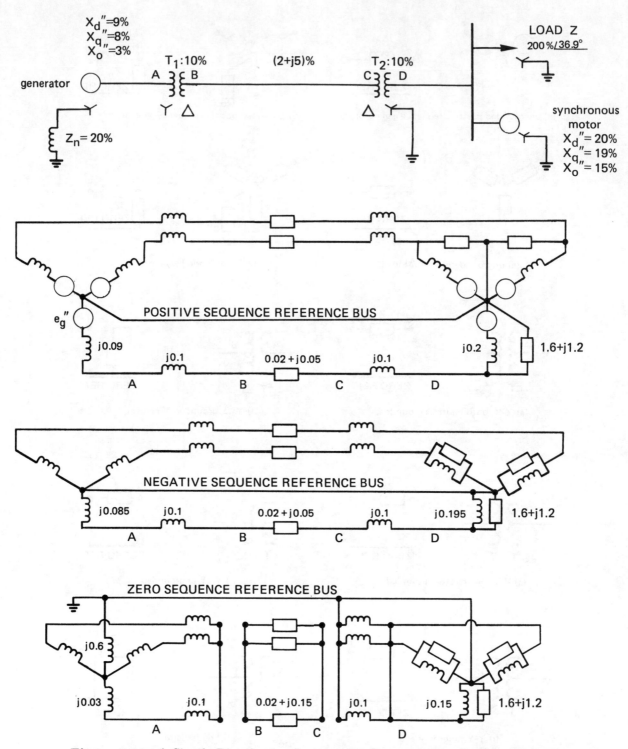

Figure 5.16 A Simple Distribution System with Corresponding Sequence Networks

$$V_{a2} = \frac{1}{3}(V_{an} + \mathbf{a}^2 V_{bn} + \mathbf{a}V_{cn}) \qquad 5.128$$

$$I_{a0} = \frac{1}{3}(1 + \mathbf{a}^2 + \mathbf{a})I_a = 0 \qquad 5.129$$

Symmetrical three-phase faults can be calculated since it is known that the load is balanced and that $I_b = \mathbf{a}^2 I_a$ and $I_c = \mathbf{a}I_a$.

$$I_{a1} = \frac{1}{3}(1 + \mathbf{a}^3 + \mathbf{a}^3)I_a = I_a \qquad 5.130$$

$$I_{a2} = \frac{1}{3}(1 + \mathbf{a}^4 + \mathbf{a}^2)I_a = 0 \qquad 5.131$$

Thus, balanced (positive sequence only) analysis can be done.

Example 5.20

For the power system shown in figure 5.16, determine the fault current for a three-phase fault at the load end (point C) of the transmission line. The load bus is at 90% voltage, the motor is running at 0.8 leading power factor, and the current is 0.5 P.U.

Since only the fault current is required, it is sufficient to obtain the Thevenin equivalent circuit using the appropriate subtransient reactances, just prior to faulting.

With the 90% voltage at the load bus taken as reference, the three-phase load takes a current of

$$\left(\frac{0.9}{2}\right) \angle -36.9° = 0.45 \angle -36.9°$$

The motor is operating with a power factor of 0.8 leading, with a current of $0.5 \angle 36.9°$. The total bus current is the sum of the load and motor currents.

$$0.45 \angle -36.9° + 0.5 \angle 36.9° = 0.76 \angle 2.3°$$
$$= 0.76 + j0.03$$

The voltage rise across the 10% transformer reactance is

$$j0.1 \times (0.76 + j0.03) = -0.003 + j0.076$$

Therefore, the pre-fault voltage is

$$0.9 + (-0.003 + j0.076) = 0.9 \angle 4.8° \text{ P.U.}$$

In figure 5.16(b), the impedance to the left of point C is

$$Z_L = j0.09 + j0.1 + 0.02 + j0.05 = 0.02 + j0.24$$

To the right of point C,

$$Z_R = j0.1 + \cfrac{1}{\cfrac{1}{j0.2} + \cfrac{1}{1.6 + j1.2}} = 0.014 + j0.288$$

The parallel combination is

$$Z_1 = \frac{Z_L Z_R}{Z_L + Z_R} = 0.131 \angle 86.1°$$

The fault current is the Thevenin equivalent pre-fault voltage divided by the impedance seen from the point of the fault to the neutral.

$$|I_{\text{fault}}| = \frac{0.9}{0.131} = 6.87 \text{ P.U.}$$

(When the load has no effect, it is necessary to treat the symmetrical fault as was done in example 5.19.)

Single-phase faults to ground require finding the Thevenin equivalent network for the fault assuming that any type of fault can occur there. The pre-fault circuit must be found for each of the sequence circuits. Equations 5.123 through 5.128 can be used to determine the fault currents.

The Thevenin equivalent circuits are found by determining the pre-fault voltage at the fault, and the impedance from fault point to reference bus for each of the sequence circuits.

For a single-phase fault from the a phase to ground, for example, the currents through the three phases of Thevenin equivalents are I_a, $I_b = 0$, and $I_c = 0$. I_a is unknown and must be determined.

From equations 5.123 through 5.125, the single phase to ground currents are:

$$I_{a0} = \frac{I_a}{3} \qquad\qquad 5.132$$

$$I_{a1} = \frac{I_a}{3} \qquad\qquad 5.133$$

$$I_{a2} = \frac{I_a}{3} \qquad\qquad 5.134$$

This makes I_{a0}, I_{a1}, and I_{a2} equal, so the three circuits shown in figure 5.17(a) are in series. With line a shorted, $V_a = 0$, so the sum of the voltages V_{a1}, V_{a2}, and V_{a0} is zero. The resulting circuit must be as in figure 5.17(b). From this it is seen that the fault current is

$$I_{a1} = \frac{e_{\text{Th},a}}{Z_0 + Z_1 + Z_2} \qquad\qquad 5.135$$

$$I_a = 3I_{a1} \qquad\qquad 5.136$$

Example 5.21

Determine the current in a fault from line to ground at point C for the circuit and conditions in example 5.20.

The Thevenin equivalent for the positive-sequence circuit has been calculated.

$$e_{a,\text{Th}} = 0.9 \angle 4.8°$$
$$Z_1 = 0.131 \angle 86.1°$$

The Thevenin equivalent impedance for the negative sequence to the left of point C is

$$0.02 + j0.05 + j0.1 + j0.85 = 0.02 + j0.235 = 0.236 \angle 85.1°$$

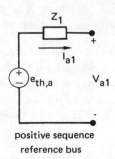

positive sequence
reference bus

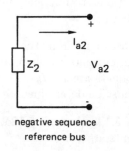

negative sequence
reference bus

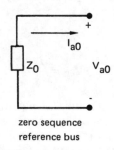

zero sequence
reference bus

Figure 5.17(a)　Thevenin Equivalent
Networks for Sequence Networks as Seen
from the Fault Point

To the right, $j0.1$ is in series with the parallel combination of $2\angle36.9°$ and $j0.195$ on a per-unit basis. This parallel combination is $0.184\angle85.7°$. When added to $j0.1$, the total to the right is $0.284\angle87.3°$. The Thevenin impedance is the sum of the parallel impedances to the left and the right.

$$Z_2 = \cfrac{1}{\cfrac{1}{0.284\angle87.3°} + \cfrac{1}{0.236\angle85.1°}}$$
$$= 0.129\angle86.0°$$

In figure 5.16(c) the impedance from point C to the zero-sequence reference is infinite because there is no zero-sequence path to ground: $Z_0 = \infty$.

Thus, a single-line to ground fault at C causes no fault current.

A *line-to-line fault* will require working with all three phases of the Thevenin equivalent network, as indicated in figure 5.18. The application of equations 5.123 and 5.128 requires the following equations:

$$I_a = 0 \qquad\qquad 5.137$$
$$I_c = -I_b \qquad\qquad 5.138$$
$$V_b = V_c \qquad\qquad 5.139$$
$$3I_{a0} = 0 + I_b - I_b = 0 \qquad\qquad 5.140$$
$$3I_{a1} = 0 + \mathbf{a}I_b - \mathbf{a}^2I_b = (\mathbf{a} - \mathbf{a}^2)I_b \qquad 5.141$$
$$3I_{a2} = 0 + \mathbf{a}^2I_b - \mathbf{a}I_b = (\mathbf{a}^2 - \mathbf{a})I_b \qquad 5.142$$

Therefore,

$$I_{a2} = -I_{a1} \qquad\qquad 5.143$$

Also,

$$3V_{a0} = V_a + V_b + V_b = V_a + 2V_b \qquad 5.144$$

$$3V_{a1} = V_a + \mathbf{a}V_b + \mathbf{a}^2V_b = V_a + (\mathbf{a} + \mathbf{a}^2)V_b \qquad 5.145$$

$$3V_{a2} = V_a + \mathbf{a}^2V_b + \mathbf{a}V_b = V_a + (\mathbf{a}^2 + \mathbf{a})V_b \qquad 5.146$$

Therefore,

$$V_{a1} = V_{a2} \qquad\qquad 5.147$$

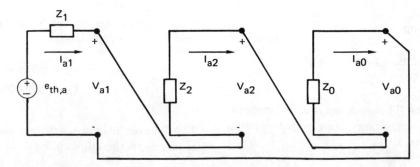

Figure 5.17(b)　Phase *a* Circuit for a Line to Ground Fault on Phase *a*

These conditions are fulfilled by the interconnection of the Thevenin equivalent sequence circuits shown in figure 5.18(b). For this equivalent circuit,

$$I_{a1} = \frac{e_{\mathrm{Th},a}}{Z_1 + Z_2} \qquad 5.148$$

$$I_b = I_{a0} + \mathbf{a}^2 I_{a1} + \mathbf{a} I_{a2} \qquad 5.149$$

With $I_{a2} = -I_{a1}$ and $I_{a0} = 0$,

$$I_b = I_{\mathrm{fault}} = -j\sqrt{3} I_{a1} \qquad 5.150$$

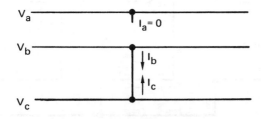

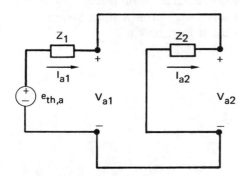

Figure 5.18 Line-to-Line Fault and Equivalent Circuit for Sequence Currents

Example 5.22

Determine the fault current for a line-to-line fault at point C of the circuit shown in example 5.20 under the load conditions of example 5.20. Also, determine the line-to-neutral voltages at point C.

The Thevenin equivalent circuits were found for point C in examples 5.20 and 5.21.

$$e_{\mathrm{Th},a} = 0.9\angle 4.8°$$
$$Z_1 = 0.131\angle 86.1°$$
$$Z_2 = 0.129\angle 86.0°$$

Then, from equation 5.148

$$I_{a1} = 3.46\angle -81.3°$$
$$V_{a1} = I_{a1} Z_2 = 0.446\angle 4.7°$$

Furthermore, it is known that $I_{a0} = 0$, $I_{a2} = -I_{a1}$, $V_{a0} = 0$, and $V_{a2} = V_{a1}$. Therefore,

$$I_b = -j\sqrt{3}I_{a1} = 5.99\angle -171.3°$$
$$V_a = V_{a0} + V_{a1} + V_{a2} = 2V_{a1}$$
$$= 0.893\angle 4.7°$$
$$V_b = V_{a0} + \mathbf{a}V_{a1} + \mathbf{a}^2 V_{a2} = (\mathbf{a} + \mathbf{a}^2)V_{a1} = -V_{a1}$$
$$= 0.446\angle 184.7° = 0.446\angle -175.3°$$

It is possible to calculate the actual (per-unit) line currents with these values.

A two-phase to ground fault from figure 5.19 follows the relations

$$V_b = V_c = 0 \qquad 5.151$$
$$I_a = 0 \qquad 5.152$$

From equations 5.123 through 5.128,

$$V_{a0} = V_{a1} = V_{a2}$$
$$= \frac{V_a}{3} \qquad 5.153$$

Using the Thevenin equivalent circuits,

$$I_{a0} = \frac{-V_{a1}}{Z_0} \qquad 5.154$$

$$I_{a2} = \frac{-V_{a1}}{Z_2} \qquad 5.155$$

$$I_{a1} = \frac{e_{\mathrm{Th},a} - V_{a1}}{Z_1} \qquad 5.156$$

Thus,

$$I_a = \frac{e_{\mathrm{Th},a}}{Z_1}$$
$$- V_{a1}\left(\frac{1}{Z_0} + \frac{1}{Z_1} + \frac{1}{Z_2}\right)$$
$$= 0 \qquad 5.157$$

$$V_{a1} = \frac{e_{\mathrm{Th},a}Z_0 Z_2}{Z_0 Z_1 + Z_1 Z_2 + Z_0 Z_2} \qquad 5.158$$

Now substitute $e_{\mathrm{Th},a} - I_{a1}Z_1$ for V_{a1}.

$$I_{a1} = \frac{e_{\mathrm{Th},a}}{Z_1 + \dfrac{Z_0 Z_2}{Z_0 + Z_2}} \qquad 5.159$$

$$I_{a2} = \frac{-e_{\mathrm{Th},a}Z_0}{Z_0 Z_1 + Z_0 Z_2 + Z_1 Z_2} \qquad 5.160$$

$$I_{a0} = \frac{-e_{\mathrm{Th},a}Z_2}{Z_0 Z_1 + Z_0 Z_2 + Z_1 Z_2} \qquad 5.161$$

This implies that the three Thevenin equivalent circuits are in parallel as shown in figure 5.19.

The fault currents are found from equations 5.162, 5.163, and 5.164.

$$I_a = 0 \qquad\qquad 5.162$$

$$I_b = I_{a0} + \mathbf{a}^2 I_{a1} + \mathbf{a}I_{a2} \qquad\qquad 5.163$$

$$I_c = I_{a0} + \mathbf{a}I_{a1} + \mathbf{a}^2 I_{a2} \qquad\qquad 5.164$$

Since $V_a = 3V_{a1}$, the line voltage can also be found.

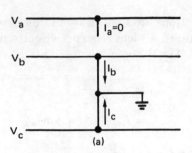

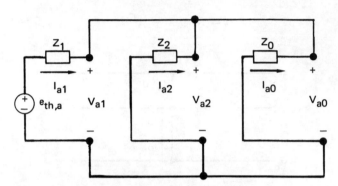

Figure 5.19   Double Line-to-Ground Fault with Corresponding Sequence Circuit for Phase a

PRACTICE PROBLEMS

Warm-ups

1. A 230 kV (line-to-line), three-phase system delivers 200 MVA to a balanced load with a 0.9 power factor (lagging). Determine the line current and the reactance necessary to achieve unity power factor.

2. A 13.8 kV single-phase transformer is subjected to an open-circuit test with the following results: $P_{in} = 900$ W, $I_{in} = 0.2$ A, and $V_{out} = 460$ V (secondary). Determine the values of the equivalent circuit parameters which can be found from this test.

3. A 32 kVA single-phase transformer is rated at 1.5 kV on the primary. A 60 Hz short-circuit test on the primary indicates $P_{in} = 1$ kW, $I_{in} = 20$ A, $I_{secondary} = 100$ A, and $V_{primary} = 80$ V. Determine the winding resistances and the leakage inductances.

4. A three-phase synchronous motor is rated 10 kVA, 440 V and has a synchronous reactance of 0.8 P.U. It is used in a 100 kVA, 440 V system. Determine the actual synchronous reactance and the per-unit value of the synchronous reactance in the system base.

5. A single-phase transformer is rated 6.8 kV : 115 kV at 500 kVA. A short-circuit test on the high-voltage side at rated current indicates $P_{sc} = 435$ W and $V_{sc} = 2.5$ kV. Determine the per-unit impedance of the transformer.

6. A 50 kVA single-phase transformer has a 440 V secondary at rated kVA and unity power factor. Determine the half-load output voltage at the same power factor for per-unit resistance of 1.4% and reactance of 2.4%. Also determine half-load to full-load and no-load to full-load voltage regulations.

7. Repeat problem 6 for a 0.4 lagging power factor.

8. A three-phase, 100 kVA transformer with phase impedance $0.03 + j0.04$ P.U. is operating at full load with a power factor of 0.8 lagging. Determine the change in voltage regulation if the power factor is corrected to unity. Assume rated current in both cases.

9. An open-delta, three-phase system consisting of two single-phase transformers drives a balanced three-phase load of 50 kVA and 440 V with unity power factor. Determine the necessary ratings for the two transformers.

10. Two transformers in parallel are rated at 2.5 MVA; one with 5% reactance and the other with 6% reactance.

If tap changes are 2.5%, what current will circulate with one tap difference between the two transformers?

Concentrates

1. A three-phase, 11.5 kV generator drives a 500 kW, 0.866 lagging power factor load. Determine (a) the line current, (b) the necessary generator rating, and (c) the compensating reactance required to bring the power factor to 1.0.

2. A transformer is rated 50 MVA, 115 kV : 13.8 kV. Open- and short-circuit tests are carried out on the secondary side: $P_{oc} = 250$ kW, $I_{oc} = 350$ A, $P_{sc} = 300$ kW, and $V_{sc} = 1.5$ kV. (a) Determine the equivalent circuit parameters. (b) Obtain the ABCD parameters. (c) For a resistive load of 0.5 ohm referred to the primary, determine the primary input impedance.

3. Two identical transformers are cascaded. Each has $Z_p = 0.01 + j0.02$ and $Y_c = 0.0002 - j0.001$. The secondary winding of the second transformer is loaded with a resistance of 0.1 ohm (referred to the primary of the first transformer). Calculate the input impedance seen at the primary of the first transformer.

4. A three-winding, single-phase transformer has the following winding impedances referred to the primary: $Z_p = 49.5 + j110$; $a_{ps}^2 Z_s = 70.5 + j90$; $a_{pt}^2 Z_t = 50.5 + j70$ ohms. The winding ratings are: primary: 50 kVA, 6.8 kV; secondary: 30 kVA, 440 V; and tertiary: 20 kVA, 1380 V. Determine the percent impedances in the secondary and tertiary bases.

5. The feeder from a distribution transformer is 500 feet long and consists of three AWG # 4/0 wires having a resistance of 0.053 ohm/1000 ft and a current capacity of 225 A. The three-phase motor load is 120 kW at 440 V with a power factor of 0.6 lagging. The transformer is rated 250 kVA, 440 V and has 1% resistance and 5% reactance. Determine the percent change in load voltage when power factor correction is made to allow 225 A phase currents to flow.

6. (a) Repeat example 5.19 for a motor PF = 1.0. (b) Repeat example 5.19 for a motor PF = 0.8 leading. (c) Draw conclusions concerning the effect of power factor on three-phase faults.

7. For the power system of figure 5.16, determine the fault current for a three-phase fault at the primary of

transformer T_1 (point A). The load bus is at 90% voltage, and the motor is running at 50% current with a power factor of 0.8 leading.

8. Determine the fault current for a single-phase fault-to-ground with conditions the same as in problem 7.

9. A three-phase, wye-connected auto-transformer connects a 345 kV substation to a 500 kV substation. It must be able to transfer 750 MVA in either direction. Determine the voltage, current, and MVA ratings of both common and series windings of the auto-transformer.

10. A 500 hp, 2400 volt, 1200 RPM squirrel cage motor is supplied from the secondary of a 2500 kVA, 13.8 kV delta to 2.4 kV wye transformers having an impedance of 5.5% and an X/R ratio of 7. The impedance of the 13.8 kV system is $5 + j20\%$ on a 100 MVA, 13.8 kV base. Determine the percent voltage drop if the motor is line started.

Timed

1. A 250 kVA, three-phase distribution transformer with a 440 V secondary and a 4% reactance supplies an induction motor rated at 440 V. (a) How large an induction motor (in horsepower) can be started directly across the line if the maximum voltage dip is limited to 3%? Assume 90% efficiency and a running power factor of 85%. (b) If the motor is delta-connected for running but wye-connected for starting, determine the maximum horsepower to meet the 3% restriction on line voltage dip. (c) If a three-phase auto-transformer with 40% taps is available for starting, what would be the largest motor which could be started without changing from delta to wye?

2. An open-delta, three-phase 440 V transformer consists of two single-phase transformers. The *c-a* leg is missing. The transformers serve a 50 kVA balanced three-phase load with a 0.8 lagging power factor. Determine the required transformer ratings if (a) a 10 kW resistive single-phase load is added to the *a-b* leg, and (b) if a 10 kW resistive load is added to the *b-c* leg.

3. Two parallel transformers, one rated 25 MVA and the other at 15 MVA, are serving a load of 30 MVA with a 0.8 lagging power factor. The 25 MVA transformer has a 10% reactance, and the 15 MVA transformer has a 6% reactance (referred to their own bases). (a) What fraction of the load current will each carry? (b) If the secondary voltage is 35 kV and the 25 MVA transformer is on a 5% overvoltage tap, determine the resulting

circulating current. What is the total current for each transformer?

4. Determine the fault current for a line-to-line fault at point A of figure 5.16 with the load bus at 90% voltage. The motor is running at 50% current with an 80% leading power factor. Also determine the line-to-neutral voltages at point A.

5. A power system has a line-to-line short at bus number 3 as shown. Find the voltages at busses 1 and 2.

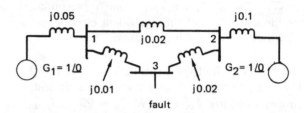

6. A three-phase transformer is fed from a three-phase, 138 kV (line-to-line) system with an impedance of $j3.1$ ohms per line. The transformer impedance base is referred to the 138 kV windings. The winding parameters from the nameplate are: H winding: 138 kV line-to-line, connected in a grounded wye; X winding: 34.5 kV line-to-line, connected in a grounded wye; T winding: 13.8 kV line-to-line, connected in delta. The impedance and kVA ratings from winding to winding are: H to X: $j9.77\%$ at 60 MVA (20 MVA/phase); H to T: $j7.46\%$ at 21 MVA (7 MVA/phase); and X to T: $j288\%$ at 21 MVA (7 MVA/phase). For faults occurring on the X winding, determine the fault current for (a) a three-phase fault, and (b) a single-phase fault to ground.

7. Two three-phase transformers, each rated 100 MVA and 138 kV : 12.47 kV, are operated in parallel in a substation. The leakage reactance is based on their own ratings and are 4% and 6% for transformers 1 and 2 respectively. The winding resistances are negligible. Transformer 1 has a tap-changing accessory which allows the turns ratio to be modified ±10% in 32 equal steps. Transformer 2 has a phase-shifting accessory that can adjust the phase angle continuously over a range of ±3 degrees. Determine the tap-changer and phase-angle settings needed to allow the two transformers to carry a load of 200 MVA at 0.8 power factor lagging without either transformer being overloaded.

8. A 30/40/50 MVA substation transformer (115 kV–12 kV) has $Z = 7.5\%$ and $X/R = 15$. The transformer is connected to a 100 MVA, 12.47 kV line with 2% R and 10% X, which in turn is connected to a 34 MW load with an 85% lagging power factor. The load voltage is measured at 12.16 kV. Find the voltage regulation at the load.

9. A transmission line with 8% reactance on a 200 MVA base connects two substations. At the first substation the voltage is $1.03\angle 5°$ P.U. and at the second the voltage is $0.98\angle -2.5°$ P.U. Find the power and volt-amperes reactive flowing out of the first substation.

10. A 500 V : 100 V transformer can be connected as an auto-transformer to obtain a 600 V : 500 V ratio (600 V input). There are four possible ways of connecting the auto-transformer, two of which are incorrect. (a) For a wrong connection, what is the output voltage? (b) What are the effects of the wrong connection? (c) Show either correct connection.

6 TRANSMISSION LINES

Nomenclature

A	area, or ABCD parameter	m^2, or volts/volt
b	normalized susceptance	–
B	ABCD parameter, or susceptance	ohms, or mhos
C	ABCD parameter, or capacitance	mhos, or farads
d_s	geometric mean distance	ft
D	ABCD parameter, or center-to-center spacing	amps/amp, or ft
f	frequency	Hz
GMR	geometric mean radius	ft
H	magnetic field intensity	amps/m
I	RMS phasor current	amps
I.D.	inner dimension, or diameter	ft, or m
J	current density	amps/m^2
K_c	capacitive correction factor	–
K_L	inductive correction factor	–
l	line length	miles
L	inductance	henrys
L_l	inductance per unit length	henrys/m
M	mutual inductance	henrys
O.D.	outside dimension, or diameter	ft, or m
P	power	watts
PF	power factor	–
q	charge per unit length	coulombs/m
Q	reactive power	volt-amps reactive
r	radial distance	m or ft
R	resistance	ohms
R_ℓ	resistance per unit length	ohms/m
S	apparent power	volt-amperes
V	RMS phasor voltage	–
x	distance, or normalized reactance	m, or ft
X	reactance	ohms
y	normalized admittance, or admittance per unit length	– mhos/m
Y	admittance	mhos
z	normalized impedance, or impedance per unit length	– ohms/m
z_l	apparent normalized impedance	–
z_o	characteristic impedance	ohms
Z	impedance	ohms

Symbols

α	attenuation constant	nepers/m
β	phase constant	radians/m
γ	propagation constant	radians/m
δ	skin depth	m
ϵ	permittivity	farads/m
μ	permeability	henrys/m
ρ	resistivity	ohm-m
ψ	flux linkage	webers
ω	radian frequency	rad/sec

Subscripts

a	line a
b	line b
c	line c
C	capacitance
cp	compensation point
e	external
eff	effective
fl	full load
i	internal
l	line, or per-unit length
L	inductive, or load
n	neutral
nl	no load
o	characteristic value
O	DC value
R	receiving
S	sending

1 INTRODUCTION

Power distribution and communication systems both use transmission lines. The behavior of such lines is an important aspect of the overall system performance, and it must be taken into account in system design or analysis. The properties of interest in transmission lines are resistance and inductance along the line, and the capacitances from line-to-line and line-to-ground. The *shunt* or *line-to-line conductance* is usually not significant and is omitted in this chapter.

A variety of length units are used in practice, including mils, inches, feet, miles, meters, and kilometers. Conversions are straightforward, and table 6.1 can be used when necessary.

Table 6.1
Length Conversion Factors

multiply	by	to obtain
mils	10^{-3}	inches
mils	2.54×10^{-3}	centimeters
mils	2.54×10^{-5}	meters
inches	2.54	centimeters
miles	5280	feet
miles	1609.3	meters
feet	0.3048	meters

The transmission line resistance, inductance, and capacitance are universally specified as *per-unit-length* quantities. The resistance is a frequency-dependent quantity because of the *skin effect*, increasing as the square root of the frequency. This, together with the *temperature effect*, presents a problem in the design of conductors for power systems. In communication systems at high frequencies, the usual length of a transmission line is short enough to neglect resistance.

Inductance is made up of two components. The *internal inductance* decreases with frequency because of the skin effect. The *external inductance* is dependent on the geometric arrangement of the conductors within the transmission line. The capacitance also is a function of the geometric arrangement of the conductors.

2 SKIN EFFECT

The *resistance per unit length* of a conductor carrying a constant DC current is the resistivity divided by the cross-sectional area. Resistance per unit length in this chapter is R_l, but it often is given with no identification.

$$R_{lo} = \frac{\rho}{A} \qquad 6.1$$

The *internal inductance* of a circular wire carrying DC current is independent of the radius. For nonmagnetic materials it can be calculated[1] from basic field principles as

$$L_{loi} = \frac{\mu}{8\pi} = 0.5 \times 10^{-7} \text{ H/m} \qquad 6.2$$

L_{li} is the internal inductance per unit length, and L_{loi} is the DC value of the internal inductance.

The *skin effect* is an electromagnetic phenomenon which determines the time needed for an external field to penetrate a conductor. This time is governed by the resistivity and permeability of the conducting material and the dimensions of the conductor. When the electric field fails to fully penetrate the conductor, the current density is higher near the surface than at the middle. This produces an effective increase of the resistance and a decrease in the internal inductance. For a flat plate conductor, the current density depends on the *skin depth*, δ.

$$J = J_{\text{surface}} e^{(-x/\delta)(1+j)} \qquad 6.3$$

$$\delta = \frac{1}{\sqrt{\dfrac{\pi f \mu}{\rho}}} \qquad 6.4$$

The impedance per unit length per unit width is

$$Z_{lw} = \frac{\rho}{\delta}(1+j) = (1+j)\sqrt{\pi f \mu \rho} \qquad 6.5$$

Equation 6.5 is interpreted to mean that the effective thickness of the plate is δ. (Obviously, this will be true only if the actual thickness of the plate is somewhat greater than δ.)

For round wires, equation 6.5 is useful only if the skin depth is much less than the radius of the wire (e.g., $r/\delta > 5.5$ for 10% or less error).

Example 6.1

The skin depth for copper is given by $\delta = 0.066/\sqrt{f}$ (meters).

---

[1] Ramo, Whinnery, and Van Duzer, *Fields and Waves in Communication Electronics*, John Wiley and Sons, 1965, p. 305.

Determine the wire radius (in mils) for which the skin depth equals the radius for frequencies of (a) 60 Hz, (b) 10 kHz, and (c) 1 MHz. (A mil is 0.001 inches.)

First convert meters to mils.

$$\delta \text{ in mils} = \delta(m) \times \frac{10^5}{2.54}$$

$$= 0.066 \times \frac{10^5}{2.54\sqrt{f}}$$

Now substitute the frequencies into the expression for δ.

$$(a)\, f = 60\,\text{Hz}, \quad \delta = 336\,\text{mils}$$
$$(b)\, f = 10\,\text{kHz}, \quad \delta = 26\,\text{mils}$$
$$(c)\, f = 1\,\text{MHz}, \quad \delta = 2.6\,\text{mils}$$

At 60 Hz, the skin depth of copper is 336 mils (from example 6.1). The largest copper wire is AWG #0000 (larger sizes being stranded). The radius of AWG #0000 is 230 mils, so there can be some appearance of the skin effect phenomenon even at 60 Hz.

Example 6.2

AWG #20 wire has a diameter of 32 mils. Its DC resistance is 10.15 ohms per 1000 feet at 25°C. Determine its resistance per 1000 feet at 25°C at a frequency of 1 MHz.

At this frequency, the skin depth is 2.6 mils (from example 6.1). The radius is 16 mils. Thus, equation 6.5 could be used for finding ρ/δ. However, the effective cross sectional area can be found by subtracting the area of a circle with a radius one skin depth less than the wire radius from the area of the wire.

$$A_{\text{effective}} = \pi r^2 - \pi(r - \delta)^2 \qquad 6.6$$

As the resistivity remains the same, the ratio of the resistance is inversely proportional to the area.

$$\frac{R_{1\,\text{MHz}}}{R_O} = \frac{A}{A_{\text{eff}}} = \frac{\pi r^2}{\pi r^2 \left[1 - \left(1 - \frac{\delta}{r}\right)^2\right]} \qquad 6.7$$

$$R_{1\,\text{Mhz}} = \frac{R_O}{1 - \left(1 - \frac{\delta}{r}\right)^2} = \frac{10.15}{1 - \left(1 - \frac{2.6}{16}\right)^2} = 34.0\,\Omega$$

For coaxial cables at high frequencies, the resistance and internal inductance of the line can be found from equation 6.5. The width of the conductor is taken as 2π times the radius for the inner and outer conductors. The thickness is taken as the skin depth.

Calculation of impedance of round wires at intermediate frequencies is a complicated process involving Bessel function tables. It is usually sufficient to use normalized curves such as those shown in figure 6.1.

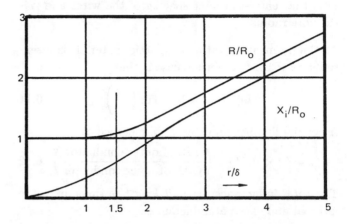

Figure 6.1 Normalized Resistance and Internal Reactance for Circular Solid Wires

Over the range of $r/\delta > 2$, the resistance is approximated by

$$\frac{R}{R_0} \approx 0.25 + 0.5\frac{r}{\delta} \qquad 6.8$$

The inductive reactance is approximated by equation 6.9 for $r/\delta > 2.5$.

$$\frac{X_i}{R_0} \approx 0.5025\frac{r}{\delta} \qquad 6.9$$

Example 6.3

Find the internal impedance per 1000 feet for AWG #1 solid copper wire operated at 60 hz. The DC resistance is 0.124 ohms per 1000 feet at 25°C. The wire diameter is 289 mils.

From example 6.1, $\delta = 336$ mils. Then,

$$\frac{r}{\delta} = \frac{289}{(2)(336)} = 0.43$$

From figure 6.1,

$$\frac{R}{R_0} = 1$$
$$\frac{X_i}{R_0} = 0.12$$

Therefore,

$$Z_L = R_0(1 + j0.12) = 0.124 + j0.0149 \,\frac{\text{ohms}}{1000\,\text{ft}}$$

3   EXTERNAL INDUCTANCE

A pair of parallel solid round wires carrying the same current in opposite directions has an external inductance in henrys per meter of approximately

$$L_{l,e} = \frac{\mu}{\pi} \ln\left(\frac{D}{r}\right) \qquad 6.10$$

D is the center-to-center spacing of the wires and r is the wire radius.

As the permeability of nonmagnetic materials is essentially $4\pi \times 10^{-7}$, the usual case is that

$$L_{l,e} = (4 \times 10^{-7}) \ln\left(\frac{D}{r}\right) \qquad 6.11$$

A coaxial line has an external inductance of

$$L_{l,e} = (2 \times 10^{-7}) \ln\left(\frac{\text{I.D. of outer conductor}}{\text{O.D. of inner conductor}}\right) \qquad 6.12$$

The *total inductance per unit length* is the sum of the internal and external inductances.

$$L_l = L_i + L_e \qquad 6.13$$

Example 6.4

Determine the total inductive reactance per meter at 10 MHz for a coaxial line (copper, $\rho = 1.8 \times 10^{-8}$ ohm-m) with an inner conductor diameter of 0.2 cm and an outer conductor inner radius of 1 cm.

The internal impedance of the line is $0.148(1 + j)$ ohm, (as determined in warm-up problem #1), so the internal reactance is the imaginary part of 0.148 ohm/m. From equation 6.12, the external inductance is

$$L_{l,e} = (2 \times 10^{-7}) \ln\left(\frac{2}{0.2}\right) = 4.605 \times 10^{-7} \frac{\text{H}}{\text{m}}$$

$$X_e = 2\pi f L_e = 28.935 \frac{\text{ohms}}{\text{m}}$$

The total inductive reactance per unit length is given by equation 6.13.

$$X_L = 0.148 + 28.935 = 29.08 \frac{\text{ohms}}{\text{m}}$$

4   CAPACITANCE

The capacitance in farads/meter of a pair of parallel solid round wires is approximately

$$C_l = \frac{2\pi\epsilon}{\ln\left[\dfrac{(D - r_1)(D - r_2)}{r_1 r_2}\right]} \qquad 6.14$$

D is the center-to-center spacing of the wires. r_1 and r_2 are the radii of the two wires. The permittivity is composed of the permittivity of free space ($\epsilon_o = 1/(36\pi \times 10^9)$) and the permittivity ratio ($\epsilon_r = 1$ for air).

$$\epsilon = \epsilon_r \epsilon_o \qquad 6.15$$

For a coaxial line, the capacitance in farads/meter is

$$C_l = \frac{2\pi\epsilon}{\ln\left(\dfrac{\text{I.D. of outer conductor}}{\text{O.D. of inner conductor}}\right)} \qquad 6.16$$

Example 6.5

Determine the capacitive reactance at 1 MHz per 1000 feet for a pair of AWG #20 wires separated by 1 cm (center-to-center).

AWG #20 wire has a diameter of 32 mils (a radius of 0.04064 cm). For air, $\epsilon = \epsilon_o = 1/(36\pi \times 10^9)$.

$$2\pi\epsilon = \frac{10^{-9}}{18}$$

$$C_l = \frac{10^{-9}}{18 \ln \dfrac{(1 - 0.04)^2}{0.04^2}} = 8.740 \times 10^{-12} \text{F/m}$$

Converting farads per meter to farads per 1000 feet,

$$C = 2.664 \times 10^{-9} \frac{\text{F}}{1000\,\text{ft}}$$

Then,

$$X_C = \frac{1}{2\pi f C} = 59.74 \frac{\text{ohms}}{1000\,\text{ft}}$$

5   POWER TRANSMISSION LINES

Power transmission lines differ from the lines previously discussed in several ways. They are typically made of aluminum strands with a core of steel or aluminum alloy wires for higher tensile strength. Also, they occur in sets of three or four lines for three-phase power transmission.

Calculating the line parameters of resistance, inductance, and capacitance is complicated and unnecessary in most cases since these parameters are tabled for standard conductors. Table 6.2 gives the electrical properties of the common standard conductors which have code names for easier identification. The inductive and capacitive reactances are per-conductor values based on a 1 foot center-to-center spacing. For a single-phase, two-wire system with 1 foot spacing, the inductive and capacitive reactances per mile would be doubled.

Table 6.2
Electrical Characteristics of the Multilayer, Steel-Reinforced Bare Aluminum Conductors

CODE NAME	aluminum area in (c. mils)	strands (Al/St)	O.D. (in)	R_{DC} 20°C (mΩ/ft)	R_{AC}, 60 Hz 20°C (Ω/mi)	R_{AC}, 60 Hz 50°C (Ω/mi)	GMR (ft)	1-ft spacing reactance per conductor at 60 Hz X_L (Ω/mi)	1-ft spacing reactance per conductor at 60 Hz X_C (MΩ-mi)
Waxwing	266,800	18/1	0.609	0.0646	0.3488	0.3831	0.0198	0.476	0.1090
Partridge	266,800	26/7	0.642	0.0640	0.3452	0.3792	0.0217	0.465	0.1074
Ostrich	300,000	26/7	0.680	0.0569	0.3070	0.3372	0.0229	0.458	0.1057
Merlin	336,400	18/1	0.684	0.0512	0.2767	0.3037	0.0222	0.462	0.1055
Linnet	336,400	26/7	0.721	0.0507	0.2737	0.3006	0.0243	0.451	0.1040
Oriole	336,400	30/7	0.741	0.0504	0.2719	0.2987	0.0255	0.445	0.1032
Chickadee	397,500	18/1	0.743	0.0433	0.2342	0.2572	0.0241	0.452	0.1031
Ibis	397,500	26/7	0.783	0.0430	0.2323	0.2551	0.0264	0.441	0.1015
Pelican	477,000	18/1	0.814	0.0361	0.1957	0.2148	0.0264	0.441	0.1004
Flicker	477,000	24/7	0.846	0.0359	0.1943	0.2134	0.0284	0.432	0.0992
Hawk	477,000	26/7	0.858	0.0357	0.1931	0.2120	0.0289	0.430	0.0988
Hen	477,000	30/7	0.883	0.0355	0.1919	0.2107	0.0304	0.424	0.0980
Osprey	556,500	18/1	0.879	0.0309	0.1679	0.1843	0.0284	0.432	0.0981
Parakeet	556,500	24/7	0.914	0.0308	0.1669	0.1832	0.0306	0.423	0.0969
Dove	556,500	26/7	0.927	0.0307	0.1663	0.1826	0.0314	0.420	0.0965
Rook	636,000	24/7	0.977	0.0269	0.1461	0.1603	0.0327	0.415	0.0950
Grosbeak	636,000	26/7	0.990	0.0268	0.1454	0.1596	0.0335	0.412	0.0946
Drake	795,000	26/7	1.108	0.0215	0.1172	0.1284	0.0373	0.399	0.0912
Tern	795,000	45/7	1.063	0.0217	0.1188	0.1302	0.0352	0.406	0.0925
Rail	954,000	45/7	1.165	0.0181	0.0977	0.1092	0.0386	0.395	0.0897
Cardinal	954,000	54/7	1.196	0.0180	0.0988	0.1082	0.0402	0.390	0.0890
Ortolan	1,033,500	45/7	1.213	0.0167	0.0924	0.1011	0.0402	0.390	0.0885
Bluejay	1,113,000	45/7	1.259	0.0155	0.0861	0.0941	0.0415	0.386	0.0874
Finch	1,113,000	54/19	1.293	0.0155	0.0856	0.0937	0.0436	0.380	0.0866
Bittern	1,272,000	45/7	1.382	0.0136	0.0762	0.0832	0.0444	0.378	0.0855
Bobolink	1,431,000	45/7	1.427	0.0121	0.0684	0.0746	0.0470	0.371	0.0837
Plover	1,431,000	54/19	1.465	0.0120	0.0673	0.0735	0.0494	0.365	0.0829
Lapwing	1,590,000	45/7	1.502	0.0109	0.0623	0.0678	0.0498	0.364	0.0822
Falcon	1,590,000	54/19	1.545	0.0108	0.0612	0.0667	0.0523	0.358	0.0814
Bluebird	2,156,000	84/19	1.762	0.0080	0.0476	0.0515	0.0586	0.344	0.0776

Used with permission from Aluminum Electrical Conductor Handbook, by the Aluminum Association, 1971.

The GMR is the *geometric mean radius* of the conductor. It takes into account the layers and the number of individual conductors per layer. The GMR is the radius adjusted for the internal inductance of the conductor. Therefore, the entire inductance (internal and external) can be found from one calculation.

For a solid round wire (when skin effect is not important), the GMR is the actual radius multiplied by $e^{-0.25}$. For example, Bluebird conductor in table 6.2 has an outside diameter of 1.762 inches, so its radius is 0.881 inches. If it were a solid wire, the GMR would have been $0.881 \times e^{-0.25} = 0.6861$ inches or 0.0572 feet. This is close to the tabulated GMR 0.0586 feet.

The inductive reactance in ohms per mile is given by equation 6.17, where D is the line-to-line spacing.

$$X_L = (2.022 \times 10^{-3})f \times \ln\left(\frac{D}{\text{GMR}}\right) \qquad 6.17$$

Equation 6.17 can be broken into two parts.

$$X_L = (2.022 \times 10^{-3})f \left[\ln\left(\frac{1}{\text{GMR}}\right) + \ln(D) \right] \quad 6.18$$

A multiplicative correction factor to X_L of table 6.2 is needed for line-to-line spacings of other than one foot.

$$K_L = 1 + \frac{\ln(D)}{\ln\left(\dfrac{1}{\text{GMR}}\right)} \qquad 6.19$$

Both D and GMR must be in feet to use equations 6.17 through 6.19.

Example 6.6

Determine the inductive reactance of Bluebird conductor when the line-to-line spacing is 20 feet.

The correction factor is

$$K_L = 1 + \frac{\ln 20}{\ln\left(\dfrac{1}{0.0586}\right)} = 2.056$$

The one foot inductive reactance from table 6.2 is 0.344. Therefore,

$$X_L = 2.056 \times 0.344 = 0.707 \,\Omega/\text{mile}$$

The capacitive reactance times miles per conductor in air is approximately

$$X_C = \frac{1.781 \times 10^6}{f} \ln\left(\frac{D}{r}\right) \qquad 6.20$$

X_C from table 6.2 must be corrected for values other than $D = 1$ by the multiplicative factor K_C.

$$K_C = 1 + \frac{\ln D}{\ln\left(\dfrac{1}{r}\right)} \qquad 6.21$$

r is the actual conductor radius, not the GMR. D and r both must be in feet.

The value of X_C from table 6.2 can be used with K_C to find the actual capacitive reactance

$$X_C = X_{C,\text{table}} \times K_C \qquad 6.22$$

Example 6.7

Find the total shunt reactance per mile of a Bluebird conductor pair with 15 feet center-to-center separation.

The table value is 0.0776 $M\Omega$-mi. The correction factor is

$$K_C = 1 + \frac{\ln 15}{\ln\left(\dfrac{24}{1.762}\right)} = 2.0369$$

The reactance per line is

$$X_C = (2.0369)(0.0776) = (0.1581) \, M\Omega\text{-mi}$$

With multiple conductors, the reactances add.

$$X_{\text{total}} = (2)(0.1581) = 0.3162 \, M\Omega\text{-mi}$$

6   THREE-PHASE TRANSMISSION LINES

For three-wire, three-phase transmission lines, the sum of the currents is instantaneously zero, as is the sum of the line-to-line voltages. This permits certain simplifications.

A. Inductance

Inductance is determined from geometry, considering the flux linking each conductor. The flux linking two conductors a and b separated by a distance r_{ab} is

$$\psi_{ab} = \frac{\mu_o I_b}{2\pi} \int_{r_{ab}}^{\infty} \frac{dr}{r} \qquad 6.23$$

Equation 6.23 is an infinite integral, and a limiting process can be used. Rather than integrating to infinity, integrate to a large value, say A.

$$\psi_{ab} = \frac{\mu_o I_b}{2\pi} \int_{r_{ab}}^{A} \frac{dr}{r} = \frac{\mu_o I_b}{2\pi} \ln\left(\frac{A}{r_{ab}}\right) \qquad 6.24$$

For a three-wire system such as is shown in figure 6.2, the flux linking conductor a is

$$\psi_a = \frac{\mu_o}{2\pi} \left[I_a \ln\left(\frac{A}{\text{GMR}}\right) + I_b \ln\left(\frac{A}{r_{ab}}\right) \right.$$
$$\left. + I_c \ln\left(\frac{A}{r_{ac}}\right) \right] \qquad 6.25$$

Since $\ln(a/b) = \ln a - \ln b$, equation 6.25 can be rearranged.

$$\psi_a = \frac{\mu_o}{2\pi} \left[(I_a + I_b + I_c)\ln(A) - I_a \ln(\text{GMR}) \right.$$
$$\left. - I_b \ln(r_{ab}) - I_c \ln(r_{ac}) \right] \qquad 6.26$$

Since $I_a + I_b + I_c = 0$, the first term of equation 6.26 vanishes as long as the radius A is large. Then, the distance to all the conductors is approximately the same. Then,

$$\psi_a = \frac{\mu_o}{2\pi}\left[-I_a \ln(\text{GMR}) - I_b \ln(r_{ab}) - I_c \ln(r_{ac})\right] \quad 6.27$$

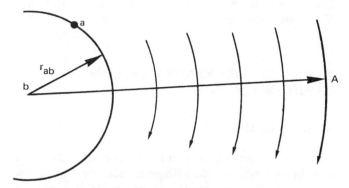

Figure 6.2 Flux Linking Conductor a Due to Conductor b, Limited to a Maximum Radius A

The *geometric mean* of the three distances separating the conductors is

$$d_s = \sqrt[3]{r_{ab}r_{bc}r_{ca}} \quad 6.28$$

Furthermore,

$$\ln(r_{ab}) = \ln\frac{r_{ab}r_{bc}r_{ca}}{r_{bc}r_{ca}} = \ln\frac{d_s^3}{r_{bc}r_{ca}}$$

$$= \ln\frac{d_s^2}{r_{bc}r_{ca}} + \ln d_s \quad 6.29$$

Equation 6.27 can be rearranged,

$$\psi_a = \frac{\mu_o}{2\pi}\left[-I_a \ln(\text{GMR}) - I_b \ln(d_s) - I_c \ln(d_s)\right.$$

$$\left. - I_b \ln\left(\frac{d_s^2}{r_{bc}r_{ca}}\right) - I_c \ln\left(\frac{d_s^2}{r_{bc}r_{ab}}\right)\right] \quad 6.30$$

Also, $I_b + I_c = -I_a$. So the first three terms of the right-hand side of equation 6.30 can be combined.

$$\psi_a = \left[\frac{\mu_o}{2\pi}\ln\left(\frac{d_s}{\text{GMR}}\right)\right]I_a + \left[\frac{\mu_o}{2\pi}\ln\left(\frac{r_{ac}r_{bc}}{d_s^2}\right)\right]I_b$$

$$+ \left[\frac{\mu_o}{2\pi}\ln\left(\frac{r_{ab}r_{bc}}{d_s^2}\right)\right]I_c \quad 6.31$$

Equation 6.31 is the sum of the self and mutual inductive terms.

$$\psi_a = L_a I_a + M_{ab}I_b + M_{ac}I_c \quad 6.32$$

When the three wires are identical (as they typically are), the three self inductances are the same.

$$L = \frac{\mu_o}{2\pi}\ln\left(\frac{d_s}{\text{GMR}}\right)$$

$$= (2 \times 10^{-7})\ln\left(\frac{d_s}{\text{GMR}}\right) \quad 6.33$$

The mutual inductances are calculated from equation 6.34.

$$M = \frac{\mu_o}{2\pi}\ln\left(\frac{r_{bc}r_{ac}}{d_s^2}\right)$$

$$= (2 \times 10^{-7})\ln\left(\frac{r_{bc}r_{ac}}{d_s^2}\right) \quad 6.34$$

When the three conductors are arranged as an equilateral triangle, the geometric mean distance is the same as each of the sides, and $\ln(1) = 0$. Then, the mutual inductance terms vanish. The self-inductance found in equation 6.33 is called *phase inductance*. It is the same as the per-conductor value given in table 6.2. The reactances are not doubled as was done in the two-conductor, single-phase system.

Example 6.8

A three-conductor transmission line has Waxwing conductors arranged in a plane, with each of the conductors separated by 1 foot. Find the self and mutual inductances per meter of conductor.

$$d_s = \sqrt[3]{1 \times 1 \times 2} = \sqrt[3]{2}$$

From table 6.2, GMR = 0.0198.

$$L = (2 \times 10^{-7})\ln\left(\frac{\sqrt[3]{2}}{0.0198}\right) = 8.306 \times 10^{-7}\,\text{H/m}$$

$$M_{12} = (2 \times 10^{-7})\ln\left(\frac{1 \times 2}{2^{2/3}}\right) = 4.62 \times 10^{-8}\,\text{H/m}$$

$$M_{13} = (2 \times 10^{-7})\ln\left(\frac{1 \times 1}{2^{2/3}}\right) = -9.24 \times 10^{-8}\,\text{H/m}$$

$$M_{23} = M_{12}$$

B. Capacitance

The *capacitance* of a three-phase, three-conductor system is also determined from the geometry. For two conductors with line-charges q_a and q_b coulombs per meter, the voltage drop from line a to line b is determined by the line integral of equation 6.35.

$$V_{ab} = \frac{q_a}{2\pi\epsilon} \int_{r_a}^{r_{ab}} \frac{dr}{r} + \frac{q_b}{2\pi\epsilon} \int_{r_{ab}}^{r_a} \frac{dr}{r} \qquad 6.35$$

r_{ab} is the center-to-center separation of line a from line b. The integration of equation 6.35 is

$$V_{ab} = \frac{q_a}{2\pi\epsilon} \ln\left(\frac{r_{ab}}{r_a}\right) + \frac{q_b}{2\pi\epsilon} \ln\left(\frac{r_a}{r_{ab}}\right) \qquad 6.36$$

If a third line with different distances is included, it will contribute a drop. The total voltage drop from line a to line b becomes

$$V_{ab} = \frac{q_a}{2\pi\epsilon} \ln\left(\frac{r_{ab}}{r_a}\right) + \frac{q_b}{2\pi\epsilon} \ln\left(\frac{r_a}{r_{ab}}\right)$$
$$+ \frac{q_c}{2\pi\epsilon} \ln\left(\frac{r_{cb}}{r_{ac}}\right) \qquad 6.37$$

Consideration of unequal spacing is complicated and lengthy. In most cases, the approximation of equal spacing is adequate. Thus, the last term of equation 6.37 will vanish because $\ln(1) = 0$.

The voltage drop from line a to line c is

$$V_{ac} = \frac{q_a}{2\pi\epsilon} \ln\left(\frac{d}{r}\right) + \frac{q_c}{2\pi\epsilon} \ln\left(\frac{r}{d}\right) \qquad 6.38$$

r is now the radius of each of the three identical conductors, and d is the separation between each pair. Equation 6.36 can be rewritten in the same form as equation 6.38.

$$V_{ab} = \frac{q_a}{2\pi\epsilon} \ln\left(\frac{d}{r}\right) + \frac{q_b}{2\pi\epsilon} \ln\left(\frac{r}{d}\right) \qquad 6.39$$

Adding equations 6.38 and 6.39,

$$V_{ab} + V_{ac} = \frac{2q_a}{2\pi\epsilon} \ln\left(\frac{d}{r}\right) + \frac{q_b + q_c}{2\pi\epsilon} \ln\left(\frac{r}{d}\right) \qquad 6.40$$

However, the sum of the charges is approximately zero. (It is exactly zero only if the lines are equilaterally spaced and the ground is an infinite distance away.) Therefore, $q_b + q_c \approx -q_a$.

$$V_{ab} + V_{ac} = \frac{3q_a}{2\pi\epsilon} \ln\left(\frac{d}{r}\right) \qquad 6.41$$

The phasor relations indicate that $V_{ab} + V_{ac} = 3V_{an}$, where V_{an} is the voltage drop to a hypothetical neutral. Therefore, the charge is

$$q_a = \frac{2\pi\epsilon}{\ln\left(\frac{d}{r}\right)} V_{an} = C_{an} V_{an} \qquad 6.42$$

C_{an} is the phase capacitance, which is the per-conductor capacitance listed in table 6.2. The capacitive reactance is $1/2\pi fC$, so with air as the dielectric, $\epsilon = 10^{-9}/36\pi$. The capacitive reactance times miles at 60 Hz is

$$X_C = 29{,}668 \ln\left(\frac{d}{r}\right) \ \Omega\text{-mi} \qquad 6.43$$

The actual capacitive reactance is obtained by dividing X_C by the number of miles of conductor.

7 POWER TRANSMISSION LINE REPRESENTATION

A. Introduction

Three-phase transmission lines are normally operated as balanced as possible. However, the spacing is not equilateral. Transposing the positions of the phases as the transmission line travels across the countryside will tend to balance the inductance and capacitance per phase, but this is seldom done. Nevertheless, calculations which assume equilateral spacing and transposition closely approximate the equilateral cases in the previous sections, as long as the geometric mean distance between pairs d_s from equation 6.28 is used for the effective spacing.

Under these assumptions, it is possible to analyze balanced three-phase systems by considering only one phase and the hypothetical neutral (which contributes no resistance, inductance, or capacitance to the system). Seriously unbalanced lines, however, must be evaluated with *symmetrical components*.

As three-phase systems are analyzed using one phase only, it is necessary to perform a delta-wye transformation to obtain the equivalent impedance-to-neutral when the actual load is a delta configuration.

B. Delta-Wye Transformation

The *delta-wye transformation* for unbalanced loads is given by equation 6.44.

$$Z_{an} = \frac{Z_{ab} Z_{ca}}{Z_{ab} + Z_{bc} + Z_{ca}} \qquad 6.44$$

For a balanced load, all the delta-connected impedances are equal. The equivalent line-to-neutral impedance is one-third of the line-to-line impedance.

$$Z_{\ln} = \frac{Z_{ll}}{3} \qquad 6.45$$

C. Short Transmission Lines

Short transmission lines at 60 Hz are under 50 miles in length. In such lines, the shunt reactance is not important because it is much greater than the usual impedances of the loads (and incomparably larger than the line impedance). In such cases, it is necessary to determine only the resistance and inductance for the line

$$Z_l = \left(\frac{R}{\text{mile}} + \frac{jX_L}{\text{mile}}(\text{corrected}) \right) \times (\text{length in miles})$$

6.46

A circuit model for a short transmission line is shown in figure 6.3. The ABCD parameters for this circuit result in equations 6.47 and 6.48.

$$V_S = V_R + ZI_R \qquad 6.47$$
$$I_S = I_R \qquad 6.48$$

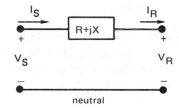

Figure 6.3 One Phase of
a Short Transmission Line

One of the principal concerns in transmission line design is the effect of power factor on the voltage regulation. The *voltage regulation* (or simply *regulation*) in percent is

$$\text{regulation} = \frac{|V_{R,\text{nl}}| - |V_{R,\text{fl}}|}{|V_{R,\text{fl}}|} \times 100 \qquad 6.49$$

$V_{R,\text{nl}}$ is the voltage at the receiving end of the transmission line when the load is disconnected (i.e., no load current is flowing); $V_{R,\text{fl}}$ is the voltage at the receiving end of the transmission line when full line current is flowing.

Example 6.9

A 25 mile Flicker conductor transmission line has 7 feet between the conductors. The full-load current is 225 amps per phase. The line-to-neutral voltage at the sending end is 11 kV per phase, and the power factor of the load is 0.9 lagging. Determine the voltage regulation in percent. Assume $T = 20°C$.

From table 6.2, the resistance per mile is 0.1943 ohm; X_L at 1 foot spacing is 0.432 ohm/mi, and GMR = 0.0284 foot. The correction factor for 7 foot spacing is

$$K_L = 1 + \frac{\ln 7}{\ln \left(\dfrac{1}{0.0284} \right)} = 1.5464$$

$$X_L = 1.5464 \times 0.432 \text{ ohm/mi} = 0.669 \text{ ohm/mi}$$

For 25 miles of line,

$$Z = (0.1943 + j0.669) \times 25 = 4.86 + j16.7$$
$$= 17.4\angle 73.8°$$

With a power factor of 0.9 lagging, the current at the receiving end lags the receiving end voltage by 25.8°. If the receiving end voltage is taken as a reference, then

$$I_R = 225\angle -25.8°$$

The line voltage drop is

$$I_R Z = 225\angle -25.8° \times 17.4\angle 73.8° = 3915\angle 48°$$

From equation 6.47,

$$V_R\angle 0° = V_S\angle \theta - I_R Z = 11\angle \theta - 3.915\angle 48° \text{ kV}$$

θ is the angle of the sending end voltage with respect to the receiving end voltage, V_R. This equation is rewritten in rectangular form.

$$V_R = 11\cos\theta + j11\sin\theta - 2.62 - j2.91$$

Since there is no imaginary part on the left-hand side of this equation, the right-hand side's imaginary part must be zero. Then

$$\sin\theta = \frac{2.91}{11}$$
$$\cos\theta = 0.9644$$
$$V_R = 11 \times 0.9644 - 2.62 = 7.99 \text{ kV}$$
$$= V_{R,\text{fl}}$$

The no-load voltage at the receiving end is the same as the voltage at the sending end.

$$V_{R,\text{nl}} = 11 \text{ kV}$$

Therefore,

$$\text{regulation} = \frac{11 - 7.99}{7.99} \times 100 = 37.7\%$$

D. Medium-Length Transmission Lines

Medium-length transmission lines are generally considered to be from 50 to 150 miles in length at 60 Hz. (It is sometimes permissible to consider a line up to 200 miles long to be a medium-length transmission line.) The capacitive reactance cannot be ignored, as significant amounts of shunt current are lost in the transmission process.

There are two circuit models for medium-length lines; the tee and pi approximations as shown in figure 6.4. The *tee approximation* is made by splitting the series impedance in half with the shunt susceptance in the downward leg.

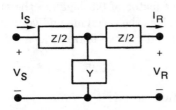

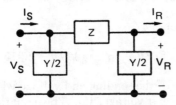

Figure 6.4   One Phase of a Medium-Length Transmission Line: (a) Tee Equivalent Circuit, (b) Pi Equivalent Circuit

$$Y = \frac{-1}{jX_C} = \frac{j}{X_C} = jB_C \qquad 6.50$$

The *pi approximation* is made by splitting the shunt susceptance with the series impedance between the two downward legs. The ABCD parameters for each of these circuit models are given by the following equations.

- tee approximation

$$A = 1 + \frac{YZ}{2} \qquad 6.51$$

$$B = Z\left(1 + \frac{YZ}{4}\right) \qquad 6.52$$

$$C = Y \qquad 6.53$$

$$D = 1 + \frac{YZ}{2} \qquad 6.54$$

- pi approximation

$$A = 1 + \frac{YZ}{2} \qquad 6.55$$

$$B = Z \qquad 6.56$$

$$C = Y\left(1 + \frac{YZ}{4}\right) \qquad 6.57$$

$$D = 1 + \frac{YZ}{2} \qquad 6.58$$

The two approximations are not the same, except when $YZ/4$ is insignificant compared to 1. The pi approximation results in calculations for regulation that are slightly simpler.

$$V_S = \left(1 + \frac{YZ}{2}\right)V_R + ZI_R \qquad 6.59$$

$$I_S = Y\left(1 + \frac{YZ}{4}\right)V_R + \left(1 + \frac{YZ}{2}\right)I_R \qquad 6.60$$

The regulation can be found by using equation 6.49.

Example 6.10

A transmission line delivers 7500 kW at 8500 kVA and 190 kV to a balanced load. The line is 100 miles long, and the conductors are Flicker with a separation of 7 feet. Assume the operating temperature is 50°C. Determine the regulation.

The power factor is $7500/8500 = 0.88$, assumed to be lagging. The phase voltage is $190/\sqrt{3} = 109.7$ kV, and the phase kVA is $8500/3 = 2833$. The phase current is then $(2833/109.7)\angle -28.07°$.

The inductive reactance per mile is the same as in example 6.9.

$$X_L = 0.669 \text{ ohm/mile} \times 100 \text{ miles} = 66.9 \text{ ohms}$$

$$R = 0.2134 \text{ ohm/mile} \times 100 \text{ miles}$$

$$= 21.34 \text{ ohms}$$

$$X_C = 0.0992 \text{ megohm-mile}$$

$$K_C = 1.58$$

From equation 6.21,

$$K_C = 1 + \frac{\ln 7}{\ln\left(\dfrac{24}{0.864}\right)}$$

$$X_C = 1.58 \times (0.0992 \times 10^6)/100 \text{ miles}$$

$$= 1569 \text{ ohms}$$

$$Y = \frac{j}{X_c} = j6.37 \times 10^{-4} \text{ mhos}$$

$$Z = 21.34 + j66.9$$

$$1 + \frac{YZ}{2} = 0.9787 + j0.0136 \approx 0.9787$$

$$IZ = 1.86 \text{ kV}\angle 44.69° = (1.32 + j1.31) \text{ kV}$$

From equation 6.59,

$$V_S = 0.9787 \times 109.7 + 1.32 + j1.31 = 108.7 \text{ kV}$$

At no load with $I_R = 0$,

$$V_{R,\text{nl}} = \frac{108.7}{0.9787} = 111.1 \text{ kV}$$

Since $V_{R,\text{fl}}$ was given as 109.7,

$$\text{regulation} = \frac{111.1 - 109.7}{109.7} \times 100 = 1.29\%$$

E. Long Transmission Lines

Long transmission lines must be evaluated with distributed parameters. Consider the section of transmission line in figure 6.5: z is the impedance per unit length, and y is the shunt admittance per unit length. (Both z and y are complex quantities.) The impedance for the section is, then, $z\,dx$, and the admittance is $y\,dx$.

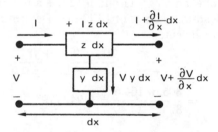

Figure 6.5 Section of a Long Transmission Line

Kirchhoff's voltage law can be written around figure 6.5.

$$V = Iz\,dx + V + \frac{\partial V}{\partial x}dx \qquad 6.61$$

This reduces to

$$\frac{\partial V}{\partial x} = -Iz \qquad 6.62$$

Similarly, Kirchhoff's current law equation can be written for the section.

$$I = Vy\,dx + I + \frac{\partial I}{\partial x}dx \qquad 6.63$$

This reduces to

$$\frac{\partial I}{\partial x} = -Vy \qquad 6.64$$

Differentiating equations 6.62 and 6.64,

$$\frac{\partial^2 V}{\partial x^2} = yzV \qquad 6.65$$

$$\frac{\partial^2 I}{\partial x^2} = yzI \qquad 6.66$$

These differential equations have solutions of the form

$$V = V_1 e^{-\gamma x} + V_2 e^{\gamma x} \qquad 6.67$$

The *propagation constant* is

$$\gamma = \sqrt{yz} \qquad 6.68$$

$$y = jB_C \qquad 6.69$$

$$B_C = \frac{1}{X_C} \qquad 6.70$$

$$z = R + jX_L \qquad 6.71$$

Then,

$$yz = -B_C X_L + jrB_C \qquad 6.72$$

$$\gamma = \alpha + j\beta \qquad 6.73$$

$$|\gamma| = \sqrt{B_C}(R^2 + X_L^2)^{1/4} \qquad 6.74$$

α is the *attenuation constant*. Common units are nepers per mile.

$$\alpha = |\gamma| \cos\left[\frac{1}{2} \arctan\left(\frac{-r}{X_L}\right)\right] \qquad 6.75$$

β is the *phase constant*. Common units are radians per mile.

$$\beta = |\gamma| \sin\left[\frac{1}{2} \arctan\left(\frac{-r}{X_L}\right)\right] \qquad 6.76$$

The voltage along the transmission line is

$$V = V_1 e^{-\alpha x} e^{-j\beta x} + V_2 e^{\alpha x} e^{j\beta x} \qquad 6.77$$

From equation 6.62,

$$I = \sqrt{\frac{y}{z}}V_1 e^{-\alpha x} e^{-j\beta x} - \sqrt{\frac{y}{z}}V_2 e^{\alpha x} e^{j\beta x} \qquad 6.78$$

These equations represent *traveling waves*. The $e^{-\alpha x}e^{-j\beta x}$ terms are waves traveling in the positive x-direction, and terms with $e^{\alpha x}e^{j\beta x}$ are waves traveling in the negative x-direction.

The *characteristic impedance* is

$$z_o = \sqrt{\frac{z}{y}} \qquad 6.79$$

Using the line parameters,

$$\frac{z}{y} = \frac{r + jX_L}{jB_C} = X_L X_C - jX_c r \qquad 6.80$$

$$|z_o| = \sqrt{X_C}(r^2 + X_L^2)^{1/4} \qquad 6.81$$

$$\angle z_o = \frac{1}{2} \arctan\left(\frac{-r}{X_L}\right) \qquad 6.82$$

The relative values of V_1 and V_2 (or I_1 and I_2) are determined from conditions at the sending and receiving ends of the line.

$$I_1 z_o = V_1 \qquad\qquad 6.83$$

$$I_2 z_o = -V_2 \qquad\qquad 6.84$$

At the receiving end,

$$I_R = I_1 + I_2 \qquad\qquad 6.85$$

$$V_R = V_1 + V_2 \qquad\qquad 6.86$$

The origin ($x = 0$) is ordinarily taken at the receiving end of the line. Since $V_R/I_R = Z_L$, equations 6.85 and 6.86 can be combined.

$$V_1 + V_2 = Z_L(I_1 + I_2) \qquad\qquad 6.87$$

Introducing equations 6.83 and 6.84,

$$V_2 = \left(\frac{Z_L - z_o}{Z_L + z_o}\right) V_1 \qquad\qquad 6.88$$

The ratio of V_2 to V_1 is the *reflection coefficient*, ρ.

$$\rho = \frac{Z_L - z_o}{Z_L + z_o} \qquad\qquad 6.89$$

Example 6.11

For Waxwing conductors with 3 foot spacing at 20°C, determine (a) the propogation constant, γ, in polar form, (b) the attenuation constant, α, and (c) the phase constant, β.

From table 6.2, the O.D. = 0.609 inch, so the radius = 0.0254 foot. GMR = 0.0198 ft. $R = 0.3488$ ohm/mi, $X_L = 0.476$ ohm/mi, and $X_C = 0.109$ Mohm-mi.

First, correct the reactance.

$$X_L = \left[1 + \frac{\ln 3}{\ln\left(\dfrac{1}{0.0198}\right)}\right] \times 0.476 = 0.6093 \ \Omega/\text{mi}$$

$$X_C = \left[1 + \frac{\ln 3}{\ln\left(\dfrac{1}{0.0254}\right)}\right] \times 0.109 = 0.1416 \ \text{Mohm-mi}$$

$$B_C = 7.06 \times 10^{-6} \ \text{mhos/mi}$$

$$yz = (j7.06 \times 10^{-6})(0.3488 + j0.6093)$$

$$= 4.96 \times 10^{-6} \angle 150°$$

The propagation constant is

$$\gamma = \sqrt{yz} = 2.23 \times 10^{-3} \angle 75°$$

$$\gamma = 5.72 \times 10^{-4} + j(2.15 \times 10^{-3})$$

The attenuation and phase constants are:

$$\alpha = 5.72 \times 10^{-4} \ \text{nepers/mile}$$

$$\beta = 2.15 \times 10^{-3} \ \text{radians/mile}$$

Example 6.12

Find the characteristic impedance for the line of example 6.11.

$$z_o^2 = \frac{z}{jB_C} = \frac{0.3488 + j0.6093}{j(7.06 \times 10^{-6})}$$

$$= 9.94 \times 10^4 \angle -29.8°$$

$$z_o = 315.3 \angle -14.9°$$

Example 6.13

The line of examples 6.11 and 6.12 has a line-to-neutral (phase) load impedance of 10 ohms. Determine the reflection coefficient.

$$\rho = \frac{Z_L - z_o}{Z_L + z_o} = \frac{10 - 304.6 + j81.1}{10 + 304.6 - j81.1}$$

$$= \frac{305.6 \angle 164.6}{325 \angle -14.40} = 0.94 \angle 179°$$

The ABCD parameters can be found from V_1, V_2, I_1, and I_2. From equations 6.83 through 6.86,

$$V_1 = \frac{V_R + z_o I_R}{2} \qquad\qquad 6.90$$

$$V_2 = \frac{V_R - z_o I_R}{2} \qquad\qquad 6.91$$

Substituting into equations 6.77 and 6.78,

$$V = \left(\frac{V_R + I_R z_o}{2}\right) e^{-\gamma x} + \left(\frac{V_R - z_o I_R}{2}\right) e^{\gamma x} \qquad 6.92$$

$$I = \left(\frac{\dfrac{V_R}{z_o} + I_R}{2}\right) e^{-\gamma x} - \left(\frac{\dfrac{V_R}{z_o} - I_R}{2}\right) e^{\gamma x} \qquad 6.93$$

Terms can be combined to obtain equations 6.94 and 6.95.

$$V = V_R\left(\frac{e^{\gamma x} + e^{-\gamma x}}{2}\right) - I_R z_o\left(\frac{e^{\gamma x} - e^{-\gamma x}}{2}\right) \qquad 6.94$$

$$I = \frac{V_R}{z_o}\left(\frac{-e^{\gamma x} + e^{-\gamma x}}{2}\right) + I_R\left(\frac{e^{\gamma x} + e^{-\gamma x}}{2}\right) \qquad 6.95$$

The hyperbolic functions are defined by equations 6.96 and 6.97.

$$\cosh A = \frac{e^A + e^{-A}}{2} \qquad 6.96$$

$$\sinh A = \frac{e^A - e^{-A}}{2} \qquad 6.97$$

When A is the complex number $a + jb$,

$$\cosh(a + jb) = \cosh a \cos b + j(\sinh a \sin b) \quad 6.98$$

$$\sinh(a + jb) = \sinh a \cos b + j(\cosh a \sin b) \quad 6.99$$

The distance x is defined as being positive to the right, the direction of power flow. Therefore, with V_R and I_R as the dependent variables in equations 6.94 and 6.95, distances toward the sending end are negative, and at the sending end $x = -l$. Thus, equations 6.94 and 6.95 become

$$V_S = (\cosh \gamma l)V_R + (z_o \sinh \gamma l)I_R \qquad 6.100$$

$$I_S = \left(\frac{1}{z_o}\right)(\sinh \gamma l)V_R + (\cosh \gamma l)I_R$$
$$6.101$$

This defines the ABCD parameters.

$$A = \cosh \gamma l \qquad 6.102$$

$$B = z_o \sinh \gamma l \qquad 6.103$$

$$C = \frac{1}{z_o} \sinh \gamma l \qquad 6.104$$

$$D = \cosh \gamma l \qquad 6.105$$

Example 6.14

The line of examples 6.11 through 6.13 has a length of 1000 miles. Determine the sending-end voltage and current, the power loss along the line, and the voltage regulation ($V_R = 110$ kV, $I_R = 200$ amps, and power factor = 0.8).

From example 6.11,

$$\alpha l = 0.572 \text{ neper}$$

$$\beta l = 2.15 \text{ radians} = 123.2°$$

$$\cosh \alpha l = 1.168$$

$$\sinh \alpha l = 0.6037$$

$$\cos \beta l = -0.5474$$

$$\sin \beta l = 0.8369$$

From equation 6.98,

$$\cosh \gamma l = 0.8149\angle 141.6°$$

From equation 6.99,

$$\sinh \gamma l = 1.032\angle 108.7°$$

From example 6.12, $z_o = 315.3\angle -14.9°$, and for PF = 0.8, $I_R = 200\angle -36.9°$. Then, from equation 6.100

$$V_S = 115.5\angle 107.5°$$

The voltage regulation is

$$\frac{\dfrac{115.5}{0.8149} - 110}{110} \times 100 = 28.8\%$$

From equation 6.101,

$$I_S = 516.9\angle 117.7° \text{ amps}$$

Then,

$$S_S = P_S + jQ_S$$
$$= V_S I_S^* = 58{,}747 - j10{,}586 \text{ kVA}$$
$$S_R = 110 \times 200\angle 36.9° = 17{,}600 + j13{,}200 \text{ kVA}$$

Then, the power loss is

$$P = P_S - P_R = 41{,}147 \text{ kW}$$

8 HIGH-FREQUENCY TRANSMISSION LINES

At high frequencies the wavelengths become shorter, so even a few feet of line must be treated as a long line. However, the resistance is negligible, considerably simplifying the calculations. With $R = 0$, the series impedance per unit length is $j\omega L_C$. The simplified equations are:

$$z = j\omega L \qquad 6.106$$

$$y = j\omega C \qquad 6.107$$

$$\gamma = \sqrt{yz} = j\omega\sqrt{LC}$$
$$= j\beta \qquad 6.108$$

$$z_o = \sqrt{\frac{z}{y}} = \sqrt{\frac{L}{C}} \qquad 6.109$$

$$\cosh j\beta x = \cos \beta x \qquad 6.110$$

$$\sinh j\beta x = j\sin \beta x \qquad 6.111$$

At high frequencies, it is desired to compensate a line to minimize reflected waves. This is done by using a capacitive compensation to make the load impedance appear to be the characteristic impedance. The sending voltage and current are found from equations 6.77, 6.90, and 6.107 through 6.111.

$$V_S = V_1(e^{j\beta l} + \rho e^{-j\beta l}) \qquad 6.112$$

$$I_S = \frac{V_1}{z_o}(e^{j\beta l} - \rho e^{-j\beta l}) \qquad 6.113$$

Then,

$$V_S = V_1\left[(1+\rho)\cos\beta l + j(1-\rho)\sin\beta l\right] \qquad 6.114$$

$$I_S = \frac{V_1}{z_o}\left[(1-\rho)\cos\beta l + j(1+\rho)\sin\beta l\right] \qquad 6.115$$

Then, *input impedance* is V_S/I_S, which becomes

$$Z_i = z_o\left[\frac{(1+\rho)\cos\beta l + j(1-\rho)\sin\beta l}{(1-\rho)\cos\beta l + j(1+\rho)\sin\beta l}\right] \qquad 6.116$$

Using the definition of the reflection coefficient, ρ, the input impedance is

$$Z_i = z_o\left(\frac{Z_L\cos\beta l + jz_o\sin\beta l}{z_o\cos\beta l + jZ_L\sin\beta l}\right) \qquad 6.117$$

Example 6.15

A transmission line has a characteristic impedance of $z_o = 50$ ohms. It has a 100 ohm terminating resistance. Find (a) the reflection coefficient, and (b) the input impedance if the line length is an even number of wavelengths ($\beta l = 2\pi n$).

$$\rho = \frac{Z_L - z_o}{Z_L + z_o}$$

$$= \frac{100 - 50}{100 + 50} = 0.333$$

$$\cos\beta l = 1$$

$$\sin\beta l = 0$$

$$Z_{in} = z_o\frac{Z_L}{z_o}$$

$$= Z_L = 100 \text{ ohms}$$

The voltage at any point along the line is a function of the distance from the load because of the two traveling waves. One of the measurable variables along the line is the ratio of the maximum (RMS) voltage to the minimum voltage, known as the voltage standing wave ratio, VSWR. Where the magnitudes of the two waves add,

the total voltage is a maximum. Where they subtract, it is a minimum.

$$\text{VSWR} = \frac{|V_1| + |V_2|}{|V_1| - |V_2|} = \frac{1 + |\rho|}{1 - |\rho|} \qquad 6.118$$

Example 6.16

Find the VSWR for the line of example 6.15.

$$\text{VSWR} = \frac{1 + 0.333}{1 - 0.333} = 2$$

The current along the line will also exhibit a standing wave pattern. The current is maximum where the voltage magnitudes add.

$$I_{max} = \frac{V_1(1 + |\rho|)}{z_o} \qquad 6.119$$

The current is minimum where they subtract.

$$I_{min} = \frac{V_1(1 - |\rho|)}{z_o} \qquad 6.120$$

From equations 6.112 and 6.113, the maximum current occurs where the voltage minimum occurs and the conditions for the minimum current occur when the voltage amplitudes add. Then, the impedance will have a maximum at the point of maximum voltage.

$$Z_{max} = z_o \times \text{VSWR} \qquad 6.121$$

$$Z_{min} = \frac{z_o}{\text{VSWR}} \qquad 6.122$$

9   THE SMITH CHART

The analysis and design of lossless transmission lines is greatly aided by the use of Smith charts.[2] The chart is a polar representation of normalized (with respect to z_o) and constant resistance and reactance curves, as well as the reflection coefficient, ρ. The apparent normalized line impedance is obtained by finding the *normalized load impedance*.

$$z_L = \frac{R_L + jX_L}{z_o} = r + jx \qquad 6.123$$

The z_L point is located on the chart at the intersection of the r and x lines. A circle centered at $1 + j0$ on the chart then is drawn through that value of $r + jx$. This circle is known as the *impedance locus circle*. Next, a radial line is drawn from the center through the z_L point. Finally, the angle βl is determined, and the normalized impedance a distance l from the load can be determined by following the circle from the load point clockwise through an angle $2\beta l$.

[2] P. H. Smith, *An Improved Transmission-Line Calculator*, Electronics, 17, 130, Jan. 1944.

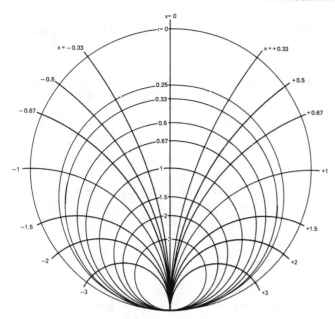

Figure 6.6   Smith Chart

Example 6.17

A transmission line of length $\beta l = 6$ radians (344°) and $z_o = 50$ ohms has a termination load of $75 + j25$ ohms. Determine the input impedance.

step 1: Calculate the normalized load impedance.

$$\frac{R_L + jX_L}{z_o} = \frac{75 + j25}{50} = 1.5 + j0.5$$

step 2: Plot point $z_L = 1.5 + j0.5$ on the Smith chart.

step 3: Draw a circle with center at $1.0 + j0$ (point C) which passes through point z_L.

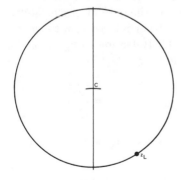

step 4: Extend a line from a point C through point z_L to the edge of the Smith chart, point L (load point).

step 5: Calculate the angular separation between the source and load points.

$$2 \times 344 = 688°$$

Since this is more than one revolution, subtract the full revolution.

$$688 - 360 = 328°$$

Move clockwise 328° from the load to the generator point.

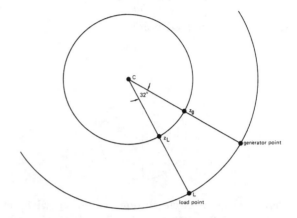

step 6: Read the impedance seen from the generator at the intersection of the generator line and the impedance circle, point z_g.

$$z_g = 1.1 + j0.6$$

step 7: Calculate the input load impedance.

$$z_g = 50(1.1 + j0.6) = 55 + j30$$

The intersection of the $r = 1$ circle and $x = 0$ is the origin of the Smith chart. At that point, the impedance circle locus has zero radius, and the impedance will not change with length of line. This is the *matched condition*, which can be achieved by placing a *compensating reactance* a particular distance away from the load (toward the source).

The compensation technique is to set up the impedance circle locus as usual and then traverse the line toward the source a distance which will put the circle on the $r = 1$ circle. At that point, the impedance will be

$$z_{cp} = 1 + jx \qquad 6.124$$

This can be adjusted to $z = 1$ by adding a series reactance of $-x$. As the most satisfactory compensating reactance is a capacitance, the compensation point should be chosen on the right side ($x > 0$) of the Smith chart.

Example 6.18

A transmission line with $z_o = 50$ ohms is terminated with a load impedance of $100 + j100$ ohms. Find the electrical angle (βl) where a compensating capacitor can be inserted in the line to cause the impedance seen at that point to be 50 ohms. Determine the reactance of the capacitor.

step 1: The normalized load impedance is

$$\frac{100 + j100}{50} = 2 + j2$$

step 2: Plot the load point z_L.

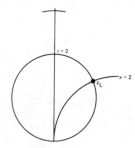

step 3: Draw a circle through point z_L with center at $1.0 + j0$ (point C). This is the impedance locus circle.

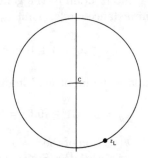

step 4: Determine the compensation point P, as the intersection of the impedance circle and the $r = 1$ circle. The intersection is at $1 + j1.6$.

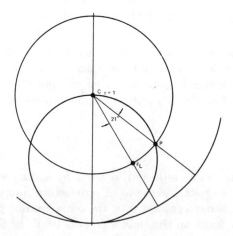

step 5: Extend lines from point C through z_L and P, and measure the angular difference between them as 21°.

step 6: Since the direction toward the source is clockwise, it is necessary to travel $360 - 21 = 339°$ around the circle toward the source. The electrical angle along the line is one-half that amount or 169.5°. The distance from the load to the compensation point is 169.5 electrical degrees or 169.5/360 wavelengths. Therefore, $l = 0.471\lambda$. At this point, the normalized reactance is about 1.6. The compensating reactance is

$$x_c = -1.6 \times 50 = -80 \text{ ohms}$$

The coordinates of the Smith chart are $p\angle\phi$ where p is the magnitude of ρ and ϕ is the angle of ρ. Other information that can be obtained from the Smith chart is $|\rho|$ and the VSWR.

$$\rho = p\angle\phi = u + jv \qquad 6.125$$

On the Smith chart, the loci of r and x values are circles described by equations 6.126 and 6.127.

$$\left(u - \frac{v}{1+r}\right)^2 + v^2 = \frac{1}{(1+r)^2} \qquad 6.126$$

$$(u - 1)^2 + \left(v - \frac{1}{x}\right)^2 = \frac{1}{x^2} \qquad 6.127$$

Equation 6.126 shows that the circle of $r = 0$ has a unit radius $(p = 1)$. Therefore, p is the ratio of the distance from the origin to the point $z = r + jx$. The VSWR is

$$\text{VSWR} = \frac{1 + |\rho|}{1 - |\rho|} = \frac{1 + p}{1 - p} \qquad 6.128$$

The circle of radius p intersects the $x = 0$ line in two places. $(1 + p)$ is the distance from the intersection to the near side of the $r = 0$ circle, and $(1 - p)$ is the distance from either intersection to the opposite side of the $r = 0$ circle. Referring to figure 6.7, the length $\overline{OZ}$ divided by the length $\overline{OA}$ is p.

$$p = \frac{\overline{OZ}}{\overline{OA}} \qquad 6.129$$

The lengths $\overline{AB}$ and $\overline{BC}$ give the VSWR.

$$\text{VSWR} = \frac{\overline{AB}}{\overline{BC}} \qquad 6.130$$

The reciprocal of a complex value can be found from the Smith chart by following a circle 180° through the value to a point diametrically opposed to the original point. This method can be used to determine an admittance from an impedance, and vice versa. Thus, for a coaxial line where a shunt capacitance is more suitable than a series capacitance for compensation, the load point for admittance is on the same P circle as the impedance 180° from the load point.

Example 6.19

The transmission line of example 6.18 is to be compensated with a shunt capacitance. Determine the value and position (in wavelengths) of the compensating capacitance.

The Smith chart is the same as example 6.18, except the load is 180° away ($y = g + jb = 0.25 - j0.25$). The compensation point is where the circle intersects $g = 1$ (also $r = 1$).

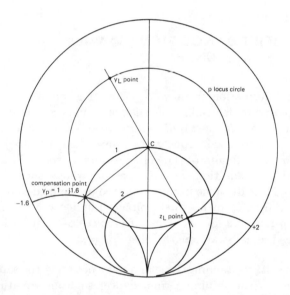

However, the compensation point using a capacitance must be a point where the susceptance b is negative. Therefore, the compensation point is at the intersection of the $g = 1$ circle with an angle of 264°. The admittance at this point is $y = 1 - j1.6$.

The electrical angle along the line is 264/2. Therefore,

$$l = \left(\frac{264}{720}\right)\lambda = 0.367\lambda$$

The susceptance is

$$B = \frac{b}{z_o} = \frac{-1.6}{50}$$

$$X_C = \frac{-50}{1.6} = -31.25 \text{ ohms}$$

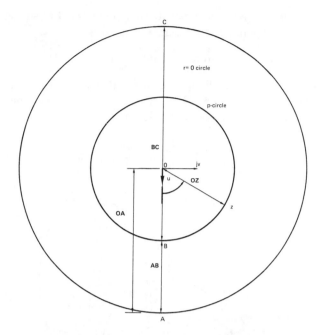

Figure 6.7 Lengths for Determining the Reflection Coefficient and the VSWR

Shorted and open transmission lines with lengths of $\lambda/4$ can be used as shunt compensation. Such a line has essentially zero resistance, so the impedance and admittance are on the $r = 0$ circle. The values of z or b can then be set by the length of the line. For a shorted line, the termination reactance is zero. This is at the top of the Smith chart where $r = 0$ and $x = 0$ coincide. The susceptance at this point is infinite (the line is shorted). On the other hand, an open line has infinite reactance at the load end. This point is at the bottom of the Smith chart (180° from $x = 0$).

In the case of shorted and open lines, it is possible to place the compensation at the first clockwise intersection from the load. This will result in different compensation points for both examples, as inductive compensation (positive reactance or negative susceptance) is available.

Example 6.20

Determine lengths of a 72 ohm line which, when placed (a) in a series, and (b) in parallel, will compensate the 50 ohm line in examples 6.18 and 6.19. (λ is the wavelength of the signal.)

(a) The angle to the first intersection of the $r = 1$ circle is 83°. So, the distance is

$$l = \left(\frac{83}{720}\right)\lambda = 0.115\lambda$$

The normalized impedance at that point is -1.6, so the actual compensating reactance needed is

$$+1.6 \times 50 = +80 \text{ ohms}$$

With a 72 ohm line, the normalized reactance is $80/72 = -1.11$ ohms. This requires a series reactance of $+1.11$ ohms for compensation at that point.

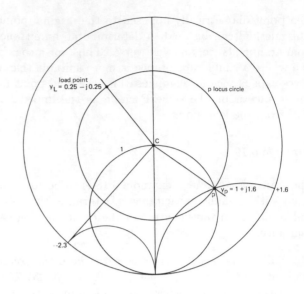

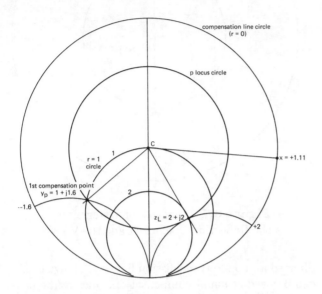

The shortest compensating line is one that will begin at the top of a Smith chart, because length increases in a clockwise direction. For a reactance of 1.11, the angle will be 94° (just below the $x = +1$ intersection with the $r = 0$ circle). So, the compensating line length is

$$l_{\text{cp}} = \left(\frac{94}{720}\right)\lambda = 0.131\lambda$$

This is a shorted length because the termination is at $x = 0$.

(b) From example 6.19, the first clockwise intersection from the load point $Y_L = 0.25 - j0.25$ occurs at $y = 1 + j1.6$. It is again capacitive ($b = 1.6$), and the angle is 160°. Therefore, the length is

$$l = \left(\frac{160}{720}\right)\lambda = 0.222\lambda$$

Since $b = 1.6$, $B = 1.6/50 = 0.032$. Then, in the 72 ohm normalization $b_C = 0.032 \times 72 = -2.3$.

This is a negative value. The shortest line will be at the bottom of the Smith chart at its termination. Since $b = \infty$, this is a shorted line. The point at $g = 0$ and $b = -2.3$ is between the $x = -2$ and $x = -3$ intersection with the $r = 0$ circle ($g = 0$). The angle from the bottom is measured as approximately 45°. Thus,

$$l_{\text{cp}} = \left(\frac{45}{720}\right)\lambda = 0.0625\lambda$$

10   IMPEDANCE FROM VSWR MEASUREMENTS

The information from voltage-standing-waves can be employed to determine the value of an unknown impedance. The *voltage-standing-wave ratio*, VSWR, and the distance from the impedance (toward the generator) of the first minimum of the voltage-standing-wave, L, are needed. In addition, with the load impedance short circuited, the distance from the shorted impedance toward the generator of the first voltage-standing-wave minimum must be found. This latter piece of information gives the half-wavelength, where $\beta l = \pi$.

The voltage-standing-wave minimum occurs at the same point as the minimum impedance, so from equation 6.117,

$$\frac{Z_i}{z_o} = \frac{Z_u/z_o + j \tan \beta L}{1 + j\left(\dfrac{Z_u}{z_o}\right)\tan \beta L}$$

and from equation 6.122,

$$\frac{Z_{\min}}{z_o} = \frac{1}{\text{VSWR}} = \frac{Z_i}{z_o}$$

Between these two expressions, the unknown impedance, Z_u, can be determined in terms of the minimum Z_i ($= Z_{\min}$), z_o, and L.

$$Z_u = z_o \frac{1 - j(\text{VSWR})\tan \beta L}{\text{VSWR} - j \tan \beta L} \qquad 6.131$$

Example 6.21

An unknown impedance is connected to a transmission line with an intrinsic impedance of 50 ohms. With the unknown impedance shorted, the distance from the shorted impedance to the first minimum of the voltage-standing-wave occurs two meters toward the generator.

With the short circuit removed, the VSWR is measured as 2.5, and the first minimum of the voltage-standing-wave is found at a distance $L = 0.75$ meter (toward the generator).

Determine the resistance and reactance of the unknown load.

Since $2\beta = \pi$, and $L = 0.75$, $\beta L = 0.375\pi$ and $\tan(0.375\pi) = +2.414$. Then, from equation 6.131,

$$Z_u = 50\frac{1 - j(2.5)(2.414)}{2.5 - j2.414}$$
$$= 88\angle-36.6° = 70.7 - j52.5$$

Example 6.22

Repeat example 6.21, but with $L = 1.25$ meters.

As $2\beta = \pi$, and $L = 1.25$, $\beta L = 0.625\pi$ and $\tan(0.625\pi) = -2.414$. Then, from equation 6.131,

$$Z_u = 50\frac{1 + j(2.5)(2.414)}{2.5 + j2.414}$$
$$= 88\angle+36.6° = 70.7 + j52.5$$

11 THE ELEMENTAL ANTENNA

High-frequency transmission lines (and wave guides) are used to connect transmitters and receivers to antennas. Transmission lines radiate very little because their conductors are separated by distances that are small compared to the wavelength of the signals. Antennas differ from transmission lines in that their dimensions are comparable to the signal wavelength.

The elemental antenna is a short section of uniform electric current, which is the basis for several definitions to be introduced later.

The development of the relationship between an element of antenna current and the resulting electromagnetic field is based on the notion of retarded magnetic potential, using phasor notation to account for the time variation of the current. The resulting expression for the magnetic field intensity, **H**, in spherical coordinates is

$$H_\phi = \frac{I\,dz\,e^{-j\beta r}}{4\pi}\sin\theta\left(j\frac{\beta}{r} + \frac{1}{r^2}\right)\left(\frac{A}{m}\right) \qquad 6.132$$

β is $\omega\sqrt{\mu_o\epsilon_o}$, ω is the radian frequency of the signal, μ_o is the *permeability* of free space, and ϵ_o the *permittivity* of free space. Referring to figure 6.8, θ is the polar angle, ϕ is the azimuthal angle, dz is on the polar axis, r is the distance from the element $I\,dz$ to the field point, and I is the RMS phasor current. The magnetic field **H** is entirely in the azimuthal direction, i.e., perpendicular to r and parallel to the plane $\theta = 90°$, and so is indicated in equation 6.132 as H_ϕ.

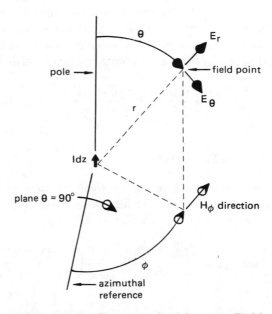

Figure 6.8 Geometry for Magnetic Field at a Point Due to a Current Element for Equations 6.132 through 6.135

The electric field, **E**, is obtained using Maxwell's curl of **H** equation in empty space.

$$\text{curl}\,\mathbf{H} = j\omega\epsilon_o\mathbf{E} \qquad 6.133$$

Carrying out the operation indicated in equation 6.133, components of the electric field intensity in the r-direction and the θ-direction can be obtained.

$$E_r = \frac{I\,dz\,e^{-j\beta r}}{2\pi}\cos\theta\left(\frac{1}{cr^2} + \frac{1}{j\beta r^3}\right)\left(\frac{V}{m}\right) \qquad 6.134$$

$c = 1/\sqrt{\mu_o\epsilon_o}$, the speed of light, and

$$E_\theta = \frac{I\,dz\,e^{-j\beta r}}{4\pi}\sin\theta\left(j\frac{\beta}{r} + \frac{1}{r^2} + \frac{1}{j\beta r^3}\right)\left(\frac{V}{m}\right) \qquad 6.135$$

A. Radiated Power

From Poynting's Theorem, $\mathbf{S} = \mathbf{E} \times \mathbf{H}$, the power density can be obtained. The radiated power is in the r-direction, and includes only $E_\theta H_\phi$. At large distances from the antenna, only the $1/r$ components contribute significant power, so the resulting expression for the radiated power density at the field point is

$$p_r = \frac{(I\,dz)^2 e^{-2j\beta r}}{(4\pi)^2} \sqrt{\frac{\mu_o}{\epsilon_o}} \sin^2\theta \left(j^2 \frac{\beta^2}{r^2}\right) \left(\frac{W}{m^2}\right) \quad 6.136$$

The total average power radiated by the antenna is obtained by integrating equation 6.136 over a sphere of radius r. As the power density is not a function of ϕ, the element of surface area would be $(2\pi r \sin\theta) r\, d\theta$, which is integrated with equation 6.136 from $\theta = 0$ to π. The term $e^{-2j\beta r}$ is merely a time delay, and vanishes when average power is considered, so the resulting integral of average radiated power can be calculated to be

$$P_r = \frac{(I\,dz)^2}{6\pi} \beta^2 \sqrt{\frac{\mu_o}{\epsilon_o}} \left(\frac{W}{m^2}\right) \quad 6.137$$

B. Antenna Directivity

Directivity is a measure of the *focus* of an antenna. The maximum power density for the elemental antenna, from equation 6.136, occurs at $\theta = 90°$.

$$\text{maximum } p_r = \frac{(Idz)^2 e^{-2j\beta r}}{(4\pi)^2} \sqrt{\frac{\mu_o}{\epsilon_o}} \left(j^2 \frac{\beta^2}{r^2}\right) \left(\frac{W}{m^2}\right)$$
$$6.138$$

An *isotropic* antenna radiates power equally in all directions, so that its radiated power is the power density at some radius times the spherical surface area at that radius. For the elemental antenna, the equivalent isotropic power is then the maximum power as given in equation 6.138 multiplied by $4\pi r^2$ (the area of the sphere at that radius). Removing the exponential term (which only contributes time delay),

$$P_{\text{isotropic}} = \frac{(Idz)^2}{4\pi} \beta^2 \sqrt{\frac{\mu_o}{\epsilon_o}} \left(\frac{W}{m^2}\right) \quad 6.139$$

Directivity is defined as the ratio of this isotropic power to the radiated power, i.e., equation 6.139 divided by equation 6.137.

The directivity for the elemental antenna is

$$g_D = \frac{P_{\text{isotropic}}}{P_{\text{rad}}} = 1.5 \quad 6.140$$

In decibels,

$$G_D = 10 \log_{10} g_D$$

So, the elemental antenna has a directivity of 1.761 dB.

C. Radiation Resistance

The *radiation resistance* is defined as the ratio of the total radiated power to the square of the RMS input current.

$$R_{\text{rad}} = \frac{P_r}{I_{\text{in}}^2} \quad 6.141$$

Thus, for the elemental antenna with radiated power given in equation 6.137, the radiation resistance is

$$R_{\text{rad}} = \frac{P_r}{I^2} = \frac{\beta^2 (dz)^2}{6\pi} \sqrt{\frac{\mu_o}{\epsilon_o}} \quad 6.142$$

In MKS units,

$$\sqrt{\frac{\mu_o}{\epsilon_o}} = \sqrt{\frac{4\pi(10^{-7})}{\left[\dfrac{1}{36\pi(10^9)}\right]}} = 120\pi \approx 377$$

D. Ohmic Resistance

The magnetic field calculations above were based on a very thin antenna wire, as the distance from the antenna was large. However, an antenna is a wire of finite diameter, typically copper, operating at a sufficiently high frequency such that the skin depth effect dominates.

For example, consider a meter length of AWG #20 copper wire operating at a frequency of 1 MHz and a temperature of 20°C. From example 6.1, such an antenna has a skin depth of 2.6 mils and a radius of 16 mils. Following example 6.2 (but at 20°C), the DC resistance is 10.15 ohms per 1000 feet, or 0.033 ohm/m. Thus, from equation 6.7 the resistance of one meter of this conductor is

$$R = \frac{0.033}{1 - \left(1 - \dfrac{\delta}{r}\right)^2} = \frac{0.033}{1 - \left(1 - \dfrac{2.6}{16}\right)^2} = 0.111 \text{ ohm}$$
$$6.143$$

The wavelength, λ, for 1 MHz is 300 meters, and $\beta = 2\pi/\lambda = 0.0209$.

E. Antenna Efficiency

Antenna *efficiency* is defined as the ratio of radiated power to input power, which is related to the radiation and ohmic resistances as

$$\text{efficiency} = e = \frac{R_{\text{rad}}}{R_{\text{ohmic}} + R_{\text{rad}}} \times 100\% \quad 6.144$$

For the example of the previous section with $R_{\text{ohmic}} = 0.111$ ohm, the radiation resistance from equation 6.142 is

$$R_{\text{rad}} = \frac{(0.0209)^2(1)^2}{6\pi} \times 120\pi = 0.00877 \text{ ohm}$$

The efficiency of such an elemental antenna is then

$$e = \frac{0.00877}{0.111 + 0.00877} \times 100\% = 7.32\%$$

F. Antenna Gain

The gain of an antenna is the ratio of the product of the radiated power and the directivity of the input power. For the elemental antenna, it is simply the antenna efficiency multiplied by the directivity.

$$g = e \times g_D \qquad 6.145$$

For the example above, the antenna gain is $0.0732 \times 1.5 = 0.11$, or in decibels it is $10\log_{10} 0.11 = -9.6$ dB.

12   SHORT DIPOLE ANTENNA

A. Radiated Power

A *dipole* antenna consists of two equal branches connected to a transmission line, as indicated in figure 6.9. Each branch has a length of h. For the short dipole antenna, the length h is much less than the wavelength of the signal.

Two identical elementary currents $I_z dz$ are indicated in the drawing; they are identical because of the symmetrical positions at $+z$ and $-z$. The current is assumed to be a standing wave in each branch, linearly diminishing to zero at $z = +h$ and $z = -h$. The current is then

$$I_z = \begin{cases} I_{\text{in}}\left(1 - \dfrac{z}{h}\right) & \text{for } 0 < z < +h \\[2mm] I_{\text{in}}\left(1 + \dfrac{z}{h}\right) & \text{for } -h < z < 0 \end{cases} \qquad 6.146$$

I_{in} is the RMS phasor current entering the upper branch of the antenna and leaving the lower branch. Because z ranges over a very small distance, it is possible to assume all the current is concentrated at the origin.

Therefore, equation 6.132 can be applied to determine the magnetic field at the field point as

$$H_\phi = \frac{e^{-j\beta r}}{4\pi} \sin\theta \left(j\frac{\beta}{r} + \frac{1}{r^2}\right) \int_{-h}^{+h} I \, dz \left(\frac{A}{m}\right)$$

but,

$$\int_{-h}^{+h} I \, dz = \int_{-h}^{0} I_{\text{in}}\left(1 + \frac{z}{h}\right) dz + \int_{0}^{+h} I_{\text{in}}\left(1 - \frac{z}{h}\right) dz$$
$$= I_{\text{in}} h$$

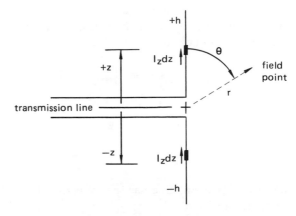

Figure 6.9   Dipole Antenna

The resulting magnetic field is

$$H_\phi = I_{\text{in}} \frac{h e^{-j\beta r}}{4\pi} \sin\theta \left(j\frac{\beta}{r} + \frac{1}{r^2}\right)\left(\frac{A}{m}\right)$$

It can be seen that the only difference between this and the results obtained from the elementary antenna is the replacement of $I dz$ by $h I_{\text{in}}$. Thus, the short dipole antenna power density can be determined from equation 6.136 to be

$$p_r = I_{\text{in}}^2 \frac{h^2 e^{-2j\beta r}}{(4\pi)^2} \sqrt{\frac{\mu_o}{\varepsilon_o}} \sin^2\theta \left(j^2\frac{\beta^2}{r^2}\right)\left(\frac{W}{m^2}\right) \qquad 6.147$$

The total average power radiated by the antenna is obtained by integrating equation 6.147 over a sphere of radius r as before. The average radiated power is calculated to be

$$P_r = I_{\text{in}}^2 \frac{\beta^2 h^2}{6\pi} \sqrt{\frac{\mu_o}{\varepsilon_o}} \left(\frac{W}{m^2}\right) \qquad 6.148$$

B. Antenna Directivity

The maximum power density for the short dipole antenna, from equation 6.147, occurs at $\theta = 90°$.

$$\text{maximum } p_r = I_{\text{in}}^2 \frac{h^2 e^{-2j\beta r}}{(4\pi)^2} \sqrt{\frac{\mu_o}{\varepsilon_o}} \left(j^2\frac{\beta^2}{r^2}\right)\left(\frac{W}{m^2}\right)$$
$$6.149$$

For the short dipole antenna, the equivalent isotropic power is the maximum power as given in equation 6.149 integrated over a sphere of radius r. Multiply equation 6.149 by the area of the sphere $(4\pi r^2)$ and remove the exponential term (which only contributes time delay for the short dipole).

$$P_{\text{isotropic}} = I_{\text{in}}^2 \frac{\beta^2 h^2}{4\pi} \sqrt{\frac{\mu_o}{\varepsilon_o}} \left(\frac{W}{m^2} \right) \qquad 6.150$$

Directivity was defined in equation 6.140 as the ratio of the isotropic power to the radiated power. Directivity for the short dipole antenna can be obtained by dividing equation 6.150 by equation 6.148.

$$g_D = \frac{P_{\text{isotropic}}}{P_{\text{rad}}} = 1.5$$

In decibels,

$$G_D = 10 \log_{10} g_D = 1.76 \text{ dB}$$

So, the short dipole antenna has the same directivity as the elemental antenna.

C. Radiation Resistance

The radiation resistance has been defined as the ratio of the total radiated power to the square of the RMS input current. Thus, for the short dipole antenna with radiated power given in equation 6.148, the radiation resistance is

$$R_{\text{rad}} = \frac{P_r}{I_{\text{in}}^2} = \frac{\beta^2 h^2}{6\pi} \sqrt{\frac{\mu_o}{\varepsilon_o}} \left(\frac{W}{m^2} \right) \qquad 6.151$$

In comparing equations 6.151 and 6.142, the radiation resistance of the elemental antenna is seen to be proportional to $(dz)^2$, while the short dipole radiation resistance is proportional to h^2. The length of the elemental antenna is dz, and the length of the short dipole is $2h$. So, for purposes of calculating the radiation resistance, the effective length of the short dipole is only half the actual length. This is due to the nonuniform distribution of current in the short dipole, which will arise again in the determination of the ohmic resistance.

D. Ohmic Resistance

Where I_z is as given in equation 6.146, the power lost in the ohmic resistance of the short dipole is

$$P_{\text{ohmic}} = \int_{-h}^{+h} I_z^2 R_l \, dz$$

$$= I_z^2 R_l \int_{-h}^{0} \left(1 + \frac{z}{h} \right)^2 dz + I_{\text{in}}^2 R_l \int_{0}^{+h} \left(1 - \frac{z}{h} \right)^2 dz$$

$$= I_{\text{in}}^2 R_l \frac{2}{3} h \qquad 6.152$$

R_l is the resistance per unit length of the antenna wire, including the skin effect. Since $P = I^2 R$, the ohmic resistance is

$$R_{\text{ohmic}} = \frac{P_{\text{ohmic}}}{I_{\text{in}}^2} = \frac{2h}{3} R_l \qquad 6.153$$

For the example of the previous section, with a frequency of 1 MHz and an antenna length of 1 meter ($h = 0.5$ m), the ohmic resistance of equation 6.153 is only one-third of that for the elemental antenna as given in equation 6.143, which is the resistance per unit length of AWG #20 copper wire at 1 MHz and 20°C. Thus, the effective length of the short dipole antenna is one-third the actual length for purposes of calculating the ohmic resistance.

E. Antenna Efficiency

With $\beta = 0.0209$, $h = 0.5$ m, and $\sqrt{\mu_0/\varepsilon_0} = 120\pi$, the radiation resistance using equation 6.151 is 0.00218 ohm. From equation 6.143, $R_l = 0.111$. So, from equation 6.153 the ohmic resistance is 0.111/3. The antenna efficiency from equation 6.144 is

$$\text{short dipole efficiency} = \frac{0.00218}{0.00218 + \dfrac{0.111}{3}} \times 100\%$$

$$= 5.6\%$$

Even though the ohmic resistance of the short dipole is one third that of the elemental antenna, the efficiency is lower because the effective length for radiation resistance is one-half, and it is squared in the calculation.

F. Antenna Gain

The antenna gain is given in equation 6.145. For the example considered here, the antenna gain for the short dipole of 1 meter length and 1 MHz signal frequency is $0.056 \times 1.5 = 0.084$. In decibels it is $10 \log_{10} 0.084 = -10.8$ dB.

13 DIPOLE ANTENNAS

When the length of the dipole antenna is comparable to the wavelength of the signal, a further complication arises; the difference in the distances to the field point from the two elemental currents, $I_{z1} dz$ and $I_{z2} dz$, can no longer be ignored. This is illustrated in figure 6.10(a). However, h is much less than r, so the distances r_1 and r_2 can be reasonably approximated as

$$r_1 = r - z \cos \theta \qquad 6.154a$$
$$r_2 = r + z \cos \theta \qquad 6.154b$$

In addition, the current is assumed to be a standing sinusoidal wave having a maximum RMS value of I_m, as indicated in figure 6.10(b). The currents are then

$$I_z = \begin{cases} I_m \sin \beta(h - z) & \text{for } 0 < z < +h \\ I_m \sin \beta(h + z) & \text{for } -h < z < 0 \end{cases} \quad 6.155$$

The antenna input current is seen to be $I_{in} = I_m \sin \beta h$.

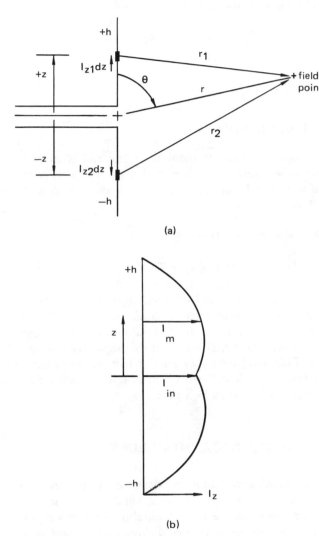

(a)

(b)

Figure 6.10 (a) Dipole Antenna with Length Comparable to Wavelength λ, (b) Assumed Current Standing Wave Distribution

In order to obtain a solution, it is assumed that the angle θ is the same from all points on the antenna to the field point. As r is much greater than h, the only effect of the differences in r_1 and r_2 from r occur in the exponential term of equation 6.132. When that equation, incorporating I_z from equation 6.155 and the distances from equation 6.154, is carried out, and after discarding

all but the radiating portion (l/r dependence),

$$H_\phi = jI_m \frac{e^{-j\beta r}}{2\pi r \sin \theta}$$

$$\times [\cos(\beta h \cos \theta) - \cos \beta h] \left(\frac{A}{m}\right) \quad 6.156$$

As before, the θ-component of electric field involved in radiation is

$$E_\theta = \sqrt{\frac{\mu_o}{\varepsilon_o}} H_\phi = 120\pi H_\phi$$

and the power density of $E_\theta H_\phi = 120\pi H_\phi^2$.

$$p(r, \theta) = -I_m^2 \frac{120\pi e^{-j2\beta r}}{(2\pi r)^2 \sin^2 \theta}$$

$$\times [\cos(\beta h \cos \theta) - \cos \beta h]^2 \left(\frac{W}{m^2}\right) \quad 6.157$$

In order to determine the power, radiation resistance, and the other parameters of interest, it is necessary to integrate equation 6.157 from $\theta = 0$ to $\theta = \pi$. This results in cosine and sine integrals which must be evaluated by computation, or from tables. This is carried out in texts on antenna theory.[3,4]

To simplify the calculations, but still demonstrate the principles, the standard antenna calculations will be made for the half-wavelength antenna or half-wave dipole where $2h = \lambda/2$.

A. Radiated Power

Equation 6.157 is to be integrated over a sphere. Using an elementary volume of $(2\pi r \sin \theta)r d\theta$, the integration is carried out for the half-wave dipole with $2h = \lambda/2$ and $\cos \beta h = 0$. So, with the time-delay term suppressed, the integration becomes

$$P_r = \int_0^\pi I_m^2 \frac{120\pi}{(2\pi r)^2 \sin^2 \theta}$$

$$\times \cos 2 \left[\left(\frac{\pi}{2}\right) \cos \theta\right] 2\pi r^2 \sin \theta d\theta (W)$$

which simplifies to

$$P_r = 60 I_m^2 \int_0^\pi \frac{\cos 2 \left[\left(\frac{\pi}{2}\right) \cos \theta\right]}{\sin \theta} d\theta (W) \quad 6.158$$

Equation 6.158 can be evaluated numerically to obtain

$$P_r = 60 I_m^2 (1.218) = 73.08 \, I_m^2 (W)$$

[3] Kraus, J. D., *Antennas*, 2d ed., McGraw-Hill, 1988, pp. 218 ff.
[4] Stutzman, W. L., and Thiele, G. A., *Antenna Theory and Design*, John Wiley, 1981, pp. 197 ff.

B. Antenna Directivity

From equation 6.157, the maximum power density occurs at $\theta = \pi/2$, which for the half-wave dipole results in a value of

$$p_{max} = \frac{120\pi I_m^2}{(2\pi r)^2}$$

Then, the corresponding isotropic power is found by multiplying this quantity by $4\pi r^2$, to obtain

$$P_{isotropic} = 120 I_m^2$$

The directivity is then

$$g_D = \frac{P_{isotropic}}{P_r} = \frac{120}{73.08} = 1.64$$

In decibels it is $10\log_{10} 1.64 = 2.15$ dB.

C. Radiation Resistance

The radiation resistance is given by

$$R_r = \frac{P_r}{I_m^2} = 73.08$$

D. Ohmic Resistance

The ohmic resistance is found from the $I^2 r$ power loss in the antenna to be

$$P_{ohmic} = R_l \int_{-h}^{h} I_z^2 dz \qquad 6.159$$

I_z is obtained from equation 6.155 to yield

$$P_{ohmic} = R_l \int_{-h}^{0} I_m^2 \sin^2 \beta(h+z) dz$$

$$+ R_l \int_{0}^{h} I_m^2 \sin^2 \beta(h+z) dz \qquad 6.160$$

Then,
$$P_{ohmic} = h R_l I_m^2 = R_{ohmic} I_{in}^2 \qquad 6.161$$

But $I_{in} = I_m \sin\beta h$, so

$$R_{ohmic} = \frac{h R_l}{\sin^2 \beta h} \qquad 6.162$$

But, as $\beta h = \pi/2$ and $I_{in} = I_m$, $R_{ohmic} = h R_l$.

E. Antenna Efficiency

Returning to the example of the previous two sections with a frequency of 1 MHz, the wavelength is 300 meters. So, for a half-wavelength dipole, $h = 75$ m, and with the resistance per unit length of AWG #20 copper

wire 0.111 ohm/meter at 1 MHz, the ohmic resistance is 8.325 ohms. Then with a radiation resistance of 73.08 ohms, the antenna efficiency becomes

$$e = \frac{73.08}{73.08 + 8.325} \times 100\% = 89.8\%$$

F. Antenna Gain

The antenna gain is the directivity multiplied by the fractional antenna efficiency.

$$g = e \times g_D = 0.899 \times 1.64$$
$$= 1.47$$

In decibels, it is 1.67 dB.

G. Input Impedance

The input impedance consists of the radiation resistance, the ohmic resistance, and a reactance. For the half-wavelength antenna, the reactance is 42.5 ohms (inductive)[5].

Then, Z_{in} can be written as

$$Z_{in} = R_{ohmic} + 73.08 + j42.5 \qquad 6.163$$

However, at about one half-wavelength, the antenna reactance is a very sensitive function of the length. By shortening the antenna by a small amount, it is possible to cause the reactive part of the impedance to go to zero. This will have only a small effect on the radiation resistance, which will remain at about 73 ohms. Such an antenna is said to be tuned.

14   COMMUNICATIONS LINKS

A communication link consists of a transmitting antenna, a receiving antenna, and a transmission medium between the two. Assuming the receiving antenna is positioned to receive the maximum power available from the transmitting antenna, the power density at the receiving antenna is

$$p_r = \frac{P_t g_t}{4\pi r^2} \left(\frac{W}{m^2}\right) \qquad 6.164$$

P_t is the input power to the transmitting antenna, g_t is the transmitting antenna's gain, and r is the distance separating the two antennas.

[5] Kraus, J. D., *Antennas, 2d ed.*, McGraw-Hill, 1988, pp. 213–422.

In free space, the power density is the square of the electric field intensity divided by z_o, the intrinsic impedance of space (120π ohms).

$$p_r = E^2 \sqrt{\frac{\epsilon_o}{\mu_o}} = \frac{E^2}{120\pi} \qquad 6.165$$

With a receiving linear antenna oriented in the direction of the electric field intensity, and having length L, the voltage received is $EL = V_r$. The receiving antenna has an input impedance $R_{in} + jX_{in}$, and it feeds a receiver (amplifier) with an impedance $R_a + jX_a$. The current that flows is then

$$I = \frac{V_r}{(R_{in} + R_a) + j(X_{in} + X_a)} \qquad 6.166$$

There are three resistances involved in R_{in} and R_a: R_a itself and $R_{in} = R_{ohmic} + R_{rad}$. Thus, there are three places the received power is distributed.

The power to the amplifier is I^2R, or

$$P_{amp} = \frac{R_a V_r^2}{(R_{in} + R_a)^2 + (X_{in} + X_a)^2} \qquad 6.167$$

Since $V_r^2 = E^2 L^2$, from equation 6.165 $V_r^2 = 120\pi L^2 p_r$. Therefore,

$$P_{amp} = \frac{R_a 120\pi L^2 p_r}{(R_{in} + R_a)^2 + (X_{in} + X_a)^2} \qquad 6.168$$

The power lost in the ohmic resistance of the antenna is

$$P_{loss} = \frac{R_{ohmic} 120\pi L^2 p_r}{(R_{in} + R_a)^2 + (X_{in} + X_a)^2} \qquad 6.169$$

The power lost in the radiation resistance is re-radiated, and is said to be *scattered*.

$$P_{scattered} = \frac{R_{rad} 120\pi L^2 p_r}{(R_{in} + R_a)^2 + (X_{in} + X_a)^2} \qquad 6.170$$

A. Antenna Apertures

From equations 6.168 through 6.170, effective, loss, and scattering *apertures* of the antenna can be defined as

$$A_{eff} = \frac{R_a 120\pi L^2}{(R_{in} + R_a)^2 + (X_{in} + X_a)^2} \qquad 6.171$$

$$A_{scat} = \frac{R_{rad} 120\pi L^2}{(R_{in} + R_a)^2 + (X_{in} + X_a)^2} \qquad 6.172$$

$$A_{loss} = \frac{R_{ohmic} 120\pi L^2}{(R_{in} + R_a)^2 + (X_{in} + X_a)^2} \qquad 6.173$$

Here, L is the effective length of the antenna.

The above development is specific to the linear antennas previously discussed, but the results are quite general. The apertures described in equations 6.171 through 6.173 have their counterparts in all antennas.

In general, the following relationships between apertures, the received power, and the expenditures of that power are valid for any receiving antenna.

$$P_{amplifier} = p_r A_{eff} \qquad 6.174$$

$$P_{scattered} = p_r A_{scat} \qquad 6.175$$

$$P_{loss} = p_r A_{loss} \qquad 6.176$$

$$p_{r-total} = (A_{eff} + A_{scat} + A_{loss})p_r = A_{total}p_r \qquad 6.177$$

It can be seen that the effectiveness of the receiver is improved if the antenna is *tuned* by making its input impedance the complex conjugate of the antenna input impedance. In that way, half the received power is transmitted to the receiver, while the remainder is divided between the ohmic and scattering losses.

For a lossless antenna with conjugate receiver impedance, the maximum amplifier input power can be determined from equation 6.168.

$$P_{amp-m} = \frac{120\pi L^2 p_r}{4R_r} = \sqrt{\frac{\mu_o}{\epsilon_o}} \frac{L^2 p_r}{4R_{rad}} \qquad 6.178$$

L is the effective length of the antenna.

B. Relationship of Directivity and Effective Aperture

Directivity has been defined as the ratio of the isotropic power to the radiated power of an antenna, which is also the ratio of the maximum power density to the average power density. For the short dipole antenna $P_{isotropic}$ is given in equation 6.150, while the relationship of the total radiated power and the radiation resistance is given in equation 6.151. Using these two relations, a new expression for the directivity can be obtained.

$$g_D = \frac{P_{isotropic}}{P_{rad}} = \frac{\beta^2 h^2}{4\pi R_{rad}} \sqrt{\frac{\mu_o}{\epsilon_o}} \qquad 6.179$$

Viewing the antenna as a receiver, with the maximum power at the receiver given in equation 6.178, and recognizing the effective length of the short dipole as $L = h$,

$$A_{eff-m} = \frac{P_{amp-m}}{p_r} = \sqrt{\frac{\mu_o}{\epsilon_o}} \frac{h^2}{4R_{rad}} \qquad 6.180$$

This is substituted into equation 6.179 along with the identity that $\beta = 2\pi/\lambda$, to obtain

$$g_D = \frac{4\pi}{\lambda^2} A_{\text{eff-m}} \qquad 6.181$$

The relationship between directivity and maximum effective aperture has been developed for the case of the short dipole antenna, but it is true for antennas in general.

C. Beam Solid Angle Ω_A

The only antennas treated here have been simple dipoles. There are a wide variety of antennas not mentioned, but having the same parameters as those discussed. One further parameter worth discussing is the *beam angle*, which is a solid angle often defined as the region where the antenna power density is 50% or more of the maximum power density. Alternatively, it is defined as the solid angle which would contain the total radiated power if the power density was uniformly equal to the maximum density.

$$P_r = p_{r-\max} \Omega_A r^2 \qquad 6.182$$

The directivity is the ratio of the solid angle of a sphere to the beam solid angle.

$$g_D = \frac{4\pi}{\Omega_A} \qquad 6.183$$

D. Communication Link Transmission

The power density at the receiving antenna is given in equation 6.164. The power to the receiving amplifier supplied by a lossless transmitting antenna is then

$$P_{\text{amp}} = \frac{g_{DT} A_{R-\text{eff}-m}}{4\pi r^2} P_t \qquad 6.184$$

P_T is the input power to the transmitting antenna, g_{DT} is the directivity of the transmitting antenna, $A_{R-\text{eff}-m}$ is the maximum effective aperture of the receiving antenna, and r is the distance between the antennas.

Using equation 6.181, this can be manipulated to

$$P_{\text{amp}} = \frac{A_{T-\text{eff}-m} A_{R-\text{eff}-m}}{\lambda^2 r^2} P_t \qquad 6.185$$

The above equation is known as the FRIIS Transmission formula. Alternatively,

$$P_{\text{amp}} = \left(\frac{\lambda}{4\pi r}\right)^2 g_{DT} g_{DR} P_t \qquad 6.186$$

When the efficiencies become important, equations 6.184 through 6.186 must be modified by factors of the antenna efficiencies.

E. Communication Link Examples

Example 6.23

A dish-type antenna has a circular beam with an angle of $2n$ degrees between the half-power points. Determine the antenna directivity.

The angle between the center of the beam and a half-power point is n degrees. At an angle of θ degrees, an element of area on the surface of a sphere of radius R can be found to be

$$dA = (2\pi R \sin\theta) R d\theta \qquad 6.187$$

The surface area within the beam half-power angles is then,

$$A = \int_0^n 2\pi R^2 \sin\theta d\theta = 2\pi R^2 (1 - \cos n) \qquad 6.188$$

The solid angle is then defined as the area divided by the radius squared.

$$\Omega_A = \frac{A}{R^2} = 2\pi(1 - \cos n) \qquad 6.189$$

From equation 6.183,

$$g_D = \frac{4\pi}{\Omega_A} = \frac{2}{1 - \cos n} \qquad 6.190$$

Example 6.24

Two dish-type antennas with $n = 1$ degree are used in a communication link with a signal whose wavelength is 10 centimeters. If the transmitter supplies 1 watt to the transmitting antenna, what is the maximum separation if the receiver is to obtain an effective power of 10 microwatts?

First, the antennas have gains of 13,132, so using equation 6.186,

$$10^{-5} = \left(\frac{0.1}{4\pi r}\right)^2 13{,}132 \times 13{,}132 \times 1$$

Therefore, $r = 33.0$ kilometers or 20.5 miles.

APPENDIX A

Copper-Wire Table, Standard Annealed Copper
American Wire Gage

gage no.	diameter mils at 20°C	cross section at 20°C circular mils	ohms per sq. in	ohms per 1,000 ft at 20°C	lb. per 1,000 ft
0000	460.0	211,600.0	0.1622	0.04901	640.5
000	409.6	167,800.0	0.1318	0.06180	507.9
00	364.8	133,100.0	0.1045	0.07793	402.8
0	324.9	105,500.0	0.08289	0.09827	319.5
1	289.3	83,690.0	0.06573	0.1239	253.3
2	257.6	66,370.0	0.05213	0.1563	200.9
3	229.4	52,640.0	0.04134	0.1970	159.3
4	204.3	41,740.0	0.03278	0.2485	126.4
5	181.9	33,100.0	0.02600	0.3133	100.2
6	162.0	26,250.0	0.02062	0.3951	79.46
7	144.3	20,820.0	0.01635	0.4982	63.02
8	128.5	16,510.0	0.01297	0.6282	49.98
9	114.4	13,090.0	0.01028	0.7921	39.63
10	101.9	10,380.0	0.008155	0.9989	31.43
11	90.74	8,234.0	0.006467	1.260	24.92
12	80.81	6,530.0	0.005129	1.588	19.77
13	71.96	5,178.0	0.004067	2.003	15.68
14	64.08	4,107.0	0.003225	2.525	12.43
15	57.07	3,257.0	0.002558	3.184	9.858
16	50.82	2,583.0	0.002028	4.016	7.818
17	45.26	2,048.0	0.001609	5.064	6.200
18	40.30	1,624.0	0.001276	6.385	4.917
19	35.89	1,288.0	0.001012	8.051	3.899
20	31.96	1,022.0	0.0008023	10.15	3.092
21	28.46	810.1	0.0006363	12.80	2.452
22	25.35	642.4	0.0005046	16.14	1.945
23	22.57	509.5	0.0004002	20.36	1.542
24	20.10	404.0	0.0003173	25.67	1.223
25	17.90	320.4	0.0002517	32.37	0.9699
26	15.94	254.1	0.0001996	40.81	0.7692
27	14.20	201.5	0.0001583	51.47	0.6100
28	12.64	159.8	0.0001255	64.90	0.4837
29	11.26	126.7	0.00009953	81.83	0.3836
30	10.03	100.5	0.00007894	103.2	0.3042
31	8.928	79.7	0.00006260	130.1	0.2413
32	7.950	63.21	0.00004964	164.1	0.1913
33	7.080	50.13	0.00003937	206.9	0.1517
34	6.305	39.75	0.00003122	260.9	0.1203
35	5.615	31.52	0.00002476	329.0	0.09542
36	5.000	25.00	0.00001964	414.8	0.07568
37	4.453	19.83	0.00001557	523.1	0.06001
38	3.965	15.72	0.00001235	659.6	0.04759
39	3.531	12.47	0.000009793	831.8	0.03774
40	3.145	9.888	0.000007766	1049.0	0.02993

PROFESSIONAL PUBLICATIONS ● Belmont, CA

APPENDIX B

Approximate Dielectric Constants
(20°C and 1 atmosphere)

material	ϵ_r
acetone	21.3
air	1.00059
alcohol	16–31
amber	2.9
asbestos paper	2.7
asphalt	2.7
bakelite	3.5–10
benzene	2.284
carbon dioxide	1.001
carbon tetrachloride	2.238
castor oil	4.7
diamond	16.5
glass	5–10
glycerine	56.2
hydrogen	1.003
lucite	3.4
marble	8.3
methanol	22
mica	2.5–8
mineral oil	2.24
mylar	2.8–3.5
olive oil	3.11
paper	2.0–2.6
paper (kraft)	3.5
paraffin	1.9–2.5
polyethylene	2.25
polystyrene	2.6
porcelain	5.7–6.8
quartz	5
rock	≈ 5
rubber	2.5–5.0
shellac	2.7–3.7
silicon oil	2.2–2.7
slate	6.6–7.4
sulfur	3.6–4.2
teflon	2.0–2.2
vacuum	1.000
water	80.37
wood	2.5–7.7

PROFESSIONAL PUBLICATIONS ● Belmont, CA

APPENDIX C

The Smith Chart

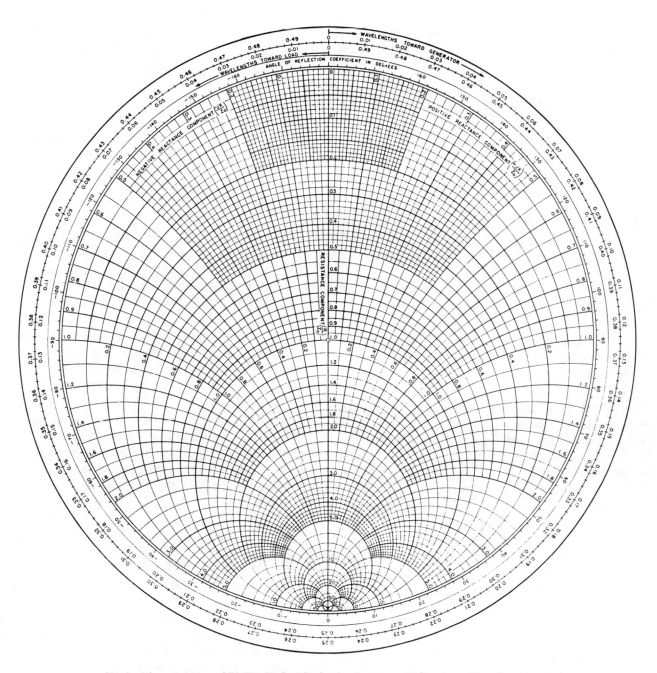

Used with permission of Phillip H. Smith, Analog Instruments Company, New Providence, New Jersey. Permission granted by Anita M. Smith, executrix of Phillip H. Smith, deceased.

PRACTICE PROBLEMS

Warmups

1. A coaxial line with an inner conductor diameter of 0.20 cm and an outer conductor with an inner radius of 1 cm is operated at 10 MHz. Determine the internal impedance per meter of length ($\rho_{cu} = 1.8 \times 10^{-8}$ ohm-meters).

2. Find the internal impedance per 1000 feet of AWG #1 solid copper wire at 10 kHz.

3. Determine the total inductive reactance per 1000 feet for a pair of AWG #20 wires separated by 1 cm (center-to-center) and operating at 1 MHz.

4. Determine the capacitance per meter for a coaxial line with an inner conductor diameter of 0.2 cm and inner diameter of the outer conductor of 2 cm.

5. Determine the inductive reactance of a Bluebird conductor when the line-to-line spacing is 10 feet.

6. Find the capacitive reactance of a 2 mile single-phase Bluebird line operating at 60 Hz with a spacing of 20 feet between centers.

7. Verify that equation 6.43 gives the correct values of per-conductor capacitance for several conductors in table 6.2. (Hint: You must convert the given conductor diameter to the conductor radius in feet. Assume that the spacing is 1 foot.)

8. An unbalanced three-phase load is delta-connected with $R_{ab} = 2$ ohms, $R_{bc} = 2.5$ ohms, and $R_{ca} = 1.8$ ohms. Determine the wye equivalent resistances.

9. In applying capacitance reactance for power factor correction, compare capacitor ratings for wye-connection and delta-connection. (Ratings are in volt-amperes reactive and rated voltage.)

10. Determine the radiation efficiency of a 1 meter length of AWG #20 wire at 20°C, for a signal frequency of 60 Hz.

Concentrates

1. A three-conductor transmission line has the conductors arranged as three of the vertices of a square which is 10 feet on a side. The conductors are Partridge, and the frequency is 60 Hz. Determine the per-phase inductive reactance, X_L, and the mutual reactances between phases X_{ab}, X_{ac}, and X_{bc}. (The order is not specified.)

2. A balanced three-phase power system has line impedances $3 + j4$ and line-to-line load impedances of $20 + j15$ ohms. The load power is 30 kW. (a) Determine the source voltage, power, and volt-amperes. (b) Determine the source voltage, power, and volt-amperes if the load power factor is corrected to 1.

3. A 25 mile length of transmission line consisting of Flicker conductors has spacing of 7 feet between the conductors. The full load current is 225 amps per phase, the voltage at the sending end of the line is 11 kV per phase (line-to-neutral), and the power factor of the load is 0.9 leading. Determine the voltage regulation. Assume ambient $T = 20°C$.

4. For Falcon conductors spaced 20 feet apart at $T = 50°C$, determine (a) the propagation constant, γ, (b) the attenuation constant, α, (c) the phase constant, β, (d) the characteristic impedance, and (e) the reflection coefficient for a load of 100 ohms per phase.

5. A transmission line with a characteristic impedance of 50 ohms and a termination resistance of 70 ohms has $\beta l = \pi/2$. Determine (a) the reflection coefficient, (b) the input impedance, and (c) the VSWR.

6. A transmission line with $z_o = 50$ ohms is terminated with a load impedance of $25 - j25$ ohms. Find the electrical angle (βl) where a compensating capacitor can be inserted in the line to cause the impedance seen at that point to be 50 ohms. Also, determine the reactance of the capacitor.

7. Given a coaxial cable with $z_o = 50$ ohms, determine the length of cable (in wavelengths) necessary to produce (a) a reactance of -25 ohms, and (b) a reactance of $+50$ ohms. (The stubs may be either open or shorted.)

8. An impedance is attached to a 50 ohm transmission line. At a frequency of 100 MHz, the voltage standing wave has its first minimum at a distance of 1 meter towards the generator from the load, and the VSWR = 2. Determine the resistive and reactive parts of this impedance.

9. Determine the efficiency of a half-wave dipole antenna made up of AWG #20 wire and operating at 20°C, for a frequency of 100 MHz.

10. An antenna has a directivity of 20 and an efficiency

of 90%. Determine its gain in decibels, and its beam solid angle in steradians.

Timed

1. A transmission line delivers 7500 kW at 8500 kVA, and 86.6 kV to a balanced load. The line is 100 miles long, and the conductors are Flicker with a separation of 7 feet. Assume the operating temperature is 50°C. Determine the regulation.

2. A transmission line with $z_o = 72$ ohms and $\beta = 0.5$ rad/m is 7.5 meters long. It is terminated with a pure 50 ohm resistance. Determine the impedance seen from the source.

3. A transmission line with $z_o = 100$ ohms is terminated with $Z_L = 25 + j25$ ohms. This is to be compensated with a 72 ohm line section. Determine the length of the compensating line section and specify its placement from the load on the 100 ohm line for (a) series compensation, and (b) parallel compensation.

4. A transmission line with $z_o = 50$ ohms and $Z_L = 25 - j25$ ohms is to be compensated with a shunt capacitor. Specify the capacitance of the capacitor and its placement from the load in wavelengths.

5. A 200 mile transmission line carrying three-phase power at 60 Hz has approximate pi parameters of $Z_{\text{series}} = 40 + j160$ and $Y_{\text{shunt}} = 0.001063j$. The receiving end voltage is 220 kV with a load power of 500 MW at a power factor of 0.9 lagging. Determine the sending end voltage.

6. A 60 Hz, three-phase transmission line is 100 miles long. It delivers 75 MW at 132 kV with a lagging power factor of 0.92. The line constants are 0.1 ohm/mile, 0.6 inductive ohm/mile, and 0.15 capacitive MΩ-mile. Using the pi approximation, determine the sending end voltage.

7. A pair of parabolic reflector antennas are operating at 10 GHz. They have circular beams with a half-power angle of 1.5 degrees. One of these is carried on a synchronous satellite which is in stationary orbit at a height of 40,000 km. What power must the satellite antenna transmit if the earth receiving amplifier is to receive 10^{-9} milliwatts.

8. A half-wave dipole antenna at 50 MHz is transmitting 100 W. At what distance is the maximum field intensity 100 microvolts/m?

9. A pair of horn antennas have physical apertures of 0.2 m^2, and may be assumed to be 100% efficient. Operating at 3 GHz and transmitting 1 watt of power, what separation will result in a received power of 10^{-3} microwatts?

10. A half-wavelength antenna has been shortened slightly in an attempt to tune it. Its efficiency may be assumed to be 100%. The radiation resistance is to be assumed to be 73 ohms. Standing wave measurements are made on a 75 ohm transmission line. With the transmission line short-circuited, the first standing-wave voltage null is found at 3 meters from the short; with the antenna attached (the short circuit removed) the first minimum of the voltage standing wave is found at 1.6 meters from the antenna. Specify the length of transmission line from the antenna to the receiving amplifier and the amplifier impedance required to achieve a match with the complex conjugate impedance as seen from the antenna.

7 ROTATING MACHINES

Nomenclature

a	equivalent turns ratio	–
a_p	number of parallel current paths	–
A_r	area of rotor winding	m²
b	susceptance	mhos
B	magnetic flux density	teslas
E_g	EMF or generated voltage	volts
EMF	electromotive force	volts
f	frequency	hertz
f_v	coefficient of viscous friction	N-m-sec/ rad
g	gap length, or conductance	m, or mhos
I	DC or RMS current	amps
I^*	complex conjugate of current	amps
J	mass moment of inertia	N-m-sec²/ rad
K_T	torque constant	–
K_v	voltage constant	–
K_w	winding constant	–
K_Φ	flux constant	–
l	a length	m
L_a	armature inductance	henrys
MMF	magnetomotive force	amp-turns
n	rotary speed	rpm
N_f	number of shunt field winding turns	–
N_r	number of rotor winding turns	–
N_s	number of series field winding turns	–
p	number of poles	–
P	power	watts
PF	power factor	–
q	number of phases	–
Q	reactive power	volt-amps reactive
r	resistance	ohms
R	resistance, or radius	ohms, or m
s	slip	–
S	apparent power	volt-amps
t	time	sec
T	torque	N-m
V	voltage	volts
x	reactance	ohms
X_s	synchronous reactance	ohms
Y	admittance	mhos
z_c	the number of armature conductors	–
Z	impedance	ohms

Symbols

δ	torque angle	degrees or rad
θ	an angle	degrees or rad
μ_o	$4\pi \times 10^{-7}$	webers/ amp-turn-meter
Φ	magnetic flux	webers
ψ	power factor angle	degrees or rad
Ω	mechanical angular velocity	rad/sec
ω	electrical angular frequency	rad/sec

Subscripts

a	armature
an	from line a to neutral
aa'	from point a to point a'
b.r.	blocked rotor
c	core, or conductors
e	equivalent
f	field
g	gap
in	input
L	load
ll	line to line
ln	line to neutral
m	magnetizing or mechanical

n	normalized, normalizing
o	operating point
p	parallel
r	rotor, or rotational
R	rated
s	stator, shaft, or synchronous
T	torque
v	voltage

1   REVIEW OF THE FUNDAMENTALS OF DC MACHINES

A circuit model of DC motors and generators is shown in figure 7.1.

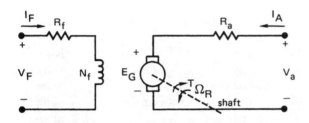

Figure 7.1   DC Motor Circuit Model
(reverse T and I_a for a generator)

The *generated voltage*, E_G, is proportional to the shaft speed, Ω_R, and the field flux, Φ. The proportionality constant, K_v, is approximately equal to the number of series turns between the commutator brushes. (When the speed is given in revolutions per minute, the constant K_v must be multiplied by $30/\pi$.)

$$E_G = K_v \Phi \Omega_R \qquad 7.1$$
$$K_v \approx N_r \qquad 7.2$$

Electro-mechanical energy conversion can be described by electrical and mechanical expressions. Mechanical power is usually given in horsepower, while torque is usually given in foot-pounds. Appendix B gives important conversion factors for energy and power variables.

$$P_{\text{shaft}} = E_G I_A = T\Omega_R \qquad 7.3$$

From equations 7.1 and 7.3, the relationship between armature current and torque is

$$T = K_T \Phi I_A \qquad 7.4$$

In the MKS system, $K_T = K_v \approx N_r$. If mixed units are used, then adjustments have to be made to K_T. If the torque is given in foot-pounds, then

$$(T \text{ in N-m}) \times 0.7376 = T \text{ in ft-lb} \qquad 7.5$$

The relationship between the flux, Φ, and the magneto-motive force, MMF, is frequently given by a *saturation curve* of generated voltage, E_G, and the field current, I_F, for constant shaft speed. The saturation curve is a scaled version of the *magnetization curve*. It is obtained by exciting the field winding with a known current and then measuring the *open circuit armature voltage* (E_G in figure 7.1 with no load current flowing).

In this chapter, the saturation curve is scaled by dividing the generated voltage, E_G, by speed, Ω_R, and normalizing it to some convenient value of E_G/Ω_R. That value of E_G is known as E_N.

The horizontal axis of the magnetization curve is frequently the field current, I_F. This is scaled from the MMF. In the case of only one current contributing to the field, MMF is $N_f I_F$. In this chapter, the horizontal axis is scaled to some convenient value of I_F known as I_N.

The resulting characteristic is a *normalized magnetization curve* for the machine, usually referred to as the *saturation curve* when it is given in terms of the field current and generated voltage at constant speed. The normalized curve is used in this section to demonstrate the basic principles, since it avoids confusion about the changes of speed and the compounding problem presented in the next section. A normalized saturation curve is shown in figure 7.2.

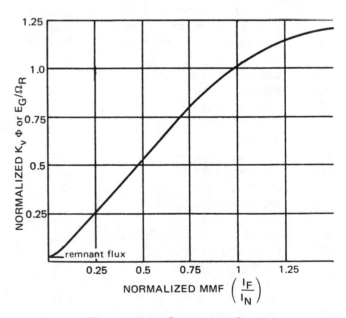

Figure 7.2   Saturation Curve

Actual curves for various load currents are better than the saturation curve for determining the behavior of DC machines. Actual curves show the effects of *armature reaction*, a phenomenon which occurs because the armature current produces a component of field flux which affects the stator poles, thereby reducing the field flux. This effect is compensated for in most machines by the use of *interpoles* in the *quadrature plane* of the field. (The rotor field component is in the quadrature plane, i.e., at 90° to the stator field.) The armature current passes through the windings of the compensating poles to cancel the flux contributed by the armature current. This chapter assumes that compensating interpoles are used so that the saturation curve is adequate.

2 DC GENERATORS

A. Introduction

A voltage applied to the field winding of a DC machine results in current and flux. The flux produces a voltage potential in each rotor (armature) conductor that passes through the flux. The result is summarized by equation 7.1. The field winding can be excited externally (resulting in *separately excited generation*) or by use of the generated voltage itself to supply the field current (resulting in *self-excited generation*).

The self-excited system results in decreased terminal voltage, V_A, as the load current increases. This is due to the armature resistance, R_a. The terminal voltage reduction is particularly detrimental in the self-excited generator because it results in a decrease in field current, causing an even lower generated voltage.

It is sometimes necessary to supply a series field winding. The load (armature) current supplies additional MMF to counter the IR loss in terminal voltage. This technique is called *compounding*, as it uses both shunt and series windings.

B. Separately Excited DC Generators

Separately excited DC generators can be analyzed with equations 7.1 through 7.5. A magnetization curve or mathematical relationship between flux and field current also is required.

Example 7.1

A separately excited DC generator is rated at 10 kW at 240 volts DC while turning at 2800 rpm with an input power of 15 horsepower. It is proposed to operate this machine at 3600 rpm and at an output of 300 volts. Determine the ratings of the machine under the new conditions. Determine the field current necessary if the generated voltage at 2800 rpm is given by

$$E_G = 100I_F - I_F^2$$

The armature resistance is $R_a = 0.5$ ohm.

With a speed increase, the generated voltage will increase proportionally. This assumes a constant flux in equation 7.1.

$$V_R = 240 \times \frac{3600}{2800} = 308 \text{ volts}$$

The current rating will not change since it is set by the wire size. Then, the power rating will increase since power is proportional to voltage.

$$P_R = 10 \text{ kW} \times \frac{3600}{2800} = 12.9 \text{ kW}$$

The input horsepower will increase, but not quite proportionally, since part of the input power must supply $I_A^2 R_a$ losses. The $I_A^2 R_a$ loss is determined from the load current at full load.

$$I_A = \frac{10,000}{240} = 41.67 \text{ amps}$$

The armature resistance loss is

$$I_A^2 R_a = (41.67)^2(0.5) = 868.2 \text{ watts}$$
$$\frac{(868.2)(1.341)}{1000} = 1.16 \text{ hp}$$

Assuming that the remainder of the 15 hp input was due to $T\Omega_R$, the increase in speed will increase this part of the input proportionally.

$$(15 - 1.16) \times \left(\frac{3600}{2800}\right) = 17.79 \text{ hp}$$

This must be added to the $I_A^2 R_a$ loss.

$$\text{hp rating } = 17.79 + 1.16 = 18.95 \text{ hp}$$

At full load with a terminal voltage of 300 volts and a current of 41.67 amps through R_a, the generated voltage must be

$$E_G = 300 + (41.67)(0.5) = 320.8 \text{ volts}$$

I_F is then found by solving the quadratic equation

$$100I_F = \left(\frac{2800}{3600}\right)320.8 + I_F^2$$
$$I_F = 2.56 \text{ amps}$$

C. Self-Excited Generators

Self-excited DC generators use the terminal voltage, V_A, to supply the field current. The circuit model for a self-excited generator is shown in figure 7.3.

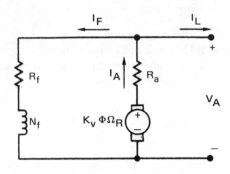

Figure 7.3 Self-Excited DC Generator Circuit Model

Self-excited DC generators behave according to equation 7.6.

$$E_G = K_v \Phi \Omega_R = V_A + R_a I_L + R_a I_F \qquad 7.6$$

To determine I_F, the values of V_A, I_A, and speed must be known. The unknowns in equation 7.6 are then E_G and I_F, as V_A, I_L, and R_a are given. The usual way to solve for two unknowns is to find another equation. However, the only other information available is the saturation curve.

The saturation curve can be used if equation 7.6 is normalized the same way that the magnetization curve is normalized. As equation 7.6 gives voltage, it is normalized to a value of $K_v \Phi \Omega_R / \Omega_R$ taken from the magnetization curve data. This implies that some reference point on the curve corresponds to given ratios of E_G/Ω and MMF/N_f. This reference point is given the normalized values of $E_G/\Omega = 1.0$ and $MMF_n = 1.0$.

If the curve is a saturation curve at a given speed, then this normalization is not necessary, and equation 7.6 can be used directly, provided the curve speed is the same as the problem speed. Otherwise a factor of $\Omega_{R,design}/\Omega_{R,data}$ must be used on the saturation curve's vertical axis.

Then, divide equation 7.6 by the normalizing voltage, E_N.

$$\frac{E_G}{E_N} = \frac{K_v \Phi \Omega_R}{E_N} = \frac{\dfrac{E_G}{\Omega_R}}{\dfrac{E_N}{\Omega_R}}$$

$$= \frac{K_v \Phi}{\dfrac{E_N}{\Omega_R}}$$

$$= \frac{V_A + R_a I_L}{E_N} + R_a \left(\frac{I_N}{E_N}\right) \times \frac{I_F}{I_N} \qquad 7.7$$

The variables of the normalized magnetization curve become E_G/E_N and I_F/I_N. Equation 7.7 is a straight line. The intersection of the straight line with the magnetization curve gives the required value of I_F/I_N from which I_F is obtained. Then, the required value of R_f is V_A/I_F.

$$R_{f,required} = \frac{V_{A,specified}}{I_F} \qquad 7.8$$

Example 7.2

A DC generator rated 10 kW, 15 hp, 260 volts, and 2800 rpm has the normalized magnetization curve shown. The normalization reference point is at $E_G = 260$ volts and $I_F = 2.75$ amps. Determine the value of R_f necessary to self-excite the generator with $V_A = 240$ volts at $I_L = 41.67$ amps if $R_a = 0.48$ ohms.

Use equation 7.7.

$$\frac{E_G}{260} = \frac{240 + 0.48(41.67)}{260} + 0.48 \left(\frac{2.75}{260}\right) \frac{I_F}{I_N}$$

$$= 1 + (5.077 \times 10^{-3}) \frac{I_F}{I_N}$$

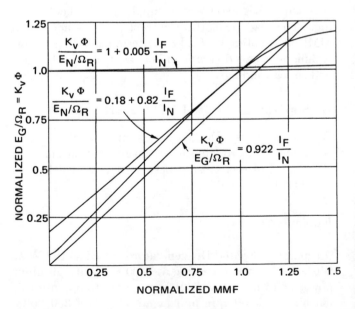

The slope of this curve is small, so it approximately intersects the magnetization curve at the point (1,1). For a more accurate determination of I_F/I_N, take a straight line approximation to the magnetization curve at that point.

$$\frac{E_G}{260} = 0.18 + 0.82 \left(\frac{I_F}{2.75}\right)$$

This intersects the straight line of the previous equation at $I_F = 2.767$ amps. Then,

$$R_f = \frac{240}{2.767} = 86.7 \,\text{ohms}$$

Regulation of the generator is determined from equation 7.6 when the load current is zero.

$$E_G = I_F(R_a + R_f) \qquad 7.9$$

Normalizing this to the magnetization curve coordinates,

$$\frac{E_G}{E_N} = \left(\frac{I_N}{E_N}\right)(R_a + R_f)\left(\frac{I_F}{I_N}\right) \qquad 7.10$$

This is a straight line through the origin. The intersection of this line with the saturation curve will give the no-load value of E_G/E_N.

$$V_A E_G - I_F R_a \qquad 7.11$$

$$V_A = \frac{E_G}{1 + \dfrac{R_a}{R_f}} \qquad 7.12$$

Example 7.3

Find the voltage regulation of the self-excited generator in example 7.2.

Substituting into equation 7.10,

$$\frac{E_G}{260} = \frac{2.75}{260}(0.48 + 86.7)\frac{I_F}{2.75}$$

This intersects the magnetization curve at $E_G/260 = 1.15$ and $I_F = 2.75 \times 1.22$.

Then,

$$V_A = 299 - 0.48 \times 2.75 \times 1.22 = 297.4 \text{ volts}$$

So, the voltage regulation is

$$\frac{297.4 - 240}{240} \times 100 = 23.9\%$$

D. Compound Generators

Compound generators are a solution to the voltage regulation problem which can occur for small armature resistances. The load or armature current is used to provide extra MMF to the field to counter the effects of the armature resistance and reaction. The range of

usefulness of this method is limited. The characteristics of the generator (i.e., the plot of terminal voltage versus load current) will remain nonlinear. This is illustrated in figure 7.4(c).

There are two possible arrangements of the compound winding in the circuit model. Figure 7.4(a) shows the *short shunt connection*, where the main field winding shunts the generator. Only the load current flows through the series compounding winding. Figure 7.4(b) shows the *long-shunt connection*, where the shunt winding comes after the series connection. The compensating winding carries the armature current in this case.

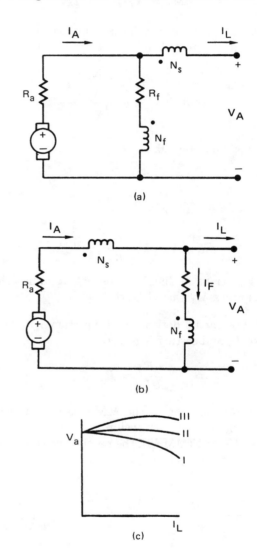

Figure 7.4 Compounded Generators:
(a) Short Shunt, (b) Long Shunt,
(c) the Voltage Current Characteristic:
I Uncompounded, II Flat-Compounded,
III Overcompounded

Compounding is used when loads are sensitive to voltage changes. (Incandescent lights are sensitive in this manner.) Flat compounding can be used when the load is near the generator, and there is negligible loss in the feeder lines. Over-compounding is necessary when there is a long feeder line with significant resistance.

This chapter uses the short-shunt configuration for computational ease. The results for the long-shunt connection are essentially the same.

With two different currents contributing to the MMF, it is necessary to modify the horizontal coordinates of the saturation curve. The MMF for a short shunt is

$$\text{MMF} = N_f I_F + N_s I_L \qquad 7.13$$

The normalization of the horizontal coordinate of the magnetization curve then becomes

$$\text{MMF}_n = \frac{I_F + \left(\dfrac{N_s}{N_f}\right) I_L}{I_N} \qquad 7.14$$

Repeating equation 7.6,

$$E_G = V_A + I_L R_a + I_F R_a \qquad 7.15$$

However, $I_F = V_A / R_f$, so

$$E_G = V_A \left(1 + \frac{R_a}{R_f}\right) + I_L R_a \qquad 7.16$$

If the V_A versus I_L line is to be flat at some value of I_L, it is necessary that $\partial V_A / \partial I_L = 0$.

Taking the partial derivative of equation 7.16 with respect to I_L and setting it to zero,

$$\frac{\partial E_G}{\partial I_L} = R_a \qquad 7.17$$

E_g is also a function of the MMF as shown in figure 7.2. Describing this relationship as an unspecified function $E_G / E_N = \mathbf{F}(\text{MMF}_n)$, the partial derivative of that function with respect to I_L is

$$\frac{\partial(\text{MMF}_n)}{\partial I_L} = \frac{\partial}{\partial I_L} \frac{\left[I_F + \left(\dfrac{N_s}{N_f}\right) I_L\right]}{I_N} = \frac{1}{I_N}\left(\frac{N_s}{N_f}\right)$$

$$7.18$$

$$\begin{aligned}\frac{\partial \mathbf{F}(\text{MMF}_n)}{\partial I_L} &= \frac{\partial \left(\dfrac{E_G}{E_N}\right)}{\partial I_L} \\ &= \frac{\partial \mathbf{F}(\text{MMF}_n)}{\partial(\text{MMF}_n)} \times \frac{\partial(\text{MMF}_n)}{\partial I_L} \\ &= \frac{1}{I_N}\left(\frac{N_s}{N_f}\right) \frac{\partial \mathbf{F}(\text{MMF}_n)}{\partial(\text{MMF}_n)} \qquad 7.19 \end{aligned}$$

However, the partial derivative of $\mathbf{F}(\text{MMF}_n)$ with respect to MMF_n is the slope of the magnetization curve at the operating point (call this point $\mathbf{OPS}$). Then

$$\frac{\partial E_g}{\partial I_L} = \frac{N_s}{N_f} \times \frac{E_N}{I_N} \times \mathbf{OPS} = R_a \qquad 7.20$$

Equation 7.20 requires trial and error because the operating point and I_F are both unknown. However, if the design follows a selection of R_f which gives the proper output voltage at the specified load current, the range of the operating point will be known. The approximate value of I_F will be somewhat less than would be required for self-excited operation. The operating point should be near the no-load voltage for the separately excited case when the field current has been adjusted to give the required voltage at the specified load current. Then, using that approximate value of $\mathbf{OPS}$ in equation 7.20, the ratio of N_s to N_f is

$$\frac{N_s}{N_f} = \frac{R_a \times \dfrac{I_N}{E_N}}{\mathbf{OPS}} \qquad 7.21$$

From this value, it is possible to determine normalized MMF_n from the expression following equation 7.13.

$$I_F = I_N \times \text{MMF}_n - \left(\frac{N_s}{N_f}\right) I_L \qquad 7.22$$

Equation 7.22 can be substituted into equation 7.7.

$$\frac{E_G}{E_N} = \frac{V_A + R_a I_L \left(1 - \dfrac{N_s}{N_f}\right)}{E_N} + R_a \left(\frac{I_N}{E_N}\right) \text{MMF}_n$$

$$7.23$$

Equation 7.23 will determine the necessary operating point. If it is sufficiently close to the operating point chosen, I_F and R_f can be determined. If the operating point is inadequate, use the operating point found as a new starting point. Repeat the process from equation 7.20 onward and iterate until the results are adequate.

Example 7.4

The generator of example 7.2 is to be flat-compounded. The flatness in the characteristic will occur at a load current of 30 amps for a voltage $V_A = 240$ volts. Determine the N_s / N_f ratio and the value of R_f to accomplish this.

Begin by assuming that the field current will be about 2.5 amps. The $I_A R_a$ drop is then

$$32.5 \times 0.48 = 15.6$$

The operating point will be

$$E_G = 240 + 15.6 = 255.6$$

The normalized value of E_G/E_N is $255.6/260 = 0.98$.

From the magnetization curve in example 7.2, the slope at the normalization point is about 0.83. From equation 7.21,

$$\frac{N_s}{N_f} = \frac{2.75 \times 0.48}{260 \times 0.83} = 0.006$$

Substituting into equation 7.23,

$$\frac{E_G}{260} = 0.978 + 0.005 \, \text{MMF}_n$$

This expression gives 0.983 for $\text{MMF}_n \approx 1$, close to the normalized value of $E_G/E_N = 0.98$. (If this hadn't been the case, it would have been necessary to iterate, starting with the value of E_G/E_N, finding the slope and the value of N_s/N_f, and recalculating the operating point.) Thus, from equation 7.22,

$$I_F = 2.75 - 0.006 \times 30 = 2.57 \text{ amps}$$
$$R_f = \frac{240}{2.57} = 93.4 \text{ ohms}$$

E. Advanced DC Generator Fundamentals

The power flow path to generators is from the mechanical source through the shaft. The input power is used to overcome mechanical or rotational losses as well as to produce the electric power, $E_G I_A$. The mechanical losses can be assumed constant unless otherwise stated. From the shaft, power goes to the load, giving up parts to the $I_A^2 R_a$ and $I_F^2 R_f$ losses. The power equation can be written as the difference between mechanical and rotational terms.

$$P_m - P_r = T\Omega = E_G I_A$$
$$= I_A^2 R_a + I_F^2 R_f + V_A I_L \qquad 7.24$$

The usual practice is to use unnormalized saturation curves. In such cases, the normalizing factors, E_N and I_N, are both 1. Also, speed is normally given in rpm with n being the variable for speed. Some machines have more than two poles. The voltage equation which incorporates these changes is

$$E_G = \left(\frac{pZ_c}{60a_p}\right)\Phi n = K_v \Phi n \qquad 7.25$$

$$K_v = \frac{pZ_c}{60a_p} \qquad 7.26$$

p is the number of poles, Z_c is the number of armature conductors, and a_p is the number of parallel paths.

The torque equation is

$$T = \left(\frac{pZ_c}{2\pi a_p}\right)\Phi I_A = K_T \Phi I_A \qquad 7.27$$

$$K_T = \frac{pZ_c}{2\pi a_p} = \frac{60K_v}{2\pi} \qquad 7.28$$

3 DC MOTORS

A. Introduction

The major difference between DC generators and DC motors is that generators are more efficiently constructed. The fundamental behavior is the same, and the same equations apply. The directions of the armature current and the shaft torque are reversed, corresponding to the reversed power flow.

Motor equivalent circuits are the same as for generators, although a DC series motor is somewhat different. Motor performance is given by speed-torque curves.

Representative speed-torque curves for the four types of motors are shown in figure 7.5.

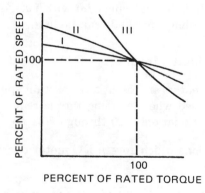

Figure 7.5 Speed-Torque Characteristics for DC Motors: I Shunt and Separately Excited, II Compounded, III Series

B. Separately Excited Motors

Separately excited DC motors are usually small servo type motors of fractional horsepower. The field is maintained at a constant current, and the speed and direction of rotation are controlled by the armature voltage (which is often a power amplifier). The application of the equations for steady-state operation is straight forward.

Analysis of transient behavior requires knowing the inductances of the motor windings, L_f and L_a, and knowing the mass moment of inertia, J, of the shaft system and load. In motor control systems, the load must be specified in terms of torque-speed relations, accounting for both friction and windage effects. This results, for the simplest case, in the load being described by the differential equation 7.29.

$$T_L = J\frac{d\Omega_R}{dt} + f_v\Omega_R \qquad 7.29$$

For constant separate excitation, Φ and K_v are constant terms. The system equations are:

$$V_A = I_A R_a + K_v\Phi\Omega_R \qquad 7.30$$

$$T = K_T\Phi I_A = \frac{K_T\Phi}{R_a}(V_A - K_v\Phi\Omega_R) \qquad 7.31$$

$$T = J\frac{d\Omega_R}{dt} + f_v\Omega_R \qquad 7.32$$

By eliminating T from the last two equations,

$$\frac{d\Omega_R}{dt} = \alpha\Omega_R + \frac{K_T\Phi}{R_a J}V_A \qquad 7.33$$

$$\alpha = \frac{f_v}{J} + \frac{K_T K_v\Phi^2}{J R_a} \qquad 7.34$$

This first-order differential equation can be used in sinusoidal analysis by replacing d/dt with $j\omega$. For transient analysis, ordinary methods can be used.

C. DC Shunt Motors

DC shunt motors are essentially constant speed devices. They are used where starting torque requirements are moderate. Equations 7.29 through 7.31 apply.

The equations which govern DC motor operation are:

$$P_{\text{in}} = V_A I_A + V_A I_F = V_A I_A + \frac{V_A^2}{R_F} \qquad 7.35$$

$$V_A I_A = E_G I_A + I_A^2 R_a \qquad 7.36$$

$$E_G I_A = T\Omega_R = \text{rotational losses} + T_L\Omega_R \qquad 7.37$$

Example 7.5

A 10 hp, 1200 rpm shunt motor has a rated voltage of 230 volts and current of 42 amps. Its field current is 2 amps and its armature resistance is 0.25 ohm. The motor actually is operating at 1200 rpm with a viscous load torque of 35 ft-lbf when the resistance of the armature circuit is suddenly increased to 1.0 ohm. What are the input and output powers after the motor reaches steady state? Assume that the rotational losses remain constant.

For convenience, first convert to MKS units. The initial speed is

$$\Omega_R = 1200 \times \frac{2\pi}{60} = 125.66 \text{ rad/sec}$$

The load torque is

$$T = 35 \times \frac{746}{550} = 47.47 \text{ N-m}$$

Thus, the load power is

$$P_L = 125.66 \times 47.47 = 5965 \text{ W}$$

The rated input power is

$$230 \times 42 = 9660 \text{ W}$$

Of this, 2 amps flow in the field circuit, accounting for 460 watts. This leaves 9200 watts for the armature circuit, along with 40 amps of armature current.

The armature resistance accounts for a 10 volt (40 amps $\times$ 0.25 Ω) drop, leaving 220 volts for the back EMF, E_g. The armature $I_a^2 R_a$ loss is 400 watts (10 volts $\times$ 40 amps) leaving an air-gap power of

$$220 \times 40 = 8800 \text{ W}$$

The power lost in the mechanical system is the difference between the air-gap power and the load power.

$$\text{mechanical loss} = P_{\text{in}} - P_L$$
$$= 8800 - (10 \times 746) = 1340 \text{ W}$$

The full load power flow diagram is

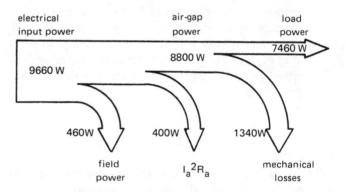

After the change of armature resistance, R_a, the field power will remain at 460 watts, and the mechanical loss will remain at 1340 watts. The $I_A^2 R_a$ loss and the load power will adjust to satisfy the system equations.

The air-gap power will be $E_G I_A$. I_A is found from the armature circuit relation.

$$230 = I_A R_a + E_G$$

$$I_A = \frac{230 - E_G}{1}$$

$$E_G I_A = 230E_G - E_G^2$$

For constant I_F, $K_v\Phi = K$, so

$$E_G = K\Omega_R$$

The original situation had $E_G = 220$ volts when Ω_R was 125.66 rad/sec. K was, then

$$K = \frac{220}{125.66} = 1.75$$

The air-gap power expressed in terms of the speed is

$$E_G I_A = 230K\Omega_R - K^2\Omega_R^2 = 402.5\Omega_R - 3.065\Omega_R^2$$

The load power can be expressed as $f_v\Omega_R^2$ for a viscous load. From $T = f_v\Omega_R$ (equation 7.31 with constant speed) and the original conditions of torque and rotational speed,

$$f_v = \frac{T}{\Omega_R} = \frac{47.47}{125.66}$$
$$= 0.378 \frac{\text{N-sec}}{\text{m-rad}}$$

The load power is, then, $0.378\Omega_R^2$. Thus, the air-gap power must balance the mechanical loss and the load power.

$$E_G I_A = 1340 + 0.378\Omega_R^2$$

Simultaneously solving the electrical and the mechanical expressions for air-gap power produces two solutions of 113.5 and 3.42 rad/sec. As the machine is slowing from 125.66 rad/sec, it will first satisfy the 113.5 solution. Then,

$$E_G = 1.75 \times 113.5 = 198.6 \text{ volts}$$
$$I_A = \frac{230 - 198.6}{1} = 31.4 \text{ amps}$$

The total input current is

$$I_{\text{IN}} = 31.4 + I_F = 31.4 + 2$$
$$= 33.4 \text{ amps}$$

The input power is

$$P_{\text{in}} = 33.4 \times 230 = 7682 \text{ W}$$

The output power is $f_v\Omega_R^2$, so

$$P_{\text{out}} = 0.378 \times (113.5)^2 = 4869 \text{ W}$$

Starting DC shunt motors larger than 2 hp is a problem. Current limiting is provided by *starting resistors* in series with the armature winding. The *controller* consists of a number of relays and armature series resistances. A review of relay and switch symbols is given in figure 7.6.

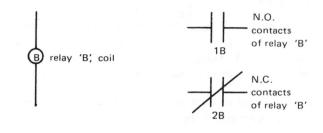

Figure 7.6 Relay and Push Button Symbols

A controller for a large DC compound motor is shown in figure 7.7. The resistors R_1 and R_2 are the *current limiting resistors*. Until the start button is pressed, the contacts 1A and 2A are open; when the start button is pushed, the main relay M is energized, which closes contact M_a. M_a maintains the energizing of coil M. Line voltage is then supplied to the armature circuit. R_1 and R_2 limit the current as the motor starts.

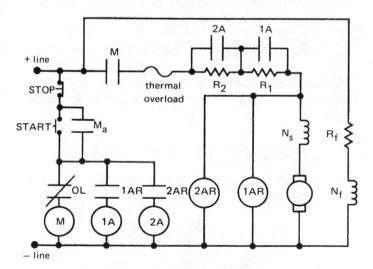

Figure 7.7 Counter-EMF Magnetic Controller

As the armature counter-EMF and motor speed build up to about half their rated values, *accelerating relays* 1AR and 2AR (which are *light-current relays*) remain un-energized. 1AR pulls in at half the rated voltage, which in turn energizes the coil of 1A to close contact 1A and bypass resistance R_1. This allows more current to flow in the armature circuit and further accelerates the motor.

Relay 2AR pulls in at about 80% of rated voltage and speed. This energizes relay coil 2A through contact 2AR, which in turn closes contact 2A and shunts R_2. Full line voltage is then applied to the armature.

The *overload contact* is a *thermal limit switch* sensing the temperature of the armature.

Example 7.6

A 250 volt DC motor starter consists of series resistors in the armature circuit and a corresponding set of relays. The rated armature current is 125 amps, and the armature circuit resistance is 0.15 ohm. The starting current is to be limited to 150% of rated current, and each relay coil is to pull in when the current reaches 100% of rated current. Determine the values and wattage of the resistors, as well as the pull-in voltage of each relay coil.

The first resistor is found from the line voltage applied at zero speed.

$$R_{x1} + R_a = \frac{250}{125 \times 1.5} = 1.333$$

$$R_{x1} = 1.333 - 0.15 = 1.1833 \text{ ohms}$$

For a current of 125 amps to flow through R_{x1}, the armature voltage must be

$$V_A = 250 - (125 \times 1.1833) = 102.1 \text{ volts}$$

This is the pull-in voltage of the first relay and is also the armature voltage at the time the first relay closes. E_G at this time is

$$E_G = 250 - (125 \times 1.33) = 83.75$$

For 150% of rated current at this value of E_G (which will occur when contact relay CR1 closes),

$$R_{x2} + R_a = \frac{250 - 83.75}{1.5 \times 125} = 0.89$$

$$R_{x2} = 0.89 - 0.15 = 0.74$$

To have 125 amps through R_{x2},

$$E_G = 250 - (125 \times 0.89) = 138.75$$
$$V_A = 250 - (125 \times 0.74)$$
$$= 157.5 = V_{\text{CR2}}$$

V_{CR2} is the pull-in voltage of the second relay.

To obtain 150% current with this value of E_G, the resistance must be

$$R_{x3} + R_a = \frac{250 - 138.75}{1.5 \times 125} = 0.59$$

$$R_{x3} = 0.59 - 0.15 = 0.44$$

Continuing in a similar fashion,

$$E_G = 250 - (125 \times 0.59) = 176.25$$
$$V_A = 250 - (125 \times 0.44)$$
$$= 195 = V_{\text{CR3}}$$

$$R_{x4} + R_a = \frac{250 - 176.25}{1.5 \times 125} = 0.39$$
$$R_{x4} = 0.39 - 0.15 = 0.24$$
$$E_G = 250 - (125 \times 0.39) = 201.25$$
$$V_A = 250 - (125 \times 0.24)$$
$$= 220 = V_{\text{CR4}}$$

$$R_{x5} + R_a = \frac{250 - 201.25}{1.5 \times 125} = 0.26$$
$$R_{x5} = 0.26 - 0.15 = 0.11$$
$$E_G = 250 - (125 \times 0.26) = 217.5$$
$$V_A = 250 - (125 \times 0.11)$$
$$= 236 = V_{\text{CR5}}$$

$$R_{x6} + R_a = \frac{250 - 217.5}{1.5 \times 125} = 0.17$$
$$R_{x6} = 0.17 - 0.15 = 0.02$$
$$E_G = 250 - (125 \times 0.17) = 228.75$$
$$V_A = 250 - (125 \times 0.02)$$
$$= 247.5 = V_{\text{CR6}}$$

For this value of E_G, it is impossible to obtain 150% rated current, so no further external resistance is needed.

The required resistances are:

$$R_6 = R_{x6} = 0.02 \text{ ohm}$$
$$R_5 = R_{x5} - R_{x6} = 0.11 - 0.02$$
$$= 0.09 \text{ ohm}$$
$$R_4 = R_{x4} - R_{x5} = 0.24 - 0.11$$
$$= 0.13 \text{ ohm}$$
$$R_3 = R_{x3} - R_{x4} = 0.44 - 0.24$$
$$= 0.20 \text{ ohm}$$
$$R_2 = R_{x2} - R_{x3} = 0.74 - 0.44$$
$$= 0.30 \text{ ohm}$$
$$R_1 = R_{x1} - R_{x2} = 1.18 - 0.74$$
$$= 0.44 \text{ ohm}$$

D. Compound DC Motors

Compound DC motors can be constructed so that the MMF of the series winding either aids (*cumulative compounding*) the shunt winding MMF or opposes (*differential compounding*) the shunt winding's MMF. The shunt winding can be in parallel with both the armature and the series winding as shown in figure 7.8 (*long shunt*), or it may parallel only the armature, with the series winding being between the parallel combination and the power source (*short shunt*). Cumulative compounding is used to increase starting torque, and differential compounding is used to improve speed regulation. Referring to figure 7.8(b), the MMF for the differentially-compounded motor is

$$\text{MMF} = N_f I_F - N_s I_A \qquad 7.38$$

The speed is derived from equation 7.6.

$$\Omega_R = \frac{V_A - (R_s + R_a)I_A}{K_v \Phi} \qquad 7.39$$

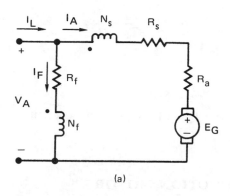

(a)

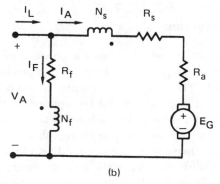

(b)

Figure 7.8 Compounded DC Motors (Long Shunt): (a) Cumulative Compounding, (b) Differential Compounding

Small performance changes around an operating point, indicated by the subscript o, can be predicted from equation 7.40.

$$\Omega_{Ro} + \Delta\Omega_R = \frac{V_A - (R_a + R_s)(I_{Ao} + \Delta I_A)}{K_v(\Phi_o + \Delta\Phi)} \qquad 7.40$$

At constant speed, $\Delta\Omega_R = 0$, so it is necessary that

$$K_v \Omega_o \Delta\Phi = -(R_a + R_s)\Delta I_A \qquad 7.41$$

Equation 7.41 can be written in partial differential form as

$$\frac{\partial}{\partial I_A}|K_v \Omega_o \Phi| = -(R_a + R_s) \qquad 7.42$$

Recognizing that $E_G = K_v \Omega_R \Phi$, the chain rule for partial differential equations results in

$$\frac{\partial}{\partial I_A}(K_v \Omega_R \Phi) = \frac{\partial}{\partial \text{MMF}} E_G \times \frac{\partial}{\partial I_A} \text{MMF} \qquad 7.43$$

Equations 7.42 and 7.43 can be considered equivalent when Ω_R is constant. From equation 7.38,

$$\frac{\partial}{\partial I_A}\text{MMF} = -N_s \qquad 7.44$$

For an unnormalized curve where the MMF is given in terms of the field current, I_F, the actual variable on the abscissa is $I_F \pm (N_s/N_f)I_A$. (The negative sign is used with differential compounding.) For such a curve,

$$\left(\frac{\partial E_G}{\partial \text{MMF}}\right)_{\text{unnormalized}} = \frac{\text{OPS}}{N_f} \qquad 7.45$$

For a normalized curve, with E_N and $N_f I_N$ as the operating point,

$$\left(\frac{\partial E_G}{\partial \text{MMF}}\right)_{\text{normalized}} = \frac{N_f I_N}{E_N}\left(\frac{\partial E_{GN}}{\partial \text{MMF}_n}\right) \qquad 7.46$$

$$\frac{\partial E_{GN}}{\partial \text{MMF}_n} = \text{OPS} \qquad 7.47$$

For constant speed around an operating point, with an unnormalized saturation curve,

$$\left(\frac{N_s}{N_f}\right) = \frac{R_a + R_s}{\text{OPS}} \qquad 7.48$$

With a normalized saturation curve,

$$\left(\frac{N_s}{N_f}\right) = \frac{(R_a + R_s)\dfrac{I_N}{E_N}}{\text{OPS}} \qquad 7.49$$

This is the same as equation 7.21 with the exception that R_s was previously ignored.

The only difference in the design between a generator and a motor is that the generator is short-shunted while the motor is long-shunted. In the motor, a *flat speed-torque* characteristic is obtained for differential compounding at the operating point. With cumulative compounding, a higher starting torque is obtained with a resulting degradation of speed regulation. Therefore, differential compounding will result in lower starting torque.

E. DC Series Motors

DC series motors (traction motors) are used in variable speed applications where the starting torque is of primary importance. Such motors always are connected directly to the load shaft. The series motor has only a series field winding, which carries the armature current. Thus, the field is approximately proportional to the armature current, since $MMF = N_s I_a$. The basic relationships for this motor are:

$$V_A = E_G + I_A R_a \qquad 7.50$$

$$E_G = K_v \Phi \Omega_R \qquad 7.51$$

$$MMF = N_s I_A \qquad 7.52$$

$$T = K_T \Phi I_A \qquad 7.53$$

In MKS units $K_T = K_v$. (When n in rpm is used rather than Ω_R in radians/sec, then $K_T = 60 K_v / 2\pi$. Torque is still in newton-meters, however.)

A reasonable approach is to take the slope of the magnetization curve through the origin and the maximum current point and set

$$L_a = \frac{K_v \Phi_{max}}{I_{A,max}} \qquad 7.54$$

Then,

$$K_v \Phi = L_a I_A \qquad 7.55$$

When working entirely in MKS units, the equations are:

$$E_G = L_a I_A \Omega_R \qquad 7.56$$

$$T = L_a I_A^2 \qquad 7.57$$

$$V_A = (L_a \Omega_R + R_a) I_A \qquad 7.58$$

$$\Omega_R = \frac{V_A}{L_a I_A} - \frac{R_a}{L_a} \qquad 7.59$$

The rated power output is

$$P_R = T_R \Omega_{r,R} \qquad 7.60$$

P_R will be given for particular Ω_R and $I_{A,R}$. From these values, the armature inductance is

$$L_a = \frac{P_R}{\Omega_R I_{A,R}^2} \qquad 7.61$$

The starting torque can be estimated from equations 7.57 through 7.59.

Example 7.7

A DC series motor is rated at 900 volts, 200 amps, and 225 hp at a speed of 1500 rpm. The armature circuit resistance, including the resistance of the series winding, is 0.001 ohm. If the starting current is limited to six times the rated current, determine the starting torque in foot-pounds.

The torque at 200 amps is

$$T_{200} = \frac{225 \text{ hp} \times 746 \text{ W/hp}}{1500 \text{ rpm} \times \dfrac{2\pi \text{ rad/sec}}{60 \text{ rpm}}} \times \frac{550 \text{ ft-lb}}{746 \text{ N-m}}$$

$$= \frac{16{,}500}{1500\pi} \times 225 = 788 \text{ ft-lb}$$

However,

$$T = K I_A^2$$

So,

$$\frac{T_2}{T_1} = \frac{I_2^2}{I_1^2}$$

Since

$$I_2 = 6 I_1$$

$$T_2 = (6)^2 \times 788 = 28{,}368 \text{ ft-lb}$$

4   INDUCTION MOTORS

A. Introduction

The three-phase induction motor consists of a set of three field windings on the stator, which produces a rotating field. The rotating field turns through 360 electrical degrees in each cycle for the applied voltage. The number of field poles is determined by the layout of stator windings. For one complete three-phase winding set on a stator, there are two poles produced by the stator currents. With two complete sets of windings, as indicated in figure 7.9, there are four poles generated by the stator currents. For purposes of systems analysis, the stator presents a balanced load to the line, and is therefore considered to be wye-connected.

In figure 7.9, phase sequence *acb* results in a field rotating in the clockwise direction. Phase sequence *abc* produces a counterclockwise field rotation. (In actual motors, the phase windings are distributed over a whole pole-span, but they are shown concentrated in figure 7.9 to clarify the concept.)

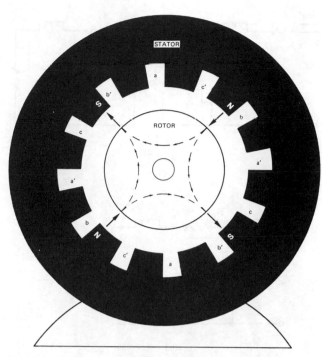

Figure 7.9 A Four-Pole, Three-Phase Winding

As the poles move through a complete set of windings during an electrical cycle, the mechanical speed of rotation of the stator field is the synchronous speed, Ω_s, which is the electrical frequency, ω, divided by the number of complete sets of windings, $p/2$.

$$\Omega_s = \frac{2\omega}{p} \qquad 7.62$$

The rotor has windings with the same distribution as the stator. They may be brought out via slip rings and brushes if speed control is necessary, or internally shorted in a squirrel-cage configuration. The source of excitation for the rotor currents is the rotating stator field, which induces rotor EMF by the relative motion between the stator field and the rotor. The EMF induced in a conductor by a constant amplitude, sinusoidally-distributed flux moving with respect to the conductor at a frequency Ω_c is

$$\text{RMS EMF per conductor} = \frac{\Omega_c K_w \Phi}{\sqrt{2}} \qquad 7.63$$

Φ in equation 7.63 is the *flux per pole* from the rotating field, and K_w is the *winding factor* for the particular conductor's winding. K_w depends on the spatial distribution of windings. It is specific to particular machines but it varies from 0.65 to 0.86. Thus, for a stator field with N_s conductors per phase with stationary windings so that the relative speed is the synchronous speed of the rotating field ($\Omega_{cs} = \omega$), the RMS EMF per phase of the stator is

$$E_s = \frac{N_s \omega K_{ws} \Phi_p}{\sqrt{2}} = \sqrt{2}\pi f N_s K_{ws} \Phi_p$$
$$= 4.44 f N_s K_{ws} \Phi_p \qquad 7.64$$

For the rotor, the relative speed of the field is

$$\omega_r = \frac{p}{2}(\Omega_s - \Omega_r) = \omega - \frac{p\Omega_r}{2} \qquad 7.65$$

Slip is defined as

$$s = \frac{\Omega_s - \Omega_r}{\Omega_s} \qquad 7.66$$

Equation 7.66 can be substituted into equation 7.65.

$$\omega_r = \frac{p}{2}\Omega_s \times s = s\omega \qquad 7.67$$

Then, the RMS EMF induced in a rotor phase with N_r conductors is

$$E_r = 4.44 s f N_r K_{wr} \Phi_p \qquad 7.68$$

The frequency of the rotor current is sf. The RMS rotor EMF can be rewritten in terms of the stator EMF.

$$E_r = \left(\frac{s N_r K_{wr}}{N_s K_{ws}}\right) E_s \qquad 7.69$$

The rotor circuit has an impedance of $r_r + js\omega L_r$, so the resulting per-phase RMS voltage is

$$E_r = s\left(\frac{N_r K_{wr}}{N_s K_{ws}}\right) E_s = r_r I_r + js\omega L_r I_r \qquad 7.70$$

Dividing equation 7.70 by the slip s and defining a, the RMS rotor current is

$$I_r = \frac{\dfrac{E_s}{a}}{\dfrac{r}{s} + jx} \qquad 7.71$$

$$a = \frac{N_s K_{ws}}{N_r K_{wr}} \qquad 7.72$$

$x = \omega L_r$ is the reactance of the rotor at the stator frequency. Thus, a circuit represented by equation 7.72 is equivalent to a transformer secondary with a turns ratio of a.

The power loss per phase in the rotor circuit is

$$P_r = I_r^2 r_r \qquad 7.73$$

From an analysis of the stator circuit, the power loss per phase appears to be $I_r^2 r_r / s$. This is greater than equation 7.73. The difference is the gap power, P_g.

$$P_g\Big|_{\text{phase}} = I_r^2 r_r\left(\frac{1}{s} - 1\right) = I_r^2 r_r\left(\frac{1-s}{s}\right) \qquad 7.74$$

For q phases, the gap power transferred to the shaft is

$$P_g = qI_a^2 r_r \left(\frac{1-s}{s}\right)$$

$$= T\Omega_r = T\Omega_s(1-s) \qquad 7.75$$

The resistance in ohms associated with the mechanical load is R_m.

$$R_m = r_r \left(\frac{1-s}{s}\right) \qquad 7.76$$

An induction motor can be represented on a per-phase basis as a transformer with the equivalent circuit shown in figure 7.10. (The per-phase basis assumes a wye connection unless otherwise stated.) The power flow through the circuit must be multiplied by q, the number of phases, in order to obtain the total power.

An analysis of figure 7.10(b) shows that the absolute value of the current i_r' is

$$|i_r'| = \sqrt{\frac{|V_{an}|^2}{\left(r_s + a^2 \frac{r_r}{s}\right)^2 + x_e^2}} \qquad 7.77$$

From equation 7.75, the torque per phase is

$$\frac{T}{q} = \frac{|V_{an}|^2}{\Omega_s}\left[\frac{\dfrac{a^2 r_r}{s}}{\left(r_s + \dfrac{a^2 r_r}{s}\right)^2 + x_e^2}\right] \qquad 7.78$$

The starting torque is obtained when $s = 1$.

$$T_{\text{starting}} = \frac{q|V_{an}|^2}{\Omega_s}\left[\frac{a^2 r_r}{(r_s + a^2 r_r)^2 + x_e^2}\right] \qquad 7.79$$

The slip at which maximum torque occurs can be found by differentiating equation 7.78 with respect to s and setting the result to zero.

$$s_{\text{maximum torque}} = \frac{a^2 r_r}{\sqrt{r_s^2 + x_e^2}} \qquad 7.80$$

Substituting equation 7.80 into equation 7.78 gives the maximum torque.

$$T_{\text{max}} = \frac{q|V_{an}|^2}{2\Omega_s}\left(\frac{1}{r_s + \sqrt{r_s^2 + x_e^2}}\right) \qquad 7.81$$

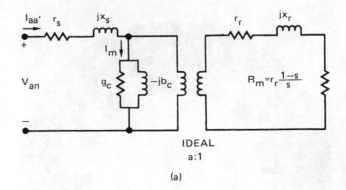

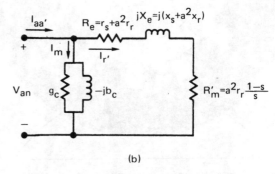

Figure 7.10 Phase Models of an Induction Motor: (a) Physical Model, (b) Practical Approximate Model

An external three-phase resistance can be connected to the rotor to adjust the point of maximum torque. It will also adjust the running speed for a given load and the starting torque. This is illustrated in figure 7.12 for shorted rotor windings and for an external resistance R. The slope of the torque-speed curve where $\Omega_r = \Omega_s$ $(s = 0)$ is

$$\left.\frac{dT}{ds}\right|_{s=0} = \frac{-q|V_{an}|^2}{\Omega_s}\left[\frac{1}{a^2(r_r + R)}\right] \qquad 7.82$$

Power can be evaluated with the aid of figure 7.10(b). The input power per phase is

$$P_{\text{in, phase}} = |V_{an}|^2 g_c + |I_r'|^2\left(r_s + \frac{a^2 r_r}{s}\right) \qquad 7.83$$

Of the power in equation 7.83, the output power transferred to the shaft is

$$P_{\text{shaft, phase}} = \frac{P_L + P_{\text{rotational}}}{q}$$

$$= |I_r'|^2 a^2 r_r\left(\frac{1-s}{s}\right) \qquad 7.84$$

Part of the shaft power goes into rotational losses, and part goes to the load.

$$P_{\text{shaft}} = P_{\text{rotational losses}} + P_{\text{output}} \qquad 7.85$$

Example 7.8

A 10 hp (nameplate), 6 pole, 60 Hz, three-phase induction motor delivers 9.9 hp with an input of 9200 watts. The core loss is 450 watts, the stator copper loss is 650 watts, and the rotational loss is 150 watts. What is the motor speed?

From equation 7.83 for three phases,

$$P_{in} - P_c - P_s = 9200 - 450 - 650 = 8100 \text{ watts}$$

From equation 7.84 for three phases,

$$P_L + P_{rotational} = 9.9 \times 746 + 150 = 7535.4 \text{ watts}$$

Then,

$$1 - s = \frac{7535.4}{8100} = 0.93$$

The synchronous speed is

$$n_s = \frac{60f}{\frac{p}{2}} = \frac{(60)(60)}{\frac{6}{2}}$$
$$= 1200 \text{ rpm}$$

The motor speed is derived from equation 7.66.

$$n_r = n_s(1 - s) = 1116 \text{ rpm}$$

B. Induction Motor Tests

Induction motor tests are comparable to open- and short-circuit tests on transformers. A no-load test will determine the core characteristics and the rotational losses. When unloaded, slip is usually a fraction of a percent, and the *transformed rotor resistance* $a^2 r_r/s$ becomes very large. Figure 7.10(b) would then be dominated by the core conductance and susceptance.

$$g_c = \frac{P_{no\text{-}load}}{|V_{an}|^2 q} \qquad 7.86$$

$$b_c = \frac{Q_{no\text{-}load}}{|V_{an}|^2 q} \qquad 7.87$$

The no-load test does not distinguish between core losses and rotational losses. It is common to combine those losses into the g_c term, or to let g_c be zero and combine the losses with the rotational losses. There will be a slight difference in results depending on which assumption is made.

The equivalent of a transformer short-circuit test is the *blocked rotor test*. The rotor is locked to prevent rotation. Under this condition, $s = 1$. The effective input impedance is

$$Z = (r_s + a^2 r_r) + j(x_s + a^2 x_r) \qquad 7.88$$

The primary use of the blocked rotor test is to determine X_e.

$$R_e = r_s + a^2 r_r = \frac{P_{b.r.}}{I_{an}^2 q} \qquad 7.89$$

$$X_e = x_s + a^2 x_r = \frac{Q_{b.r.}}{I_{an}^2 q} \qquad 7.90$$

To obtain $a^2 r_r$ and r_s from R_e, a third test must be made. The third test is to measure input and output power and the slip. By combining the third test results with the results of the no-load and blocked rotor tests, it is possible to calculate $a^2 r_r$ from equation 7.84.

Example 7.9

The results of a no-load test on a three-phase induction motor are: $V_{ll} = 440$ RMS volts, line current = 4 amps, and input power = 900 watts. Determine g_c and b_c.

$$V_{an}^2 = \frac{(440)^2}{3}$$
$$= \left(\frac{440}{\sqrt{3}}\right)^2 = 64,533$$

From equation 7.86,

$$g_c = \frac{\frac{900}{3}}{64,533} = 0.00465 \text{ mho}$$

The apparent power per phase is

$$\frac{|S|}{q} = \frac{440}{\sqrt{3}} \times 4 = 1016 \text{ VA per phase}$$

$$\frac{P}{q} = 300 \text{ watts per phase}$$

$$\frac{Q}{q} = \sqrt{1016^2 - 300^2} = 971 \text{ VAR per phase}$$

From equation 7.87, on a per-phase basis (i.e., $q = 1$),

$$b_c = \frac{Q}{|V_{an}|^2} = -0.0150 \text{ mho reactive}$$

Example 7.10

The motor of example 7.9 is subjected to a blocked rotor test with the following results: $P_{in} = 100$ watts, $I_{line} = 12$ amps, and $V_{ll} = 20$ volts. Determine R_e and X_e.

From equation 7.89,

$$R_e = \frac{100}{(12)^2(3)} = 0.231 \text{ ohm}$$

$$\frac{|S|}{q} = \frac{V_{ll}}{\sqrt{3}} \times I_{\text{line}}$$

$$= \left(\frac{20}{\sqrt{3}}\right) \times 12 = 138.6 \text{ VA per phase}$$

$$\frac{Q}{q} = \sqrt{138.6^2 - 33.3^2} = 134.5 \text{ VAR}$$

From equation 7.90,

$$X_e = \frac{134.5}{(12)^2} = 0.934 \text{ ohm reactive}$$

Example 7.11

The motor of examples 7.9 and 7.10 has a slip of 0.05 with an output power of 10 hp and an input power of 9.3 kW. Determine r_s and $a^2 r_r$.

The output power is given by equation 7.84 since the rotational losses in the no-load test were lumped into g_c. Then, from equation 7.75 or 7.84,

$$I_r'^2 a^2 r_r = \frac{0.05}{0.95}\left(\frac{10 \times 746}{3}\right) = 130.88$$

$$I_r'^2\left(\frac{a^2 r_r}{s}\right) = \frac{130.88}{0.05} = 2617.5 \text{ watts}$$

The input power per phase is 3.1 kW. From example 7.9, $g_c V_{an}^2 = 300$ watts. Then,

$$P_{\text{in}} - g_c V_{an}^2 - I_r'^2 r_s = I_r'^2\left(\frac{a^2 r_r}{s}\right)$$

$$I_r'^2 r_s = 3100 - 300 - 2617.5 = 182.5$$

Then,

$$\frac{I_r'^2 a^2 r_r}{I_r'^2 r_s} = \frac{a^2 r_r}{r_s} = \frac{130.88}{182.5}$$

$$= 0.717$$

From example 7.10, $R_e = 0.231$.

$$R_e = r_s + a^2 r_r = r_s(1 + 0.717)$$

$$= 0.231$$

So,

$$r_s = 0.135 \text{ ohm}$$

$$a^2 r_r = 0.135 \times 0.717 = 0.0968 \text{ ohm}$$

C. Starting Induction Motors

Large starting currents can cause serious drops on branch and feeder circuits. These drops can interfere with other equipment, or the currents drawn may trip over-current protection devices. Some loads cannot be started abruptly, so the starting torque may have to be limited as well.

One control method starts with the windings in a wye configuration and switches them to a delta configuration after starting.

Example 7.12

Assume the motor of examples 7.9 through 7.11 is delta connected for all the previous measurements. Determine the starting torque for delta and wye connections if the synchronous speed is 900 rpm.

The actual impedance attached from line to line is three times the phase impedance. At starting, the admittance for the line-to-neutral phase was found to be

$$g_c = 0.00465$$

$$b_c = 0.0150$$

$$R_e = 0.231$$

$$X_e = 0.934$$

$$Y_{\text{in}} = g_c - jb_c + \frac{1}{R_e + jX_e} = 0.254 - j1.024$$

Therefore,

$$Z_{l-n} = Z_{\text{in}} = \frac{1}{Y_{\text{in}}}$$

$$= 0.228 + j0.920$$

$$Z_{l-l} = 3\, Z_{l-n} = 0.685 + j2.760$$

If this is the actual line-to-line phase impedance, and if it is initially connected in a wye configuration, the currents flowing will be one-third those calculated previously. The initial current flowing would then be

$$I_{\text{line}} = \frac{254}{|Z_{l-l}|} = 89.3 \text{ amps}$$

With a delta connection, the line current would be 268 amps. The starting torque given by equation 7.79 is only one-third of the delta-connected torque.

From equation 7.79 with a delta connection,

$$T_{\text{starting}} = \frac{3(440)^2}{\left(\dfrac{2\pi}{60}\right)900} \times \frac{3a^2 r_r}{(3R_e)^2 + (3X_e)^2}$$

$$= \frac{3(440)^2}{30\pi} \times \frac{a^2 r_r}{3(R_e^2 + X_e^2)}$$

With a wye connection, $V_{\text{ln}} = 254$ volts.

$$T_{\text{starting}} = \frac{3(254)^2}{30\pi} \times \frac{3a^2 r_r}{(3R_e)^2 + (3X_e)^2}$$

$$= \frac{V_{\text{phase}}^2}{30\pi} \times \frac{a^2 r_r}{R_e^2 + X_e^2}$$

Since it is known that $a^2 r_r = 0.096$, $R_e = 0.231$, and $X_e = 0.934$,

$$\frac{a^2 r_r}{R_e^2 + X_e^2} = 0.1037$$

$$T_{\text{starting}} = \frac{V_{\text{phase}}^2}{30\pi} \times 0.1037 = 0.0011 V_{\text{phase}}^2$$

For $V_{\text{phase}} = V_{\text{ll}} = 440$, $T_{\text{starting}} = 213$ N-m.

For $V_{\text{phase}} = V_{\text{ln}} = 440/\sqrt{3}$, $T_{\text{starting}} = 71$ N-m.

Another type of controller is the *auto-transformer* with voltage taps that are switched from a low voltage at starting to higher values as the speed increases. There are three auto-transformers in wye connections. All are switched together. They effectively control the line-to-neutral voltage. The lowest voltage tap is for starting, and one or more taps may involve intermediate voltage levels.

Example 7.13

The maximum line current for the motor of examples 7.9 through 7.11 has been set at 180 amps. The motor is to be controlled by a tapped three-phase auto-transformer (wye connection). On starting, the initial current is to be limited to 180 amps when the voltage of tap no. 1 is applied. When the current drops to 150 amps the voltage is to be switched to tap no. 2 to supply sufficient voltage to draw 180 amps.

Subsequent taps will be switched in when the current drops to 140, 130, 120 amps, etc. At each switching, the initial current is limited to 180 amps. Switching will continue until the voltage necessary to draw 180 amps is greater than the line-to-neutral voltage of 254 volts.

Specify the tap voltage and the switching speed in rpm for a synchronous speed of 900 rpm. Ignore g_c and b_c in the calculations.

The voltage necessary is found from figure 7.10(b).

$$V_{\text{tap}} = I_{\text{max}} \sqrt{\left(\frac{a^2 r_r}{s} + r_s\right)^2 + X_e^2}$$

$$= 180 \sqrt{\left(\frac{0.096}{s} + 0.135\right)^2 + 0.934^2}$$

The slip at which the next tap-change should occur is found from the same equation.

$$s_{\text{switching}} = \frac{a^2 r_r}{-r_s + \sqrt{\left(\dfrac{V_{\text{tap}}}{I_{\text{min}}}\right)^2 - X_e^2}}$$

$$= \frac{0.096}{-0.135 + \sqrt{\left(\dfrac{V_{\text{tap}}}{I_{\text{min}}}\right)^2 - 0.934^2}}$$

The switching speed is

$$n_{\text{switching}} = n_s (1 - s_{\text{switching}})$$

Initially, the machine is stopped, so $s = 1$.

$$V_1 = 180 \sqrt{(0.096 + 0.135)^2 + 0.934^2} = 173 \text{ volts}$$

$$s_1 = \frac{0.096}{-0.135 + \sqrt{\left(\dfrac{173}{150}\right)^2 - 0.934^2}} = 0.177$$

$$n_1 = 900(1 - 0.177) = 741 \text{ rpm}$$

$$V_2 = 180 \sqrt{\left(\frac{0.096}{0.177} + 0.135\right)^2 + 0.934^2} = 208 \text{ volts}$$

$$s_2 = \frac{0.096}{-0.135 + \sqrt{\left(\dfrac{208}{140}\right)^2 - 0.934^2}} = 0.094$$

$$n_2 = 900(1 - 0.094) = 815 \text{ rpm}$$

$$V_3 = 180 \sqrt{\left(\frac{0.096}{0.094} + 0.135\right)^2 + 0.934^2} = 268 \text{ volts}$$

This exceeds the line-to-neutral voltage, so the final voltage is 254 volts.

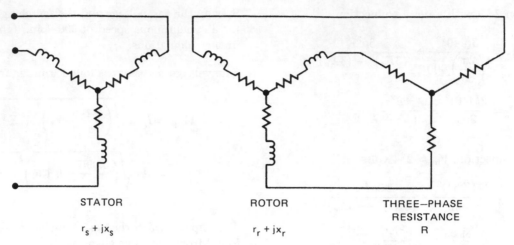

Figure 7.11 External Rotor Resistance Connections

A third method of current control uses series resistors, as were used to start DC motors. In such cases, the series resistance is added to R_e. Calculations are similar to those of example 7.13. The procedure for torque limitation is not covered here.

Wound rotor induction motors can be controlled by an external three-phase rotor resistance. This resistance can be controlled by relays to limit starting current torque as well as running speed. In this case, $a^2 r_r$ is replaced by $a^2(r_r + R_r)$ where R_r is the external rotor resistance. Figures 7.11 and 7.12 illustrate this method of control.

5   SYNCHRONOUS MACHINES

A. Introduction

Three-phase synchronous machines have stator windings constructed similarly to three-phase induction machines. However, while the stator is the field for an induction machine, the stator of a synchronous machine is known as an *armature*. The synchronous machine field is supplied by a DC current in the rotor. The rotor will consist of *salient poles* for machines operating at 1800 rpm or less. A *cylindrical rotor* is used for higher speeds.

The number of poles in the machine is related to the ratio of electrical frequency to mechanical frequency.

$$p = 2\left(\frac{\text{electrical frequency}}{\text{mechanical frequency}}\right) = \frac{2\omega}{\Omega}$$

$$= \frac{2 \times 60 f}{n} \qquad\qquad 7.91$$

In synchronous machines, the slip is zero, so rotor currents are not induced by the stator currents. Instead,

the *rotor flux* is produced by a DC current, and the flux rotates with the rotor at the synchronous speed.

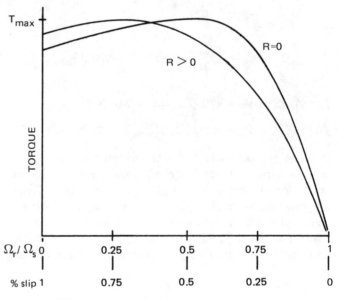

Figure 7.12   Performance Curves for the Two External Rotor Resistances

The rotor flux is approximated by equation 7.92.

$$\Phi_r = \left(\frac{\mu_o N_r A_r}{gp}\right) I_R \qquad\qquad 7.92$$

N_r is the number of rotor winding turns per pole of the rotor, g is the *gap length* (the separation of the rotor from the stator), p is the number of poles, A_R is the rotor cross-section area, and I_R is the rotor current.

It is assumed that the flux produced by the rotor current is sinusoidally distributed around the rotor or stator. With respect to the rotor, the flux density will complete its positive portion (i.e., the region where the flux density is radially outward) over the span of one pole. Using the electrical angle, the flux density can be written as a sinusoid. (The point of maximum outward flux density is assumed to occur at $\theta_r = 0$.)

$$B(\theta_r) = B_{\max} \cos \theta_r \qquad 7.93$$

If the length of the rotor is l_r and the radius is R_r, the flux from one pole is found by integrating equation 7.93 over the electrical angle $-\pi/2$ to $+\pi/2$.

$$\Phi_r = 2l_r R_r B_{\max} = B_{\max} A_r \qquad 7.94$$

A_r in equation 7.94 is the cross-sectional area of the rotor.

$$A_r = 2l_r R_r \qquad 7.95$$

The rotor turns at the synchronous speed Ω_s, so the rotor flux density seen from the stator will appear as

$$B_r(\theta_s, t) = B_{\max} \cos(\omega_s t - \theta_s) \qquad 7.96$$

Equation 7.96 assumes counterclockwise rotation of the rotor. θ_s is the angle of the stator at which the flux density is observed, measured counterclockwise from where the rotor flux density is maximum at $t = 0$ (i.e., *the stator reference point*).

The EMF induced in a single stator turn located θ_s counterclockwise from the stator reference can be determined.

$$\mathrm{EMF}_{\mathrm{turn}} = 2B_{\max} A_r \omega_s \sin(\omega_s t - \theta_s) \qquad 7.97$$

A three-phase stator winding has the phases separated by 120 *electrical degrees* ($240/p$ *mechanical degrees*), and the windings are distributed symmetrically about their respective centers. Thus, the voltages induced will not be simply the number of turns per phase multiplied by equation 7.97, but will be somewhat less because the individual components of each phase voltage are slightly out-of-phase. For a perfectly sinusoidal distribution of the phase windings, the effective number of turns is the actual number times $2/\pi$, or about $0.64N_s$, where N_s is the number of turns per phase.

Then, the RMS EMF induced in the stator winding due to the rotor current is

$$\mathrm{EMF}_{\mathrm{RMS}} = \sqrt{2}K_\Phi I_R \sin(\omega_s t - \theta_s) \qquad 7.98$$

θ_s takes on values of $0°$, $120°$, and $240°$ for an *abc* sequence for phases a, b, and c, respectively, and K_ϕ is an ideal constant given by equation 7.99.

$$K_\Phi = \frac{\sqrt{2}N_r N_s \mu_o A_r \omega_s}{\pi g p} \qquad 7.99$$

Then, the RMS value of the induced voltage per phase is

$$\mathrm{EMF}_{\mathrm{RMS}} = E_g = K_\Phi I_R \qquad 7.100$$

Equation 7.100 shows that the generated voltage is proportional to the rotor current. This is valid because the air-gap between the rotor and stator is fairly large, minimizing the effects of nonlinear rotor and stator reluctances. Equation 7.100 cannot be used when I_R is very large, and magnetization data must then be considered.

When a balanced three-phase current flows in the stator windings, a rotating component of flux is induced, as in the induction machine. In contrast with the induction machine, however, there is no induced voltage in the rotor windings because the rotor is traveling at synchronous speed in unison with the stator field. The stator currents induce an EMF in the stator winding because of the self-inductance of the windings. This EMF corresponds to the rotating stator field, which can be thought of as being in-phase with the stator current. The stator EMF will lead the stator current by approximately 90 degrees, because the inductive reactance is much larger than the winding resistance.

In addition, there is a component of stator-generated EMF which is due to leakage inductance, not linking the other stator winding via the air-gap. Thus, the total self-generated EMF of the stator winding can be accounted for by an equivalent stator reactance called the *synchronous reactance*, X_s. Winding resistance is typically ignored except when power flow and efficiency are being evaluated.

The RMS terminal voltage of the stator winding is, therefore, made up of two phasor components. One is due to the stator current and the other is due to the rotor current. (The subscript a in equation 7.101 designates *armature*, as the stator is considered to be the armature in a synchronous machine.)

$$\mathbf{V}_a = jX_s \mathbf{I}_a + \mathbf{E}_g \qquad 7.101$$

Equation 7.101 is a phasor equation, and the power into the machine is positive whenever $\mathbf{I}_a$ is within $90°$ of $\mathbf{V}_a$. In such a case, the machine will be acting as a motor. If, on the other hand, the current is more than $90°$ out of phase (positive or negative), the power flow is out of the terminals, and the machine will be acting as a generator.

Equation 7.101 represents a machine acting as a motor with the per-phase circuit shown in figure 7.13. The generated voltage, $\mathbf{E}_g$, lags behind the applied voltage, $\mathbf{V}_a$, according to the phasor diagram figure 7.13(b).

Taking the applied phase voltage as the phasor reference, the generated RMS voltage due to the rotor current is

$$E_g = |E_g| \cos \delta - j |E_g| \sin \delta \qquad 7.102$$

δ is the angle by which $\mathbf{E}_g$ lags $\mathbf{V}_a$. Then, equation 7.101 can be rewritten to give the RMS stator current.

$$I_a = \frac{|E_g| \sin \delta}{X_s} - j \frac{V_a - |E_g| \cos \delta}{X_s} \qquad 7.103$$

The input power to the motor is the terminal voltage $|V_a|$ multiplied by the real part of the terminal current (when the terminal voltage is taken as reference). The apparent power per phase is the terminal voltage multiplied by the complex conjugate of the RMS terminal current. Therefore, for a motor,

$$S = P + jQ = q(V_a I_a^*)$$
$$= q \left(\frac{V_a |E_g| \sin \delta}{X_s} + j \frac{V_a^2 - V_a |E_g| \cos \delta}{X_s} \right) \qquad 7.104$$

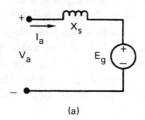

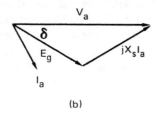

Figure 7.13   (a) Per-Phase Equivalent Circuit for a Synchronous Motor, (b) Phasor Diagram

Assuming all circuit losses are negligible, all electrical power is converted to *air-gap mechanical power*, which is the product of the *air-gap torque* and the mechanical angular velocity: $T\Omega$. Then, motor torque can be written as equation 7.105 where Ω_s is the mechanical synchronous speed.

$$T_{\text{air-gap}} = q \left(\frac{V_a |E_g| \sin \delta}{X_s \Omega_s} \right) \qquad 7.105$$

For a synchronous generator, the generated voltage causes a current to flow out of the terminals. So, equation 7.101 is modified to account for the reversal of the current direction. The RMS EMF is

$$\mathbf{E}_g = j \mathbf{I}_a X_s + \mathbf{V}_a \qquad 7.106$$

Equation 7.106 represents the per-phase circuit indicated in figure 7.14. Taking the terminal voltage V_a as the reference and letting δ be the angle by which $\mathbf{E}_g$ leads $\mathbf{V}_a$, the EMF voltage for a generator is

$$E_g = |E_g| \cos \delta + j |E_g| \sin \delta \qquad 7.107$$

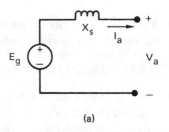

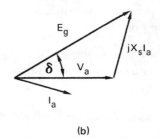

Figure 7.14   (a) Per-Phase Equivalent Circuit for a Synchronous Generator, (b) Phasor Diagram

The apparent power for a generator can be calculated from equations 7.106 and 7.107. The air-gap power will be the product of the generated voltage and the complex conjugate of the terminal current, $E_g I_a^*$.

$$\frac{S}{q} = \frac{|E_g| V_a \sin \delta}{X_s} + j \frac{|E_g|^2 - |E_g| V_a \cos \delta}{X_s} \qquad 7.108$$

The air-gap torque is the ratio of power to angular velocity, so equation 7.105 can be used for the generator.

The ideal torque-power angle relationship is a sinusoidal curve as shown in figure 7.15. However, there is a second harmonic of the angle in salient pole machines. This harmonic produces higher torque and a non-sinusoidal relationship. Only pure sinusoidal performance is covered in this chapter.

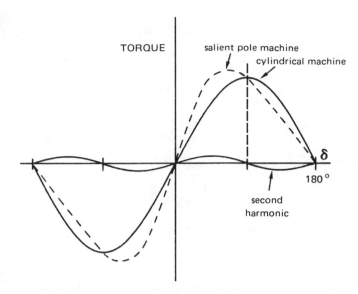

Figure 7.15 Torque-Power Angle Relationship

B. Synchronous Motors

Synchronous motors generally are started under no-load by means of a squirrel-cage rotor winding. This brings the motor up to nearly synchronous speed, at which time the rotor current is applied. After several oscillations, the rotor locks to the synchronous speed.

Under constant rotor current, the rotor generated voltage remains constant. The terminal voltage is usually constant (except when line regulation is a factor), so the power is related to the sine of the power angle, δ. In many synchronous motor problems, the information given includes input power, input volt-amperes, power factor, and a suitable number of other parameters. For this reason, it is convenient to express the apparent power in several forms.

Two additional angles are defined in figure 7.16. θ is the usual *power factor angle*. It is positive when the terminal current leads the terminal voltage. ψ is the power factor angle between the generated voltage, $\mathbf{E}_g$, and the terminal current, $\mathbf{I}_a$. It also is positive when the current leads the voltage.

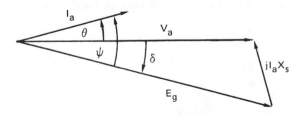

Figure 7.16 Angles for a Synchronous Motor

The apparent power can be rewritten from equation 7.104 as

$$S = P + jQ = V_a I_a^*$$
$$= E_g I_a^* + jI_a I_a^* X_s \qquad 7.109$$

Using the angles defined in figure 7.16,

$$S = V_a\,|I_a|\cos\theta + jV_a\,|I_a|\sin\theta \qquad 7.110$$
$$S = |E_g|\,|I_a|\cos\psi$$
$$\quad - j(|E_g|\,|I_a|\sin\psi - |I_a|^2\,X_s) \qquad 7.111$$

Using figure 7.16 and equations 7.110 and 7.111, the following useful formulas can be derived.

$$|I_a X_s|^2 = \frac{|E_g|^2}{2} + \left(\frac{Q}{q}\right) X_s$$

$$\pm \sqrt{\frac{|E_g|^4}{4} + \frac{|E_g|^2 QX_s}{q} + \frac{(PX_s)^2}{q^2}} \qquad 7.112$$

$$|E_g|^2 = |I_a|^2\,X_s^2 + \frac{P^2 + Q^2}{q^2\,|I_a|^2} - \frac{2QX_s}{q} \qquad 7.113$$

$$X_s = \frac{Q}{q\,|I_a|^2} \pm \sqrt{\frac{|E_g|^2}{|I_a|^2} - \frac{P^2}{q^2\,|I_a|^4}} \qquad 7.114$$

$$|I_a X_s|^2 = |V_a|^2 + |E_g|^2 - 2\,|E_g|\,|V_a|\cos\delta \qquad 7.115$$

Example 7.14

A three-phase synchronous machine with a line-to-line voltage of 440 volts and a synchronous speed of 900 rpm operates with a power of 9 kW and a lagging power factor of 0.8. The synchronous reactance per phase is 10 ohms. Determine (a) the torque angle, δ, in electrical degrees and (b) the generated voltage, E_g. If the overall machine efficiency is 90%, determine (c) the torque delivered to the output shaft.

The per-phase current is

$$I_a = \frac{\left[\dfrac{(9)(1000)}{3}\right]}{254 \times 0.8}\angle-\arccos(0.8) = 14.76\angle-36.9°$$

The per-phase reactive power is

$$\frac{Q}{q} = \frac{9\,\text{kW}}{3 \times (0.8)}\sin 36.9° = 2.25\,\text{kVAR}$$

From equation 7.113,

$$E_g = 203.3\,\text{volts}$$

From equation 7.115,

$$\cos\delta = \frac{-I_a^2 X_s^2 + V_a^2 + E_g^2}{2E_g V_a} = 0.8140$$

$$\delta = \arccos(0.8140) = -35.51°$$

The input power less the losses is the load power.

$$P_L = 0.9 \times 9\,\text{kW} = 8.1\,\text{kW}$$

$$T = \frac{P_L}{\Omega} = \frac{8100}{900 \times \frac{2\pi}{60}}$$

$$= 85.9\,\text{N-m}$$

C. Power Factor Correction

Power factor correction of synchronous motors can be accomplished by adjusting the rotor current, since E_g is approximately proportional to the rotor current. The phasor diagram of figure 7.16 illustrates the principle involved.

For a leading PF (positive θ), the voltages are related by figure 7.16.

$$|E_g|^2 = |V_a|^2 + 2|V_a|\,|I_a X_s|\sin\theta + |I_a X_s|^2 \qquad 7.116$$

The input power per phase is

$$\frac{P}{q} = |V_a|\,|I_a|\cos\theta \qquad 7.117$$

Substituting equation 7.117 into equation 7.116, the generated voltage with a leading power factor angle is

$$|E_g|^2 = |V_a|^2 + 2X_s\left(\frac{P}{q}\right)\tan\theta$$

$$+ \frac{X_s^2\left(\frac{P}{q}\right)^2}{|V_a|^2}\sec^2\theta \qquad 7.118$$

$$\sec^2\theta = \frac{1}{\text{PF}^2} \qquad 7.119$$

Example 7.15

The machine in example 7.14 was operating with a rotor current of 5 amps. It is desired to continue carrying the same load, but provide 5 kVAR of power factor correction to the line. Determine the required rotor current to accomplish this.

E_g can be calculated from equation 7.118. θ is taken as negative so that $\tan\theta$ is negative and $\sec^2\theta$ is positive.

From example 7.14,

$$E_g = 203.3\,\text{volts}$$

From equation 7.100,

$$E_g = K_r I_R$$

So,

$$K_r = \frac{203.3}{5} = 40.66$$

The power factor is corrected to

$$\text{PF} = \frac{9}{\sqrt{5^2 + 9^2}} = 0.874$$

$$\theta = \arccos(0.874) = 29.06°$$

Again, from equation 7.118,

$$|E_g|^2 = 116 \times 10^3$$

$$|E_g| = 341\,\text{volts}$$

Then, equation 7.100 gives the rotor current

$$I_R = \frac{341}{40.66} = 8.39\,\text{amps}$$

D. Synchronous Capacitors

Synchronous capacitors are machines run with no load (only rotational losses). The power angle, δ, is nearly zero, so that $\mathbf{V}_a$ and $\mathbf{E}_g$ are in phase. At no-load, the vector relationship is

$$j\mathbf{I}_a X_s = \mathbf{V}_a - \mathbf{E}_g \qquad 7.120$$

$\mathbf{I}_a$ leads $\mathbf{V}_a$ by 90° if $|E_g| > |V_a|$. $\mathbf{I}_a$ lags $\mathbf{V}_a$ by 90° if $|E_g| < |V_a|$.

The magnitude of I_a is given by equation 7.120. Thus, for a synchronous capacitor,

$$\frac{Q}{q} = V_a I_a = \frac{E_g V_a - V_a^2}{X_s} \qquad 7.121$$

Example 7.16

The machine of examples 7.14 and 7.15 is to provide 10 kVAR power factor correction when running lightly loaded. Determine the rotor current necessary.

From equation 7.121,

$$E_g = \frac{X_s\left(\frac{Q}{3}\right)}{V_a} + V_a$$

$$= \frac{10 \times 10^4}{3 \times 254} + 254 = 385.23\,\text{volts}$$

Then, by direct proportion,

$$I_R = 5 \times \frac{385.23}{203.3} = 9.47 \text{ amps}$$

E. Synchronous Generators

Synchronous generators are typically tied into a grid. It is assumed that the terminal voltage and phase are set by the grid. The individual machines are then constrained to that operating frequency and phase. If the prime mover of the synchronous generator does not continuously supply positive torque, the machine will behave as a motor. Thus, the power angle, δ, must be positive, with E_g leading V_a.

If current direction is reversed from that of the motor, all of the previous equations can be used with that change.

Figure 7.17 shows the current and voltage vectors for a synchronous generator. The *over-excited case* ($|E_g| \cos \delta > V_a$) is where there is a lagging power factor. The *under-excited case* ($|E_g| \cos \delta < V_a$) is where the power factor is leading. For simplicity, describe $\mathbf{E}_g$ as a vector $E \cos \delta + jE \sin \delta$ and $\mathbf{I}_a$ as $I \cos \theta + jI \sin \theta$, where $E = |E_g|$ and $I = |\mathbf{I}_a|$.

As the terminal voltage, $\mathbf{V}_a$, is set by the grid, it is given a phase angle of zero. For generator action, $\mathbf{E}_g$ must lead $\mathbf{V}_a$. Then, the following equations describe the terminal quantities.

$$\mathbf{I}_a = \frac{E_g - V_a}{jX_s} = \frac{E \sin \delta}{X_s} + j\frac{V_a - E \cos \delta}{X_s} \qquad 7.122$$

$$(IX)^2 = V_a^2 + E^2 - 2V_a E \cos \delta \qquad 7.123$$

$$E^2 = V_a^2 + (IX)^2 - 2V_a IX \sin \theta \qquad 7.124$$

As E is roughly proportional to the rotor current, I_R, it is possible to use equations 7.123 and 7.124 when there are changes in power, power factor, or rotor current. Since $S = P + jQ = V_a I_a^*$,

$$P = V_a I \cos \theta = \left(\frac{V_a E}{X_s}\right) \sin \delta \qquad 7.125$$

$$Q = -V_a I \sin \theta = \frac{V_a E \cos \delta - V_a^2}{X_s} \qquad 7.126$$

Example 7.17

A 50 MVA, 13.8 kV, 3600 rpm, 60 Hz grid-connected generator with $X_s = 3.8$ ohms reactive (per phase) operates with a lagging PF of 0.88 at rated voltage and current. Determine the new power factor if the rotor current is decreased by 5%.

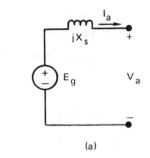

(a)

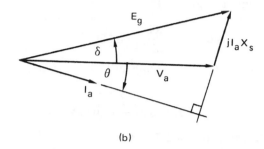

(b)

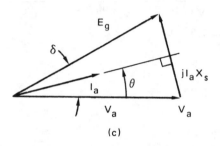

(c)

Figure 7.17 Synchronous Generator (a) Per-Phase Model, (b) Over-Excited with Lagging Power Factor, (c) Under-Excited with Leading Power Factor

As the prime-mover's power has not changed, it is assumed that the output power is constant. Thus, $E \sin \delta = IX_s \cos \theta$ is constant. The phase voltage is

$$\frac{13.8}{\sqrt{3}} = 7.967$$

The rated current per phase is

$$I_{\text{phase}} = \frac{\dfrac{50 \times 10^6}{3}}{7.967 \times 10^3} = 2092$$

$$I = 1841 - j994$$

$$IX \sin \theta = 3776$$

Using equation 7.124, and noting that θ and $\sin \theta$ are negative,

$$E^2 = (7.967)^2 + (3.8 \times 2.092)^2 - 2 \times 7.967 \times (-3.776)$$
$$= 186.8 \text{ (in thousands)}$$

Therefore,
$$E_1 = 13.67\,\text{kV}$$

Since the generated voltage is proportional to I_R,
$$E_2 = 0.95E_1 = 12.99\,\text{kV}$$

The output power is constant so $E\sin\delta$ is constant.
$$I_1 X\cos\theta_1 = 6.996\,\text{kV}$$

Referring to figure 7.17,
$$\sin\delta_2 = \frac{7.00}{12.99} = 0.539$$
$$\cos\delta_2 = \cos[\arcsin(0.539)] = 0.842$$

Therefore,
$$E_2\cos\delta_2 = 10.94\,\text{kV}$$

Then, from equation 7.123
$$I_2 X = \sqrt{(7.967)^2 + (12.99)^2 - 2\times 7.967\times 10.94}$$
$$= 7.61\,\text{kV}$$
$$\cos\theta_2 = \text{PF}_2 = \frac{7.00}{7.61}$$
$$= 0.92$$

Example 7.18

The machine of example 7.17 is to provide a lagging reactive power of 40 MVAR and still operate at its rated current. What rotor current is necessary to accomplish this?

In this case, I remains constant, and the input power varies.
$$\sin\theta_2 = \frac{40}{50} = 0.8$$

From equation 7.124,
$$E^2 = (7.967)^2 + (3.8\cdot 2.092)^2 + 2\cdot 7.967\cdot 2.092\cdot 3.8\cdot 0.8$$
$$E = 15.1\,\text{kV}$$

Since the original value from example 7.17 was 13.67 kV, the percent increase in rotor current is
$$\frac{15.1 - 13.67}{13.67}\times 100 = 10.48\%$$

E is proportional to rotor current, and it will also experience a 10.5% increase.

Example 7.19

The machine in example 7.17 has its prime-mover power decreased by 25%. If the rotor current decreases by the same percentage, what will the new power and power factor be?

Since the input power drops by 25%, the output power also will drop by 25%. However, V_a will remain constant at 13.8 kV.
$$I_2 X_s\cos\theta_2 = 0.75(I_1 X_s\cos\theta_1) = 5.247\,\text{kV}$$

Similarly,
$$E_2\sin\delta_2 = 0.75(E_1\sin\delta_1)$$

However, the rotor current drops by 25%. Therefore,
$$E_2 = 0.75E_1$$

From equation 7.100, E is proportional to I_R.
$$\sin\delta_2 = \sin\delta_1$$
$$\cos\delta_2 = \cos\delta_1$$

Since $E_1 = 13.67$, equation 7.125 results in
$$I_1 X_s\cos\theta_1 = E_1\sin\delta_1$$
$$7.00 = 13.67\sin\delta_1$$
$$\delta_1 = 30.8°$$
$$\cos\delta_1 = 0.86$$
$$E_1\cos\delta_1 = 13.67\times 0.86 = 11.76$$
$$E_2\cos\delta_2 = 0.75\times 11.76 = 8.82$$

Thus from equation 7.123,
$$I_2 X_s = \sqrt{(7.967)^2 + (0.75\cdot 13.67)^2 - 2\cdot 7.967\cdot 8.82}$$
$$= 5.30\,\text{kV}$$

However,
$$I_2 X_s\cos\theta_2 = 5.247$$

Therefore,
$$\text{PF}_2 = \cos\theta_2 = \frac{5.247}{5.30}$$
$$= 0.99$$

F. Power Factor Relations for Synchronous Generators

A three-phase generator operating with constant load torque has
$$T_{\text{shaft}} = \frac{3V_a E_g}{X_s\Omega_s}\sin\delta \qquad 7.127$$

E_g is approximately proportional to the rotor current, I_R. Figure 7.18(a) shows the phasor diagram with a dashed line for the locus of $\mathbf{E}_g$ at constant load.

The line perpendicular to $\mathbf{V}_a$ is the minimum value of $jX_s\mathbf{I}_a$. The angle θ is the power factor angle. From trigonometric considerations,

$$\text{PF} = \cos\theta$$

$$= \frac{jX_sI_{a,\min}}{jX_sI_a} = \frac{I_{a,\min}}{I_a} \qquad 7.128$$

With E_g proportional to I_R, it is common to plot I_a versus I_R at constant load. This results in the so-called *vee curve* as shown in figure 7.18(b).

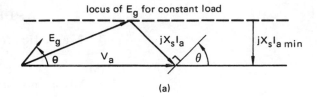

(a)

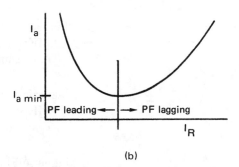

(b)

Figure 7.18 (a) Generator Vector Diagram, (b) Vee Curve for Constant Load

APPENDIX A

Magnetic Conversion Factors

MMF
    MKS: amp-turns
    cgs: 1 gilbert $= 0.4\pi$ amp-turns

Flux, ϕ
    MKS: webers (volt-sec)
    cgs: 1 maxwell $= 1$ line $= 10^{-8}$ webers

Magnetic Field Intensity, H
    MKS: amp-turns/m
    cgs: 1 oersted $= (250/\pi)$ amp-turns/m

Magnetic Flux Density, B
    MKS: 1 tesla $= 1$ weber/m^2
    cgs: 1 gauss $= 10^{-4}$ teslas

APPENDIX B
Power and Energy Conversions

multiply	by	to get
BTU	3.929×10^{-4}	hp-hr
BTU	778.3	ft-lbf
BTU	2.930×10^{-4}	kW-hr
BTU	1.0×10^{-5}	therm
BTU/hr	0.2161	ft-lbf/sec
BTU/hr	3.929×10^{-4}	hp
BTU/hr	0.2930	watts
ft-lbf	1.285×10^{-3}	BTU
ft-lbf	3.766×10^{-7}	kW-hr
ft-lbf	5.051×10^{-7}	hp-hr
ft-lbf/sec	4.6272	BTU/hr
ft-lbf/sec	1.818×10^{-3}	hp
ft-lbf/sec	1.356×10^{-3}	kW
hp	2545.0	BTU/hr
hp	550	ft-lbf/sec
hp	0.7457	kW
hp-hr	2545.0	BTU
hp-hr	1.976×10^{6}	ft-lbf
hp-hr	0.7457	kW-hr
kW	1.341	hp
kW	3412.9	BTU/hr
kW	737.6	ft-lbf/sec

PRACTICE PROBLEMS

Warmups

1. A DC generator has a 250 volt output at 50 amp load current and 260 volts at no-load. Estimate the armature resistance.

2. A separately excited DC generator is rated 10 kW, 240 volts, 2800 rpm for input power of 15 hp. No-load voltage is 260 volts. Determine the armature resistance and the mechanical power loss at 2800 rpm.

3. A DC shunt motor is running at rated values when the line voltage suddenly drops by 10%. Determine the resulting changes in input power and motor speed if (a) the load torque is constant, and (b) the load torque is proportional to speed.

4. A DC series motor is running at rated speed, voltage, and current when the load torque increases by 25%. Determine the effects of this change, assuming that the line voltage remains the same.

5. A four-pole, 60 Hz, three-phase induction motor is drawing 25 kW under load and running at 1700 rpm. The no-load losses have been measured at 850 watts. The stator copper losses have been measured at 550 watts. Determine the hp and efficiency of this machine.

6. A no-load test on a 440 volt (line-to-line) three-phase induction motor yields 1200 watts and 2500 VA. Determine the per-phase parameters r_c ($1/g_c$) and x_Φ ($-1/b_c$).

7. A blocked rotor test on a 440 volt (line-to-line) three-phase induction motor finds a line current of 21 amps, $P_{\text{in}} = 375$ watts, and power factor of 0.37. Find the per-phase values of R_e and X_e.

8. The motor of problems 6 and 7 is run at a slip of 0.03 with an output of 20 hp and an input power of 16.95 kW. Determine r_s and $a^2 r_r$ on a per-phase basis.

9. A three-phase, 440 volt (line-to-line) synchronous motor has $X_s = 20$ ohms and armature resistance of 0.5 ohm per phase. The synchronous speed is 1800 rpm. The no-load field current is adjusted for minimum armature current at $I_f = 5$ amps DC, at which time the input power is 900 watts. This machine is to provide 10 kVAR of power factor correction when running lightly loaded. Determine the rotor current required.

10. A 50 MVA, 13.8 kV, 3600 rpm synchronous motor with per-phase reactance X_s of 2.8 ohms at 60 Hz is operating into a grid. The power factor is 0.88 lagging at rated voltage and current. Determine the new power factor if the rotor current is increased by 10%.

Concentrates

1. A DC generator is rated 10 kW, 240 volts, and 2800 rpm with 15 hp. Its armature resistance is 1.0 ohm, and it has a no-load voltage of 260 volts at $I_F = 2.75$ amps and 2800 rpm. Using the magnetization curve of figure 7.2 normalized to the no-load values, determine the shunt field resistance necessary to self-excite this generator at rated voltage and current at 2800 rpm.

2. A DC generator rated 10 kW, 240 volts, 2800 rpm, and 15 hp has an armature resistance of 1.0 ohm, a self-excitation shunt field resistance of 77 ohms, and follows the normalized magnetization curve of figure 7.2 for $E_N = 260$ volts and $\text{MMF}_n = 2.75$ amps. Find the voltage regulation if the normalized $K_V \Phi$ is 1.2 for normalized MMF > 1.5 in figure 7.2.

3. A DC shunt motor takes 200 volts and 25 amps at 1000 rpm and 5 hp. It has an armature resistance of 0.5 ohm and a shunt field resistance of 100 ohms. Determine (a) the rotational losses, and (b) the no-load speed at 200 volts.

4. A 500 volt, 330 amp, 200 hp DC series motor is rated at 900 rpm. With $R_a = 0.08$ ohm, find the starting current and torque at rated voltage.

5. A 440 volt (line-to-line) induction motor has per-phase parameters $X_e = 0.712$ ohm, $R_e = 0.283$ ohm, and $r_s = 0.126$ ohm when it is connected in a delta configuration. The synchronous speed is 1200 rpm. Determine the starting current and the starting torque if (a) the stator windings are connected in the delta configuration, and (b) if the stator windings are connected in the wye configuration.

6. A 440 volt (line-to-line) induction motor with synchronous speed of 1200 rpm has per-phase parameters $X_e = 0.943$ ohm, $R_e = 0.231$ ohm, and $r_s = 0.135$ ohm. The starting of this motor is to be controlled through a tapped auto-transformer which is in a wye-connection. The maximum line current is to be limited to 150 amps, and switching to the next higher tap is to take place when the line current drops to 100 amps. Specify the voltage taps and the switching speeds for the starting transformer.

7. A three-phase, 440 volt (line-to-line), 60 Hz, 3600 rpm synchronous motor with per-phase X_s of 20 ohms

and armature resistance of 0.5 ohm has its field current adjusted for minimum armature current at $I_f = 5$ amps, at which time the input power is 900 watts. For a rated current $I_a = 12$ amps, determine (a) the power angle, (b) the power factor angle, (c) the load power, (d) the efficiency, and (e) the new rotor current necessary to provide 5 kVAR of power factor correction at these operating conditions.

8. A 50 hp, 50 kVA, 1746 rpm squirrel-cage induction motor has phase windings rated at 440 volts. Under a light-running test conducted at a line voltage of 440 V, the results are $P + jQ = 3,342 + j19,750$ volt-amperes including all three phases. If the starting current under full line voltage (440 volts line-to-line) is found to be $367\angle -68°$ A, determine the starting torque under the same conditions.

9. A 440 volt, three-phase synchronous motor is driving a 45 horsepower pump. Assume the motor is 93% efficient. The rotor current is adjusted until the line current is at a minimum, occurring when the rotor current is 8.2 DCA. Determine the rotor current necessary to provide a leading power factor of 0.8.

10. A processing plant draws 1000 kW at 0.8 lagging power factor. The air-handling equipment is driven by a 100 hp induction motor which must be replaced. This induction motor requires 80 kW at a lagging power factor of 0.85. If the replacement is a 100 hp synchronous motor of the same efficiency as the existing induction motor, and rated to operate at a power factor of 0.8, what improvement of the plant power factor can be achieved?

Timed

1. A DC generator rated 10 kW, 240 volts, 2800 rpm, and 15 hp has $R_a = 0.48$ ohm. It has a no-load armature voltage of 260 volts at 2800 rpm when a separate excitation of 2.75 amps is applied to the field winding. Use the magnetization curve of figure 7.2 normalized to the no-load values to determine the ratio of series field turns to shunt field turns, N_s/N_f, and the shunt field resistance necessary for flat compounding at rated values.

2. A DC shunt motor is rated 7.5 hp, 60 amps, 1000 rpm, and 120 volts. Assume the load torque is constant, regardless of speed. Give circuit diagrams with parameter values to change the speed to (a) 1150 rpm, and (b) 750 rpm.

3. Design a starting circuit for the armature of a 250 volt, 125 amp, 1150 rpm, 35 hp DC shunt motor,

assuming the line voltage of 250 volts is always connected to the shunt field winding circuit. The armature resistance is 0.15 ohm, and the maximum starting current is 250% of rated current.

4. A 200 hp squirrel-cage (induction) motor is rated at 460 volts (line-to-line) and 1760 rpm. At no-load, it requires 4.5 kW at 460 volts, 135 amps, and turns at 1798 rpm. Under locked rotor conditions, it draws 240 amps at 78 volts and produces a torque of 17.2 ft-lb. The motor is started on an 80% tap. Determine (a) the starting torque, and (b) the starting current.

5. A 60 Hz, 480 volt, 50 kVA alternator with rated field current of 20 amps is driven at a constant 30 hp while the following measurements are taken:

I_R (amps)	2	3	5.5	8	12	15	20	23	27
I_a (amps)	60	45	30	21	15	18	32	45	60

For the same load, determine (a) the rotor current necessary for an 80% lagging power factor, (b) the rotor current necessary to obtain $I_a = 40$ amps with a leading power factor, and (c) the resulting power factor.

6. A 50 hp, 50 kVA, 1746 rpm squirrel-cage induction motor has phase windings rated at 440 volts. With the windings connected in a delta configuration, a blocked rotor test is carried out under rated current conditions. The result in per-phase volt-amperes is found to be $P + jQ = 1097 + j2769$ volt-amperes. This machine is fed from a three-phase, 50 kVA, 440 volt transformer which has 10% reactance per phase (on its own base). Determine the dip in the line voltage when the motor is started. The transformer can be modeled as a wye connection of 254 volt sources in series with the 10% reactance on each phase.

7. A 50 hp, 50 kVA, 1746 rpm squirrel-cage induction motor has phase windings rated at 440 volts and is connected in delta. A blocked rotor test yields per-phase impedance of $0.255 + j0.643$ ohm. A light running test yields per-phase impedance of $1.614 + j9.531$ ohms. If the windings were connected in a wye with 762 volts line-to-line, or 440 volts line-to-neutral, determine (a) the new starting line current, and (b) the ratings of the machine in this configuration.

8. A 50 hp, 50 kVA, 1746 rpm squirrel-cage induction motor has phase windings rated at 440 volts and is normally connected in delta. Its starting current is 367 A and 38.3 kW per phase when started directly across the line, and under those conditions it develops a starting torque of 190 lb-ft. A blocked rotor test at rated

current results in a voltage of 78.6 V and a power of 1.1 kW per phase. A light-running test at rated voltage results in a current of 25.9 A and 1.1 kW per phase. If this machine is started in the wye configuration, determine its starting current and torque.

9. A large three-phase alternator is providing rated volt-amperes to an even larger power system at a lagging power factor of 0.9. If the rotor current is reduced by 10%, determine the percentage change in the output current and the new power factor if the synchronous reactance is 100% and the armature resistance is negligible.

10. A three-phase alternator is providing power which is 40% of rated kVA to a large power system. It is under-excited so that its open-circuit voltage would be 75% of the system voltage. The output power is scheduled to be doubled. Assume 100% synchronous reactance and negligible Y resistance. (a) Determine the per-unit line current and the power factor of the alternator when running as above, at $P = 40\%$ of rated kVA. (b) Determine the minimum change of rotor current necessary to obtain $P = 80\%$ of rated kVA without exceeding any generator ratings. (c) Under the excitation determined in part (b), determine the new generator line current and power factor.

FUNDAMENTAL SEMICONDUCTOR CIRCUITS

8

Nomenclature

A_{cm}	common-mode voltage gain	volts/volt
A_{dm}	differential-mode voltage gain	volts/volt
A_i	current gain	–
A_v	voltage gain	–
C	Celsius	–
$C_{b'c}$	base-collector capacitance	farad
$C_{b'e}$	base-collector capacitance	farad
C_d	diffusion capacitance	farad
C_{gd}	gate-drain capacitance	farad
C_{gs}	gate-source capacitance	farad
C_t	transition capacitance	farad
C_{ISS}	gate-channel capacitance	farad
C_{OSS}	output capacitance (FET)	farad
CMMR	common-mode rejection ratio	dB
f	functions	–
f_{cf}	corner frequency	hertz
GBW	gain-bandwidth product	rad/s
g_m	transconductance	mhos
h_{ie}	common-emitter input impedance	ohms
h_{fe}	common-emitter current gain	–
h_{oe}	common-emitter output admittance	mhos
h_{re}	common-emitter feedback voltage gain	–
i_b	small-signal base current	amps
i_c	small-signal collector current	amps
i_d	small-signal drain current	amps
i_e	small-signal emitter current	amps
i_{in}	small-signal input current	amps
i_l	small-signal load current	amps
i_o	small-signal output current	amps
i_B	variable base current	amps

i_C	variable collector current	amps
i_D	variable drain current	amps
i_E	variable emitter current	amps
i_{PN}	variable P-N junction current	amps
I_{BQ}	bias level of base current	amps
I_{CO}	collector reverse saturation current	amps
I_{CQ}	bias level of collector current	amps
I_{DQ}	bias level of drain current	amps
I_{DSS}	drain saturation current level with $v_{GS} = 0$	amps
I_{EO}	emitter reverse saturation current	amps
I_{EQ}	bias value of emitter current	amps
I_{REF}	reference current	amps
I_S	constant current source current	amps
k	(with T) Boltzmann's constant	1.38×10^{-23} joules/K
k	a constant	–
k_Q	ratio: $I_{CQ,max}/I_{CQ,min}$	–
K	Kelvin scale	–
K	enhancement FET constant	amps/V^2
q	charge of an electron	1.609×10^{-19} coulombs
Q	quiescent or operating point	–
r	small-signal	ohms
$r_{bb'}$	small-signal base bulk resistance	ohms
$r_{b'e}$	small-signal base spreading resistance	ohms
r_c	small-signal collector resistance	ohms
r_d	small-signal diode resistance	ohms

PROFESSIONAL PUBLICATIONS ● Belmont, CA

r_{ds}	small-signal drain-source resistance	ohms
r_m	transresistance	ohms
r_o	small-signal output resistance	ohms
R	resistance	ohms
S_v	stability factor	mA/°C
T	temperature	Kelvins
v_{bc}	small-signal base-collector voltage drop	volts
v_{be}	small-signal base-emitter voltage drop	volts
v_{ce}	small-signal collector-emitter voltage drop	volts
v_{dm}	differential-mode voltage	volts
v_{ds}	small-signal drain-source voltage drop	volts
v_e	small-signal emitter voltage	volts
v_{ec}	small-signal emitter-collector voltage drop	volts
v_{gs}	small-signal gate-source collector voltage drop	volts
v_s	small-signal voltage	volts
v_{sd}	small-signal source-drain voltage drop	volts
v_{sg}	small-signal source-gate voltage drop	volts
v_{BC}	variable base-collector voltage drop	volts
v_{BE}	variable base-emitter voltage drop	volts
v_{CE}	variable collector-emitter voltage drop	volts
v_{CM}	common-mode voltage	volts
v_D	generic FET channel voltage	volts
v_{DG}	variable drain-gate voltage drop	volts
v_{DS}	variable drain-source voltage drop	volts
v_{EB}	variable emitter-base voltage drop	volts
v_{EC}	variable emitter-collector voltage drop	volts
v_G	generic FET control voltage	volts
v_{GD}	variable gate-drain voltage drop	volts
v_{GS}	variable gate-source voltage drop	volts
v_{NP}	variable N-P voltage drop	volts
v_{PN}	variable P-N voltage drop	volts
v_{SD}	variable source-drain voltage drop	volts
v_{SG}	variable source-gate voltage drop	volts

V_A	Early voltage (constant)	volts
V_{BR}	avalanche breakdown level	volts
V_{CEQ}	collector-emitter bias voltage level	volts
V_{DD}	drain supply voltage	volts
V_{DSQ}	drain-source bias voltage level	volts
V_F	P-N junction forward threshold voltage level	volts
V_{GG}	gate supply voltage	volts
$V_{GS(off)}$	turn-off gate-source voltage level for depletion-mode N-channel FET	volts
$V_{GS(on)}$	enhancement N-channel FET threshold voltage level	volts
V_{GSQ}	gate-source bias voltage level	volts
V_P	pinchoff voltage	volts
V_R	P-N junction reverse threshold voltage level	volts
V_S	constant source voltage	volts
$V_{SG(off)}$	turn-off source-gate voltage level for depletion-mode P-channel FET	volts
$V_{SG(on)}$	enhancement P-channel FET threshold voltage level	volts
V_T	thermal voltage	volts
Y	admittance	mhos
Z_{in}	input impedance	ohms
Z_{out}	output impedance	ohms

Symbols

α_o	common-base DC current gain	–
α_N	normal common-base current gain	–
α_R	reverse common-base current gain	–
β	common-emitter current gain	–
η	a number	–
ω	frequency	rad/s
ω_{cf}	cut-off frequency	rad/s

1   P-N JUNCTION CHARACTERISTICS

A. Introduction

Pure silicon and germanium are poor conductors. They must be doped with a column III element such as boron to become *P-type* (*P* for positive charge carriers) semiconductors, or with a column V element such as phosphorous to become *N-type* (*N* for negative charge carriers) semiconductors.

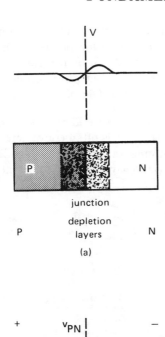

junction

depletion
P layers N

(a)

When doping is changed from P-type to N-type over a short distance, a *semiconductor junction* (*P-N junction*) is formed. Near this junction (where the effective doping level is zero), a thermally generated *depletion region* exists. Here the charge carriers have diffused away from the junction, and an internal *voltage barrier* exists. See figure 8.1(a).

With the application of an external voltage, the voltage barrier will lower or raise depending on the polarity of the applied voltage. Correspondingly, the depletion layer will narrow or widen depending on the polarity of the externally applied voltage. This is indicated in figure 8.1(b) or (c).

When the applied junction voltage exceeds the potential barrier, the current flow is approximately exponential with the voltage. The potential barrier is about 0.6 volt for silicon and 0.3 volt for germanium. This corresponds to a *forward-bias condition* as indicated in figure 8.1(b).

In the *reverse-bias condition*, figure 8.1(c), the current is approximately constant due to *minority carrier conduction* (holes in the N-region and electrons in the P-region) which thermally diffuse to the depletion region and are swept across. These minority carriers are produced from a thermal phenomenon which is constant at a given temperature.

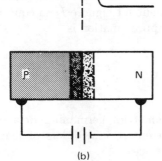

(b)

Thus, minority carrier generation is constant, resulting in a constant reverse current. This reverse current is called the *reverse saturation current*, I_S. Over a certain range of applied voltages, the behavior of the junction can be described approximately by the *ideal P-N diode equation*.

$$i_{PN} = I_S \left(e^{qv_{PN}/kT} - 1 \right) \qquad 8.1$$

The subscript *PN* indicates that the positive (forward-bias) direction of current is from the P-side to the N-side of the junction, and the positive polarity of voltage is to drop from the P-side to the N-side.

B. Avalanche Breakdown of P-N Junctions

The width of the depletion region is inversely proportional to the doping level and proportional to the $1/x$ power of the applied voltage v_{NP}.

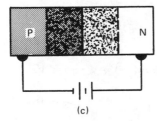

(c)

$$\text{depletion width} \propto \frac{1}{\text{doping level}} v_{NP}^{1/x} \quad (2 < x < 3) \quad 8.2$$

Figure 8.1 P-N Junction Voltage Barrier and Depletion Regions: (a) No Applied Voltage, (b) Voltage Dropping From P to N, and (c) Voltage Rising From P to N

When the junction is reverse biased, the depletion region widens. At the same time, however, an electric field proportional to the voltage and inversely proportional to the depletion-layer width is established. When the field becomes strong enough, some of the electrons in the atomic bonds are ionized and a current flows.

If the depletion region is very narrow due to intentional high doping density on both sides of the junction, the ionization of the bonding electrons is the major source of current. This ionization is called the *Zener effect*. If the depletion regions are lightly doped so that a wider depletion layer occurs, the ionized electrons collide with other atoms and thereby cause the release of more charge carriers. This is called the *avalanche effect*.

For any P-N junction with reverse bias, there is *reverse breakdown*, usually called *Zener breakdown*. Actual Zener breakdown occurs only for very high doping densities on both sides of the junction and in the range of only a few volts. *Avalanche breakdown* is the more usual mechanism in semiconductors. By changing the doping level, avalanche breakdown can be made to occur over a wide range of voltages. This effect is used in *voltage regulator diodes*, usually misnamed *zener diodes*.

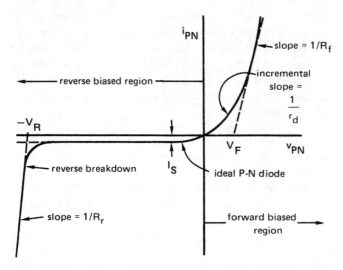

Figure 8.2 Low-Frequency Characteristic of P-N Junctions

C. Low-Frequency Characteristics of P-N Junctions

For frequencies low enough that capacitive effects of the junction can be ignored, and in forward-biased operation, i.e., with positive v_{PN}, the voltage-current relationship is similar to the ideal diode equation as given in equation 8.1. The difference in practical junctions is that the equation becomes

$$i_{PN} = I_S(e^{v_{PN}/\eta V_T} - 1) \qquad 8.3$$

V_T is the thermal voltage,

$$V_T = \frac{kT}{q} = 86.21 \times 10^{-6}T \qquad 8.4$$

T is in Kelvin ($273\,\text{Kelvin} = 0°\text{C}$). The number η depends on the manufacturing process. For discrete silicon diodes η is about 2, but in integrated circuits, where a P-N diode is realized using a junction transistor, the value of η is close to 1.0. Germanium diodes have $\eta = 1$.

When a junction is operated at very nearly constant voltage and current, it is often convenient to characterize it by its incremental resistance which is obtained by taking the partial derivative of the current with respect to the voltage.

$$\frac{1}{r_d} = \frac{\partial i_{PN}}{\partial v_{PN}} = \frac{1}{\eta V_T}I_S e^{v_{PN}/\eta V_T} \qquad 8.5$$

But, $I_S e^{v_{PN}/\eta V_T} = i_{PN} - I_S$, so

$$\frac{1}{r_d} = \frac{i_{PN} - I_S}{\eta V_T} \qquad 8.6$$

A silicon diode with $\eta = 2$ operating at room temperature where $V_T = 0.025$ volt, and with $v_{PN} = 0.70$ volt has its current on the order of 10^6 times I_S, so it is certainly reasonable to ignore I_S in equation 8.6. Thus, a reasonable approximation is

$$r_d = \eta\frac{V_T}{i_{PN}} \qquad 8.7$$

For similar geometries, germanium devices have reverse saturation current several orders of magnitude larger than silicon devices.

D. Piecewise Linear Model

When the voltage applied to the P-N junction varies widely, it is often convenient to use a piecewise linear model. The piecewise linear model is obtained by finding the best linear approximations to the diode characteristic in the three regions separated by the voltages $-V_R$ and V_F.

- Forward-bias region, $v_{PN} > V_F$

$$i_{PN} = \frac{v_{PN} - V_F}{R_f} \qquad 8.8$$

- Off-region, $-V_R < v_{PN} < V_F$

$$i_{PN} = 0 \qquad 8.9$$

- Reverse breakdown region, $v_{PN} < -V_R$

$$i_{PN} = \frac{v_{PN} + V_R}{R_r} \qquad 8.10$$

A single equivalent circuit can be obtained from the definition of an ideal diode. An *ideal diode* will allow any amount of current to flow in the forward direction without forward voltage drop. An ideal diode also will not allow any reverse current to flow, regardless of how much reverse voltage is applied.

Using ideal diodes, the piecewise linear model is obtained.

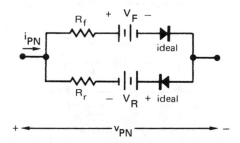

Figure 8.3 Piecewise Linear Model of Real Diode

E. Temperature Effects

Both the *thermal voltage*, V_T, and the reverse *saturation* current, I_S, are functions of temperature. V_T is directly proportional to temperature: $V_T = kT/q = 86.21 \times 10^{-6}T$. For silicon, the reverse saturation current doubles for an increase of 10°C.

With constant forward-biased current, the junction voltage, v_{PN}, decreases at a rate of 2.2 millivolts per degree Celsius.

$$\frac{\partial v_{PN}}{\partial T} = -2.2 \text{ mV/°C} \qquad 8.11$$

This relationship is useful in designing temperature measurement circuits.

F. Junction Capacitances

The main frequency effects in P-N junctions are two capacitances. The depletion region is related to the depletion-layer width. The *diffusion capacitance* is related to charge storage associated with forward current when the forward voltage is large (primarily in bipolar junction transistors).

The *depletion-region capacitance* is called the *transition capacitance*, C_t, and the *diffusion capacitance* is C_d. The transition capacitance is inversely proportional to the depletion-layer width and, therefore, is inversely proportional to $v_{NP}^{1/x}$ where x is again between two and three. Transition capacitance is the primary capacitance for reverse-bias operation. Both capacitances appear in parallel with the terminals.

The diffusion capacitance is proportional to the forward current, while the incremental resistance is inversely

proportional to the forward current. Therefore, the product $r_d C_d$ is constant and is known as the junction cutoff frequency.

$$r_d C_d = \omega_{cf} \qquad 8.12$$

2 JUNCTION FIELD EFFECT TRANSISTORS (JFETs)

A. Introduction

Field effect transistors (FETs) are divided into junction (JFET) and metal-oxide-semiconductor (MOSFET) types. These are further divided according to the doping in the channel as P-channel or N-channel. JFETs are commonly discrete devices (i.e., not part of an integrated circuit), and the P-channel configuration is rarely used because of inferior performance. On the other hand, most MOSFETs are used in integrated circuits, where N-channel devices (NMOS) predominate. However, both N-channel and P-channel configurations are used in so-called *complementary* circuits (CMOS).

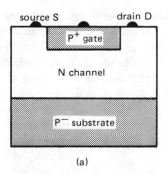

(a)

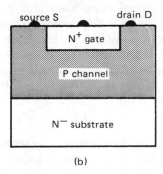

(b)

Figure 8.4 Junction FET
(a) N-Channel, (b) P-Channel

B. N-Channel JFET Characteristics

The fabrication of JFETs is indicated schematically in figure 8.4 with a lightly doped substrate, a moderately doped channel, and a heavily doped gate. Figure 8.4(a) shows the N-channel JFET, and figure 8.4(b) shows the P-channel JFET. Electrical connections are made to the regions indicated as *source*, *gate*, and *drain*.

The P-N junction between the gate and the channel is normally reverse biased. Transistor action is obtained by modulating the P-N junction's depletion region. As shown in figure 8.5(a), the depletion region exists primarily in the channel side of the P-N junction, because the gate region is much more heavily doped than the channel, and the depletion region on the gate side is negligible.

Figure 8.5(a) shows the N-channel JFET with the source, gate, and drain tied together, so that the P-N junction has zero bias.

With zero voltage between gate and source, and various values of drain-source voltage, the characteristic curve is indicated in figure 8.5(d) which shows three regions of operation. Beginning at $v_{\text{DS}} = 0$ the curve rises linearly at first, but next reaches a saturation region where the current changes along a small slope until an avalanche region is reached.

The first region of operation, which begins as a constant slope, is variously called the unsaturated region, the ohmic region, or the triode region. Because this region of operation extends to a point where it meets the flattest part, in this text it is referred to as the triode region. In this triode region of operation, the depletion layer expands toward the substrate as indicated in figure 8.5(b). In this region, the characteristic can be approximated by the following expression:

$$i_D = I_{\text{DSS}} \left(2\frac{v_{\text{DS}}}{V_P} - \frac{v_{\text{DS}}^2}{V_P^2} \right) \qquad 8.13$$

In the second region, where the characteristic is nearly flat, the depletion layer has essentially reached the substrate, but leaves a current path as indicated in figure 8.5(c). This region is variously called the saturation region, or the pinchoff region. In this text it is referred to as the *pinchoff region*. With $v_{\text{GS}} = 0$, a first-order approximation is

$$i_D = I_{\text{DSS}} \qquad 8.14$$

It can be seen that these two regions meet at $v_{\text{DS}} = V_P$, where V_P is called the *pinchoff voltage*.

Operation in the avalanche region is to be avoided as damage may occur because of overheating.

With reverse-bias voltage on the P-N junction obtained by applying a negative v_{GS}, saturation is reached at lower values of i_D. With a sufficiently large negative bias, the channel can be completely pinched off. The value of the negative bias necessary to turn the channel off is $-V_P$, where again V_P is the pinchoff voltage.

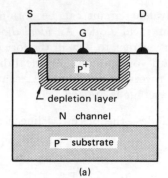

(a)

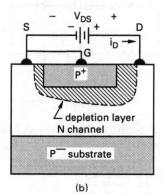

(b)

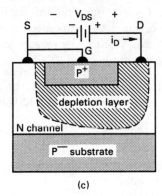

(c)

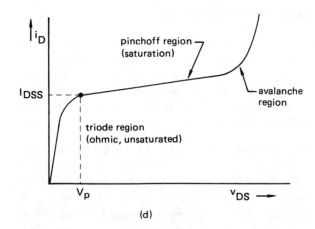

(d)

Figure 8.5 Zero-Biased N-Channel JFET (a) Depletion Layer with $v_{\text{DS}} = 0$, (b) Depletion Layer with Positive v_{DS} in Triode Operating Region, (c) Depletion Layer in Pinchoff Region of Operation, (d) Characteristic Curve

This is illustrated in figure 8.6(b), where a set of characteristics for an N-channel JFET with $V_P = 6$ volts is shown.

In the triode region, the drain current can be approximated by

triode:

$$i_D = I_{DSS}\left[2\left(1 - \frac{v_{GS}}{V_{GS(off)}}\right)\frac{v_{DS}}{V_P} - \frac{v_{DS}^2}{V_P^2}\right] \quad 8.15$$

$V_{GS(off)} = -V_P$, and v_{GS} is negative.

In the pinchoff region, the first-order approximation does not take into account the slope of the curves, and is given as

pinchoff—first approximation:

$$i_D = I_{DSS}\left(1 - \frac{v_{GS}}{V_{GS(off)}}\right)^2 \quad 8.16a$$

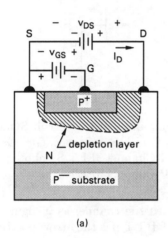

(a)

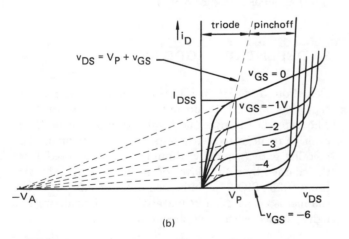

(b)

Figure 8.6 N-Channel JFET with $V_P = 6$ volts
(a) Wiring Configuration,
(b) Family of Characteristic Curves

A second approximation takes into account the slope of the pinchoff region characteristics, which tend to meet at a point $i_D = 0$ and $v_{DS} = -V_A$. V_A is dependent upon doping levels, and generally ranges between 50 and 200 volts, with 100 volts as a typical value.

pinchoff—second approximation:

$$i_D = I_{DSS}\left(1 - \frac{v_{GS}}{V_{GS(off)}}\right)^2\left(1 + \frac{v_{DS}}{V_A}\right) \quad 8.16b$$

In normal operation of an N-channel JFET, v_{GS} is negative, v_{DS} drops from drain to source, and i_D flows from drain to source.

The boundary between the triode and pinchoff regions is usually found by equating equations 8.15 and 8.16a, which gives the boundary as $v_{DS} = V_P + v_{GS}$. This is indicated by the dashed curve in figure 8.6(b).

C. P-Channel JFET Characteristics

For the P-channel JFET junction to be reverse biased, the gate lead must be made positive with respect to the source lead. To make the drain-end of the channel more strongly reverse biased than the source-end, it is necessary that the drain voltage be more negative than the source voltage. Thus, the characteristic curves could be shown in the third quadrant by using the same definitions as with the N-channel device. However, it is more common to redefine directions for the P-channel device using terms v_{SG} and v_{SD}, and defining i_D as flowing from source to drain. This results in a characteristic of the same form as figure 8.6, but with v_{SG} negative, v_{SD} dropping from source to drain, and i_D flowing from source to drain.

This results in equations where $V_{SG(off)} = -V_P$ and v_{SG} is negative, as follows:

triode: $i_D = I_{DSS}\left[2\left(1 - \frac{v_{SG}}{V_{SG(off)}}\right)\frac{v_{SD}}{V_P} - \frac{v_{SD}^2}{V_P^2}\right] \quad 8.17$

pinchoff—first approximation:

$$i_D = I_{DSS}\left(1 - \frac{v_{SG}}{V_{SG(off)}}\right)^2 \quad 8.18a$$

pinchoff—second approximation:

$$i_D = I_{DSS}\left(1 - \frac{v_{SG}}{V_{SG(off)}}\right)^2\left(1 + \frac{v_{SD}}{V_A}\right) \quad 8.18b$$

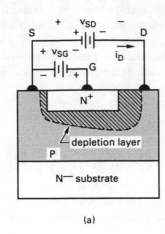

(a)

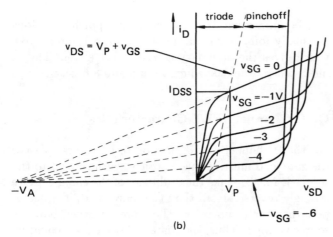

(b)

Figure 8.7 P-Channel JFET with $V_P = 6$ volts
(a) Wiring Configuration,
(b) Family of Characteristic Curves

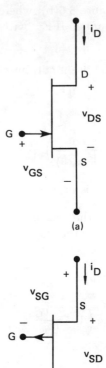

(a)

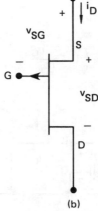

(b)

Figure 8.8 JFET Circuit Symbols:
D = Drain, S = Source, G = Gate.
(a) N-Channel JFET Symbol,
(b) P-Channel JFET Symbol

D. JFET Circuit Symbols

There is no real distinction between the source and drain ends of the channel in a field-effect transistor, unless there is a body lead to the substrate that is internally connected to the source. The schematic representation of a JFET has the channel indicated as a line with source and drain connections perpendicularly connected on one side, and a gate lead perpendicularly connected on the other side. The gate lead is usually shown closer to the source than to the drain.

Current practice is to show a circuit on paper with the most positive voltage at the top of the page and the most negative voltage at the bottom of the page, with bias currents flowing from top to bottom. In that case, the N-channel JFET should be drawn with the drain above the source as indicated in figure 8.8(a), and the P-channel JFET drawn with the source above the drain as indicated in figure 8.8(b). The arrow on the gate lead is the indicator of the channel, pointing in the P to N direction. Thus, an N-channel JFET has the arrow

pointing toward the channel as in figure 8.8(a), while the P-channel JFET has the arrow pointing away from the channel as in figure 8.8(b).

3 DEPLETION MODE MOSFETS

Depletion-Mode MOSFETs behave much as JFETs, except that MOSFETs do not rely on a P-N junction. The difference is that the insulated gate can be made positive or negative with respect to the channel. The metallic gate (typically aluminum) is separated from the channel by a thin layer of insulation (typically silicon dioxide, SiO_2), as indicated in figure 8.9, where the channel is either N or P material, and the substrate is lightly doped with the opposite impurity.

There are two significant differences between the depletion-mode MOSFET and the JFET. First, the MOSFET can operate in a so-called *enhancement* mode, which will be discussed in a following section. Second, the MOSFET is susceptible to catastrophic breakdown of the insulation between the gate and the channel.

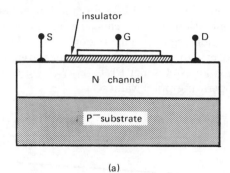

(a)

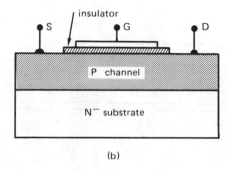

(b)

Figure 8.9 Depletion-Mode MOSFET:
(a) N-Channel, (b) P-Channel

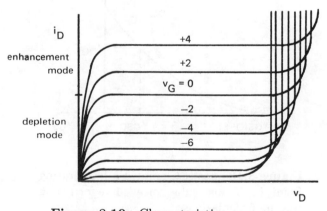

Figure 8.10 Characteristics
of a Depletion-Mode MOSFET
For N-Channel, $v_G = v_{GS}$ and $v_D = v_{DS}$
For P-Channel, $v_G = v_{SG}$ and $v_D = v_{SD}$

Regardless of whether the gate-drain voltage is positive or negative, the N-channel depletion-mode MOSFET is described by equations 8.15 and 8.16, as is the N-channel JFET. Similarly, the P-channel MOSFET is described by equations 8.17 and 8.18, as is the P-channel JFET. The difference here is that both polarities of the gate-source voltage are permitted, resulting in characteristics as shown in figure 8.10.

Static electricity constitutes a serious hazard to MOSFETs, while JFETs are much less sensitive. JFETs can be damaged by static electricity, but because of the P-N junction they can recover from minor shocks. The MOSFET has a thin insulating layer between the gate and channel which can be permanently damaged by relatively small electrostatic charges which can create conductive paths through the insulation. For this reason, MOSFET devices must be handled with extreme care to avoid such damage.

Circuit symbols for depletion-mode MOSFETs are shown in figure 8.11. The space between the gate and channel symbolizes the insulation. The substrate or body lead in integrated logic circuits is typically connected to one of the power-supply voltages, i.e., for N-channel MOSFETs the body is connected to the most negative voltage available, and for P-channel MOSFETs the body is connected to the most positive voltage available. In analog applications, the body lead is commonly

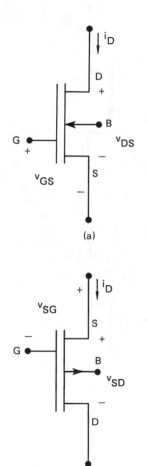

Figure 8.11 Depletion-Mode MOSFET
Circuit Symbols: D = Drain, S = Source,
G = Gate, B = Body or Substrate.
(a) N-Channel MOSFET Symbol,
(b) P-Channel MOSFET Symbol

connected to the source lead. Note that the only P-N junction involved with the MOSFET is between the channel and the substrate, so the arrow indicating the type of channel is associated with the substrate or body (B) lead. For N-channel the substrate is P-type, so the arrow points toward the N-type channel, while for a P-channel the body arrow points away from the channel.

While the symbols shown in figure 8.11 are logically consistent, when the body lead is connected to the source some authors use other symbols which are simpler to draw, but lack the graphic qualities of these. Alternative symbols are shown in figure 8.12 where the direction of the arrow indicates the positive direction for i_D rather than the channel polarity. Thus, the N-channel MOSFET has the arrow pointing out of the source, while the P-channel MOSFET has the arrow pointing into the source.

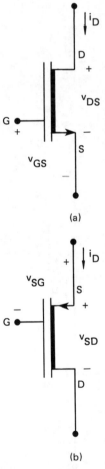

(a)

(b)

Figure 8.12 Alternative Depletion-Mode MOSFET Circuit Symbols When the Body is Connected to the Source: D = Drain, S = Source, G = Gate.
(a) N-Channel MOSFET Symbol,
(b) P-Channel MOSFET Symbol

4 DEPLETION MODE FET BIASING

A. Introduction

Biasing depletion-mode FETs for small-signal amplifier (linear) operation requires choosing an *operating point* in the pinchoff region.[1] This region is bounded by pinchoff and avalanche breakdown as indicated in figure 8.13(a) for an N-channel FET. V_{DSQ}, V_{GSQ}, and I_{DQ} are the operating point values.

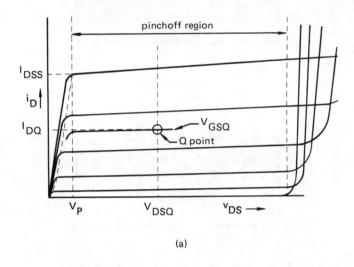

(a)

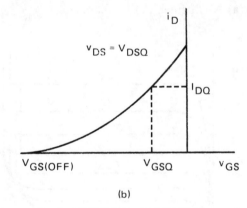

(b)

Figure 8.13 (a) N-Channel FET Operating Point, (b) Transconductance Curve for Use in Biasing Calculations

The most useful curve for determining biasing is the *transconductance curve* shown in figure 8.13(b). This curve is obtained by plotting i_D versus v_{GS} with $v_{DS} = V_{DSQ}$.

B. Simple Self-Biasing Circuits

Figure 8.14 shows a simple self-biasing circuit using an N-channel depletion-mode MOSFET. Because no gate

[1] The operating point also is known as the *quiescent point* or *Q-point*.

current flows, the resistance R_g carries no current, and the gate terminal is at ground potential. The source terminal will be at the voltage $i_D R_s$, so that $v_{GS} = -i_D R_s$.

On the transconductance curve, the relation between i_D and v_{GS} is that of a straight line through the origin with slope $-1/R_s$. The design then requires that

$$R_s = \frac{-V_{GSQ}}{I_{DQ}} \qquad 8.19$$

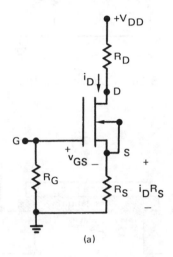

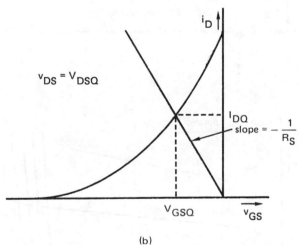

(b)

Figure 8.14 (a) Simple Self-Biasing Circuit FET, (b) Use of Transconductance Curve

The design is completed by determining R_d. Kirchhoff's voltage law used around the drain-source circuit yields

$$V_{DD} = i_D R_d + v_{DS} + i_D R_s \qquad 8.20$$

But $i_D R_s = -v_{GS}$, so at the operating point

$$R_d = \frac{V_{DD} - V_{DSQ} + V_{GSQ}}{I_{DQ}} \qquad 8.21$$

The bias method used here is adequate for only one circuit, when device-to-device variations are not important. When many similar circuits are produced, it is necessary to ensure that the operating points of v_{DS} and I_{DS} be maintained within narrow limitations.

C. Biasing For Device Characteristic Variations

Nominally identical FETs exhibit a range of characteristic curves such as those shown in figure 8.15(a) and 8.15(b). From these extreme cases the corresponding transconductance curves can be plotted on a single graph as in figure 8.15(c).

The most important parameter to control for small-signal FET amplifiers is the *incremental transconductance*, g_m. The incremental transconductance is the slope of the transconductance curve at the operating point.

Variations in transconductance can be minimized by choosing operating points on the two limiting curves of figure 8.15(c) that have the same slope.

The *bias load line* can be calculated from the two-point form for the two operating points.

$$\frac{i_D - I_{DQ2}}{v_{GS} - V_{GSQ2}} = \frac{I_{DQ1} - I_{DQ2}}{V_{GSQ1} - V_{GSQ2}} \qquad 8.22$$

Solving for v_{GS},

$$v_{GS} = V_{GG} - R_s i_D \qquad 8.23$$

$$V_{GG} = V_{GSQ2} + \left(\frac{V_{GSQ2} - V_{GSQ1}}{I_{DQ1} - I_{DQ2}} \right) I_{DSQ2} \qquad 8.24$$

$$R_s = \frac{V_{GSQ2} - V_{GSQ1}}{I_{DQ1} - I_{DQ2}} \qquad 8.25$$

(Note that V_{GSQ1} and V_{GSQ2} are both negative numbers, so all terms in equations 8.24 and 8.25 are positive.)

Because V_{GG} is positive, it can be supplied by the drain voltage source. A biasing circuit that will accomplish this is shown in figure 8.16.

Using Kirchhoff's voltage law around the drain-source circuit,

$$R_d = \frac{V_{DD} - V_{DSQ}}{I_{DQ}} - R_s \qquad 8.26$$

$$I_{DQ} = \frac{1}{2}(I_{DQ1} + I_{DQ2}) \qquad 8.27$$

The gate resistance, R_g, should be large and $V_{GG} = V_{DD}/k$, so

$$k = \frac{V_{DD}}{V_{GG}} \qquad 8.28$$

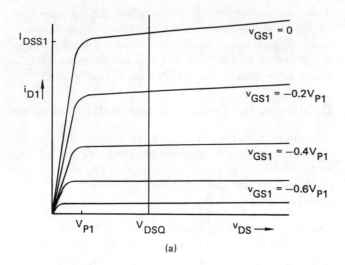

(a)

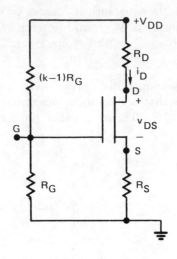

Figure 8.16　Constant g_m Bias Circuit

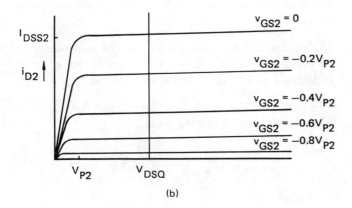

(b)

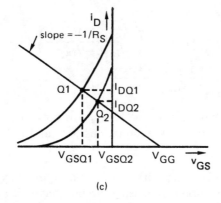

(c)

Figure 8.15　Two Depletion-Mode FETs of the Same Type: (a) Highest-Gain Version, (b) Lowest-Gain Version, (c) Transconductance Curves

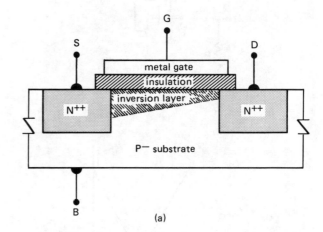

(a)

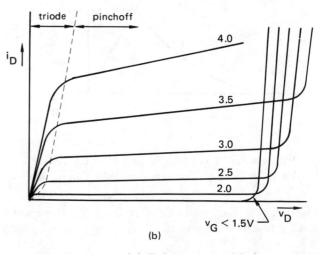

(b)

Figure 8.17　(a) Enhancement-Mode N-Channel MOSFET, (b) Characteristics of an Enhancement-Mode MOSFET with Threshold Voltage of 1.5 Volts: For N-Channel, $v_G = v_{GS}$ and $v_D = v_{DS}$; for P-Channel, $v_G = v_{SG}$ and $v_D = v_{SD}$

5　ENHANCEMENT-MODE MOSFETS

An *enhancement-mode* device is schematically shown in figure 8.17(a) for an N-channel version, where the source and drain are heavily doped N-type and the channel is lightly doped P-type.

With a zero applied gate-source voltage, the channel is essentially an open circuit, so no current will flow unless avalanche is induced by a sufficiently high drain-source voltage.

A. N-Channel Enhancement-Mode Characteristics

An N-channel enhancement-mode MOSFET requires positive v_{GS} to turn it on. v_{GS} must exceed a certain threshold voltage, $V_{GS(on)}$, to be activated. With sufficient v_{GS} applied, electrons are drawn to the lower side of the insulation from the N^{++} regions of the source and drain to create an *inversion layer* adjacent to the gate as seen in figure 8.17(a). This inversion layer provides the channel where electrons are available to conduct current. The higher v_{GS} is made, the wider the channel, and thus the lower the resistance in the triode region of operation. The channel becomes pinched off when the gate to source voltage is insufficient to exceed the threshold voltage, $V_{GS(on)}$. Thus, the characteristics for the enhancement-mode transistor are quite similar to that of a depletion-mode transistor.

The threshold voltage, $V_{GS(on)}$, depends on the doping levels of the N^{++} regions and the P^- region of the channel, as well as the thickness of the insulating material and its dielectric constant. The value of $V_{GS(on)}$ ranges from 1 to 3 volts.

In triode operation, the N-channel enhancement-mode transistor is described by

triode: $\quad i_D = K[2(v_{GS} - V_{GS(on)})v_{DS} - v_{DS}^2]$ $\quad$ 8.29

In pinchoff operation, ignoring the slope of the characteristics, the first-order N-channel approximation is

pinchoff—first-order: $\quad i_D = K(v_{GS} - V_{GS(on)})^2$ $\quad$ 8.30a

Or, accounting for the slope with the voltage V_A as before, the second-order N-channel approximation is

pinchoff—second-order:

$$i_D = K(v_{GS} - V_{GS(on)})^2 \left(1 + \frac{v_{DS}}{V_A}\right) \qquad 8.30b$$

Equating equations 8.29 and 8.30a, the locus of the boundary between the triode and pinchoff regions can be found where

$$v_{DS} = v_{GS} - V_{GS(on)} \quad \text{or} \quad v_{GD} = V_{GS(on)} \quad 8.31$$

The constant K is related to the physical properties of the semiconductor as well as to the channel dimensions. The length of the channel is L, and corresponds to the distance between the two N^{++} regions in figure 8.17(a).

The channel width, W, is in the direction into the paper, and corresponds to the width of the two N^{++} regions in that direction. The insulation thickness is d_{ins}, so that $K = \mu_n \epsilon W/(2L d_{ins})$, where μ_n is the electron mobility and ϵ is the dielectric constant of the insulator. This is also written as

$$K = k'\left(\frac{W}{L}\right) = \frac{\mu_n \epsilon}{2d_{ins}}\left(\frac{W}{L}\right) \qquad 8.32$$

For NMOS with a 0.1 μmeter oxide thickness, $k' = 10\ \mu A/V^2$.

It is possible to recast equations 8.29 and 8.30 to the form seen in the depletion-mode transistors by extracting $V_{GS(on)}^2$ from each as follows:

triode:

$$i_D =$$

$$KV_{GS(on)}^2 \left[2\left(\frac{v_{GS}}{V_{GS(on)}} - 1\right)\frac{v_{DS}}{V_{GS(on)}} - \left(\frac{v_{DS}}{V_{GS(on)}}\right)^2\right]$$

$$8.33$$

pinchoff—first-order:

$$i_D = KV_{GS(on)}^2 \left(\frac{v_{GS}}{V_{GS(on)}} - 1\right)^2 \qquad 8.34a$$

pinchoff—second-order:

$$i_D = KV_{GS(on)}^2 \left(\frac{v_{GS}}{V_{GS(on)}} - 1\right)^2 \left(1 + \frac{v_{DS}}{V_A}\right) \qquad 8.34b$$

It is seen that $KV_{GS(on)}^2$ takes on the role of I_{DSS}.

B. P-Channel Enhancement-Mode Characteristics

For the P-channel enhancement-mode transistor, the order of subscripts is interchanged, so that in triode operation the P-channel enhancement-mode transistor is described by

triode: $\quad i_D = K[2(v_{SG} - V_{SG(on)})v_{SD} - v_{SD}^2]$ $\quad$ 8.35

pinchoff—first-order: $\quad i_D = K(v_{SG} - V_{SG(on)})^2$ $\quad$ 8.36a

pinchoff—second-order:

$$i_D = K(v_{SG} - V_{SG(on)})^2 \left(1 + \frac{v_{SD}}{V_A}\right) \qquad 8.36b$$

The value of K is as given in equation 8.32, but with μ_n replaced by μ_p.

Equations 8.35 and 8.36 can also be put into the form of equations 8.33 and 8.34 by factoring out $V_{SG(on)}^2$.

C. Enhancement-Mode MOSFET Circuit Symbols

The line indicating the channel for the enhancement-mode MOSFET is broken to distinguish it from the depletion-mode symbol. The N- and P-channel symbols are indicated in figure 8.18(a) and (b). The alternative forms, which are sometimes used when the body contact is connected to the source, are indicated in figure 8.18(c) and (d).

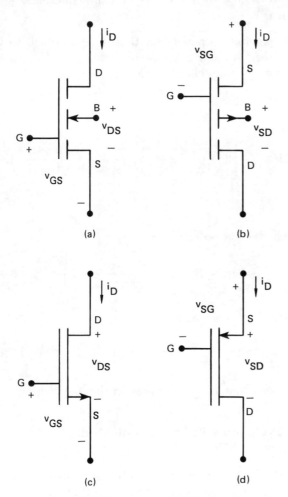

(a)   (b)

(c)   (d)

Figure 8.18 Enhancement-Mode MOSFET Circuit Symbols: D=Drain, S=Source, G=Gate, B=Body or Substrate. (a) N-Channel MOSFET Symbol, (b) P-Channel MOSFET Symbol, (c) Alternative N-Channel MOSFET Symbol, (d) Alternative P-Channel MOSFET Symbol

D. Voltage Feedback Biasing

One common method of biasing the enhancement-mode MOSFET is shown in figure 8.19. The bias level of the gate is tied directly to the average value of the drain-source voltage. This is known as *voltage feedback*

biasing. With the characteristic shown in figure 8.19(b), the locus of points with a gate-source voltage equal to the drain-source voltage is the bias curve. The Kirchhoff voltage law equation for the drain-source circuit is

$$v_{DS} = V_{DD} - i_D R_d \qquad 8.37$$

The DC load line intercepts are $v_{DS} = V_{DD}$ and $i_D = V_{DD}/R_d$. The intersection of these two curves gives the operating point, marked by Q on the characteristic curves.

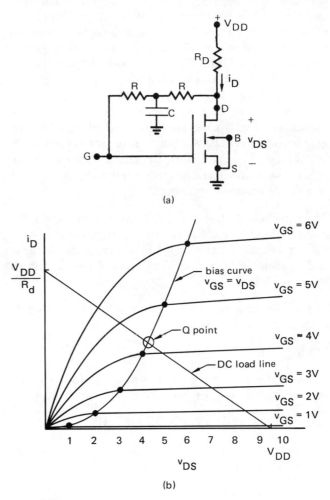

(a)

(b)

Figure 8.19 (a) Voltage Feedback Biasing, (b) Load Line and the Bias Curve

E. Current Feedback Biasing

Another common biasing method for enhancement-mode MOSFETs is shown in figure 8.20(a). This is similar to bias stabilization of depletion-mode FETs. The transconductance curve is used (as with depletion-mode FETs) and is shown in figure 8.20(b).

The application of Kirchhoff's voltage law around the gate-source circuit generates the bias load-line equation.

$$v_{GS} = \frac{V_{DD}}{k} - i_D R_s \qquad 8.38$$

The load line can be plotted on the transconductance curve. The intersection of these two curves is the operating point Q as indicated in figure 8.20(b).

If I_{DQ} is specified and k is selected as 2 for maximum input resistance, the two-point method can be used for a design problem. (V_{GSQ} is found from the transconductance curve.)

$$\frac{v_{GS} - V_{GSQ}}{i_D - I_{DQ}} = \frac{\dfrac{V_{DD}}{k} - V_{GSQ}}{0 - I_{DQ}} \qquad 8.39$$

$$v_{GS} = \frac{V_{DD}}{k} - i_D \left(\frac{\dfrac{V_{DD}}{k} - V_{GSQ}}{I_{DQ}} \right) \qquad 8.40$$

$$R_s = \frac{\dfrac{V_{DD}}{k} - V_{GSQ}}{I_{DQ}} \qquad 8.41$$

The design is completed by using Kirchhoff's voltage law around the drain-source circuit.

$$R_d = \frac{V_{DD} - V_{DSQ} - I_{DQ}R_s}{I_{DQ}} \qquad 8.42$$

F. Vertical Enhancement-Mode MOSFET (VMOS)

All of the MOS devices discussed up to this point are limited to drain currents in the milliampere range. This limitation is due primarily to the relatively long channel. An alternative uses three diffusions: an N^+ drain, a P^- depletion channel, and an N^+ source. After the diffusions are completed, a V-shaped notch is cut in the wafer. An oxide layer placed in the groove is followed by a metal gate layer as in figure 8.21.

Because the channel can be made short, VMOS transistors are capable of high currents in the range of tens of amperes or more.

The characteristics and control of VMOS circuitry essentially are the same as for the enhancement-mode N-channel MOSFET.

G. Double-Diffused MOSFETs (DMOS)

DMOS is another geometry which has been developed for power applications, and has largely replaced VMOS except in high-frequency applications. DMOS is quite similar to VMOS in principle, but has eliminated the mechanically difficult groove by essentially flattening out the V as indicated in figure 8.22. The processing involves two diffusions in an originally lightly doped N-material. The first diffusion establishes two lightly doped P^- regions which will provide the channel. The

second diffusion provides the two N^+ source regions and the N^+ drain region seen at the bottom of the drawing of figure 8.22. The N^+ region provides the path to the drain, which is necessary to provide a sufficiently high avalanche breakdown voltage to allow high-voltage

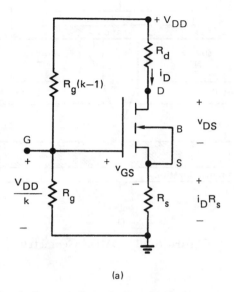

(a)

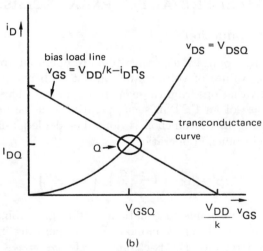

(b)

Figure 8.20 (a) Current Feedback Biasing, (b) Transconductance Characteristic

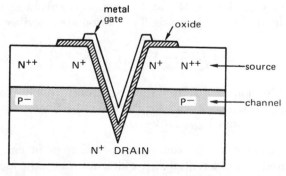

Figure 8.21 Vertical MOSFET (VMOS)

applications. Again, this device has characteristics and control requirements similar to other enhancement-mode MOSFETs.

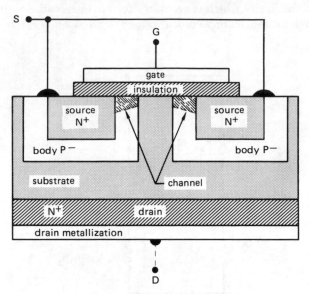

Figure 8.22 DMOS Geometry

6 SMALL-SIGNAL FET PARAMETERS

A. Introduction

In many applications, transistors are operated around biasing values of voltage and current with only small signal variations from those operating points. The general case of an FET in such operation has the device operating in the pinchoff region. For depletion-mode FETs in pinchoff operation,

$$i_D = I_{\text{DSS}} \left(1 + \frac{v_G}{V_P} \right)^2 \left(1 + \frac{v_D}{V_A} \right) \qquad 8.43$$

which corresponds to equations 8.16b and 8.18b, and is valid for all depletion-mode FETs operating in the pinchoff region. For N-channel FETs $v_G = v_{\text{GS}}$ and $v_D = v_{\text{DS}}$, while for P-channel FETs $v_G = v_{\text{SG}}$ and $v_D = v_{\text{SD}}$. v_G is normally negative.

From equations 8.34b and 8.36b, the enhancement-mode MOSFETs in pinchoff operation are described by

$$i_D = K(v_G - V_{\text{G(on)}})^2 \left(1 + \frac{v_D}{V_A} \right) \qquad 8.44$$

For N-channel FETs $v_G = v_{\text{GS}}$, $v_D = v_{\text{DS}}$, and $V_{\text{G(on)}} = V_{\text{GS(on)}}$, while for P-channel FETs $v_G = v_{\text{SG}}$, $v_D = v_{\text{SD}}$, and $V_{\text{G(on)}} = V_{\text{SG(on)}}$.

The change in i_D around an operating point can be treated as a differential di_D, which in turn can be found from the partial differentials of i_D.

$$di_D = \frac{\partial i_D}{\partial v_G} dv_G + \frac{\partial i_D}{\partial v_D} dv_D \qquad 8.45$$

dv_G is the small variation of v_G around its bias value, and dv_D is the corresponding small variation of v_D around its operating or bias value.

The *transconductance* g_m can be defined as

$$g_m = \frac{\partial i_D}{\partial v_G} \qquad 8.46$$

From equations 8.18b and 8.36b, for a particular operating point designated by the values $i_D = I_{\text{DQ}}$, $v_G = V_{\text{GQ}}$ and $v_D = V_{\text{DQ}}$, for N-channel FETs $v_G = v_{\text{GS}}$ and $v_D = v_{\text{DS}}$, and for P-channel FETs $v_G = v_{\text{SG}}$ and $v_D = v_{\text{SD}}$, the transconductance can be determined by

depletion mode: $\quad g_m = \dfrac{\partial i_D}{\partial v_G} = 2\dfrac{I_{\text{DQ}}}{V_{\text{GQ}} + V_P} \qquad 8.47$

enhancement mode: $\quad g_m = \dfrac{\partial i_D}{\partial v_G} = 2\dfrac{I_{\text{DQ}}}{V_{\text{GQ}} - V_{\text{G(on)}}} \qquad 8.48$

The drain resistance, r_{ds}, is defined from

$$\frac{1}{r_{\text{ds}}} = \frac{\partial i_D}{\partial v_D} \qquad 8.49$$

From equations 8.18b or 8.36b, for an operating point,

$$\frac{1}{r_{\text{ds}}} = \frac{I_{\text{DQ}}}{V_A + V_{\text{DQ}}} \qquad 8.50$$

As V_P, $V_{\text{GS(on)}}$, and V_A are all positive quantities, for the transistor to operate in the pinchoff region, in all cases both g_m and r_{DS} are positive.

The small variations di_D, dv_G, and dv_D are actually small-signal variations about the quiescent Q-values or bias values. As such, these signals are designated by lower case variables and subscripts: $di_D = i_d$, $dv_G = v_g$, and $dv_D = v_d$. Then, equation 8.45 can be rewritten integrating g_m and r_{ds}.

$$i_d = g_m v_g + \frac{v_d}{r_{\text{ds}}} \qquad 8.51$$

As before, for N-channel FETs $v_g = v_{\text{gs}}$ and $v_d = v_{\text{ds}}$, while for P-channel FETs $v_g = v_{\text{sg}}$ and $v_d = v_{\text{sd}}$.

B. Low-Frequency Small-Signal FET Equivalent Circuit

For small variations about the operating point, each variable X can be expressed as $X = X_Q + x$. X_Q is the operating point value, and x is a small variation in that value. Thus,

$$v_{DS} = V_{DSQ} + v_{ds} \qquad 8.52$$

$$v_{GS} = V_{GSQ} + v_{gs} \qquad 8.53$$

$$i_D = I_{DQ} + i_d \qquad 8.54$$

As the operating values are constant and satisfy Kirchhoff's laws independently of the small variations, they can be omitted from the equations, and the variations can be considered alone. The small variations must satisfy Kirchhoff's laws by themselves.

Thus small-signal circuits do not need to include any constant (DC) sources. For any FET, the small-signal equivalent circuit is shown in figure 8.23. To find the total voltages or currents for any specific form, the operating point values must be added to the small-signal values.

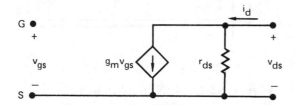

Figure 8.23 Low-Frequency, Small-Signal FET Equivalent Circuit

C. High-Frequency Small-Signal FET Equivalent Circuit

The intrinsic design of FETs gives rise to capacitances which are important at high frequencies. These capacitances are associated with depletion and inversion layers in general and are dependent upon the voltages associated with them.

The *gate-channel capacitance*, C_{ISS}, and the *output capacitance*, C_{OSS}, both are distributed capacitances. They are difficult to assign to either lead along the channel. The input capacitance C_{ISS} usually is divided between the drain and the source to produce capacitances C_{gs} and C_{gd}. The output capacitance includes the *gate-channel capacitance*, C_{ISS}, and the *channel-substrate capacitance*. Because the substrate often is connected to the source, the channel-substrate capacitance usually is attributed to the drain. The resulting high-frequency, small-signal equivalent circuit is shown in figure 8.24.

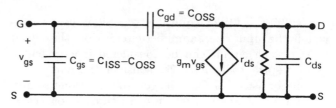

Figure 8.24 High-Frequency, Small-Signal FET Equivalent Circuit

D. Miller Effective Impedance

When two nodes are connected by an admittance Y as indicated in figure 8.25(a), where the second node's voltage is a constant multiplied by the first node's voltage, certain simplifications of the circuit can be made with the objective of separating the two parts of the circuit.

As v_2 in figure 8.25(a) is Av_1, the current i_{12} is found to be

$$i_{12} = Y(v_1 - Av_1) = Y(1 - A)v_1 \qquad 8.55$$

And the current flowing from node 2 to node 1 is simply $i_{21} = -i_{12}$, so that the current i_{21} can be written in terms of v_2, where $v_1 = v_2/A$.

$$i_{21} = Y\left(v_2 - \frac{v_2}{A}\right) = Y\left(1 - \frac{1}{A}\right)v_1 \qquad 8.56$$

These relationships permit the replacement of the connecting admittance Y by two separate admittances to ground as indicated in figure 8.25(b). The reciprocal of the input admittance $Y(1 - A)$ is commonly called the *Miller effective impedance*, and when Y is a capacitance it is called the *Miller effective capacitance*.

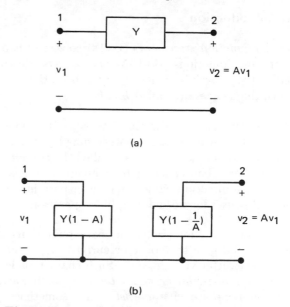

Figure 8.25 Development of the Miller Effective Impedance: (a) Original Connection, (b) Miller Equivalent Circuit

Typically, the Miller equivalent circuit is used to decouple the input of an amplifier circuit from the output so that each circuit can be considered independently. With respect to figure 8.24, consider the part of the circuit consisting of C_{gd}, the controlled current source, and the resistance r_{ds}, which constitutes a connection as in figure 8.25. With $A = -g_m r_{ds}$ and $Y = sC_{gd}$,

$$Y(1 - A) = sC_{gd}(1 + g_m r_{ds}) \qquad 8.57$$

and

$$Y\left(1 - \frac{1}{A}\right) = sC_{gd}\left(1 + \frac{1}{g_m r_{ds}}\right) \qquad 8.58$$

so that figure 8.24(b) can be simplified to that shown in figure 8.26.

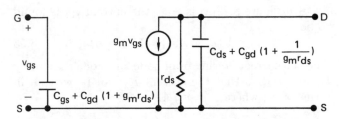

Figure 8.26 Application of Miller's Theorem to Simplify the High-Frequency FET Equivalent Circuit

7   BIPOLAR JUNCTION TRANSISTORS

A. Introduction

Bipolar junction transistors (BJTs) consist of two opposing P-N junctions. Both *N-type carriers* (electrons) and *P-type carriers* (holes) carry substantial current. Thus, the devices are called *bipolar*.

The bipolar junction transistor consists of two opposing P-N junctions with a narrow intervening layer called the *base*. The other two regions are called the *collector* and the *emitter*. The collector-base junction is biased in the reverse direction while the emitter-base junction is forward-biased for normal amplifier operation.

The bipolar junction transistor characteristics most often provided give the collector current, i_C, versus the collector-emitter voltage, v_{CE}. Such characteristics are called *common-emitter characteristics*. Alternatively, the *common-base characteristics* are sometimes provided. These are curves of constant i_E (emitter current) with coordinates of the collector-base voltage and the collector current.

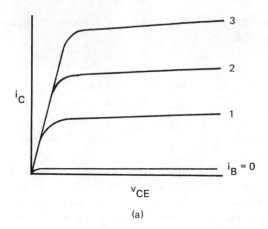

(a)

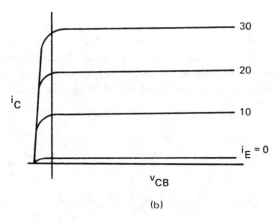

(b)

Figure 8.27 NPN Bipolar Junction Transistor Characteristics: (a) Common Emitter, (b) Common Base

The common-base characteristics can be modeled as two coupled P-N junctions, described by the Ebers and Moll equations as follows:

$$i_E = I_{EO}(e^{v_{BE}/V_T} - 1) - \alpha_R I_{CO}(e^{v_{BC}/V_T} - 1) \quad 8.59$$

P-N collector current

$$i_C = -\alpha_N I_{EO}(e^{v_{BE}/V_T} - 1) + I_{CO}(e^{v_{BC}/V_T} - 1) \quad 8.60$$

V_T is the thermal voltage kT/q, α_N is the normal current gain, and α_R is the reverse current gain which is typically about 80 percent of the normal current gain. The directions of the current are positive in the P to N direction, and the voltage drops are from the P-side to the N-side of the junction. Therefore, the device modelled by equations 8.59 and 8.60 is an NPN BJT.

In normal operation the collector-base junction is reverse biased so that the exponential associated with that junction is negligible. The emitter-base junction is forward biased so that the exponential there dominates the expression. Equations 8.59 and 8.60 are reasonably approximated by

$$i_E = I_{EO}e^{v_{BE}/V_T} + \alpha_R I_{CO} \qquad 8.61$$

P-N collector current

$$i_C = -\alpha_N I_{EO} e^{v_{BE}/V_T} - I_{CO} \qquad 8.62$$

As I_{CO} is negligible in equation 8.61, and since i_C is defined in the direction from N to P, equations 8.61 and 8.62 can be rewritten as an approximation which is adequate in almost all situations where the normal biasing is in effect.

$$i_E = I_{EO} e^{v_{BE}/V_T} \qquad 8.63$$
$$i_C = \alpha_N i_E + I_{CO} \qquad 8.64$$

i_E flows from P to N, and i_C flows from N to P.

B. Bipolar Junction Transistor Biasing

The operating point Q for a bipolar junction transistor will depend on the DC component of current into the third terminal (i.e., the base for common-emitter characteristics and the emitter for common-base characteristics). The operating point is established by the combination of the load line (found from the voltage source and resistance connected to the output terminals of the bipolar junction transistor) and the current into the third terminal.

Most bias designs use the common-base DC parameters. In this case, the collector current is proportional to the emitter current.

$$i_C = I_{CO} + \alpha_o i_E \qquad 8.65$$

I_{CO} is the collector current (see figure 8.28(b)) when the emitter current is zero, and α_o is the average change in collector current divided by the corresponding change in emitter current.[2] Plotting the load line on the characteristic curve, the operating point is the intersection of the load line with the characteristic curve corresponding to the DC value of the emitter current.

The collector current in the common-base configuration is independent of the supply voltage, V_S, and the Thevenin equivalent resistance, R_s, is a function of i_E only. The load line sets the relationship between the collector current and the collector-base voltage. The general biasing circuit is shown in figure 8.28.

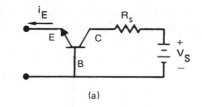

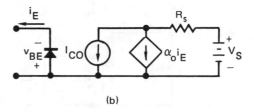

Figure 8.28 (a) Common-Base Biasing Circuit, (b) Equivalent Circuit

The bipolar junction transistor has three terminals, and the currents in each terminal must add to zero to satisfy Kirchhoff's current law. With directions as given in figure 8.29, the current equation is

$$i_B + i_C = i_E \qquad 8.66$$

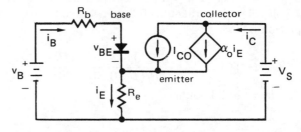

Figure 8.29 Circuit for Biasing Calculations

From equation 8.65,

$$i_E = \frac{i_C - I_{CO}}{\alpha_o} \qquad 8.67$$

Therefore,

$$i_B = \left(\frac{1 - \alpha_o}{\alpha_o}\right) i_C - \frac{I_{CO}}{\alpha_o} \qquad 8.68$$

Writing Kirchhoff's voltage law around the left-hand loop of figure 8.29,

$$i_C = \frac{\alpha_o(v_B - v_{BE}) + I_{CO}(R_b + R_e)}{R_b(1 - \alpha_o) + R_e} \qquad 8.69$$

Example 8.1

A silicon NPN transistor is used in the circuit shown. The transistor parameters are $I_{CO} = 10^{-6}$ amp, $\alpha_o = 0.95$, $R_e = 50$ ohms, and $V_{CC} = 24$ volts. Assume $R_1 = R_2$. Determine the value of R_b to provide an operating point of $I_{CQ} = 10$ mA. (R_b is the parallel combination of R_1 and R_2 and is written as $R_b = R_1 \parallel R_2$.)

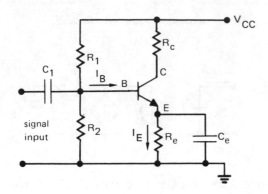

The Thevenin equivalent DC circuit for the bias is R_b in series with 12 volts.

$$V_B = 24\left(\frac{R_2}{R_1 + R_2}\right)$$

For a silicon transistor, the base-emitter voltage is the diode drop for a forward-biased diode, on the order of 0.6 to 1.5 volts. The value usually is taken as 0.6 or 0.7 volt. Here it is taken as 0.6 volt. Solve for R_b from equation 8.69.

$$I_{CQ} = 0.01 = \frac{0.95(12 - 0.6) + 10^{-6}(R_b + 50)}{R_b(1 - 0.95) + 50}$$

$$R_b = 20.7 \text{ kilohms}$$

C. Stability

Bipolar junction transistor stability is fundamental in the design of reliable electronic circuits. The base-emitter junction voltage decreases at a rate of 2.5 mV per degree Celsius. This is the major problem in temperature stabilization of silicon transistors.

$$\frac{\partial v_{BE}}{\partial T} = -2.5 \text{ mV/}^\circ\text{C} \qquad 8.70$$

The resulting change in collector current is

$$\frac{\partial i_C}{\partial T} = \frac{\partial i_C}{\partial v_{BE}} \times \frac{\partial v_{BE}}{\partial T} = -0.0025\frac{\partial i_C}{\partial v_{BE}} \qquad 8.71$$

Using equation 8.69,

$$\frac{\partial i_C}{\partial v_{BE}} = \frac{-\alpha_o}{R_b(1 - \alpha_o) + R_e} \qquad 8.72$$

As α_o is approximately 1, a reasonable approximation is

$$S_v = \frac{\partial i_C}{\partial T} = \frac{2.5}{R_e}\alpha_o \text{ mA/}^\circ\text{C} \qquad 8.73$$

S_v is called the *voltage stability factor*. It is necessary to examine $R_b(1 - \alpha_o)$ to determine whether it is negligible compared to R_e. If it is not, then

$$S_v = \frac{2.5\alpha_o}{(1 - \alpha_o)R_b + R_e} \text{ mA/}^\circ\text{C} \qquad 8.74$$

Example 8.2

Determine the change in collector current for a temperature change of 125°C for the circuit of example 8.1.

From equation 8.74,

$$S_v = \frac{2.5\alpha_o}{0.05 \times 20{,}700 + 50} = \frac{2.5(0.95)}{1085} = 0.0022 \text{ mA/}^\circ\text{C}$$

In this case, $R_b(1 - \alpha_o)$ is not negligible. (See values from example 8.1.)

$$\Delta i_C = S_v \times 125 = 0.275 \text{ mA}$$

D. Parameter Variation

Interchangeability of bipolar junction transistors is a major concern for circuits which are to be produced in quantity. Parameter variations among transistors of the same type must be taken into account. The primary parameter is the large signal current gain α_o.

The usual method is to design the circuit so that the collector current will remain within a specified range when α_o varies. For a known maximum and minimum α_o, use equation 8.69 but neglect I_{CO} because of its insignificant contribution.

$$i_{C,\text{max}} = \frac{\alpha_{\text{max}}(v_B - v_{BE})}{(1 - \alpha_{\text{max}})R_b + R_e} \qquad 8.75$$

$$i_{C,\text{min}} = \frac{\alpha_{\text{min}}(v_B - v_{BE})}{(1 - \alpha_{\text{min}})R_b + R_e} \qquad 8.76$$

To maintain a specified range independent of the voltages, take the ratio of equations 8.75 and 8.76.

$$k_Q > \frac{i_{C,\text{max}}}{i_{C,\text{min}}} = \frac{(1 - \alpha_{\text{min}})R_b + R_e}{(1 - \alpha_{\text{max}})R_b + R_e} \times \frac{\alpha_{\text{max}}}{\alpha_{\text{min}}} \qquad 8.77$$

k_Q is the specified maximum ratio of the maximum i_C to the minimum at the operating point. This results in the following inequality relating R_b and R_e.

$$\left[\frac{1 - \alpha_{\text{min}}}{\alpha_{\text{min}}} - k_Q\left(\frac{1 - \alpha_{\text{max}}}{\alpha_{\text{max}}}\right)\right]R_b < \left(\frac{k_Q}{\alpha_{\text{max}}} - \frac{1}{\alpha_{\text{min}}}\right)R_e$$

$$8.78$$

Example 8.3

The transistor of example 8.1 has a range of α_o from 0.94 to 0.96. Determine whether R_b and R_e will keep I_{CQ} between 9 and 11 mA. If not, redesign the circuit to meet the specifications.

$$k_Q = \frac{11}{9}$$

From equation 8.78,

$$\left(\frac{1-0.94}{0.94} - \frac{11}{9} \times \frac{1-0.96}{0.96}\right)R_b < \left(\frac{11}{9 \times 0.96} - \frac{1}{0.94}\right)R_e$$

$$R_e > 0.0616 R_b$$

This is not satisfied when $R_e = 50$ and $R_b = 20.7k$. The minimum and maximum values of I_{CQ} calculated from equation 8.69 are 8.29 and 12.46 mA, respectively (neglecting I_{CO}).

The procedure is to set R_e within the restriction (greater than $0.0616 R_b$), so set $R_e = 0.0625 R_b$. This results in minimum and maximum currents of $87.48/R_b$ and $106.8/R_b$, respectively, which satisfies the range ratio of 11 : 9. Then, setting the low value at 9 mA,

$$R_b = \frac{87.48}{0.009} = 9720\ \Omega$$

$$R_e = 607.5\ \Omega$$

E. Small-Signal Parameters

Common-emitter characteristic curves such as those shown in figure 8.30 relate variables i_C, v_{CE}, and i_B. There are three nonlinear functional relationships between the characteristic curves. However, i_B is an independent variable, so only the two functions are needed.

$$i_C = \mathbf{f}_1(v_{CE}, i_B) \qquad 8.79$$

$$v_{CE} = \mathbf{f}_2(i_C, i_B) \qquad 8.80$$

These equations are expanded for small variations. The small change in i_C is represented by i_c and is a variation around the operating Q-point. Small changes in v_{CE} are similarly represented as v_{ce}, and small changes in i_B around its bias value are represented by i_b. Thus, for small variations around an operating point,

$$I_{CQ} + i_c = \mathbf{f}_1 + \frac{\partial i_C}{\partial v_{CE}}v_{ce} + \frac{\partial i_C}{\partial i_B}i_b \qquad 8.81$$

$$V_{CEQ} + v_{ce} = \mathbf{f}_2 + \frac{\partial v_{CE}}{\partial i_C}i_c + \frac{\partial v_{CE}}{\partial i_B}i_b \qquad 8.82$$

However, $V_{CEQ} = \mathbf{f}_2$ and $I_{CQ} = \mathbf{f}_1$, so

$$i_c = \frac{\partial i_C}{\partial v_{CE}}v_{ce} + \frac{\partial i_C}{\partial i_B}i_b \qquad 8.83$$

$$v_{ce} = \frac{\partial v_{CE}}{\partial i_C}i_c + \frac{\partial v_{CE}}{\partial i_B}i_b \qquad 8.84$$

These partial derivatives are found from the characteristics. The partial derivative of i_C with respect to v_{CE} is the slope of the constant i_B curve through the operating point.

$$\frac{1}{r_c} = \frac{\partial i_C}{\partial v_{CE}} = \lim_{\Delta v_{CE} \to 0} \frac{\Delta i_C}{\Delta v_{CE}}\bigg|_{i_B = \text{constant}} \qquad 8.85$$

The slope of the constant i_B curve is easily evaluated if the curve goes through the actual operating point. In most cases, the operating point will fall between curves, and the slope must be estimated.

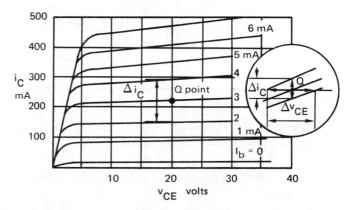

Figure 8.30 Characteristics for Determining Small-Signal Parameters

Equation 8.60 indicates a dependence of the collector current on the collector voltage, but that effect is negligible. The main reason collector current is dependent upon collector voltage is that the length of the base is modulated by the collector-base voltage. As a result, the base length decreases with increasing collector-base voltage, thus fewer minority carriers injected from the emitter are intercepted by the base, thereby increasing the collector current. This phenomenon was explained by James Early, and is known as the *Early effect*. Early showed that the slopes of the common-emitter-connected BJT all met at a single point $(-V_A)$ on the negative v_{CE} axis, as indicated in figure 8.31 for various characteristics with constant v_{BE} values.

For a fixed value of v_{BE}, the slope of the characteristic through a particular operating point in the active region can be seen to be

$$\text{slope} = \frac{I_{CQ}}{V_{CEQ} + V_A} = \frac{1}{r_c} \qquad 8.86$$

V_A is the Early voltage, typically in the range from 50 to 200 volts.

The partial derivative of i_C with respect to i_B is more difficult to evaluate. It is estimated by determining the change in i_C from one constant i_B curve below the operating point to the next constant i_B curve above the operating point, both values taken for the operating point value of v_{CE}. Then, this difference is divided by the difference in i_B on two constant i_B curves.

$$\beta = \frac{\partial i_C}{\partial i_B} = \lim_{\Delta i_B \to 0} \frac{\Delta i_C}{\Delta i_B}\Big|_{v_{CE}=\text{ constant}} \qquad 8.87$$

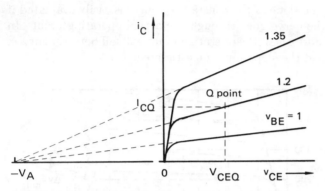

Figure 8.31 Constant v_{BE} Characteristics for NPN BJT Showing the Early Voltage

The small-signal output circuit equation is derived from equation 8.83.

$$i_c = \frac{1}{r_c}v_{ce} + \beta i_b \qquad 8.88$$

The equivalent circuit derived from equation 8.88 is shown in figure 8.32.

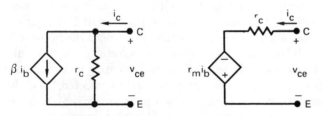

Figure 8.32 Bipolar Junction Transistor Output Circuits

r_c can be evaluated in another way from the characteristic curves.

$$r_c = \frac{\partial v_{CE}}{\partial i_C} = \lim_{\Delta i_C \to 0} \frac{\Delta v_{CE}}{\Delta i_C}\Big|_{i_B=\text{ constant}} \qquad 8.89$$

This is the inverse of the slope of the constant i_B characteristic curve at the operating point. The partial derivative of v_{CE} with respect to i_C is not easily evaluated for transistors, because of the curve flatness. If the characteristic curves had a steeper slope, as do those shown in the inset of figure 8.30, then the derivative would be taken along the constant i_B value at the operating point.

$$r_m = \frac{\partial v_{CE}}{\partial i_B} = \lim_{\Delta i_B \to 0} \frac{\Delta v_{CE}}{\Delta i_B}\Big|_{i_C=\text{ constant}} \qquad 8.90$$

Equations 8.89 and 8.90 can be combined with equation 8.84.

$$v_{ce} = r_c i_c - r_m i_b \qquad 8.91$$

This is represented by the output circuit shown in figure 8.32(b). Note that this is the Thevenin equivalent of the circuit in figure 8.32(a) with $r_m = \beta r_c$. This is valid only if very close values of i_B are available on the curves. Usually, the most satisfactory method of obtaining the circuit shown in figure 8.32(b) is by a Thevenin transformation of the circuit of figure 8.32(a) which has readily obtainable parameters.

Example 8.4

For the operating point given in figure 8.30, obtain the small-signal output parameters.

The slope of the characteristic through the operating point is approximately

$$\frac{1}{r_c} = \frac{0.03}{30} = \frac{1}{1000}$$
$$r_c = 1000 \text{ ohms}$$

Using the Δi_C shown, Δi_B is $4 - 2 = 2$ mA. The corresponding values of i_C are 280 and 140, so $\Delta i_C = 140$ mA. Thus,

$$\beta = \frac{140}{2} = 70$$

Input characteristics normally are not given on manufacturer specification sheets, but the parameters are. The output parameters are hybrid parameters for the common-emitter configuration. Thus, equation 8.88 could be written as

$$i_c = h_{\text{fe}}i_b + h_{\text{oe}}v_{ce} \qquad 8.92$$

The corresponding input parameters are h_{ie} and h_{re}. h_{re} is negligibly small in the common-emitter parameters, so only h_{ie} is necessary. This may be given in the specifications for a specific collector current. It also can be constructed from the *base-spreading resistance*, $r_{\text{bb}'}$, and $r_{\text{b}'\text{e}}$ (also called r'_e).

$$r_{\text{b}'\text{e}} = \frac{r_e}{1-\alpha} \qquad 8.93$$

$$r_e = \frac{kT}{q i_E}$$

$$= \frac{0.026}{i_E} \text{ at } 300 \text{ K} \qquad 8.94$$

Then,

$$h_{\text{ie}} = r_{\text{bb}'} + r_{\text{b}'\text{e}} \qquad 8.95$$

$$v_{\text{be}} = h_{\text{ie}}i_b \qquad 8.96$$

h_{ie} can vary from about 150 ohms to 750 ohms. 500 ohms is a common estimate when the actual value is unknown.

Equation 8.63 can be used to find the relationship between the emitter current and the base-emitter voltage.

$$\frac{1}{r_e} = \frac{\partial i_E}{\partial v_{BE}} = \frac{1}{V_T} I_{eo} e^{v_{BE}/V_T} = \frac{I_{EQ}}{V_T} \qquad 8.97$$

I_{EQ} is the bias value of the emitter current.

8 SMALL-SIGNAL AMPLIFIERS

A. Introduction

The analysis of small-signal amplifiers is largely a circuit analysis problem. A number of amplifier parameters, including input impedance, output impedance, voltage gain, current gain, transconductance, and transresistance can be involved. (The *transconductance* is the ratio of the output current to the input voltage, and the *transresistance* is the ratio of the output voltage to the input current.)

In small-signal analysis, the circuit should be analyzed for its *mid-frequency behavior*. In this case, the reactive elements are not significant (except in the case of tuned circuits), and the capacitors are treated as short circuits. In particular, coupling capacitances along the signal path isolate bias circuits from one stage to the next. These are treated as short circuits, with the provision that the actual impedance must be negligible compared with the impedance seen from the terminals of the capacitance.

Self-biasing resistors also often have bypass capacitors to reduce feedback. These too are considered short circuits to small signals. This results in the shunting resistance being shorted out of the circuit and vanishing from the small-signal equivalent circuit.

Small-signal analysis involves small variations about the operating point. All DC bias values disappear from the small-signal equivalent circuit with DC sources being replaced by their internal resistance. Thus, the small-signal equivalent circuit for the amplifier of figure 8.33(a) is shown in figure 8.33(b).

B. Common-Emitter Bipolar Junction Transistor Amplifier

A common-emitter bipolar junction transistor amplifier is shown in figure 8.33 in mid-frequency, small-signal form. The parameters of interest are input and output impedances (resistances in this case) and the voltage and current gains.

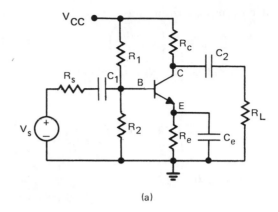

(a)

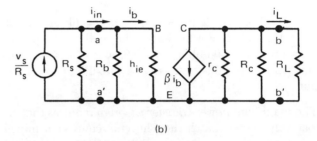

(b)

Figure 8.33 Common-Emitter Amplifier: (a) Circuit Diagram, (b) Small-Signal Equivalent Circuit

Input impedance is the impedance seen from the input terminals (a-a' in figure 8.33(b)). For the common-emitter bipolar junction transistor amplifier shown, the input impedance is

$$Z_{in} = h_{ie} \parallel R_1 \parallel R_2 = h_{ie} \parallel R_b \qquad 8.98$$

$R_1 \parallel R_2$ means the parallel combination of R_1 and R_2.

Output impedance is the impedance seen from the output terminals (b-b' in figure 8.33(b)) looking back into the output of the amplifier.

$$Z_{out} = r_c \parallel R_c \qquad 8.99$$

Voltage gain can be defined as the ratio of the voltage input terminals (i.e., $A_v = v_{bb'}/v_{aa'}$ in figure 8.33(b)). The input current to the base of the transistor is

$$i_b = \frac{v_{aa'}}{h_{ie}} \qquad 8.100$$

The output voltage is

$$v_{bb'} = -\beta i_b (r_c \parallel R_c \parallel R_L) \qquad 8.101$$

The voltage gain is

$$A_v = \frac{v_{bb'}}{v_{aa'}} = -\frac{\beta(r_c \parallel R_c \parallel R_L)}{h_{ie}} \qquad 8.102$$

The input current at terminals a-a' is

$$i_{\text{in}} = i_b \left(1 + \frac{h_{\text{ie}}}{R_b} \right) \qquad 8.103$$

The output current is $v_{\text{bb}'}/R_L$, so the current gain is

$$\begin{aligned} A_i &= \frac{i_l}{i_{\text{in}}} \\ &= -\frac{\beta(r_c \parallel R_c \parallel R_L)}{R_L} \div \left(1 + \frac{h_{\text{ie}}}{R_b} \right) \qquad 8.104 \end{aligned}$$

$$R_b = R_1 \parallel R_2 \qquad 8.105$$

Example 8.5

The amplifier in figure 8.33 has $R_1 = 6\text{k}$, $R_2 = 3\text{k}$, $R_c = 1\text{k}$, and $R_s = 1$ kilohms. The small-signal parameters are $h_{\text{ie}} = 500$ ohms, $\beta = 50$, and $r_c = 5$ kilohms. Determine the value of R_L to give maximum power output.

The *maximum power transfer theorem* requires that the load impedance match the Thevenin equivalent impedance of the driving source. This is the output impedance of the amplifier. From equation 8.99, the output impedance is

$$\begin{aligned} Z_{\text{out}} &= r_c \parallel R_c = 1\text{k} \parallel 5\text{k} \\ &= 833 \text{ ohms} \end{aligned}$$

Thus, the load resistance must be 833 ohms.

C. Common-Collector Amplifier

The *common-collector amplifier (emitter follower)* is shown in figure 8.34(a). This circuit has a voltage gain of slightly less than 1. It is used primarily for current gain or impedance matching between a low impedance load and a high impedance source. Its small-signal equivalent circuit requires a more complicated analysis than does the common-emitter circuit.

The input voltage is the base-collector voltage, since the collector is effectively grounded to small signals and the output current is the emitter current. A new set of parameters can be defined.

$$v_{\text{bc}} = h_{\text{ie}} i_b + v_{\text{ec}} \qquad 8.106$$

$$i_e = (1 + \beta) i_b + \frac{v_{\text{ec}}}{r_c} \qquad 8.107$$

Equations 8.106 and 8.107 define the common-collector *hybrid parameters* for a bipolar junction transistor.

$$h_{\text{ic}} = h_{\text{ie}} \qquad 8.108$$

$$h_{\text{rc}} = 1 \qquad 8.109$$

$$h_{\text{fc}} = 1 + \beta = 1 + h_{\text{fe}} \qquad 8.110$$

$$h_{\text{oc}} = \frac{1}{r_c} = h_{\text{oe}} \qquad 8.111$$

These parameters could be used to establish a different equivalent circuit, but there is no advantage in doing so. Therefore, the circuit of figure 8.34(b) is used.

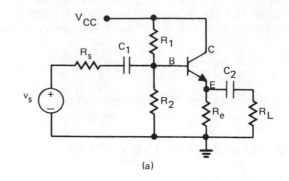

(a)

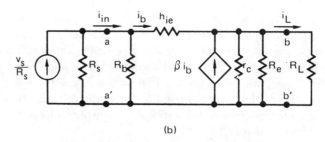

(b)

Figure 8.34 Common-Collector (Emitter Follower) Amplifier: (a) Circuit, (b) Small-Signal Equivalent Circuit

At terminals a-a',

$$Z_{\text{in}} = R_b \parallel [h_{\text{ie}} + (1 + \beta)(r_c \parallel R_e \parallel R_L)] \qquad 8.112$$

At terminals b-b',

$$Z_{\text{out}} = R_e \parallel r_c \parallel \left(\frac{h_{\text{ie}} + R_b \parallel R_s}{1 + \beta} \right) \qquad 8.113$$

From terminals a-a' to the load, the current and voltage gains are:

$$A_i = \frac{(1 + \beta)R_b}{Z_{\text{in}}} \times \frac{r_c \parallel R_c \parallel R_L}{R_L} \qquad 8.114$$

$$A_v = \cfrac{1}{1 + \cfrac{h_{\text{ie}}}{(1 + \beta)(r_c \parallel R_c \parallel R_L)}} \qquad 8.115$$

D. The Common-Base Amplifiers

The *common-base amplifier* is shown in figure 8.35. It has a current gain of less than 1. The change in configuration allows the use of a different set of parameters, the *common-base hybrid parameters*. These are calculated from the common-emitter parameters.

$$h_{ib} = \frac{h_{ie}}{1 + \beta + \left(\dfrac{h_{ie}}{r_c}\right)} \approx \frac{h_{ie}}{1 + \beta} \qquad 8.116$$

$$h_{rb} = \frac{h_{ie}}{(1 + \beta)r_c + h_{ie}} \approx 0 \qquad 8.117$$

$$h_{fb} = \frac{-(\beta r_c + h_{ie})}{(1 + \beta)r_c + h_{ie}} \approx \frac{-\beta}{1 + \beta} = -\alpha \qquad 8.118$$

$$h_{ob} = \frac{1}{(1 + \beta)r_c + h_{ie}} \approx \frac{1}{(1 + \beta)r_c} \qquad 8.119$$

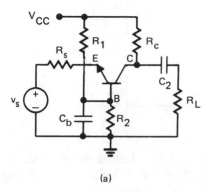

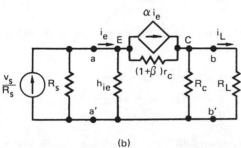

Figure 8.35 Common-Base Amplifier:
(a) Circuit, (b) Small-Signal Equivalent Circuit

Analysis of the small-signal equivalent circuit produces the following equations for terminals a-a'.

$$Z_{in} = \frac{h_{ie}(r_c + R_L \parallel R_c)}{(1 + \beta)r_c + h_{ie} + R_L \parallel R_c} \approx \frac{h_{ie}}{1 + \beta} \qquad 8.120$$

For terminals b-b',

$$Z_{out} = R_c \parallel \left[(r_c \parallel h_{ie}) + R_s + \frac{\beta r_c R_s}{h_{ie} + r_c}\right]$$

$$\approx R_c \parallel [h_{ie} + (1 + \beta)R_s] \qquad 8.121$$

From terminals a-a' to the load,

$$A_i = \frac{R_c}{R_c + R_L} \times \frac{(1 + \beta)r_c + h_{ie}}{(1 + \beta)r_c + h_{ie} + R_L \parallel R_c}$$

$$\approx \frac{\alpha R_c}{R_c + R_L} \qquad 8.122$$

$$A_v = \frac{R_c \parallel R_L}{h_{ie}} \times \frac{\beta r_c + h_{ie}}{r_c + h_{ie}} \approx \frac{\beta(R_c \parallel R_L)}{h_{ie}} \qquad 8.123$$

E. Common-Source FET Amplifiers

FET amplifiers are simpler to analyze because of the electrical isolation of the gate from the source-drain. For any common-source circuit such as the one shown in figure 8.36, the amplifier parameters are:

$$Z_{in} = R_3 + R_1 \parallel R_2 \qquad 8.124$$

$$Z_{out} = R_d \parallel r_{ds} \qquad 8.125$$

$$A_i = -\frac{g_m Z_{in}}{R_L}(r_{ds} \parallel R_d \parallel R_L) \qquad 8.126$$

$$A_v = -g_m(r_{ds} \parallel R_d \parallel R_L) \qquad 8.127$$

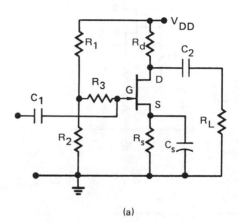

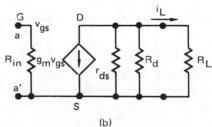

Figure 8.36 Common-Source FET Amplifier: (a) Circuit Diagram, (b) Small-Signal Equivalent Circuit

F. Common-Drain FET Amplifiers

For the common-drain (*source follower*) amplifier shown in figure 8.37, the amplifier parameters are:

$$Z_{in} = \frac{(R_1 + R_{s2})\left(R_{s1} + \dfrac{R_o}{1 + g_m R_o}\right) + R_1 R_{s2}}{R_{s1} + \dfrac{R_{s2} + R_o}{1 + g_m R_o}} \qquad 8.128$$

where $R_o = R_L \parallel r_{ds}$

Let $R_s = R_{s1} + R_{s2}$

$$Z_{out} = \frac{1}{g_m + \dfrac{1}{r_{ds}} + \dfrac{1}{R_{s1} + R_{s2} \parallel R_1}} \qquad 8.129$$

$$A_v = \cfrac{1}{1 + \cfrac{R_o R_1 + R_s(R_1 + R_{s1} \parallel R_{s2})}{R_o\left[R_{s2} + g_m R_s(R_1 + R_{s1} \parallel R_{s2})\right]}} \quad 8.130$$

$$A_i = \frac{r_{ds}}{r_{ds} + R_L}$$

$$\times \frac{R_{s2} + g_m R_s(R_1 + R_{s1} \parallel R_{s2})}{R_s + R_o(1 + g_m R_{s1})} \quad 8.131$$

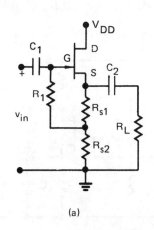

(a)

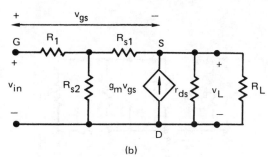

(b)

Figure 8.37 Common-Drain FET Amplifier: (a) Circuit Diagram, (b) Small-Signal Equivalent Circuit

9　MOSFETS WITH ACTIVE LOADS

A. Introduction

In integrated circuits it is quite expensive to use resistances because they take up a lot of space. For this reason other transistors are used for loads whenever possible. In this section some simple FET circuits are presented to demonstrate some of the possibilities.

B. Enhancement-Mode MOSFET With Enhancement Load

A typical circuit is shown in figure 8.38 where the upper FET is the load and the lower FET is the amplifying transistor. When used as a load, an enhancement-mode

MOSFET normally has the gate and drain connected, so that its operation is always in pinchoff as $v_{GD} = 0 < V_{GS(on)}$. Both v_{GS} and v_{DS} of the upper FET are the difference between the supply voltage and the drain-source voltage of the lower transistor.

$$v_{GS2} = v_{DS2} = V_{DD} - v_{DS1} \quad (8.132)$$

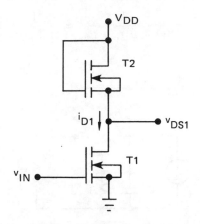

Figure 8.38 Enhancement-Mode N-MOSFET with Enhancement-Mode N-MOSFET Load

Taking the simplest model for the pinchoff region of the upper FET,

$$\begin{aligned} i_{D2} &= K_2(v_{GS2} - V_{GS(on)2})^2 \\ &= K_2(V_{DD} - v_{DS1} - V_{GS(on)2})^2 \end{aligned}$$
$$8.133$$

However, as i_{D2} and i_{D1} are identical, i_{D2} versus v_{DS1} from equation 8.133 can be plotted as a *load curve* (serving the same purpose as a load line) on the characteristic of the lower FET, as indicated in figure 8.39.

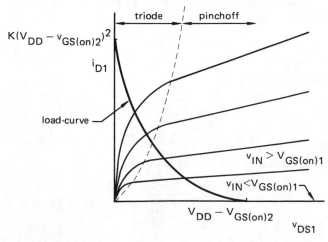

Figure 8.39 Characteristic Curves for the Lower FET in Figure 8.38 with Load Curve of the Upper FET

As can be seen in figure 8.39, for positive values of v_{IN}, there are three possible modes of operation for the lower FET. First, when v_{IN} is less the $V_{GS(on)1}$ the lower FET is off, so the current is zero. Then from equation 8.133 $v_{DS1} = V_{DD} - V_{GS(on)2}$.

Second, when v_{IN} first exceeds $V_{GS(on)1}$ the lower FET is in pinchoff, so that

$$i_{D1} = K_1 (v_{IN} - V_{GS(on)1})^2 \qquad 8.134$$

which when equated with equation 8.133 yields

in pinchoff:

$$v_{DS1} = V_{DD} - V_{GS(on)2} - \sqrt{\frac{K_1}{K_2}} (v_{IN} - V_{GS(on)1}) \quad 8.135$$

As v_{IN} continues to increase, the lower FET will go from pinchoff to triode operation where $v_{IN} - V_{GS(on)1} = v_{DS1}$, which from equation 8.135 occurs where

$$v_{DS1} = \frac{V_{DD} - V_{GS(on)2}}{1 + \left(\dfrac{K_1}{K_2}\right)^{1/2}} \qquad 8.136$$

In triode operation the relationship of v_{DS1} to v_{IN} becomes quite nonlinear, and is not included here, except in the graphical form as indicated in figure 8.40 which is a transfer characteristic of v_{DS1} versus v_{IN}. This transfer characteristic can be seen to be linear from $v_{IN} - V_{GS(on)1} = 0$ to where $v_{IN} - V_{GS(on)1}$ is equal to v_{DS1} in equation 8.136. This linear range can be adjusted with the ratio K_1/K_2.

C. Enhancement-Mode MOSFET With Depletion Load

It is possible to fabricate both depletion- and enhancement-mode MOSFETs on the same chip. The circuit indicated in figure 8.41 shows T1 (an enhancement-mode N-MOSFET) as the amplifying device, and T2 (a depletion-mode N-MOSFET) as the load. In this case, the gate and source of T2 are connected together so that $v_{GS} = 0$ in all cases. T1 is characterized by its constant K and threshold voltage $V_{GS(on)}$, while T2 is characterized by I_{DSS} and V_P.

Again, $v_{DS2} = V_{DD} - v_{DS1}$, so to describe T2 in terms of v_{DS1} and i_{D1},

T2 in triode:

$$i_D = I_{DSS} \left[2 \left(\frac{V_{DD} - v_{DS1}}{V_P} \right) - \left(\frac{V_{DD} - v_{DS1}}{V_P} \right)^2 \right]$$

$$8.137$$

T2 in pinchoff: $\qquad i_D = I_{DSS} \qquad 8.138$

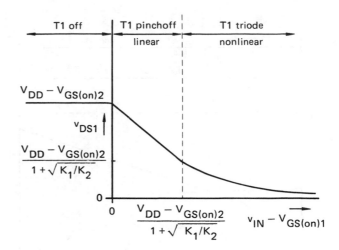

Figure 8.40 Transfer Characteristic of the Circuit of Figure 8.38

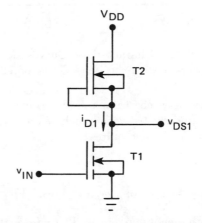

Figure 8.41 Enhancement-Mode N-MOSFET with Depletion-Mode N-MOSFET Load

In order to plot the load curve of this FET load, recognize that it is the depletion-mode device's $v_{GS} = 0$ characteristic, but reversed so that it extends to the left from $v_{DS1} = V_{DD}$ as indicated in figure 8.42, rather than to the right from v_{DS1}.

With v_{IN} less than $V_{GS(on)}$, i_{D1} is zero, and so the voltage on the load curve is V_{DD}. With T1 off, T2 will be in triode operation, where it will remain until v_{IN} is large enough to lower v_{DS1} to $V_{DD} - V_P$. While in this range (A), T1 is in pinchoff.

In the next range (B), both T1 and T2 are operating in pinchoff.

When v_{IN} is raised sufficiently high, it will drive T1 into triode operation. This is the region (C) furthest to the left on the drawing, where T1 is in triode and T2 in pinchoff.

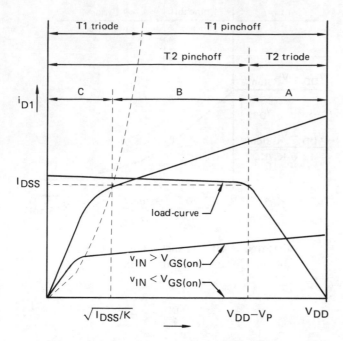

Figure 8.42 Characteristic Curves for T1 in Figure 8.41 with the Load Curve of T2

The boundary between regions A and B can be found from T2's boundary between pinchoff and triode, or where $v_{DS2} = V_P$. This requires

A-B boundary:

$$V_{DD} - v_{DS1} = V_P \quad \text{or} \quad v_{DS1} = V_{DD} - V_P \qquad 8.139$$

And equating pinchoff current in T1 to I_{DSS}, the value of v_{IN} at that point is

A-B boundary:

$$v_{IN} - V_{GS(on)} = \sqrt{\frac{I_{DSS}}{K}} \qquad 8.140$$

Because first-order approximations are being used here for the equations describing operation of FETs, over the next ranges i_D doesn't change, nor does v_{IN}. The boundary between regions C and B can be found from the boundary between pinchoff and triode operation for T1, or where $v_{DS1} = v_{IN} - V_{GS(on)}$. Since $v_{IN} - V_{GS(on)}$ remains the same as in equation 8.140, the C-B boundary is

C-B boundary:

$$v_{IN} - V_{GS(on)} = \sqrt{\frac{I_{DSS}}{K}} \qquad 8.141$$

C-B boundary:

$$v_{DS1} = \sqrt{\frac{I_{DSS}}{K}} \qquad 8.142$$

The expressions relating v_{DS1} to v_{IN} are all nonlinear to some extent, and so are not developed here. However, the values of those voltages at the boundaries between the regions of operation can be used to sketch the transfer characteristic as indicated in figure 8.43.

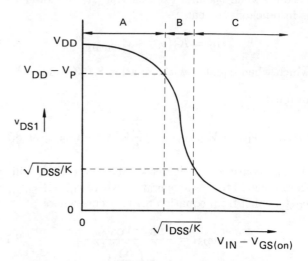

Figure 8.43 Transfer Characteristic of Enhancement N-MOSFET Amplifier with Depletion N-MOSFET Load

The exact location of $\sqrt{I_{DSS}/K}$ is not given in figure 8.43 as it corresponds to both boundaries of region-B operation, again the result of the use of first-order approximations. When the finite slopes of the FET characteristics are included, the two boundaries become different.

10 FET DIFFERENTIAL AMPLIFIERS

A. Introduction

The *differential amplifier* is a basic building block in many applications, including both amplifiers and logic circuits. Their large scale behavior is examined in the following sections. Small-signal analysis is then possible using ordinary circuit analysis techniques with the small-signal models of the transistors.

B. Depletion-Mode FET Differential Amplifier

The behavior of the JFET differential amplifier and a depletion-mode MOSFET version will be the same. Thus, the analysis of one is the same as the other. Here, the JFET version will be used where the biasing is provided by the current source I_S as indicated in figure 8.44.

Assume pinchoff operation and that the two JFETs are identical. Again using the first-order approximation for pinchoff operation,

$$i_D = I_{\text{DSS}} \left(1 + \frac{v_{\text{GS}}}{V_P} \right)^2 \qquad 8.143$$

Recall that V_P is a positive number, while v_{GS} will be negative in normal operation.

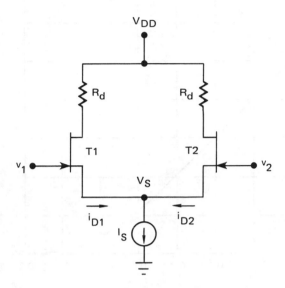

Figure 8.44 JFET Differential-Amplifier Circuit

Assume that voltages v_1 and v_2 are composed of a common mode and a differential mode.

Common-Mode Voltage:

$$v_{\text{CM}} = \frac{(v_1 + v_2)}{2} \qquad 8.144$$

Differential-Mode Voltage:

$$v_{\text{dm}} = v_1 - v_2 \qquad 8.145$$

so that $v_1 = v_{\text{CM}} + v_{\text{dm}}/2$ and $v_2 = v_{\text{CM}} - v_{\text{dm}}/2$.

Note that $v_{\text{GS}1} = v_{\text{CM}} + v_{\text{dm}}/2 - V_S$ and $V_{\text{GS}2} = v_{\text{CM}} - v_{\text{dm}}/2 - V_S$, so that

$$i_{D1} = I_{\text{DSS}} \left(1 + \frac{v_{\text{CM}} - v_S}{V_P} + \frac{v_{\text{dm}}}{2V_P} \right)^2 \qquad 8.146$$

and

$$i_{D2} = I_{\text{DSS}} \left(1 + \frac{v_{\text{CM}} - v_S}{V_P} - \frac{v_{\text{dm}}}{2V_P} \right)^2 \qquad 8.147$$

By adding equations 8.146 and 8.147, carrying out the squares in the right hand side, and noting that $i_{D1} + i_{D2} = I_S$,

$$\left(1 + \frac{v_{\text{CM}} - v_S}{V_P} \right)^2 = \frac{I_S}{2I_{\text{DSS}}} - \left(\frac{v_{\text{dm}}}{V_P} \right)^2 \qquad 8.148$$

Substituting equation 8.148 into equations 8.146 and 8.147,

$$\frac{i_{D1}}{I_S} = \frac{1}{2} + \sqrt{\frac{I_{\text{DSS}}}{2I_S}} \frac{v_{\text{dm}}}{V_P} \sqrt{1 - \frac{I_{\text{DSS}}}{2I_S} \left(\frac{v_{\text{dm}}}{V_P} \right)^2} \qquad 8.149$$

and

$$\frac{i_{D2}}{I_S} = \frac{1}{2} - \sqrt{\frac{I_{\text{DSS}}}{2I_S}} \frac{v_{\text{dm}}}{V_P} \sqrt{1 - \frac{I_{\text{DSS}}}{2I_S} \left(\frac{v_{\text{dm}}}{V_P} \right)^2} \qquad 8.150$$

The normalized curves of currents as a function of the normalized differential-mode voltage are shown in figure 8.45.

The differential-mode output voltage is the difference between the two drain voltages. Since $v_{D1} = V_{\text{DD}} - R_d i_{D1}$ and $v_{D2} = V_{\text{DD}} - R_d i_{D2}$,

$$v_{\text{out-dm}} = v_{D2} - v_{D1} = R_d(i_{D1} - i_{D2})$$
$$= 2R_d I_S \sqrt{\frac{I_{\text{DSS}}}{I_S}} \frac{v_{\text{dm}}}{V_P} \sqrt{1 - \frac{I_{\text{DSS}}}{2I_S} \left(\frac{v_{\text{dm}}}{V_P} \right)^2} \qquad 8.151$$

C. Enhancement-Mode Differential Amplifier

An N-MOSFET differential-amplifier circuit is shown in figure 8.46. Here again, if the first-order approximation for the drain currents is used,

$$i_D = K(v_{\text{GS}} - V_{\text{GS(on)}})^2 \qquad 8.152$$

with $v_{\text{GS}1} = v_{\text{CM}} + v_{\text{dm}}/2 - v_S$ and $v_{\text{GS}2} = v_{\text{CM}} - v_{\text{dm}}/2 - v_S$. Thus,

$$i_{D1} = K(v_{\text{CM}} - V_{\text{GS(on)}} - v_S + v_{\text{dm}}/2)^2 \qquad 8.153$$

and

$$i_{D2} = K(v_{\text{CM}} - V_{\text{GS(on)}} - v_S - v_{\text{dm}}/2)^2 \qquad 8.154$$

and, as $i_{D1} + i_{D2} = I_S$, adding equations 8.153 and 8.154 while completing the squares,

$$v_{\text{CM}} - V_{\text{GS(on)}} - v_S = \sqrt{\frac{I_S}{2K} - \left(\frac{v_{\text{dm}}}{2} \right)^2} \qquad 8.155$$

By substituting into equations 8.153 and 8.154, the currents can be obtained.

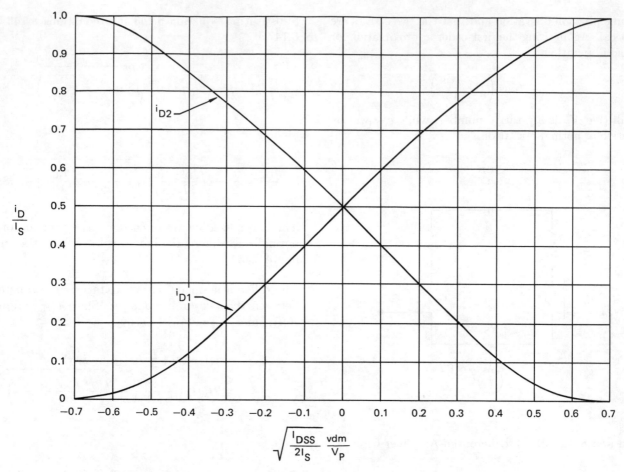

Figure 8.45 Plots of the Normalized Drain Currents for the Differential
Amplifier as Functions of the Differential-Mode Input Voltage

$$\frac{i_{D1}}{I_S} = \frac{1}{2} + \sqrt{\frac{K}{2I_S}}\frac{v_{\mathrm{dm}}}{2}\sqrt{1 - \frac{K}{2I_S}\left(\frac{v_{\mathrm{dm}}}{2}\right)^2} \qquad 8.156$$

and

$$\frac{i_{D2}}{I_S} = \frac{1}{2} - \sqrt{\frac{K}{2I_S}}\frac{v_{\mathrm{dm}}}{2}\sqrt{1 - \frac{K}{2I_S}\left(\frac{v_{\mathrm{dm}}}{2}\right)^2} \qquad 8.157$$

It can be seen that this is of the same form as the currents in the previous section, so the results there can be used by making the substitution of

$$\sqrt{\frac{K}{2I_S}}\frac{v_{\mathrm{dm}}}{2} \text{ for } \sqrt{\frac{I_{\mathrm{DSS}}}{2I_S}}\frac{v_{\mathrm{dm}}}{V_P} \qquad 8.158$$

and the graph of figure 8.45 is then equally applicable to the enhancement-mode differential amplifier.

D. Small-Signal Analysis of an FET Differential Amplifier

As enhancement- and depletion-mode FETs have similar large-scale behavior, and since both have the same

equivalent circuit, it is necessary to analyze only one small-signal circuit. Figure 8.47 represents the small-signal equivalent circuit, including the shunt resistance of the current source and r_{ds} for the FETs.

Three node equations are necessary to determine the behavior of this circuit. Writing in terms of the voltages v_{D1}, v_{D2}, and v_S, a matrix equation can be obtained with v_1 and v_2 the input variables.

$$\begin{bmatrix} -g_m v_1 \\ -g_m v_2 \\ g_m(v_1 + v_2) \end{bmatrix} =$$

$$\begin{bmatrix} \frac{1}{R_c} + \frac{1}{r_{\mathrm{ds}}} & 0 & -g_m + \frac{1}{r_{\mathrm{ds}}} \\ 0 & \frac{1}{R_c} + \frac{1}{r_{\mathrm{ds}}} & -g_m + \frac{1}{r_{\mathrm{ds}}} \\ -\frac{1}{r_{\mathrm{ds}}} & -\frac{1}{r_{\mathrm{ds}}} & \frac{1}{R_c} + 2\left(g_m + \frac{1}{r_{\mathrm{ds}}}\right) \end{bmatrix} \times \begin{bmatrix} v_{D1} \\ v_{D2} \\ v_S \end{bmatrix}$$

$$8.159$$

Solving these equations for v_1 and v_2,

$$v_{D1} = A_{11}v_1 + A_{12}v_2$$

and
$$\qquad\qquad\qquad\qquad\qquad\qquad\qquad 8.160$$

$$v_{D2} = A_{12}v_1 + A_{11}v_2$$

where

$$A_{11} = -\frac{g_m R_d r_{ds}}{R_d + r_{ds}} \frac{R_d + r_{ds} + R_s(1 + g_m r_{ds})}{R_d + r_{ds} + 2R_s(1 + g_m r_{ds})} \qquad 8.161$$

and

$$A_{12} = -\frac{g_m R_d r_{ds}}{R_d + r_{ds}} \frac{R_s(1 + g_m r_{ds})}{R_d + r_{ds} + 2R_s(1 + g_m r_{ds})} \qquad 8.162$$

As $v_{D2} - v_{D1} = A_{dm}(v_1 - v_2)$, and $0.5(v_{D1} + v_{D2}) = 0.5A_{cm}(v_1 + v_2)$, the gains can be found to be

$$A_{dm} = \frac{g_m R_d r_{ds}}{R_d + r_{ds}} \qquad 8.163$$

and

$$A_{cm} = \frac{g_m R_d r_{ds}}{R_d + r_{ds} + 2R_s(1 + g_m r_{ds})} \qquad 8.164$$

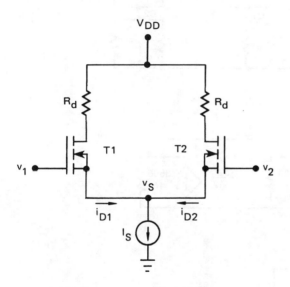

Figure 8.46 N-MOSFET
Differential-Amplifier Circuit

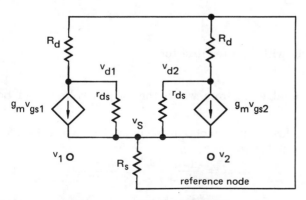

Figure 8.47 N-MOSFET
Differential-Amplifier Equivalent Circuit

The *common-mode rejection ratio (CMRR)* is defined as the ratio of the common-mode gain to the differential-mode gain, which then becomes

$$\text{CMRR} = \frac{R_d + r_{ds}}{R_d + r_{ds} + 2R_s(1 + g_m r_{ds})} \qquad 8.165$$

11 MULTIPLE TRANSISTOR CIRCUITS

A. Introduction

To analyze a multiple transistor circuit, it is necessary to obtain the small-signal equivalent circuit. Even with only two transistors, it may be wise to perform an entirely numerical analysis, as symbolic analysis can be unwieldly. The method of analysis may require matrix analysis, or network reduction techniques can be used. The latter is easier to check for errors, since it is a step-by-step process, and each step can be checked separately.

There are several two-transistor amplifiers which are sufficiently common to consider as special cases.

B. BJT Differential Amplifier With Emitter Resistor

The differential amplifier is a very common circuit with some variation in construction. Figure 8.48(a) shows the configuration of one version of this amplifier. The two input signals supposedly are equal and opposite, but there may be an undesirable common signal which is the same for both inputs. This common signal is the so-called *common mode*. The differential amplifier has very good common-mode rejection, particularly when capacitively coupled.

Two different gains are present in the differential amplifier: one involving the differential part of the signal, designated A_{dm}, and the other dealing with the common-mode signal and designated A_{cm}. The *common-mode rejection ratio* is the ratio A_{cm}/A_{dm}. It is often expressed in decibels.

$$\text{CMRR} = 20 \log \left| \frac{A_{cm}}{A_{dm}} \right| \qquad 8.166$$

Example 8.6

The transistors of figure 8.48(a) have $h_{fe} = 50$, $r_c = 3$ kilohms, and $h_{ie} = 500$ ohms. The circuit values are $R_b = 100$, $R_e = 300$, $R_{c1} = 500$, and $R_{c2} = R_L = 1000$ ohms. Determine (a) the voltage gain for differential signals, (b) the voltage gain for common-mode signals, and (c) the common-mode rejection ratio.

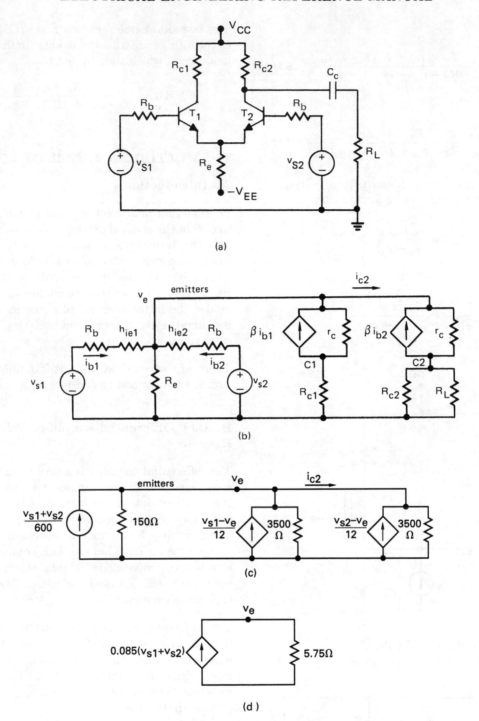

Figure 8.48   Differential Amplifier with Emitter Resistor

The small-signal equivalent network is shown in figure 8.48(b). To use network reduction, it is necessary to preserve voltages and currents that are involved with the control of controlled sources. The currents i_{b1} and i_{b2} are the source controls, and they can be written in terms of v_e and input voltages v_{s1} and v_{s2}. Therefore, it is necessary to preserve these quantities to account for the controlled source variables. The strategy is to determine v_e, from which the output voltage will be found.

The first step is to obtain the numerical expressions for βi_{b1} and βi_{b2}.

$$\beta i_{b1} = \frac{\beta(v_{s1} - v_e)}{R_b + h_{ie}} = \frac{v_{s1} - v_e}{12} \qquad 8.167$$

$$\beta i_{b2} = \frac{v_{s2} - v_e}{12} \qquad 8.168$$

The next step is to reduce the base-emitter circuits to a single Norton equivalent. The controlled sources are converted to Thevenin equivalent circuits and then combined with the series resistances. They are then converted to Norton equivalents as shown in figure 8.48(c).

At this point, control for the controlled current sources involving $-v_e/14$ is dependent on a 14 ohm resistance from the emitter to ground. This leaves only the terms $v_{s1}/14$ and $v_{s2}/14$ which can be combined with the current source. Then, combining the 14 ohm resistances, the 3500 ohm resistances, and the 150 ohm resistance in parallel, the equivalent circuit seen from v_e to ground is given in figure 8.48(d).

The equivalent circuit is described by

$$v_e = 0.4888(v_{s1} + v_{s2})$$

Referring to figure 8.48(c),

$$i_{c2} = 0.4888(v_{s1} + v_{s2})\left(\frac{1}{12} + \frac{1}{3500}\right) - \frac{v_{s2}}{12}$$

$$= 0.0409\,v_{s1} - 0.0425\,v_{s2}$$

Since $R_L = R_{c2}$, the current to the load is $i_{c2}/2$. So the output voltage is

$$v_o = 20.45\,v_{s1} - 21.25\,v_{s2}$$

In *differential operation*, $v_{s2} = -v_{s1}$, and $v_{in} = v_{s1} - v_{s2}$, so the *differential gain* is

$$A_{dm} = \frac{v_o}{v_{in}} = \frac{20.45 + 21.25}{2} = 20.85$$

For common mode, the input $v_{in} = v_{s1} = v_{s2}$. The *common-mode gain* is

$$A_{cm} = 20.45 - 21.25 = -0.800$$

The common-mode rejection ratio is $A_{cm}/A_{dm} = -0.038$ or $CMRR_{dB} = -28.4$ dB.

C. BJT Differential Amplifier With Ideal Current Source Bias

Differential amplifiers are widely used in integrated circuits. While discrete amplifiers use resistors and voltage sources in their biasing schemes, integrated circuits are biased with current sources.

Here, the first assumption is that the BJTs are identical, and the current source as shown in figure 8.49 is ideal, which then sets the emitter voltage at a level of -0.6 volt when the input is zero. Also, the bias levels of the

collector voltages are at $V_{CC} - R_c(0.5\alpha I_S)$. This causes the collector-emitter voltage to be

$$V_{CEQ} = V_{CC} - R_c(0.5\alpha I_S) + 0.6 \qquad 8.169$$

which must be greater than 0.2 volt for the transistors to be operating in the linear region. It also sets the emitter resistance of the BJTs at $r_\epsilon = 0.026/(0.5 I_S)$.

Obviously, the restriction placed on equation 8.169 limits the range of common-mode voltages the amplifier can accept.

From a straightforward circuit analysis, with zero common-mode signal so that $v_{IN} = v_{in}$, it can be found that the differential-output voltage is

$$v_{o-diff} = v_{c2} - v_{c1} = \frac{\alpha R_c}{r_\epsilon}v_{in} = \frac{0.5\alpha I_S R_c}{0.026}v_{in} \qquad 8.170$$

Similarly, the input impedance can be found to be

$$r_{in} = 2(1+\beta)r_\epsilon = 2(1+\beta)\frac{0.026}{0.5 I_S} \qquad 8.171$$

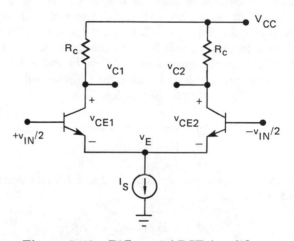

Figure 8.49 Differential BJT Amplifier with Current Source Biasing

Example 8.7

For an amplifier such as that shown in figure 8.49, with $V_{CC} = 12$ volts and $I_S = 2$ mA, determine the maximum value of R_c to allow a common-mode voltage range from -6 to $+6$ volts. What is the differential-mode voltage range?

With a common-mode voltage of $+6$ volts, the emitter bias voltage level will be 5.4 volts. Equation 8.169 can then be modified such that

$$V_{CEQ} = V_{CC} - R_c(0.5\alpha I_S) - 5.4$$

or, taking $\alpha = 1$,

$$V_{CEQ} = 12 - R_c(0.5 \times 0.002) - 5.4 > 0.2$$

so that $0.001 R_c < 6.4$ or $R_c < 6.4k$.

Now, with $R_c = 6.4$ kilohms, the BJTs are biased to their saturation level. As a result, any input signal will result in one of the BJTs saturating, and when the common-mode voltage is +6 volts, the range of input signals to keep the BJTs in their linear range is zero. Therefore, this is not a useful design.

Example 8.8

For an amplifier as shown in figure 8.49, select a value of R_c so that the differential output can vary over the range of +2 to −2 volts when the common-mode voltage is +6 volts. What are the implications of having a common-mode voltage of −6 volts for this design?

With a common-mode voltage of 6 volts, the emitter average voltage will be at 5.4 volts, and the minimum permissible v_{CE} that will allow linear operation is 0.2 volt. Assume a sufficiently high gain so that the emitter voltage will change by only a few millivolts for a signal sufficient to change the output by one volt from its quiescent value. So with the differential-mode output having a swing from −2 to +2 volts, each collector will have to vary ±1 volt from its quiescent value. If the left BJT is at the minimum of 5.6 volts and is 2 volts lower than the right BJT, the right collector will be at 7.6 volts. The two voltage relations are:

$$\text{left: } V_{CC} - 5.6 = i_{C1}R_c$$
$$\text{right: } V_{CC} - 7.6 = i_{C2}R_c$$

But, the two collector currents add to αI_S, and assume that $\alpha = 1$, so

$$I_S R_c = 2V_{CC} - 5.6 - 7.6$$

or

$$R_c = \frac{2V_{CC} - 13.2}{I_S} = 5.4 \text{ kilohms}$$

With a common-mode voltage of −6 volts, the emitter voltage will be −6.6 volts. The collector voltages will remain as above, so the collector-emitter voltages are:

$$V_{CE1} = 5.6 + 6.6 = 12.2 \text{ volts}$$
$$V_{CE2} = 7.6 + 6.6 = 14.2 \text{ volts}$$

This is no problem as long as the BJTs do not suffer from avalanche breakdown.

Example 8.9

For the design of the previous example, determine the differential-voltage gain and input resistance of the amplifier. Assume the value of α is 0.98.

First, find the value of r_ϵ: $0.026/I_{EQ} = 0.026/0.001 = 26$ ohms. Then, from equation 8.170,

$$A_v = \frac{\alpha R_c}{r_\epsilon} = \frac{0.98 \times 5400}{26} = 203$$

Thus, the assumption that the emitter voltage remained at essentially the common-mode voltage level appears valid, as the input swing for an output swing of 2 volts is 2/203 or 10 mV.

The input resistance can be obtained by noting that

$$\beta = \frac{\alpha}{1-\alpha} = \frac{0.98}{1-0.98} = 49$$

Then, from equation 8.171,

$$r_{in} = 2(1+\beta)r_\epsilon = 2(50)26 = 2.6 \text{ kilohms}$$

An approximate large-scale solution for the differential amplifier with an ideal current source providing emitter biasing can be obtained using equation 8.63 by recognizing that $v_{BE1} = v_1 - v_E$ and $v_{BE2} = v_2 - v_E$.

$$i_{E1} = I_{EO}e^{(v_1-v_E)/V_T}$$
$$i_{E2} = I_{EO}e^{(v_2-v_E)/V_T}$$

so that

$$\frac{i_{E1}}{i_{E2}} = \frac{e^{v_1/V_T}}{e^{v_2/V_T}} = e^{(v_1-v_2)/V_T}$$

but, as $i_{E1} + i_{E2} = I_S$,

$$i_{E2} = \frac{I_S}{1 + e^{(v_1-v_2)/V_T}} \quad\quad 8.172$$

By a similar development,

$$i_{E1} = \frac{I_S}{1 + e^{(v_2-v_1)/V_T}} \quad\quad 8.173$$

These currents can be normalized with respect to I_S, and the voltages normalized to the thermal voltage. With these plotted, the characteristic shown in figure 8.50 can be obtained.

It can be seen that the linear region of operation is restricted to emitter currents ranging from about 45 percent to 55 percent of I_S. This corresponds to an input voltage difference of less than $0.25V_T$ volts which corresponds to about 6 mV. This restriction must be added to those which involve the DC power supply voltages and the common-mode voltage.

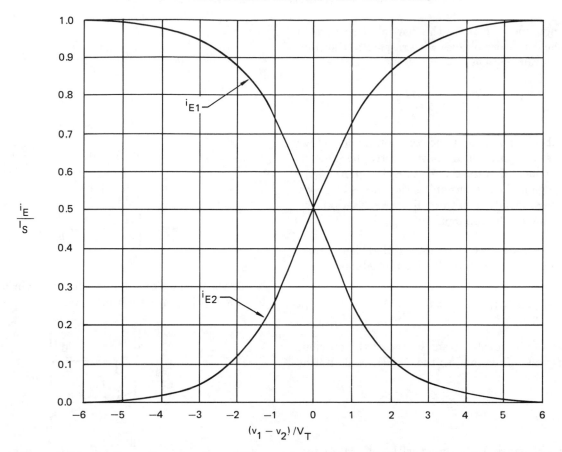

Figure 8.50 Emitter Currents for a Differential Amplifier with Ideal Current Source Biasing

It can be seen that the differential amplifier can be used for a binary output device by having the input voltage difference change by as little as $12V_T$ which again corresponds to a voltage change of only about 300 mV.

D. BJT Differential Amplifier With Real Current Source Bias

Any practical current source has an equivalent shunt resistance which is indicated as R_s in figure 8.51. It is also presumed that the current source and its shunt resistance are connected to a negative voltage, which is typically the same magnitude as V_{CC}, and is taken as $-V_{CC}$ here. In this case, the circuit analysis is somewhat more difficult.

Using the large-scale model where $i_E r_\epsilon = v_{BE} - 0.6$, analysis of the two emitter currents and the emitter node equation are combined to solve first for the emitter currents. The left BJT current is

$$i_{E1} = \frac{2R_s}{2R_s + r_\epsilon}\left(\frac{V_{CC} + v_{CM} - 0.6}{2R_s} + \frac{I_S}{2}\right) + \frac{v_{dm}}{r_\epsilon}$$
$$8.174$$

and the right BJT emitter current is

$$i_{E2} = \frac{2R_s}{2R_s + r_\epsilon}\left(\frac{V_{CC} + v_{CM} - 0.6}{2R_s} + \frac{I_S}{2}\right) - \frac{v_{dm}}{r_\epsilon}$$
$$8.175$$

The collector voltages can be found from $v_C = V_{CC} - \alpha R_c i_E$, and the differential-output voltage $v_{C2} - v_{C1}$ is

$$v_{\text{out-differential}} = \frac{\alpha R_c}{r_\epsilon} v_{dm} \qquad 8.176$$

from which differential mode gain is $A_{dm} = \alpha R_c / r_\epsilon$. Note that the differential gain is unaffected by the resistance R_s.

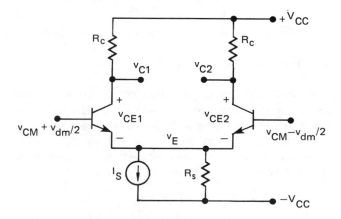

Figure 8.51 Differential BJT Amplifier with Real Current Source Biasing, Shunt Resistance, and Negative Power Supply Included

The common mode gain is the partial derivative of either of the collector voltages with respect to the common-mode input voltage.

$$A_{\text{cm}} = \frac{\partial v_{C2}}{\partial v_{\text{CM}}} = \frac{\alpha R_c}{2R_s + r_\epsilon} \qquad 8.177$$

It should also be noted that the common-mode voltage and the power supply voltage have exactly the same effect on the collector voltages. As a result, a power supply voltage gain (i.e., determining the effects of power supply voltage variation on the collector voltages) is exactly the same as the common-mode gain.

The common-mode rejection ratio (CMRR) is the ratio of the common-mode gain to the differential-mode gain.

$$\text{CMRR} = \frac{A_{\text{cm}}}{A_{\text{dm}}} = \frac{r_\epsilon}{2R_s} \qquad 8.178$$

In the case of the ideal current source of the previous section, R_s was infinite and so the CMRR could not be calculated.

Example 8.10

The circuit of figure 8.51 is used with the parameters and values of example 8.8, where R_c was determined to be 5.4 kilohms and $\alpha = 0.98$, with $V_{\text{CC}} = 12$ volts. For a current source of 2 mA and a shunt resistance of 20 kilohms, determine the common-mode rejection ratio in decibels.

It was previously determined that $r_\epsilon = 26$ ohms. With $R_s = 20$ kilohms, the common-mode rejection ratio is $26/(2 \times 20\text{k})$, or in decibels

$$\text{CMRR}\Big|_{\text{dB}} = 20 \log \frac{26}{2 \times 20\text{k}} = -64\,\text{dB}$$

E. Darlington Circuits

The *Darlington circuit* is a pair of transistors (known as *compound transistors*) connected as shown in figure 8.52(a). The increase in gain is approximately the product of the current gains of the two transistors. The input impedance is increased as well. An analysis shows that the three-terminal equations for the pair are:

$$v_{\text{be}'} = i_{b1}\left[h_{\text{ie}1} + \frac{(1+\beta_1)h_{\text{ie}2}}{1 + \dfrac{h_{\text{ie}2}}{r_{\text{cl}}}}\right] + v_{\text{ce}2}\left[\frac{h_{\text{ie}2}}{r_{\text{cl}} + h_{\text{ie}2}}\right]$$

$$8.179$$

$$i_{e'} = i_{b1}\left[\frac{(1+\beta_1)(1+\beta_2)}{1 + \dfrac{h_{\text{ie}2}}{r_{c1}}}\right]$$
$$+ v_{\text{ce}2}\left[\frac{1}{r_{c2}} + \frac{1+\beta_2}{r_{c1} + h_{\text{ie}2}}\right] \qquad 8.180$$

$$i_c = i_{b1}\left[\frac{\beta_1 + \beta_2 + \beta_1\beta_2 - \left(\dfrac{h_{\text{ie}2}}{r_{c1}}\right)}{1 + \dfrac{h_{\text{ie}2}}{r_{c1}}}\right]$$
$$+ v_{\text{ce}2}\left[\frac{1}{r_{c2}} + \frac{1+\beta_2}{r_{c1} + h_{\text{ie}2}}\right] \qquad 8.181$$

These three equations define effective hybrid parameters for Darlington pair transistors.

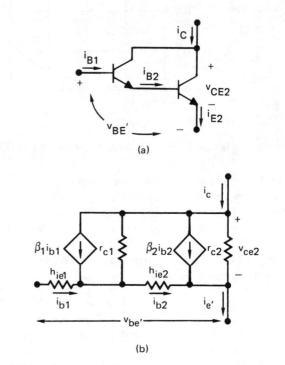

Figure 8.52　Darlington Pair: (a) Circuit, (b) Small-Signal Equivalent

F. Cascode Amplifiers

The *cascode amplifier* is another two-transistor circuit that finds use in high-frequency circuits because of its superior high-frequency response. The input transistor (the lower one in figure 8.53(a)) is a common-emitter stage, while the output transistor is in the common-base configuration. (The same configurations, with appropriate bias circuits, can be created with FETs or vacuum tubes. For FETs, the input stage is common source, and the output is common gate. For

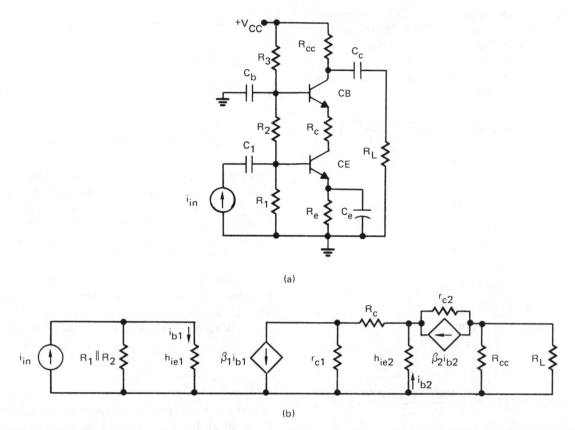

Figure 8.53   Cascode Bipolar Junction Transistor: (a) Circuit Diagram, (b) Small-Signal Equivalent Circuit

vacuum triode systems, the input stage is common cathode and the output is common-grounded grid.) The small-signal equivalent circuit for the bipolar junction transistor cascode amplifier is shown in figure 8.53(b). For good high-frequency response, R_c should be zero.

The behavior of a cascode amplifier is not significantly different from that of a common-emitter amplifier at mid-frequencies. Its advantage is at high frequencies where improved bandwidth is achieved.

12   FREQUENCY EFFECTS

A. Low-Frequency Behavior

Low-frequency behavior usually is influenced by emitter resistance and the bypass capacitors. This is due to the relatively small resistance and the subsequent need for large capacitance. The impedance of a parallel resistance and capacitance is

$$Z_{R|C} = \frac{R}{1 + j\omega RC} \qquad 8.182$$

The *corner frequency* is

$$\omega_{\text{cf}} = \frac{1}{RC} \qquad 8.183$$

The amplitude of the impedance at the corner frequency is $R/\sqrt{2}$ or 3 dB down from the DC value. For $\omega > 10/RC$, the impedance is essentially the capacitive reactance, inversely proportional to the frequency. The self-biasing resistances for common-emitter and common-source circuits are of primary interest here, because significant impedance in that position causes degenerative feedback, which lowers the gain of the amplifier.

Coupling capacitors and bypass capacitors both diminish the gain when their impedances become comparable to the resistance they see. The resistance typically seen consists of a parallel combination of the biasing resistances with the input impedance of the amplifying device as well as the output impedance of the previous stage.

Example 8.11

Determine the low-frequency limit for the emitter impedance of example 8.5. The bypass capacitance C_e is 100μF and $R_e = 100$ ohms.

The resistance seen by the capacitance is not simply the emitter resistance R_e. It must be found by determining the effective output impedance as if the amplifier was an emitter follower. The collector circuit is essentially isolated by the large value of r_c. For the common-collector amplifier, equation 8.113 indicates that the output impedance included the series connection of h_{ie} and the resistance connected to it, all divided by $1 + \beta$. The term of r_c in parallel should be increased to include the load and collector resistances.

$$R_{\|C} = R_e \| (r_c + R_L \| R_c) \| \left(\frac{h_{ie} + R_b \| R_s}{1 + \beta} \right)$$

For this example,

$$R_b = 3k R_{\|C} 6k = 2000 \text{ ohms}$$
$$R_s = 1000 \text{ ohms}$$
$$h_{ie} = 500 \text{ ohms}$$
$$\beta = 50$$
$$r_c = 5000 \text{ ohms}$$
$$R_c = 1000 \text{ ohms}$$
$$R_L = 833 \text{ ohms}$$

Then,

$$R_{\|C} = 100 \| 5454 \| \left(\frac{500 + 1000 \| 2000}{51} \right)$$
$$= 18.6 \text{ ohms}$$

Thus,

$$\frac{1}{RC} = 538 \text{ rad/s} = 85.6 \text{ hertz}$$

B. High-Frequency Bipolar Junction Transistor Behavior

Analysis of *high-frequency limitations of bipolar junction transistors* requires a high-frequency equivalent circuit. The most general model is the *Hybrid pi* model shown in figure 8.54. The circuit can be simplified by using the *Miller effective impedance*. This causes the capacitance $C_{b'c}$ to be multiplied by $(g_m R'_c + 1)$ and connected from the active base b' to the emitter as shown in figure 8.55.

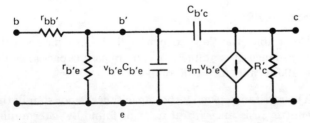

Figure 8.54 Hybrid Pi Equivalent Circuit for Bipolar Junction Transistor at High Frequency

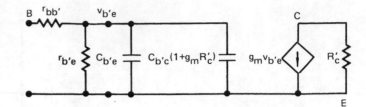

Figure 8.55 Hybrid Pi Equivalent Circuit with Miller Capacitance

The *high-frequency cutoff* is found by determining the resistance seen by the parallel combination of $C_{b'e}$ and $C_{b'c}(1 + g_m R'_c)$. For a common-emitter amplifier, this is the parallel combination of the base biasing resistors and the output impedance of the previous stage (R_s) in series with $r_{bb'}$ all in parallel with $r_{b'e}$.

For a cascode amplifier, the Miller effective capacitance is reduced because the first transistor (CE) sees R_c and the low input impedance of the second transistor (CC). With $R_c = 0$, the Miller capacitance is essentially $C_{b'c} + C_{b'e}$.

Example 8.12

Determine the high-frequency cutoff for an emitter follower which has the following parameters:

$$C_{b'e} = 1000 \text{ pF}$$
$$C_{b'c} = 25 \text{ pF}$$
$$r_{bb'} = 100 \text{ ohms}$$
$$r_{b'e} = 100 \text{ ohms}$$
$$\beta = 50$$
$$R_s \| R_b = 200 \text{ ohms}$$
$$R_e = 100 \text{ ohms}$$

As the collector is grounded, the resistance seen from the controlled source is approximately R_e. (See figure 8.34.)

$$g_m = \frac{\beta}{r_{b'e}} = 0.5$$

The Miller capacitance is

$$25(1 + 50) = 1275 \text{ pF}$$
$$C_{total} = 1275 + C_{b'e} = 2275 \text{ pF}$$

The resistance seen from the capacitance is

$$R_{\|C} = r_{b'e} \| (r_{bb'} + R_s \| R_b + (1 + \beta)R_e)$$
$$= 100 \| (150 + 200 + 51 \times 100) = 98.2 \text{ ohms}$$

Then, the cutoff frequency is

$$\omega_{cf} = \frac{1}{98.2 \times 2275 \times 10^{-12}} = 4.48 \times 10^6 \text{ rad/s}$$
$$f_{cf} = \frac{4.48 \times 10^6}{(2\pi)(1000)} = 713 \text{ kHz}$$

13   CURRENT SOURCES

A. Introduction

Integrated circuits use current biasing, as it is more readily controlled in the fabrication process than are precision resistances. Precision resistors also require much more room than the transistors used in current sources.

B. Enhancement MOSFET Current Mirror

In the circuits of figure 8.56, the left FETs (T1) are carrying the reference current, which is established via a resistance and another power supply or ground. In a CMOS circuit both types of current mirrors would be used with a single reference current, so the two circuits would be joined on the left by a single precision resistance.

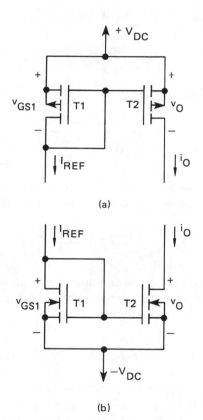

(a)

(b)

Figure 8.56 MOSFET Current Mirror Circuits: (a) P-MOSFET Version, (b) N-MOSFET Version

As both FETs are fabricated in the same process, their threshold voltages will be identical, but different dimensions on the channel width will result in different values of the constant K $(K = k'W/L$, see equation 8.4).

Because of the design $v_{DS1} = v_{GS1}$, which will be a small and constant voltage, the first-order approximation can

be used to obtain the drain current for the reference FET.

$$I_{REF} = K_1(v_{GS1} - V_{GS(on)})^2 \qquad 8.184$$

For the mirror FET (T2) the drain-source voltage is quite variable, so the second-order approximation is used.

$$i_O = K_2(v_{GS2} - V_{GS(on)})^2 \left(1 + \frac{v_O}{V_A}\right)$$
$$= K_2(v_{GS1} - V_{GS(on)})^2 \left(1 + \frac{v_O}{V_A}\right) \qquad 8.185$$

Combining these two equations produces

$$i_O = \frac{K_2}{K_1} I_{REF} \left(1 + \frac{v_O}{V_A}\right) \qquad 8.186$$

Or, in terms of channel dimensions,

$$i_O = \frac{W_2/L_2}{W_1/L_1} I_{REF} \left(1 + \frac{v_O}{V_A}\right) \qquad 8.187$$

In practice, a number of current mirrors can be created using only one reference source, as the mirror circuits draw no current from the reference circuit so that there is no loading from the mirrors. Each mirror can be designed for the needed amount of current by scaling the channel dimensions during fabrication, where in general only the width W would be variable, as L involves the avalanche specification.

It can be seen from equation 8.187 that the mirror current source has a shunt resistance given by

$$r_o = \frac{W_1/L_1}{W_2/L_2} \frac{V_A}{I_{REF}} \qquad 8.188$$

Example 8.13

Specify the relative dimensions of the channels of two FETs used to obtain a 1 mA current mirror source using a reference current of 0.1 mA. If the Early voltage $V_A = 100$ volts, determine the output resistance of the current source.

Assuming the channel lengths of the two FETs are equal, the width of the output FET must be ten times the width of the reference FET.

With a current source of 1 mA, the output resistance of the current source is $100\,\text{V}/1\,\text{mA} = 100$ kilohms.

C. BJT Current Mirror Circuits

An NPN version of a BJT current mirror is shown in figure 8.57. The reference BJT (T1), is connected in the diode configuration, which in the linear region of operation behaves as a diode with a constant $\eta = 1$. (See equation 8.3.)

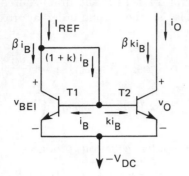

Figure 8.57　BJT Current Mirror Circuit

In this circuit the output BJT (T2) has a base-emitter junction area that is k times larger than that of the reference BJT (T1). This means that with the same base-emitter voltage, T2 will have k times as much base current as T1. Identical current gains for the two transistors are assumed.

Using the simplest transistor model, with $i_C = \beta i_B$, the reference current divides as indicated between the collector and base circuits as shown in figure 8.57. Thus,

$$I_{\text{REF}} = (1 + k + \beta)i_B \qquad 8.189$$

As $i_O = \beta k i_B$,

$$i_O = \frac{k I_{\text{REF}}}{1 + \dfrac{(1 + k)}{\beta}} \qquad 8.190$$

In this case the output resistance of the current source (T2) is its Early voltage divided by the output current, or

$$r_o = \frac{V_A}{k I_{\text{REF}}}\left(1 + \frac{1 + k}{\beta}\right) \qquad 8.191$$

Two difficulties are seen in this circuit. One is the output resistance and the other is the loading of the reference transistor by the mirror BJT. These both suffer uncertainty due to lack of reliability of the current gain β. An improved circuit would have less loading effect.

D. Improved BJT Current Mirror Circuits

In order to reduce the loading of the reference BJT by the base of the mirror BJT, a third transistor can be used to buffer the base circuits. This is shown in figure 8.58 (T3).

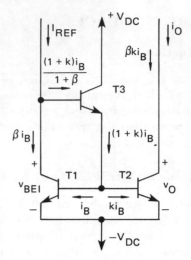

Figure 8.58　Improved Version of the BJT Current Mirror Circuit

In analyzing the currents of figure 8.58,

$$I_{\text{REF}} = \left(\beta + \frac{k + 1}{\beta + 1}\right) i_B \qquad 8.192$$

As $i_O = \beta k i_B$,

$$i_O = \frac{k I_{\text{REF}}}{1 + \dfrac{1 + k}{\beta(\beta + 1)}} \qquad 8.193$$

Again, the output resistance of the current source (T2) is its Early voltage divided by the output current, or

$$r_o = \frac{V_A}{k I_{\text{REF}}}\left(1 + \frac{1 + K}{\beta(\beta + 1)}\right) \qquad 8.194$$

In order to demonstrate an improvement in the reliance of the current mirror output on the current gain, consider the following example.

Example 8.14

For BJT current mirrors as in figures 8.57 and 8.58, with $\beta = 50$ and $k = 1$, determine the percentage the output current differs from the reference current in each case.

From equation 8.190, $i_O = 0.962 I_{\text{REF}}$, which is a 3.8 percent difference of the output from the reference.

From equation 8.193, $i_O = 0.999 I_{\text{REF}}$. Thus, the circuit of figure 8.58 is a great improvement over that of figure 8.57 since it provides greater independence of the output current from the current gain β.

APPENDIX 8.A
Semiconductor Device Symbols and Characteristics

(A) P-N Diode

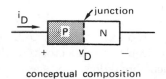

conceptual composition

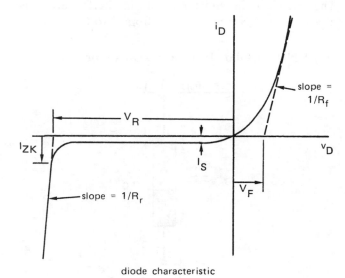

diode characteristic

circuit symbol

PIECEWISE LINEAR MODEL

$v_D > V_F : i_D = (v_D - V_F)/R_f$

$V_F > v_D > -V_R : i_D = 0$

$-V_R > v_D : i_D = (v_D + V_R)/R_r$

IDEAL DIODE

$V_F = 0, R_f = 0, R_r = 0, V_R \longrightarrow \infty$

COMPOSITE PIECEWISE LINEAR MODEL

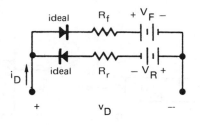

PERFECT DIODE EQUATION:

$$i_D = I_S(e^{q v_D /kT} - 1)$$

$$\frac{q}{kT} = 40 \text{ at } T = 293 \text{ K}$$

INCREMENTAL RESISTANCE

$$r_f = \frac{kT}{qI_d} \text{ for } v_D > V_F$$

BIAS DEFINITION

positive bias: $v_D > 0$ (drops from P to N)

forward bias: positive bias

negative bias: $v_D < 0$ (drops from N to P)

reverse bias: negative bias

FORWARD VOLTAGE V_F

Germanium: $V_F = 0.2$ volt

Silicon: $V_F = 0.6$ volt

(B) Zener or Avalanche Diode

The same as the P-N diode but with voltage and current directions reversed.

Changes in nomenclature: $v_R = v_Z$, $R_r = R_z$

The avalanche or zener diode will have its nominal voltage when carrying rated current.

Keep-alive current I_{ZK} is the minimum for which characteristic is linear.

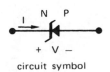

circuit symbol

(C) NPN Bipolar Junction Transistor (BJT)

COMMON EMITTER

COMMON BASE

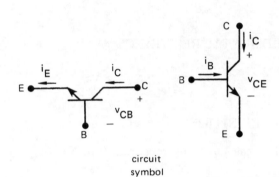

conceptual
composition

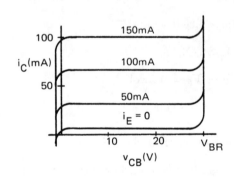

circuit
symbol

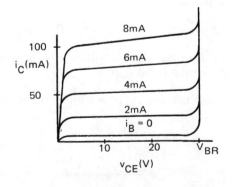

form of
characteristic

Nomenclature

B	base
C	collector
E	emitter
i_B	base current
i_C	collector current
i_E	emitter current
v_{CB}	voltage drop from collector to base
v_{CE}	voltage drop from collector to emitter
V_{BR}	avalanche breakdown voltage

The arrow on the emitter points in the forward bias direction for P-N junction (i.e., from P to N).

(D) PNP Bipolar Junction Transistor (BJT)

COMMON EMITTER

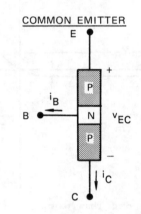

COMMON BASE

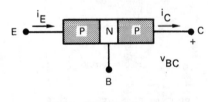

conceptual
composition

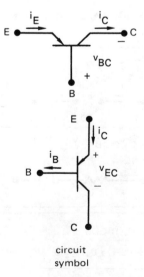

circuit
symbol

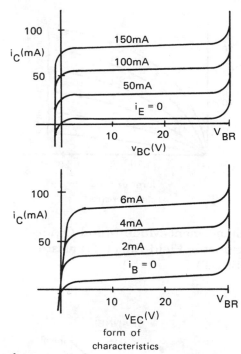

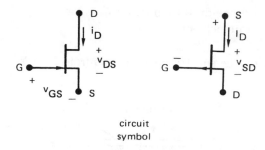

circuit
symbol

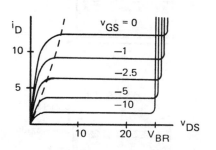

form of
characteristics

Nomenclature

B	base
C	collector
E	emitter
i_B	base current
i_C	collector current
i_E	emitter current
v_{CB}	voltage drop from collector to base
v_{CE}	voltage drop from collector to emitter
V_{BR}	avalanche breakdown voltage

The arrow points in the forward bias direction for a P-N junction (i.e., from P to N).

(E) Junction Field-Effect Transistors (JFET)

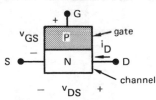

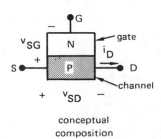

conceptual
composition

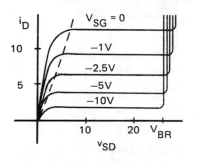

form of
characteristics

Nomenclature

B	body or substrate
D	drain
G	gate
S	source
i_D	drain current
v_{DS}	voltage drop from drain to source
v_{GS}	voltage drop from gate to source
V_{BR}	avalanche breakdown voltage

Arrow on the gate lead points from P to N.

(F) Depletion-Mode Metal-Oxide-Semiconductor Field-Effect Transistors (D-MOSFET)

Nomenclature

B	body or substrate
D	drain
G	gate
S	source
v_{DS}	voltage drop from drain to source

v_{GS} voltage drop from gate to source
i_D drain current

Arrow on the substrate lead points from P to N.

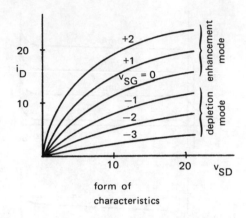

form of
characteristics

(G) Enhancement-Mode Metal-Oxide-Semiconductor Field-Effect Transistors (E-MOSFET)

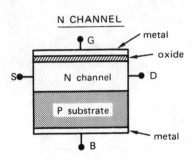

N CHANNEL

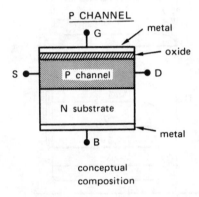

illustration goes here

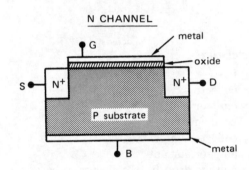

conceptual
composition

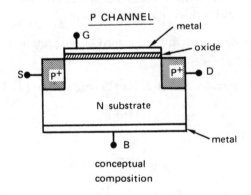

conceptual
composition

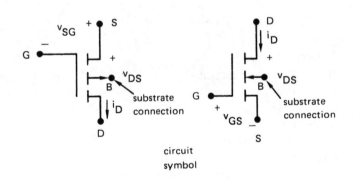

circuit
symbol

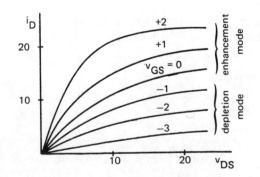

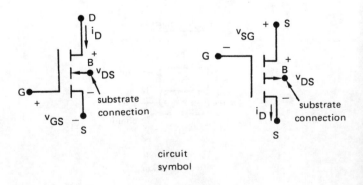

circuit
symbol

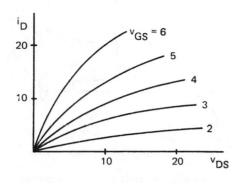

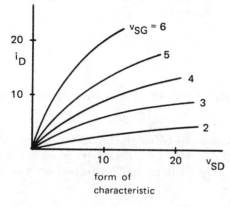

form of
characteristic

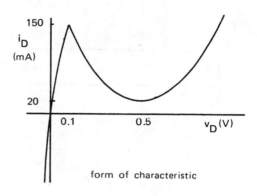

form of characteristic

(I) Unijunction Transistor (UJT) or Double-Base Diode

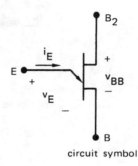

circuit symbol

Nomenclature

B body or substrate
D drain
G gate
S source
v_{DS} voltage drop from drain to source
v_{GS} voltage drop from gate to source
i_D drain current

Arrow on substrate lead points from P to N.

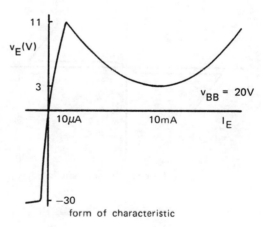

form of characteristic

(H) Tunnel Diode

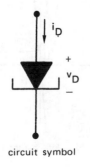

circuit symbol

(J) Current Regulator Diode (see JFET)

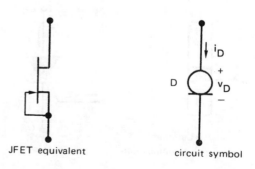

JFET equivalent   circuit symbol

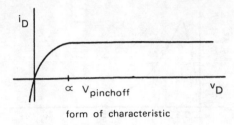

form of characteristic

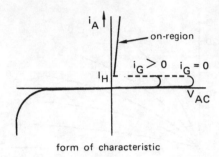

form of characteristic

I_H is the minimum current needed to maintain operation in the on-region.

(K) DIAC

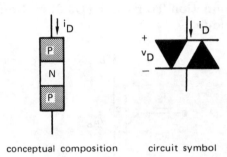

conceptual composition   circuit symbol

(M) Gate Turn-off Silicon-Controlled Rectifier (GTO)

This device is an SCR which can be turned off by reversing the voltage on the gate (making V_{gc} negative). The ordinary SCR cannot be turned off via the gate lead.

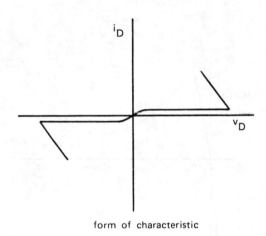

form of characteristic

(N) TRIAC

This device is essentially two SCRs connected back-to-back, with the gate leads internally connected.

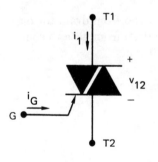

circuit symbol

(L) Silicon-Controlled Rectifier (SCR)

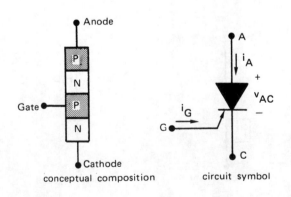

conceptual composition   circuit symbol

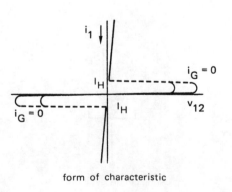

form of characteristic

PRACTICE PROBLEMS

Warmups

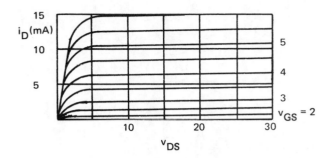

1. At a room temperature of 293 K, a particular diode has a voltage of 0.60 volt when the current is 1.00 mA, and 0.70 volt when the current is 9.83 mA. Determine the value of the exponent parameter η in the diode equation for this device.

2. At a room temperature of 293 K, a diode has a voltage of 0.600 volt when it is driven by a current source of 1 mA. If the temperature drops 50 degrees, determine the new voltage across the diode.

3. At a room temperature of 293 K, a diode with $\eta = 2$ has a voltage of 0.6 volt at a current of 1.0 mA. It is forward biased in a series circuit with a 10 volt battery and a 1000 ohm resistance. A quick estimate of the current, assuming a drop of 0.7 volt across the diode, is 9.3 mA. Determine the error in this estimate.

4. At a room temperature of 293 K, a diode with $\eta = 1.5$ is rated at 10 mW. At a current of 1 mA it has a voltage drop of 0.6 volt. Determine its rated current.

5. An N-channel JFET has $I_{DSS} = 5$ mA and $V_P = 5$ volts. Estimate i_D at $v_{GS} = -3$ volts.

6. An N-channel JFET has the characteristics shown. As a small-signal amplifier, it is to be self-biased at $v_{DS} = 15$ volts and $i_D = 5$ mA with a drain supply of 30 volts. Specify R_d and R_s.

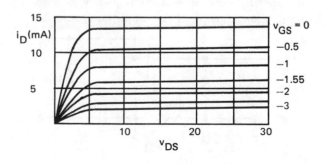

7. An N-channel enhancement-mode MOSFET with the given characteristics is to be operated as a source-follower with $R_d = 500$ ohms, $R_s = 1000$ ohms, and a DC supply of 16 volts. Determine a bias circuit which will permit an input voltage swing of at least one volt peak-to-peak.

8. The circuit in example 8.1 has $R_1 = R_2 = 10$ kilohms. Determine (a) the value of R_e necessary for an operating point of 15 mA when $\alpha_o = 0.95$, $I_{CO} = 10^{-6}$, and $V_{CC} = 24$ volts, and (b) the change in collector current for a temperature change of 125°C and with the parameters found in part (a). Assume silicon construction.

9. A bipolar junction transistor characteristic is shown in figure 8.30. For an operating point of $V_{CC} = 10$ volts and $i_B = 1$ mA, obtain the output circuit for a small-signal amplifier of the form shown in figure 8.32(a).

10. An amplifier in the configuration of figure 8.33 has $R_1 = 6$ kilohms, $R_2 = 3$ kilohms, $R_c = 1$ kilohms, $R_s = 1$ kilohms, and $R_L = 833$ ohms. With $h_{ie} = 500$ ohms, $\beta = 50$, and $r_c = 5$ kilohms, determine the amplifier voltage gain.

Concentrates

1. The two JFETs shown have $I_{DSS} = 8$ mA and $V_P = 2.5$ volts. Determine the range of the input voltage v_{IN} over which the output follows the input.

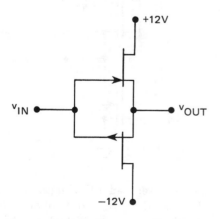

2. The JFETs shown have $V_P = 1$ volt and $V_A = 150$ volts. T1 has $I_{DSS} = 12$ mA and T2 has $I_{DSS} = 8$ mA. Find the small-signal gain and output resistance.

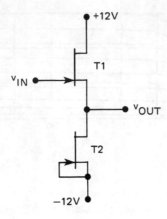

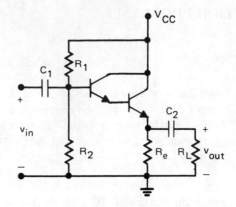

3. An N-channel JFET has the circuit shown and $C_{OSS} = 2$ pF, $g_m = 2 \times 10^{-3}$ mhos, and $C_{ISS} = 6$ pF. Determine the voltage gain and high-frequency cutoff (in hertz) if (a) $R_s = 5000$ ohms, and (b) $R_s = 50$ ohms.

6. For the amplifier shown, $h_{ie} = 500$ ohms, $\beta = 39$, and $r_c = 5$ kilohms. Determine (a) the low frequency cutoff due to the emitter bypass capacitance, assuming that C_1 is a short circuit, (b) the value of C_1 necessary to ensure that the associated corner frequency is less than 10 hertz regardless of whether or not the emitter resistance is bypassed, and (c) the frequency response of the voltage gain, assuming C_2 can be neglected. The transistor has high-frequency parameters of $C_{b'c} = 25$ pF, $C_{b'e} = 1000$ pF, $r_{bb'} = 400$ ohms, and $r_{b'e} = 100$ ohms.

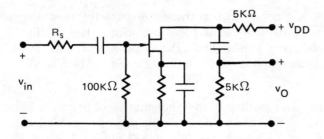

4. The N-MOSFETs have $K = 0.001$, $V_{GS(on)} = 1$ volt, and $V_A = 125$ volts. Find the range of v_{IN} over which T2 remains in pinchoff operation. Set the operating point for v_{IN} at the middle point of that range and determine the small-signal gain at that operating point.

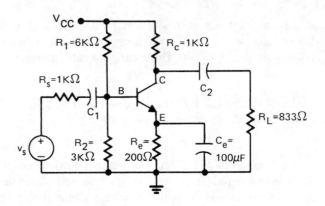

7. For the amplifier circuit and MOSFET characteristics shown, determine the frequency response for v_o/v_{IN}. Show a graph of the amplitude response.

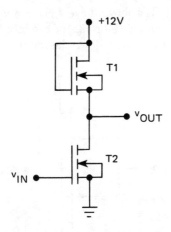

5. An emitter follower uses a Darlington pair as shown. $R_1 = R_2 = 220$ kilohms, $R_e = 10$ kilohms, $R_L = 1$ kilohms, $h_{ie1} = 750$ ohms, $h_{ie2} = 250$ ohms, $h_{fe1} = 100$, $h_{fe2} = 50$, $r_{c1} = 20$ kilohms, and $r_{c2} = 10$ kilohms.

Determine the current gain of this amplifier.

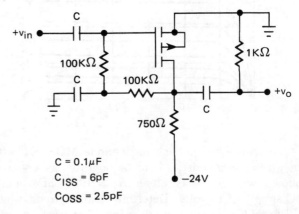

$C = 0.1 \mu F$
$C_{ISS} = 6 pF$
$C_{OSS} = 2.5 pF$

(a)

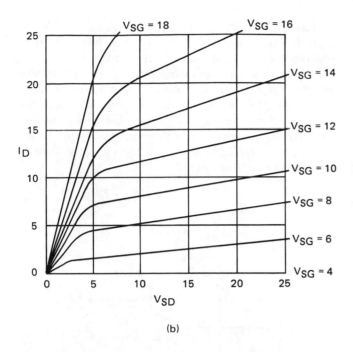

(b)

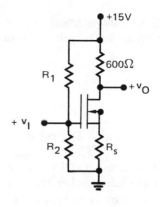

(a) Circuit Diagram
(b) Drain-Source Characteristic

9. The JFETs shown have $I_{DSS} = 5$ mA and $V_P = 2$ volts. (a) Determine the range of common-mode input voltage that can occur if the output voltage is to have its maximum swing. (b) Determine the differential-mode gain for small differential inputs.

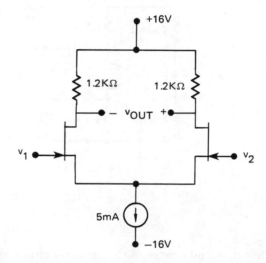

8. Design the bias circuit for the amplifier shown. The desired Q-point is 10 mA and $v_{DS} = 5$ volts, with an input impedance of 100 kilohms. Determine the amplifier gain if the resistance R_s is capacitively bypassed.

10. The FETs in the current mirror shown have $K = 0.0005$ A/V^2, $V_{GS(on)} = 1.5$ volts, and $V_A = 120$ volts. (a) Determine the value of the reference current I_{REF}. (b) Determine the range of the voltage v_O, over which

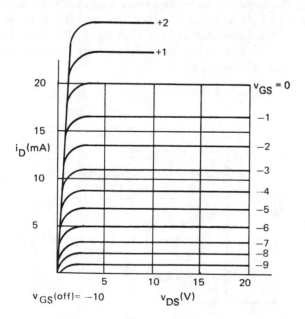

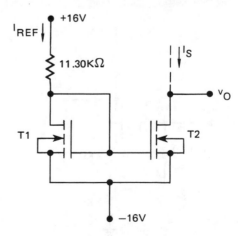

T2 remains in pinchoff operation. (c) Determine I_s over the range of current source voltages found in (b).

Timed

1. The BJT shown has a DC value of $\alpha_o = 0.95$ and a small-signal value of $\beta = 49$. The base-spreading resistance is 350 ohms. (a) Determine the operating point. (b) Determine whether the bias stability with respect to temperature is adequate. (c) Determine the small-signal gain at frequencies where the capacitive impedance is negligible.

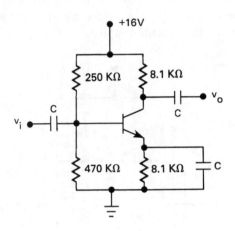

3. The BJT shown has a DC value of $\alpha_o = 0.95$ and a small-signal value of $\beta = 49$. The base-spreading resistance is 350 ohms. The Early voltage is 100 volts. (a) Determine the operating point. (b) Determine whether the bias stability with respect to temperature is adequate. (c) Determine the small-signal gain at frequencies where the capacitive impedances are negligible. (d) How much error is there in an approximation that $A_v = -R_c/R_e$, where R_e is the unbypassed resistance in the emitter circuit?

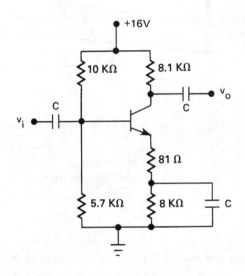

2. The transconductance of the amplifier circuit shown is 5000 μmhos and the drain-source resistance is 20 kilohms. Neglect the capacitor effects and determine (a) the voltage gain, and (b) the input impedance.

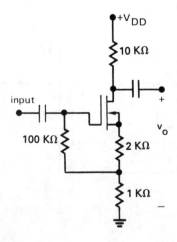

4. The amplifier shown has the following transistor parameters. For T_1: $h_{fe} = 100$, $h_{ie} = 750$ ohms, and $r_c = 20$ kilohms; for T_2: $h_{fe} = 50$, $h_{ie} = 250$ ohms, and $r_c = 10$ kilohms. Determine the mid-frequency voltage gain for this amplifier.

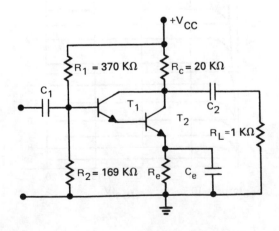

5. The circuit shown can produce v_o to be the product of kv_1I_2 when the voltage v_1 is small compared with the thermal voltage, which is taken as 25 mV for this problem. Making necessary approximations, design the value of R to obtain the value of k to be 10,000.

7. The JFET shown has $I_{DSS} = 4$ mA, an Early voltage of 100 volts, and $V_P = 2.5$ volts. The quiescent value of the drain current is to be 2.0 mA. (a) Determine R such that the quiescent drain current is 2 mA. (b) Determine the transconductance at that current.

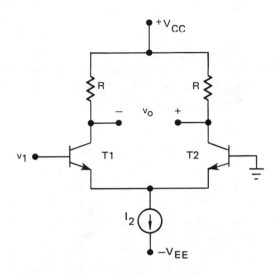

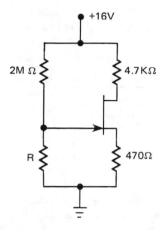

6. For the circuit shown, i_{IN} is given by

$$i_{IN} = 175 + 40\cos 1000t \ \mu A$$

The transistor has $\beta = 49$ and an Early voltage of 50 volts. (a) Determine the output voltage v_o. (b) Determine the power delivered by the 16 volt power supply. (c) Determine the power supplied by the 10 mA current source.

8. The differential amplifier circuit shown has a single-ended output. The transistors T1 and T2 are matched with $\beta = 49$ and $V_A = 50$ volts. The current source has a shunt resistance of 100K ohms. There is a -4 VDC common-mode voltage at the input. (a) Find the differential-mode gain. (b) Find the CMRR. (c) Find the input resistance.

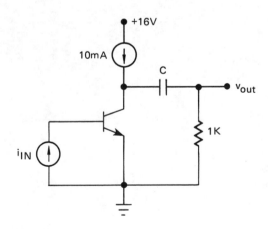

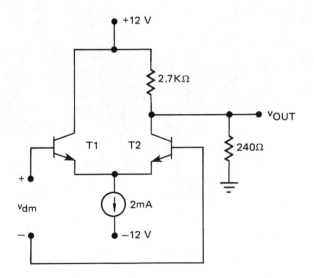

9. The differential amplifier shown has a current mirror load used to increase the gain such that the single-ended output will provide gain comparable to the differential output circuit. Assume that all transistors have $\beta = 49$ and $V_A = 50$ volts. The current source has a shunt resistance of 100K ohms, and there is a -4 volt common-mode voltage at the input. (a) Determine the differential-mode gain. (b) Determine the CMRR.

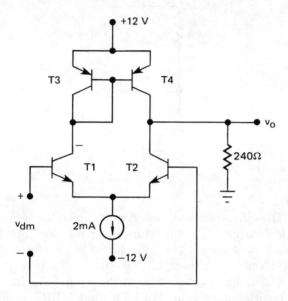

10. The Wilson current mirror shown has the same independence of the current gain β as the improved current mirror of figure 8.57, but because of the transistor in the emitter circuit of T1 the output resistance is also increased. (a) Using only the βi_b current source as the circuit model of the three identical BJTs, show that

$$i_S = \frac{I_{REF}}{1 + \dfrac{2}{\beta^2 + 2\beta}}$$

(b) Using a small-signal BJT model including input and output resistance as well as transconductance, determine the output resistance of the Wilson current mirror in terms of i_S, V_A, r_{bb}, and β.

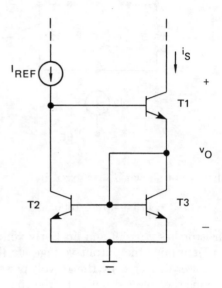

9 AMPLIFIER APPLICATIONS

Nomenclature

$A(f)$	gain spectrum	–
A_{cm}	common-mode voltage gain	volts/volt
A_{dm}	differential mode voltage gain	volts/volt
A_I	current gain	amps/amp
A_o	DC gain of operational amplifier	volts/volt
A_p	power gain	watts/watt
A_v	voltage gain	volts/volt
CMRR	common-mode rejection ratio	–
e^2	spectral voltage squared density	volts^2/hertz
f	frequency	hertz
f_b	operational amplifier open-loop corner frequency	hertz
f_t	operational amplifier gain-bandwidth product	hertz
F	noise figure	–
g_m	transconductance	mhos
G	power gain	watts/watt
GBW	gain-bandwidth product	hertz
$G(f)$	power gain spectrum	–
I	Laplace transformed current	amps/s
I	Laplace transformed current(constant)	amps/s
I^2	spectral current squared density	amps^2/hertz
I_b	bias current	amps
I_{off}	offset current	amps

k	a constant	–
k	(with T) Boltzmann's constant	1.38×10^{-23} joules/K
N	noise power	watts
p	spectral power density	watts/hertz
q	electron charge	6.09×10^{-19} coulombs
S	signal power	watts
$S(f)$	spectral voltage squared density	volts/hertz
S_R	maximum slewing rate	volts/s
T	temperature	Kelvin
v_{IN}	input voltage	volts
v_O	output voltage	volts
v_{OUT}	output voltage	volts
V	Laplace transformed voltage	volt/s
V	voltage (constant)	volts
Y	admittance	mhos
Z_{in}	input impedance	ohms
Z_{out}	output impedance	ohms

Symbols

μ	micro	10^{-6}
μ	voltage gain	volts/volts
ω_b	opamp open-loop corner frequency	rad/s
ω_{cf}	corner frequency	rad/s
ω_{hf}	high-frequency pole	rad/s
ω_n	notch frequency	rad/s
ω_p	pole frequency	rad/s
ω_t	GBW product of operational amplifier	rad/s
ω_z	zero frequency	rad/s

1 IDEAL OPERATIONAL AMPLIFIERS

A. Introduction

Amplifiers are characterized by parameters such as voltage gain, (A_v), current gain, (A_I), power gain, $(A_p$ or $G_p)$, input impedance, (Z_{in}), and output impedance, (Z_{out}). These parameters are functions of frequency, and in practical operational amplifiers have tolerances up to 50 percent of their nominal values.

Operational amplifiers (opamps) are used in circuits where their parameter variations must not affect the correct operation of the circuit. In many applications this function can be achieved quite independently of the opamp parameters, and the opamp can be treated as ideal. In order to determine the behavior of an ideal opamp, first consider the characteristics of practical opamps.

B. Amplifier Definitions

The opamp symbol is a triangle, as indicated in figure 9.1, having an output terminal at the right, with the single-ended output voltage, v_O, defined as positive at that terminal. The input terminals are to the left and constitute a differential input. The upper input terminal in this drawing is marked with a minus sign and is called the *negative terminal*. The lower input terminal is marked with a positive sign and is called the *positive input terminal*. The input voltages are both defined as positive with respect to ground. Depending on the application, the upper terminal may be the positive input terminal and the lower negative, but the arrangement shown is the one most commonly used.

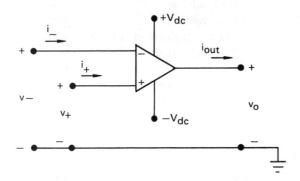

Figure 9.1 Opamp Symbol and Definitions

The output voltage is proportional to the difference between the two input voltages, provided the input voltages are small enough. The voltage gain when the output is open-circuited, $(i_{\text{OUT}} = 0)$ is given by

$$v_{\text{OUT}} = A_{\text{vo}}(v_+ - v_-)$$
$$= -A_{\text{vo}}(v_- - v_+)$$
$$= A_{\text{vo}}v_+ - A_{\text{vo}}v_- \qquad 9.1$$

A_{vo} is the open-circuit DC voltage gain of the opamp. Typical values of A_{vo} are from 10^5 to 10^7.

The two voltages $+V_{\text{DC}}$ and $-V_{\text{DC}}$ are often called the *rail voltages* and are the constant power supply voltages needed to operate the electronic devices in the opamp. The range of v_O is between these two values, but not within about three volts of either. Thus for a 16 volt value of V_{DC}, the range of v_{OUT} is from about -13 to $+13$ volts.

For the opamp to be in linear operation, it can be seen from equation 9.1 that the input voltage difference is restrained to be between the values of the rail voltages divided by the gain, or

$$|v_+ - v_-| < \frac{V_{\text{DC}} - 3}{A_{\text{vo}}} \qquad 9.2$$

For 16 volt power supply voltages and A_{vo} of 10^5, the result is a maximum input voltage difference of 0.13 millivolts.

The input impedance of practical opamps is typically more than 10^5 ohms, so the input currents, in order for the opamp to remain in linear operation, will have to be less than the maximum input voltage difference (equation 9.2) divided by 10^5. For the example cited above with a maximum of 0.13 mV, the maximum input current will be 1.3 nanoamps.

Thus, for the analysis of any practical opamp, the input current is assumed to be negligible.

In addition, for analysis of ideal opamps in linear operation, the input voltage is also assumed to be zero. For analysis, including frequency effects and other limitations, the assumption that the input voltage difference is zero must be abandoned. However, for most cases the ideal opamp can be assumed, or at least must be assumed to find the low-frequency behavior of a circuit using one or more opamps.

The assumptions that are made to carry out analysis of opamp circuits in the ideal case are:

- The opamp is in its linear range.
- The input voltage difference is zero.
- The input current is zero.

A further restriction is that any feedback of the output signal is to be negative. This requires that a single feedback link go to the negative input terminal, and for dual feedback there be stronger feedback to the negative terminal than to the positive terminal. The reasons for this cannot be understood until the frequency response of opamps is examined in a later section.

C. Inverting Amplifier Analysis

The inverting opamp circuit is shown in figure 9.2 without the rail voltages which are not normally shown for the analysis of the circuit operation. The analysis of this circuit is accomplished by writing Kirchhoff's current law (KCL) at the node marked "a", where initially the voltage of that node is called v_A. Thus,

$$\frac{v_{\text{IN}} - v_A}{R_1} + \frac{v_O - v_A}{R_2} = i_n \qquad 9.3$$

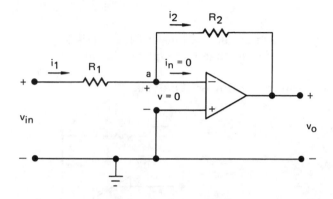

Figure 9.2   Inverting Opamp Circuit

Then review the assumptions and adjust accordingly.

- The opamp is in its linear range.

- The input voltage difference is zero.

This means that with v_+ already zero because of its grounded condition, v_- is also zero. Thus, v_A is also zero in equation 9.3.

- The input current is zero.

Here the input current has been called i_n, which is zero in equation 9.3.

Then, take these results to equation 9.3 which becomes

$$\frac{v_{\text{IN}}}{R_1} + \frac{v_{\text{OUT}}}{R_2} = 0 \quad \text{or} \quad v_{\text{OUT}} = -\frac{R_2}{R_1} v_{\text{IN}} \qquad 9.4$$

The assumptions of zero voltage difference and zero current at the input of the opamp results in the assumption that as long as the opamp remains in its linear operating range, the output of the opamp will adjust its voltage so that the current flowing through R_2 (i.e., i_2) will be exactly the same as that flowing through R_1 (i.e., i_1).

Note that the current i_2 is provided by the output of the opamp. A question about the validity of v_O under current flow may be raised. The answer is that the

output current adjusted itself to match the circuit input current i_1, so that any voltage drop across an internal resistance on the inside of the opamp is irrelevant.

The linear range of input signal level can be determined from the rail voltages. For the output voltage to reach the rail voltage minus 3 for the inverting amplifier with a gain $-R_2/R_1$, the magnitude of the input voltage to reach saturation is

$$\left| v_{\text{IN}|\text{max}} \right| = \frac{(V_{\text{DC}} - 3)}{A}$$

$$= \frac{R_1}{R_2}(V_{\text{DC}} - 3) \qquad 9.5$$

For larger input voltage magnitudes the output will remain at essentially the positive or negative $V_{\text{dc}} - 3$. This is called *saturation*. Then the transfer characteristic for the inverting amplifier is as shown in figure 9.3.

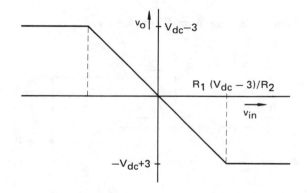

Figure 9.3   Transfer Characteristic of an Inverting Opamp Circuit

D. Inverting Amplifier With Multiple Inputs

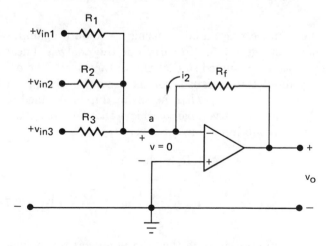

Figure 9.4   Inverting Opamp with Multiple Inputs

An inverting opamp with three inputs is shown in figure 9.4, which is again analyzed via KCL at the node of the negative or inverting terminal of the opamp. With the node "a" voltage still zero, and the opamp input current zero, the result is

$$\frac{v_{IN1}}{R_1} + \frac{v_{IN2}}{R_2} + \frac{v_{IN3}}{R_3} + \frac{v_O}{R_f} = 0$$

or

$$v_O = -\frac{R_f}{R_1}v_{IN1} - \frac{R_f}{R_2}v_{IN2} - \frac{R_f}{R_3}v_{IN3} \qquad 9.6$$

This result can be extended to any number of inputs.

Note that the impedance seen from any of the input sources is the attached resistance, as the negative terminal of the opamp is at virtual ground.

E. Noninverting Amplifier Analysis

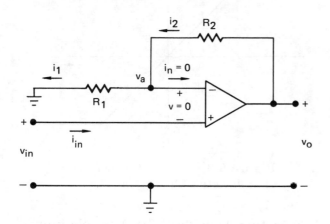

Figure 9.5 Noninverting Opamp Circuit

The noninverting amplifier using a standard opamp is shown in figure 9.5. To carry out the analysis, linear operation is assumed so that the voltage across the opamp input terminals is zero. Therefore, v_A is v_{IN}. It follows that i_1 is v_{IN}/R_1. As the opamp is assumed to adjust the output voltage to achieve that level of i_1, and as the opamp input current is zero, $i_2 = i_1$.

Then $v_{OUT} = i_2(R_2 + R_1)$, which is

$$v_{OUT} = \frac{v_{IN}}{R_1}(R_2 + R_1) = v_{IN}\left(\frac{R_2}{R_1} + 1\right) \qquad 9.7$$

It is of interest to note that the noninverting configuration of the opamp has infinite input impedance, since $i_{IN} = -i_N = 0$.

The transfer characteristic for noninverting opamp circuit is shown in figure 9.6.

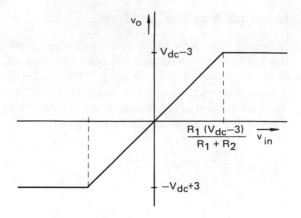

Figure 9.6 Transfer Characteristic of a Noninverting Opamp Circuit

F. Voltage Follower Analysis

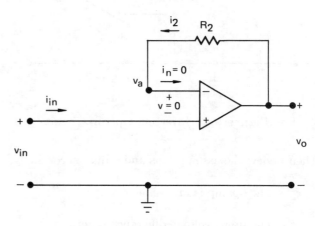

Figure 9.7 Voltage Follower Opamp Circuit

The *voltage follower* circuit is shown in figure 9.7. Its output voltage closely follows its input voltage. The operation is quite simple: assume linear operation and require the output voltage to adjust itself so the voltage difference between the two input terminals is zero. The positive terminal has v_{IN} applied, so the opamp's job is to adjust the output voltage until the negative terminal also has the voltage v_{IN}. When this is achieved,

$$v_O = v_{IN} \qquad 9.8$$

As no current flows through R_2, it is always replaced with a short circuit, i.e., zero ohms. The voltage follower can be thought of as a special case of the noninverting amplifier with R_1 infinite and R_2 zero.

2 OPAMP FREQUENCY CHARACTERISTICS

A. Introduction

Internally compensated opamps have frequency characteristics as indicated in figure 9.8. This is a low-pass filter characteristic having a low-frequency pole at f_b, a DC gain of A_o, and a high-frequency pole at a frequency f_{hf} that is equal to or higher than $A_o f_b$.

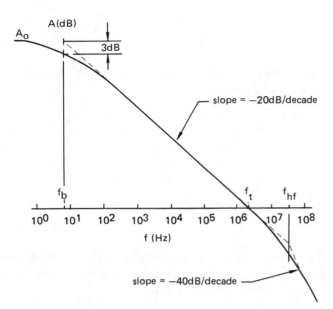

Figure 9.8 Frequency Characteristic of a Typical Opamp Gain

Where $\omega = 2\pi f$ and s is the complex frequency variable, the characteristic shown in figure 9.8 can be written in a second-order form as

$$A_2(s) = \frac{A_o}{\left(1 + \dfrac{s}{\omega_b}\right)\left(1 + \dfrac{s}{\omega_{\text{hf}}}\right)} \qquad 9.9$$

In most cases a first-order approximation ignoring f_{hf} (or ω_{hf}) is adequate to determine the most significant frequency effects. This is written as

$$A = A_1(s) = \frac{A_o}{1 + \dfrac{s}{\omega_b}} \qquad 9.10$$

The low-frequency gain is A_o, and the frequency where the gain is $0.707 A_o$, or $-3\,\text{dB}$ from the low-frequency gain, is ω_b. The product of the low-frequency gain and the $-3\,\text{dB}$ frequency is defined as the gain-bandwidth product of a DC amplifier. For the opamp, GBW = $A_o \omega_b$. As $V_{\text{out}} = A(V_+ - V_-)$,

$$V_+ - V_- = \frac{V_{\text{out}}}{A} = V_o\left(\frac{1}{A_o} + \frac{s}{A_o\omega_b}\right) \qquad 9.11$$

But $A_o\omega_b = \omega_t$, so

$$V_+ - V_- = V_o\left(\frac{1}{A_o} + \frac{s}{\omega_t}\right) \qquad 9.12$$

B. Frequency Response of the Inverting Opamp Circuit

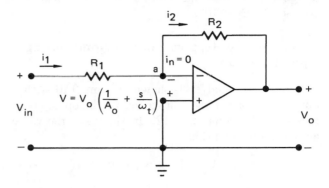

Figure 9.9 Inverting Opamp Circuit with Frequency Effects

The circuit in figure 9.9 is analyzed using KCL at node "a."

$$\frac{V_{\text{in}} + V}{R_1} + \frac{V_o + V}{R_2} = 0 \qquad 9.13$$

But $V = V_+ - V_-$, so substituting from equation 9.12,

$$\frac{V_o}{V_{\text{in}}} = -\frac{R_2}{R_1}\frac{1}{1 + \left(\dfrac{1}{A_o} + \dfrac{s}{\omega_t}\right)\left(1 + \dfrac{R_2}{R_1}\right)} \qquad 9.14$$

As long as R_2 and R_1 are within two orders of magnitude of each other, the term in the denominator involving A_o is negligible compared to 1, so that equation 9.14 can be rewritten as

$$\frac{V_o}{V_{\text{in}}} = -\frac{R_2}{R_1}\frac{1}{1 + \dfrac{s}{\omega_t}\left(1 + \dfrac{R_2}{R_1}\right)} \qquad 9.15$$

This is also a low-pass filter characteristic, with a DC gain of $-R_2/R_1$, and a corner or 3 dB frequency

$$\omega_{\text{cf}} = \frac{\omega_t}{1 + \dfrac{R_2}{R_1}} \qquad 9.16$$

It can be seen that the gain of the inverting amplifier can be made independent of the opamp parameters as long as the ratio R_2/R_1 remains much less than A_o. On the other hand, the bandwidth of the resulting amplifier is dependent on the term ω_t, which is sometimes called the *gain-bandwidth product* of the opamp. For this circuit, however, the gain-bandwidth product is R_2/R_1 multiplied by ω_{cf}, or

$$\text{gain-bandwidth product} = \text{GBW}$$
$$= \frac{R_2\omega_t}{R_2 + R_1} \qquad 9.17$$

The gain-bandwidth product approaches ω_t only if R_1 is much less than R_2. In practice, opamp circuits are designed to have a maximum gain of less than about 100, so $R_2/R_1 < 100$.

C. Frequency Response of the Noninverting Opamp Circuit

The circuit for analysis is shown in figure 9.10, with the approximation for the voltage across the opamp terminals given by equation 9.12. In this case, the voltage across the terminals is

$$V_+ - V_- = V_{\text{in}} - V_a \qquad 9.18$$

By voltage division, it is seen that

$$V_a = \frac{R_1 V_O}{R_1 + R_2} \qquad 9.19$$

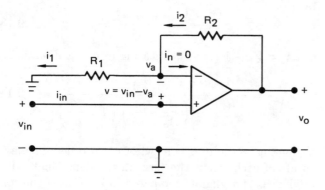

Figure 9.10   Noninverting Opamp Circuit

Combining equations 9.12, 9.18 and 9.19 to solve for V_o in terms of V_{in},

$$\frac{V_o}{V_{\text{in}}} = \frac{1 + \frac{R_2}{R_1}}{1 + \frac{1}{A_o}\left(1 + \frac{R_2}{R_1}\right) + \frac{s}{\omega_t}\left(1 + \frac{R_2}{R_1}\right)} \qquad 9.20$$

Again the denominator term involving A_o can be neglected under the condition that R_2/R_1 is much less than A_o. Then the result is a low-pass characteristic that has a DC gain of $1 + R_2/R_1$ and a corner frequency

$$\omega_{\text{cf}} = \frac{\omega_t}{1 + \frac{R_2}{R_1}} \qquad 9.21$$

So in this case the gain-bandwidth product is ω_t.

$$\text{GBW} = \omega_t$$

D. Effects of Secondary Amplifier Pole

As indicated in figure 9.8, there is a second pole, shown there as occurring at f_{hf}. For the inverting amplifier

of figure 9.9, the node "a" equation can be written together with the expression for the amplifier gain including the secondary pole in the form

$$V_o = V\frac{A_o}{\left(\dfrac{s}{\omega_b} + 1\right)\left(\dfrac{s}{\omega_{\text{hf}}} + 1\right)} \qquad 9.22$$

Eliminating V from equations 9.22 and 9.13, the voltage relationship becomes

$$\frac{V_o}{V_{\text{in}}} =$$

$$\frac{\dfrac{-R_2}{R_1}}{1 + \left(1 + \dfrac{R_2}{R_1}\right)\left[\dfrac{s^2}{A_o\omega_b\omega_{\text{hf}}} + \dfrac{s}{A_o}\left(\dfrac{1}{\omega_b} + \dfrac{1}{\omega_{\text{hf}}}\right) + \dfrac{1}{A_o}\right]} \qquad 9.23$$

The zeros of the denominator of equation 9.23 can be solved by the quadratic equation. By substitution of ω_t for $A_o\omega_b$, the poles are found to be

$$s = -\frac{\omega_{\text{hf}} + \omega_b}{2}$$

$$\pm\sqrt{\frac{(\omega_{\text{hf}} - \omega_b)^2}{4} - \frac{\omega_t\omega_{\text{hf}}}{1 + \dfrac{R_2}{R_1}}} \qquad 9.24$$

In many opamps, ω_{hf} is very nearly equal to ω_t, and ω_b is negligible when compared to ω_t, so for these cases the following approximation of the locus of the poles is valid.

$$s = \frac{\omega_t}{2}\left(-1 \pm \sqrt{1 - \frac{4}{1 + \dfrac{R_2}{R_1}}}\right) \qquad 9.25$$

Note that when $R_2/R_1 = 3$, there are two real poles at $s = -\omega_t/2$, and for smaller ratios of R_2/R_1 there are complex conjugate poles. In general, the single pole model of the opamp is valid only for ratios of R_2/R_1 that are greater than about 10.

For opamps with $\omega_{\text{hf}} \cong \omega_t$, equation 9.23 can be written as

$$\frac{V_o}{V_{\text{in}}} = \frac{\dfrac{-R_2}{R_1}}{\left(1 + \dfrac{s}{\omega_1}\right)\left(1 + \dfrac{s}{\omega_2}\right)} \qquad 9.26$$

$\omega_1 = (\omega_t/2)(1 - \sqrt{1 - (4/(1 + R_2/R_1))})$ and $\omega_2 = (\omega_t/2)(1 + \sqrt{1 - (4/(1 + R_2/R_1))})$. In order that equation 9.26 behave as a single pole system up to the first pole frequency, it is necessary that the second pole have little effect at that frequency. The worst effect is phase shift, so that if $\omega_1 = 0.1\omega_2$, there would be 5.7 degrees of excess phase shift over the single pole system, but only 0.5 percent change in amplitude due to the second pole. Setting $\omega_1 = 0.1\omega_2$, the minimum ratio R_2/R_1 is 11.1. This results in 5.7 degrees of excess phase shift at the first pole frequency which occurs at about one-eleventh of the gain bandwidth product, ω_t, with the second pole at ten-elevenths of ω_t.

E. Noninverting Amplifier With Positive Feedback

As previously mentioned, opamps are normally used with negative feedback, with a portion of the output voltage fed back to the input. There are applications where the feedback must be to the positive input terminal, particularly where some hysteresis is required, such as in comparator circuits.

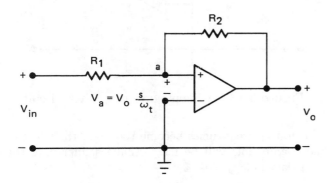

Figure 9.11   Noninverting Opamp Circuit with Positive Feedback

Figure 9.11 shows the simplest noninverting circuit with positive feedback. The voltage, v_A, is indicated in the figure, so a Kirchhoff current law equation at node "a", under the Laplace transformation yields the relationship

$$V_o(s) = \frac{R_2}{R_1} \frac{1}{\tau s - 1} V_{\text{in}}(s) \qquad 9.27$$

In equation 9.27, $\tau = \left(1 + \frac{R_2}{R_1}\right)\frac{1}{\omega_t}$.

It can be seen that the DC gain is $-R_2/R_1$, which will be valid whenever the opamp is operating in a steady DC condition in its linear region. To understand the behavior of this circuit, assume the circuit is at rest with zero input and zero output, when a small thermal noise spike occurs in the input resistor, R_1. This spike can be treated as an impulse voltage input or $v_{\text{IN}} = k\delta(t)$. The Laplace transform of the delta function is a constant, so $V_{\text{in}}(s) = k$. Then, the Laplace transformed output voltage is obtained from equation 9.27.

$$V_{\text{out}}(s) = \frac{R_2}{R_1} \frac{1}{\tau s - 1} k$$

The inverse Laplace transformation of $V_{\text{out}}(s)$ is $v_{\text{OUT}}(t)$.

$$v_{\text{OUT}}(t) = \frac{kR_2}{\tau R_1} e^{+t/\tau} \qquad 9.28$$

Equation 9.28 shows the fundamental effect of the positive feedback: the output grows exponentially. Of course it can only grow to the saturation level of the opamp. From this equation it can be seen that the polarity of the growing output is the same as that of the input delta function, i.e., the output is proportional to $+k$. It follows that the growing exponential response to any input will be in the same polarity as the input. This then results in a transfer function having a negative slope in the active region which will end in saturation as indicated in figure 9.12

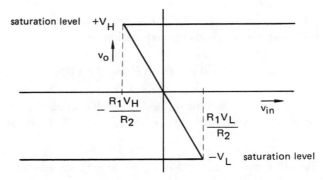

Figure 9.12   Transfer Characteristic of a Noninverting Amplifier with Positive Feedback

In figure 9.12 the positive saturation voltage is V_H and the negative saturation voltage is $-V_L$. Operation on the negative sloped part of the characteristic is unstable, and as such the output will progress quickly to the appropriate saturation level. Then, if the output is at the negative saturation level and the input is less than $R_1 V_L/R_2$, the output will remain at the negative saturation level. If an input voltage exceeding $R_1 V_L/R_2$ is applied, the opamp will enter its linear range long enough for the output to reach the positive saturation level, whereupon it will remain until a voltage less than $-R_1 V_H/R_2$ is applied. Thus, this circuit exhibits hysteresis with an input width of

$$\Delta v_{\text{IN}} = \frac{R_1(V_L + V_H)}{R_2} \qquad 9.29$$

Example 9.1

The opamp shown has a gain-bandwidth product of 10^6 hertz and has saturation output levels of 11 and -11 volts. The initial situation has the output at -11 volts. Select R_2 so that an input exceeding 3.100 volts will initiate a switch to +11 volts. With this value of R_2, determine the input hysteresis width. Also determine the switching time if the input is 3.110 volts. (Assume 4 significant figure accuracy.)

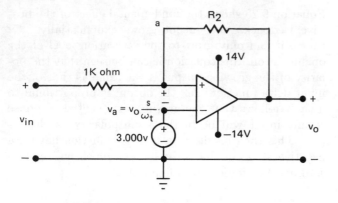

A DC analysis (i.e., $s = 0$), assuming the opamp is in the linear range, yields the expression

$$v_{\text{OUT}} - 3.000 = -0.001 R_2 (v_{\text{IN}} - 3.000)$$

As we require the output be -11 volts when the input is 3.100 volts, we obtain $0.001 R_2 = 140$, or $R_2 = 140 \text{k ohms}$.

The value of v_{IN} to cause v_{OUT} to be $+11$ volts is then found from

$$11 - 3 = -140(v_{\text{IN}} - 3.000) \quad \text{or} \quad v_{\text{IN}} = 2.943$$

The hysteresis is then $3.100 - 2.943 = 0.157$ volts or 157 mV.

For the response to a 3.110 volt step input, nodal analysis at node "a" yields the Laplace expression

$$\frac{141}{\omega_t}[sV_{\text{out}}(s) - v_{\text{OUT}}(0)] - V_{\text{out}}(s) = \frac{12.4}{s}$$

As $v_O(0) = -11$ and $\omega_t = 2\pi \times 10^6$, the partial fraction expansion yields

$$V_o(s) = -\frac{12.4}{s} + \frac{1.4}{s - 4.456 \times 10^4}$$

The inverse transformation yields

$$v_O(t) = -12.4 + 1.4 e^{44,560 t}$$

Solving for the time when v_{OUT} reaches $+11$ volts,

$$t_{\text{switching}} = \frac{1}{44,560} \log_e \frac{11 + 12.4}{1.4} = 63.2 \ \mu\text{s}$$

3　SIMPLE INVERTING ACTIVE FILTERS

A. Introduction

The circuit shown in figure 9.13 can be analyzed under the ideal opamp assumption, which yields a transfer function

$$\frac{V_o}{V_{\text{in}}} = -Y_1 Z_2 \qquad 9.30$$

The admittance, Y_1, and the impedance, Z_2, are functions of s. The poles and zeros of the transfer function are the poles and zeros of Y_1 and Z_2.

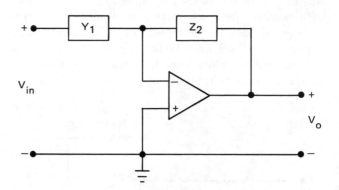

Figure 9.13　A Simple Inverting Filter Circuit

If operating frequencies become too high, the effect of the opamp pole will be felt. Analysis, including the secondary pole, yields

$$\frac{V_o}{V_{\text{in}}} = \frac{-Y_1 Z_2}{1 + \dfrac{s}{\omega_t}\left(1 + \dfrac{s}{\omega_{\text{hf}}}\right)(1 + Y_1 Z_2)} \qquad 9.31$$

This is an important check on the range of validity of the approximation given in equation 9.14.

B. Inverting Integrator

An *inverting integrator* is obtained by placing resistance in the Y_1 location and capacitance in the Z_2 location, yielding the low-frequency approximation to an integrator with $Y_1 = 1/R$ and $Z_2 = 1/sC$. The resulting equation is

$$\frac{V_o}{V_{\text{in}}} = -\frac{1}{sRC} \qquad 9.32$$

Example 9.2

An inverting integrator is designed using an opamp with a gain-bandwidth product of 1 mHz and a high-frequency pole at 1.5 mHz. The resistance used is 1 meg-ohm and the capacitance is 1 μF. Determine the range of use that results in less than 1 degree of phase error due to the opamp poles.

$Y_1 Z_2 = 1/s$, so by substituting into equation 9.31 and using the fact that $\omega_t \gg 1$,

$$\frac{V_o}{V_{\text{in}}} = \frac{-\dfrac{1}{s}}{1 + \dfrac{s}{\omega_t}\left(1 + \dfrac{s}{\omega_{\text{hf}}}\right)}$$

The angle in question is that of the denominator when $j\omega$ is substituted for s. The angle has a tangent of

$$\tan \angle \,(\text{denominator}) = \frac{\omega_{\text{hf}}\,\omega}{(\omega_{\text{hf}}\,\omega_t - \omega^2)} \leq \tan 1^\circ$$
$$= 1.746 \times 10^{-2}$$

As $\omega_{\text{hf}} = 1.5\,\omega_t$, solve for the frequency where the angle is 1 degree in terms of ω_t: $\omega/\omega_t = 1.745 \times 10^{-2}$.

Then, as the gain-bandwidth product is 1 mHz, the frequency at which the denominator contributes 1 degree of phase shift is 17.45 kHz.

Example 9.3

For the system of example 9.2, determine the location of the amplifier poles.

Letting $x = s/\omega_t$, the denominator is expanded to

$$1 + x + \frac{x^2}{1.5}$$

which has complex zeros at $s = (-0.75 \pm j0.968)\omega_t$, which are also the poles due to the opamp.

Example 9.4

An inverting integrator is connected in a closed-loop system with a noninverting amplifier with positive feedback, which was discussed in a previous section. This amplifier has two levels of output and exhibits hysteresis in its transfer characteristic as indicated in figure 9.12. In this version of the noninverting amplifier with positive feedback, which will hereafter be called the *switching amplifier*, the output is limited by matched zener diodes to a maximum level of 10 volts. Except for the brief time needed to switch that amplifier from one output level to the other, the output is always at +10 or −10 volts, and will be seen to be a square wave of that magnitude.

The integrator integrates the square-wave, producing a ramp, which continues until the output becomes large enough to cause the positive feedback amplifier to switch to the opposite output level. The output of the integrator is a triangular wave. The frequency of the

system is determined by the time necessary for the integrator output to change from one threshold of the switching amplifier to the other.

Determine the range of the resistance R if it is to be adjustable to provide frequencies from 10 Hz to 10 kHz.

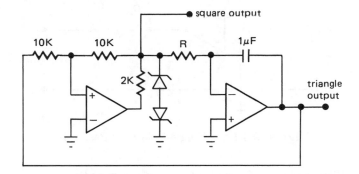

Assume the switching amplifier output is +10 volts, and that this voltage is being integrated by the integrator so that the triangular output is decreasing. It will decrease only until the voltage reaches the −10 volt level, at which time it will cause the switching amplifier to switch to its negative level of −10 volts.

The integrator now begins to integrate the −10 volt output, but its beginning value is known, i.e., the initial voltage is −10 volts, which is across the capacitor. The capacitor now begins to charge toward the positive output of +10 volts, at a rate of $10/R$ coulombs/s.

$$v_{\text{TRIANGLE}} = \int_0^t \frac{10}{CR}\,dt - 10 = 10\left(\frac{t}{CR} - 1\right)$$

The capacitor will continue to charge at this constant rate until it reaches +10 volts, at which time the switching amplifier will enter its linear region of operation proceeding to switch to the positive 10 volt level of output.

The time at which the triangle output reaches 10 volts is found by

$$10 = 10\left(\frac{t}{CR} - 1\right) \quad \text{or} \quad t = 2CR$$

The time for the triangle to change from −10 volts to +10 volts constitutes half of a cycle, so the frequency of operation of the function generator is

$$f = \frac{1}{2t} = \frac{1}{4CR} = \frac{250}{R}\ \text{kHz}$$

For a frequency of 10 hertz, R should be 25 k ohms, and for 10 kHz, R should be 25 ohms. Thus, R should be made up from a 25 ohm fixed resistance in series with a 25 kilohm rheostat.

C. Single-Pole Low-Pass Filter

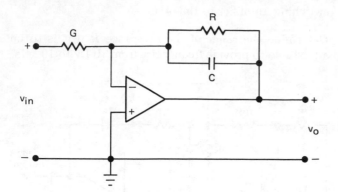

Figure 9.14 A Single-Pole Low-Pass Filter

The circuit of figure 9.14 can be compared to that of figure 9.13 to produce the low-frequency approximation.

$$\frac{V_o}{V_{\text{in}}} = -YZ = -\frac{GR}{sCR + 1} \qquad 9.33$$

This is a low-pass transfer function with a pole at $s = -1/RC$. To determine the effects of the amplifier poles on this circuit, substitute for YZ in equation 9.31 from 9.33, then by setting the frequency to be $1/RC$, determine the difference in phase angle from -45 degrees, using parameters of the opamp. If the difference is only a few degrees, the circuit approximation is adequate.

D. Single-Pole High-Pass Filter

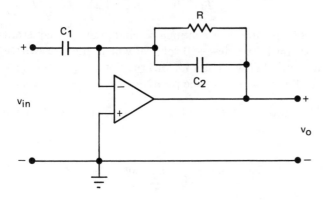

Figure 9.15 A Single-Pole High-Pass Filter

For the circuit of figure 9.15: $Y = sC_1$ and $Z = R/(sC_2R + 1)$, so its low-frequency approximation is

$$\frac{V_{\text{out}}}{V_{\text{in}}} = -YZ = -\frac{sC_1R}{sC_2R + 1} \qquad 9.34$$

This is a high-pass transfer function with a pole at $s = -1/RC$ and a zero at $s = 0$. To determine the effects of the amplifier poles on this circuit, substitute

for YZ in equation 9.31 from 9.34. Then by setting the frequency to be $1/RC$, determine the difference in phase angle from $+45$ degrees, using parameters of the opamp. If the difference is less than a few degrees, the circuit approximation is adequate.

E. Lead-Lag or Lag-Lead Filter

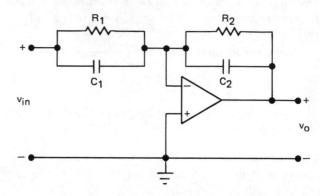

Figure 9.16 A Simple Lead-Lag or Lag-Lead Filter

For the circuit of figure 9.16,

$$Y = sC_1 + \frac{1}{R_1} \quad \text{and} \quad Z = \frac{R_2}{sC_2R_2 + 1}$$

Its low-frequency approximation is

$$\frac{V_o}{V_{\text{in}}} = -YZ = -\frac{R_2}{R_1}\frac{sC_1R_1 + 1}{sC_2R_2 + 1} \qquad 9.35$$

or

$$\frac{V_o}{V_{\text{in}}} = -\frac{C_1}{C_2}\frac{s + \omega_z}{s + \omega_p} \qquad 9.36$$

$\omega_z = 1/R_1C_1$, and $\omega_p = 1/R_2C_2$.

This transfer function has 180 degrees of phase angle at both zero and infinite frequency. At zero frequency it has a gain of R_2/R_1, and at infinite frequency it has a gain of C_1/C_2. It provides a maximum phase shift at a frequency,

$$\omega_{\text{max phase shift}} = \omega_{\text{mp}}$$

$$= \frac{1}{\sqrt{R_1C_1}}\frac{1}{\sqrt{R_2C_2}}$$

$$= \sqrt{\omega_p\omega_z} \qquad 9.37$$

The maximum phase shift can then be determined to be

$$\text{maximum phase shift} = \arctan\sqrt{\frac{\omega_p}{\omega_z}} - \arctan\sqrt{\frac{\omega_z}{\omega_p}}$$

$$9.38$$

And at the maximum phase shift it has a gain

$$\text{gain}\Big|_{\omega=\omega_{\text{mp}}} = \sqrt{\frac{C_1 R_2}{C_2 R_1}} \qquad 9.39$$

The ratio of the infinite frequency gain to the DC gain is

$$\frac{\text{gain}_{\text{hf}}}{\text{gain}_{\text{DC}}} = \frac{R_1 C_1}{R_2 C_2} = \frac{\omega_p}{\omega_z} \qquad 9.40$$

The *lead-lag filter* is often used for phase-angle correction in linear servo systems. At very low frequencies, the transfer function has a gain of R_2/R_1 and a phase angle of -180 degrees. It has $R_1 C_1 > R_2 C_2$ so that the numerator term has more influence at lower frequencies where the gain increases and the phase angle becomes less negative (or more positive). At higher frequencies the denominator influence becomes important, so that the effects of the numerator are counteracted, causing the phase angle to return toward -180 degrees and the gain to approach C_1/C_2.

The *lag-lead filter* has $R_1 C_1 < R_2 C_2$ and has a DC gain of R_2/R_1. Its DC or very low-frequency phase angle is to be considered to be $+180$ degrees, however. At low frequencies the denominator becomes effective in decreasing the transfer function gain and decreasing the phase angle, and at higher frequencies the numerator counters the denominator phase shift.

F. Differentiator

An ideal *differentiator* is approximated via equation 9.30 with $Y = sC_1$ and $Z = R_2$ as indicated in figure 9.17(a). However, the differentiation places a zero at zero frequency, and this interacts with the opamp poles in such a way that oscillations generally occur. This problem is eliminated by the addition of two components, R_1 and C_2, which are shown in figure 9.17(b).

For figure 9.17(b),

$$Y = \frac{sC_1}{sC_1 R_1 + 1} \quad \text{and} \quad Z = \frac{R_2}{sC_2 R_2 + 1}$$

Then the low-frequency approximation becomes

$$\frac{V_{\text{out}}}{V_{\text{in}}} = \frac{-sC_1 R_2}{(1 + sC_1 R_1)(1 + sC_2 R_2)} \qquad 9.41$$

The poles are at $1/C_1 R_1$ and $1/C_2 R_2$.

In order to eliminate the oscillations it is necessary that these poles be at much lower frequencies than the gain-bandwidth frequency.

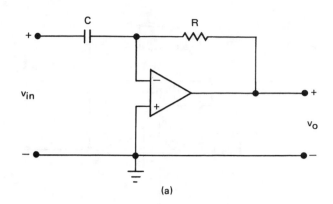

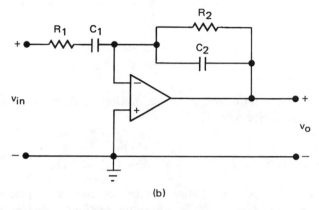

Figure 9.17 Inverting Differentiators: (a) Ideal Circuit, (b) Practical Differentiator with Bandwidth Limiting

4 BIQUAD FILTERS USING INVERTING OPAMPS

A. Introduction

A *biquadratic (biquad)* function has quadratic expressions in s in both the numerator and the denominator, being of the form

$$F = \frac{a_2 s^2 + a_1 s + a_0}{b_2 s^2 + b_1 s + b_0} \qquad 9.42$$

This is a fundamental building block in higher-order filters, and can be realized using an inverting operational amplifier in the topology indicated in figure 9.18.

Analysis of figure 9.18 is by Kirchhoff's current law applied at nodes "a" and "b", taking V_{in} and V_{out} as the other node voltages. At node "a" the equation is found to be

$$V_{\text{in}} Y_1 + V_o Y_4 = V_a(Y_1 + Y_2 + Y_3 + Y_4) - V_b Y_3 \qquad 9.43$$

And at node b,

$$V_o Y_5 = V_b(Y_3 + Y_5) - V_a Y_3 \qquad 9.44$$

Recognize that $V_b = -V_o/A$, where A is the transfer function of the opamp.

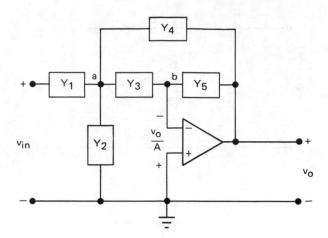

Figure 9.18　Topology for Biquadratic Inverting Filters

In the low-frequency approximation, $1/A = 0$ so that $V_b = 0$. Solve for V_o/V_{in}.

$$\frac{V_o}{V_{in}} = \frac{-Y_1 Y_3}{Y_5(Y_1 + Y_2 + Y_3 + Y_4) + Y_3 Y_4} \qquad 9.45$$

Equation 9.45 will be used to develop the low-frequency approximations in the following sections. However, when the frequency limitations of the opamp become important, it is necessary to carry A through the analysis, which then yields

$$\frac{V_o}{V_{in}} = \frac{-Y_1 Y_3}{D} \qquad 9.46$$

$$D = (1 + \frac{1}{A})Y_5(Y_1 + Y_2 + Y_3 + Y_4) + Y_3 Y_4$$
$$+ \frac{Y_3}{A}(Y_1 + Y_2 + 2Y_3 + Y_4)$$

In equation 9.46, $1/A = [(s/\omega_t) + (1/A_o)][(s/\omega_{hf}) + 1]$.

In general, the active filters using operational amplifiers are limited to resistive and capacitive components, so the simplest Y's will be of the form of $1/R = G$ and sC.

B.　Two-Pole Low-Pass Filters

A low-pass filter will have the form

$$\frac{V_o}{V_{in}} = \frac{-K}{s^2 + bs + c} \qquad 9.47$$

Comparing equations 9.47 and 9.45, and first attempting to realize the filter with only one resistance or one capacitance in each box in figure 9.18, note that Y_1 and Y_3 must be resistive in order to have a constant in the numerator. Then, to have an s^2 term in the denominator, it is necessary that Y_5 and either Y_2 or Y_4 be

capacitors. Following from that, in order that there be a constant term in the denominator, it is necessary that Y_3 and Y_4 both be resistive. This fully specifies the types of components.

$$Y_1 = G_1, \ Y_2 = sC_2, \ Y_3 = G_3, \ Y_4 = G_4 \text{ and } Y_5 = sC_5$$

The resulting circuit is shown in figure 9.19.

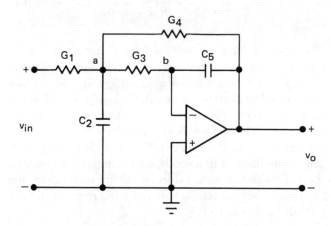

Figure 9.19　A Biquadratic Inverting Low-Pass Filter

The transfer function for the filter of figure 9.19 can be written as

$$\frac{V_o}{V_{in}} = \frac{-G_1 R_4}{R_3 R_4 C_2 C_5 s^2 + R_3 R_4 C_5 (G_1 + G_3 + G_4)s + 1} \qquad 9.48$$

$R_3 = 1/G_3$, $R_4 = 1/G_4$, and $R_1 = 1/G_1$. Equation 9.48 is in a standard form corresponding to

$$\frac{V_o}{V_{in}} = \frac{-K}{\left(\dfrac{s}{\omega_o}\right)^2 + \dfrac{s}{Q\omega_o} + 1} \qquad 9.49$$

By comparing terms in these two equations, it can be seen that

$$K = G_1 R_4 \qquad 9.50$$

$$\frac{1}{\omega_o^2} = C_2 C_5 R_3 R_4 \qquad 9.51$$

$$\frac{1}{Q\omega_o} = C_5[R_3(1 + K) + R_4] \qquad 9.52$$

Then with some manipulations, the following relationships can be obtained.

$$R_3 = \frac{1}{2Q\omega_o C_5(1 + K)}\left[1 \pm \sqrt{1 - 4Q^2 \frac{C_5}{C_2}(1 + K)}\right] \qquad 9.53$$

$$R_4 = \frac{1}{2Q\omega_o C_5}\left[1 \pm \sqrt{1 - 4Q^2 \frac{C_5}{C_2}(1 + K)}\right] \qquad 9.54$$

From these it is clear that a necessary requirement is

$$C_2 \geq 4Q^2(1+K)C_5 \qquad 9.55$$

Example 9.5

Using the topology of figure 9.19, design a second-order low-pass filter with repeated poles at 15 kHz (critical damping) and having a DC gain of -1.

Referring to equation 9.49, $K = 1, Q = 0.5$ (for critical damping), and $\omega_o = 2\pi \times 15,000 = 94.25$ krad/s.

Taking the equality in equation 9.55: $C_2 = 2C_5$, from equation 9.53: $R_3 = 1/188,500C_5$, and from equation 9.54: $R_4 = 2R_3$.

As it is desirable to have these resistances at least 1k ohm, it is necessary that C_5 be less than 5.3 nF, so choose $C_5 = 2$ nF, which results in $C_2 = 4$ nF, $R_3 = 2650$, and $R_4 = 5300$ ohms. Also, for $K = 1$, $R_1 = R_4$, so all parameters are specified.

C. Two-Pole High-Pass Filters

A *high-pass filter* has zeros at zero frequency, so the transfer function from equation 9.45 calls for capacitors in Y_1 and Y_3. In order to have the s^2 term in the denominator, either Y_4 or Y_5 must be capacitors, but not both, as there is to be a constant term in the denominator. As Y_3 is a capacitance, Y_5 is designated a resistance to provide a current path around the opamp to prevent DC offset problems. Therefore, Y_4 is to be a capacitance. The transfer function then becomes

$$\frac{V_o}{V_{\text{in}}} = \frac{C_1 C_3 R_2 R_5 s^2}{s^2 C_3 C_4 R_2 R_5 + s(C_1 + C_3 + C_4)R_2 + 1} \qquad 9.56$$

As in equation 9.49, the denominator is in the standard form, so

$$\omega_o^2 = \frac{1}{C_3 C_4 R_2 R_5} \quad \text{and} \quad Q\omega_o = \frac{1}{(C_1 + C_3 + C_4)R_2}$$

and the expressions for the resistances can be found to be

$$R_2 = \frac{1}{(C_1 + C_3 + C_4)Q\omega_o} \qquad 9.57$$

and

$$R_5 = \frac{Q(C_1 + C_3 + C_4)}{C_3 C_4 \omega_o} \qquad 9.58$$

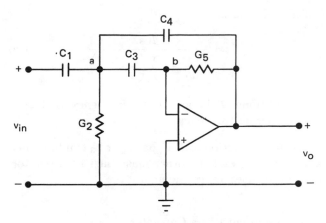

Figure 9.20 A Biquadratic Inverting High-Pass Filter

The circuit for the second-order high-pass filter is shown in figure 9.20. It is to be noted that the high-frequency gain of this circuit is C_1/C_4.

Example 9.6

A high-pass filter is to have a corner frequency of 1000 Hz, a Q of 5, and a high-frequency gain of 1. Any resistors connected to the output terminal of the opamp are to exceed 1000 ohms.

As the high-frequency gain is to be unity, it is necessary that $C_1 = C_4$. Unless there is some difficulty that develops later, all capacitors are assumed to have the same value.

Then from equation 9.58, because R_5 must exceed 1000 ohms, the capacitance must be

$$C = \frac{3Q}{R_5 \omega_o} < \frac{15}{10^3 \times 2\pi \times 10^3} = 2.4 \; \mu\text{F}$$

The capacitances are taken as 0.1 μF, so from equation 9.58,

$$R_5 = \frac{15}{10^{-7} \times 2\pi \times 10^3} = 23.9\text{k ohms}$$

From equation 9.57,

$$R_2 = \frac{1}{3 \times 10^{-7} \times 5 \times 2\pi \times 10^3} = 106 \text{ ohms}$$

Using these resistances and three 0.1 μF capacitors, the design will meet all the given specifications.

D. Two-Pole Band-Pass Filters

A *two-pole band-pass filter*, if overdamped, has a transfer function of the form

$$\frac{V_o}{V_{\text{in}}} = \frac{Ks}{s^2\left(\dfrac{1}{\omega_1 \omega_2}\right) + s\left(\dfrac{1}{\omega_1} + \dfrac{1}{\omega_2}\right) + 1} \qquad 9.59$$

or in factored form

$$\frac{V_o}{V_{\text{in}}} = \frac{Ks}{\left(\dfrac{s}{\omega_1} + 1\right)\left(\dfrac{s}{\omega_2} + 1\right)} \qquad 9.60$$

ω_1 is traditionally the low corner frequency, and ω_2 is the high corner frequency.

Referring to equation 9.45, either Y_1 or Y_3 can be chosen as a capacitor, and the other must then be a resistor. Here, both cases are considered.

• Design with Y_1 a Capacitor

With $Y_1 = sC_1$ and $Y_3 = G_3$, it can be seen in equation 9.45 that it is necessary that $Y_4 = G_4$ and $Y_5 = sC_5$. Furthermore, Y_2 is unnecessary, and so it is set to zero (an open circuit). The resulting circuit is shown in figure 9.21.

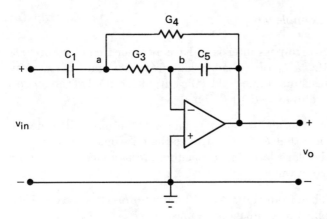

Figure 9.21 One Biquadratic Inverting Band-Pass Filter

The resulting transfer function is

$$\frac{V_o}{V_{\text{in}}} = \frac{-sC_1R_4}{s^2C_1C_5R_3R_4 + s(R_3 + R_4)C_5 + 1} \qquad 9.61$$

Solving for the corner frequencies,

$$s_{1,2} = -\frac{1}{2}\frac{R_3 + R_4}{R_3R_4C_1}\left[1 \pm \sqrt{1 - \frac{4R_3R_4}{(R_3 + R_4)^2}\frac{C_1}{C_5}}\right] \qquad 9.62$$

Equation 9.62 permits real or complex roots. When real roots are required, the expression can be simplified by setting $C_1 = C_5 = C$ to give roots determined simply as

$$s_{1,2} = -\frac{1}{R_3C}, \ -\frac{1}{R_4C} \qquad 9.63$$

Either resistor could be chosen to be identified with ω_1, the lower corner frequency.

The midband gain is found by setting the frequency to the geometrical mean of ω_1 and ω_2, or

$$\omega_{\text{midband}} = \sqrt{\omega_1\omega_2}$$

First using equation 9.59, and then substituting parameters from equation 9.61 results in

$$\left.\frac{V_o}{V_{\text{in}}}\right|_{\text{midband}}$$

$$= \frac{-j\sqrt{\omega_1\omega_2}}{j^2(\omega_1\omega_2)\left(\dfrac{1}{\omega_1\omega_2}\right) + j\sqrt{\omega_1\omega_2}\left(\dfrac{1}{\omega_1} + \dfrac{1}{\omega_2}\right) + 1}$$

$$= \frac{-1}{\left(\dfrac{1}{\omega_1} + \dfrac{1}{\omega_2}\right)} = \frac{C_1R_4}{(R_3 + R_4)C_5} \qquad 9.64$$

Note that with $C_1 = C_5$, unity gain cannot be achieved.

• Design with Y_1 a Resistance

With $Y_1 = G_1$, according to equation 9.45, it is necessary that $Y_3 = sC_3$. Then it follows that Y_5 must be G_5 in order to have a constant term in the denominator, and Y_4 must be sC_4 to have an s^2 term. That makes Y_2 unnecessary again, so it can be removed. The resulting circuit is as shown in figure 9.22.

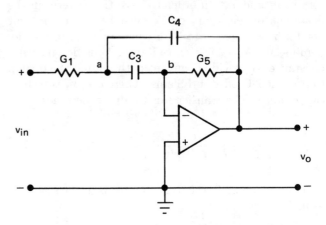

Figure 9.22 Another Biquadratic Inverting Band-Pass Filter

The resulting transfer function is

$$\frac{V_o}{V_{\text{in}}} = \frac{-sC_3R_5}{s^2C_3C_4R_1R_5 + s(C_3 + C_4)R_1 + 1} \qquad 9.65$$

Solving for the corner frequencies results in

$$s_{1,2} = -\frac{1}{2}\frac{C_3 + C_4}{C_3C_4R_5}\left[1 \pm \sqrt{1 - \frac{4C_3C_4}{(C_3 + C_4)^2}\frac{R_5}{R_1}}\right] \qquad 9.66$$

For the choice of making $C_3 = C_4 = C$, the corner frequencies are

$$s_{1,2} = -\frac{1}{CR_5}\left[1 \pm \sqrt{1 - \frac{R_5}{R_1}}\right] \qquad 9.67$$

which again can be real or complex roots.

The midfrequency gain, determined as in the first case, is

$$\left.\frac{V_o}{V_{\text{in}}}\right|_{\text{midband}}$$

$$= \frac{-j\sqrt{\omega_1\omega_2}}{j^2(\omega_1\omega_2)\left(\frac{1}{\omega_1\omega_2}\right) + j\sqrt{\omega_1\omega_2}\left(\frac{1}{\omega_1} + \frac{1}{\omega_2}\right) + 1}$$

$$= \frac{-1}{\left(\frac{1}{\omega_1} + \frac{1}{\omega_2}\right)} = \frac{C_3 R_5}{(C_3 + C_4)R_1} \qquad 9.68$$

This second band-pass filter is generally preferred over the first for generating complex roots because of the simplicity of its dependence on the resistances.

5 BIQUAD INVERTING FILTERS WITH FEEDFORWARD

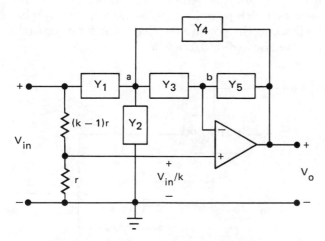

Figure 9.23 Topology for Biquadratic Inverting Filters with Feedforward

The use of *feedforward*, as indicated in figure 9.23, allows the possibility of complex zeros in a transfer function using only one operational amplifier. The feedforward is provided by applying a fraction of the input signal at the noninverting terminal of the opamp. Otherwise the topology is the same as the simple biquad inverting filter of figure 9.18.

With the low-frequency approximation for the opamp, its input voltage is essentially zero, so that the voltage

at node "b" is V_{in}/k. The Kirchhoff current law applied at node "a" then yields

$$Y_1 V_{\text{in}} = (Y_1 + Y_2 + Y_3 + Y_4)V_a - \frac{Y_3 V_{\text{in}}}{k} - Y_4 V_o \quad 9.69$$

and at node "b",

$$Y_3 V_a - \frac{Y_3 V_{\text{in}}}{k} + Y_5 V_o - \frac{Y_5 V_{\text{in}}}{k} = 0 \qquad 9.70$$

Eliminating V_a between these two equations results in the transfer function

$$\frac{V_o}{V_{\text{in}}} = -\frac{kY_1Y_3 + Y_3^2 - (Y_3 + Y_5)(Y_1 + Y_2 + Y_3 + Y_4)}{k[Y_5(Y_1 + Y_2 + Y_3 + Y_4) + Y_3Y_4]}$$
$$9.71$$

The denominator is to yield a quadratic in any case, which requires that Y_3 and Y_4 both be the same type, either both resistances or both capacitances. If they are both resistances, Y_5 and either or both of Y_1 and Y_2 must be capacitances. If Y_3 and Y_4 are both capacitances, Y_5 and either or both of Y_1 and Y_2 must be resistances. Then the numerator of equation 9.71 (together with the leading negative sign) can be simplified to

$$Y_3Y_4 + [Y_3[Y_2 + Y_5 - (k-1)Y_1] + Y_4Y_5] + Y_5(Y_1 + Y_2)$$

But Y_2 is redundant and can be set to zero. This requires that Y_1 and Y_5 be of the opposite type from Y_3 and Y_4. Thus, the transfer function can be written in the following form:

$$\frac{V_o}{V_{\text{in}}} = \frac{Y_3Y_4 + [Y_3[Y_5 - (k-1)Y_1] + Y_4Y_5] + Y_5Y_1}{k[Y_3Y_4 + Y_5(Y_3 + Y_4) + Y_5Y_1]}$$
$$9.72$$

The primary use of this topology is to provide a *notch filter*, which requires that the numerator coefficient of s should be zero. This requires

$$k - 1 = \frac{Y_5}{Y_1}\left(1 + \frac{Y_4}{Y_3}\right) \qquad 9.73$$

A. Notch Filter With Y_1 Resistive

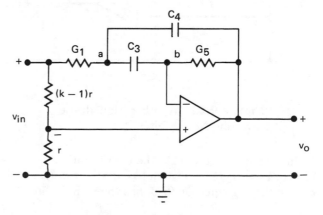

Figure 9.24 One Biquadratic Inverting Notch Filter

As indicated in figure 9.24, the circuit with $Y_1 = G_1$, $Y_2 = 0$, $Y_3 = sC_3$, $Y_4 = sC_4$, and $Y_5 = G_5$, together with $k-1$ satisfying equation 9.73, results in the transfer function

$$\frac{V_o}{V_{\text{in}}} = \frac{s^2 C_3 C_4 R_1 R_5 + 1}{k[s^2 C_3 C_4 R_1 R_5 + R_1(sC_3 + sC_4) + 1]} \quad 9.74$$

which has a high- and low-frequency gain of $1/k$, and has infinite loss at the notch frequency defined as

$$\omega_n^2 = \frac{1}{C_3 C_4 R_1 R_5} \quad 9.75$$

The denominator has a quality factor Q of

$$Q = \sqrt{\frac{R_5}{R_1}} \frac{\sqrt{C_3 C_4}}{C_3 + C_4} \quad 9.76$$

The bandwidth is defined as the difference in the two frequencies where the amplitude of the transfer function is 0.707 times the high- and low-frequency gain of $1/k$. The difference is found to be

$$\omega_2 - \omega_1 = \frac{\omega_n}{Q} \quad 9.77$$

In the cases where the two capacitors are made equal, the bandwidth becomes

$$\frac{\omega_2 - \omega_1}{\omega_n} = 2\sqrt{\frac{R_1}{R_5}} \quad 9.78$$

B. Notch Filter with Y_1 Capacitive

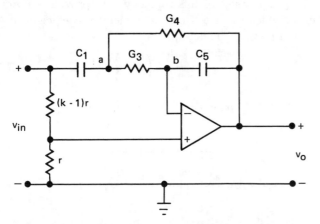

Figure 9.25 Another Biquadratic Inverting Notch Filter

As indicated in figure 9.25, the circuit with $Y_1 = sC_1$, $Y_2 = 0$, $Y_3 = G_3$, $Y_4 = G_4$, and $Y_5 = sC_5$, together with $k-1$ satisfying equation 9.73, has the transfer function

$$\frac{V_o}{V_{\text{in}}} = \frac{s^2 C_1 C_5 R_3 R_4 + 1}{k[s^2 C_1 C_5 R_3 R_4 + sC_5(R_3 + R_4) + 1]} \quad 9.79$$

which again has high- and low-frequency gains of $1/k$, and has infinite loss at the notch frequency found from

$$\omega_n^2 = \frac{1}{C_1 C_5 R_3 R_4} \quad 9.80$$

The denominator has a quality factor Q of

$$Q = \sqrt{\frac{C_1}{C_5}} \frac{\sqrt{R_3 R_4}}{R_3 + R_4} \quad 9.81$$

The bandwidth is defined in equation 9.77.

For this circuit, making the two capacitors equal does not yield a simple expression for the bandwidth, so for easier design the previous circuit is preferred.

6 ACTIVE FILTERS WITH NONINVERTING OPERATIONAL AMPLIFIERS

A. Introduction

Sallen and Key[1] developed numerous *active filters* using the positive feedback scheme where a noninverting op-amp configuration has the usual resistive negative feedback to the inverting terminal, and includes a three-port network which provides both feedforward and positive feedback to the positive terminal. The filter configuration is indicated in figure 9.26.

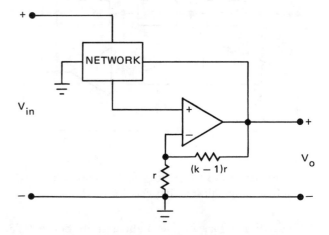

Figure 9.26 Configuration of a Positive Feedback Active Filter

The filters considered here can be treated with an analysis of the circuit shown in figure 9.27. There it is to

---

[1] Sallen, R. P. and Key, E. L., *A Practical Method of Designing RC Active Filters*, IRE Transactions of Circuit Theory, CT-2, May 1955, pp. 74-85.

be noted that the voltage at node "b", for the low-frequency approximation which presumes negligible input voltage to the opamp, has a value of V_o/k, which is established by the voltage divider at the output.

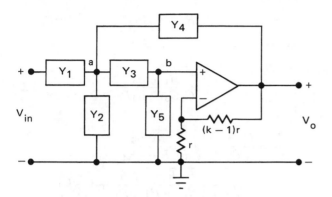

Figure 9.27 Topology Sallen and Key Active Filters

The Kirchhoff current law equation written for node "b" results in

$$V_a Y_3 = \frac{V_o Y_3}{k} + \frac{V_o Y_5}{k} \qquad 9.82$$

The equation for node "a" is

$$V_{in} Y_1 + V_o \left(Y_4 + \frac{Y_3}{k} \right) = V_a (Y_1 + Y_2 + Y_3 + Y_4) \quad 9.83$$

Solving for V_a in equation 9.82, and substituting the result into equation 9.83 yields the transfer function

$$\frac{V_o}{V_{in}}$$

$$= \frac{k Y_1 Y_3}{Y_5 (Y_1 + Y_2 + Y_3 + Y_4) + Y_3 [Y_1 + Y_2 + (1-k) Y_4]}$$

$$9.84$$

In the following sections, low-pass, high-pass, and band-pass versions of the Sallen and Key circuits will be examined.

B. Low-Pass Sallen and Key Circuits

For a low-pass circuit using only resistance and capacitance in the admittances of figure 9.27, it can be seen from equation 9.84 that both Y_1 and Y_3 must be resistive to be assured that there is no s in the numerator, which therefore requires that Y_5 and either Y_2 or Y_4 be capacitors, in order to have an s^2 in the denominator. It can also be seen that Y_2 is redundant, and so it is removed for simplicity. Then the admittances are: $Y_1 = G_1$, $Y_2 = 0$, $Y_3 = G_3$, $Y_4 = sC_4$, and $Y_5 = sC_5$. These are shown in figure 9.28.

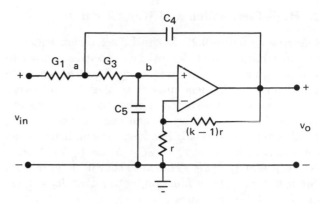

Figure 9.28 Sallen and Key Low-Pass Filter

The resulting transfer function, noting that $R_1 = 1/G_1$ and $R_3 = 1/G_3$, becomes

$$\frac{V_o}{V_{in}}$$

$$= \frac{k}{s^2 C_4 C_5 R_1 R_3 + s[C_5(R_1 + R_3) + (1-k)C_4 R_1] + 1}$$

$$9.85$$

which has

$$\omega_o^2 = \frac{1}{C_4 C_5 R_1 R_3} \qquad 9.86$$

and

$$\frac{1}{Q} = \sqrt{\frac{C_5}{C_4}} \frac{R_1 + R_3}{\sqrt{R_1 R_3}} - (k-1)\sqrt{\frac{C_4}{C_5}} \frac{R_1}{R_3} \qquad 9.87$$

Example 9.7

A second-order noninverting low-pass filter is to be designed to have a corner frequency of 10,000 rad/s and to be critically damped.

Using the circuit of figure 9.28, the design will be attemped to have equal capacitances. The value of k must be at least one, and typically ranges between 1 and 4/3. For simplicity here, $k = 1$ is chosen (this makes the amplifier a voltage follower, and r may also be removed).

Critical damping requires $Q = 1/2$, so with equal capacitances and $k = 1$, equation 9.87 becomes

$$Q = \frac{\sqrt{R_1 R_3}}{R_1 + R_3} = 0.5$$

This relationship is solved to show $R_1 = R_3$.

Choosing $R_1 = R_3 = 10$ kilohms, equation 9.86 then yields the values of $C_4 = C_5 = 0.01$ μF.

C. High-Pass Sallen and Key Circuits

Referring to figure 9.27, it can be seen from equation 9.84 that both Y_1 and Y_3 must be capacitive in order to provide two zeros at zero frequency. These also provide the s^2 term in the denominator so that Y_5 and either Y_2 or Y_4 must be resistance in order to have a constant term in the denominator. It can also be seen that Y_2 is redundant, and so it is removed for simplicity. Then the admittances are: $Y_1 = sC_1$, $Y_2 = 0$, $Y_3 = sC_3$, $Y_4 = G_4$, and $Y_5 = G_5$. These are shown in figure 9.29. The resulting transfer function, noting that $R_4 = 1/G_4$ and $R_5 = 1/G_5$, becomes

$$\frac{V_o}{V_{in}} =$$

$$\frac{s^2 kC_1C_3R_4R_5}{s^2C_1C_3R_4R_5 + s[(C_1+C_3)R_4 + (1-k)C_3R_5] + 1} \quad 9.88$$

which has

$$\omega_o^2 = \frac{1}{C_1C_3R_4R_5} \quad 9.89$$

and

$$\frac{1}{Q} = \sqrt{\frac{R_4}{R_5}}\frac{C_1+C_3}{\sqrt{C_1C_3}} - (k-1)\sqrt{\frac{C_3}{C_1}\frac{R_5}{R_4}} \quad 9.90$$

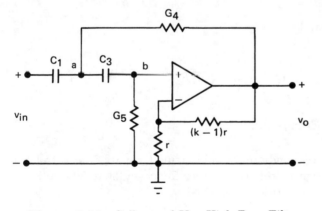

Figure 9.29 Sallen and Key High-Pass Filter

D. Sallen and Key Band-Pass Circuits

The band-pass filter requires one zero at zero frequency (an s term in the numerator) and a full quadratic in s in the denominator. Referring to figure 9.27, it can be seen from equation 9.84 that Y_1 and Y_3 must be different types of elements, one capacitive and the other resistive. The denominator requires that Y_5 be the opposite type of component from Y_3, which results in two possible topologies for a band-pass filter. Figure 9.30 shows one implementation. There the admittances are: $Y_1 = G_1, Y_2 = sC_2, Y_3 = sC_3, Y_4 = G_4$, and $Y_5 = G_5$. The resulting transfer function, noting that $R_1 = 1/G_1$, $R_4 = 1/G_4$, and $R_5 = 1/G_5$, becomes

$$\frac{V_o}{V_{in}} = \frac{a_1 s}{b_2 s^2 + b_1 s + b_0} \quad 9.91$$

$$a_1 = kC_3R_4R_5$$
$$b_2 = C_2C_3R_1R_4R_5$$
$$b_1 = (C_2+C_3)R_1R_4 + C_3R_5[R_4 + (1-k)R_1]$$
$$b_0 = R_1 + R_4$$

which has

$$\omega_o^2 = \frac{R_1+R_4}{C_2C_3R_1R_4R_5} \quad 9.92$$

and

$$\frac{1}{Q} = \frac{C_2+C_3}{\sqrt{C_2C_3}}\sqrt{\frac{1}{R_5}\frac{R_1R_4}{R_1+R_4}} + \sqrt{\frac{C_3R_5}{C_2R_4}\frac{R_1}{R_1+R_4}}\left(\frac{R_4}{R_1}+1-k\right) \quad 9.93$$

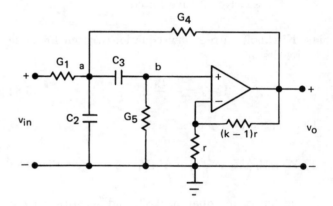

Figure 9.30 A Sallen and Key Band-Pass Filter

Example 9.8

A noninverting band-pass filter is to be designed to have a low corner frequency at 10 rad/s, and a high corner frequency at 1000 rad/s.

The solution is based on the Sallen and Key circuit of figure 9.30. With the given pole locations, the denominator is to be

$$\left(\frac{s}{10}+1\right)\left(\frac{s}{1000}+1\right)$$

As the circuit has six parameters and only two specifications are given, some parameter values can be assumed. Assume that R_1 and R_4 are equal, that $k = 1$, and that the two capacitors are equal: $C_2 = C_3 = C$.

Then the transfer function from equation 9.91 becomes

$$\frac{V_o}{V_{in}} = \frac{\dfrac{sCR_5}{2}}{\dfrac{s^2C^2R_4R_5}{2} + sC\left(\dfrac{R_4+R_5}{2}\right)+1}$$

The denominator can be factored into: $(sCR_4 + 1) \times (sCR_5/2 + 1)$, so that each corner frequency can be identified with a resistance and capacitance as

$$CR_4 = \frac{1}{1000} \quad \text{and} \quad \frac{CR_5}{2} = \frac{1}{10}$$

Arbitrarily selecting $C = 1 \ \mu\text{F}$, the resistances become $R_4 = 1$ kilohm, and $R_5 = 200$ kilohms.

7 LARGE SIGNALS AND OPERATIONAL AMPLIFIERS

A. Introduction

An understanding of the internal organization of an operational amplifier is necessary to deal with its large signal characteristics. Figure 9.31 shows a functional arrangement, omitting the output stage which is a unity gain buffer.

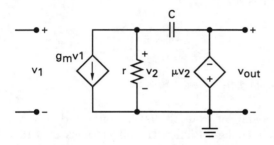

Figure 9.31 A Functional Equivalent of an Opamp

A straightforward nodal analysis yields the transfer function

$$\frac{V_o}{V_1} = \frac{\mu g_m r}{sCr(1 + \mu) + 1} \qquad 9.94$$

which yields the usual opamp parameters of

$$A_o = \mu g_m r \quad \text{and} \quad \omega_b = \frac{1}{Cr(1 + \mu)}$$

As $\omega_t = A_o \omega_b$, and $\mu \gg 1$,

$$\omega_t = \frac{\mu g_m r}{Cr(1 + \mu)} \cong \frac{g_m}{C} \qquad 9.95$$

B. Saturation and Maximum Slew Rate

When a large signal is applied to the opamp input terminals, the transconductance amplifier saturates, so that instead of $g_m v_1$, the output is limited to a maximum current, I. The Laplace transform of the current, I, is I/s, so the expression for the output voltage becomes

$$V_o = \frac{\mu r I}{s[sCr(1 + \mu) + 1]} = \frac{\mu r I}{s(\frac{s}{\omega_b} + 1)}$$

$$= \mu r I \left(\frac{1}{s} - \frac{1}{s + \omega_b}\right) \qquad 9.96$$

The inverse Laplace transform gives the time function

$$v_O(t) = \mu r I(1 - e^{-\omega_b t}) \cong \mu r I \omega_b t \qquad 9.97$$

The maximum slew rate, S_R, can be found by taking the time derivative of equation 9.97, which occurs at $t = 0$ for the exact solution. So,

$$S_R = \mu r I \omega_b = \frac{\mu r I}{Cr(1 + \mu)} \cong \frac{I}{C} \qquad 9.98$$

C. Slew Limiting in the Voltage Follower

The voltage follower is the favored circuit used to examine slew-rate limiting, as the output voltage eventually matches the input voltage, at least within a few microvolts. Figure 9.32 shows the voltage follower, where the opamp is modeled as in figure 9.31.

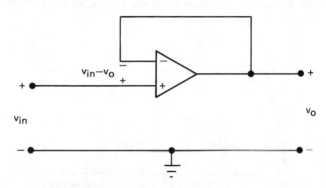

Figure 9.32 Voltage Follower Circuit

The internal transconductance amplifier saturates whenever

$$g_m(v_{\text{IN}} - v_O) > I = CS_R$$

or when

$$v_{\text{IN}} - v_O > \frac{CS_R}{g_m} = \frac{S_R}{\omega_t} \qquad 9.99$$

Suppose v_{IN} has been zero for a long time and at $t = 0$ it changes to a constant value V_S. If the opamp responds without saturating the transconductance amplifier, the output will try to follow via an exponential rise, or

$$V_o = \frac{A_o \omega_b}{s + \omega_b} V_1 \qquad 9.100$$

but

$$V_1 = V_{in} - V_o = \frac{1}{s}V_S - \frac{A_o\omega_b}{s+\omega_b}V_1$$

so

$$V_1 = \frac{1}{s}\frac{s+\omega_b}{s+(1+A_o)\omega_b}V_S = \frac{1}{s}\frac{s+\omega_b}{s+\omega_t}V_S$$

Then, substituting from equation 9.100 with $\omega_t = A_o\omega_b$

$$V_o = \frac{1}{s}\frac{\omega_t}{s+\omega_t}V_S = V_S\left(\frac{1}{s} - \frac{1}{s+\omega_t}\right) \qquad 9.101$$

and the inverse Laplace transformation gives

$$v_O(t) = V_S(1 - e^{-\omega_t t}) \qquad 9.102$$

However, the transconductance amplifier is saturated during the first part of this transient time interval, so that equation 9.97 is in effect during that time. The result is that the opamp is slewing at essentially a slope of S_R until the transconductance amplifier comes out of saturation. Figure 9.33 shows the maximum slew rate response and the idealized response of equation 9.102.

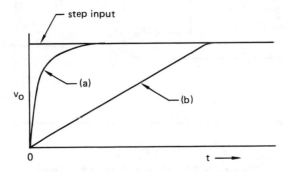

Figure 9.33   (a) Ideal Exponential Step Response, (b) Slew-Rate Limited Response

Example 9.9

A particular opamp is connected in a voltage follower configuration and has a maximum slew rate $S_R = 650,000$ volts/s, and a gain-bandwidth product of 1 MHz. Compare the idealized response to a unit step to that of the slew-rate limited case.

$\omega_t = 2\pi \times 10^6 = 6.28 \times 10^6$, so the idealized response from equation 9.102 is

$$v_{IDEAL} = 1 - e^{-6,280,000t}$$

This has an initial slope of 6,280,000 volts/s, and a time constant of 1/6,280,000 or 0.159 μs, while the actual response will be $S_R t$, or

$$v_{ACTUAL} = 650,000t$$

The actual voltage slope is only 10 percent of the initial slope of the ideal exponential response. The slewing voltage will reach 1 volt at $t = 1/650,000$ or 1.54 μs. This corresponds to 10 time constants, so the actual case is compared on the basis that the ideal reaches 95 percent of the final value in 3 time constants, about 30 percent of the actual time to reach saturation.

D. Sinusoidal Full Output Frequency Limit

The maximum slew rate, S_R, is the limit of the slope of the output voltage of an opamp. The linear range of the output is one to three volts less than the rail voltages. So an undistorted sine wave output voltage can have a peak value up to that maximum amount, which we shall designate as V_{max}.

A sinusoid $V_{max}\sin\omega t$ has a slope of $\omega V_{max}\cos\omega t$, which has its maximum of ωV_{max} at $t = 0$. The largest this slope can reach is S_R, so the maximum full output frequency is S_R/V_{max}, or

$$\omega_{max} = \frac{S_R}{V_{max}} \qquad 9.103$$

Example 9.10

The opamp of example 9.9 is connected in a voltage follower configuration with rail voltages of ±15 volts. Assume the extent of the linear range of the output is ±12 volts. Determine the maximum frequency of a full-output sinusoid.

From equation 9.103, the maximum full-output sinusoidal frequency is $650,000/12 = 54$ krad/s or 8.6 kHz.

8   DC AND OTHER OPERATIONAL AMPLIFIER PROBLEMS

A.   Introduction

Practical opamps have biasing effects which can cause serious problems in certain applications. These can be accounted for with three separate sources: an effective offset of the output when the input is zero, a bias current associated with the inverting terminal, and a bias current associated with the noninverting terminal.

The *offset voltage* is normally associated with the noninverting input terminal, and modeled as a DC source there. The bias currents are modeled with current sources at the appropriate input terminal. The entire model is as indicated in figure 9.34.

Also of interest is the *common-mode gain* of the amplifier. For instance, if the two input terminals are shorted together and a voltage applied, there will be a finite output response. This leads to the *common-mode rejection ratio* and related subjects, to be covered in following sections.

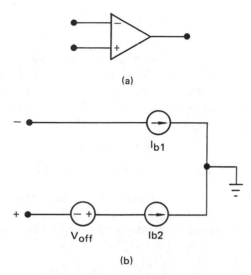

(a)

(b)

Figure 9.34 Opamp Offset Model
(a) Circuit Symbol, (b) Input Circuit Model

B. Offset Voltage

The effects of the offset voltage can be determined by grounding the input of the opamp in the voltage follower configuration as indicated in figure 9.35. Referring to figure 9.34(b), it can be seen that the output voltage will be simply the offset voltage as referred to the noninverting input terminal. In commercially available opamps there is a method of nulling out the offset voltage, often with a 10K potentiometer connected between two pins, with the wiper arm connected to one of the rail voltages. This potentiometer is then adjusted until the output is zero with the input grounded.

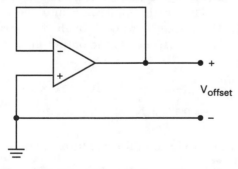

Figure 9.35 Offset Voltage Measurement

The following presumes that the offset voltage has been nulled.

C. Inverting Input Bias Current

With the offset voltage nulled, it is possible to identify the two offset currents by careful measurement. With the noninverting terminal grounded and the opamp connected in the noninverting arrangement as indicated in figure 9.36, the inverting input is held at virtual ground potential by the large DC gain, so the output becomes $-i_{B1}R_{b1}$, or

$$i_{B1} = -\frac{v_{B1}}{R_{b1}} \qquad 9.104$$

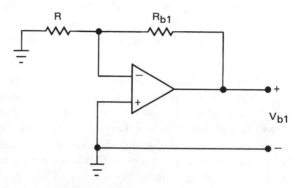

Figure 9.36 Inverting Input
Bias Current Measurement

D. Noninverting Input Bias Current

Again with the offset voltage nulled, the circuit shown in figure 9.37 will identify the noninverting terminal bias current. With this connection, the output voltage becomes $-R_{b2}i_{B2}$, or

$$i_{B2} = -\frac{v_{B2}}{R_{b2}} \qquad 9.105$$

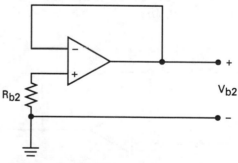

Figure 9.37 Noninverting Input
Bias Current Measurement

E. Reduction of Bias Current Effects

The two bias currents are in the same direction and fairly close in amplitude, but may differ by up to about 10 percent of their values. Manufacturers specify two currents that make up the individual bias currents. The average of the two is normally called the bias current.

$$i_B = \frac{i_{B1} + i_{B2}}{2} \qquad 9.106$$

The difference between the two bias currents is called the offset current.

$$i_{\text{OFFSET}} = |i_{B1} - i_{B2}| \qquad 9.107$$

A typical value for the bias current in bipolar opamps is 100 nA, and for the offset current it is 10 nA, or about 10 percent of the bias current. But, the manufacturer cannot guarantee which bias current will be larger, so the design techniques used to reduce the bias current effects are limited.

The fundamental requirement to reduce the effects of the bias currents is to provide a DC path to each input terminal. The alternative is to have a capacitive path only, which will result in the current source charging the capacitance. This is of great importance in data acquisition systems where analog-to-digital converters use integrators involving capacitors in the feedback path, and current sources for inputs. These can result in significant errors.

The secondary method is to match the resistances at the two terminals, on the assumption that the bias current (equation 9.106) is predominant, and that the offset current is random in direction (from one component to the next), but of small magnitude. Then from figure 9.38, with the current directions taken as into the terminals, the lower terminal voltage is $-i_{B2}R_3$ and the high DC gain of the amplifier causes the upper terminal to have the same voltage. Therefore, the current through R_1 is to the right and is $i_{B2}R_3/R_1$. At the positive terminal the current source, i_{B1}, draws that current in, so to satisfy that source, the current through R_2 is to the left and is $i_{B1} - i_{B2}R_3/R_1$.

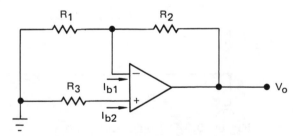

Figure 9.38   Circuit Showing the Effect of Offset Currents on the Output Voltage

The output voltage is then the voltage at the input terminals (assumed identical at both because of the high DC gain of the amplifier) added to the voltage across R_2, or

$$v_O = -i_{B2}R_3 + R_2 \left(i_{B1} - i_{B2}\frac{R_3}{R_1} \right)$$

$$= -i_{B2}R_3 \frac{R_1 + R_2}{R_1} + i_{B1}R_2$$

$$= R_2 \left(i_{B1} - i_{B2}R_3 \frac{R_1 + R_2}{R_1 R_2} \right) \qquad 9.108$$

If i_{B1} and i_{B2} were identical, this expression could be made zero by making R_3 equal to the parallel combination of R_2 and R_1. But, $i_{B1} = i_B + i_{\text{OFFSET}/2}$ and $i_{B2} = i_B - i_{\text{OFFSET}/2}$, so making the resistances equivalent results in an output voltage of

$$v_O = R_2 i_{\text{OFFSET}} \qquad 9.109$$

Equation 9.109 is true for $R_3 = (R_1 R_2)/(R_1 + R_2)$.

If, as is the tendency, R_3 were zero, the output voltage due to the bias currents would be $R_2 i_{B1}$, which would be about an order of magnitude larger than the case above.

Example 9.11

An inverting opamp (bipolar) integrator has a 0.1 μF feedback capacitor and a 10 kilohm input resistor. What output error rate will occur if (a) there is no resistance between the noninverting terminal and ground, and (b) if there is a 10 kilohm resistance between that terminal and ground?

(a) With the noninverting terminal at ground, there will be no error due to i_{B2}. However, all of i_{B1} will have to flow through the capacitor, or

$$\frac{dv}{dt} = \frac{1}{C}(-i_{B1}) = 10^7(-100 \times 10^{-9}) = -1 \text{ volt/s}$$

(b) With the 10 kilohm resistor connected to the noninverting terminal, there will be a voltage at that terminal due to i_{B2} of $-10\text{k}i_{B2}$. This voltage also appears at the inverting terminal causing a current of $10\text{k}i_{B2}/10\text{k} = i_{B2}$ to flow to the right in the input resistance. Then with i_{B1} flowing into the inverting terminal, the current flowing through the capacitance is $i_{B2} - i_{B1} = -i_{\text{OFFSET}}$, so the output error rate will be

$$\frac{dv}{dt} = \frac{1}{C}(-i_{\text{OFFSET}}) = 10^7(-10 \times 10^{-9})$$

$$= -0.1 \text{ volt/s}$$

F. Common-Mode Gain in Differential Amplifiers

If the terminals of a practical opamp are connected together and an input signal voltage is applied to them, there will be an output voltage signal proportional to the input signal voltage, so that the common-mode gain will be the ratio of the output to the input signal voltage. This gain will be small compared to the differential gain, typically 10^{-5} to 10^{-4} times the DC gain. The ratio of the differential-mode gain to the common-mode gain is called the *common-mode rejection ratio* (CMRR).

$$\text{CMRR} = \frac{A_{\text{differential mode}}}{A_{\text{common mode}}} \qquad 9.110$$

CMRR is usually given in decibels.

Common-mode amplification can be a serious problem in instrument amplifiers and other precision applications. For inverting amplifiers the problem is not serious, as the terminals are virtually shorted because of the high amplifier gain, and the non inverting terminal is very nearly grounded (even when there is a resistance connected to eliminate bias current problems). As a result, there is little common-mode signal to be amplified.

The serious problems arise in noninverting amplifiers where the input terminals are at the input level so that the entire input signal is common mode, and in inverting amplifiers using feedforward to the noninverting terminal as discussed in section 9.5, where the common-mode input is the feedforward portion of the input signal.

9   AMPLIFIER NOISE

A. Shot Noise

Shot noise is associated with current. It is an aggregate phenomenon involving numerous individual charge carriers. A current can be calculated from the number of charge carriers passing a cross section per second.

$$I = q\frac{dn}{dt} \qquad 9.111$$

q is the charge per carrier, and dn/dt is the number of carriers passing the cross section per second. Because of the random thermal velocities of the charge carriers, dn/dt fluctuates. If the average current is I, and the total current is $i(t)$, then the noise current, $i_n(t)$, can be defined as

$$i_n(t) = i(t) - I \qquad 9.112$$

The resulting spectral density of the noise current, assuming that the arrival of individual charge carriers is entirely random, is constant. That is, all frequencies are equally likely. This can be considered as an aggregate of delta functions in time, which in the frequency domain becomes a constant (Fourier transformation). The amplitude of the spectral density is, then, proportional to the average current.

$$S_{\text{shot}}(f) = 2qI \qquad 9.113$$

S_{shot} is the noise shot power spectral density in W/Hz. It is constant over all frequencies and is known as *white noise*.

B. Johnson or Thermal Noise

Thermal noise is the result of random motions of electrons in conductors, which give rise to random voltage fluctuations across the conductor. This noise voltage exists whether or not there is any current flow in the conductor. The energy of individual electrons in a conductor is random, but the average energy per electron is kT. The resulting noise voltage is related to energy and is expressed in terms of its RMS value squared.

$$S_{\text{thermal}}(f) = 4kTR \qquad 9.114$$

S is the spectral power density of the thermal noise in watts/hertz. R is the resistance of the conductor in ohms.

C. Low-Frequency Noise

Parameter variations due to changes in temperature over time, aging of components, and any other slow changes (drift) in a system produce low-frequency noise. The spectral distribution is typically $1/f$ in nature, so it would be expected to become infinite at zero frequency. In reality, there is no zero-frequency component actually present.

The lowest frequency which must be considered is $1/T_{\text{on}}$, where T_{on} is the length of time since the device was first (most recently) turned on.

D. Interference Noise

Interference noise is caused by power lines and unrelated signals. These are called *cross-talk* when due to signals. The frequency spectra depend on the source. Power line noise is typically at the power frequency and the frequencies of its harmonics, but it also can involve random bursts from switching of power equipment.

Interference noise can be filtered out if it is not within the spectrum of the desired signal.

E. Bandwidth Limiting of Noise

All the noise sources except interference are described by a power frequency spectral distribution. The output noise power from an amplifier will be related to the frequency characteristics of the amplifier and the spectral distribution of the noise sources.

Figure 9.39 shows a thermal noise source (taken as its Thevenin equivalent voltage source) and an amplifier that has a voltage gain frequency response of $A(f)$. The corresponding power gain response is

$$G(f) = |A(f)|^2 \qquad 9.115$$

For a voltage-squared input spectrum, S_n, from the noise voltage source, the output voltage-squared value will be

$$\overline{v_O}^2 = \int_{f=0}^{\infty} S_n(f) G(f) \, df \qquad 9.116$$

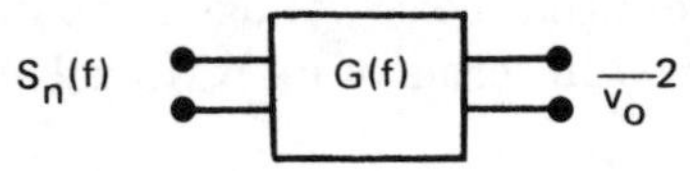

Figure 9.39   Noise into an Amplifier

A single-pole, low-pass amplifier corresponding to $A(f)$ will have

$$A(f) = \frac{A_v}{1 + j\left(\dfrac{f}{f_{\text{cf}}}\right)} \qquad 9.117$$

$$G(f) = \frac{A_v^2}{1 + \left(\dfrac{f}{f_{\text{cf}}}\right)^2} \qquad 9.118$$

Thus, for thermal or shot noise with a constant $S(f) = S$,

$$\overline{v_O}^2 = S \int_0^{\infty} \frac{A_v^2}{1 + \left(\dfrac{f}{f_{\text{cf}}}\right)^2} \, df$$

$$= A_v S f_{\text{cf}} \arctan\left(\frac{f}{f_{\text{cf}}}\right)\Big|_{f=0}^{f=\infty}$$

$$= A_v S f_{\text{cf}} \frac{\pi}{2} \qquad 9.119$$

For a two-pole, low-pass filter with critical damping,[2]

---

[2] The power gain G is related to the voltage gain by $G = R_{\text{in}} A_v^2 / R_{\text{out}}$. Here, it is assumed that $R_{\text{out}} = R_{\text{in}}$.

$$A(f) = \frac{A_v}{\left[1 + j\left(\dfrac{f}{f_{\text{cf}}}\right)\right]^2} \qquad 9.120$$

$$G(f) = \frac{A_v^2}{\left[1 + \left(\dfrac{f}{f_{\text{cf}}}\right)^2\right]^2} \qquad 9.121$$

Again, for thermal or shot noise with $S(f) = S$,

$$\overline{v_O}^2 = S \int_0^{\infty} \frac{A_v^2}{\left[1 + \left(\dfrac{f}{f_{\text{cf}}}\right)^2\right]^2} \, df = A_v S f_{\text{cf}} \frac{\pi}{4} \qquad 9.122$$

Bandwidth usually is defined by 3 dB points, so that the bandwidth for the single-pole, low-pass amplifier is f_{cf}. The bandwidth for the two-pole filter is not f_{cf}, however. The filter gain is 6 dB down at that frequency, and the 3 dB frequency is at

$$0.6436 f_{\text{cf}} = f_{\text{cf}}'$$

Scaling this upwards,

$$\frac{\pi}{4} f_{\text{cf}} = 1.22 f_{\text{cf}}'$$

The *noise bandwidth* is defined as the frequency range Δf over which the mid-frequency gain A_v holds such that a square filter amplifier would pass noise energy $S A_v^2 \Delta f$. The resulting bandwidths are summarized in table 9.1.

F. Noise Source Equivalent Circuits

A white noise source may produce thermal noise and shot noise, one of which is dependent on temperature and the other on current level. The thermal noise source

Table 9.1
Signal and Noise Bandwidths for Amplifiers

	signal bandwidth (-3 dB)	noise bandwidth (Δf)
1-pole, low pass	f_{cf}	$1.571 f_{\text{cf}}$
2-pole, low-pass (critical damping)	$0.6436 f_{\text{cf}} = \text{BW}_s$	$0.785 f_{\text{cf}} = 1.22 \text{BW}_s$
2-pole, band-pass	$f_2 - f_1$	$1.571(f_2 - f_1)$
4-pole, band-pass (critical damping)	$0.6436 f_2 - 1.554 f_1 = \text{BW}_s$	1.22BW_s

is identified as an RMS voltage source, and the shot noise source appears as an RMS current source. The general circuit for a resistive device with both noise sources is shown in figure 9.40. Because the two sources are statistically independent, squaring variables does not produce any cross-products. Thus, the noise input spectrum is the independent sum of the two. From equations 9.113 and 9.114,

$$S(f) = S_{\text{thermal}}(f) + S_{\text{shot}}(f)$$
$$= e_{tn}^2 + Ri_{sn}^2 \qquad 9.123$$
$$e_{tn}^2 = 4kTR \qquad 9.124$$
$$Ri_{sn}^2 = 2qIR \qquad 9.125$$

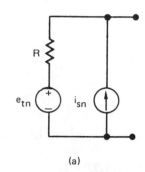

(a)

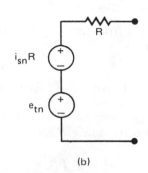

(b)

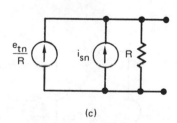

(c)

Figure 9.40 Noise Source Equivalent Circuits:
(a) Thermal and Shot Noise Sources,
(b) Thevenin Equivalent Circuit,
(c) The Norton Equivalent Circuit

G. Optimum Source Resistance

From the usual thermal point of view, it is possible to minimize the effects of internal noise sources in an amplifier by proper selection of the source resistance. Consider the noiseless amplifier shown in figure 9.41(a).

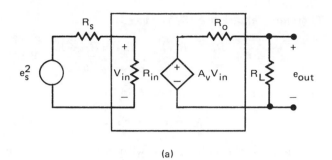

(a)

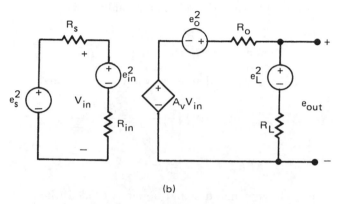

(b)

Figure 9.41 (a) Noiseless Amplifier,
(b) Amplifier with Internal Noise Sources

Circuit analysis of the noiseless amplifier results in

$$e_{\text{out}} = \frac{A_v R_{\text{in}} R_L e_s}{(R_{\text{in}} + R_s)(R_L + R_o)} \qquad 9.126$$

$$e_{\text{out}}^2 = \frac{A_v^2 R_{\text{in}}^2 R_L^2 e_s^2}{(R_{\text{in}} + R_s)^2 (R_L + R_o)^2} \qquad 9.127$$

$$e_s^2 = 4kTR_s \qquad 9.128$$

Circuit analysis of the same amplifier with internal noise sources due to the internal resistances and the load resistance results in

$$e_{\text{out}}^2 = \frac{A_v^2 R_L^2 (R_{\text{in}}^2 e_s^2 + R_s^2 e_{\text{in}}^2)}{(R_{\text{in}} + R_s)^2 (R_L + R_o)^2} + \frac{R_o^2 e_L^2 + R_L^2 e_o^2}{(R_o + R_L)^2} \qquad 9.129$$

The *noise figure* is a measure of the noise generated within an amplifier as compared to the noise that would exist without internally generated noise. It is the ratio of equation 9.129 to equation 9.127.

$$F = 1 + \frac{R_s^2 e_{\text{in}}^2}{R_{\text{in}}^2 e_s^2} + \frac{(R_s + R_{\text{in}})^2 (R_o^2 e_L^2 + R_L^2 e_o^2)}{A_v^2 R_L^2 R_{\text{in}}^2 e_s^2} \qquad 9.130$$

However,

$$e_{\text{in}}^2 = 4kTR_{\text{in}} \qquad 9.131$$
$$e_o^2 = 4kTR_o \qquad 9.132$$
$$e_L^2 = 4kTR_L \qquad 9.133$$

Therefore, equation 9.130 reduces to

$$F_{\text{thermal}} = 1 + \frac{R_s}{R_{\text{in}}} + \frac{R_o(R_s + R_{\text{in}})^2(R_L + R_o)}{A_v^2 R_s R_{\text{in}}^2 R_L} \quad 9.134$$

Use equation 9.135 to express F in dB.

$$F_{\text{dB}} = 10\log(F) \quad 9.135$$

To minimize the noise figure with respect to R_s, the partial derivative of F with respect to R_s is set to zero. As the second partial derivative of F with respect to R_s is always positive, the minimum condition is

$$R_s^2\big|_{\min F} = \frac{R_{\text{in}}^2 R_o(R_L + R_o)}{A_v^2 R_{\text{in}} R_L + R_o(R_L + R_o)} \quad 9.136$$

The noise power is defined as the voltage squared.

$$N = V_{\text{on}}^2 = e_s^2(BW)\left[\frac{A_v^2 R_{\text{in}}^2 R_L^2}{(R_{\text{in}} + R_s)^2(R_o + R_L)^2}\right] \quad 9.137$$

$$e_s^2 = 4kTR_s \quad 9.138$$

The voltage squared signal power is needed when calculating the ratio of signal power to noise power, S/N.

$$S = V_{oS}^2 = V_s^2\left[\frac{A_v^2 R_{\text{in}}^2 R_L^2}{(R_{\text{in}} + R_s)^2(R_o + R_L)^2}\right] \quad 9.139$$

The ratio of signal power to noise power frequently is given in decibels.

$$\left(\frac{S}{N}\right)_{\text{dB}} = 10\log\left(\frac{S}{N}\right) \quad 9.140$$

The noise and signal powers actually are powers only if they appear across a 1 ohm resistance. The actual powers are obtained by dividing S and N by the actual load resistance. However, this does not change the ratio.

Example 9.12

The circuit in figure 9.41 has $R_{\text{in}} = R_o = R_L = 1000$ ohms, and $A_v = 10$. (a) Determine the optimum R_s and the resulting noise figure in decibels. (b) With a noise bandwidth of 10^4 hertz and a temperature of 300K, determine the noise power voltage at the load. (c) For an RMS input signal of 10^{-5} volts, determine the ratio of output signal to noise.

From equation 9.136, using k as shorthand notation for 1000,

$$R_s^2 = \frac{(1k)^2(1k)(1k + 1k)}{(10)^2(1k) + (1k)(1k + 1k)} = \frac{2k^4}{102k^2}$$

$$R_s = 140 \text{ ohms}$$

From equation 9.134,

$$F = 1 + \frac{140}{1k} + \frac{(1k)(140 + 1k)^2(1k + 1k)}{(10)^2(1k)(1k)^2(1k)}$$

$$= 1.166$$

$$F_{\text{dB}} = 10\log(1.166) = 0.67 \text{ dB}$$

Now use equations 9.138 and 9.137.

$$e_S^2 = 4(1.38 \times 10^{-23})(300)(140) = 2.32 \times 10^{-18}$$
$$N = V_{\text{on}}^2$$
$$= (2.32 \times 10^{-18})(10^4)\left[\frac{(100)(1k)^2(1k)^2}{(140 + 1k)^2(1k + 1k)^2}\right]$$
$$= 4.46 \times 10^{-13}\text{volts}^2$$

For the signal input of 10^{-5} volts, $V_s^2 = 10^{-10}$. From equation 9.139,

$$S = 1.92 \times 10^{-9}$$

Therefore, the ratio of signal to noise power is

$$\frac{S}{N} = \frac{1.92 \times 10^{-9}}{4.46 \times 10^{-13}} = 4.31 \times 10^3$$

$$\left(\frac{S}{N}\right)_{\text{dB}} = 10\log 4310 = 31.3 \text{ dB}$$

H. Noise Power to an Input Resistance

The RMS voltage squared per unit bandwidth (in volts^2/hertz) for a resistance R_s is

$$e_n^2 = 4kTR_s \quad 9.141$$

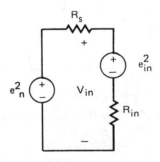

Figure 9.42 Input Noise Circuit

If this voltage is modeled as in figure 9.42, the voltage across R_{in} for $e_{\text{in}} = 0$ is

$$v_{\text{in}} = \frac{e_n R_{\text{in}}}{R_s + R_{\text{in}}} \quad 9.142$$

With power being voltage squared divided by resistance, the power in watts/hertz from the resistance R_s that reaches R_{in} is

$$P_{in} = \frac{e_n^2 R_{in}^2}{(R_s + R_{in})^2 R_{in}} = \frac{(4kTR_s)R_{in}}{(R_s + R_{in})^2} \qquad 9.143$$

In many cases, the input resistance is set equal to the source resistance in order to maximize power transfer. Unfortunately, this also maximizes the noise power input from the source resistance.

The total noise power at the input, including the noise from the input resistance, is minimized by making the source resistance as small as possible. With maximum signal power to the input, the corresponding maximum effect of the input resistance noise occurs as

$$P_{in} = kT \qquad 9.144$$

I. Noise Figures For Cascaded Amplifiers

The noise output from an amplifier with only an input resistance noise is

$$P_o = FGP_{in} \qquad 9.145$$

Equation 9.144 is used for p_{in} as impedance matching is assumed. This is illustrated for the input stage of figure 9.43.

The excess noise accounting for F is modeled by an equivalent internal noise source at the input (indicated by p_1 in figure 9.43). Then,

$$P_{o1} = G_1(kT + P_1) = F_1 G_1 kT \qquad 9.146$$
$$P_1 = kT(F_1 - 1) \qquad 9.147$$

Similarly for the other matched stages,

$$P_2 = kT(F_2 - 1) \qquad 9.148$$
$$P_3 = kT(F_3 - 1) \qquad 9.149$$
$$P_n = kT(F_n - 1) \qquad 9.150$$

Then,

$$
\begin{aligned}
P_{o2} &= G_2(P_{o1} + P_2) \\
&= G_2 G_1 F_1 kT + G_2(F_2 - 1)kT
\end{aligned}
\qquad 9.151
$$

Since the noiseless contribution is $G_2 G_1 kT = G_2 G_1 P_{in}$, and the noise figure is $P_{o2}/G_2 G_1 P_{in}$, an overall noise figure for the first two stages is

$$F_{12} = F_1 + \frac{F_2 - 1}{G_1} \qquad 9.152$$

Similarly for P_{o3} and F_{13},

$$
\begin{aligned}
P_{o3} &= G_3(P_{o2} + P_3) \\
&= G_3 G_2 G_1 F_1 kT + G_3 G_2(F_2 - 1)kT \\
&\quad + G_3(F_3 - 1)kT
\end{aligned}
\qquad 9.153
$$
$$F_{13} = F_1 + \frac{F_2 - 1}{G_1} + \frac{F_3 - 1}{G_1 G_2} \qquad 9.154$$

It can be seen that the noise figure is obtained by referring all subsequent noise sources to the input of the first stage, dividing by the gains of all previous amplifiers, and then dividing by the input power density. Then, each succeeding stage's contribution to the noise figure is found from

$$F_n = \frac{F_n - 1}{G_1 G_2 G_3 \dots G_{n-1}} \qquad 9.155$$

Equation 9.155 is applicable to any amplifier components involved in a cascaded system, including antennas, modulators, and mixers. The equivalent noise power density of any source is kT.

J. Noise Temperature and Noise Resistance

Another method of representing device noise effects is by an equivalent *noise temperature*, T_e. The output of the input stage is $G_1(kT_o + kT_e)$ where T_o commonly is taken as room temperature (290K). Then,

$$F_1 = 1 + \frac{T_e}{T_o} \qquad 9.156$$
$$F_n - 1 = \frac{T_{en}}{T_o} \qquad 9.157$$

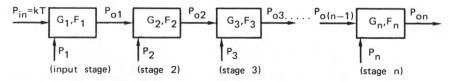

Figure 9.43 Cascaded Amplifier Stages

Yet another method of representing a device's contribution to noise is by an equivalent *noise resistance*, Ri_{sn}. For this method, the input resistance of the nth stage must be known.

$$F_1 = 1 + \frac{R_e}{R_s} \qquad 9.158$$

$$F_n - 1 = \frac{R_{\text{en}}}{R_{\text{sn}}} \qquad 9.159$$

PRACTICE PROBLEMS

Warmups

1. An operational amplifier has a gain-bandwidth product of $2\pi \times 10^7$ rad/s. The amplifier is used in an inverting mode with a feedback resistance of 25,000 ohms and an input resistance of 5000 ohms. Determine the gain and bandwidth of the circuit.

2. The circuit shown is to provide $V_0 = V_1 + V_2$. Determine R_1 and R_2.

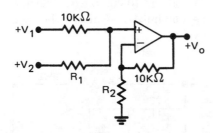

3. A low-pass active filter using one operational amplifier is to have a low-frequency gain of 10 and a bandwidth of 30 hertz. The only capacitor available is 0.1 μF. Specify the resistances and show the circuit.

4. Design a two-pole high-pass active filter as in figure 9.20. $G_2 \geq 10^{-4}$ mhos and all capacitors must be powers of 10 less than 10 μF. The poles are to be at 100 rad/s, and the circuit is to be critically damped.

5. An amplifier has an input resistance of 50 ohms, a bandwidth of 100 kHz, and a noise figure of 6 dB. Determine (a) the power gain at 293K if the output noise power is 10^{-12} watts and the input source is matched, (b) the equivalent noise temperature, and (c) the equivalent noise resistance.

6. For the circuit shown, determine the output voltage as a function of the input voltage.

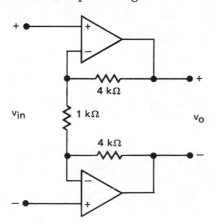

7. The opamp circuit shown is to provide a characteristic similar to a relay, so that when the output has been negative and the input current increases, the output will go to the positive limit of +12 volts when the input voltage exceeds +1 volt, and will remain there until the input voltage becomes less than −1 volt, at which time the output will go to the negative limit of −12 volts. Select an appropriate value for R.

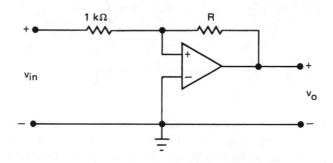

8. An opamp has a maximum slew rate of 0.75 volt/μs. Determine the maximum frequency that this amplifier can provide for a sinusoidal output of 10 volts peak to peak.

9. The circuit shown simulates an inductance with a series resistance. Determine their values, and the frequency range over which this is a reasonable approximation.

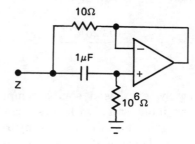

10. The circuit shown is a capacitance multiplier. Determine the capacitance as seen from the terminal, and the frequency range over which the approximation is valid.

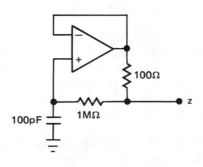

Concentrates

1. For the circuit shown, find v_O as a function of V_1, v_2, v_3, and v_4.

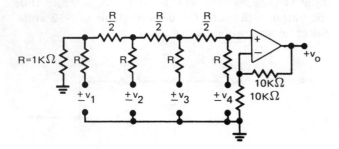

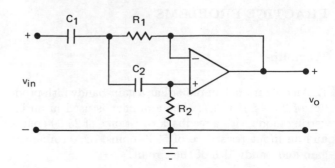

2. A capacitive displacement transducer has a capacitance given by $C_x = A/(0.225d)$ where d is the displacement to be measured. A voltage proportional to the displacement is obtained using the accompanying circuit.

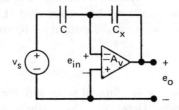

Show that for v_s being a constant amplitude sinusoid at approximately 50 kHz and $A \approx 10^8$,

$$e_o = \frac{-(0.225)Cv_sd}{A}$$

3. A receiving system has a source temperature of 293 degrees, an input stage with noise bandwidth of 1 GHz, a power gain of 12 dB, a noise figure of 6 dB, and an input impedance of 72 ohms. The mixer stage has a gain of -6 dB and a noise figure of 12 dB. The I.F. stage has a power gain of 40 dB and a noise figure of 6 dB, with a noise bandwidth of 10 MHz. Determine (a) the overall noise figure, and (b) the minimum detectable input voltage within the input band-pass to permit a signal-to-noise ratio of 1.

4. The transfer function of the circuit shown has a low-frequency asymptote slope of $+40$ dB/decade. The high-frequency asymptote has a slope of 0 dB/decade and a value of 0 dB. The two asymptotes intersect at a frequency of 1000 rad/s. At the intersection frequency, the transfer function is 0 dB.

Given that $C_1 = C_2 = 0.1$ μF, design the circuit by specifying the values of the resistances.

5. The circuit shown is a band-pass filter with corner frequencies at 100 and 10,000 rad/s. The mid-frequency gain is to be maximized. Specify the resistance and capacitance values.

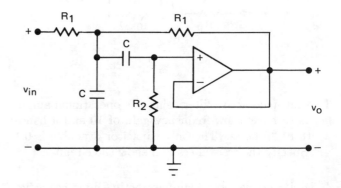

6. For the circuit shown, determine the equivalent component corresponding to the input impedance.

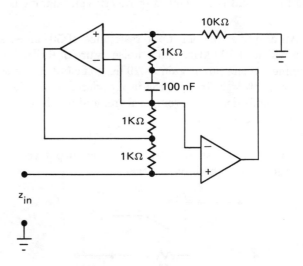

7. The circuit shown is to be used to provide an all-pass transfer function of

$$G(s) = \frac{s^2 - 2000s + 10^6}{s^2 + 2000s + 10^6}$$

Given that $C = 1$ μF, specify R_1, R_2, and k to achieve $G(s)$.

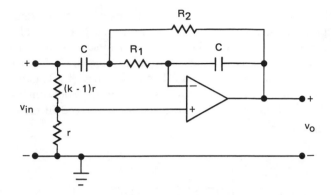

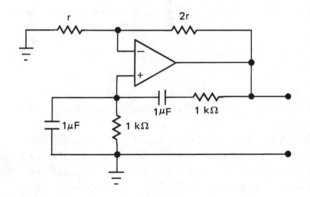

8. The Wien bridge oscillator shown will oscillate at the frequency where the complex loop gain to the positive opamp terminal is $1 + j0$. Determine the frequency of oscillation.

9. The voltage across a diode driven by a constant current source decreases by 2 mV for every degree C of increase that the temperature changes. The silicon diodes in the circuit shown are identical and driven by identical current sources. The reference diode is maintained at a constant temperature, while the sensing diode experiences the temperature to be probed. The difference amplifier is to provide a voltage output which is to be zero when the temperature of the two diodes are identical, and -10 volts when the probe temperature is 100°C higher than the reference temperature.

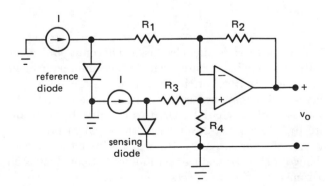

In order to minimize offset current problems in the opamp, each input terminal should see the same resistance, so that the parallel combinations of R_1 with R_2, and R_3 with R_4, should be made equal.

Taking $R_1 = 1$ kilohm, specify the resistances R_2, R_3, and R_4 to achieve these specifications.

10. The sample-and-hold circuit shown is operated at 40 kHz. The opamp manufacturer lists the input bias current at 100 nA and the offset current at 10 nA. Determine the maximum average error in the output voltage due to these currents.

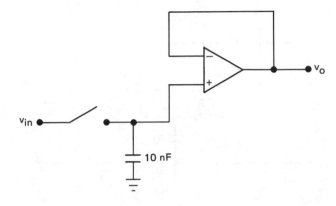

Timed

1. The circuit shown is a notch filter providing at least 30 dB of attenuation at 400 hertz. It has a 3 dB bandwidth of 80 hertz. 0.01 μF capacitors are available. Determine R_1, R_2, and R_3.

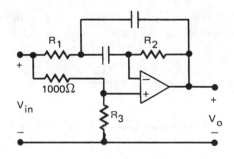

2.

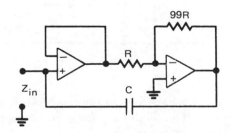

The operational amplifiers shown are identical and have a raw GBW of 10^7 rad/s. Determine the input impedance Z_{in} and the frequency range over which it appears essentially as a capacitance.

3. The transfer function of the circuit shown has a DC gain of 0 dB with a low-frequency asymptote slope of 0 dB/decade. The high-frequency asymptote has a slope of -40 dB/decade, and the two asymptotes intersect at a frequency of 10,000 rad/s. At the intersection frequency, the transfer function is 3 dB below the DC value.

Given that C_2 is 0.1 μF, design the circuit by specifying the values of the resistances and C_1.

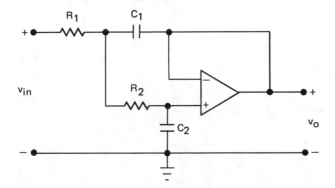

4. The transfer function of the circuit shown has a DC gain of 6 dB with a low-frequency asymptote slope of 0 dB/decade. The high-frequency asymptote has a slope of -40 dB/decade, and the two asymptotes intersect at a frequency of 10,000 rad/s. At the intersection frequency, the transfer function is at the DC value.

Given that $C_1 = C_2$ and that $R_1 = R_2$, design the circuit by specifying their values and the value of k.

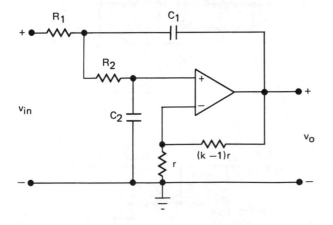

5. Identical diodes are used in the temperature sensing circuit shown. The diodes are ideal and have current given by the diode equation:

$$i_{\text{DIODE}} = I_o(e^{-qv/kT} - 1)$$

Assume that I_o does not vary significantly over the temperature range of interest. (Note, this is not a valid assumption for real diodes.) (a) Determine the output voltage as a function of input voltage, the reference temperature, and the sensed temperature. (b) The circuit can be made more sensitive by adding more sensing diodes. Show whether these should be placed in series or in parallel.

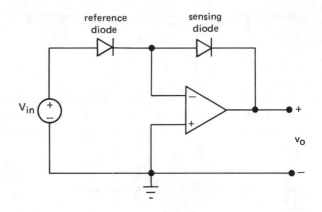

6. The amplifier shown is to be used to drive the input of a 12 bit analog-to-digital converter with a range of 0–5 volts. The opamp has an input bias current of 100 nA and an offset current of 10 nA. Specify the maximum values of the resistances for an amplifier gain of two so that the error due to the bias and offset currents does not exceed one-half of the least significant bit.

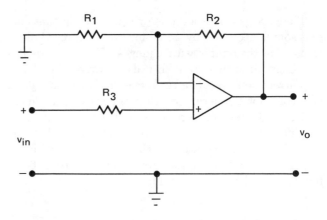

7. A transducer is modeled as a Thevenin equivalent voltage source, e, in series with a Thevenin equivalent resistance, r. Both of these parameters are subject to change with temperature. In order to desensitize the output to these changes, the two circuits shown below are to be examined. Which of these, when providing a gain of approximately 10, will be least sensitive to changes of the transducer parameters? Show how each parameter variation affects the output.

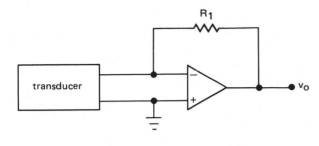

circuit A

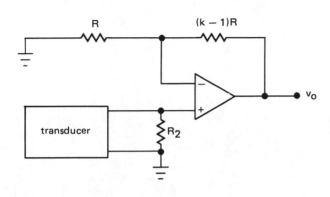

circuit B

8.

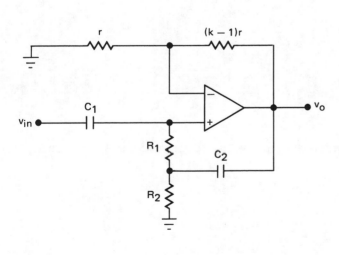

For the circuit shown, assume an ideal operational amplifier. For parts (a), (b), and (c), $k = 1$.

(a) Find the input impedance. (b) Find the voltage transfer function. (c) For an input of $V_{dc} + V_{ac}\sin(t/\sqrt{C_1 C_2 R_1 R_2})$, find the input current and the output voltage. (d) What changes occur in the above for $k = 2$?

9. For the circuit shown, $-1 < a < 1$. Obtain the function e/a, and plot e as a function of a.

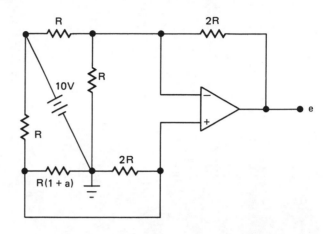

10. For the circuit shown, the opamp, transistor, and resistors R_1, R_2, and R_3 are to provide a firm current source of 10 mA for a range of load voltage from -6 to $+4$ volts. Assume the opamp is ideal and the transistor has infinite beta. Specify suitable resistances and determine the required power rating for the transistor.

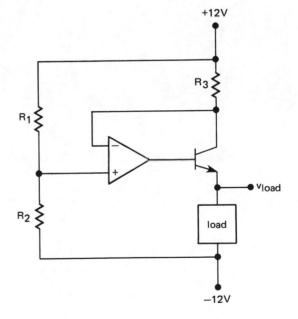

WAVESHAPING, LOGIC, AND DATA CONVERSION

10

Nomenclature

A_v	amplifier voltage gain	
C	capacitance	farads
G_c	i-v slope in region C	mhos
h_{IE}	base-emitter input impedance	ohms
H	high voltage level	volts
i_B	variable base current	amps
i_C	variable collector current	amps
i_D	variable diode current	amps
i_G	variable gate current	amps
i_{IN}	variable input current	amps
i_L	variable load current	amps
i_T	variable terminal current	amps
i_Z	variable zener diode current	amps
I_B	region B intercept current (constant)	amps
I_{BR}	breakdown current level (constant)	amps
I_C	region C intercept current (constant)	amps
I_{COE}	reverse collector saturation current with open emitter (constant)	amps
I_{GT}	gate turn-on current level (constant)	amps
I_H	Hold current level (constant)	amps
I_{ZK}	zener diode keep-alive current (constant)	amps
k	voltage division ratio	–
L	inductance	henrys
P	average power	watts
P_D	diode average power	watts
P_R	resistor average power	watts
R_a	region A v-i slope	ohms
R_b	bases resistance	ohms
R_f	P-N junction forward resistance	ohms
R_r	P-N junction reverse breakdown resistance	ohms
R_s	source resistance	ohms
R_{Th}	Thevenin equivalent resistance	ohms
R_z	zener diode resistance	ohms
t_d	delay time	s
t_f	full time	s
t_r	rise time	s
t_s	storage time	s
t_{tr}	triggering time	s
T	period of time	s
v_{BE}	variable base-emitter voltage drop	volts
v_C	variable control voltage	volts
v_{CE}	variable collector-base voltage drop	volts
v_D	variable diode voltage drop	volts
v_{DS}	variable drain-source voltage drop	volts
v_F	variable P-N forward voltage drop	volts
v_{GS}	variable gate-source voltage drop	volts
v_{IN}	variable input voltage	volts
v_L	variable load voltage	volts
v_N	variable node voltage	volts
v_O	variable output voltage	volts
v_S	variable source voltage	volts
v_T	variable terminal voltage	volts
v_Z	variable zener diode voltage	volts
V_B	region B intercept voltage (constant)	volts
V_{BB}	base supply voltage	volts
V_{BEF}	base-emitter forward voltage drop (constant)	volts

V_{BO}	breakover voltage level (constant)	volts
V_{BR}	breakdown voltage level (constant)	volts
V_{CC}	collector supply voltage	volts
V_{DD}	drain supply voltage (constant)	volts
V_F	forward threshold voltage level (constant)	volts
$V_{GS(off)}$	gate to source turn-off voltage level (constant)	volts
$V_{GS(on)}$	gate to source threshold voltage level for enhancement mode MOSFET (constant)	volts
V_{GT}	gate turn-on voltage level (constant)	volts
V_H	holding voltage level (constant)	volts
V_H	high voltage level (constant)	volts
V_I	ionization voltage level (constant)	volts
V_L	low voltage level (constant)	volts
V_R	reference voltage (constant)	volts
V_S	source voltage (constant)	volts
V_{Th}	Thevenin equivalent voltage (constant)	volts
V_{ZO}	zener diode equivalent voltage source (constant)	volts

Symbols

β	current gain	–
τ	time constant	s
ω	frequency	rad/s

1   DIODE CIRCUITS

A. Introduction

Semiconductor diodes have characteristics similar to figure 10.1. For *silicon diodes*, the *forward voltage drop* is approximately 0.7 volts, while for *germanium diodes*, it varies from 0.15 to 0.25 volts. The *reverse voltage breakdown* value is set by the doping levels. (Decreasing doping results in a higher breakdown voltage.) This latter property is used in the regulation of DC voltage supplies.

The linear equivalent circuit is derived from the curve parameters shown in figure 10.1(a) and results in the circuit shown in figure 10.1(b).

In figure 10.1(b), the lower branch is active when v_D is less than $-V_{BR}$. The upper branch is active when v_D is greater than $+V_F$. When v_D is between $-V_{BR}$ and $+V_F$, the device behaves like an open circuit.

B. Precision Diodes

The equivalent circuit shown in figure 10.1(b) is useful when voltages are greater than V_F, but it is of limited use for voltages near or smaller than V_F. For such low-voltage situations, an equivalent circuit more closely approximating the ideal diode is needed. Such an approximation is obtained by using an operational amplifier in conjunction with one or two ordinary diodes.

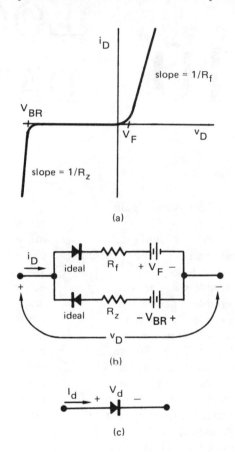

Figure 10.1   (a) Semiconductor Diode Characteristics, (b) Equivalent Circuit, (c) Circuit Symbol

Consider the circuit of figure 10.2(a). A diode has been inserted between the amplifier output and the feedback point of a voltage follower. For negative inputs, the amplifier output will be saturated at the maximum negative output. The diode will then be reverse-biased. (The diode must have a reverse breakdown greater in magnitude than the maximum negative amplifier voltage.) Then, the output will be disconnected from the input, and the load will see only R_o.

When the input voltage is positive, the amplifier output will be $A_v v_{IN}$ until the diode becomes forward-biased (i.e., until v_D exceeds V_F). In this case, the output voltage v_L follows the input with the small offset V_F/A_v.

(A_v is the raw gain of the operational amplifier.) This circuit also suppresses the forward resistance and any source resistance associated with v_{IN}.

For comparison, consider the diode circuit of figure 10.2(b). Circuit analysis yields equation 10.1.

$$v_L = \frac{R_o v_{IN}}{R_f + R_o} - \frac{R_o V_F}{R_f + R_o} \qquad 10.1$$

R_f and V_F are the linear equivalent circuit parameters for the diode from figure 10.1 when the diode is forward-biased. In cases where the source (v_{IN}) has an associated resistance, it should be added to R_f.

The equivalent circuit for when the diode of figure 10.2(a) is forward-biased is shown in figure 10.2(c). The circuit analysis yields equation 10.2.

$$v_L = \frac{R_o v_{IN}}{R_o + \dfrac{R_o + R_f}{A_v}} - \frac{R_o \dfrac{V_F}{A_v}}{R_o + \dfrac{R_o + R_f}{A_v}} \qquad 10.2$$

Since A_v is on the order of 10^5, the effective source resistance is $(R_o + R_f) \times 10^{-5}$. This is negligible in the denominator compared to R_o. Thus, the resulting transfer equation is

$$v_L = v_{IN} - \frac{V_F}{A_v} \qquad 10.3$$

A *precision diode* will be seen from the output as the voltage v_{IN} in series with a small battery. The equivalent V_F for this precision diode would be about 7 microvolts for a silicon diode, assuming an operational amplifier raw gain of 10^5. The result is a rectifier that can be used for millivolt level signals. Precision and ordinary diode transfer characteristics are shown in figure 10.2(d).

C. Clipper-Limiter Circuits

The diode can perform clipping and limiting functions when used in the circuits shown in figure 10.3. Analysis of the first circuit produces equations 10.4 and 10.5.

$$v_L = \frac{R(V_R - V_F)}{R_f + R} + \frac{R_f v_{IN}}{R_f + R} \text{ for } v_{IN} < V_R - V_F \qquad 10.4$$

$$= v_{IN} \text{ for } v_{IN} > V_R - V_F \qquad 10.5$$

The resulting voltage characteristics are shown. V_R can be either negative or positive. If it is negative, the circuit is considered a *clipper*. If V_R is positive, the circuit is a *limiter*.

The circuit of figure 10.3(b) is sensitive to negative values of V_S. The ideal circuit would have $R_f/(R_f+R)$ be equal to zero, and the effect of V_F would be negligible. This can be accomplished in practice by using the precision diode circuit. This is illustrated in figures 10.3(c) and 10.3(d). R_f and V_F are negligible for the last two circuits.

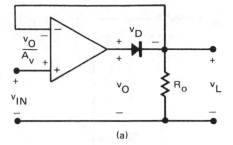

(a)

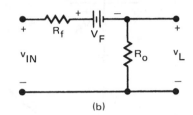

(b)

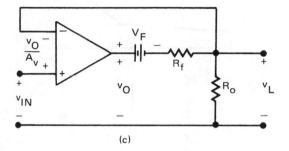

(c)

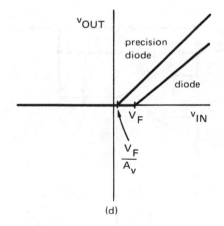

(d)

Figure 10.2 (a) Simple Precision Diode Circuit, (b) Ordinary Diode Circuit, (c) Equivalent Circuit of Forward-Biased Precision Diode, (d) Input-Output Voltage Characteristics

Example 10.1

Design a circuit that will linearly transfer any signal with an amplitude of 1.0 volt or less, but will clip all larger signals at an amplitude of 1.0 volt. Assume that supply voltages of +12 volts and −12 volts are available.

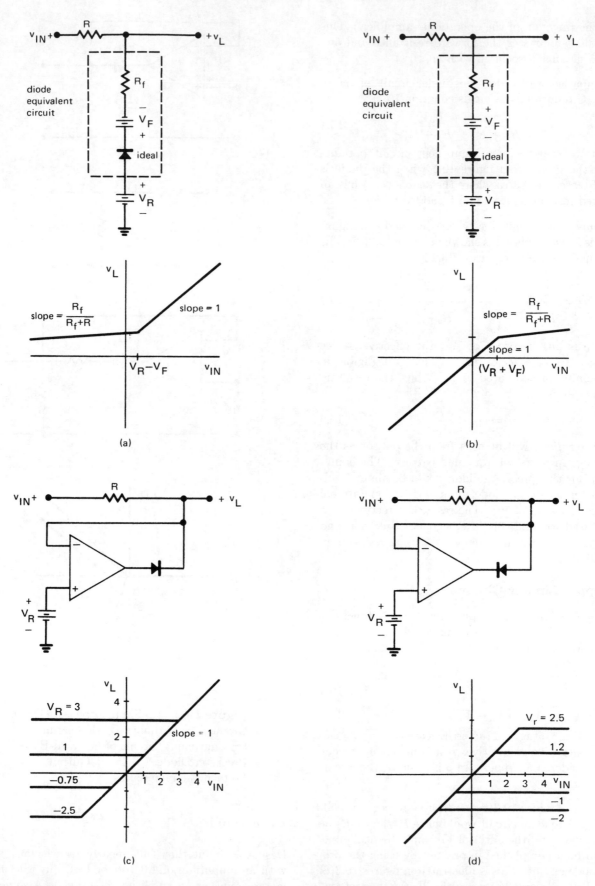

Figure 10.3   (a) Limiter Circuit, (b) Clipper Circuit, (c) Precision Limiter/Clipper Circuit, (d) Precision Limiter/Clipper Circuit Sensitive to Negative Voltages

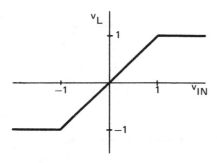

The characteristic required is shown. For v_{IN} greater than 1 volt, a precision diode must be connected from the output terminal to a 1.0-volt source. For the negative voltage situation, a precision diode must be connected from a -1.0-volt source to the output terminal. The 1-volt sources are derived from the two 12-volt supply sources. The resulting circuit is shown.

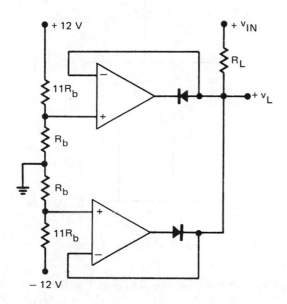

Example 10.2

Obtain the voltage transfer characteristic for the circuit shown.

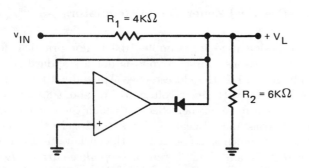

The Thevenin equivalent of the input sources, R_1 and R_2, are needed. R_1 and R_2 form a voltage divider, and the Thevenin equivalent voltage source is

$$\frac{R_2 v_{IN}}{R_1 + R_2} = 0.6 v_{IN}$$

The Thevenin equivalent resistance is the parallel combination of R_1 and R_2.

$$R_{Th} = \frac{R_1 R_2}{R_1 + R_2}$$

The effective source voltage is $0.6 v_{IN}$, so the slope is 0.6. So with the precision diode circuit inactive, the output v_L is $0.6 v_{IN}$. With v_{IN} negative the output is negative, which makes the opamp output positive, so the diode is reverse biased and the precision diode circuit is inactive.

The limiting action of the precision diode prevents v_L from becoming positive, so that with v_{IN} positive, v_L is zero. The voltage transfer charateristic is then as shown.

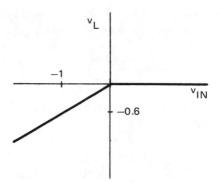

2 VOLTAGE REGULATORS

A. Introduction

The characteristics shown in figure 10.3(b) are similar to voltage regulator characteristics. The accompanying circuit approximates the equivalent circuit of a zener diode (e.g., the characteristics of an ordinary diode operated with reverse voltage). A zener or voltage-regulator diode is manufactured to have a specific breakdown voltage. These devices are used in circuits where the source voltage, v_{IN}, is kept between limits, and the load current must be able to vary from zero to some maximum value.

Zener diodes are used to control the output voltage within a narrow range of values regardless of changes in the source load voltages.

B. Ideal Zener Voltage Regulators

An ideal zener voltage regulator is shown in figure 10.4. (An ideal zener diode maintains a constant output voltage as long as the zener current is positive.)

The source resistance R_s that will result in the lowest power loss while maintaining a constant load voltage during changes of v_{IN} and i_L is needed. Design problems also include finding the power rating of the zener diode, since it absorbs the difference between the maximum and minimum load currents.

From figure 10.4(b), it can be seen that voltage control will not occur unless the ideal diode is forward-biased. This requires that v_L be equal to V_{ZO}, so that $v_{IN} - i_L R_s$ must be at least V_{ZO} (i.e., $i_{IN} = i_L + i_Z$, and i_Z must be zero or positive for the ideal diode to be forward-biased). To ensure regulation, it is necessary that with the maximum possible load current, the ideal diode passes no negative current. This requires

$$R_s \le \frac{v_{IN,min} - V_{ZO}}{i_{L,max}} \qquad 10.6$$

To minimize the power loss in R_s, the equality is taken. The current through the zener diode is

$$i_Z = \frac{v_{IN} - v_Z}{R_s} - i_L \qquad 10.7$$

Equation 10.7 is maximum when v_{IN} is maximum and when i_L is minimum. Then,

$$i_{Z,max} = \frac{v_{IN,max} - V_{ZO}}{R_s} = \left(\frac{v_{IN,max} - V_{ZO}}{v_{IN,min} - V_{ZO}} \right) i_{L,max} \qquad 10.8$$

The zener maximum power is $V_{ZO} i_{z,max}$. The source resistance must absorb a maximum power of $i_{Z,max}^2 R_s$, and the resistor must be rated accordingly.

Example 10.3

Design a voltage regulator circuit to maintain an output voltage of 5 volts when the source voltage varies between 11 and 14 volts, and the load current varies between 0 and 5 amps.

$$R_s = \frac{11 - 5}{5} = 1.2 \, \text{ohms}$$

$$i_{Z,max} = \left(\frac{14 - 5}{11 - 5} \right) 5 = 7.5 \, \text{amps}$$

$$P_D = 5 \times 7.5 = 37.5 \, \text{watts}$$

Therefore, the diode must be rated at 40 watts. Also,

$$P_R = \frac{(14 - 5)^2}{1.2} = 67.5 \, \text{watts}$$

Therefore, the resistor must be rated at 75 watts.

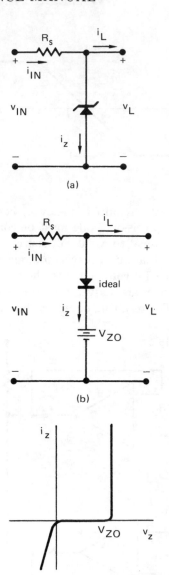

Figure 10.4 (a) Simple Zener Diode Voltage Regulator (b) Equivalent Circuit with Ideal Zener Diode, (c) The Voltage-Current Characteristics of an Ideal Zener Diode

C. Practical Zener Diode Regulators

The design used in example 10.3 is not practical for several reasons. First, 5 volts is not a standard zener voltage. (5.1 is the closest. See Appendix 10.A.) Second, diodes with zener voltage below about 8 volts are controlled by the zener effect in which there is considerable resistance. Because of this, the zener voltage can rise by as much as 20 percent of the basic value at rated current. In addition, there is a *keep-alive current*, I_{ZK}, necessary to maintain the zener characteristics in the linear region of operation. Therefore, the zener current must always be maintained at that level or higher.

As might be expected, the zener diode regulator in a practical circuit may not approximate the ideal case at all. To demonstrate this, consider the equivalent circuit and practical characteristics shown in figure 10.5.

For the circuit of 10.5(a),

$$v_{\text{IN}} - v_L = (i_L + i_Z)R_s \qquad 10.9$$
$$v_L = V_{\text{ZO}} + i_Z R_z \qquad 10.10$$

Solving simultaneously,

$$v_L = \frac{V_{\text{ZO}}R_s + v_{\text{IN}}R_z}{R_s + R_z} - i_L\left(\frac{R_s R_z}{R_s + R_z}\right) \qquad 10.11$$

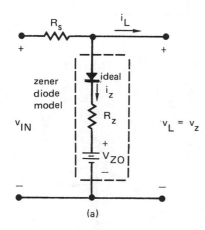

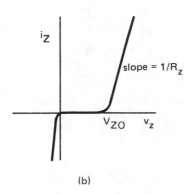

Figure 10.5 (a) Practical Voltage Regulator, (b) Characteristics of Practical Zener Diode

Example 10.4

Determine the regulation for the circuit shown in figure 10.5. Assume $V_{\text{ZO}} = 5.1$ volts, $I_{\text{ZK}} = 0.5$ amp, and $R_z = 0.1$ ohm. From example 10.3, $v_{\text{IN,max}} = 14$ volts, $v_{\text{IN,min}} = 11$ volts, and $i_{L,\text{max}} = 5$ amps.

Because it is necessary to maintain i_Z at 0.5 amp or greater, v_L must be maintained at

$$v_L = V_{\text{ZO}} + I_{\text{ZK}}R_z = 5.15 \text{ volts}$$

This minimum voltage can be designed to occur only when the source voltage is at its minimum and the load current is at its maximum. The 5.15 value for v_L will occur when

$$5.15 = \frac{5.1R_s}{0.1 + R_s} + \frac{11 \times 0.1}{0.1 + R_s} - 5\left(\frac{0.1 \times R_s}{0.1 + R_s}\right)$$
$$R_s = 1.06 \text{ ohms}$$

Substituting into equation 10.11,

$$v_L = 4.66 + 0.086v_{\text{IN}} - 0.091i_L$$

For $V_s = 11$ volts, the load voltage reaches its limiting value at $i_L = 5$ amps. For increased i_L, the zener diode will come out of the breakdown region and cease to regulate. In that case, the load voltage will be simply

$$v_L = v_{\text{IN}} - i_L R_s \qquad 10.12$$

The resulting characteristic of v_L versus i_L for the two extremes of source voltage are shown.

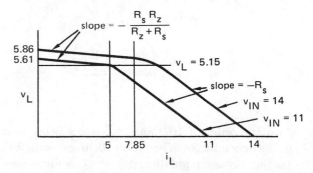

The maximum regulation will occur between the maximum source voltage at no load current, and the minimum source voltage at the maximum load current. The minimum load voltage was found to be 5.15 volts. The maximum load voltage occurs at $v_{\text{IN}} = 14$ and $i_L = 0$. From equation 10.12, the load voltage is 5.86 volts. Thus, the voltage regulation at maximum is

$$\text{regulation} = \frac{5.86 - 5.15}{5.15} \times 100 = 13.8\%$$

D. Operational Amplifier Voltage Regulators

A zener diode regulator is useful when current demands are low but not when there are currents as large as were used in example 10.4. It is possible to use a precision diode circuit to improve the voltage regulation. In general, the operational amplifier will have a limited current capacity. The most usual method to reduce the voltage regulation and still to supply larger currents is to use the zener diode as a reference source for the amplifier. This, in turn, serves a *pass transistor* as illustrated in figure 10.6. This circuit has the added advantage that the zener voltage need not be the same as the desired regulated voltage. The *regulated voltage* can be adjusted by varying the ratio of the resistances R_1 and R_2.

The input to the operational amplifier is $v_Z - kv_L$ where

$$k = \frac{R_2}{R_1 + R_2} \qquad 10.13$$

So, the output is $A_v(v_Z - kv_L)$. The transistor is modeled by its base-emitter P-N voltage drop, and the current gain β is indicated. The circuit analysis of the output circuit results in

$$A_v(v_Z - kv_L) = i_B R_b + v_{BE} + (\beta+1)R_s i_B + v_L \qquad 10.14$$

$$v_L = (1+\beta)i_B(R_1 + R_2) - i_L(R_1 + R_2) \qquad 10.15$$

$$v_L = \frac{A_v v_Z - v_{BE} - i_L\left(R_s + \frac{R_b}{\beta+1}\right)}{kA_v + 1 + \frac{k}{R_2}\left(R_s + \frac{R_b}{\beta+1}\right)}$$

$$\approx \frac{v_Z}{k} - i_L\frac{\left(R_s + \frac{R_b}{\beta+1}\right)}{kA_v} \qquad 10.16$$

A_v is on the order of 10^5, and the terms other than kA_v in the denominator of equation 10.16 are negligible. The output voltage regulation due to i_L is diminished by a factor of $1/kA_v$ from that of the zener regulator of figure 10.5.

The variation of v_Z with v_{IN} is important. With v_{IN} at its minimum value, R_r is set to allow the minimum keep-alive current to flow.

$$R_r = \frac{v_{IN,min} - V_{ZO}}{I_{ZK}} - R_z \qquad 10.17$$

Then

$$v_{Z,min} = V_{ZO} + I_{ZK}R_z \qquad 10.18$$

$$v_{Z,max} = \frac{v_{IN,max}R_z + V_{ZO}R_r}{R_r + R_z} \qquad 10.19$$

Example 10.5

Design an operational amplifier voltage regulator to meet the specifications of $v_L = 5$ volts, $v_{IN,max} = 14$ volts, $v_{IN,min} = 11$ volts, and $i_{L,max} = 5$ amps. Determine the worst-case voltage regulation for the circuit.

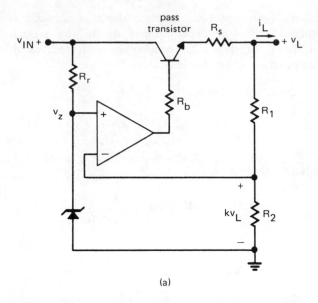
(a)

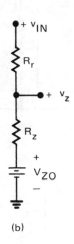
(b)

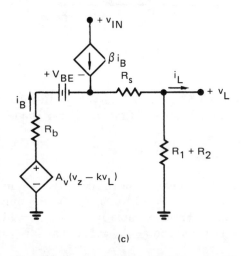
(c)

Figure 10.6 (a) Simple Operational Amplifier Voltage Regulator Circuit, (b) Reference Voltage Circuit, (c) Equivalent Output Circuit

Refer to Appendix A. Begin by choosing a 500 mW zener diode with a voltage of 4.7 volts at the keep-alive current of 1.0 mA and a zener resistance of 19 ohms. From equation 10.18,

$$v_{Z,\min} = 4.7 \text{ volts}$$

$$4.7 = V_{ZO} + I_{ZK}R_z = V_{ZO} + 0.001(19)$$

So,

$$V_{ZO} = 4.681 \text{ volts}$$

At $i_Z = 1.0\,\text{mA}$ and $v_{IN} = 11$ volts, the necessary value for R_r is

$$R_r = \frac{11 - 4.7}{0.001} - 19 = 6281 \text{ ohms}$$

From equation 10.19,

$$v_{Z,\max} = \frac{14 \times 19}{6281 + 19} + \frac{4.681 \times 6281}{6281 + 19} = 4.709 \text{ volts}$$

Next, choose a current limiting resistance that will permit the transistor some voltage (say one volt from collector to emitter) when the maximum load current and minimum source voltage exist simultaneously.

$$R_s = \frac{v_{IN,\min} - V_L - 1}{i_{L,\,\max}} = \frac{11 - 5 - 1}{5} = 1 \text{ ohm}$$

The transistor has to absorb 4 volts at 5 amps when the source voltage is maximum. Therefore, it must be rated at least 20 watts and 5 amps. Such a transistor will have an input resistance h_{IE} of about 2 ohms and a base-emitter turn-on voltage of about 0.7 volt. (At rated current, the base-emitter voltage is about 1.2 volts, and the current gain is about 15 for a high-current transistor. Thus, the base current is about $I_{rated}/15$. h_{IE} is $(v_{BE} - 0.7)/i_B = 1.5$ ohms.)

As R_b includes both the transistor input resistance and the operational amplifier output resistance, choose R_b as 50 ohms total.

With an amplifier gain on the order of 10^5, all the terms in equation 10.16 are negligible except v_Z/k. So,

$$v_L = \frac{v_Z}{k}$$

Setting $v_L = 5$ at the minimum V_{IN},

$$k = \frac{4.7}{5} = 0.940$$

Setting $R_1 + R_2 = 10$ kilohms, $R_2 = 9.4$ kilohms, and $R_1 = 600$ kilohms. This completes the design.

The voltage regulation is

$$\text{regulation} = \frac{v_{Z,\max} - v_{Z,\min}}{v_{IN,\min}} \times 100$$

$$= \frac{4.709 - 4.7}{4.7} \times 100 = 0.19\%$$

This is an improvement over example 10.4. Additional improvement is possible by using either an additional higher-voltage zener diode to supply the reference zener, or a higher-voltage reference zener and a voltage divider.

3 LOGIC CIRCUITS

A. Introduction

In the previous chapter, the analysis of small signal behavior for amplifying devices assumed that small input signals would not cause radical departures from the operating point. The devices were assumed to operate within their linear regions. Departure from these conditions can lead to saturation or cutoff.

With *saturation*, the input becomes large enough to drive the output current to a limiting value (usually a voltage so low that the output current is limited only by the external resistance of the circuit). *Cutoff* is also a limiting case, but it usually refers to the output current. In the cutoff state, the output appears as an open circuit.

Figure 10.7 shows the saturation and cutoff regions for an NPN transistor. The transistor saturates when the base current is 0.6 mA or more; it is cutoff for a base current of 0.0 mA. The required saturation current to the base is a function of the load line. When the base-emitter voltage is negative (or less than about 0.6 volt), the transistor is cut off.

Most logic circuits use these two extreme regions to define the two states needed in binary logic.

B. Logical NOT Inverter Circuits

The common emitter amplifier is the basic NOT (*inverter* or *complement*) circuit, as shown in figure 10.8. (This is essentially the same as figure 10.7.) If the applied voltage to the base is low (less than 0.6 volt), the transistor will be cut off, and no collector current will flow. In this condition, the collector voltage will be high. If the base-emitter voltage is high, the transistor will be saturated. The collector-emitter voltage will be low (on the order of 0.2 volt). Such inverter circuits are designed so that the input to one stage is the output of an identical stage.

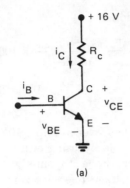

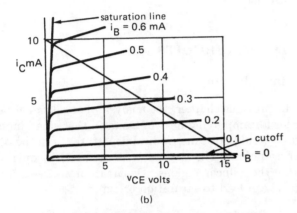

Figure 10.7   Saturation
and Cutoff Characteristics

The analysis of the circuit shown in figure 10.8 is straightforward. The base current of transistor T2 must be sufficient to saturate T2 when T1 is cut off (i.e., when v_{CE} is high). When T1 is saturated (i.e., when v_{CE} is less than 0.2 volt), the base voltage of T2 will be less than 0.7 volt, and T2 will be cut off.

In practical problems, single transistors must be able to drive multiple inputs to subsequent stages. Thus, when a transistor is cut off, it may have to provide saturation input current to a number of other identical circuits. This is a major design consideration. The ability of a transistor to drive other circuits is called its *fan-out*.

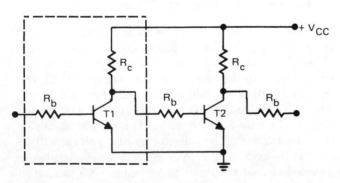

Figure 10.8   NOT Gates

Example 10.6

A transistor inverter has the following parameters: $i_{B,\text{sat}} > 0.5$ mA and $v_{BE,\text{sat}} < 1.5$ volts. $V_{CC} = 5$ volts and $R_c = 500$ ohms. How many identical inverter circuits can be driven by such an inverter drive if the input base resistance is 1000 ohms?

Using the circuit of figure 10.8, the source current must divide among N identical circuits. All identical circuits must be provided with the minimum base current of 0.5 mA when the base voltage is 1.5 volts. Then, recognizing that $i_C = N i_{B,\text{min}}$ and using Kirchhoff's voltage law across R_b and R_c,

$$V_{CC} = N i_{B,\text{min}} R_c + i_{B,\text{min}} R_b + v_{BE,\text{sat}}$$
$$5 = N(0.0005 \times 500) + (0.0005 \times 1000) + 1.5$$

The fan-out is $N = 12$.

When the inverter is on, the collector-emitter voltage is typically 0.2 volt. Both the collector-base and the emitter-base junctions will be forward-biased. The collector must be able to accept the reverse saturation currents of the load transistors without coming out of saturation. This also limits the number of loads a gate can handle and thereby affects the fan-out.

Logic gates are typically specified in terms of the current they can source (i.e., the current they can supply when the collector voltage is high) and the current they can sink (i.e., the current the collector or output circuit can accept from loads in the low state without being pulled out of saturation).

Within a particular logic family, this may be specified in terms of *unit loads*. A *unit source current* is the input current drawn by a single gate when the input voltage is high. A *sinking unit load* is the current drawn from the input of a single gate when the input voltage is low. The fan-out will be the minimum of these two unit loads that the output of the gate can provide.

Consider figure 10.9. Gate 1 sources a unit load by providing base current to gate 2. With the base of T1 low (v_{BE} less than 0.6 volt), T1 will be cut off. Since this appears as an open circuit to the output, the current to the base of T2 will be

$$i_{B2} = \frac{V_{CC} - v_{BE2}}{R_c + R_b} \qquad 10.20$$

Therefore, i_{B2} is the current being sourced by gate 1.

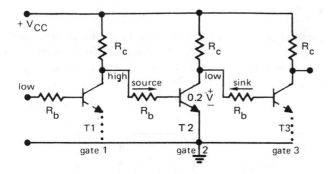

Figure 10.9 Unit Loads for Fan-Out

Gate 2 will be saturated so that the collector-emitter voltage is 0.2 volt. This results in the transistor of gate 3 being cut off. The base will draw reverse saturation current from the reverse-biased collector-base junction. This unit sink current is over and above the current being drawn from V_{CC}. If I_{COE3} is the unit sink current, the collector current of T2 is

$$i_{C2} = \frac{V_{CC} - 0.2}{R_c} + I_{COE3} \qquad 10.21$$

For multiple loads, the output circuits are as shown in figure 10.10. Figure 10.10(a) illustrates the sourcing configuration. With $v_{BE} \approx V_{BEF} \approx 0.7$ volt, the current necessary to maintain saturation will require the common emitter characteristic as in figure 10.7 to determine $i_{B,min}$ for a required i_C. The necessary value of i_C will depend on the number of loads to be sinked, so that the design problem is based on both the sinked and sourced currents. The minimum value of i_B to hold the transistor in saturation will depend on the number of loads being served. The saturation collector current is

$$i_{C,sat} = \frac{V_{CC} - v_{CE,sat}}{R_c} + NI_{COE} \qquad 10.22$$

N is the number of load gates, $v_{CE,sat} \approx 0.2$, and I_{COE} is the reverse collector current with the emitter terminal open. The load line is shown in figure 10.11. $i_{B,min}$ can be obtained from this figure.

For the sourcing condition of figure 10.10(a), the current through R_c will include the reverse saturation current I_{COE} for the cutoff transistor.

$$\frac{V_{CC} - v_{CE}}{R_c} - i_C = N\left(\frac{v_{CE} - V_{BEF}}{R_b}\right) \qquad 10.23$$

$V_{BEF} \approx 0.7$ volt to hold the load transistors in saturation, and $i_C = I_{COE}$. The load-line equation is obtained from equation 10.23.

$$i_C + \left(\frac{1}{R_c} + \frac{N}{R_b}\right)v_{CE} = \frac{V_{CC}}{R_c} + \left(\frac{N}{R_b}\right)V_{BEF} \qquad 10.24$$

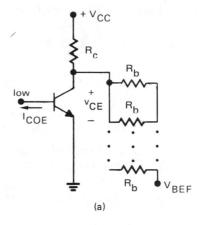

(a)

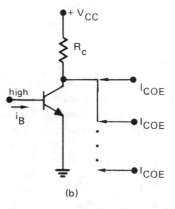

(b)

Figure 10.10 Circuits for Calculating (a) Source Loads, (b) Sink Loads

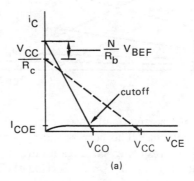

(a)

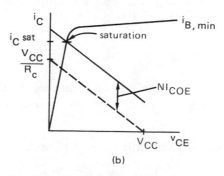

(b)

Figure 10.11 Load Lines for (a) Cutoff, and (b) Saturation of an Inverter Gate with N Similar Loads

For each load input, the base current is

$$i_B = \frac{v_{CE} - V_{BEF}}{R_b} \qquad 10.25$$

The simultaneous solution of equations 10.24 and 10.25 is

$$i_C + \left(N + \frac{R_b}{R_c}\right) i_B = \frac{V_{CC} - V_{BEF}}{R_c} \qquad 10.26$$

The design problem is formulated by setting $i_C = I_{COE}$ and solving for N, using equations 10.22 and 10.26,

$$N \le \frac{R_c i_{C,\text{sat}} + v_{CE,\text{sat}} - V_{CC}}{R_c I_{COE}} \qquad 10.27$$

$$N + \frac{R_b}{R_c} \le \frac{V_{CC} - V_{BEF} - R_c I_{COE}}{i_{B,\min} R_c} \qquad 10.28$$

Example 10.7

A transistor with the characteristic shown has $R_c = 1000$ ohms and $V_{CC} = 5$ volts. Determine the base current necessary for a fan-out of five.

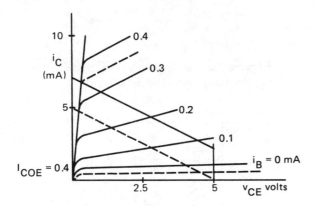

$I_{COE} = 0.4$ mA, so five load gates require $5 \times 0.4 = 2$ mA excess current. The corresponding load line is shown solid. The resulting saturation current requires a base current of about 0.35 mA. Setting the value at 0.4 mA gives a margin of error for temperature effects and variations between transistors.

Example 10.8

Determine the maximum base resistance R_b that will allow sourcing five gates from the transistor gate of example 10.7.

From equation 10.28 with $V_{BEF} = 0.7$ volt,

$$5 + \frac{R_b}{1000} \le \frac{5 - 0.7 - 0.4}{1000 \times 0.0004}$$
$$R_{b,\max} = 4750 \text{ ohms}$$

C. Gate Delays

Switching between saturation and cutoff does not occur instantaneously. Figure 10.12 shows the relationships between input and output voltages versus time for purposes of identifying the delays.

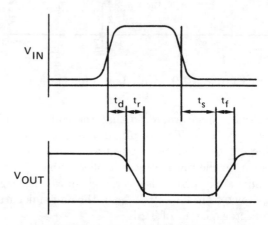

Figure 10.12　Gate Delays

When the input voltage rises in an inverter gate, it will take some time for the base-emitter voltage to reach the *turn-on level*. This is called the *delay time*, t_d. It may range from 1 to 10 nanoseconds (ns). Typical values are 2 to 4 ns.

The time for the collector output voltage to fall to the saturation level is called the *rise time*, t_r. It can range from 1 to 100 ns, with a typical value being 50 ns.

When the transistor is driven well into saturation, an excess charge is stored. There will be a time after the base current has dropped before this excess charge is cleared out of the base region. Then, the collector current will drop, allowing the output voltage to rise. The delay is called the *storage time*, t_s. It can range from 5 to 50 ns.

Once the excess charge has been removed from the base, the collector current can begin to fall. The time interval involved here is called the *fall time*, t_f. It can range from 1 to 100 ns, with a typical value of 30 ns.

The minimum time for a transistor to pass a pulse (i.e., the time from saturation to cutoff and back to saturation, or from cutoff to saturation and back to cutoff) is the sum of these delays. Therefore, the maximum frequency at which a transistor can operate is the reciprocal of the sum of the delays.

D. Logical Functions

Logical functions can be obtained by combining gates. The basic logical functions are inversion (NOT), logical sum (OR), and logical product (AND). Logical variables and functions have two possible values: true (1) and false (0).

Definition of functions is often done with a *truth table*. The function value is given for each combination of the input variables.

Table 10.1
$C = \overline{A}$ truth table
(NOT)

A	C
0	1
1	0

Table 10.2
$C = A + B$ truth table
(OR)

A	B	C
0	0	0
0	1	1
1	0	1
1	1	1

Table 10.3
$C = A \cdot B$ truth table
(AND)

A	B	C
0	0	0
0	1	0
1	0	0
1	1	1

The most commonly occurring functions in transistor-transistor logic are the negations of AND (NAND) and OR (NOR).

Table 10.4
$C = \overline{A + B}$
(NOR)

A	B	C
0	0	1
0	1	0
1	0	0
1	1	0

Table 10.5
$C = \overline{A \cdot B}$
(NAND)

A	B	C
0	0	1
0	1	1
1	0	1
1	1	0

A basic operation in Boolean algebra is De Morgan's theorem, which can be written in two ways.

$$\overline{A \cdot B} = \overline{A} + \overline{B} \qquad 10.29$$

$$\overline{A + B} = \overline{A} \cdot \overline{B} \qquad 10.30$$

Two useful identities are repeated here.

$$A \cdot \overline{A} = 0 \qquad 10.31$$

$$A + \overline{A} = 1 \qquad 10.32$$

The symbols for logic gates are shown in figure 10.13.

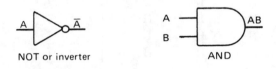

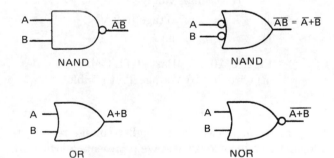

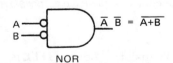

Figure 10.13 Gate Symbols

4 TYPES OF LOGIC

A. Resistor-Transistor Logic (RTL)

The inverter illustrated in figure 10.8 is an RTL circuit, the basic building block for RTL logic. RTL was popular prior to large-scale integration. The fundamental logical gate is the NOR circuit shown in figure 10.14.

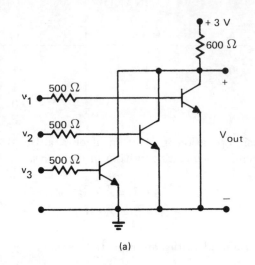

(a)

v_1	v_2	v_3	v_{OUT}
L	L	L	H
L	L	H	L
L	H	L	L
L	H	H	L
H	L	L	L
H	L	H	L
H	H	L	L
H	H	H	L

H = high voltage (3 V)

L = low voltage (0.2 V)

(b)

Figure 10.14   Basic RTL Gate:
(a) Circuit, (b) Voltage Truth Table

For *positive logic* (where the higher voltage represents a logical 1 and the lower voltage represents a logical 0), the circuit and truth table correspond to a NOR circuit. However, the choice of positive or *negative logic* is often arbitrary.

B. Diode-Transistor Logic (DTL)

The basic DTL circuit is shown in figure 10.15. The output voltage, v_C, is high, H, in all cases except when both inputs are low, L. A desirable feature of this circuit is that when sourcing a load, no current is required from the output. On the other hand, the current necessary to sink the load when the output voltage is low is much more than is required by the RTL circuit. The two diodes in series with the base ensure that the transistor will not be turned on when only one input is low.

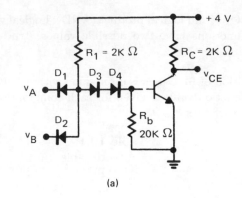

(a)

v_A	v_B	v_C
L	L	H
L	H	H
H	L	H
H	H	L

(b)

Figure 10.15   Basic DTL Gate:
(a) Circuit, (b) Voltage Truth Table

C. Transistor-Transistor Logic (TTL)

A very fast logic which uses saturation switching (from cutoff to saturation) is the TTL family. The basic two-input circuit is shown in figure 10.16 with an accompanying truth table. For positive logic, the basic gate is the NAND gate.

The basic TTL NAND gate can be considered a modification of the DTL NAND gate, with the input diodes replaced by a multi-emitter transistor's base-emitter junctions. The two series diodes are replaced by the base-collector junction of the multi-emitter input transistor and by the base-emitter junction of T2.

The behavior of the input transistor is unique because bias voltages are not provided. When one or both of the input voltages are low, L, the base voltage of T2 is pulled down to about 0.8 volt. This voltage attempts to forward bias the string of P-N junctions (T1: B-C, T2: B-E, and T3: B-E) to ground. However, the three junctions required $3 \times 0.7 = 2.1$ volts in order to turn on. So, T3 is cutoff and v_C is high, H.

When both of the input voltages are high, H, the base of T1 is connected to V_{CC}. This is high enough to bias all three junctions to ground in the forward direction. T2 acts as an emitter-follower, providing the current necessary to hold v_C down to 0.2 volt, L.

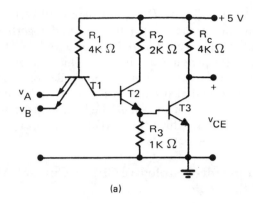

(a)

v_A	v_B	v_C
L	L	H
L	H	H
H	L	H
H	H	L

(b)

Figure 10.16 Basic TTL Gate:
(a) Circuit Diagram, (b) Voltage Truth Table

Example 10.9

Determine the power required of the two-input TTL stage shown in figure 10.16 if the fan-out is five.

When v_C is low (0.2 volt), the base-emitter voltage of T3 is 0.7 volt. The base-emitter voltage of T2 and the base-collector voltage of T1 will also be 0.7 volt. T2 can be assumed to be saturated since its base current will be 0.725 mA and its saturated current is 2.05 mA. Then, the DC equivalent circuit shown can be used. The current through R_c will be 1.2 mA.

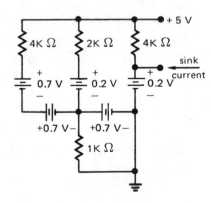

The unit load must be found from the figure shown. The low-output power will be determined after the power for high-output is found.

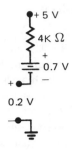

With output high at T3, both T2 and T3 are assumed off. (Even with the emitter of T2 at ground, the voltage at the base of T1 would have to be 1.4 volts to turn on T2, allowing for a 0.7 volt drop across the two P-N junctions.) At least one of the inputs must be low (0.2 volt). With the base-emitter voltage of T1 at 0.7 volt, the voltage across R_1 is 4.1 volts. Then, the unit current is

$$\frac{4.1}{4000} = 1.025 \text{ mA}$$

As the reverse collector saturation current will be on the order of 0.1 μA, the power drain due to collector saturation will be negligible compared to the 1.025 mA through R_1.

The power in the high output state (lost in this stage) is

$$(V_{CC} - v_{IN})i_{R_1} = (5 - 0.2) \times 1.025 = 4.92 \text{ mW}$$

Returning to the low output state, the unit current from one load will be divided among the sourcing gates that have low outputs. The worst case of sinking is when there is only one sourcing gate (i.e., only one low input). This single gate would receive all of the 1.025 mA unit current.

Thus, for five loads in the low-output state, the maximum sink current would be

$$5 \times 1.025 = 5.125 \text{ mA}$$

The power dissipated for this sinked current is

$$0.2 \times 5.125 = 1.025 \text{ mW}$$

The power directly from V_{CC} to the stage (exclusive of the sinked current) is

$$5 \times (0.725 + 2.05 + 1.2) = 19.875 \text{ mW}$$

The total power dissipated in the stage for low output is

$$19.875 + 1.025 = 20.90 \text{ mW}$$

D. TTL Output Circuits

A method of improving the fall time of TTL gates is to reduce the collector current in the saturated state. Circuits that accomplish this are shown in figure 10.17. The addition of T4 and D disconnects the collector of T3 from V_{CC} when T4 is off. For T3 to be saturated, T2 must also be saturated, which brings the base of T4 to 0.8 volt. This would be just enough to turn on T4 if the diode D were not present.

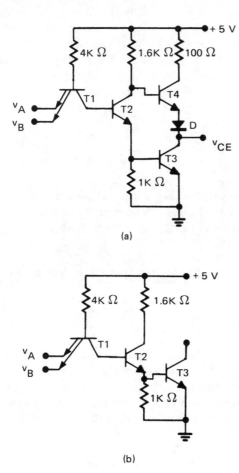

(a)

(b)

Figure 10.17   Alternative TTL Output Circuits: (a) Totem-Pole, (b) Open Collector

However, the presence of diode D requires that 1.4 volts be applied to the base of T4 in order for it to be turned on. Thus, with T3 saturated, T4 is cut off, disconnecting T3 from V_{CC} and requiring a collector current only from the loads. When T3 is cut off, T2 is also cut off and the 1.6K resistance turns on T4 and so pulls up the collector voltage of T3. This *totem-pole configuration* results in a lower storage time and faster turn-off of the output transistor.[1]

---

[1] Totem-pole is also known as *active pull-up*.

The *open-collector TTL gate* (figure 10.17(b)) is used when two or more outputs are wired together. This results in a so-called *wired OR* or *wired AND* configuration. (Depending on the viewpoint, it may be called either.) When two or more open-collector outputs are connected and tied through an externally added resistor to V_{CC}, all outputs must be high in order to obtain a high output. This is an AND function from the point of view of the outputs.

E. Emitter-Coupled (ECL) or Current-Mode (CML) Logic

The speed limitation on TTL is due to the saturation switching. A faster configuration is based on the emitter-coupled amplifier which does not drive the transistor into saturation. The basic circuit is shown in figure 10.18. Outputs are taken through emitter-followers. This accounts for the difference in input and output voltages in the truth table. ECL can be up to an order of magnitude faster than TTL, but it is much more demanding of power than TTL.

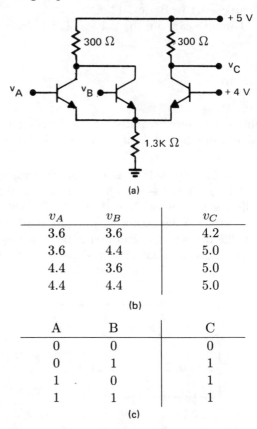

(a)

v_A	v_B	v_C
3.6	3.6	4.2
3.6	4.4	5.0
4.4	3.6	5.0
4.4	4.4	5.0

(b)

A	B	C
0	0	0
0	1	1
1	0	1
1	1	1

(c)

Figure 10.18   ECL or CML Basic Gate: (a) Circuit Diagram, (b) Actual Voltage Table, (c) Truth Table

F. PMOS Logic Gates

PMOS stands for P-channel enhancement mode MOS-FET. This logic is used in integrated circuit form, requiring no external resistors or capacitors. The attraction of this logic is its simplicity of manufacture. It requires only one diffusion into N-type substrates, plus three additional steps for oxidation and metalization. PMOS is slower than bipolar junction transistor technology, but much higher gate densities are possible.

A PMOS inverter is shown in figure 10.19. The upper transistor is biased to an on condition. It serves as a somewhat nonlinear resistance. The substrate lead is connected to the most positive voltage available (the ground in figure 10.19) to insure that the P-N junctions between drain and substrate and between source and substrate remain reverse-biased.

Negative logic is typically used in PMOS logic. For negative logic, $-V_{DD}$ corresponds to a logical 1 and zero volts corresponds to a logical 0.

When the gate of T1 is at ground, the output voltage approaches V_{DD}, a negative voltage corresponding to a logical 1. When the gate voltage has a logical 1, (i.e., a negative voltage applied), the source is positive with respect to the gate. T1 will then be turned on, and the output voltage will approach ground, the logical 0 level.

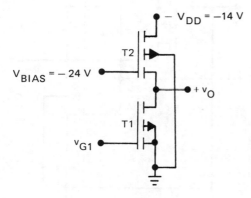

Figure 10.19 PMOS Inverter with Negative Logic

The circuit of figure 10.20 shows a basic PMOS NOR gate. The output is pulled toward ground if either gate v_A or v_B is brought to the 1 voltage (i.e., a negative voltage).

G. CMOS Logic Gates

It is possible (and, in some cases, worthwhile) to employ N-channel enhancement mode MOSFETs in a manner similar to the PMOS. That is called NMOS and is faster than PMOS because electron mobility is about 2.5 times hole mobility. NMOS is more difficult and expensive to manufacture, so it has not been as popular as PMOS.

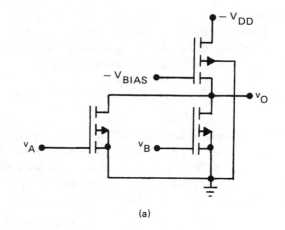

v_A	v_B	v_C
0	0	$-V_{DD}$
0	$-V_{DD}$	0
$-V_{DD}$	0	0
$-V_{DD}$	$-V_{DD}$	0

(b)

Figure 10.20 (a) PMOS NOR Gate with Negative Logic, (b) Truth Table

A combined technology using both N-channel and P-channel enhancement mode MOSFETs has proven to be very useful. It requires much less power than the single-types of logic circuits, yet is relatively simple to fabricate. This is called CMOS for complimentary MOSFETs. CMOS requires only a single supply voltage, which allows for additional simplifications.

The basic CMOS inverter is shown in figure 10.21. The upper transistor is P-channel with its substrate lead connected to the most positive voltage available (+10 volts). The lower transistor is N-channel with its substrate lead connected to the most negative voltage available. These connections are made to maintain reverse bias on the substrate P-N junctions.

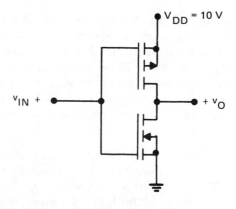

Figure 10.21 CMOS Inverter

When the input voltage is high (near +10 volts), the upper transistor has $v_{GS} = 0$ (or at least lower than the turn-on voltage of the upper transistor), so the upper transistor is off. The lower transistor will be turned on by the high v_{GS}, and the output terminal will be grounded through the lower transistor's on-resistance.

When the input voltage is low (near zero volts), the lower transistor is turned off, breaking the path to ground from the output terminal. The upper transistor has a large negative v_{GS} and, being a P-channel device, is turned on. The output terminal is connected to V_{DD} through its internal on-resistance.

The only power supply current that flows is that which is necessary to charge up the gate-channel capacitances of load stages (corresponding to the input). Some current also leaks through the off-transistor. Thus, power consumption is extremely low, but the speed is limited by the time constants associated with the charging of the off-transistor gates.

Typical CMOS logic gate construction is shown in figure 10.22. For NOR logic (figure 10.22(a)), a positive input at either terminal will turn on one of the lower transistors while turning off one of the upper. Thus, if either or both of the inputs are high, the output is low. Both inputs must be low to turn on the series combination and to connect V_{DD} to the output.

In figure 10.22(b), a low voltage at either input terminal turns on at least one of the parallel upper transistors while turning off a least one of the series lower transistors. Thus, a low voltage on either or both of the inputs results in a high output voltage. A high voltage on both inputs is necessary to turn off both upper transistors and to connect the output to ground through the two lower transistors.

5 ANALOG SWITCHES

A. Introduction

Electronic analog switches can be controlled by logic inputs (i.e., TTL levels of less than 0.8 volt low and about 3 volts high). They are designed to switch analog signals up to about 75 percent of the DC supply voltage in electronic circuits. Switches are used in sample-and-hold circuits, programmable amplifiers, digital-to-analog converters, and a wide variety of other applications.

Analog switches make use of the on-resistance of FET channels. The channel resistances range from a few ohms for VMOS, to 10 or 100 ohms for JFETs and enhancement mode MOSFETs. Because channel resistances vary with current in the channel, it is

almost always necessary that the series or load resistance be high in comparison with the nominal channel resistance.

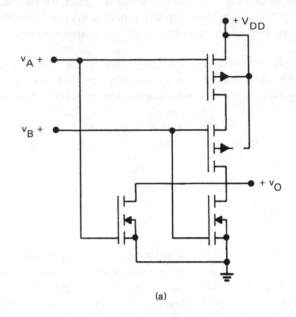

(a)

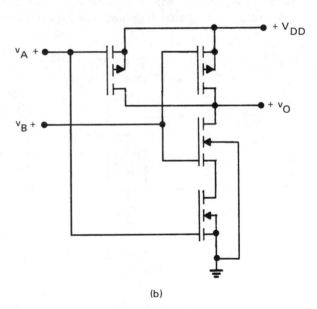

(b)

Figure 10.22 Positive Logic CMOS (a) NOR Gate, (b) NAND Gate

B. CMOS Transmission Switch

The CMOS switch is similar to CMOS logic. The basic switch is shown in figure 10.23. CMOS logic level voltages are expected as the control signal, v_C. With v_C low and no signal present, both parallel switch channels are turned off. Either terminal of the channel can be the source (or drain), depending on the polarity of the signal being switched.

Consider the off case first. v_C is virtually zero. This level is applied to the lower transistor in the switching pair. For this transistor to turn on, both terminals must be negative by at least the turn-on voltage of v_{GS} (from 1 to 3 volts). Thus, to maintain an off condition, it is necessary that at least one terminal be connected to a potential more positive than about -1 volt. This requirement is satisfied whenever one terminal is connected to a resistive load.

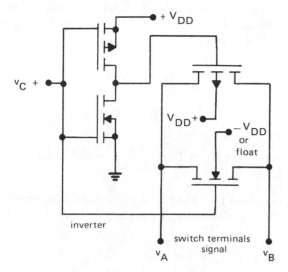

Figure 10.23 CMOS Switch

With the lower transistor still in the off condition, consider the upper (P-channel) transistor in the switching pair. The gate is near the V_{DD} level, so the channel cannot be activated unless both terminals are more positive than the supply voltage, V_{DD}.

When the control signal is high (i.e., nearly V_{DD}), the lower channel will be turned on for negative signals and for signals up to within a few volts of V_{DD}. The upper channel will turn on when the switching voltages are sufficiently positive to make v_{GS} sufficiently negative to turn on the P-channel. With proper design, the upper channel will compensate for the turning off of the lower channel for positive switching voltages. With proper design of the load impedance, the difference between v_A and v_B will be small when the switch is on.

The limitations on switching are in the switched voltage when the switch is off.

C. JFET Analog Switches

Another analog switch device is developed from depletion mode P-channel JFETs. These are on when the gate-source voltage is zero. A positive voltage from gate to channel is required to turn them off. This is the *pinchoff voltage* $V_{GS(off)}$. A major problem with

an ordinary JFET is that it cannot have any forward bias on the gate-channel P-N junction. For this reason, JFETs for switching application use a reverse diode in series with the gate lead as indicated in figure 10.24.

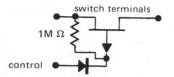

Figure 10.24 JFET Analog Switch

An advantage of JFET logic over CMOS logic is in a lower on-resistance. Disadvantages are complexity and price. The JFET is restricted to switching voltages that are at least a few volts less than the control voltage, as larger voltages can turn the switch on. Also, a sufficiently large signal while the switch is in the on condition can turn the channel off.

D. VMOS Analog Switches

VMOS transistors are available as an N-channel (enhancement mode) switch. It is, therefore, used only for switching of positive voltages (e.g., when the drain is more positive than the source). VMOS transistors will operate with negative voltages only to about $-V_{GS(on)}$. Their advantage is the lower channel resistance, typically a few ohms, in the on condition.

6 POWER SWITCHING

A. Introduction

Power switching up to approximately 50 amps can be accomplished with power bipolar junction transistors. VMOS transistors can be used up to approximately 10 amps. Switching is typically DC, but bridges for control of AC power also require switching. Hold-off voltages for these devices may be as high as 500 volts with power dissipation up to 200 watts. The power controlled can be many times the device power ratings, as the higher currents occur at the lowest voltage level.

Control of AC power is still accomplished, in a large part, by the use of SCRs (silicon controlled rectifiers) and TRIACs (bi-directional SCRs). These are designed for high (up to 1000 volts) hold-off voltages and currents up to 80 amps.

The GTO (Gate Turn-Off) is a special SCR which can be turned off from the gate. However, it requires substantial voltage (about 70 volts) to do so. It does not rely on the current falling below an extinguishing value, as is done in SCRs. The turn-on voltage for GTOs, as for SCRs and TRIACs, is about 2.5 volts.

B. Thyristor Control

The SCR has the turn-on characteristic shown in figure 10.25. With no gate current, the characteristic indicates the maximum voltage V_{BO} which can be applied without the SCR breaking down. With increasing gate current i_G, the breakdown voltage decreases.

At $i_G = I_{GT}$, the SCR behaves essentially like a P-N diode. I_{GT} is a DC current. It occurs in conjunction with a gate-cathode (main terminal 1) voltage from 1 to 2.5 volts. The precise value of I_{GT} is specified by the manufacturer, as is V_{GT}.

The minimum value of DC current that will turn on the SCR is I_{GT}. This is not sufficient in a pulse condition. In general, I_{GT} must be applied for about 50 μs to fully turn on the SCR. For pulse turn-on, the total charge must approximately equal the current-time product ($I_{GT} \times 50$ μs). This current will not flow unless the voltage exceeds V_{GT}.

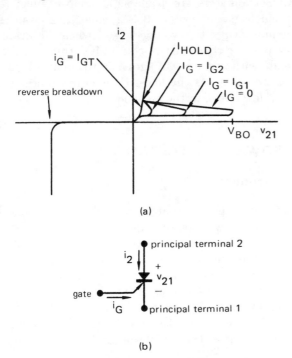

(a)

(b)

Figure 10.25 (a) SCR and GTO Characteristics, (b) SCR Symbol

Example 10.10

An SCR has $I_{GT} = 80$ mA. Determine the current necessary to turn on this SCR in 1 μs. Also determine the voltage on a 0.1 μF capacitor necessary to deliver this charge to the gate in 1 μs.

The total charge needed in 1 μs will be $I_{GT} \times 50$ μs or

$$0.08 \times 50 = 4 \ \mu C$$

The average current will be

$$\frac{4 \ \mu C}{1 \ \mu s} = 4 \ \text{amp}$$

The capacitor will discharge from its initial value to V_{GT} (assumed to be 2.5 volts). Thus, the charge to be delivered will obey

$$C(V_{INITIAL} - 2.5) = 4 \ \mu C$$

With $C = 0.1$ μF, the initial voltage is

$$V_{INITIAL} = 2.5 + \frac{4 \ \mu C}{0.1 \ \mu F} = 42.5 \ \text{volts}$$

This assumes that the RC time constant will permit the capacitor to discharge in 1 μs.

C. UJT Trigger for SCRs

A common trigger device for SCRs is the *unijunction transistor*, UJT, sometimes called a *double-base diode*. This device can hold a capacitor voltage off until it reaches a critical value. Above the critical value, the UJT releases all stored charge. The charge can be directed to the gate of an SCR as shown in the typical circuit of figure 10.26.

The UJT is structurally similar to the N-channel JFET. The difference is that the P-region is concentrated near base B2. The lead to the P-region is called the *emitter*. While the P-N junction is reverse-biased, the voltage v_1 is simply the *voltage division* of the voltage v_{21}. The voltage division ranges from 0.5 to 0.8, with a typical value of 0.75. The resistance $R_1 + R_2$ varies from 1 kilohm to 10 kilohms, with a typical value of 5 kilohms.

When the junction voltage ($v_E - v_1$) exceeds the usual P-N forward voltage of about 0.7 volt, the E-B1 combination becomes a forward-biased diode. The resulting v_{EB1} versus i_F characteristics are shown in figure 10.26(d). I_{HOLD} is the minimum current necessary to keep the UJT in the triggered or on condition. Below this value, it will revert to the untriggered state as indicated by the arrows.

Note the similarity of the circuit symbol to an N-JFET symbol. The difference is the bent P-lead, which points toward base B1.

Example 10.11

A UJT circuit as shown in figure 10.26(a) has $R_1 = 4$ kilohms, $R_2 = 1$ kilohm, $R_{b1} = 1$ kilohm, and $V_{BB} = 24$ volts. Determine the triggering voltage.

The UJT will trigger when

$$v_E \geq V_{BB} \left(\frac{R_1 + R_{b1}}{R_1 + R_{b1} + R_2} \right) + 0.7$$

The 0.7 volt accounts for the usual diode drop.

$$V_{TRIGGER} = 24 \left(\frac{4+1}{4+1+1} \right) + 0.7 = 20.7 \text{ volts}$$

Example 10.12

The SCR of example 10.10 had $I_{GT} = 80$ mA and $V_{GT} = 2.5$ volts. Determine the capacitance necessary to trigger the SCR if the UJT trigger voltage is 21.2 volts.

The trigger voltage is 21.2 volts, but 0.7 volt is used for the P-N junction. 2.5 volts are needed to drive the gate of the SCR. This leaves 18.0 volts for charge transfer. The charge needed was found in example 10.10 to be 4 μC, so

$$C_e = \frac{4 \times 10^{-6}}{18.0} = 0.22 \ \mu\text{F}$$

D. DIAC Trigger for TRIAC Circuits

A TRIAC is essentially a bi-directional SCR which can be triggered for either polarity of v_{21} (with a corresponding polarity of V_{GT}). The DIAC is a TRIAC with no gate. It is designed to have a controlled *breakover voltage*, V_{BO}. The forward and reverse breakdown voltages are nearly equal. Typical breakover voltages range from 25 to 40 volts. Fallback voltages are approximately 9 volts, as indicated in figure 10.27.

The firing angle of the TRIAC in figure 10.27(b) is controlled by resistance R_c. The capacitance C_c releases charge through the DIAC to swing the capacitor voltage from the breakover value to the *dynamic on-resistance* region (9 volts in figure 10.27(a)). Once the DIAC has triggered, the capacitor voltage will remain at the dynamic on value until the AC source voltage has reversed. The waveforms for the control voltage v_C across the capacitor are shown together with other waveforms in figure 10.28.

Before the TRIAC is triggered, only the current through R_c and C_c flows in the load. As R_c is much greater than the load impedance, it can be assumed that all of v_{IN} appears across R_c and C_c. At the beginning of

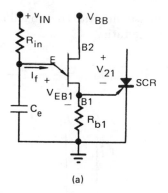

(a)

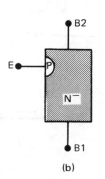

(b)

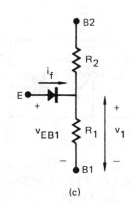

(c)

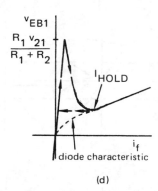

(d)

Figure 10.26 (a) UJT-SCR Trigger Circuit, (b) UJT Construction, (c) Equivalent UJT Circuit Prior to Triggering, (d) Characteristic

each half-cycle, the capacitor has the dynamic DIAC voltage across it from the last half-cycle firing. Therefore, it is opposite in polarity from the next firing (V_{BO}) value. The phasor voltage, V_c, across the capacitor in a steady-state phasor analysis tries to follow V_{in} according to the impedance ratio.

$$V_{c,\text{steady state}} = \frac{V_{in}}{j\omega RC + 1} \qquad 10.33$$

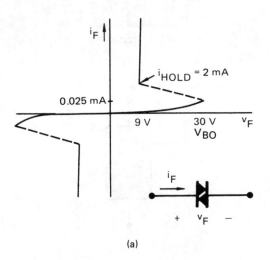

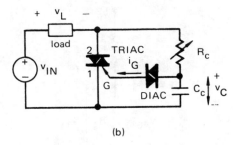

(b)

Figure 10.27 (a) DIAC Symbol and Characteristic, (b) Simple Triggering Circuit for AC Power

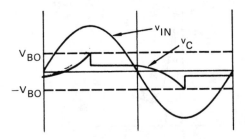

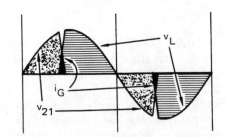

Figure 10.28 DIAC Trigger Waveforms

However, the initial charge on the capacitor must go through a transient to approach the value specified by equation 10.33. Details of this transient are shown in figure 10.28. The firing angle will be where V_c reaches the breakover voltage, V_{BO}. The worst case for the firing angle is when the transient has completely died out before the breakover voltage is reached.

When the transient can be ignored (e.g., when it is completed before v_C reaches V_{BO}), the trigger point can be found from the expression of V_c in steady state.

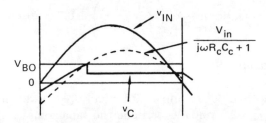

Figure 10.29 Dashed Sinusoid Showing the Steady-State Curve that the Capacitor Voltage Attempts to Follow Before the DIAC Breaks Over

$$\begin{aligned} v_{C,\text{sinusoidal steady state}} &= \frac{v_{IN,\text{max}}}{\sqrt{\omega^2 R_c^2 C_c^2 + 1}} \\ &\quad \times \sin[\omega t_{tr} - \arctan(\omega R_c C_c)] \\ &= V_{BO} \qquad 10.34 \end{aligned}$$

Equation 10.34 is a first approximation to actual performance. Complete analysis of the circuit, including initial conditions, yields equation 10.35.

$$\begin{aligned} v_C(t) &= v_{C,\text{sinusoidal steady state}} \\ &\quad + \left[v_C(0) + \frac{v_{IN,\text{max}}\omega R_c C_c}{\omega^2 R_c^2 C_c^2 + 1} \right] e^{-t/R_c C_c} \\ &\qquad\qquad\qquad\qquad\qquad\qquad 10.35 \end{aligned}$$

$v_C(t)$ must reach V_{BO} at $t = t_{tr}$. Equation 10.35 must be solved by iteration beginning with the approximation of equation 10.34. The angles in equations 10.34 and 10.35 are in radians.

Example 10.13

The circuit of figure 10.27(b) uses a DIAC with the characteristics of figure 10.27(a). $R_c = 10$k and $C_c = 0.1$ μF. V_s has an RMS value of 120 volts, and the frequency is 60 Hz. Determine the firing angle.

Since the frequency is 60 Hz, $\omega = 377$. Then,

$$\omega R_c C_c = 0.377$$

and

$$\omega^2 R_c^2 C_c^2 + 1 = 1.142$$

$$v_{C,\text{sinusoidal steady state}} = \frac{120\sqrt{2}}{\sqrt{1.142}} \sin[377t_{\text{tr}} - \arctan(0.377)]$$
$$= 30$$

Solving for t_{tr},

$$377t_{\text{tr}} - 0.361 = \arcsin(0.189) = 0.190 \text{ radians}$$

$$t_{\text{tr}} = 0.0015 \text{ s}$$

This is a first approximation to the TRIAC triggering or firing time. It is used as a starting value in equation 10.35.

$$30 = \frac{120\sqrt{2}}{\sqrt{1.142}} \sin(377t_{\text{tr}} - 0.361)$$
$$+ \left[-9 + \frac{120\sqrt{2}(0.377)}{1.142} \right] e^{-10^3 t_{\text{tr}}}$$

The iteration formula is

$$\sin(377t_{\text{tr}} - 0.361) = 0.189 - 0.2961 e^{-10^3 t_{\text{tr}}}$$

Substituting $t_{\text{tr}} = 0.0015$ in the right-hand side, solve for t_{tr} from the left-hand side. Repeat until there is no change in t_{tr}.

$$t_{\text{tr1}} = 1.28 \text{ ms}$$
$$t_{\text{tr2}} = 1.24 \text{ ms}$$
$$t_{\text{tr3}} = 1.23 \text{ ms}$$
$$t_{\text{tr4}} = 1.23 \text{ ms}$$

The firing angle is

$$377 \times 0.00123 = 0.464 \text{ radians} = 26.6°$$

For any sizeable phase shift ($\arctan \omega R_c C_c$), the initial value of the capacitance voltage is usually negligible.

7 NON-SINUSOIDAL WAVEFORM GENERATION

A. Introduction

Waveform generation can be accomplished with a variety of circuits. These circuits use feedback switching devices and energy storage elements (i.e., capacitors or inductors) to shape waves. A few representative circuits are described here.

B. UJT Relaxation Oscillator

The unijunction transistor can be used in a simple circuit to generate the non-sinusoidal waveform shown in

figure 10.30. The UJT base acts as a voltage divider until the emitter exceeds the voltage divider voltage. Then the UJT's emitter-B2 base junction becomes forward-biased and behaves more or less like an ordinary junction diode. With the UJT described by two resistors (R_1 attached to B1 base and R_2 to B2 base) in the off state, the voltage divider provides a triggering voltage of

$$\frac{V_{\text{BB}}R_1}{R_1 + R_2 + R_{\text{bb}}} \qquad 10.36$$

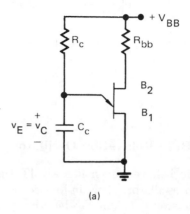

(a)

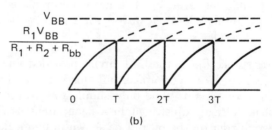

(b)

Figure 10.30 (a) UJT Relaxation Oscillator, (b) Oscillation Waveform across C_c

The UJT is off until the emitter voltage reaches the value given by equation 10.36, after which time the E-B1 terminals appear as a forward-biased P-N junction. This allows the capacitor to discharge. Once the capacitor is discharged, the UJT turns off, and the capacitor can again begin to charge toward V_{BB} according to equation 10.37.

$$v_C = V_{\text{BB}}(1 - e^{-t/\tau}) = v_E \qquad 10.37$$
$$\tau = R_c C_c \qquad 10.38$$

Example 10.14

A UJT oscillator with $R_1 = 10$ kilohms and $R_2 = 5$ kilohms is to produce a sawtooth wave with a maximum of 10 volts and a period of 10 ms. A 24 volt DC supply is available. Specify the values of the parameters for the circuit of figure 10.30(a) to accomplish this.

The maximum of 10 volts sets the divider ratio.

$$\frac{R_1 V_{BB}}{R_1 + R_2 + R_{bb}} = \frac{(10k)(24)}{10k + 5k + R_{bb}}$$

$$R_{bb} = 9 \text{ kilohms}$$

The time that the voltage on the capacitor reaches 10 volts can be derived from equation 10.37.

$$10 = 24(1 - e^{-t/\tau})$$

$$t = \tau \ln\left(\frac{24}{24 - 10}\right) = 0.54\tau$$

However, the period, T, was set at 10 ms, so,

$$\tau = \frac{10}{0.54} = 19 \text{ ms}$$

$$R_c C_c = \tau = 0.019$$

C. Neon Bulb Relaxation Oscillator

A neon bulb behaves much like a UJT when used in a relaxation oscillator. The un-ionized neon bulb acts like an open circuit. It ionizes when the applied voltage is at a sufficient level, V_I. If a neon bulb replaced the UJT in figure 10.30(a), a wave with a peak voltage of V_I would result.

The resistance R_{bb} in the UJT circuit provided a means of adjusting the peak voltage, but this is not available for the neon circuit. The only simple way to provide for peak voltage adjustment in a neon circuit is by a voltage divider on the output as shown in figure 10.31. The output voltage is

$$v_O = \frac{v_C R_1}{R_1 + R_2} \tag{10.39}$$

To determine the timing and peak voltage, it is convenient to obtain the Thevenin equivalent circuit as seen from the parallel combination of the capacitance and the bulb.

$$V_{Th} = V_S \left(\frac{R_1 + R_2}{R_1 + R_2 + R_s}\right) \tag{10.40}$$

$$R_{Th} = R_s \left(\frac{R_1 + R_2}{R_1 + R_2 + R_s}\right) \tag{10.41}$$

The Thevenin equivalent circuit is shown in figure 10.31(b). In order to achieve oscillatory behavior, V_{Th} must be greater than V_I.

Prior to ionization, the capacitance voltage is

$$v_C = V_{Th}(1 - e^{-t/\tau}) \tag{10.42}$$

$$\tau = R_{Th} C_c \tag{10.43}$$

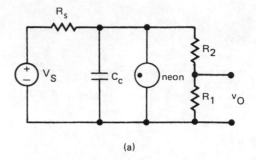

(a)

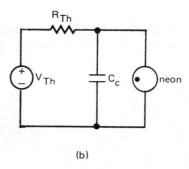

(b)

Figure 10.31   (a) Neon Relaxation Oscillator, (b) Thevenin Equivalent Circuit

When the capacitance voltage reaches V_I, the neon ionizes and discharges the capacitance. The oscillation period is

$$T = R_{Th} C_c \ln\left(\frac{V_{Th}}{V_{Th} - V_I}\right) \tag{10.44}$$

The resulting waveform is shown in figure 10.32.

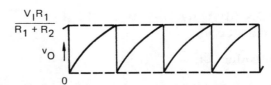

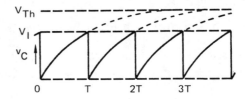

Figure 10.32   Capacitance and Output Waveforms for Neon Relaxation Oscillator

D. Negative Resistance Oscillator

A *negative resistance device* is one in which the v-i curve has a negative slope as indicated in figure 10.33(a). (Voltage is limited to single values for any current.)

The DIAC has such characteristics, and the UJT with a fixed voltage at base B2 can also be considered to be a negative resistance device. The principal parameters defined in figure 10.33(b) are used to define a piecewise linear model of the device in figure 10.33(c).

In order to obtain oscillatory behavior, it is necessary that the load line of the device-capacitance parallel combination intersect the negative resistance region.

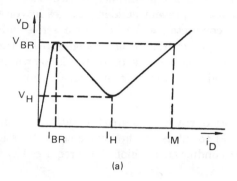

(a)

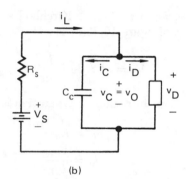

(b)

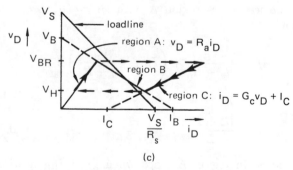

(c)

Figure 10.33 (a) Negative Resistance Characteristic, (b) Oscillator Circuit, (c) Load Line and Piecewise Linear Model of Device

The intersection of the source load line with the negative resistance region (region B) implies that $V_S/R_s < I_B$ and $V_S > V_B$. From a straight-line approximation to the negative resistance region with the points (V_{BR}, I_{BR}) and (V_H, I_H),

$$V_B = \frac{V_{BR}I_H - V_H I_{BR}}{I_H - I_{BR}} < V_S \qquad 10.45$$

$$I_B = \frac{V_{BR}I_H - V_H I_{BR}}{V_{BR} - V_H} > \frac{V_S}{R_s} \qquad 10.46$$

When these conditions are met, the circuit will oscillate. If the source load line intersects the characteristic at a positive slope, the circuit will stabilize at that point and will fail to oscillate.

With the device operating in region A, the source voltage charges the capacitance toward V_S. Since the device and capacitance voltages are identical, the device voltage and current are increasing. The charging current for the capacitance is the difference between the source current, which lies on the load line at the voltage $v_C = V_S$, and the device characteristic at the same voltage.

From figure 10.33(c), the charging current is positive in region A, so the capacitance voltage increases exponentially until V_{BR} is reached. Then the device and capacitance voltage cannot increase further. To remain on the negative resistance portion of the characteristic, the capacitance voltage must decrease. This requires that the capacitance current be negative, which is impossible in the negative resistance region.

The source current must lie on the load line, and the device current must lie on the negative resistance region at V_{BR}. The capacitance voltage cannot change instantaneously, so a jump from positive to negative voltage is physically impossible. Therefore, the capacitance current must remain positive.

The device accomplishes the desired voltage change by increasing its current over the source current. Being constrained to operate on its characteristic, it does this by jumping to the point (V_{BR}, I_M), in the operating region C.

With the device in region C, i_C is negative, and the capacitance will discharge toward the intersection of the straight-line representation of region C and the source load line. However, when it reaches the holding current point (V_H, I_H), it faces a similar dilemma of region B operation: to operate in region B, it is necessary to increase the voltage on the capacitance with negative capacitance current which is impossible in this region. Thus, the current jumps back to region A operation.

In cyclic operation, the device begins region A operation with $v_C = v_D = V_H$, as indicated in figure 10.33(c). The device is characterized by the slope of the characteristic.

$$R_a = \frac{V_{BR}}{I_{BR}} \qquad 10.47$$

The Kirchhoff current law equation of the circuit is

$$\frac{V_S}{R_s} = \frac{v_C}{R_s} + \frac{v_C}{R_a} + C\frac{dv_C}{dt} \qquad 10.48$$

The solution to equation 10.48 is

$$v_C = \frac{V_S R_a}{R_s + R_a} + \left(V_H - \frac{V_S R_a}{R_s + R_a}\right) e^{-at} \qquad 10.49$$

$$a = \frac{R_s + R_a}{C_c R_s R_a} \qquad 10.50$$

Region A operation terminates when v_C reaches V_{BR}. This occurs at the time T_A where

$$T_A = \frac{C_c R_s R_a}{R_s + R_a} \times \ln\left[\frac{V_S R_a - V_H(R_s + R_a)}{V_S R_a - V_{BR}(R_s + R_a)}\right] \qquad 10.51$$

In region C, the device begins operation with $v_C = v_D = V_{BR}$. In this region, the device is characterized by the straight line connecting points (V_H, I_H) and (V_{BR}, I_M). Then,

$$i_D = G_c v_D + I_C \qquad 10.52$$

$$G_c = \frac{I_M - I_H}{V_{BR} - V_H} \qquad 10.53$$

$$I_C = \frac{V_{BR} I_H - V_H I_M}{V_{BR} - V_H} \qquad 10.54$$

Kirchhoff's voltage law equation for region C operation is

$$\frac{V_S}{R_s} - I_C = v_C\left(G_c + \frac{1}{R_s}\right) + C\frac{dv_C}{dt} \qquad 10.55$$

$$v_C = \frac{V_S - I_C R_s}{1 + G_c R_s}$$
$$+ \left(V_{BR} - \frac{V_S - I_C R_s}{1 + G_c R_s}\right) e^{-c(t - T_A)} \qquad 10.56$$

$$c = \frac{1 + G_c R_s}{R_s C_c} \qquad 10.57$$

Region C operation terminates when $v_C = V_H$ at time T_C.

$$T_C - T_A = \frac{R_s C_c}{1 + G_c R_s}$$
$$\times \ln\left[\frac{V_S - I_C R_s - V_{BR}(1 + G_c R_s)}{V_S - I_C R_s - V_H(1 + G_c R_s)}\right] \qquad 10.58$$

The waveform for one cycle is shown in figure 10.34. Region A operation is from $t = 0$ to T_A. Region C operation is from T_A to T_C.

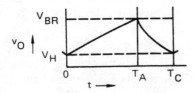

Figure 10.34 Negative Resistance Oscillator Waveform

E. Negative Conductance Oscillator

Negative conductance devices are duals of negative resistance devices. The negative conductance device has single current values for each voltage, in contrast to the negative resistance device which has single values of voltage for each current. A *tunnel diode* is an example of a negative conductance device.

While the negative resistance device uses a parallel capacitance to prevent sudden changes in voltage, the negative conductance device uses a series inductance to prevent the current from changing instantaneously. A negative conductance oscillator is shown in figure 10.35(a), with the device characteristic shown in figure 10.35(b).

Oscillatory operation requires the source load line ($V_S = i_T R + v_T$) to intersect the device characteristic in the negative conductance region. This requires $V_S/R_s > I_B$ and $V_S < V_B$.

Of course, $i_T = i_D = i_L$. Kirchhoff's voltage law for the circuit of figure 10.35(a) is

$$V_S = R_s i_L + L\frac{di_L}{dt} + v_D \qquad 10.59$$

$0 < v_D < V_{BR}$ in region A. Operation is governed by

$$v_D = R_a i_L \qquad 10.60$$

$$R_a = \frac{V_{BR}}{I_{BR}} \qquad 10.61$$

The device enters region A with $i_D = I_H$ at $t = 0$. Then,

$$i_L = \frac{V_S}{R_s + R_a} + \left(I_H - \frac{V_S}{R_s + R_a}\right) e^{-at}$$
$$= \frac{v_D}{R_a} \qquad 10.62$$

$$a = \frac{R_s + R_a}{L} \qquad 10.63$$

Operation in region A stops when $i_S = I_{BR}$ at time $t = T_A$.

$$T_A = \frac{L}{R_s + R_a} \times \ln\left[\frac{V_S - (R_s + R_a)I_H}{V_S - (R_s + R_a)I_{BR}}\right] \qquad 10.64$$

The device enters region C ($v_D > V_H$) operation at $t = T_A$ with $i_L = I_{BR}$. Operation is characterized by

$$v_D = V_C + i_L R_c \qquad 10.65$$

$$R_c = \frac{V_M - V_H}{I_{BR} - I_H} \qquad 10.66$$

$$V_C = \frac{V_H I_{BR} - V_M I_H}{I_{BR} - I_H} \qquad 10.67$$

Substituting for v_D in the KVL equation and solving for the transient, with $c = (R_s + R_c)/L$,

$$i_L = \frac{V_S - V_C}{R_s + R_c} + I_{BR} - \left(\frac{V_S - V_C}{R_s + R_c}\right)e^{-c(t-T_A)}$$

$$= \frac{1}{R_c}(v_D - V_C) \qquad 10.68$$

Region C operation terminates when $i_L = I_H$. This occurs at $t = T_B$.

$$T_B - T_A = \frac{L}{R_s + R_c} \times \ln\left[\frac{I_{BR}(R_s + R_c) + V_C - V_S}{I_H(R_s + R_c) + V_C - V_S}\right]$$

$$10.69$$

The waveform of v_D is shown in figure 10.35(c).

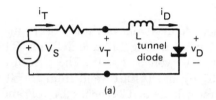

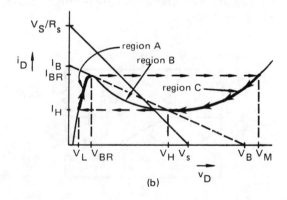

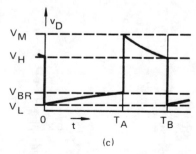

Figure 10.35 (a) Tunnel Diode Negative Conductance Oscillator, (b) Tunnel Diode Characteristic and Source Load Line, (c) Tunnel Diode Voltage Waveform

F. Operational Amplifier Square-Wave Generator

An operational amplifier can be used as a switching device for capacitor charging circuits. The amplifier provides the timing for an oscillator, and the output is a rectangular waveform.

A circuit to produce rectangular waves is shown in figure 10.36. The operational amplifier characteristics are limited by the zener diodes. The operational amplifier is used in its linear zone only during the brief switching interval when it is operating open-loop (i.e., no feedback). It has an essentially infinite slope at that time.

If v_O is $+V_H$, then v_N is V_H/k. v_C will charge toward $+V_H$ through diode D_2 and resistance R_2. As long as v_C is less than v_N, the operational amplifier input voltage, v_{IN}, remains negative. The operational amplifier will remain positively saturated with the output voltage at $+V_H$.

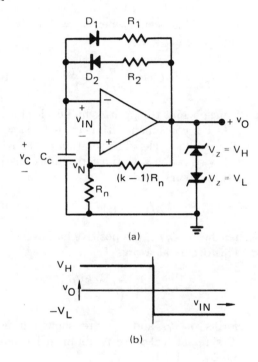

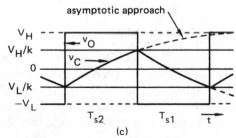

Figure 10.36 (a) Square-Wave Generator, (b) Operational Amplifier Characteristic with Zener Output Diodes, (c) Capacitor and Output Voltages

As the capacitance voltage charges past V_H/k, the operational amplifier input becomes positive, causing the output to switch to $-V_L$. The voltage at the N input also switches to $-V_L/k$, while the capacitance voltage remains essentially $+V_H/k$. The input voltage will become quite positive, holding the output at $-V_L$.

Thus, the capacitance begins charging toward $-V_L$ through the diode D_1 and the resistance R_1, beginning at a value of $+V_H/k$. Assuming time starts at $t = 0$, the capacitance charging equation is found from Kirchhoff's voltage law.

$$-V_L = v_C + R_1 C_c \frac{dv_C}{dt} \qquad 10.70$$

$$v_C = -V_L + (V_H/k + V_L)e^{-t/R_1 C_c} \qquad 10.71$$

When V_C reaches $-V_L/k$, v_{IN} will pass through zero and reverse the output to $+V_H$, terminating the negative half-cycle. This occurs when $v_C = -V_L/k$. The corresponding time is $t = T_{s1}$.

$$T_{s1} = R_1 C_c \ln\left[\frac{V_H + kV_L}{V_L(k-1)}\right] \qquad 10.72$$

The capacitance begins the positive half-cycle at a voltage of $-V_L/k$. For convenience, time is reset to $t = 0$ at the beginning of this cycle. The solution of the Kirchhoff voltage law equation with the capacitance charging toward V_H through R_2 yields

$$v_C = V_H + (-V_L/k - V_H)e^{-t/R_2 C_c} \qquad 10.73$$

When V_C reaches $+V_H/k$, the positive half-cycle is terminated. This occurs at a time T_{s2}.

$$T_{s2} = R_2 C_c \ln\left[\frac{V_L + kV_H}{V_H(k-1)}\right] \qquad 10.74$$

The waveforms of v_O and v_C are shown in figure 10.36(c). A symmetrical wave is obtained by making $R_1 = R_2$ and $V_H = V_L$.

G. Astable Multivibrators

The basic circuit of an *astable multivibrator* requires two amplifying elements. These amplifiers can be in the common-emitter configuration, or they can have a common-emitter resistance. The collector-coupled common-emitter configuration is shown in figure 10.37.

Assume transistor T2 is in the saturated condition and T1 is in the cutoff state. For these conditions to hold, v_{C1} must have a voltage greater than -0.5 volt. This, together with the collector saturation voltage (0.2 volt) of T2, does not turn on T1 since the base-emitter voltage is kept less than the turn-on voltage (taken as 0.7 volt). With these conditions, C_1 charges toward $-V_{\text{CC}} + 0.2$ for the polarities shown. When v_{C1} does reach -0.5 volt, it will initiate the regenerative switching of the

circuit. The switching process listed here occurs very quickly. The capacitor voltages do not change during switching.

- v_{C1} reaches -0.5 volt, so v_{BE1} turns on T1.

- v_{CE1} begins to drop. This drop is coupled to the base of T2 through C_2 causing v_{BE2} to drop.

- The drop in v_{BE2} causes T2 to come out of saturation and v_{CE2} to rise.

- The rise in v_{CE2} is coupled to the base of T1 through C_1, driving T1 farther into the on region.

- v_{CE1} drops farther. This turns off T2 more which turns on T1 more, and the cycle continues.

Begin a more detailed analysis at $t = 0$, the instant v_{BE1} reaches 0.7 volt, turning T1 on and T2 off. The voltage C_1 is then -0.5 volt at $t = 0$. The partial circuits of figure 10.37(b) and (c) are valid from $t = 0$ until the base-emitter voltage of T2 reaches 0.7 volt and initiates switching again. The next switching occurs a half-cycle from $t = 0$, or $T/2$.

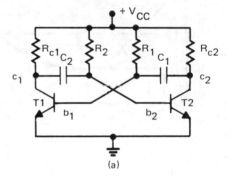

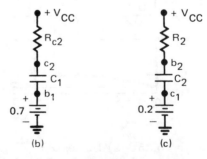

Figure 10.37 Collector-Coupled Astable Multivibrator: (a) Circuit Diagram, (b) C_1 Charging Circuit, (c) C_2 Charging Circuit

Figure 10.37(b) shows the charging path of capacitor C_1. The equation which governs this charging is

$$V_{\text{CC}} - 0.7 = v_{C1} + R_c C \frac{dv_{C1}}{dt} \qquad 10.75$$

Using $v_{C1}(0) = -0.5$ volt, the solution is

$$v_{C1} = (V_{CC} - 0.7) - (V_{CC} - 0.2)e^{-t/R_cC} \qquad 10.76$$

At $t = T/2$, v_{C1} will have the value that v_{C2} has at $t = 0$.

$$v_{C1}|_{t=T/2} = v_{C2}|_{t=0} = (V_{CC}-0.7)-(V_{CC}-0.2)e^{-T/2R_cC}$$
$$10.77$$

The behavior of the base-emitter voltage, V_{BE2}, on the cutoff transistor is desired. While T1 is saturated,

$$v_{BE2} = 0.2 - v_{C2} < 0.7$$
$$v_{C2} > -0.5$$

As C_2 is charging toward $v_{C2} = -V_{CC}+0.2$, it will turn T2 on again at $t = T/2$.

The differential equation for v_{C2} with $R_2 = R_b$ is

$$-V_{CC} + 0.2 = v_{C2} + R_bC\frac{dv_{C2}}{dt} \qquad 10.78$$

The solution is

$$v_{C2} = -(V_{CC}-0.2)+[V_{CC}-0.2+v_{C2}(0)]e^{-t/R_bC} \quad 10.79$$

At $t = T/2$, v_{C2} will have the same value that v_{C1} had at $t = 0$.

$$v_{C2}|_{t=T/2} = v_{C1}|_{t=0} = -0.5 \qquad 10.80$$

Then,

$$v_{C2}|_{t=T/2} = -0.5 = -(V_{CC} - 0.2)$$
$$+ [(V_{CC} - 0.2 + v_{C2}(0)]e^{-T/2R_bC}$$
$$10.81$$

Equations 10.77 and 10.81 contain the same two unknowns: $v_{C2}(0)$ and T. Taking the logarithms of the equations,

$$\frac{T}{2R_cC} = \ln\left[\frac{V_{CC} - 0.2}{V_{CC} - 0.7 - v_{C2}(0)}\right] \qquad 10.82$$

$$\frac{T}{2R_bC} = \ln\left[\frac{V_{CC} - 0.2 + v_{C2(0)}}{V_{CC} - 0.7}\right] \qquad 10.83$$

Equations 10.82 and 10.83 can only be solved algebraically when $R_b = R_c$. In other cases, particularly if R_b is much greater than R_c, it can be assumed that $v_{C2}(0) \approx V_{CC} - 0.2$. Then, solve for the cycle time, which will converge with successive iterations.

Example 10.15

An astable multivibrator, such as the one shown in figure 10.37(a), has $V_{CC} = 16$ volts, $R_c = 1$ kilohms,

$R_b = 5$ kilohms, and $C = 2$ nF. Determine the frequency of oscillation.

Equating equations 10.82 and 10.83 and cancelling the common T and C terms,

$$5\ln\left[\frac{15.8 + v_{C2}(0)}{15.3}\right] = \ln\left[\frac{15.8}{15.3 - v_{C2}(0)}\right]$$

Since $R_c \neq R_b$, trial and error is required. Taking the antilog of both sides,

$$v_{C2}(0) = 15.3 - 15.8e^{-5\ln([15.8+v_{C2}(0)]/15.3)}$$

Starting with $v_{C2}(0) = 15$ in the right-hand side, $v_{C2}(0)$ on the left is 14.82207. Putting that value into the right-hand side, the left side is 14.80803. A third iteration yields 14.80690, so the value of v_{C2} is taken as 14.81. From equation 10.83,

$$T = 2 \times 5000 \times 2 \times 10^{-9}\ln\left[\frac{15.8 + 14.81}{15.3}\right]$$
$$= 1.39 \times 10^{-5}$$

The frequency of oscillation is

$$f = \frac{1}{T} = 72.1 \text{ kHz}$$

H. Monostable Multivibrators

A *monostable multivibrator* (also known as a *blocking oscillator*) is one that has one stable state. There is one astable state that is typically controlled by the charging of a capacitor. Such a circuit is shown in figure 10.38. The capacitor C_1 is a *speed-up capacitor* to enhance the regenerative action whenever it begins. It does this by coupling the change at the collector of T2 to the base of T1. The timing capacitor is C_2, which charges through R_{C1} when T2 is on, and through R_2 when T2 is off and T1 is on.

The stable state is when T2 is on, being driven by V_{CC} through R_2. In the quiescent condition, B2 is at 0.7 volt, and v_2 is at about 0.2 volt, as T2 is saturated. Since T1 is off, C_2 is charged to $-V_{CC} + 0.7$ volt, and C_1 is discharged.

When a trigger pulse is applied to B1 for a sufficient time, the following processes occur.

step 1: T1 turns on and lowers the voltage v_1. Since v_1 is coupled through C_2 (which cannot change voltage instantaneously), T2 begins to turn off.

step 2: The turn-off of T2 causes the voltage v_2 to increase. This increase is coupled through C_1 to reinforce the turn-on of T1.

step 3: Steps 1 and 2 are very fast. T1 is turned on and T2 is turned off very quickly. The capacitance conditions do not have time to change significantly.

v_{C2} begins to charge toward $V_{CC} - 0.2$ since the collector of T1 is at about 0.2 volt. Charging is through R_2 with a time constant R_2C_2. Charging continues until a voltage of about 0.5 volt is reached, whereupon that (together with the 0.2 volt of v_1) will begin to turn T2 on again. This terminates the astable state. The regenerative action quickly turns T1 off and T2 on, returning to the stable state. Before another trigger is accepted, C_2 must be charged to $-V_{CC} + 0.7$ again.

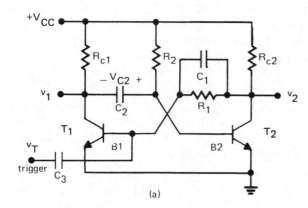

(a)

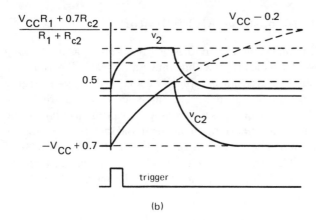

(b)

Figure 10.38 (a) Monostable Multivibrator Circuit, (b) Principal Waveforms

8 DATA CONVERSION

A. Introduction

Analog information is typically obtained as voltage, where the voltage is proportional to the variable being measured. In many systems the information is converted to digital form before it is processed further. After processing, the information may be again converted to an analog voltage, particularly in control systems. Thus, there are two conversions of interest:

conversion from analog voltage to digital form, which is analog-to-digital (A-D) conversion, and conversion from digital form to analog form, which is digital-to-analog (D-A) conversion.

The digital information format that is considered here is binary, where each digit is weighted according to its position as the appropriate power of two. Limiting the discussion to positive numbers, the value of an n-bit binary number N is

$$N = b_{n-1}2^{n-1} + b_{n-2}2^{n-2} + \cdots + b_3 2^3 + b_2 2^2 + b_1 2^1 + b_0 2^0$$

$$10.84$$

The values of the b's (binary digits or bits) may be either zero or one. This number when coded in binary would have the code

$$N = b_{n-1}b_{n-2}\ldots b_3 b_2 b_1 b_0 \qquad 10.85$$

The b's occur in a time sequence on a single line for serial information, and occur simultaneously on a set of information bus wires for parallel information. Here we are principally interested in the parallel case, although the serial case could be handled with the use of a storage register.

The digit b_{n-1} is called the most significant bit (MSB), and the digit b_0 is the least significant bit (LSB). In general, the place value of the MSB and the total number of bits in the code are necessary to define the range of a number. Each digit position has half the place value of its left-hand neighbor. For example, if the MSB position has a weighting value of 8, the next position value is 4, the next is 2, the next is 1, the next 0.5, and so on. Thus a six-bit number 100101 using these place values is

$$100101 = 1 \times 8 + 0 \times 4 + 0 \times 2 + 1 \times 1 + 0 \times 0.5$$
$$+ 1 \times 0.25$$
$$= \text{decimal value of } 9.25$$

For a positive number, the maximum value that can be represented by the code is nearly twice the MSB place value. To be exact, the maximum value is $2 \times \text{MSB} - \text{LSB}$.

B. Digital to Analog Conversion

A bus line of a digital computer is considered to be high if it is above about 1.8 volts and low if it is below about 1 volt. The high level is normally considered to be a one and can range from 1.8 to almost 5 volts. The low level is considered to be zero and can range from about 1 volt to nearly zero volts. As a result of these variations, the voltages on such bus lines cannot be used directly for conversion to analog information. Rather, they can be used to drive switches which connect to a firm 5 volts if the bus line is above 1.8 volts, or to a firm zero volts when the bus line voltage is below 1 volt. This is accomplished with electronically controlled analog switches, which are designated by S in the following drawings.

Example 10.16

One method of D-A conversion uses the weighting of a multi-input operational amplifier, which requires doubling of each successive resistor beginning with the MSB input. The ratio of the feedback resistance to the input resistance from the MSB line is chosen to provide half the maximum output voltage when the MSB is one and the rest of the bits are zero. An eight-bit D-A converter with a range of 0 to 10 volts is shown below.

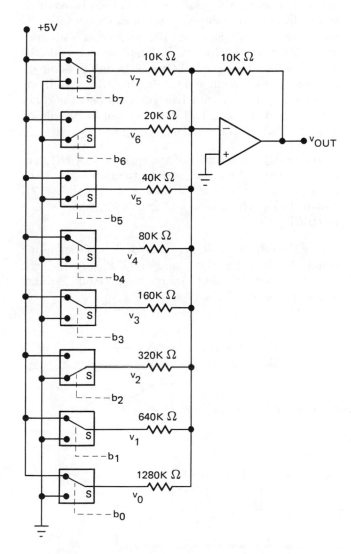

The inputs to the switches are the eight data bus lines. When a particular line is high, the switch will be in the upper position and supply +5 volts through the switch.

When the line is low, the switch will be in the lower position and apply zero volts through the switch.

From a previous analysis, the output of the opamp will become

$$v_{OUT} = -\frac{10k}{10k}v_7 - \frac{10k}{20k}v_6 - \frac{10k}{40k}v_5$$
$$- \frac{10k}{80k}v_4 - \frac{10k}{160k}v_3 - \frac{10k}{320k}v_2$$
$$- \frac{10k}{640k}v_1 - \frac{10k}{1280k}v_0$$

Using the binary values of the b's as 1 or 0, so that the corresponding switch output voltages are 5 volts when $b = 1$ and zero volts when $b = 0$, the expression for this particular D-A converter is

$$v_{OUT} = -5b_7 - 2.5b_6 - 1.25b_5 - 0.625b_4 - 0.3125b_3$$
$$- 0.15625b_2 - 0.078125b_1 - 0.0390625b_0$$

The accuracy of the conversion is one half the LSB value or 0.01953125, or about 19.5 mV.

C. Ladder Type D-A Converters

One ladder type D-A converter is based on a cell concept, where each succeeding cell has twice the weighting as the previous one. The first cell, which corresponds to the LSB, is as shown in figure 10.39(a), where the voltage, V_R, or zero occurs depending on the switch position.

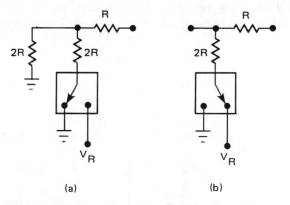

Figure 10.39 Cells of a Ladder Type D-A Converter: (a) Starting Cell, for LSB. (b) Typical Cell

A Thevenin equivalent for the starting cell in figure 10.39(a), taking the switch output as v_0, is as shown in figure 10.40(a). When that circuit is combined with the next cell, which has voltage v_1 as in figure 10.40(b), the resulting Thevenin equivalent is as shown in figure 10.40(c).

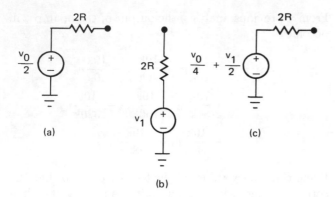

Figure 10.40   Thevenin Circuit Reduction

Then a ladder constructed of the starting cell and seven typical cells have an output given by

$$v_{\text{OUT}} = \frac{v_0}{256} + \frac{v_1}{128} + \frac{v_2}{64} + \frac{v_3}{32}$$
$$+ \frac{v_4}{16} + \frac{v_5}{8} + \frac{v_6}{4} + \frac{v_7}{2} \qquad 10.86$$

The voltages are either the reference value, V_R, or zero, depending on the individual switch settings. The output of such a ladder can then be connected to a non-inverting opamp to provide a buffered output of the ladder type D-A converter as indicated in figure 10.41.

D. Inverting Ladder Type D-A Converters

An improved ladder type D-A converter is shown in figure 10.42. The improvement is in that one side of the switches are connected directly to the inverting opamp

terminal, which is a virtual ground because the non-inverting terminal is connected to ground. This is the preferred arrangement for silicon switches, as it eliminates possible problems with the switching range. In addition, it is less sensitive to capacitive coupling from the resistors to ground, which can cause propagation delays of the switchings.

The analysis of this circuit is by inspection, beginning at the lower right resistor R in figure 10.42. Whether S_0 is as shown or in the other position, the switch is effectively grounded (either directly or by its connection to the inverting opamp terminal). Then the resistance from the right end of that lower right resistor to ground is the parallel combination of $2R$ with $2R$, which is R.

Then moving to the left end of that last R, the resistance to ground at that point is also R. Moving left to the next junction the resistance to ground is again R, and so from each node along the bottom row of resistors in figure 10.42, the resistance to ground is a parallel combination of $2R$ with $2R$. Thus, at each node the current will divide equally between the two possible directions. Then the current coming from V_R will divide at the first junction with half going up and the other half to the right. The half going up can be calculated as $V_R/2R$, which is the amount going to the right.

At the next junction, half of the current ($V_R/4R$) goes up, and the same amount goes to the right. So at the third junction, the current going up through the $2R$ resistor is $V_R/8R$, and at the fourth junction there is $V_R/16R$.

Taking the value of a bit, b_j, as zero if the corresponding switch, S_j, is to the right, and one if it is to the left, the total current flowing toward the inverting terminal of the opamp is

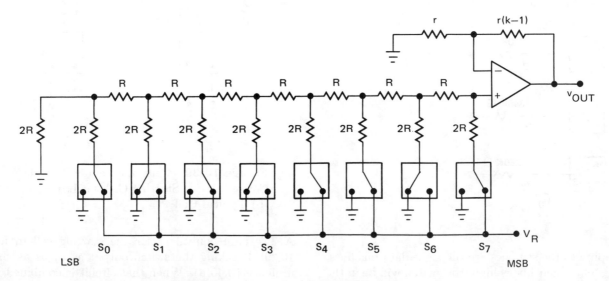

Figure 10.41   A Ladder Type D-A Converter

PROFESSIONAL PUBLICATIONS ● Belmont, CA

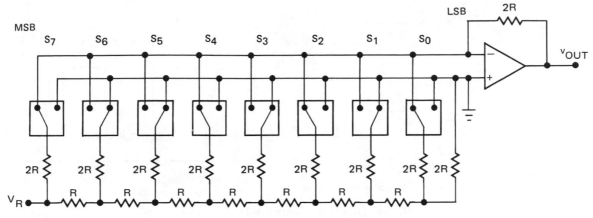

Figure 10.42   Inverting D-A Converter

$$i_{\text{IN}} = \frac{V_R}{2R}\left(b_7 + \frac{1}{2}b_6 + \frac{1}{4}b_5 + \frac{1}{8}b_4\right.$$

$$\left. + \frac{1}{16}b_3 + \frac{1}{32}b_2 + \frac{1}{64}b_1 + \frac{1}{128}b_0\right)$$

As no current can flow into the opamp terminal, this current must flow through the feedback resistance, making the output voltage

$$v_{\text{OUT}} = -2Ri_{\text{IN}}$$

$$= V_R\left(b_7 + \frac{1}{2}b_6 + \frac{1}{4}b_5 + \frac{1}{8}b_4\right.$$

$$\left. + \frac{1}{16}b_3 + \frac{1}{32}b_2 + \frac{1}{64}b_1 + \frac{1}{128}b_0\right) \quad 10.87$$

E. Multiplying D-A Converters

The noninverting ladder type D-A converter of figure 10.41 with output as described in equation 10.86 can be used as a programmable attenuator of an analog signal, which would be applied where the reference voltage V_R is shown. When used in this manner, the device is called a multiplying D-A converter.

For this eight-bit converter, the maximum output is obtained when all of the inputs are ones (the switches in figure 10.41 are to the right). This maximum output is $0.99609375 \times V_R$ (i.e., $V_R[1 - (1/256)]$ and can be adjusted in steps of $V_R/256$ down to zero. Thus, the attenuation can be adjusted via computer data byte, which is useful in a number of applications.

The circuit of figure 10.42 can also be adapted to this application by changing the feedback resistance (the resistance connected from the output of the opamp to its negative input terminal) from $2R$ to R.

Similarly, the multi-input D-A converter of example 10.16 can be adapted to a programmable attenuator by

changing the feedback resistance from 10K to 5K ohms, and by applying the input where the +5 volt reference is shown at the upper left of the circuit diagram.

F. Counting Analog to Digital (A-D) Converters

A straightforward method of converting an analog voltage to digital format is by the use of a binary counter, a D-A converter, and a comparator. A sample and hold (S/H) circuit is usually employed to avoid changing the input while conversion is taking place. Figure 10.43 shows a typical system.

The comparator behaves much like an open-loop opamp, with the exception that the output levels are compatible with the logic system (i.e., zero and 5 volt levels for TTL logic systems). The comparator output remains at the high level as long as the analog voltage remains higher than the D-A output. When the D-A output becomes higher, the comparator output goes low.

The clock input and the comparator output are inputs to an AND gate. While the comparator output is high, the AND output is simply the clock signal. The counter increases its count each time the clock changes from its low level to its high level (rising edge triggered). When the comparator output goes low, the clock is blocked, and the AND output becomes low and remains at that level until the next conversion begins.

The conversion is started by the counter being reset as the sample and hold circuit takes a new sample of the analog voltage.

The binary counter begins at zero and counts the clock signal with its output sent through a D-A converter and compared to the analog voltage. As long as the converted digital count is insufficient to exceed the analog voltage, the counter continues to count. When the converted count is sufficient to exceed the analog voltage,

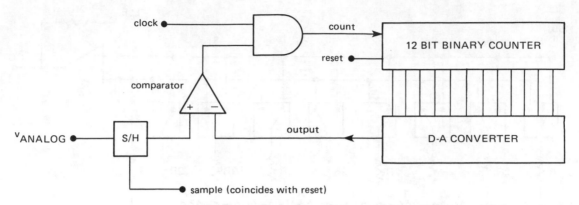

Figure 10.43   Counting A-D Converter

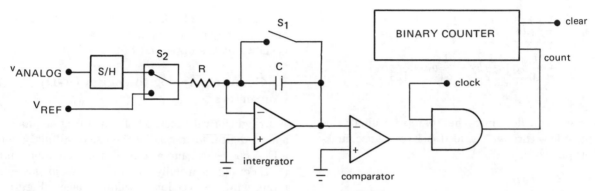

Figure 10.44   Dual-Slope A-D Converter

the counter is stopped, and the count is then the converted value of the analog voltage.

G. Dual Slope A-D Converters

Another A-D conversion system uses an integrator and a reference voltage, rather than a D-A converter. Such a system is shown in figure 10.44.

The (positive) analog voltage is sampled and held, and at the same time analog switch S_1 closes to discharge the integrating capacitor, and the counter is reset to zero. This time is designated as $t = 0$. Then the sampled analog voltage is integrated while the binary counter counts through its entire range. When it has reached its maximum count (all bits are high), the next clock input will cause the count to revert to zero (the only time the MSB switches from high to low). At this time, T_1, the capacitance is charged to an output voltage of $-v_{\text{ANALOG}}T_1$.

The transition of the MSB from high to low is used to switch the integrator input from the analog voltage to the negative reference voltage. The integrator then integrates this negative voltage, beginning at the level $-v_{\text{ANALOG}}T_1$.

$$v_O = -v_{\text{ANALOG}}T_1 + V_{\text{REF}}(t - T_1) \qquad 10.88$$

While the output of the integrator remains negative, the comparator output remains high, gating the clock signal through the AND gate to the counter. When the integrator output becomes barely positive, the comparator output will go low, preventing the clock signal from reaching the counter. Thus, the count is terminated at this time, T_2. The voltage can be found at time T_2 by setting equation 10.88 to zero.

$$v_{\text{ANALOG}} = \frac{V_{\text{REF}}(T_2 - T_1)}{T_1} \qquad 10.89$$

The fraction $T_2 - T_1/T_1$ is the ratio of the count of the counter at the time it reached zero, to the full count. Thus, the final count is proportional to the analog voltage.

$$\frac{\text{final count}}{\text{full count}} = \frac{\text{analog voltage}}{\text{reference voltage}} \qquad 10.90$$

Appendix 10.A
Table of Standard Zener Diodes
With Typical Zener Impedances and Keep-Alive Currents

nominal Zener voltage	500 milliwatt		1 watt		5 watt		50 watt	
	Z_z ohms	I_{ZK} mA	Z_z ohms	I_{ZK} mA	Z_z ohms	I_{ZK} mA	Z_z ohms	I_{ZK} mA
2.4	30	1.0						
2.7	30	1.0						
3.0	29	1.0						
3.3	28	1.0	10	1.0	3.0	1.0		
3.6	24	1.0	10	1.0	2.5	1.0		
3.9	23	1.0	9.0	1.0	2.0	1.0	0.16	5.0
4.3	22	1.0	9.0	1.0	2.0	1.0	0.16	5.0
4.7	19	1.0	8.0	1.0	2.0	1.0	0.12	5.0
5.1	17	1.0	7.0	1.0	1.5	1.0	0.12	5.0
5.6	11	1.0	5.0	1.0	1.0	1.0	0.12	5.0
6.2	7.0	1.0	2.0	1.0	1.0	1.0	0.14	5.0
6.8	5.0	1.0	1.5	1.0	1.0	1.0	0.16	5.0
7.5	5.5	0.5	1.5	1.0	1.5	1.0	0.24	5.0
8.2	6.5	0.5	4.5	0.5	1.5	1.0	0.4	5.0
8.7					2.0	1.0		
9.1	7.5	0.5	5.0	0.5	2.0	1.0	0.5	5.0
10	8.5	0.25	7.0	0.25	2.0	1.0	0.6	5.0
11	9.5	0.25	8.0	0.25	2.5	1.0	0.8	5.0
12	11.5	0.25	9.0	0.25	2.5	1.0	1.0	5.0
13	13	0.25	10	0.25	2.5	1.0	1.1	5.0
14					2.5	1.0	1.2	5.0
15	16	0.25	14	0.25	2.5	1.0	1.4	5.0
16	17	0.25	16	0.25	2.5	1.0	1.6	5.0
17					2.5	1.0	1.8	5.0
18	21	0.25	20	0.25	2.5	1.0	2.0	5.0
19					3.0	1.0	2.2	5.0
20	25	0.25	22	0.25	3.0	1.0	2.4	5.0
22	29	0.25	23	0.25	3.5	1.0	2.5	5.0
24	33	0.25	25	0.25	3.5	1.0	2.6	5.0
25					4.0	1.0	2.7	5.0
27	41	0.25	35	0.25	5.0	1.0	2.8	5.0
28					6.0	1.0		
30	49	0.25	40	0.25	8.0	1.0	3.0	5.0
33	58	0.25	45	0.25	10	1.0	3.2	5.0
36	70	0.25	50	0.25	11	1.0	3.5	5.0
39	80	0.25	60	0.25	14	1.0	4.0	5.0
43	93	0.25	70	0.25	20	1.0	4.5	5.0
45							4.5	5.0
47	105	0.25	80	0.25	25	1.0	5.0	5.0
50							5.0	5.0
51	125	0.25	95	0.25	27	1.0	5.2	5.0

Appendix 10.A
(continued)

nominal Zener voltage	500 milliwatt		1 watt		5 watt		50 watt	
	Z_z ohms	I_{ZK} mA	Z_z ohms	I_{ZK} mA	Z_z ohms	I_{ZK} mA	Z_z ohms	I_{ZK} mA
56	150	0.25	110	0.25	35	1.0	6.0	5.0
60					40	1.0		
62	185	0.25	125	0.25	42	1.0	7.0	5.0
68	230	0.25	150	0.25	44	1.0	8.0	5.0
75	270	0.25	175	0.25	45	1.0	9.0	5.0
82	330	0.25	200	0.25	65	1.0	11	5.0
87					75	1.0		
91	400	0.25	250	0.25	75	1.0	15	5.0
100			350	0.25	90	1.0	20	5.0
105							25	5.0
110					125	1.0	30	5.0
120			550	0.25	170	1.0	40	5.0
130			700	0.25	190	1.0	50	5.0
140					230	1.0	60	5.0
150			1000	0.25	330	1.0	75	5.0
160			1100	0.25	350	1.0	80	5.0
170					380	1.0		
175							85	5.0
180			1200	0.25	430	1.0	90	5.0
190					450	1.0		
200			1500	0.25	480	1.0	100	5.0

PRACTICE PROBLEMS

Warmups

1. Using a precision diode circuit, design a clipping circuit which will pass voltages above -0.25 volt and clip voltages below that value. Assume a stable power supply.

2. Design a pre-reference zener circuit using an 8.1 volt zener diode ($V_Z = 8.1$ volts at $I_Z = 8$ mA, with $R_z = 8$ ohms) from which the reference diode of example 10.5 is to be operated. (a) Design the entire reference voltage circuit, and (b) calculate the new voltage regulation for the output circuit of example 10.5.

3. For the transistor circuit shown, determine (a) the necessary base current for this gate to sink five identical gates connected to its output, and (b) the minimum size of R_b to allow this gate to source ten identical gates.

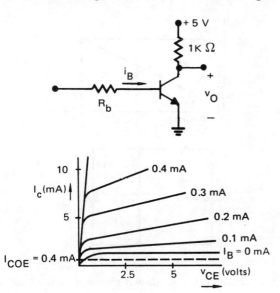

4. A particular logic gate has the following parameters: $t_d = 5$ ns, $t_r = 50$ ns, $t_s = 30$ ns, and $t_f = 30$ ns. Determine the maximum frequency at which this gate can operate.

5. Use the truth table to show that the circuit of figure 10.15 is a NOR gate for negative logic (i.e., low voltage = true, high voltage = false).

6. A particular SCR has $I_{GT} = 50$ mA. Determine the current pulse amplitude necessary to trigger it in $10~\mu s$.

7. **For the relaxation oscillator of example 10.14, determine the change of peak voltage and period for a 25 percent variation in V_{BB}.**

8. The switch in the circuit shown is alternately opened and closed for 0.5 s intervals. Determine the voltage across the capacitance, and show the waveform.

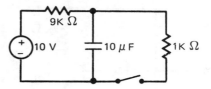

9. Refer to the circuit of figure 10.38. $V_{CC} = 12$ volts, $R_2 = 10$ kilohms, and $C_2 = 0.1~\mu F$. Determine the duration for which the monostable output, V_2, is high.

10. Use an operational amplifier to convert a rectangular wave voltage of ± 5 volts and a period of 10 ms to a triangular wave with peak output of ± 5 volts with the same period.

Concentrates

1. For the circuit shown, determine the transfer characteristic of the output voltage versus the input voltage. The diodes are ideal.

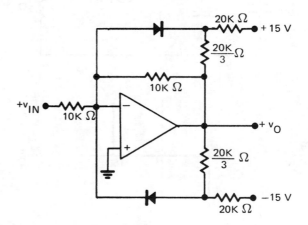

2. Design a voltage regulator that has a maximum supply voltage of 30 volts, a minimum of 24 volts, and is to supply a load voltage of 16 volts for currents up to 250 mA. (A 16-volt zener diode—rated at a current of 155 mA—with $R_z = 4$ ohms is available.) Its maximum current is 530 mA. Determine the maximum regulation of your design, and discuss methods of improving the regulation.

3. For the DTL gate shown, the diode drops may range from 0.6 to 0.8 volt, the base-emitter drop of the transistor will be between 0.7 and 0.8 volt, the transistor current gain ranges between 18 and 32, and the saturation voltage of the transistor is 0.2 volt. The resistance values are nominal, with 20 percent possible variation.

Determine the maximum reliable fan-out for production of these gates in large quantities.

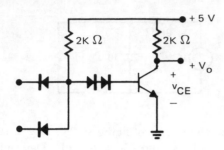

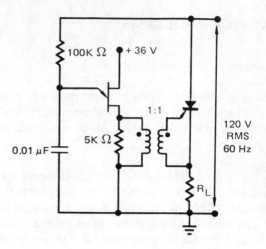

4. The circuit shown below is used to flash a camera lamp. The lamp characteristics also are shown. (a) Determine the time after the switch is closed when the lamp illuminates. (b) Determine the time after the switch is closed when the lamp extinguishes.

6. A DIAC with characteristics shown is to be used in a negative resistance oscillator circuit shown. Determine a reasonable value of R_s. Select a capacitance to provide a cyclic frequency of 400 hertz. Show the resulting waveform.

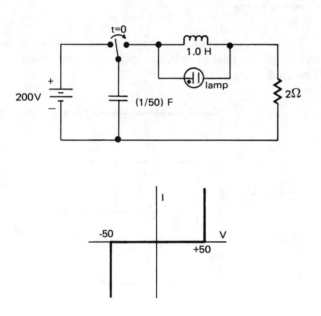

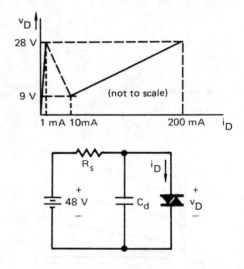

7. An A-D converter as shown uses a 3 mHz clock and a 5 volt reference voltage. Determine the maximum number of bits for the binary counter if the conversion is to be completed in less than 1 ms. Determine the value of RC to permit full range of the integrator output to -10 volts. What is the fundamental accuracy of the conversion of an analog voltage?

5. For the circuit shown, the UJT has $R_1 = 10$ kilohms and $R_2 = 5$ kilohms. Determine the firing angle of the SCR.

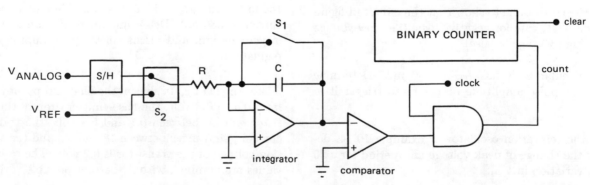

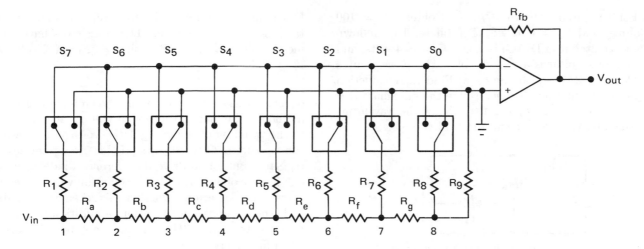

8. The circuit shown is a programmable gain amplifier. It is to be programmable up to a maximum gain of 2.00. It is to be controlled by an 8-bit binary coded data word which controls the electronic switches S_0 through S_7, with S_7 the most significant bit. Specify the values of the resistances to accomplish this.

9. A 500 mW zener diode is to be used as a reference source for a voltage regulator, and will be connected to the noninverting terminal of an operational amplifier. It has a nominal voltage of 5.6 volts, a zener resistance of 11 ohms, and a keep-alive current of 1.0 mA. If the unregulated supply varies between 30 and 40 volts, determine the resulting regulation of the output voltage.

A 1-watt zener diode having a nominal voltage of 10 volts, a zener resistance of 7.0 ohms, and a keep-alive current of 0.25 mA is proposed to provide the voltage for the reference diode above. Design the circuit and determine the regulation due to the variation of the unregulated voltage (between 30 and 40 volts).

10.

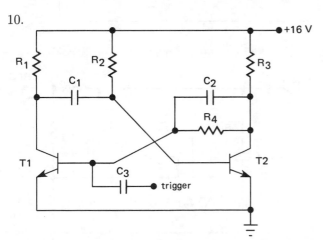

For the monostable multivibrator shown, determine the output pulse shape when it has been triggered with an appropriate signal. Determine the maximum frequency

(steady state) that the circuit can be triggered. All capacitors are 1 μF. All resistances are 1000 ohms.

Timed

1. The circuit shown is to provide full-wave rectification of a pure sinusoidal input voltage and to feed that voltage to an average reading voltmeter. The voltmeter has a 10 volt scale and is to indicate the RMS value of the input voltage v_{IN} directly (i.e., no scale change). If the amplifiers saturate at positive and negative 12 volts, what is the maximum voltage which the meter can measure accurately? Determine the values of R_1, R_2, and R_3 necessary for this circuit to operate correctly.

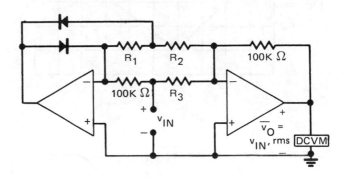

2. Determine the power requirement for the DTL circuit shown for a fan-out of five.

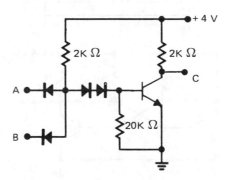

3. For the circuit shown, $R_L = 10$ ohms, $R_c = 100$ kilohms, and $C_c = 0.01$ μF. The bilateral breakdown diode (treated as a DIAC) breaks over at ± 4 volts, and the holding voltages are negligible. The power source is 220 volts RMS at 60 hertz. (a) Find the conduction angle by calculation. (b) Determine the maximum I_{GT} that the TRIAC can demand. (c) Show the waveforms for the TRIAC gate current and load voltage.

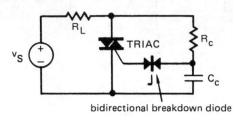

bidirectional breakdown diode

4. The tunnel diode in the circuit has the characteristics given. (a) Find the period of the astable waveform, $v_D(t)$. (b) Sketch the voltage waveform of $v_D(t)$. (c) Sketch the current waveform, $i_D(t)$.

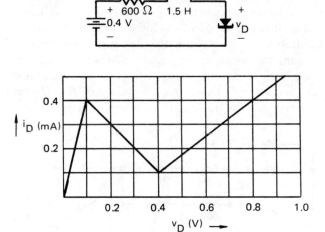

5.

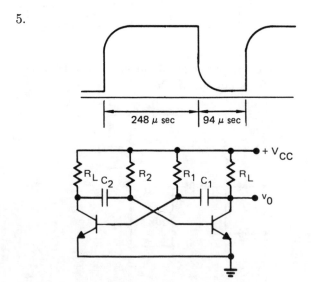

Design an astable multivibrator circuit as shown to obtain the waveform shown. The transistors have parametric values of $V_{BE} = 0.7$ volt, $V_{CE,sat} = 0.2$ volt, and $\beta_{min} = 25$.

6. The voltage regulator circuit shown is proposed as a method of eliminating the output voltage regulation due to the variation of the unregulated voltage, v_{UN}, which can vary from 30 to 40 volts. If this circuit can use its regulated output to drive its reference diode, then the only source of output voltage regulation will be the load current, i_L. Determine whether this scheme is workable, and if not, propose an alteration or addition that would make it workable. (Hint, it may not be self-starting when first energized.)

If the circuit is operable as proposed, or if it can be made to work with some modification, then it is to provide a regulated output voltage of 24 volts for a load current which can be as large as 2 amps. The 500 mW zener diode has a nominal zener voltage of 10 volts, a zener impedance of 8.5 ohms, and a keep-alive current of 0.25 mA. Determine resistance values to meet the specifications, and the resulting maximum voltage regulation of V_{REG}.

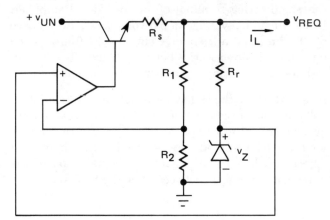

7.

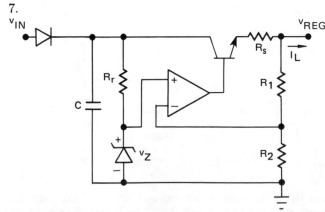

A voltage regulator with the configuration shown has $R_s = 2$ ohms, $R_1 = R_2$, and a 500 mW zener diode with a nominal zener voltage of 10 volts, a zener impedance of 8.5 ohms, and a keep-alive current of 0.25 mA. The input voltage, v_{IN}, is a rectified 60 Hz sinusoid with a peak voltage of 48 volts. The maximum load current, i_L, that can be drawn without losing regulation of the output voltage, v_{REG}, is to be 5 amps. Specify the minimum value of capacitance C that can be used. For that value of C, determine the ripple of the output voltage when $i_L = 0$.

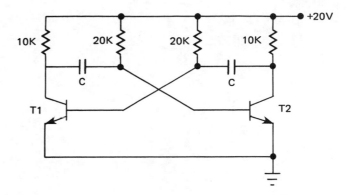

8. For the circuit shown, determine the waveform and frequency for the output taken from the collector of T_2. The capacitors are both 0.1 μF.

11 DIGITAL LOGIC

Nomenclature

D	data or delay (flip-flop)
J-K	J-K (flip-flop)
m	minterm
M	maxterm
MUX	multiplexer or data selector
PLA	programmable logic array
POS	product of sums
Q	a state variable
Q^+	the next value of Q
res	residue
ROM	read-only-memory
S	selection variable
SOP	sum of products
S-R	set-reset (flip-flop)
T	toggle (flip-flop)
X	a "don't-care" condition

1 FUNDAMENTAL LOGICAL OPERATIONS

A. Symbols, Truth Tables, and Expressions

- Logical AND

The symbol and interpretation of logic circuits are introduced with the symbol for an AND gate. Inputs are on the left; output is on the right. The *truth table* shows the input logic values on the left and the corresponding output values in the right-hand column. By convention, a logic "1" means *true*, and a logic "0" means *false*.

The logical expressions utilize the dot "·" to indicate AND, and the "+" for OR. In expressions where all variables are represented by single letters, the "·" is commonly omitted.

Figure 11.1   Symbol for Logic Element AND

Table 11.1
Truth Table for AND Gate

A	B	C	D
0	0	0	0
0	0	1	0
0	1	0	0
0	1	1	0
1	0	0	0
1	0	1	0
1	1	0	0
1	1	1	1

$$D = A \cdot B \cdot C \qquad 11.1$$
$$A \cdot B \cdot C = D \qquad 11.2$$

- Logical OR

Figure 11.2   OR Symbol

Table 11.2
Truth Table for OR Gate

A	B	C	D
0	0	0	0
0	0	1	1
0	1	0	1
0	1	1	1
1	0	0	1
1	0	1	1
1	1	0	1
1	1	1	1

Table 11.4
Truth Table for NAND Gate

A	B	C	D
0	0	0	1
0	0	1	1
0	1	0	1
0	1	1	1
1	0	0	1
1	0	1	1
1	1	0	1
1	1	1	0

$$D = A + B + C \qquad \text{11.3}$$

$$A + B + C = D \qquad \text{11.4}$$

$$D = \overline{A \cdot B \cdot C} \qquad \text{11.7}$$

$$\overline{A \cdot B \cdot C} = D \qquad \text{11.8}$$

- Logical NOT

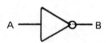

Figure 11.3　NOT Symbol

- Logical NOR

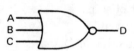

Figure 11.5　NOR Symbol

Table 11.3
Truth Table for NOT Gate

A	B
0	1
1	0

$$B = \overline{A} \qquad \text{11.5}$$

$$\overline{A} = B \qquad \text{11.6}$$

Table 11.5
Truth Table for NOR Gate

A	B	C	D
0	0	0	1
0	0	1	0
0	1	0	0
0	1	1	0
1	0	0	0
1	0	1	0
1	1	0	0
1	1	1	0

- Logical NAND

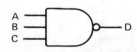

Figure 11.4　NAND Symbol

$$D = \overline{A + B + C} \qquad \text{11.9}$$

$$\overline{A + B + C} = D \qquad \text{11.10}$$

- Exclusive OR (XOR)

Figure 11.6 XOR Symbol

Table 11.6
Truth Table for XOR Gate

A	B	C
0	0	0
0	1	1
1	0	1
1	1	0

$$C = A \oplus B \qquad 11.11$$
$$A \oplus B = C \qquad 11.12$$

- Exclusive NOR (XNOR)

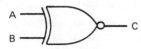

Figure 11.7 XNOR Symbol

Table 11.7
Truth Table for XNOR Gate

A	B	C
0	0	1
0	1	0
1	0	0
1	1	1

$$C = \overline{A \oplus B} = A \odot B \qquad 11.13$$
$$A \odot B = C \qquad 11.14$$

The XNOR gate is also known as a *coincidence gate* and has an alternate symbol.

Figure 11.8 Alternate XNOR Symbol

B. Logical Variables and Constants

A *logical variable* can have one of two possible values: true (1) and false (0). *Logical constants* have only one possible value: 0 or 1. Two logical variables can take on four possible combinations of values: 00, 01, 11, and 10.

Three logical variables can take on eight possible combinations of values: 000, 001, 010, 011, 100, 101, 110, and 111.

In general, N logical variables can take on a total of 2^N possible values. The position of each variable in the string of 1's and 0's remains the same. For example, when $A = 1$, $B = 0$, and $C = 1$, we can write 101 for the sequence ABC.

C. Logical Identities

- Relationships Between Constants

$$0 \cdot 0 = 0 \qquad 11.15$$
$$0 \cdot 1 = 0 \qquad 11.16$$
$$1 \cdot 1 = 1 \qquad 11.17$$
$$0 + 0 = 0 \qquad 11.18$$
$$0 + 1 = 1 \qquad 11.19$$
$$1 + 1 = 1 \qquad 11.20$$
$$\overline{0} = 1 \qquad 11.21$$
$$\overline{1} = 0 \qquad 11.22$$
$$0 \oplus 0 = 0 \qquad 11.23$$
$$0 \oplus 1 = 1 \qquad 11.24$$
$$1 \oplus 0 = 1 \qquad 11.25$$
$$1 \oplus 1 = 0 \qquad 11.26$$

- Relationships Between Variables

$$A + B = B + A \qquad 11.27$$
$$A \cdot B = B \cdot A \qquad 11.28$$
$$A + \overline{A} = 1 \qquad 11.29$$
$$A \cdot \overline{A} = 0 \qquad 11.30$$
$$A + A = A \qquad 11.31$$
$$A \cdot A = A \qquad 11.32$$
$$\overline{\overline{A}} = A \qquad 11.33$$
$$A + (B + C) = A + B + C \qquad 11.34$$
$$A \cdot (B + C) = (A \cdot B) + (A \cdot C) \qquad 11.35$$
$$A + (B \cdot C) = A + B \cdot C \qquad 11.36$$
$$A \oplus B = \overline{A} \cdot B + A \cdot \overline{B} \qquad 11.37$$
$$A \odot B = \overline{A} \cdot \overline{B} + A \cdot B \qquad 11.38$$
$$A \oplus B \oplus C = A \oplus (B \oplus C)$$
$$= (A \oplus B) \oplus C \qquad 11.39$$

D. Logical Theorems

$$A + A \cdot B = A \qquad 11.40$$

$$A + \overline{A} \cdot B = A + B \qquad 11.41$$

$$A \cdot C + \overline{A} \cdot B \cdot C = A \cdot C + B \cdot C \qquad 11.42$$

$$A \cdot B + A \cdot C + \overline{B} \cdot C = A \cdot B + \overline{B} \cdot C \qquad 11.43$$

$$A \cdot B + A \cdot \overline{B} = A \;(\text{logical adjacency})$$
$$11.44$$

$$\overline{A \cdot B \cdot C \cdot D \cdot E} = \overline{A} + \overline{B} + \overline{C} \qquad 11.45$$
$$+ \overline{D} + \overline{E} \;(\text{de Morgan})$$

$$\overline{A + B + C + D + E} = \overline{A} \cdot \overline{B} \cdot \overline{C} \cdot \overline{D} \cdot \overline{E}$$
$$11.46$$

(de Morgan)

E. Logical Functions

A single logical variable can have two possible values: *true* and *false*.

For two logical variables there are four possible combinations. These can be expressed two different but equivalent ways: as *product terms* and *sum terms*. The sum terms are equivalent to the product terms by de Morgan's theorem, equations 11.45 and 11.46.

All possible functions of two logical variables can be expressed in product term (called *minterms*) or in sum terms (called *maxterms*).

Table 11.8
Product and Sum Terms

variable combinations $A \quad B$	product terms (minterms)	sum terms $\overline{\text{(maxterms)}}$
0 0	$\overline{A} \cdot \overline{B} = m_{00}$	$\overline{A + B} = \overline{M}_{00}$
0 1	$\overline{A} \cdot B = m_{01}$	$\overline{A + \overline{B}} = \overline{M}_{01}$
1 0	$A \cdot \overline{B} = m_{10}$	$\overline{\overline{A} + B} = \overline{M}_{10}$
1 1	$A \cdot B = m_{11}$	$\overline{\overline{A} + \overline{B}} = \overline{M}_{11}$

Each of the four possible *minterms* given in table 11.8 is true only for the corresponding combination of the two variables. A function that is true for more than one combination is obtained by *ORing* the corresponding minterms. The result is called the *sum of products* or *SOP* form of the logical function expression.

Example 11.1

A function of A and B is

$$F_1 = A$$

This is independent of whether B is true or false. In minterm formulation, however, all terms that contain A must be included:

$$F_1 = A \cdot B + A \cdot \overline{B}$$

From the logical adjacency theorem, this is $F_1 = A$.

Example 11.2

The XOR or EXOR function of two variables is true when one variable is true and the other is false. This corresponds to

$$F_2 = A \oplus B = \overline{A} \cdot B + A \cdot \overline{B}$$

Example 11.3

The NEXOR or XNOR function is true when both variables are false or when both are true.

$$F_3 = A \odot B = \overline{A} \cdot \overline{B} + A \cdot B$$

The minterm formulation of a logical function may not be in its simplest form with respect to the utilization of logic gates. For example, in example 11.1 a blind implementation would require two AND gates and one OR gate, while a straight-through wire would be sufficient.

The *maxterm* formulation of a function is determined somewhat indirectly. To begin with, the *complement* or NOT of the function is formed with a sum of the corresponding maxterms. Then de Morgan's theorem is applied to obtain the function as a product-of-sums (POS) formulation.

Example 11.4

The function is F_1 as in example 11.1. To form the POS or maxterm formulation, first express the NOT of the function as a sum of the corresponding maxterms:

$$\overline{F_1} = \overline{M}_{00} + \overline{M}_{01} = \overline{(A + B)} + \overline{(A + \overline{B})}$$

Then complement the result and apply de Morgan's theorem:

$$F_1 = \overline{\overline{F_1}} = \overline{\overline{M}_{00} + \overline{M}_{01}} = \overline{\overline{M}_{00}} \cdot \overline{\overline{M}_{01}}$$

$$= M_{00} \cdot M_{01}$$

$$F_1 = \overline{\overline{(A + B)} + \overline{(A + \overline{B})}} = \overline{\overline{(A + B)}} \cdot \overline{\overline{(A + \overline{B})}}$$

$$= (A + B) \cdot (A + \overline{B})$$

Functions of three variables are sufficiently complex to justify describing them with a truth table augmented by minterms and maxterms. This is done in example 11.5.

Example 11.5

Given the truth table, obtain the function in both SOP and POS forms (called *canonical forms*).

$X\ Y\ Z$	F_4	m	M
0 0 0	0	$-$	$X + Y + Z$
0 0 1	1	$\overline{X} \cdot \overline{Y} \cdot Z$	$-$
0 1 0	1	$\overline{X} \cdot Y \cdot \overline{Z}$	$-$
0 1 1	0	$-$	$X + \overline{Y} + \overline{Z}$
1 0 0	1	$X \cdot \overline{Y} \cdot \overline{Z}$	$-$
1 0 1	0	$-$	$\overline{X} + Y + \overline{Z}$
1 1 0	0	$-$	$\overline{X} + \overline{Y} + Z$
1 1 1	1	$X \cdot Y \cdot Z$	$-$

First, fill in the columns with the corresponding minterms (where F_4 is 1) and maxterms (where F_4 is 0).

Next, combine the minterms and maxterms.

$$\text{SOP}: F_4 = \overline{X} \cdot \overline{Y} \cdot Z + \overline{X} \cdot Y \cdot \overline{Z}$$
$$+ X \cdot \overline{Y} \cdot \overline{Z} + X \cdot Y \cdot Z$$
$$\text{POS}: F_4 = (X + Y + Z) \cdot (X + \overline{Y} + \overline{Z})$$
$$\cdot (\overline{X} + Y + \overline{Z}) \cdot (\overline{X} + \overline{Y} + Z)$$

(Note that the truth table entry of maxterms was the actual maxterm instead of the NOTed maxterms in the general table for combinations of two variables.)

For the function illustrated in example 11.5, there was the same number of 1's as 0's. This was a coincidence. $A \cdot B \cdot C$ has one 1 and seven 0's, while $A + B + C$ has seven 1's and only one 0.

F. Canonical SOP Realization, NAND-NAND

The term *realization* indicates a logic diagram using gates. The SOP expression of example 11.5 indicates a two-level AND-OR realization, requiring four three-input AND gates. The outputs from those go to a four-input OR gate as indicated in figure 11.9. There it is assumed that X, Y, and Z (the original variables) and their complements $\overline{X}$, $\overline{Y}$, and $\overline{Z}$ are all available.

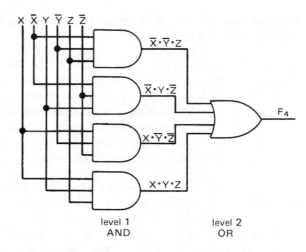

Figure 11.9 Realization of Example 11.5

In figure 11.9, the connection points are indicated with a dot; non-connecting crossings of lines do not have the dot. This is standard practice in digital circuits.

There are two levels of logic in figure 11.9. The first is the AND level, and the second is the OR level.

It is possible to achieve the same two-level SOP realization using only NAND gates. This is best seen in a truth table for the NAND gate.

Table 11.9
NAND Gate Truth Table

A	B	$\overline{A \cdot B}$	$\overline{A} + \overline{B}$
0	0	1	1
0	1	1	1
1	0	1	1
1	1	0	0

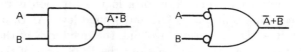

Figure 11.10 Two Equivalent Symbols for NAND Gate

De Morgan's theorem is used to prove the equivalence.

$$\overline{A \cdot B} = \overline{A} + \overline{B} \qquad 11.47$$

Thus, a NAND gate has two representations—an AND gate with a *bubble* on the output, and an OR gate with bubbles on the input as indicated in figure 11.10. The bubbles are (inverting) NOT functions.

The NAND gate at level one in figure 11.10 has the inverting bubble on the output, while the NAND gate at level two has bubbles on the input. The result is a cancellation of inversions.

A two-level NAND-NAND realization for the function of example 11.5 is shown in figure 11.11. This is canonical AND-OR realization because it is composed entirely of minterms. It is not a minimum realization since the function can be implemented using fewer gates. (In fact, the function is the XOR of X, Y, and Z. As such it can be realized using two two-input XOR gates.)

In general, logic functions are minimized for gate logic, resulting in circuit simplification. For minimized functions, it is common for one or more variables to be brought in at the second level (the OR input level). The variables must be inverted (i.e., put through a NOT gate) at this level if they are not also available in the complement.

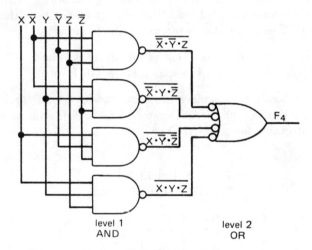

Figure 11.11   NAND-NAND Realization of the AND-OR Function

G. Canonical POS Realization, NOR-NOR

The product of sums form of a logical function is realized using a first-level OR and second-level AND, or an OR-AND realization involving the maxterms of the function. For example, consider the POS form of the function F_4 of example 11.5.

$$F_4 = (X+Y+Z)\cdot(X+\overline{Y}+\overline{Z})\cdot(\overline{X}+Y+\overline{Z})\cdot(\overline{X}+\overline{Y}+Z)$$

The NOR function equivalents from de Morgan's theorem are

$$A \text{ NOR } B = \overline{A+B} = \overline{A}\cdot\overline{B}$$

Thus, the NOR gate can be represented as an OR gate with a bubble on the output, or as an AND gate with bubbles on the input. In the OR-AND realization, as shown in figure 11.12, the first level can be thought of as an OR level, and the second as an AND level. This is apparent since the two bubbles between the output

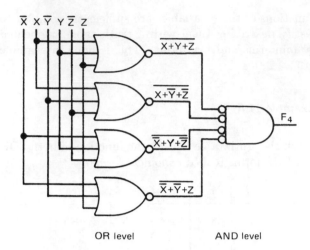

Figure 11.12   OR-AND Realization of a POS Function Using NOR-NOR Logic

of the OR stage and the input of the AND stage cancel each other.

H. Logic Levels

Mathematically, a logic 1 means true and 0 means false. However, logic circuits use voltage levels which are differentiated as high and low. For example, in TTL logic the high level is above about +2 volts and the low level is between 0 and about +1 volt. The association of one level or the other with a logic 1 is arbitrary, although most often the logic 1 is associated with the higher voltage level and the logic 0 with the lower level. This leads to difficulties in some applications where it is necessary to use the lower voltage level as the logic 1 (true) level.

For example, in the NAND-NAND logic scheme, the AND outputs are complemented, which means they are at the low level when they are true. However, the OR outputs are true at the higher voltage level, so the logic alternates between positive and negative logic. In order to avoid confusion, manufacturer truth tables are usually expressed in terms of H (high voltage level) and L (low voltage level), rather than 1 and 0. When such tables are expressed in terms of 1 and 0, the user is to take 1 as meaning the higher voltage level and 0 the lower. In such cases, the actual logical values of signals (inputs or outputs) may be reversed, depending on the situation.

A preferred alternative is to use *assertion levels*. For instances when the low voltage level is true (and the higher voltage level is false), the term *active low* indicates the assertion (true) level is low. The corresponding term for *positive logic* or high assertion level is *active high*.

The logic symbols are used to indicate the assertion levels, using the circles or "bubbles" at the low assertion level. For instance, a NOT gate can have the bubble at either the input or the output as indicated in figure 11.13, depending whether (a) the input is asserted high (bubble at output), or (b) the input is asserted low (bubble at the input).

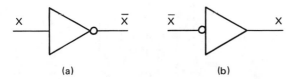

Figure 11.13   Alternative Symbols for a NOT Gate: (a) Usual Symbol When the Input Is Asserted High, (b) Symbol When the Input Is Asserted Low

The assertion level determines the function of gates more complicated than the NOT gate. For instance, the NAND gate with low assertion level inputs performs the OR function as indicated in figure 11.10 and the level 2 of figure 11.11. This can be understood through de Morgan's theorem. Similarly, the NOR gate performs the AND function when its inputs are asserted low, as indicated in the AND level symbol of figure 11.12.

The usual practice is to express a variable in its complemented form when it is asserted low. This is because with positive logic, the variable is true when its complement is low.

Like the NAND gate, the AND gate performs the OR function when its inputs are asserted low (the output is also asserted low) as indicated in figure 11.14. Similarly to the NOR gate, the OR gate performs the AND function when its inputs are asserted low (the output is also asserted low) as indicated in figure 11.15.

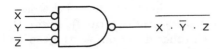

Figure 11.14   Symbol for an AND Gate When the Inputs (and Output) Are Asserted Low

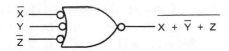

Figure 11.15   Symbol for an OR Gate When the Inputs (and Output) Are Asserted Low

2   LOGICAL MINIMIZATION TECHNIQUES

The primary method of minimizing logical functions depends on the *theorem of logical adjacency*:

$$A \cdot B + A \cdot \overline{B} = A \qquad 11.48$$

This can be generalized for functions, as well as for variables.

$$F \cdot B + F \cdot \overline{B} = F \qquad 11.49$$
$$F \cdot G + F \cdot \overline{G} = F \qquad 11.50$$

F and G are functions, and B is a simple logical variable. Equations 11.49 and 11.50 indicate a method of minimizing minterm expressions (SOP). A simple expansion of maxterm expressions shows that

$$(A + B) \cdot (A + \overline{B}) = A \cdot A + A \cdot \overline{B} + B \cdot A + B \cdot \overline{B}$$
$$= A + A \cdot (B + \overline{B}) + 0 = A$$
$$11.51$$

In general,
$$(F + G) \cdot (F + \overline{G}) = F \qquad 11.52$$

Equation 11.52 is another form of the logical adjacency theorem.

The logical adjacency theorem implies that minimization of SOP and POS functions can be achieved by inspecting the logical adjacency of minterms or maxterms. Adjacency can be determined from inspection of the truth table and by combining minterms (or maxterms) that differ by only one value.

For example, $A \cdot B \cdot C \cdot D = 1111$ differs from $\overline{A} \cdot B \cdot C \cdot D$ ($= 0111$), from $A \cdot \overline{B} \cdot C \cdot D (= 1011)$, from $A \cdot B \cdot \overline{C} \cdot D (= 1101)$, and from $A \cdot B \cdot C \cdot \overline{D} (= 1110)$, each by only the value of a binary variable. Therefore, any of the latter can be combined with $A \cdot B \cdot C \cdot D$ to eliminate the one variable that differs.

Example 11.6

A function consists of the maxterms $X + \overline{Y} + \overline{Z}$, $X + \overline{Y} + Z$, and $\overline{X} + Y + \overline{Z}$. Simplify the function.

These maxterms can be represented by the logical values of the variables when they are true.

$$X + \overline{Y} + \overline{Z} = 100$$
$$X + \overline{Y} + Z = 101$$
$$\overline{X} + Y + \overline{Z} = 010$$

The first two maxterms are adjacent since they differ in the true value of Z. Thus, they can be combined. The third term is not adjacent to either of the others, and it cannot be simplified by combining. The resulting function is

$$F = (X + \overline{Y}) \cdot (\overline{X} + Y + \overline{Z})$$

Truth tables with multiple input variables are not conducive to visualizing adjacencies. The Karnaugh map is much easier to work with.

The *Karnaugh map* is a representation of the input variable *space* for a logical function arranged to visually present the logical adjacencies. There is a one-to-one relationship between the Karnaugh map and the truth table. The binary values associated with the input values in the truth table are converted to decimal values for convenience. These values are then placed in the Karnaugh map to indicate the corresponding location of minterms and maxterms.

decimal value	inputs A B	output C
0	0 0	0
1	0 1	1
2	1 0	0
3	1 1	0

(a)

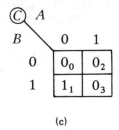

(b)

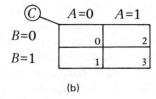

(c)

Figure 11.16   Two-Input
Karnaugh Map Development

A two-input Karnaugh map development appears in figure 11.16. Part (b) shows the Karnaugh map and the correspondence of the map locations to the truth table decimal equivalents of the binary number the input conditions require for each true condition. Part (c) shows the entries from the truth table, with ones corresponding to the minterms of the function C and 0's to the maxterm values of the function.

decimal value	inputs A B C	output F
0	0 0 0	1
1	0 0 1	1
2	0 1 0	0
3	0 1 1	0
4	1 0 0	0
5	1 0 1	1
6	1 1 0	1
7	1 1 1	0

(a)

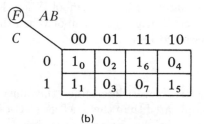

(b)

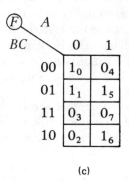

(c)

Figure 11.17   Three-Input
Karnaugh Map Development

The development of two possible three-input Karnaugh maps is shown in figure 11.17. Part (b) shows one possible arrangement with the first two variables (AB) values across the top and the last (so-called *least-significant* because of its position as a binary bit) on the side. Part (c) shows a different arrangement that contains the same information arranged with the least significant variables down the side.

In Figure 11.17(b), the decimal term 0 is adjacent to terms 1 and 2, while in (c) the 0-term is adjacent to terms 1 and 4. This seeming inconsistency can be accounted for by visualizing the maps as wrapping around a cylinder. In (b), the right and left columns meet, and adjacencies are obvious, (i.e., term 1 is adjacent to term 5 and term 4 is adjacent to term 0). In (c), the cylinder axis is in the other direction, so the top and bottom rows are adjacent.

In the three-input map, the numbering for the dimension containing two variables is not binary, as binary

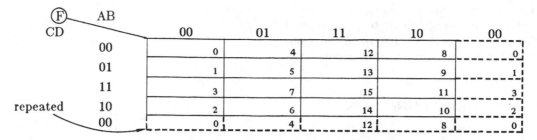

Figure 11.18 A Four-Input Karnaugh Map

G ABC

DE	000	001	011	010	110	111	101	100
00	0	4	12	8	24	28	20	16
01	1	5	13	9	25	29	21	17
11	3	7	15	11	27	31	23	19
10	2	6	14	10	26	30	22	18

Figure 11.19 Single-Layer
Five-Input Karnaugh Map

numbering does not ensure logical adjacency. The numbering technique is called *unit distance coding*, and adjacent numbers differ by only one bit. This technique of indicating adjacency is used in the Karnaugh map for four inputs shown in figure 11.18. The adjacency is made obvious by repeating the first column on the right, and by repeating the first row on the bottom.

The four-input Karnaugh map has 16 boxes, four-by-four. The columns are from left to right: $AB = 00, 01, 11, 10$, with rows from top to bottom: $CD = 00, 01, 11, 10$. This is shown in figure 11.18. In figure 11.18 the first column is repeated (dashed lines) to the right of the fourth column to emphasize logical adjacencies. Also the first row is repeated below the fourth row for the same emphasis. In practice the repeated rows are not used.

A five-input Karnaugh map can be made by doubling one dimension on the four-input map as in figure 11.19, where adjacencies are somewhat difficult to perceive. For example, the fact that term 17 and term 25 are adjacent could be easily overlooked.

Another representation of the five-input Karnaugh map conceptually folds the right-hand side of the map of figure 11.19 under the left-hand side to make a three-dimensional map as in figure 11.20. The most significant variable (A) is indicated by dividing each square diagonally with $A = 0$ in the upper half and $A = 1$

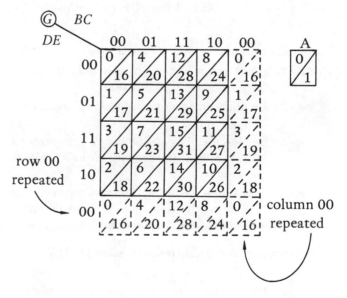

Figure 11.20 Three-Dimensional Karnaugh Map

in the lower half. This representation shows all adjacencies by the extension method of the four-input map. Terms are adjacent if they are in the same square or if they are in adjacent squares in the same relative half (adjacent squares in the upper half, or adjacent squares in the lower half).

A five-input Karnaugh map is as complex as can be used practically. A six-input map can be constructed by dividing each square of the five-dimensional map in half again, but this becomes difficult to perceive, and all adjacencies cannot be shown. For more than four inputs, it is useful to achieve minimization by computer.

3　MINIMIZATION OF FUNCTIONS

As SOP and POS expressions are nontrivial only for functions of three or more inputs, minimization methods are needed. These minimization techniques guarantee a minimum number of gates and gate leads only for POS or SOP realizations. There is no guarantee that the reduced functions found are absolute minimizations. They are good starting places for small-scale logic designs. The techniques developed here are very useful in the design of flip-flop logic for sequential circuits.

For canonical forms with three-input variables, there is a total of eight possible terms, some of which correspond to minterms for the SOP realization. The remainder corresponds to maxterms for a POS implementation. In an otherwise unrestricted situation, choose the canonical form that results in the least number of packages. This generally corresponds to the minimum number of gate input leads. Unless otherwise restricted, try minimizing both the POS and the SOP realizations of the function.

The minimization procedure is given here.

step 1: Construct a Karnaugh map from the truth table or given logic function.

step 2: Combine adjacent terms into the largest possible groupings.

step 3: Include a sufficient number of groupings to include each of the original terms in at least one group.

step 4: Check the resulting function against the original truth table. It must have the same outputs for the same inputs.

This procedure is demonstrated by example 11.7

Example 11.7

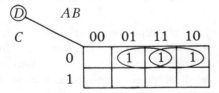

The function $\mathbf{D}\ (A, B, C)$ has the truth table shown. Obtain a minimal NAND-NAND or NOR-NOR realization of this function.

ABC	$\mathbf{D}$
0 0 0	0
0 0 1	0
0 1 0	1
0 1 1	0
1 0 0	1
1 0 1	0
1 1 0	1
1 1 1	0

The function has only three ones, so the SOP form is tried first. The minterms are entered in the map. Next, the adjacent terms are circled to indicate that they can be combined. With this function, the largest possible grouping is of two terms.

In general, terms can be grouped in twos, fours, and eights (i.e., in powers of 2), as will be demonstrated later. Each grouping has eliminated the one variable that is not common to the group. The left-hand group has $C = 0$ and $B = 1$ in common. This is true when the combination $\overline{C} \cdot B$ is true. The right-hand group has $C = 0$ and $A = 1$ in common, which is true when $\overline{C} \cdot A$ is true. Thus, the function written with these two terms is

$$\mathbf{D} = \overline{C} \cdot B + \overline{C} \cdot A$$

A	B	C	$B \cdot \overline{C}$	$A \cdot \overline{C}$	$\overline{C} \cdot B + \overline{C} \cdot A$
0	0	0	0	0	0
0	0	1	0	0	0
0	1	0	1	0	1
0	1	1	0	0	0
1	0	0	0	1	1
1	0	1	0	0	0
1	1	0	1	1	1
1	1	1	0	0	0

This can be checked with a truth table. The reduced function has the same outputs as the original function. The realization using NAND-NAND logic, assuming all inputs and their complements are available, is shown.

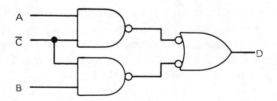

The NAND-NAND function can be further simplified, but it would no longer be a NAND-NAND function. As will be shown, the simplification is the direct result of the POS or maxterm formulation. Because the maxterm formulation is one step more abstract than the minterm formulation, the process will be described here in detail.

To obtain the maxterms, enter the zeros from the truth table as shown. The maxterms are made up of the ORed complements of the input variable values.

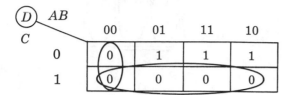

A location containing a zero has a maxterm that includes the entire map excepting that zero location. For instance, the location $ABC = 010$ has the associated maxterm $A+\overline{B}+C$. Similarly, groups of zeros have sumterms, or OR terms, associated with them. The group of two zeros at locations 000 and 001 have the sum-term $A + B$. Likewise, the row of zeros across the bottom have locations $ABC = 001$, 011, 111, and 101. These have the respective maxterms $A + B + \overline{C}$, $A + \overline{B} + \overline{C}$, $\overline{A} + \overline{B} + \overline{C}$, and $\overline{A} + B + \overline{C}$. These can be combined by logical adjacency to $\overline{C}$, which is the OR term of the row of zeros.

The Karnaugh map, then, contains two OR terms: $A + B$ and $\overline{C}$. The function **D** is the AND of these two ORs.

$$\mathbf{D} = (\overline{C}) \cdot (A + B)$$

This can be realized in NOR-NOR logic. Because the $\overline{C}$ must be introduced at the AND (second) level, it must be in complement form (i.e., NOT $\overline{C}$ or C).

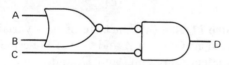

The verification of the truth table is omitted here.

4 MULTIPLE ADJACENCIES AND VEITCH DIAGRAMS

Multiple adjacencies are important in function minimization. They have been briefly introduced in the previous example. The procedure can be formalized by the extended use of the logical adjacency theorem.

$$A \cdot B \cdot \overline{C} \cdot \overline{D} + A \cdot B \cdot C \cdot \overline{D} + A \cdot B \cdot \overline{C} \cdot D$$
$$+ A \cdot B \cdot C \cdot D = A \cdot B \qquad 11.53$$

It is useful to introduce another way of viewing the Karnaugh map—in terms of the variables by name rather than by value. First, redefine the rows and columns of a three-input Karnaugh map in terms of the input values that select them. This is shown in figure 11.21(a).

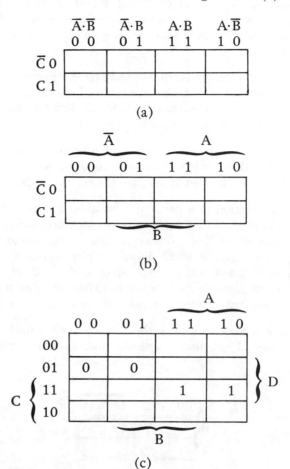

Figure 11.21 (a) Three-Inputs by Name, (b) Veitch Diagram Notation, (c) Four-Input Veitch Diagram

When $A = 0$, A is false, and the true condition is $\overline{A}$. Therefore, wherever A is noted as a 0 on the Karnaugh map, the corresponding value of $\overline{A}$ is true, or $\overline{A} = 1$. Thus, the *Veitch diagram* is simply a Karnaugh map, which tells where a variable is true. Where it is not true, the complement is true.

This is an extremely useful notion in interpreting the Karnaugh map for purposes of establishing reduced functions. In the notation indicated in figure 11.21(b), the location 011 is seen to be $\overline{A} \cdot B \cdot C$ in minterm realization, and $A + \overline{B} + \overline{C}$ in maxterm realization.

A four-input Karnaugh map with Veitch diagram enhancement is shown in figure 11.21(c). The entry of the maxterms $A + \overline{B} + C + \overline{D}$ and $A + B + C + \overline{D}$ can be seen as the zeros at $\overline{A} \cdot B \cdot \overline{C} \cdot D$ (0101) and at

$\overline{A} \cdot \overline{B} \cdot \overline{C} \cdot D$ (0001) as indicated on the diagram. These can be combined to give the term $A + C + \overline{D}$, eliminating the B term.

Likewise, the two ones at positions 1111 and 1011 can be interpreted as $A \cdot B \cdot C \cdot D$ and $A \cdot \overline{B} \cdot C \cdot D$ minterms, respectively. These terms can be combined as $A \cdot C \cdot D$. To do this, note that both terms have A, C, and D in common.

The Veitch diagram is equivalent to the Karnaugh map, but it permits the immediate entry of ones and zeros from a logical expression, and it further allows the direct writing of a logical expression.

In general, it is recommended that Karnaugh maps be constructed to include both the minterm information (values) and the Veitch diagram information (names).

As the position of a particular variable in a truth table is random, all locations of combinations of different variables in the K-V (Karnaugh-Veitch) map should be inspected. Figure 11.22 shows the placements of the groups of four terms corresponding to A, $\overline{A}$, B, $\overline{B}$, C, and $\overline{C}$ for three-input functions. (These are shown as minterms, but maxterms replace the ones and zeros.)

With four-variable inputs, the groupings (also called *encirclements*) are more numerous and, therefore, more

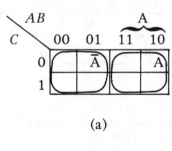

(a)

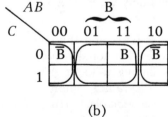

(b)

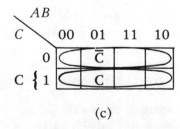

(c)

Figure 11.22   (a) A and $\overline{A}$,
(b) B and $\overline{B}$, and (c) C and $\overline{C}$

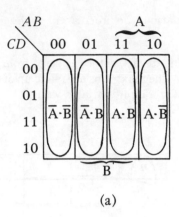

(a)

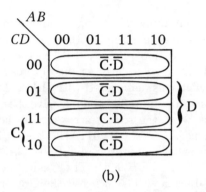

(b)

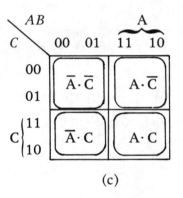

(c)

Figure 11.23   (a) Functions of Four Variables Dependent Only on A and B, (b) Functions Dependent Only on C and D, and (c) Functions Dependent Only on A and C

complex. The obvious groups of four terms are shown in figure 11.23. For the arbitrary definition of variables to this point, these correspond to (a) $\overline{A} \cdot \overline{B}$, $\overline{A} \cdot B$, $A \cdot B$, and $A \cdot \overline{B}$; (b) $\overline{C} \cdot \overline{D}$, $\overline{C} \cdot D$, $C \cdot D$, and $C \cdot \overline{D}$; and (c) $\overline{A} \cdot \overline{C}$, $\overline{A} \cdot C$, $A \cdot C$, and $A \cdot \overline{C}$. The corresponding maxterms involve the complements of the variables in each case.

The three remaining combinations (shown in figure 11.24) of four terms are more complicated, as they require wrapping a cylinder around the K-V map. Those combinations that eliminate all but A and D require a cylinder placed so that the top and bottom of the

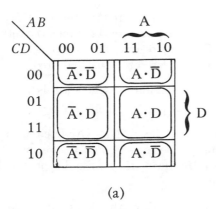

(a)

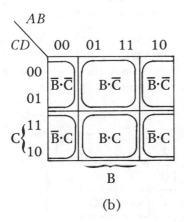

(b)

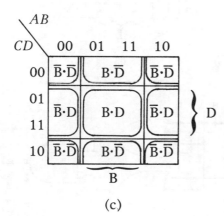

(c)

Figure 11.24 (a) Functions of Four Variables Dependent Only on A and D, (b) Functions Dependent Only on B and C, and (c) Functions Dependent Only on B and D

map are joined (a); those removing all but B and C as variables require a cylinder connecting the left and right edges (b); and those removing all but B and D as variables require a cylinder that is more like a sphere, in that it has to connect simultaneously the top and bottom and the sides. This joins the four corners (c).

For functions of four variables with eight combinable adjacencies, the results are the values given in the Veitch diagram, or their complements: A, $\overline{A}$, B, $\overline{B}$, C, $\overline{C}$, D, and $\overline{D}$.

inputs	segments	decimal
$A\ B\ C\ D$	$a\ b\ c\ d\ e\ f\ g$	number
0 0 0 0	1 1 1 1 1 1 0	0
0 0 0 1	0 1 1 0 0 0 0	1
0 0 1 0	1 1 0 1 1 0 1	2
0 0 1 1	1 1 1 1 0 0 1	3
0 1 0 0	0 1 1 0 0 1 1	4
0 1 0 1	1 0 1 1 0 1 1	5
0 1 1 0	1 0 1 1 1 1 1	6
0 1 1 1	1 1 1 0 0 0 0	7
1 0 0 0	1 1 1 1 1 1 1	8
1 0 0 1	1 1 1 1 0 1 1	9
1 0 1 0	X X X X X X X	not used
1 0 1 1	X X X X X X X	not used
1 1 0 0	X X X X X X X	not used
1 1 0 1	X X X X X X X	not used
1 1 1 0	X X X X X X X	not used
1 1 1 1	X X X X X X X	not used

(a)

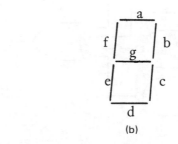

(b)

CD \ AB	00	01	11	10
00	0	4	X_{12}	8
01	1	5	X_{13}	9
11	3	7	X_{15}	X_{11}
10	2	6	X_{14}	X_{10}

(c)

Figure 11.25 Seven-Segment Display (a) Truth Table, (b) Identification of Segments, (c) Karnaugh Map with X's to Indicate Don't-Care Conditions

For four-input minterm realizations, the minimization involves the recognition of combinable terms in all possible combinations. An interesting case is where there are inputs that cannot occur and are, therefore, called *don't-care* inputs.

An example of this is a binary-coded decimal seven-segment display. There are four inputs that can vary in decimal value from 0 through 9. The input signals reflect this limitation, and the binary combinations of 1010 (10_{10}) through 1111 (15_{10}) do not occur.

The don't-care conditions can be used to minimize the logical function by assigning values of true or false as is advantageous. For example, take the segment a from the truth table of figure 11.25. Enter the ones and zeros from there into the Veitch diagram on figure 11.26.

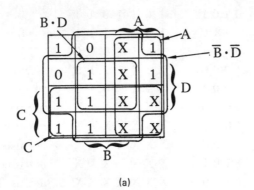

(a)

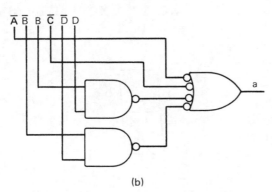

(b)

Figure 11.26 (a) Veitch Diagram for Segment a, (b) NAND-NAND Realization of Segment a Logical Function

The reduced logical expression for a must include all of the ones as minterms. This means that each must be included in at least one chosen group. The minimum function will have the minimum number of reduced minterm products. The unreduced minterms are referred to by number, and they have been enumerated in the Karnaugh map of figure 11.25(c).

If the don't-care of minterm 10 is treated as a *one*, the four-corners grouping ($\overline{B} \cdot \overline{D}$) takes care of minterms 0, 2, 8, and 10.

The don't-cares of minterms 13 and 15 are treated as ones, to group with minterms 5 and 7 in $B \cdot D$.

When the don't-cares of 11 and 14 are added to the list of ones, minterms 2, 3, and 6 can be covered by C. This also covers minterm 7, which had already been covered. There is no reason to avoid multiple coverages of one term if it reduces the complexity.

The only remaining uncovered minterm is 9. By making the one remaining don't-care, 12, a one, the total region A can be included to cover 9. In doing so, this also covers terms that were previously covered.

The resulting expression for segment a is

$$a = A + \overline{B} \cdot \overline{D} + B \cdot D + C \qquad 11.54$$

This is realized by the NAND-NAND circuit shown in figure 11.26(b). Because A and C are introduced at the second (OR) level, they must be present in their complements.

In this illustration, all of the don't-care were changed to ones. This is not necessary in every case.

An alternative to minterm realization is the *maxterm method*, resulting in *NOR-NOR logic*. For purposes of comparison, the segment a of the seven-segment display is used. The Veitch diagram is shown in figure 11.27.

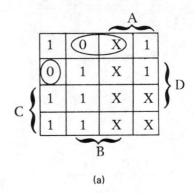

(a)

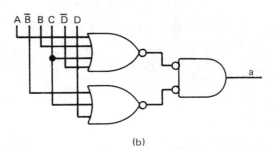

(b)

Figure 11.27 POS Realization of the Segment a (a) Veitch Diagram, (b) NOR-NOR Realization

As the two zeros cannot be combined, there will be two maxterms. The term at minterm position 1 cannot be combined with the other zero or don't-care. The maxterm is made up of the complements of the position's minterm

$$A + B + C + \overline{D}$$

The zero at position 4 can be combined with the don't-care at position 12. The resulting maxterm is the complement of the combined minterm for the group

$$\overline{B} + C + D$$

Thus, the maxterm expression is

$$a = (A + B + C + \overline{D}) \cdot (\overline{B} + C + D) \qquad 11.55$$

Only one don't-care was changed to a zero, so the remaining are treated as ones. The NOR-NOR realization of the reduced function is shown.

5 EXCLUSIVE OR AND EXCLUSIVE NOR LOGIC

A. Introduction

The XOR function is also called a *ring sum*, and it can be considered as an *odd-parity function*. For example, A XOR B, $A \oplus B$, is true only when one of the two variables is true. A XNOR B is true when both variables are the same.

When a third variable is included, as in $A \oplus B \oplus C$, the operation must be done in steps. $A \oplus B$ is taken first; then XOR is applied to that result with C. The order in which this is done is of no importance. $A \oplus B \oplus C$ will be true whenever there is an odd number of trues present.

$$A \oplus B \oplus C = (A \cdot B \cdot C) + (\overline{A} \cdot \overline{B} \cdot C) + (\overline{A} \cdot B \cdot \overline{C}) + (A \cdot \overline{B} \cdot \overline{C})$$
$$11.56$$

It also can be seen from simple permutation theory that

$$A \oplus B \oplus C = \overline{A} \oplus \overline{B} \oplus C = \overline{A} \oplus B \oplus \overline{C}$$
$$= A \oplus \overline{B} \oplus \overline{C} \qquad 11.57$$

The XNOR function is the complement of the XOR function and is an *even-parity function*.

$$A \odot B \odot C = \overline{A \oplus B \oplus C} = \overline{A} \oplus \overline{B} \oplus \overline{C}$$
$$= \overline{A} \oplus B \oplus C = A \oplus \overline{B} \oplus C$$
$$= A \oplus B \oplus \overline{C} \qquad 11.58$$

In general, an XOR function of any number of variables can have only two values. There will be either an odd or an even number of trues. For any XOR function it is necessary only to count the number of complemented variables in the expression. If this number is even, then the function is true for odd parity. If the number of complemented variables is odd, then the function is true for even parity.

For example, the function $A \oplus B \oplus C \oplus \overline{D}$ is true with even parity when an even number of the variables are true. When $ABCD$ = 0000, 0011, 0101, 1001, 1010, 1100, 0110, or 1111, the function is true.

B. XOR Maps

The Karnaugh map and Veitch diagram for exclusive-OR functions using all map variables result in a checkerboard pattern as shown in figure 11.28. The alternating ones and zeros result in even-parity functions.

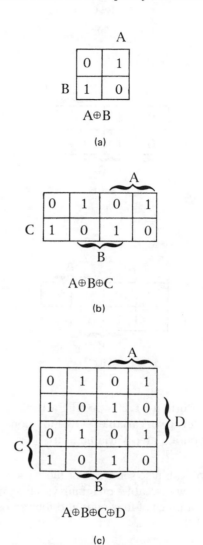

Figure 11.28 XOR Patterns for (a) Two-Variable, (b) Three-Variable, and (c) Four-Variable Maps

It is useful to recognize XOR functions with fewer than all of the map variables. For example, the XOR functions of two variables on a three-variable map are shown in figure 11.29.

The XOR function of two variables results in groupings of two ones and two zeros. Patterns for a four-variable map involving XOR of three variables also result in groupings of ones and of zeros pairs, as is shown in figure 11.30.

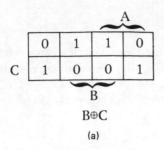

B⊕C

(a)

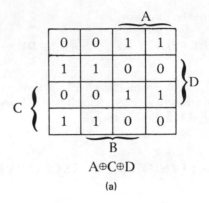

A⊕C⊕D

(a)

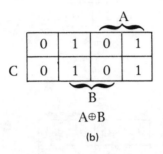

A⊕B

(b)

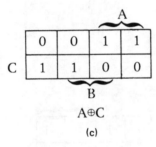

A⊕C

(c)

Figure 11.29   XOR Patterns
for Two-Variable XOR Functions
on a Three-Variable Map

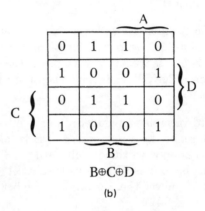

B⊕C⊕D

(b)

More complex, but to some extent more useful, are XOR
functions of two variables on a four-variable map. These
result in groupings of four ones and four zeros as shown
in figure 11.31.

C. XOR With Minterms

XOR functions can be included with minterms. Additional logic levels are produced, but with the further
minimization of gates and gate leads. If the XOR or
$\overline{\text{XOR}}$ function involves only two terms, it can be implemented with two-input XOR or $\overline{\text{XOR}}$ gates. The usefulness of these gates depends on recognizing an XOR
function from the map, or from the recognition of such a
term in the minterm expression (i.e., a term $\overline{A}\cdot B + A\cdot\overline{B}$
recognized as XOR, or $A\cdot B + \overline{A}\cdot\overline{B}$ recognized as XNOR).

An example of this is the segment a in a seven-segment
display. This has the logical function

$$a = A + (\overline{B}\cdot\overline{D} + B\cdot D) + C$$

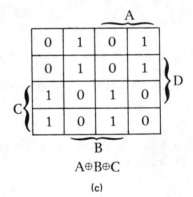

A⊕B⊕C

(c)

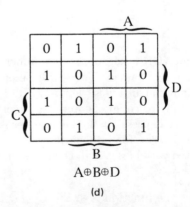

A⊕B⊕D

(d)

Figure 11.30   XOR Patterns
for Functions of Three Variables
on a Four-Variable Map

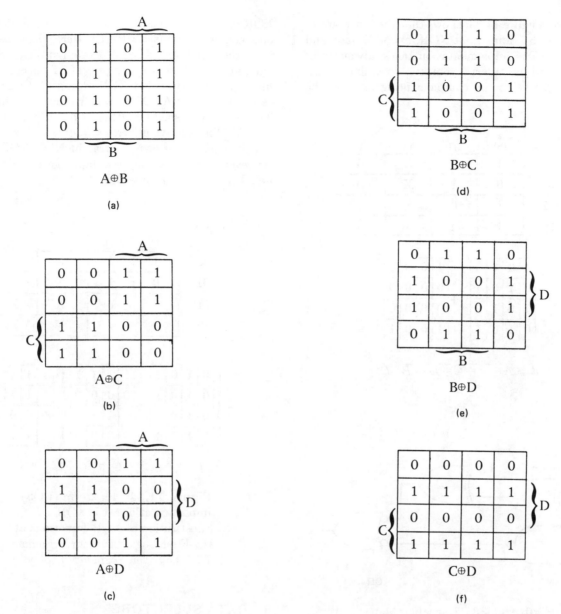

A⊕B

(a)

A⊕C

(b)

A⊕D

(c)

B⊕C

(d)

B⊕D

(e)

C⊕D

(f)

Figure 11.31 XOR Patterns for Functions of Two Variables on a Four-Variable Map

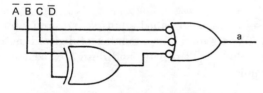

$\overline{A}\ \overline{B}\ \overline{C}\ \overline{D}$

a

Figure 11.32 Using an XOR Gate
to Further Minimize SOP Realization

This can be minimized further by using an XOR gate at the first logic level since the function must be complemented at the second logic level. Figure 11.32 illustrates this. Here $\overline{B} \oplus \overline{D}$ is an odd-parity function and is, therefore, equivalent to $B \oplus D$. Also, the required function was the even parity $\overline{B \oplus D}$. At the OR level of

NAND-NAND logic, however, it is necessary to invert the inputs.

Example 11.8

The logical function which drives segment d of the seven-segment display of figure 11.25 requires a minterm or SOP function of

$$d = A + \overline{B} \cdot C + \overline{B} \cdot \overline{D} + B \cdot \overline{C} \cdot D + C \cdot \overline{D}$$

Realization of this function requires a very awkward set of gates. An eight-input NAND is required, as there are no five-input NAND gates available. This function is not an obvious candidate for the use of an XOR realization. However, a careful study of the Karnaugh map, as indicated in figure 11.33, shows that the function can be realized with a four-input XOR, an AND gate, and an OR gate. The implementation takes advantage of the various odd-parity arrangements, providing the required odd parity to the complementing input of the OR level.

$$d$$ AB

CD	00	01	11	10
00	1	0	X	1
01	0	1	X	1
11	1	0	X	X
10	1	1	X	X

$$= \begin{array}{|c|c|c|c|} \hline 1 & 0 & 1 & 0 \\ \hline 0 & 1 & 0 & 1 \\ \hline 1 & 0 & 1 & 0 \\ \hline 0 & 1 & 0 & 1 \\ \hline \end{array} + \begin{array}{|c|c|c|c|} \hline 1 & 0 & 0 & 1 \\ \hline 0 & 0 & 0 & 0 \\ \hline 0 & 0 & 0 & 0 \\ \hline 1 & 0 & 0 & 1 \\ \hline \end{array}$$

$$d = \overline{A \oplus B \oplus C \oplus D} + \overline{B} \cdot \overline{D}$$

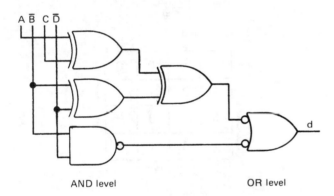

Figure 11.33   Karnaugh Map NAND-NAND Implementation (with XOR Gates Substituted) for Segment d

With this type of implementation, there is an additional level of logic added. The extra level slows down the system due to gate delays. However, it requires a smaller number of integrated circuit packages.

D. XOR With Maxterms

XOR or XNOR gates usually can be substituted for either minterms or maxterms at the expense of increased gating levels. With maxterms it is, as usual, more difficult to visualize the process. For example, consider segment f of the seven-segment display, which has a maxterm representation of

$$f = (A + B + \overline{D}) \cdot (B + \overline{C}) \cdot (\overline{C} + \overline{D}) \qquad 11.59$$

XOR or XNOR terms could be substituted, depending on recognition of basic patterns. Here, the substitution is at the OR level. Careful examination of the Karnaugh map of f, shown in figure 11.34, reveals that the function is the sum of A XOR B and C XNOR D, except for rows C AND D, which must be eliminated. This is accomplished by ORing the two XOR functions and ANDing the result with $\overline{C}$ OR $\overline{D}$. The resulting circuit is not superior to a straight NOR-NOR circuit. However, it requires only two-input gates. This might be advantageous in some cases.

$$f$$ AB

CD	00	01	11	10
00	1	1	X	1
01	0	1	X	X
11	0	0	X	X
10	0	1	X	X

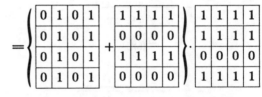

$$f = (A \oplus B + \overline{C} \oplus D) \cdot (\overline{C} + \overline{D})$$

Figure 11.34   Maxterm (POS) Implementation with XOR Substitution of XOR Functions at the First Logic Level for Segment f of the Seven-Segment Display

6   DATA SELECTORS

A. Introduction

Data selectors, also called *multiplexers* and abbreviated MUX, are useful for a number of digital functions. These include multiplexing and design of central processor units.

The simplest data selector involves one *select input* and two function inputs, as shown in figure 11.36. The select must be low to allow I_0 to pass through to the output Y, while it must be high to allow I_1 to pass through. In effect, the select is acting like a control of a two-position switch. The schematic representation of the two-input data selector is shown in figure 11.36(c). Data selectors are available commercially in single packages with four- or two-input (*quad*) versions with a common select for all.

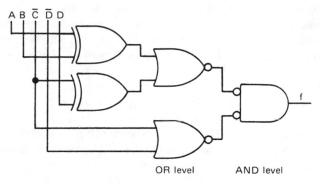

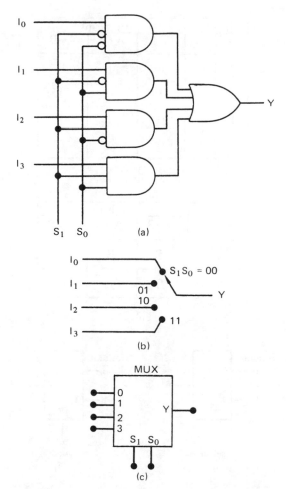

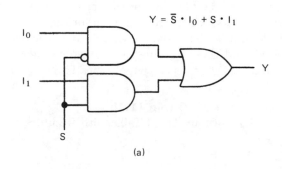

Figure 11.35 XOR Gate
Realization for Segment f

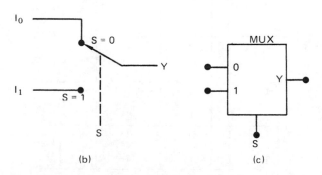

Figure 11.37 (a) Four-Input MUX Logic
Diagram, (b) Switch Analogy, (c) Symbol

Figure 11.36 (a) Two-Input Data
Selector Logic Diagram, (b) Switch
Analogy, (c) Circuit Representation

A four-input data selector has the logic diagram shown in figure 11.37. Note that the input selected corresponds to the line with the decimal number equivalent to the binary number $S_1 S_0$.

B. Two-Input MUX Logic

There are three distinct functions of two variables—AND, OR, and XOR. The AND functions are minterms with four possibilities. The OR functions are maxterms, also with four possibilities. The XOR functions include only two possibilities—the odd-parity and the even-parity functions. All possible functions of two variables are included in this set. (The *trivial functions* of true and false, and the functions that reduce

to only one variable are not included.) A two-input MUX is able to produce any of these functions of two variables.

Representatives of the three types of functions making up all nontrivial functions of two variables are shown in the truth table of figure 11.38(a). For these examples, A is used for the select input. The truth table is divided into two parts, one where $A = 0$ and the other where $A = 1$. For each value of A, the functions are compared to B, and the corresponding *residue functions* are placed in the reduced truth table of figure 11.38(b).

For example, in the AND function for $A = 0$, $\overline{A} \cdot B = B$. For $A = 1$, $\overline{A} \cdot B = 0$. These are entered as the residues in the reduced table, and they become the inputs to the two-input MUX used to generate $\overline{A} \cdot B$ shown in figure 11.38(c).

The OR function is handled similarly. When $A = 0$, the residue is $\overline{B}$ which is the function to be provided at input 0 of the MUX of figure 11.38(d). When $A = 1$, the OR residue is 1, and this is the value provided to input 1.

input		functions		
A	B	$\overline{A}\cdot B$	$A+\overline{B}$	$A\oplus B$
0	0	0	1	0
0	1	1	0	1
1	0	0	1	1
1	1	0	1	0

(a)

select input	residues		
	$\overline{A}\cdot B$	$A+\overline{B}$	$A\oplus B$
$\overline{A}$	B	$\overline{B}$	B
A	0	1	$\overline{B}$

(b)

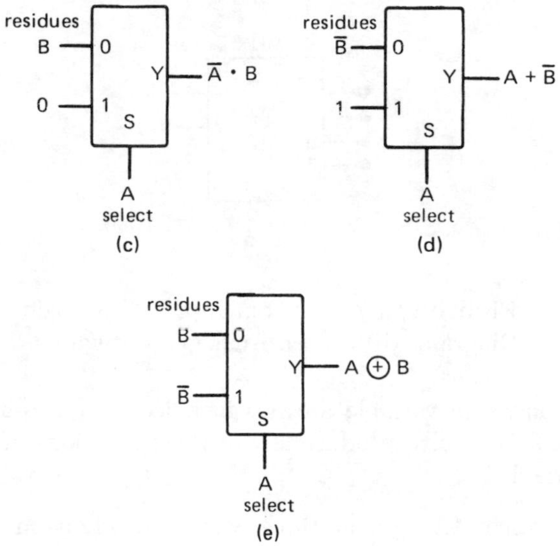

(c)   (d)

(e)

Figure 11.38   Two-Input MUX Logic

For the XOR function with $A = 0$, the function is B; while for $A = 1$, the function is $\overline{B}$. These are the residues to be applied to inputs 0 and 1, respectively, in figure 11.38(e).

These examples show that any function of two variables can be generated with a two-input MUX.

C. Four-Input MUX Logic

A four-input data selector (MUX) requires two select inputs to distinguish between the four data inputs. This permits the realization of single functions with three variables. Although there are 14 possible nontrivial functions of two variables, there are 206 distinct functions using all three variables.

An arbitrary function of three variables can be realized with a four-input data selector with a procedure similar to that of the previous section: (1) Divide the truth table in half by grouping both possibilities of the right-most variable. (2) Examine the relationship of that variable to the function in each case. (3) Find the residues, which can be 1, 0, the right-most function, or its complement. (4) The residue is the input to the data selector for the input corresponding to the decimal equivalent of the binary value of the two left-most variables.

As an illustration, consider the function given in table 11.10. The truth table is divided into groupings according to the two left-most inputs. The right-most input is compared to the function for that combination of the left-most inputs, which are then used for the two select inputs S_1 and S_0. The comparison results in the residue indicated, and these become the corresponding inputs to the data selector, as shown in figure 11.39.

Table 11.10

A Three-Input Truth Table with Residues

A	B	C	Y	res
0	0	0	0	
0	0	1	0	0
0	1	0	1	
0	1	1	0	$\overline{C}$
1	0	0	0	
1	0	1	1	C
1	1	0	1	
1	1	1	1	1

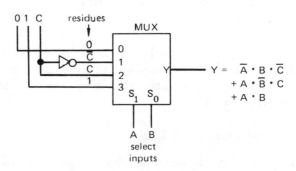

Figure 11.39   Four-Input
Data Selector Realization

This function was chosen to illustrate the four possible residues of one input variable. Four-input data selectors are available in dual packages with common select inputs. Two three-input functions can be generated for each package.

With additional logic gates, it is possible to realize functions of even more than three variables. To do this, it is necessary to extend the concept of residues. The two left-most variables in the truth table are taken as the select inputs; the two right-most variables are used to generate the residues. These are necessarily functions of two variables, which can be generated with two-input data selectors. As a four-input data selector can use only four inputs, a quad two-input package can generate all functions of two variables.

As an illustration, consider the function of four variables given in table 11.11. The truth table is divided into groups by the values of the two left-most variables, which are used for the select inputs.

Table 11.11
A Four-Input Truth Table with Residues

A	B	C	D	Y	D-res
0	0	0	0	1	
		0	1	0	$\overline{D}$
		1	0	0	
		1	1	0	0
0	1	0	0	1	
		0	1	1	1
		1	0	0	
		1	1	1	D
1	0	0	0	1	
		0	1	0	$\overline{D}$
		1	0	1	
		1	1	1	1
1	1	0	0	1	
		0	1	0	$\overline{D}$
		1	0	0	
		1	1	1	D

The two-variable groups for each value of the left-most pair are further divided into two-input function generator form. From these latter divisions, the resulting two-input functions provide the residues for the four-input data selectors with the left-most variables used as select inputs.

While this method does not necessarily produce the minimum realization, it demonstrates a method of using multiplexers to realize functions of more than three variables.

D. Eight-Input And Higher MUX Logic

An eight-input data selector requires three select inputs to account for the addressing of eight distinct inputs. This provides a method for generating functions of four variables. The left-most three variables are used for the select, and the right-most variable is used for the generation of residues. The method is the same as with two- and four-input data selectors.

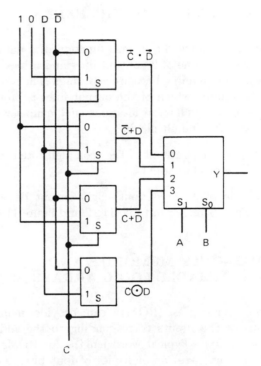

Figure 11.40 Two-Level Logic Realization with Four- and Two-Input Data Selectors

Consider table 11.11 as an example of eight-input data selector logic. A, B, and C are taken as the select inputs. The D residues are applied to the eight inputs as indicated in figure 11.41.

16-input data selectors are also available. These are capable of generating any function of five variables. However, they require about twice the space of the 8:1, dual 4:1, or quad 2:1 data selectors.

Data selector trees, illustrated by figure 11.40, can be used to implement functions of more than five variables, but other techniques (described in the following sections) are more efficient and practical.

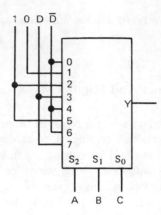

Figure 11.41 Eight-Input Data Selector Use

7   DECODERS/DEMULTIPLEXERS

A decoder performs the inverse function of a data selector. It selects one of 2^n output lines and brings that line low. All remaining lines are kept high. The n select inputs determine which of the output lines is brought low. These devices have at least one *enable* input which can be used for demultiplexing.

Decoders can generate all of the (complemented) minterms of the n select inputs.

An example of a three-input, eight-output decoder/demultiplexer is given in table 11.12 and figure 11.42.

8   READ-ONLY MEMORIES AND PROGRAMMABLE LOGIC ARRAYS

Read-only memories (ROMs) can provide multiple functions of the inputs corresponding to the address selection inputs. Typical word lengths for ROMs are eight bits (one *byte*) or multiples of four bits. Thus,

they can provide eight distinct functions of five, six, seven, or eight variables. The specified functions are *burned* into the ROM from the function truth table.

In many cases, the ROM solution is wasteful of space, particularly when the functions do not contain a majority of the possible minterms of the input variables. A special device, called a *programmable logic array*, has been developed to function as a 14- or 16-input, 8-output sum-of-products generator. These are available with provisions for 48 or 96 minterms. Some minimization may be necessary to fit complex multiple functions into the space available.

9   FLIP-FLOPS

The basic elements of sequential logic circuits are *flip-flops*, which are essentially bi-stable multivibrators. Their operation can be explained with either NAND gates or NOR gates connected with positive feedback.

The basic flip-flop is the *S-R (set-reset) flip-flop*, which can be constructed from two NAND gates or two NOR

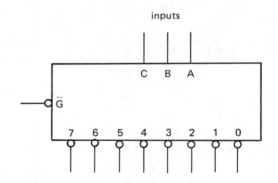

Figure 11.42 3:8 Decoder with One Enable

Table 11.12
Truth Table for a 3:8 Decoder with One Enable; X=Don't Care

inputs				outputs							
G	C	B	A	Z0	Z1	Z2	Z3	Z4	Z5	Z6	Z7
L	L	L	L	L	H	H	H	H	H	H	H
L	L	L	H	H	L	H	H	H	H	H	H
L	L	H	L	H	H	L	H	H	H	H	H
L	L	H	H	H	H	H	L	H	H	H	H
L	H	L	L	H	H	H	H	L	H	H	H
L	H	L	H	H	H	H	H	H	L	H	H
L	H	H	L	H	H	H	H	H	H	L	H
L	H	H	H	H	H	H	H	H	H	H	L
H	X	X	X	H	H	H	H	H	H	H	H

Table 11.13
Next-State Table for S-R Flip-Flop

present			next
S	R	Q	Q
0	0	0	0
0	1	0	0
1	0	0	1
1	1	0	N
0	0	1	1
0	1	1	0
1	0	1	1
1	1	1	N

N = input not permitted

gates. The other common flip-flops (T, D, and J-K flip-flops) are constructed from S-R flip-flops and other gates of the same type.

A. The Set-Reset (S-R) Flip-Flop

The S-R flip-flop is constructed from either two NAND gates, as shown in figure 11.43, or two NOR gates, as shown in figure 11.44. The *next condition* of a flip-flop depends on both the present output and the present input. Thus, a *next state table* (rather than a truth table) is appropriate. Note that for the NAND realization the inputs are the complements of set and reset.

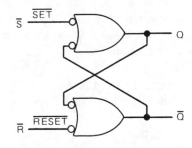

Figure 11.43 NAND Gate
Realization of an S-R Flip-Flop

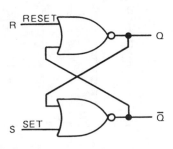

Figure 11.44 NOR Gate
Realization of an S-R Flip-Flop

B. The Toggle (T) Flip-Flop

The toggle flip-flop has only one data input (T) but is clocked so it will change states only when the clock is high. This can be realized with NAND- or NOR-based S-R flip-flops as shown in figure 11.45.

The necessary configuration is determined from the required inputs for the S-R flip-flop together with the requirements for the T flip-flop. A *transition table* from table 11.13 and the T flip-flop requirements are derived as follows:

Table 11.14
Transition Table for T Flip-Flop

present	next			
Q	Q	T	$\overline{S}$	$\overline{R}$
0	0	0	1	X
0	1	1	0	1
1	0	1	1	0
1	1	0	X	1

X = don't care

It is desired to determine the logic in NAND form for the generation of $\overline{S}$ and $\overline{R}$, the inputs to the NAND S-R flip-flop. The inputs are T, the present value of Q, and the complement $\overline{Q}$.

From table 11.14, S is true ($\overline{S}$ is false) when the present Q is false and T is true. Then,

$$S = \overline{Q} \cdot T \qquad 11.60$$

$$\overline{S} = \overline{\overline{Q} \cdot T} \qquad 11.61$$

This is the output of two input NAND gates with inputs $\overline{Q}$ and T. In addition, the clock (**Cl**) must be high for toggling to occur.

$$S = \overline{Q} \cdot T \cdot \mathbf{Cl} \qquad 11.62$$

Inverting both sides of equation 11.62,

$$\overline{S} = \overline{\overline{Q} \cdot T \cdot \mathbf{Cl}} \qquad 11.63$$

The realization of equation 11.63 requires the three-input NAND gate shown in figure 11.45(a) with inputs T, $\overline{Q}$, and **Cl**.

Similarly, the transition table shows that R is true ($\overline{R}$ is false) when the present Q is 1 and T is 1. Since the clock must be high,

$$\overline{R} = \overline{Q \cdot T \cdot \mathbf{Cl}} \qquad 11.64$$

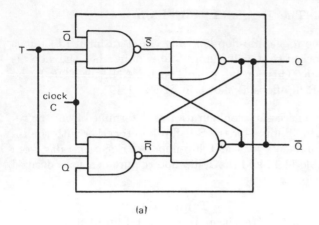

(a)

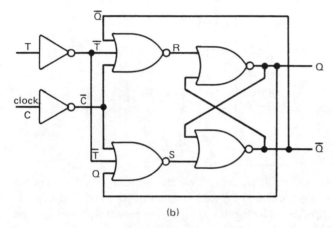

(b)

Figure 11.45 Toggle (T) Flip-Flop Configurations: (a) NAND Realization, (b) NOR Realization

This is the output of the lower input NAND gate of figure 11.45(a).

To obtain the NOR gate logic for figure 11.45(b), it is necessary to put equations 11.63 and 11.64 in OR form by applying de Morgan's theorem to the right-hand side. Then, the negation of the NOR gate will provide S and R.

$$\overline{S} = \overline{\overline{Q} \cdot T \cdot \mathbf{Cl}} = \overline{\overline{Q}} + \overline{T} + \overline{\mathbf{Cl}}$$

$$= Q + \overline{T} + \overline{\mathbf{Cl}} \qquad 11.65$$

$$S = \overline{Q + \overline{T} + \overline{\mathbf{Cl}}} \qquad 11.66$$

This is the output of a three-input NOR gate such as the lower input gate of figure 11.45(b).

The clock pulse for figure 11.45 can be high no longer than the propagation time through the gates.

C. The Data (D) Flip-Flop

The D flip-flop is clocked so that the output Q will follow the input when the clock is high, when the clock

makes its transition (*edge-triggered*) from low to high, or when the clock makes its transition from high to low. The simplest D flip-flop, called the *D-latch*, uses an *enable* (which may be a clock). The transition table, including the S-R input requirements, is shown in table 11.15.

The X's in S and R can be used because they allow the variable to be 0 or 1 as needed. If the X's are 1, the logical expressions for S and R are

$$S = E \cdot D \qquad 11.67$$

$$R = E \cdot \overline{D} \qquad 11.68$$

Table 11.15
D Flip-Flop Transition Table

		present	next		
E	D	Q	Q	S	R
1	0	0	0	0	X
1	0	1	0	0	1
1	1	0	1	1	0
1	1	1	1	X	0
0	X	X	present	0	0

D. J-K Flip-Flops

The J-K flip-flop has been developed to make use of the two disallowed inputs of the S-R flip-flop. The inputs $S, R = 1, 1$ result in a toggled output (i.e., if the present state is $Q = 1$, the next state is $Q = 0$, and vice versa). There are two types of J-K flip-flops, an ordinary J-K and a so-called *master-slave J-K*. The difference is in when the change of output occurs. This will be discussed in the next section.

Table 11.16
J-K Flip-Flop Truth Table

		present	next		
J	K	Q	Q	S	R
0	0	0	0	0	X
0	0	1	1	X	0
0	1	0	0	0	X
0	1	1	0	0	1
1	0	0	1	1	0
1	0	1	1	X	0
1	1	0	1	1	0
1	1	1	0	0	1

$$S = J \cdot \overline{Q} \qquad 11.69$$

$$R = K \cdot Q \qquad 11.70$$

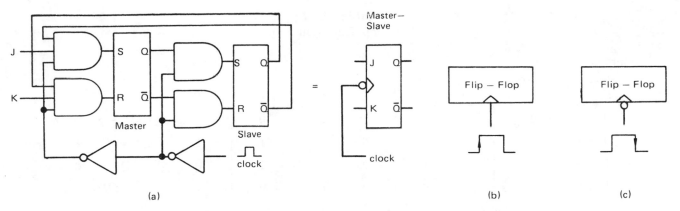

Figure 11.46 (a) Master-Slave J-K Flip-Flop,
(b) Clock Input Symbol for a Rising Edge-Triggered Flip-Flop,
(c) Clock Input Symbol for a Falling Edge-Triggered Flip-Flop

E. Clocked Flip-Flops

Flip-flops can be triggered by either the rising edge or the falling edge of the clock pulse. The master-slave J-K flip-flop consists of two parts, a *master* and a *slave*, as indicated in figure 11.46. For these, the inputs must be stable during the time the clock is high. This sets the first master stage, and the change is propagated to the slave stage on the falling edge of the clock. (There are also master-slave versions of the S-R flip-flop that operate similarly.)

The triangular symbol on the clock input indicates that the flip-flop is triggered at the rising edge of the clock input as in figure 11.46(b). An inversion (bubble) with the triangle indicates that the flip-flop is triggered by the falling edge of the clock pulse.

Ordinary flip-flops (not master-slave units) include D and J-K types, which are triggered on either the rising edge or the falling edge of the clock pulse. *T flip-flops* are not commercially available. They must be built from one of the other available types and added logic.

Other inputs usually found on flip-flops are *preset*, which *asynchronously* (immediately) sets the output state to $Q = 1$, and *clear*, which asynchronously sets the output to $Q = 0$. Both provide an overriding of the logic. They do not generally wait for the proper clock transition.

There are several important *propagation delay times* associated with flip-flops. These are usually different for high-to-low and low-to-high output transitions. They also differ depending on the origin of the initiator of the change (e.g., clock, preset, or clear). These times determine the maximum frequency at which the flip-flops can operate.

Two additional times are the *set-up time*, which is the minimum time that the inputs must be stable before the triggering clock transition, and the *hold time*, which is the minimum time the inputs must remain stable after the triggering clock transition.

10 FLIP-FLOP TRANSITION TABLES

In many design problems involving flip-flops, the *transitions* from the present states to the next output states are needed to design logic to satisfy the problem specifications. The four possible transitions are: 0 to 0, 0 to 1, 1 to 1, and 1 to 0. While 0 to 0 and 1 to 1 may not appear to be transitions, they are necessary in logic considerations. The transitions for the four principal flip-flop types are given in table 11.17.

Table 11.17 shows that a T flip-flop can be constructed from a J-K flip-flop by making $J = K = T$. It can also be constructed from a D flip-flop by making $D = T$ XOR Q, or from an S-R flip-flop by making $S = \overline{Q} \cdot T$ and $R = Q \cdot T$.

11 COUNTERS

Counters are sets of flip-flops. Usually, the flip-flops are all the same type, and they all respond to the same clock edge for triggering. A counter passes through a sequence of *states* in response to one or more inputs. The count is determined by the number of states in the sequence and is limited to 2^n, where n is the number of flip-flops in the counter.

A simple example is the *ripple counter* shown in figure 11.47(a). The input signal is applied to the clock input of the lowest flip-flop. This flip-flop is triggered by the falling edge of the signal. When the lowest flip-flop makes a high-to-low transition, the middle flip-flop is

Table 11.17
Flip-Flop Transition Table

present state Q	next state Q^+	S	R	J	K	T	D	
0 ⟶ 0		0	X	0	X	0	0	$Q^+ = S + \overline{R} \cdot Q$
0 ⟶ 1		1	0	1	X	1	1	$Q^+ = J \cdot \overline{Q} + \overline{K} \cdot Q$
1 ⟶ 0		0	1	X	1	1	0	$Q^+ = T \oplus Q$
1 ⟶ 1		X	0	X	0	0	1	$Q^+ = D$

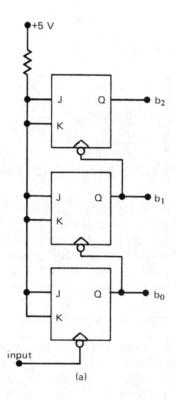

(a)

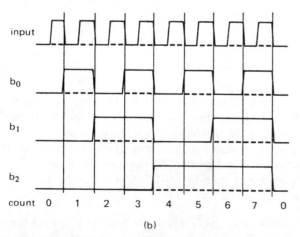

(b)

Figure 11.47 (a) A Ripple Counter
(b) Timing Diagram for One-Count
Sequence

triggered. When the middle flip-flop makes a transition from high-to-low, the upper flip-flop is triggered. Figure 11.47(b) shows the count cycle. The three flip-flops are at their low states at the beginning of the cycle.

This type of a counter is called *asynchronous*, in that the outputs of the flip-flops are not actually clocked. The count ripples through the circuit from the input flip-flop, so that delay times become important with additional flip-flops. This *propagation delay* is a fundamental limitation with this type of counter. Other means (involving common clocking) are required for count lengths of more than a few bits.

A. Synchronous Binary Counters

In order to make a counter (or similar sequential circuit) synchronous, the flip-flops all must make transitions at the same time. A *state table* for the counter can then be constructed where the *state* is the concatenation of the Q outputs of the flip-flops.

For example, the state sequence for the count indicated in figure 11.48(b) is 000, 001, 010, 011, 100, 101, 110, 111, 000, with the positional significance of $Q_2 Q_1 Q_0$ in that order.

The state table is constructed by listing the state sequence in order as the left-hand column as shown in table 11.18. The next state for each of the present states is shown in the second column. (The superscript + is used to indicate the next state.) The third column contains the set of inputs to the flip-flops, which are necessary to obtain the proper transition (see table 11.17).

The design requires determining the flip-flop inputs as logical functions of the present state values of Q_2, Q_1, and Q_0. This is normally accomplished with Karnaugh maps or Veitch diagrams. For this example, the logic is simple enough to merely inspect the state table. *T flip-flops* (which are obtained by tying the J and K inputs together on J-K flip-flops) are assumed. The resulting counter circuit is shown in figure 11.48(b).

Table 11.18

State Table for a Three-Bit Binary Counter

present state $Q_2 Q_1 Q_0$	next state $Q_2^+ Q_1^+ Q_0^+$	required T inputs $T_2 T_1 T_0$
0 0 0	0 0 1	0 0 1
0 0 1	0 1 0	0 1 1
0 1 0	0 1 1	0 0 1
0 1 1	1 0 0	1 1 1
1 0 0	1 0 1	0 0 1
1 0 1	1 1 0	0 1 1
1 1 0	1 1 1	0 0 1
1 1 1	0 0 0	1 1 1

$$T_2 = Q_0 \cdot Q_1$$
$$T_1 = Q_0$$
$$T_0 = 1$$

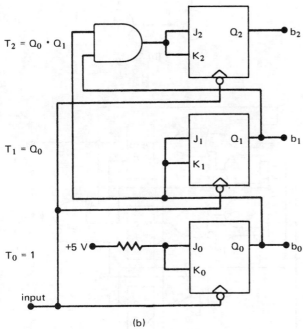

(a)

(b)

Figure 11.48 Three-Bit Binary Counter
(a) State Diagram (b) T Flip-Flop Realization

Figure 11.48(a) shows a *state diagram* for the system. This is a useful device for the design or analysis of

sequential logic circuits. The states are indicated by subscripted S's. They do not have to be given the same subscripts as the counts, although they have been in this example.

A major problem with this design is that more than one output can change at the same time. Because there are always differences in the speeds of transitions, there is a time during the transitions in which a false value of the count can occur. For example, in going from S_1 to S_2, both Q_0 and Q_1 are changing. If Q_1 reaches 1 before Q_0 reaches 0, there will be a momentary output of 011 existing. This can cause problems when the outputs are used as logical inputs to some other system.

A better design would have changes occur with only one state variable changing. This is equivalent to making state assignments so that the next state is logically adjacent to the present state. While this is not possible in all designs, it is good practice to try to do so.

B. An Adjacent-State Binary Counter

Because the binary counter uses all 2^n states, it is possible to assign all next states adjacent to the present state. This is complicated by the fact that the state variables no longer correspond to the correct count. To correct this, it is necessary to use logic to provide the correct count for each state.

Consider the previous example. Start with state assignments so that the next state is logically adjacent to the present state. These are shown in the second column of table 11.19. The first column has the state it names, which are subscripted to match the correct count.

The third column contains the next state, and the fourth column contains the required outputs in each state. In this example, D flip-flops are used. The required values of inputs for D flip-flops are in the fifth column.

Once the state table has been established, the logic to obtain flip-flop inputs and the outputs is determined by the use of Karnaugh maps as indicated in figure 11.49(a). Having determined the necessary logic, the circuit is realized in figure 11.49(c). (There is a redundant gate in this circuit diagram. $Q_1 \cdot \overline{Q_0}$ is needed for both D_1 and D_2.) A simplification of the state table is possible when using D flip-flops, because the value of D is always the next state value for the corresponding state variable. Thus, the last column is redundant when using D flip-flops.

Table 11.19
Adjacent-State Binary Counter

state	present state $Q_2Q_1Q_0$	next state $Q_2^+Q_1^+Q_0^+$	outputs $b_2b_1b_0$	D inputs $D_2D_1D_0$
S_0	0 0 0	0 0 1	0 0 0	0 0 1
S_1	0 0 1	0 1 1	0 0 1	0 1 1
S_2	0 1 1	0 1 0	0 1 0	0 1 0
S_3	0 1 0	1 1 0	0 1 1	1 1 0
S_4	1 1 0	1 1 1	1 0 0	1 1 1
S_5	1 1 1	1 0 1	1 0 1	1 0 1
S_6	1 0 1	1 0 0	1 1 0	1 0 0
S_7	1 0 0	0 0 0	1 1 1	0 0 0

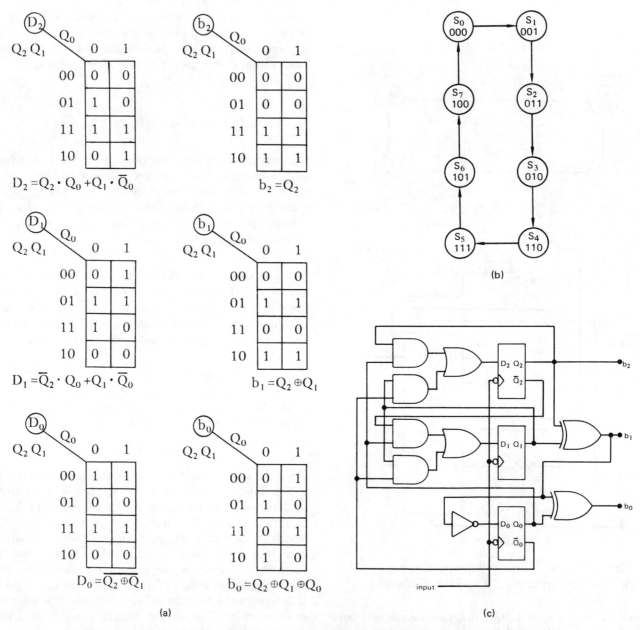

Figure 11.49   Adjacent-State Counter (a) Logic, (b) State Diagram, (c) Circuit

Even for this relatively simple circuit, the inclusion of all logic lines from the state variables to the flip-flop inputs is confusing. For that reason, these feedback lines are usually omitted, and the state variables are indicated at the various gate inputs.

C. Counter Analysis

In order to analyze a counter circuit, the most orderly approach is to construct a state table. The logical connections are used to establish the flip-flop inputs from the present state. The usual starting point is at the state with all zero state variables. Each successive state is then entered in sequence.

The following example is intended to illustrate the method. The circuit shown in figure 11.50 uses three different types of flip-flops and two AND gates. The flip-flops are all triggered by the rising edge of the input signal.

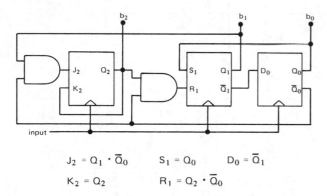

$$J_2 = Q_1 \cdot \overline{Q_0} \qquad S_1 = Q_0 \qquad D_0 = \overline{Q_1}$$

$$K_2 = Q_2 \qquad R_1 = Q_2 \cdot \overline{Q_0}$$

Figure 11.50 Sample Counter Circuit

From the counter circuit and the flip-flop logical equations given in table 11.17, the next-state variables can be written as

$$Q_2^+ = J_2 \cdot \overline{Q_2} + \overline{K_2} \cdot Q_2 = \overline{Q_2} \cdot Q_1 \cdot \overline{Q_0} + \overline{Q_2} \cdot Q_2$$
$$= \overline{Q_2} \cdot Q_1 \cdot \overline{Q_0} \qquad \qquad 11.71$$

$$Q_1^+ = S_1 + \overline{R_1} \cdot Q_1 = Q_0 + (\overline{Q_2} + Q_0) \cdot Q_1$$
$$= Q_0 + \overline{Q_2} \cdot Q_1 \qquad \qquad 11.72$$

$$Q_0^+ = D_0 = \overline{Q_1} \qquad \qquad 11.73$$

These equations are used to generate the second column of the state table. The first four rows in the table complete the count sequence since the next state from S_4 is S_0. Because there are three unused states, it is necessary to check for hang-up states.

Hang-up states are two or more unused states which have no next state in the desired sequence. Unless there

Table 11.20
State Table for Figure 11.50

	present state $Q_2Q_1Q_0$	next state $Q_2^+Q_1^+Q_0^+$
S_0	0 0 0	0 0 1
S_1	0 0 1	0 1 1
S_2	0 1 1	0 1 0
S_3	0 1 0	1 1 0
S_4	1 1 0	0 0 0
S_5	1 1 1	0 1 0
S_6	1 0 1	0 1 1
S_7	1 0 0	0 0 1

is an automatic clearing of all flip-flops when a system is turned on, it could come up into one of the unused states and be hung-up there, never getting into the desired sequence. To make sure this doesn't happen, all unused states must have a path into the desired sequence. A check of the states 5, 6, and 7 shows that they all return to the main sequence.

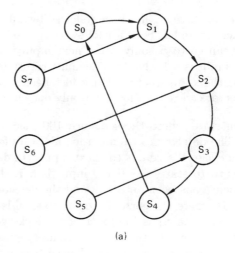

(a)

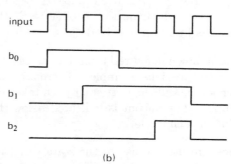

(b)

Figure 11.51 (a) State Diagram, (b) Timing Diagram for Figure 11.50

The state diagram, with the same designations as in the state table, is shown in figure 11.51(a). The timing diagram in figure 11.51(b) shows the outputs in reference to the input. The resulting sequence is 000, 001, 011, 010, 110, 000. Thus, this is a five-state counter.

12   SEQUENCE DETECTION

A *sequence detector* is a sequential circuit which determines when a specific input sequence has occurred. For example, a detector might be needed to detect the sequence 1001. The inputs will be monitored using a clock with the same frequency as the monitored data line.

The detector starts in a state S_0 where the previous input was not from a proper sequence (e.g., was a zero following two or more zeros). This will be the *start-up state*. The detector will remain in state S_0 as long as 0's are received and will go to the next state S_1 as soon as a 1 is received.

The detector will remain in state S_1 as long as 1's are received. It will go to the next state S_2 when a 0 is received.

S_2 is a state where a 1 has been followed by a 0. S_3 is a state where there has been a 1 followed by two 0's. The detector can go to S_3 only if the next input is a 0. If the next input is a 1, the detector must return to state S_1 (S_1 is the state where a 1 has just been received). The detector can remain in S_2 for only one clock cycle.

S_3 is the state where the sequence 100 has been received. The detector can remain in this state for only one clock cycle. If a zero is the next input, the detector must return to state S_0. If the input is a 1, the correct sequence has been detected, and the detector goes to state S_4 where the correct sequence signal is given. The stay in state S_4 is limited to one clock cycle. If the next input is a 0, the detector makes a transition to state S_2, since a 1 has been followed by a 0. If the next input is a 1, the transition is to state S_1, the state where a 1 has just been received.

This logic is shown in figure 11.52. This state diagram is different from previous examples. There are two paths from each state, and the particular path taken depends on the input. The output is indicated below the horizontal line in each circle.

The procedure used to assign the state variable values for each state is complicated. Three rules can be used to minimize logic costs.

> **Rule 1.** States that have the same next states for the same inputs should be given adjacent assignments.

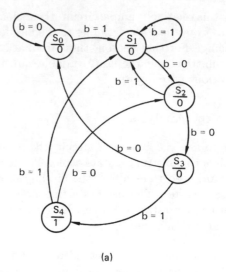

(a)

present state	next state for input b		present output
	$b = 0$	$b = 1$	
S_0	S_0	S_1	0
S_1	S_2	S_1	0
S_2	S_3	S_1	0
S_3	S_0	S_4	0
S_4	S_2	S_1	1

(b)

	S_4	S_3	S_2	S_1
S_0	1, 2	1, 3	1, 3	1, 2, 3
S_1	1, 1	2, 3	1, 2, 2, 3	
S_2	1	3		
S_3				

(c)

Figure 11.52   (a) State Diagram for Sequence 1001 Detector, (b) State Table, (c) Adjacency Weighting According to State Assignment Rules

> **Rule 2.** States that are the next state of the same state should be given adjacent assignments.

> **Rule 3.** States that have the same outputs for a given input should be given adjacent assignments.

In figure 11.52(c), the adjacency rules are entered into a matrix to evaluate the desirability of assigning various states with adjacent variable values. Each entry in the matrix shows the number of times a rule applies. Because there is only one state with an output, rule 3 is of

Table 11.21
State Table for 1001 Detector

S	present state $Q_2 Q_1 Q_0$	next state $b=0$ $Q_2^+ Q_1^+ Q_0^+$	$b=1$ $Q_2^+ Q_1^+ Q_0^+$	present output Y	flip-flop inputs J_2	K_2	J_1	K_1	J_0	K_0
S_0	0 0 0	0 0 0	0 0 1	0	0	X	0	X	b	X
S_1	0 0 1	0 1 1	0 0 1	0	0	X	$\bar{b}$	X	X	0
S_2	0 1 1	0 1 0	0 0 1	0	0	X	X	b	X	$\bar{b}$
S_3	0 1 0	0 0 0	1 0 1	0	b	X	X	1	b	X
S_4	1 0 1	0 1 1	0 0 1	1	X	1	$\bar{b}$	X	X	0

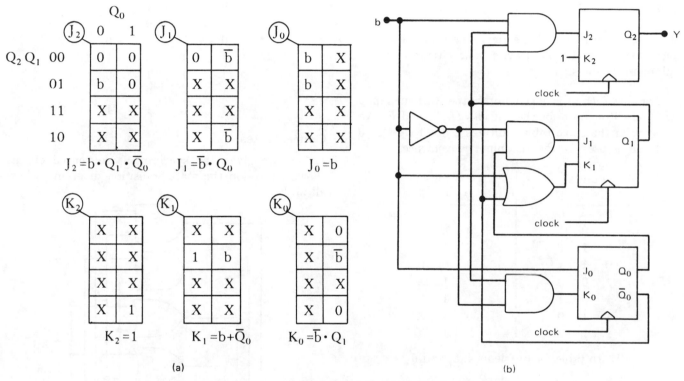

Figure 11.53 (a) J-K Logic Determination Using Karnaugh Maps, (b) Realization with Rising Edge-Triggered J-K Flip-Flops

minor importance here. It should be used only to break ties. Giving rules 1 and 2 equal weight, it is obvious that the $S_1 - S_2$ adjacency must be made.

$S_0 - S_1$ is the second priority because of rule 3. Next in priority are the groups $S_1 - S_4$ and $S_0 - S_4$. Then the three groupings of $S_0 - S_2$, $S_0 - S_3$, and $S_1 - S_3$ should be ranked. The final consideration is $S_2 - S_4$, followed

by $S_2 - S_3$ and $S_3 - S_4$.

Several combinations are approximately equal, and all groupings are not possible. Because of reset on start-up, it is common to assign 000 to state S_0. From these, one efficient grouping is as shown in table 11.21. The logic for the flip-flop inputs is obtained for J-K flip-flops using table 11.17 relationships.

PRACTICE PROBLEMS

Warmups

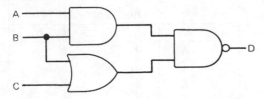

1. For the circuit shown, obtain the truth table for the function **D**.

2. Obtain the sum-of-products and the product-of-sums expressions for **D**.

3. Using identities and theorems, determine the simplest expression for **D**. Revise your circuit to use a minimum of gates.

4. For the truth table given, obtain the Karnaugh map and Veitch diagram for the outputs D, E, and F. An X entry in the truth table indicates a don't care, so 0 or 1 can be used for the maximum convenience.

inputs			outputs		
A	B	C	D	E	F
0	0	0	0	1	1
0	0	1	0	X	1
0	1	0	X	1	0
0	1	1	0	0	0
1	0	0	1	1	X
1	0	1	1	1	X
1	1	0	0	X	X
1	1	1	X	1	X

Truth table for problems 4, 5, and 6

5. For the outputs D, E, and F, obtain the minimum gate realization between the SOP (NAND-NAND) and the POS (NOR-NOR) implementations. (Hint: determine which canonical form will result in the minimum number of gates after minimization.)

6. Using 4:1 data selectors, determine the residues for each of the functions D, E, and F.

7. In the truth table for a seven-segment display (figure 11.25), notice that segment c requires only one 0, while segment e requires six 0's. From Karnaugh maps for these two segments, determine whether a minterm SOP realization or a maxterm POS realization is preferable.

8. For segment f of the seven-segment display, design an 8:1 data selector realization of the logic. Use the variable D for residues.

9. For the NOR form of the S-R flip-flop (see figure 11.44), consider that the gate with R as an input is faster than the other. Trace the signals when the inputs have been $R = 0$ and $S = 1$ for some time when R is suddenly changed to 1. Does it make any difference whether one or the other gate is faster?

10. A sequential circuit is to have the state table given. Using an S-R flip-flop, design a circuit. Q is the present state and output, A and B are inputs, and Q^+ is the next state and output.

Q	A	B	Q^+
0	0	0	0
0	0	1	1
0	1	0	1
0	1	1	0
1	0	0	1
1	0	1	0
1	1	0	0
1	1	1	1

Concentrates

1. You are given the synchronous sequential circuit shown. Develop the state transition diagram. Show all work.

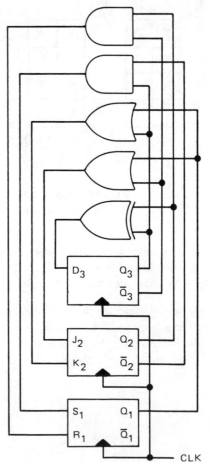

2. A one-bit full-adder cell has inputs A, B, and CI (carry in) and outputs S (sum) and CO (carry out). Each of these quantities is one bit. (a) Obtain a NAND-NAND realization of the full-adder cell. (b) Obtain a dual 4:1 data selector realization of the full-adder cell.

3. A shift register is to be designed using D flip-flops and data selectors. This shift register has four functions: shift left, shift right, broadside load, and do nothing. The register has six inputs: L (left input), R (right input), $D0$, $D1$, $D2$, and $D3$, where the D's are the broadside data bits. The register has four outputs: $Q0$, $Q1$, $Q2$, and $Q3$. The finished register and its function table are shown.

C_1	C_0	$Q3^+$	$Q2^+$	$Q1^+$	$Q0^+$
0	0	$Q3$	$Q2$	$Q1$	$Q0$
0	1	L	$Q3$	$Q2$	$Q1$
1	0	$Q2$	$Q1$	$Q0$	R
1	1	$D3$	$D2$	$D1$	$D0$

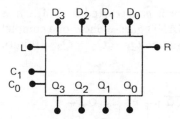

4. For the counter circuit shown, develop the state table and the state transition diagram.

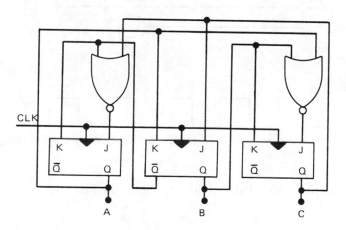

5. Using T flip-flops, design a counter that will produce the following sequence:

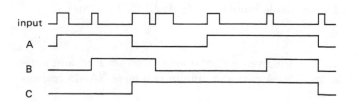

6. An M-N flip-flop is to have the following truth table:

M	N	Q	Q^+
L	L	L	H
L	L	H	H
L	H	L	L
L	H	H	L
H	L	L	L
H	L	H	L
H	H	L	H
H	H	H	H

Determine whether a D flip-flop, a J-K flip-flop, or an S-R flip-flop will result in the simplest logic to realize this M-N flip-flop. Show the resulting circuit together with its transition table.

7. A microprocessor has eight address lines A7 A6 A5 A4 A3 A2 A1 A0, and is required to address up to 16 input-output (I/O) ports. A decoder (4:16 with two active low enables), as shown below, is to be used to provide the 16 active-low outputs to select the input-output ports based on the least significant address bits A3 A2 A1 A0 when the address bits A7 A6 A5 A4 = HHHH = 1111. Using a minimum number of two-input NAND gates, design the logic to accomplish this.

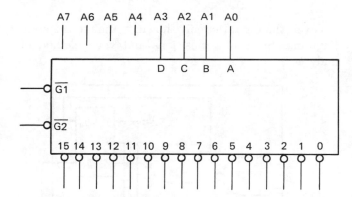

8. A clocked sequential circuit has an input that can change voltage levels only during the rising edge of the clock. Design a sequence detector which will provide a high (1) output whenever the input has remained low (0) for four consecutive clock risings. The output is to be provided at the next clock rising after such a sequence is detected.

9. Four-bit binary ripple counters are available. Each has parallel outputs DCBA, with A the LSB, a count input (CI), and a carry out (CO) which is low except when the counter output is DCBA = HHHH. They also have a level-sensitive preset input PR, which will set DCBA = HHHH when PR is high. The counters respond to the falling edge of the count input (CI).

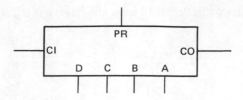

Using two such counters, and necessary logic gates, design a circuit which will provide a 1 (high) after each 100 counts.

10. The following Karnaugh map is to be implemented using an 8:1 multiplexer and NOT gates. The signals A, B, C, and D are available, but not their complements.

F	CD			
AB	00	01	11	10
00	1	0	0	0
01	0	0	0	1
11	0	0	1	0
10	0	1	0	1

Timed

1. Given the digital circuit shown, complete the state diagram. Show all transitions and your method of analysis.

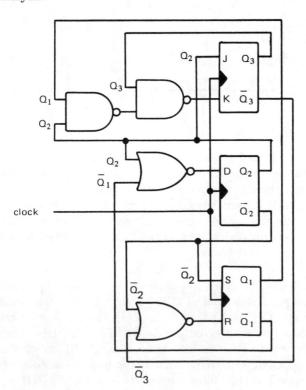

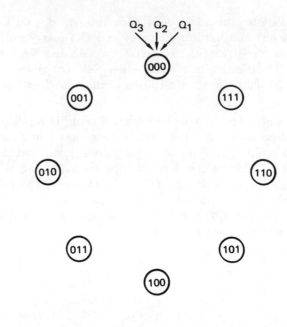

2. A counter is constructed from J-K master-slave flip-flops and NAND gates. Show one complete cycle of the counter. Assume $Q_1 = Q_2 = Q_3 = 0$ initially.

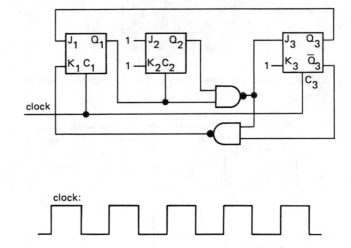

3. Using an 8:1 data selector, obtain the function

$$\mathbf{F} = \overline{A} \cdot B \cdot D + A \cdot C + \overline{A} \cdot B \cdot \overline{C} \cdot D + \overline{A} \cdot \overline{B} \cdot \overline{C} \cdot \overline{D}$$

The available inputs are A, $\overline{A}$, B, C, D, 0, and 1.

4. You are given a digital logic function $\mathbf{F}$ of four variables. Develop the minimum two-level NAND realization for $\mathbf{F}$.

W	X	Y	Z	F
0	0	0	0	0
0	0	0	1	0
0	0	1	0	0
0	0	1	1	1
0	1	0	0	0
0	1	0	1	0
0	1	1	0	1
0	1	1	1	1
1	0	0	0	0
1	0	0	1	0
1	0	1	0	0
1	0	1	1	1
1	1	0	0	1
1	1	0	1	0
1	1	1	0	1
1	1	1	1	1

5. You are to design a digital circuit using only two-input NAND gates and J-K master-slave flip-flops. You may use as many of each as is necessary, but your design should be minimized with respect to packages. (There are four two-input NAND gates in a package and two J-K flip-flops per package.)

The system is to have two inputs: a clock and x. x is a sequence of binary voltages—high = 1, low = 0. Binary data is presented in sequences of one, two, or three 1's followed by at least one 0. No other sequence will ever occur (i.e., a sequence of four ones will not occur).

There are to be two outputs, A and B. A is to go high when the first 1 of a sequence occurs, and to go low after the second or when the next 0 occurs. B is to go high when the second 1 of a sequence occurs, and stay high until the next 0 occurs.

6. An input/output port selector is to use a 4:16 decoder (with two active-low enables) as indicated, with the most significant bits of an address bus as indicated.

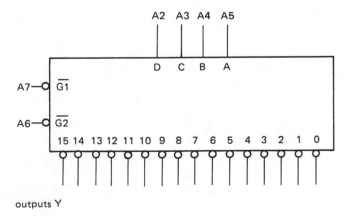

In addition, the selection of a port is to take place in coincidence with a timing strobe (asserted low) $\overline{X}$. Using only a quad two-input NAND gate package and a package with six NOT gates, design the circuit to select the address 00101101. Indicate which output from the 4:16 decoder (Y) is to be used. Inputs to the logic are to be the correct output from the decoder, address lines A1 and A0, and the strobe $\overline{X}$.

7. A three-bit ripple down-counter is preset to 7 and counts down to zero in response to input pulses. The outputs of the flip-flops are connected to a 3-input AND gate which is to produce a high output pulse during each subsequent time that the output is 7. Under all other conditions the output is to be low.

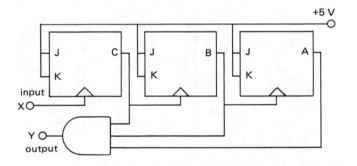

During the countdown sequence, one or more errors occur at the output. (a) Where in the sequence can such errors occur? (b) What is the form of the error(s)? (c) How can this problem be corrected?

8. A synchronous sequential circuit is to have the following next-state table. Design the circuit using only D flip-flops and EXOR gates. A is the input.

X	Y	A	X^+	Y^+
0	0	0	1	0
0	0	1	0	0
0	1	0	1	1
0	1	1	0	1
1	0	0	1	1
1	0	1	0	1
1	1	0	1	0
1	1	1	0	0

9. A shaft is encoded by having a conductor covering 0 to 180 degrees, and an insulator covering from 180 to 360 degrees. A 5-volt source is connected to the conductor. One stationary carbon brush sensor contacts the shaft at a reference position, and another contacts the shaft 90 degrees from the first in the positive direction of shaft rotation. When a brush is in contact with the metal sector of the shaft, it delivers a high (approximately 5-volt) voltage to one input of the logic circuit, and when not in contact with the metal sector it provides zero volts to that input.

A sequential logic circuit is to be clocked at a frequency more than ten times the maximum shaft rotational velocity, and is to provide input signals to an up-down counter to correspond with the number of rotations the shaft makes. The counter output is then to be the integral number of rotations the shaft makes.

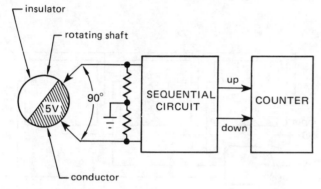

10. The following flowchart deals with three inputs variables: A, B, C, and three output variables X, Y, and Z. Using two-input NAND gates, design combinational logic to realize the flow chart functions.

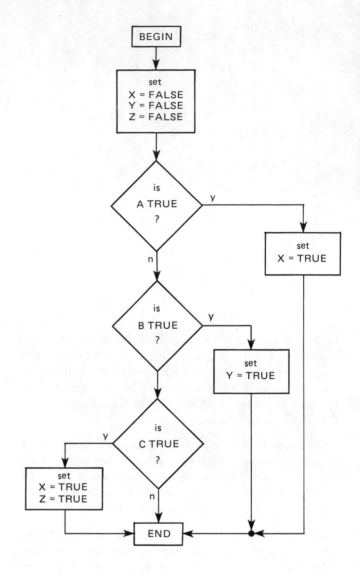

12 CONTROL SYSTEMS

Nomenclature

a	a coefficient	–
A	voltage gain	–
b	a coefficient	–
c	controlled variable of time	–
C	Laplace-transformed controlled variable	–
D	mechanical damping coefficient	N-m-sec
e	error variable of time	–
E	Laplace-transformed error variable, or electromotive force	– or V
$\mathbf{f}$	function of time	–
$\mathbf{F}$	function of s	–
$\mathbf{G}$	transfer function of s	–
$\mathbf{GH}$	loop transfer function of s	–
$\mathbf{H}$	feedback transfer function of s	–
I	current	amps
J	mass moment of inertia	N-m-sec^2
K	gain constant	–
K_v	electromotive force constant	V-sec/rad
L	inductance	henrys
p	power	watts
r	reference variable of time	–
R	Laplace-transformed reference variable	–
R	resistance	ohms
s	Laplace transform variable	1/sec
t	time	sec
$u(t)$	unit step function	–
V	voltage	V
x	an arbitrary variable	–

Symbols

ζ	damping ratio	–
θ	angular position	rad
σ	real part of s	nepers
τ	time constant	sec
Ω	mechanical angular velocity	rad/sec
ω	frequency	rad/sec

Subscripts

a	armature
c	compensation or controlled
e	error
f	field
g	generator or forward
h	feedback
m	an integer
n	an integer or natural
p	plant or pole
r	reference
s	source
ss	steady state
t	terminal, or tachometer
z	zero

1 GENERAL FEEDBACK THEORY

Linear *control systems* often are represented in block diagram form. Transfer functions typically are given as ratios of polynomials of the complex frequency s, often in factored form. The standard terminology for such a system is shown in figure 12.1(a). C is the controlled variable, R is the reference (input) variable, and E is the error (difference) between R and C for unity feedback.[1]

$$E = R - C \qquad 12.1$$

The forward block indicated by $\mathbf{G}$ in figure 12.1(a) performs the transformation

$$C = \mathbf{G}E \qquad 12.2$$

---

[1] C, E, and R are Laplace-transformed variables: $C(s) = \mathcal{L}\{c(t)\}$, $E(s) = \mathcal{L}\{e(t)\}$, and $R(s) = \mathcal{L}\{r(t)\}$. In this chapter, lower case variables are functions of t, and upper case variables are functions of s.

The equations for the output and error in terms of the reference are:

$$\frac{C}{R} = \frac{\mathbf{G}}{1+\mathbf{G}} \qquad 12.3$$

$$\frac{E}{R} = \frac{1}{1+\mathbf{G}} \qquad 12.4$$

A single-feedback system is shown in figure 12.1(b), where $\mathbf{G}$ is the forward transfer function and $\mathbf{H}$ is the feedback transfer function. Figure 12.1(a) with equations 12.3 and 12.4 are for the case where $\mathbf{H}$ is 1.0, called *unity gain* feedback. The more general case where $\mathbf{H}$ is not unity has the relationships

$$\frac{C}{R} = \frac{\mathbf{G}}{1+\mathbf{GH}} \qquad 12.5$$

$$\frac{E}{R} = \frac{1}{1+\mathbf{GH}} \qquad 12.6$$

In the most general case, the forward transfer function will consist of a controlled *plant* and a *compensator* to improve the stability or to reduce the steady-state errors of the system. In this case, the forward transfer function is expressed as a product of the *compensator transfer function*, $\mathbf{G}_c$, and the *plant transfer function*, $\mathbf{G}_p$. This is indicated in figure 12.1(c).

The terminology for control systems is based on the classical positioning control system, called a *servomechanism*. The *controlled variable* corresponds to a position (typically an angular position). Thus, when a second feedback proportional to the derivative of the output is used, it is called *velocity feedback* or *tachometer feedback*.

Since d/dt corresponds to s in Laplace-transformed variables, a unity feedback augmented with tachometer feedback will have the form

$$\mathbf{H} = 1 + k_t s \qquad 12.7$$

k_t is the *effective tachometer feedback coefficient*. For a true *tachometer*, the terminal voltage is proportional to the angular velocity, Ω. (A_t is the ratio of tachometer voltage to the angular velocity in equation 12.8.)

$$V_t = A_t \Omega \qquad 12.8$$

The classical positioning control system is shown in figure 12.2. The output device is a DC motor with separate field excitation, and the armature is driven by a power amplifier.

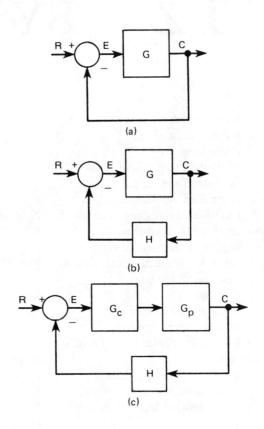

(a)

(b)

(c)

Figure 12.1   Block Diagrams:
(a) Unity Feedback, (b) General Feedback, and
(c) General System with Compensation

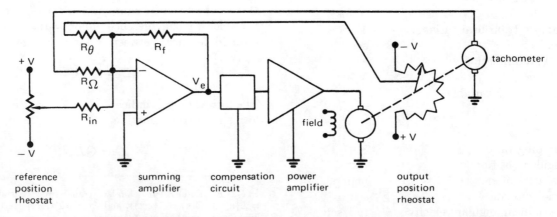

Figure 12.2   DC Motor Positional Control System with Tachometer Feedback

A separately excited DC servomotor will have its armature current limited by the back-EMF, which is proportional to the angular velocity of the shaft. The proportionality constant is K_v. It can be found transformed from the steady-state circuit equation

$$V_s = (R_a + R_s)I_a + sL_aI_a + K_v\Omega$$

$$= (R_a + R_s + sL_a)I_a + K_v\Omega \qquad 12.9$$

In steady state, $s = 0$, so the armature voltage, V_a, is

$$V_a\big|_{\text{steady state}} = I_aR_a + K_v\Omega \qquad 12.10$$

K_v can be calculated when the armature voltage, the armature current, and the shaft velocity are known.

For a *linear system*, the torque applied to the shaft by the armature current is balanced by the friction countertorque and the change in angular momentum.[2]

$$T = D\Omega + J\frac{d\Omega}{dt}$$

$$= (D + Js)\Omega \qquad 12.11$$

D is the friction coefficient and J is the mass moment of inertia of the rotating part. The *mechanical time-constant* is J/D.

$$\tau_m = \frac{J}{D} \qquad 12.12$$

$$T = D(1 + \tau_m s)\Omega \qquad 12.13$$

The electromechanical energy conversion relationship is

$$P_{\text{shaft}} = T\Omega = E_gI_a$$

$$= K_v\Omega I_a \qquad 12.14$$

Therefore, equations 12.11 and 12.13 can be written as

$$T = K_vI_a \qquad 12.15$$

or,

$$I_a = \frac{D}{K_v}(1 + \tau_m s)\Omega \qquad 12.16$$

$s = 0$ in steady state, so D can be determined from speed, armature current, voltage, and armature terminal voltage.

$$D\big|_{\text{steady state}} = \frac{K_vI_a}{\Omega} = \frac{\left(\frac{E_g}{\Omega}\right)I_a}{\Omega}$$

$$= \frac{E_gI_a}{\Omega^2} \qquad 12.17$$

and,

$$E_g\big|_{\text{steady state}} = V_a - I_aR_a \qquad 12.18$$

Substituting equation 12.16 into equation 12.9,

$$\frac{V_s}{\Omega} = [(R_s + R_a) + sL_a](1 + s\tau_m)\frac{D}{K_v} + K_v \qquad 12.19$$

or,

$$\frac{\Omega}{V_s} =$$

$$\frac{K_v}{K_v^2 + D(R_a + R_s) + sD[L_a + \tau_m(R_a + R_s)] + s^2\tau_mL_aD}$$

$$12.20$$

Equation 12.20 normally is written as

$$\frac{\Omega}{V_S} = \frac{K}{1 + \frac{2\zeta s}{\omega_n} + \frac{s^2}{\omega_n^2}} \qquad 12.21$$

$$K = \frac{K_v}{K_v^2 + D(R_a + R_s)} \qquad 12.22$$

$$\omega_n^2 = \frac{K_v^2 + D(R_a + R_s)}{DL_a\tau_m} \qquad 12.23$$

$$2\zeta = \frac{\omega_n D[L_a + \tau_m(R_a + R_s)]}{K_v^2 + D(R_a + R_s)} \qquad 12.24$$

The denominator of equation 12.21 has two s roots, p_1 and p_2, called *poles*.

$$s = p_1, p_2\big|_{\zeta>1} = -\zeta\omega_n \pm \omega_n\sqrt{\zeta^2 - 1} \qquad 12.25$$

$$s = p_1, p_2\big|_{\zeta<1} = -\zeta\omega_n \pm j\omega_n\sqrt{1 - \zeta^2} \qquad 12.26$$

$$s = p_1, p_2\big|_{\zeta=1} = -\omega_n \qquad 12.27$$

The position θ, not the velocity, is to be controlled. Since $\Omega = s\theta$, equation 12.21 can be rewritten as equations 12.28, 12.29, and 12.30 for the poles defined in equations 12.25, 12.26, and 12.27, respectively.

$$\frac{\theta}{V_s}\bigg|_{\zeta<1} = \frac{K}{s\left[1 + \left(\frac{2\zeta}{\omega_n}\right)s + \left(\frac{1}{\omega_n^2}\right)s^2\right]} \qquad 12.28$$

$$\frac{\theta}{V_s}\bigg|_{\zeta>1} = \frac{K}{s\left[1 + \left(\frac{s}{p_1}\right)\right]\left[1 + \left(\frac{s}{p_2}\right)\right]} \qquad 12.29$$

$$\frac{\theta}{V_s}\bigg|_{\zeta=1} = \frac{K}{s\left[1 + \left(\frac{s}{\omega_n}\right)\right]^2} \qquad 12.30$$

---

[2] D and J are constants in linear systems.

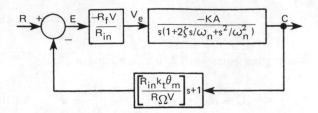

Figure 12.3　Servo Block Diagram

The position-measuring rheostats have a maximum voltage swing of $-V$ to $+V$, so the angular output and reference can be normalized to these maximum values.

$$V_r = \left(\frac{V}{\theta_m}\right)\theta_r = V\left(\frac{\theta_r}{\theta_m}\right)$$

$$= VR \qquad 12.31$$

$$V_c = -\left(\frac{V}{\theta_m}\right)\theta_c = -V\left(\frac{\theta_c}{\theta_m}\right)$$

$$= -VC \qquad 12.32$$

The operational amplifier used as a summing point adds the inputs weighted by their resistances.

$$V_e = -\left[\left(\frac{R_f V}{R_{\text{in}}\theta_m}\right)\theta_r - \left(\frac{R_f V}{R_\theta\theta_m}\right)\theta_c - \left(\frac{R_f k_t s}{R_\Omega}\right)\theta_c\right]$$

$$= -V\left[\left(\frac{R_f}{R_{\text{in}}}\right)R - \frac{R_f}{R_\theta}\left(1 + \frac{R_\theta k_t\theta_m s}{R_\Omega V}\right)C\right]$$

$$12.33$$

For unity feedback, it is necessary that $R_\theta = R_{\text{in}}$. Then, equation 12.33 would be rewritten as

$$V_e\big|_{\text{unity feedback}} = \frac{-VR_f}{R_{\text{in}}}\left[R - C\left(1 + \frac{R_{\text{in}}k_t\theta_m s}{VR_\Omega}\right)\right]$$

$$12.34$$

The locations of R, C, and E are shown in figure 12.3.

The gain of the power amplifier must be negative to cancel out the negative effect of the summing amplifier gain. The open-circuit amplifier output voltage, V_s, is $-A_p G_c V_e$. We can combine the machine's transfer function with the summing amplifier gain.

$$\theta_c = \left(\frac{A_p R_f V}{R_{\text{in}}}\right)E \times \frac{K}{s\left[1 + \left(\dfrac{2\zeta}{\omega_n}\right)s + \left(\dfrac{1}{\omega_n^2}\right)s^2\right]}$$

$$12.35$$

However, $\theta_c = C\theta_m$, so the forward transfer function is

$$G = \frac{C}{E} = \frac{A}{s\left[1 + \left(\dfrac{2\zeta}{\omega_n}\right)s + \left(\dfrac{1}{\omega_n^2}\right)s^2\right]} \qquad 12.36$$

$$A = \frac{A_p R_f VK}{R_{\text{in}}\theta_m} \qquad 12.37$$

$$H = 1 + ks \qquad 12.38$$

$$k = \frac{R_{\text{in}}k_t\theta_m}{R_\Omega V} \qquad 12.39$$

2　BLOCK DIAGRAM ALGEBRA

The previous section illustrates how an engineering system can be modeled by operational blocks. Several such blocks can be grouped together to obtain the desired response.

Complex systems of several block diagrams can be simplified by using the equivalent structures shown in table 12.1.

Example 12.1

A complex system is constructed from five blocks and two summing points as shown.

Simplify the system and determine its overall effect on an input.

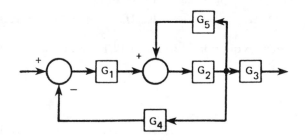

Use case 5 from table 12.1 to move the second summing point back to the first summing point.

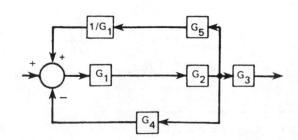

Table 12.1
Equivalent Block Diagrams

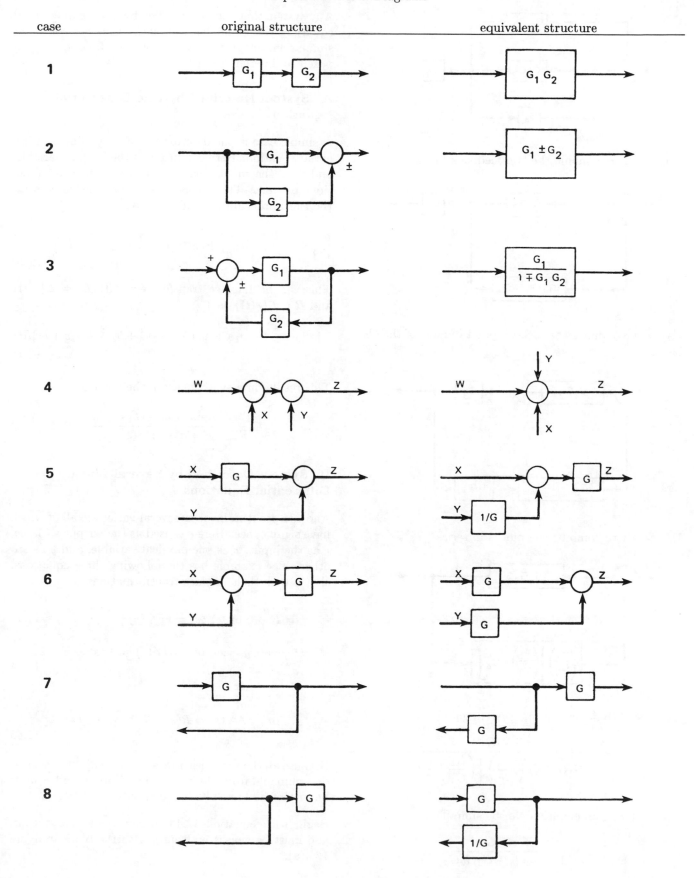

case	original structure	equivalent structure

Use case 1 to combine blocks in series.

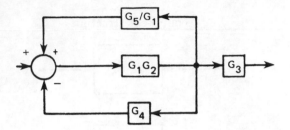

Use case 2 to combine the two feedback loops.

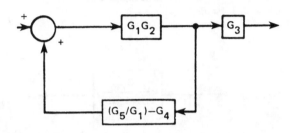

Use case 8 to move the pick-off point outside of the G_3 block.

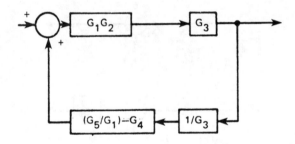

Use case 1 to combine the blocks in series.

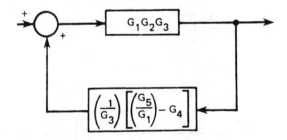

Use case 3 to determine the system gain.

$$G = \frac{G_1 G_2 G_3}{1 - (G_1 G_2 G_3)\left(\dfrac{1}{G_3}\right)\left[\left(\dfrac{G_5}{G_1}\right) - G_4\right]}$$

The system gain equation can be simplified.

$$G = \frac{G_1 G_2 G_3}{1 - G_2 G_5 + G_1 G_2 G_4}$$

3　SYSTEM MODELING

Any system that can be described by a set of differential equations can be modeled in block form. For computer simulation, integration is the preferred function, as opposed to differentiation.

A. System Described by One Differential Equation

A simple case is one described by a single linear differential equation. Here c is taken as the output variable, and r as the input variable, both of which are functions of time. The general case is demonstrated with the third-order differential equation

$$\frac{d^3c}{dt^3} + a_2\frac{d^2c}{dt^2} + a_1\frac{dc}{dt} + a_0 c = b_3\frac{d^3r}{dt^3} + b_2\frac{d^2r}{dt^2} + b_1\frac{dr}{dt} + b_0 r$$

$$12.40$$

This can be Laplace transformed, with $C = \mathcal{L}\{c(t)\}$ and $R = \mathcal{L}\{r(t)\}$ as

$$[s^3 + a_2 s^2 + a_1 s + a_0]C = [b_3 s^3 + b_2 s^2 + b_1 s + b_0]R$$

$$12.41$$

This results in a single block of the form

$$\frac{C}{R} = \frac{b_3 s^3 + b_2 s^2 + b_1 s + b_0}{s^3 + a_2 s^2 + a_1 s + a_0}$$

$$12.42$$

B. System Described by Several Linear Differential Equations

Another problem involves several variables, all of which have significance. Here c is used as the output variable, r as the input, x as one physical variable, and y as another. The example has the following three equations, shown with their Laplace transformations:

$$y = F\frac{dc}{dt} \quad \text{or} \quad Y(s) = sFC(s) \qquad 12.43$$

$$D\frac{dy}{dt} + Ey = x \quad \text{or} \quad sDY(s) = X(s) - EY(s)$$

$$12.44$$

$$c = r - A\frac{dx}{dt} - Bx$$
$$\text{or} \quad sAX(s) = R(s) - BX(s) - C(s)$$

$$12.45$$

Because all the variables have significance, they must be determinable from the model, i.e., both x and y must be intermediate results in a computer simulation.

Begin with equation 12.43, because c is the output, and must be generated from y with the block in figure 12.4(a).

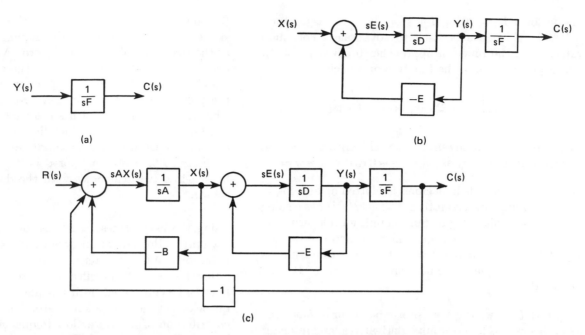

Figure 12.4 Model Development for Multi-Variable System

$Y(s)$ is generated via equation 12.44 from a block structure added as indicated in figure 12.4(b), where $sDY(s)$ is obtained from $X(s)$ and $Y(s)$.

Finally, $X(s)$ is obtained from the generation of $sAX(s)$ as indicated in equation 12.45, and added to the left of figure 12.4(b) to generate the entire model indicated in figure 12.4(c).

4 BODE DIAGRAMS

A primary design problem with control systems is assuring system stability. If the system is stable, the nature of the response of the system and the accuracy of the system also can be evaluated. A great deal can be determined from the *frequency response* of the system. The most useful form of frequency response data is the *Bode diagram*, where the amplitude (in decibels) and the phase angle (in degrees) are plotted against the logarithm of frequency.

When the open-loop transfer function, **GH**, is expressed in terms of the transform variable, s, the frequency response can be determined by replacing s with $j\omega$. The logarithmic nature of the Bode diagram is most useful when the transfer function is expressed in factored form. The transfer function can, in general, be expressed as a ratio of two polynomials.

$$\mathbf{GH}(s) =$$
$$A\left[\frac{a_n s^n + a_{n-1}s^{n-1} + a_{n-2}s^{n-2} + \cdots + a_1 s^1 + a_0}{b_m s^m + b_{m-1}s^{m-1} + b_{m-2}s^{m-2} + \cdots + b_1 s^1 + b_0}\right]$$
$$12.46$$

This is more conveniently expressed in factored form.

$$\mathbf{GH}(s) = \frac{a_0 A}{b_0}\left[\frac{\left(1 + \dfrac{s}{z_1}\right)\left(1 + \dfrac{s}{z_2}\right)\cdots\left(1 + \dfrac{s}{z_n}\right)}{\left(1 + \dfrac{s}{p_1}\right)\left(1 + \dfrac{s}{p_2}\right)\cdots\left(1 + \dfrac{s}{p_m}\right)}\right]$$
$$12.47$$

The z's are the zeros of **GH**, and the p's are the poles of **GH**. Taking the logarithm of a complex number (s is a complex variable),

$$\log \mathbf{GH} = \log|\mathbf{GH}| + j(\log e)\angle\mathbf{GH} \qquad 12.48$$

(*log* is understood to be a logarithm to the base 10, and e is the base of a natural logarithm.)

The Bode diagram measures amplitude in decibels.

$$\mathbf{GH}_{\mathrm{dB}} = 20\log|\mathbf{GH}| \qquad 12.49$$

Consider the fact that $\log(AB) = \log A + \log B$, and $\log(A/B) = \log A - \log B$, expressing $\mathbf{GH}(s)$ in decibels as in equation 12.48, the magnitude part is

$$\mathbf{GH}_{\mathrm{dB}} =$$
$$20\log\left(\frac{a_0 A}{b_0}\right) + 20\log\left|1 + \frac{s}{z_1}\right| + 20\log\left|1 + \frac{s}{z_2}\right|\cdots$$
$$-20\log\left|1 + \frac{s}{p_1}\right| - 20\log\left|1 + \frac{s}{p_2}\right|\cdots - 20\log\left|1 + \frac{s}{p_m}\right|$$
$$12.50$$

Each part of **GH** (in the factored form) can be plotted separately and the results added. However, functions of the form $\mathbf{F} = 20\log|1 + s/s_0|$ have an asymptotic behavior which can be used to advantage. For small values of s, the function approaches 0 dB ($\log 1 = 0$). For large values of s, the function approaches

$$20\log\left|\frac{s}{s_0}\right| = 20\log|s| - 20\log|s_0|$$

This intersects the low-frequency asymptote ($s = j\omega$, and the low-frequency asymptote is 0 dB) where $\omega = s_0$.

When the amplitude in decibels is plotted against $\log\omega$, the high-frequency asymptote ($20\log|s/s_0|$) has a slope of 20 dB per unit of $\log\omega$. Since a unit for a logarithmic scale is a change of the argument by a factor of 10 (i.e., $\log 1000 - \log 100 = 3 - 2 = 1$), the logarithmic unit is a *decade* of frequency. Therefore, the slope of $\mathbf{F}$ is 20 dB per decade.

Figure 12.5 shows the asymptotic behavior of functions like $\mathbf{F}$, as well as for constants, derivatives, and resonant terms. (Note that when a term is in the denominator of the transfer functions, the slope is negative.)

$$F_a = -20\log\left|1 + \frac{s}{\omega_p}\right| \qquad (a)$$

$$F_b = 20\log|K| \qquad (b)$$

$$F_c = -20\log|s| \qquad (c)$$

$$F_d = -20\log\left|1 + \frac{2\zeta s}{\omega_n} + \frac{s^2}{\omega_n^2}\right| \qquad (d)$$

The errors involved in the four types of terms plotted asymptotically in figure 12.5 are of considerable interest. For the constant and derivative types plotted in

(b) and (c), there is no error. For those representing a single pole or zero, as in (a), the maximum error is 3 dB at the frequency of the pole or zero. At a frequency of one octave (a factor of 0.5 or 2), the error is approximately 1 dB. At two octaves (a factor of 4 or 0.25) from the pole or zero, the error is approximately 0.25 dB. The errors which arise with a resonant term, as in (d), can be quite significant, depending on the value of the damping constant ζ. Representative curves are shown in figure 12.6. Note that for $\zeta = 2$, the term can be broken into two real poles with the dotted asymptote shown.

Using the asymptotes, it is possible to construct the asymptotic Bode diagram by adding the individual contributions. Where the terms involving s are in the form $20\log|1 + (s/\omega_c)|$, the effect of a pole or zero does not have to be considered until moving to the positive frequencies beyond the pole or zero. This is because the contribution of a term at low frequencies is zero. The procedure for plotting the asymptotic Bode diagram is given here.

step 1: Find the dB value of the constant term. Plot it as a horizontal line (dashed) at that height parallel to the 0 dB line.

step 2: For any s^n terms, the slope will be $20 \times n$ dB/dec. It will be a positive slope if the numerator term is a zero. It will be a negative slope if the denominator term is a pole. Such a term will cross the 0 dB axis at $\omega = 1$. When added to the constant term, the s^n term will intersect the original constant term at $\omega = 1$. This is illustrated in figure 12.7.

The sum asymptote should not be extended to the right of any pole or zero. It should be ended at the first additional pole or zero.

step 3: Upon reaching the next pole or zero, add its slope to the existing slope, beginning at the pole or zero frequency. Simple poles have negative slope, so they subtract 20 dB/dec from the slope immediately to the left. Simple zeros add 20 dB/dec to the slope immediately to the left. This is illustrated in figure 12.8 for a function $10(1 + s/20)/(1 + s/2)s$.

step 4: For a resonant term, the slope will change by 40 dB/dec from the slope just to the left of the pole or zero. It will be a positive change for a zero and a negative change for a pole.

The asymptotic Bode plot is typically within 3 dB of the actual value unless poles are close together. A decade (factor of 10) is enough to cause an error of no more than

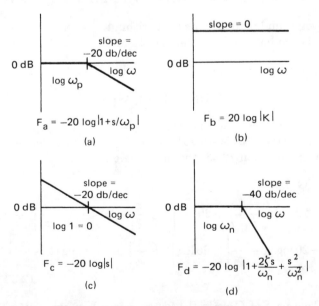

Figure 12.5 Asymptotic Behavior Plotted on a Bode Diagram

PROFESSIONAL PUBLICATIONS ● Belmont, CA

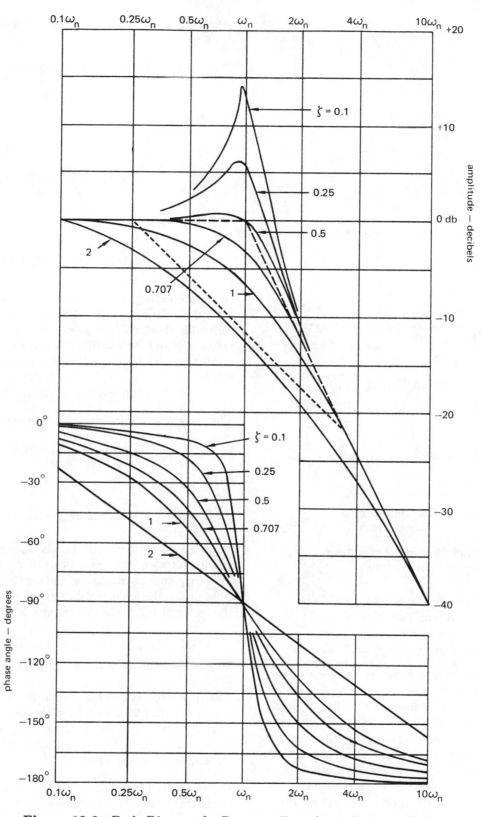

Figure 12.6 Bode Diagram for Resonant Term (as in figure 12.5) Showing the Error from the Asymptotic Value for Various Values of ζ

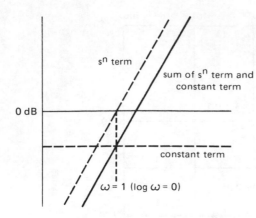

Figure 12.7 Asymptotic Addition
of Constant and s^n Terms

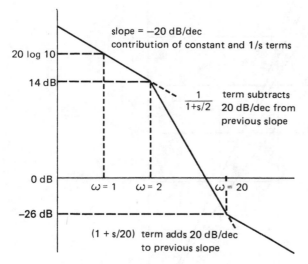

Figure 12.8 Sample Asymptotic
Bode Diagram (not to scale)

1 dB at the pole, but an *octave* (factor of 2) between poles (or between zeros) can cause an error of 4 dB at the poles. The adjacency of a pole and a zero tends to diminish the error for both to less than 3 dB.

However, there can be significant errors for resonant terms with ζ less than 0.5. For complex poles, the dependency on ζ is shown in figure 12.6. For ζ greater than 0.707, the actual values lie below the asymptotes (which intersect at ω_n, shown dashed), while for ζ less than 0.5 the actual values are above these asymptotes. The actual curve will be below the asymptotes for a pole, and above the asymptotes for a zero for simple poles and zeros, or complex poles and zeros with ζ greater then 0.707.

Example 12.2

Find the error at the natural frequency, ω_n, for a system consisting of a resonant pole pair and a zero one octave

below the resonant frequency. Use a damping factor $\zeta = 0.2$.

This can be normalized to the resonant frequency.

$$
\mathbf{GH}|_{s=j\omega_n} = K\left[\frac{1 + \dfrac{2s}{\omega_n}}{1 + \dfrac{2(0.2)s}{\omega_n} + \dfrac{s^2}{\omega_n^2}}\right]
$$

$$
= K\left[\frac{1 + j2}{1 + j0.4 - 1}\right] = K\left[\frac{1 + j2}{j0.4}\right]
$$

The amplitude is

$$
\frac{K\sqrt{5}}{0.4} = 5.59K \mapsto (20\log K + 14.95)\ \text{dB}
$$

Next, the asymptotic value at the resonant frequency is needed. Up to the zero, the value was $20\log K$. The zero adds a slope of 20 dB per decade. An octave is a factor of 2 and $\log 2 = 0.301$, while a decade is a factor of 10 and $\log 10 = 1$. Then, the change in an octave will be

$$
20 \times 0.301 = 6.02\ \text{dB}
$$

Thus, the asymptotic value at resonance will be

$$
20\log K + 6.02\ \text{dB}
$$

The error is

$$
14.95 - 6.02 = 8.93\ \text{dB}
$$

This can be verified analytically. Since the curve for $\zeta = 0.5$ intersects the asymptote at resonance, and the high-frequency asymptote of the zero is $2s/\omega_n$, it is possible to use the asymptotic zero value and the value of the resonant term with a damping of 0.5.

$$
\mathbf{GH}|_{\text{asymptotic}} = 20\log K + 20\log 2 - 20\log 1
$$
$$
= 20\log K + 6.02
$$

The *phase response* of a system is not as easily plotted as the amplitude because the asymptotes are not straight lines. Strictly speaking, they are not asymptotes at all. Rather, they are very predictable curves for simple poles and zeros.

For resonant terms, the phase graph in figure 12.6 is applicable. It contributes negative phase for denominator terms (poles) and positive phase for numerator terms (zeros). For simple poles and zeros, there is a 45 degree phase shift at the pole or zero frequency. The effect of a pole or zero on phase is much more extensive than the effect on amplitude. The phase effect is felt over several octaves of frequency above and below the pole or zero location. Table 12.2 gives values at octave intervals. Intermediate values can be calculated from equation 12.51.

$$\text{phase angle} = \arctan\left(\frac{\omega}{z}\right) = -\arctan\left(\frac{\omega}{p}\right) \quad 12.51$$

Phase angles for the entire system must be obtained by summing the contributions of all the zeros and subtracting the contributions of all poles. The phase angle is important because of feedback. The denominator of equations 12.5 and 12.6 is $1 + \mathbf{GH}$. This can go to zero for $|\mathbf{GH}| = 1$ when $\angle\mathbf{GH} = 180°$.

For absolute stability of a feedback control system, it is necessary that the gain be less than 1 (0 dB) for any frequency at which the phase is +180 degrees or −180 degrees. The usual case in control systems is for the phase to be negative, so the −180 degree phase point is of primary interest.

Satisfactory transient and steady-state performance of control systems requires the phase to be no worse than −150 degrees for the frequency at which the gain crosses the 0 dB line. The difference between the actual phase and 180 degrees at this point is called the *phase margin*. As *critical damping* is obtained with a phase angle of 45 degrees, the closed-loop system may be underdamped if the phase margin is less than 45 degrees. The phase margin and the *gain margin* are shown in figure 12.9. Empirical results indicate that the gain margin should be at least 8 to 10 dB.

5   TRANSPORTATION DELAY

A pure time delay, typically due to transit time, is expressed in a time function as $f(t-\tau)$, where τ is the time delay, delay time, propagation delay, or *transportation delay*. One example of such a delay can be visualized in a hydraulic system where a pulse of pressure is applied at one point and propagates through the fluid as a shock wave. Another example is the propagation time of an electromagnetic wave from a transmitter to a receiver. Recall that the Laplace transform of a time-delayed function is

$$\mathcal{L}\{f(t-\tau)\} = e^{-\tau s}F(s) \quad 12.52$$

So the delay function appears in the s-domain as the exponential $e^{-\tau s}$. In the frequency domain, where $s = j\omega$, this becomes $e^{-j\tau\omega}$, which has a magnitude of 1 and an angle of $-\tau\omega$ radians, which is also $-2\pi\tau f$ radians, or $-360\tau f$ degrees, or $-180\tau\omega/\pi$ degrees. So the delay function is a linear function of frequency.

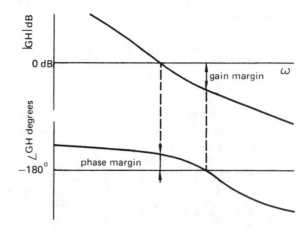

Figure 12.9   Gain and Phase Margins

The delay function can be used directly on the Bode diagram since it does not alter the gain, and subtracts the angle $180\tau\omega/\pi$ degrees. Relative stability can be estimated from gain and phase margins.

In other analysis approaches, the delay can be approximated in four progressively more complex stages. The simplest is with a first-order phase lag, which is accurate within 5 percent of the phase for $\tau\omega$ from 0 to $\pi/4$ radians (45 degrees).

$$e^{-j\tau\omega} \cong \frac{1}{1 + j\tau\omega} \quad 12.53$$

This function reaches an error in magnitude of 3 dB at 45 degrees as well.

The next approximation is completely accurate on magnitude, but has phase accuracy the same as the simpler function of equation 12.53.

Table 12.2
Phase Angles for Simple Poles and Zeros

octave	−5	−4	−3	−2	−1	0	1	2	3	4	5
ω/ω_{pz}	1/32	1/16	1/8	1/4	1/2	1	2	4	8	16	32
phase°	1.8	3.6	7.1	14.0	26.6	45	63.4	76	82.9	8 6.4	88.2

$$e^{-j\tau\omega} \cong \frac{1 - \dfrac{j\tau\omega}{2}}{1 + \dfrac{j\tau\omega}{2}} \qquad 12.54$$

A more accurate approximation uses double the number of terms of equation 12.54, but has phase accuracy within 5 percent up to $\tau\omega = 1.571$ radians or 90 degrees. It also has no error in magnitude.

$$e^{-j\tau\omega} \cong \frac{1 - \dfrac{j\tau\omega}{4}}{1 + \dfrac{j\tau\omega}{4}} \frac{1 - \dfrac{j\tau\omega}{4}}{1 + \dfrac{j\tau\omega}{4}} \qquad 12.55$$

The most accurate approximation uses a number of terms of the general form seen in equation 12.54, with the last terms squared, or

$$e^{-j\tau\omega} \cong \frac{1 - \dfrac{j\tau\omega}{4}}{1 + \dfrac{j\tau\omega}{4}} \frac{1 - \dfrac{j\tau\omega}{8}}{1 + \dfrac{j\tau\omega}{8}} \frac{1 - \dfrac{j\tau\omega}{16}}{1 + \dfrac{j\tau\omega}{16}} \frac{1 - \dfrac{j\tau\omega}{16}}{1 + \dfrac{j\tau\omega}{16}} \qquad 12.56$$

For phase, equation 12.56 is accurate within 5 percent up to $\tau\omega = 2.2$ radians or 126 degrees.

6 THE NYQUIST DIAGRAM

There are other graphic methods of representing frequency functions. In the *Nyquist diagram*, the amplitude of **GH** is plotted against the phase angle in a polar form. In such plots, the points are found by determining both the magnitude and the phase angle at each frequency. Thus, the frequency is a parameter on the curves.

The *absolute stability* of a closed-loop system can be determined from the behavior of the curve near the -1 point, as indicated in figure 12.10. The system will be absolutely stable if the curve of $|\mathbf{GH}|$ versus $\angle\mathbf{GH}$ does not encircle the -1 point on the Nyquist diagram.

The behavior of a closed-loop system will be determined largely by the nearness of the curve to the -1 point. It is useful to note that the gain and phase margins also are available on the Nyquist diagram. These are indicated in figure 12.10.

The two curves shown in figure 12.10 are typical of actual control systems with a single integrator. Curve (a) is stable by the Nyquist criterion because it does not encircle the -1 point. However, if the gain of the system is increased sufficiently, it will approach the unstable characteristic form (b) which does encircle the -1 point. Then the system would become unstable by the Nyquist criterion.

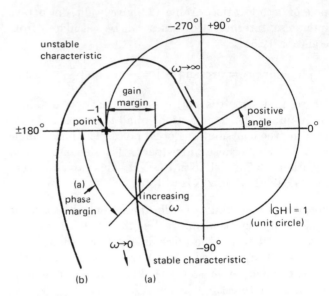

Figure 12.10 Nyquist Diagram: (a) Stable Function, (b) Unstable Function

In general, the Nyquist diagram is much more difficult to construct than the Bode diagram. It contains no more information than the Bode diagram, so it is not usually chosen as the method to present system information. When the system characteristics are presented in this form, however, it is important to be able to interpret the stability and phase margin.

7 STEADY-STATE ERRORS

A. Introduction

When the open-loop transfer function is expressed as a function of s, and the input is transformable to a function of s, the error as a function of s can be written as equation 12.57.

$$E(s) = R(s) \times \frac{1}{1 + \mathbf{GH}(s)} \qquad 12.57$$

There are three types of input for which errors are well defined: the *step input* results in a so-called position error; the *ramp* or *velocity input* results in an error in velocity; and the *parabolic input* corresponds to an error in acceleration.

The type of response is determined by the number of integrators in the open-loop system. Integrators result in denominator terms in **GH** of the form s^n, where n is the number of integrators. A *type one system* has one integrator, a *type two system* has two integrators, and a *type zero system* has no integrators in the open-loop transfer function $\mathbf{GH}(s)$. The corresponding values of n are 1, 2, and 0, respectively.

B. Position Error

The *position error* is the steady-state error when the input $r(t)$ is a step function $R \times u(t)$ of magnitude R. The Laplace transform of the unit step is $1/s$, so $E(s)$ is

$$E(s) = \frac{R}{s} \times \frac{1}{1 + \mathbf{GH}(s)} \qquad 12.58$$

Using the Laplace final value theorem, the steady-state positional error is

$$e_{ss}(t) = \lim_{s \to 0} \frac{R}{1 + \mathbf{GH}(s)} \qquad 12.59$$

For the three types of systems under consideration, $\mathbf{GH}(s)$ can be written similarly to equation 12.47.

$$\mathbf{GH}(s) = K \left[\frac{\left(1 + \dfrac{s}{z_1}\right)\left(1 + \dfrac{s}{z_2}\right) \cdots \left(1 + \dfrac{s}{z_k}\right)}{s^n \left(1 + \dfrac{s}{p_1}\right)\left(1 + \dfrac{s}{p_2}\right) \cdots \left(1 + \dfrac{s}{p_m}\right)} \right] \qquad 12.60$$

Taking the limit as s goes to zero, terms with the form $1 + s/s_j$ will become 1. Then,

$$\lim_{s \to 0} \mathbf{GH}(s) = \lim_{s \to 0} \frac{K}{s^n} \qquad 12.61$$

$$e_{ss}(t) = \lim_{s \to 0} \frac{Rs^n}{s^n + K} \qquad 12.62$$

For a type zero system ($n = 0$),

$$e_{ss}(t)|_{\text{type zero step input}} = \frac{R}{1 + K} \qquad 12.63$$

For a type one system ($n = 1$),

$$e_{ss}(t)|_{\text{type one step input}} = \frac{0}{0 + K} = 0 \qquad 12.64$$

For a type two system ($n = 2$),

$$e_{ss}(t)|_{\text{type two step input}} = \frac{0}{0 + K} = 0 \qquad 12.65$$

C. Velocity Error

The *velocity* is the standard term used for the output function when the input is a *ramp function* characterized by $r(t) = R't$. (R' is the slope of the ramp.) The Laplace transform of $R't$ is R'/s^2, so $E(s)$ is

$$E(s) = \frac{R'}{s^2[1 + \mathbf{GH}(s)]} \qquad 12.66$$

Then from the final value theorem, the steady-state error for a ramp input is

$$e_{ss}(t) = \lim_{s \to 0} s \frac{R'}{s^2[1 + \mathbf{GH}(s)]} \qquad 12.67$$

By using $\mathbf{GH}(s)$ from equation 12.61,

$$e_{ss}(t) = \lim_{s \to 0} \frac{\dfrac{R'}{s}}{1 + \dfrac{K}{s^n}}$$

$$= \lim_{s \to 0} \frac{R's^{n-1}}{s^n + K} = \lim_{s \to 0} \frac{R's^{n-1}}{K} \qquad 12.68$$

For a type zero system ($n = 0$),

$$e_{ss}(t)|_{\text{type zero ramp input}} = \lim_{s \to 0} \frac{R'}{sK} = \infty \qquad 12.69$$

For a type one system ($n = 1$),

$$e_{ss}(t)|_{\text{type one ramp input}} = \lim_{s \to 0} \frac{R'}{K} = \frac{R'}{K} \qquad 12.70$$

For a type two system ($n = 2$),

$$e_{ss}(t)|_{\text{type two ramp input}} = \lim_{s \to 0} \frac{R's}{K} = 0 \qquad 12.71$$

D. Acceleration Error

The acceleration error occurs when the input is a constant acceleration term of the form $r(t) = R''t^2$. This has a Laplace transform of $R(s) = 2R''/s^3$, so $E(s)$ is

$$E(s) = \frac{2R''}{s^3} \left[\frac{1}{1 + \mathbf{GH}(s)} \right] \qquad 12.72$$

By inserting the limiting value of $\mathbf{GH}(s)$ as s goes to zero from equation 12.61,

$$e_{ss}(t) = \lim_{s \to 0} s \frac{2R''}{s^3 \left[1 + \dfrac{K}{s^n} \right]}$$

$$= \lim_{s \to 0} \frac{2R''s^{n-2}}{s^n + K} = \lim_{s \to 0} \frac{2R''s^{n-2}}{K} \qquad 12.73$$

For a type zero system ($n = 0$),

$$e_{ss}(t)|_{\text{type zero parabolic input}} = \infty \qquad 12.74$$

For a type one system ($n = 1$),

$$e_{ss}(t)|_{\text{type one parabolic input}} = \lim_{s \to 0} \frac{2R''s^{-1}}{K} = \infty \quad 12.75$$

For a type two system ($n = 2$),

$$e_{ss}(t)|_{\text{type two parabolic input}} = \lim_{s \to 0} \frac{2R''s^0}{K}$$

$$= \frac{2R''}{K} \quad 12.76$$

Systems without an integrator in the open loop (*type zero systems*) will have a finite position error and infinite velocity and acceleration errors. An infinite steady-state error indicates that the system cannot track the particular type of function which causes the error. The system will continually lose ground against the function.

A system with one integrator in the open loop will be able to follow a position input with zero steady-state error. That is, it will track the position perfectly, given enough time. A *type one system* will follow a ramp input with a constant error proportional to the velocity input, but it cannot track an acceleration input at all.

A *type two system* can track both position and velocity with no long-run error, but it will have a constant error to an acceleration input.

Systems with more than two integrators in the open loop will have closed-loop performance able to track all derivatives up to one less than the number of integrators. However, systems with even two integrations have difficulty damping transients. Systems with more than two integrators are very difficult to stabilize.

8 THE ROUTH-HURWITZ STABILITY CRITERION

A mathematical method for determining the stability of a system from the open-loop transfer function is the *Routh-Hurwitz stability criterion*. This method is based on the concept of multiplying out the closed-loop transfer function $\mathbf{G}_c = \mathbf{G}/(1 + \mathbf{GH})$ where $\mathbf{G}$ and $\mathbf{H}$ are expressed as ratios of polynomials in s.

$$\mathbf{G} = \frac{N_g}{D_g} = K\left[\frac{(s + z_{g1})(s + z_{g2})\cdots}{s^n(s + p_{g1})(s + p_{g2})\cdots}\right] \quad 12.77$$

$\mathbf{H}$ is expressed in terms of a numerator polynomial N_h and a denominator polynomial D_h. The rationalization of $\mathbf{G}_c$ is

$$\mathbf{G}_c = \frac{N_g D_h}{D_g D_h + N_g N_h} \quad 12.78$$

$\mathbf{G}_c$ will be absolutely stable if the denominator term has no zeros with positive real parts, since positive real parts will result in positive exponentials which are unstable.

The Routh-Hurwitz criterion starts by expanding the denominator of $\mathbf{G}_c$ in equation 12.78 to

$$D = a_n s^n + a_{n-1}s^{n-1} + a_{n-2}s^{n-2} + \cdots + a_1 s + a_0 \quad 12.79$$

When this is done, note whether any of the a_j coefficients have different signs from the rest. If so, the system automatically is unstable since it will have zeros of the denominator (poles) with positive real parts. However, this criterion is not sufficient when there are no sign changes among the coefficients. There still may be poles with positive real parts.

The Routh-Hurwitz criterion is based on the development of an array generated from the coefficients of s in D. The first row is made up of a_n, a_{n-2}, a_{n-4}, etc., and the second row is generated from a_{n-1}, a_{n-3}, a_{n-5}, etc., as shown below.

$$\begin{array}{c|cccc} s^n & a_n & a_{n-2} & a_{n-4} & a_{n-6}\cdots \\ s^{n-1} & a_{n-1} & a_{n-3} & a_{n-5} & a_{n-7}\cdots \end{array}$$

The third row is generated from the first and second rows by taking the determinant of the two left-most coefficients of the first and second rows, along with the coefficients just to the right of the position where the new coefficient is to be placed, and then dividing the determinant by the negated left-most coefficient of the second row.

$$b_{n-1} = \frac{\begin{vmatrix} a_n & a_{n-2} \\ a_{n-1} & a_{n-3} \end{vmatrix}}{-a_{n-1}} \quad 12.80$$

$$b_{n-3} = \frac{\begin{vmatrix} a_n & a_{n-4} \\ a_{n-1} & a_{n-5} \end{vmatrix}}{-a_{n-1}} \quad 12.81$$

This results in the rows

$$\begin{array}{c|ccc} s^n & a_n & a_{n-2} & a_{n-4}\cdots \\ s^{n-1} & a_{n-1} & a_{n-3} & a_{n-5}\cdots \\ s^{n-2} & b_{n-1} & b_{n-3} & b_{n-5}\cdots \end{array}$$

The next row is generated from the two above it in a similar manner.

$$c_{n-3} = \frac{\begin{vmatrix} a_{n-1} & a_{n-5} \\ b_{n-1} & b_{n-5} \end{vmatrix}}{-b_{n-1}} \quad 12.82$$

This process is continued until no more rows with nonzero terms can be generated. Notice the practice of placing decreasing powers of s to the left of the rows. This is based on the theoretical justification of the process, not discussed here. Beginning with the highest power of s at the first row, $n+1$ rows will be generated by the process. The last row will have s^0 to the left.

The Routh-Hurwitz stability criterion states that there are exactly the same number of poles with positive real parts as there are sign changes in going down the first column of the array. If there are no poles with positive real parts (i.e., no sign changes), the system is absolutely stable.

Example 12.3

Evaluate the stability of the system with an open-loop transfer function.

$$\mathbf{GH}(s) = \frac{5s^2 + 3s + 2}{3s^4 + 2s^3 + 3s^2 + s}$$

The denominator of $\mathbf{G}_c$ will be

$$D = (3s^4 + 2s^3 + 3s^2 + s) + (5s^2 + 3s + 2)$$

$$= 3s^4 + 2s^3 + 8s^2 + 4s + 2$$

Then the first two rows of the array are

$$
\begin{array}{c|ccc}
s^4 & 3 & 8 & 2 \\
s^3 & 2 & 4 &
\end{array}
$$

So,

$$b_3 = \frac{\begin{vmatrix} 3 & 8 \\ 2 & 4 \end{vmatrix}}{-2} = 2$$

$$b_1 = \frac{\begin{vmatrix} 3 & 2 \\ 2 & 0 \end{vmatrix}}{-2} = 2$$

The third row is

$$
\begin{array}{c|cc}
s^2 & 2 & 2
\end{array}
$$

$$c_3 = \frac{\begin{vmatrix} 2 & 4 \\ 2 & 2 \end{vmatrix}}{-2} = 2$$

$$c_1 = \frac{\begin{vmatrix} 2 & 0 \\ 2 & 0 \end{vmatrix}}{-2} = 0$$

The fourth row is

$$
\begin{array}{c|cc}
s & 2 & 0
\end{array}
$$

$$d_3 = \frac{\begin{vmatrix} 2 & 2 \\ 2 & 0 \end{vmatrix}}{-2} = 2$$

The entire array is

$$
\begin{array}{c|ccc}
s^4 & 3 & 8 & 2 \\
s^3 & 2 & 4 & \\
s^2 & 2 & 2 & \\
s^1 & 2 & & \\
s^0 & 2 & &
\end{array}
$$

There are no sign changes in the first column, so the system is stable.

9 ROOT LOCUS

A. Introduction

In order to evaluate the relative stability and transient performance of a closed-loop system, the roots of the denominator of the closed-loop transfer function must be determined. From equation 12.78, the zeros of the closed-loop system are the zeros of $\mathbf{G}$ and the poles of $\mathbf{H}$, $N_g D_h$.

The *root locus method* deals with the term $\mathbf{GH}$, which for $\mathbf{GH} + 1 = 0$ must have a magnitude of 1 and an angle of $180° + 360n°$ (where n is an integer). When these conditions are satisfied, there is a zero-root of the denominator of $\mathbf{G}_c$. The root loci are all possible points where the angle of the function is $180° + 360n°$. This results in curves of constant angles in the complex s-plane where poles and zeros of the transfer function $\mathbf{GH}(s)$ are plotted.

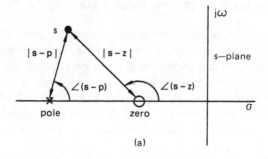

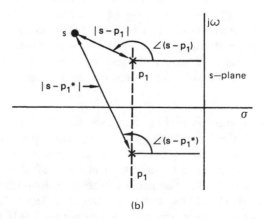

Figure 12.11 Angles and Magnitudes in the s-Plane

Consider the measurement of an angle at an arbitrary point in the s-plane for a function with one pole and one zero as shown in figure 12.11. Poles are represented by X's and zeros by O's. Angles are measured counterclockwise from the positive-real direction. The function in figure 12.11(a) is defined by its magnitude and direction.

$$\mathbf{F} = \frac{s-z}{s-p} \qquad 12.83$$

$$|\mathbf{F}| = \frac{|s-z|}{|s-p|} \qquad 12.84$$

$$\angle \mathbf{F} = \angle(s-z) - \angle(s-p) \qquad 12.85$$

Thus, $\mathbf{F}$ can be evaluated by dividing the distance from the zero to s by the distance from the pole to s. The angle for $\mathbf{F}$ is the angle $\angle z$ minus the $\angle p$.

Figure 12.11(b) shows a pair of complex poles: p_1 and its conjugate p_1^*. The function defined by this pair has magnitude and direction given by equations 12.86 and 12.87.

$$|\mathbf{F}| = \frac{1}{|s-p||s-p_1^*|} \qquad 12.86$$

$$\angle \mathbf{F} = -\angle(s-p_1) - \angle(s-p_1^*) \qquad 12.87$$

B. Root Locus on the Real Axis

The root locus on the real axis is independent of any complex poles or zeros. This is because the contribution of angle from a complex conjugate pair is zero on the real axis. Complex zeros and poles always occur in conjugate pairs. The pole or zero in the upper half-plane will contribute the negative of the angle contributed by the conjugate in the lower half-plane along the real axis.

For any point on the real axis, any zero to the right will contribute +180 degrees of phase angle, and any pole to the right will contribute −180 degrees to the phase angle at that point on the real axis. Thus, a root locus on the real axis will exist whenever there is an odd number of poles and zeros on the real axis to the right.

As the denominator D of the closed-loop transfer function is of the form $D_g D_h + K N_g N_h$ when the gain K of $\mathbf{GH}$ is zero, the roots of D are the same as the poles of $\mathbf{GH}$. As K goes to infinity, the roots of D will be the roots of $N_g N_h$ (the zeros of the open-loop transfer function $\mathbf{GH}$). Thus, the roots begin at the poles of $\mathbf{GH}$ with zero gain and end on the zeros of $\mathbf{GH}$ with infinite gain.

Example 12.4

A system consisting of an integrator with variable gain has $\mathbf{GH} = K/s$. Determine the root locus for this system, and determine the gain necessary to place the closed-loop pole at $s = -5$.

The root locus is shown. As there are no poles or zeros to the right of the origin of the s-plane, the positive real axis is not part of the root locus. There is one pole to the right of any point on the negative real axis, an odd number of poles and zeros, so the entire negative real axis is in the root locus. The locus extends from the origin to negative infinity ($\mathbf{GH}$ is zero at s equal to infinity, so there is a zero at infinity for the locus to terminate on).

To find the value of K for a closed-loop pole at $s = -5$, write the equation $\mathbf{GH} = -1$ and insert the desired value of s to solve for K.

$$\mathbf{GH} = \frac{K}{s} = -1$$
$$= \frac{K}{-5}$$

Therefore, $K = 5$.

When two poles or zeros are adjacent on the negative real axis, and there are an even number of poles and zeros to the right of the right-most of the pair, a root locus exists between the pair. That is only part of the root locus diagram because the loci begin on poles and end on zeros. The root loci must leave the negative real axis and become complex. Such root loci do so in conjugate fashion, as roots must occur in conjugate pairs when they have an imaginary part.

The simplest case is an integrator and one time constant of the form

$$\mathbf{GH} = \frac{K}{s(s-p)} \qquad 12.88$$

In this case the roots are defined by

$$s^2 - ps + K = 0 \qquad 12.89$$

The roots are

$$s = \frac{p}{2} \pm \sqrt{\left(\frac{p}{2}\right)^2 - K} \qquad 12.90$$

For $K = (p/2)^2$, there are two roots at $p/2$. Further increase in K will result in complex conjugate roots with a real part of $p/2$. Thus, the root locus will be as in figure 12.12(a). In the general case of only two poles, the root locus will leave the axis at the average of the two roots as indicated in figure 12.12(b).

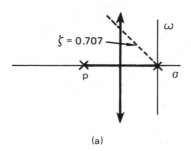

(a)

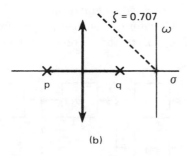

(b)

Figure 12.12 (a) Root Loci for **GH** $= K/[s(s-p)]$, (b) Root Loci for **GH** $= K/[(s-p)(s-q)]$

C. Adjacent Poles With a Zero

Two adjacent poles with a zero will result in a circular root locus centered on the zero and having a radius which is the geometric mean of the distances from the zero to the poles.

$$\text{radius of root-locus circle } = \sqrt{(z - p_1)(z - p_2)}$$

$$12.91$$

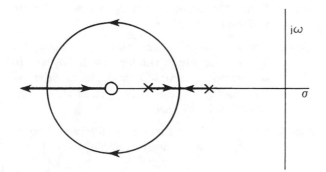

Figure 12.13 Root Loci for Two Adjacent Poles on Real Axis with a Zero

D. Higher-Order Systems

Higher-order systems can become unstable if the root locus crosses into the positive half-plane ($\sigma > 0$). Systems with two poles (i.e., a second-order system) and all zeros in the left-hand plane will be unconditionally stable. In previous sections, root loci were found on the real axis, parallel to the imaginary axis in the negative half-plane, and in a circle.

There also were a number of root loci going to infinity. The number of loci going to infinity is equal to the number of poles minus the number of zeros. When there is one excess pole, the single locus is on the negative real axis. When there are two excess poles, the loci are at +90 degrees and −90 degrees.

These results are general: the number of loci going to infinity is the difference in the number of poles and zeros. The angles at which these loci go to infinity are symmetrical with respect to the negative real axis. The cases for one and two excess poles have been covered. For three excess poles (or excess zeros), the loci are at ±60 degrees and 180 degrees. For four excess poles, the loci are at ±45 degrees and ±135 degrees.

The root loci do not immediately follow these angles to infinity for third order and higher systems. They approach them as asymptotes. The asymptotes are centered at a point on the real axis which is found from equation 12.92.

$$X_{\text{asymptote}} = \frac{\text{sum of pole values} - \text{sum of zero values}}{\text{number of poles} - \text{number of zeros}}$$

$$12.92$$

The angles are given by equation 12.93.

$$\text{angles } = \frac{2i + 1}{n_p - n_z} \times 180° \quad \text{for } i = 0, 1, \cdots, (n_p - n_z - 1)$$

$$12.93$$

In equation 12.93, n_p is the number of poles and n_z is the number of zeros.

When there are two adjacent poles on the real axis with an intervening root locus, the root locus must leave the real axis and become complex at some point between the two poles. The point where this happens is called a *break-away point*. The value of this point can be found from equation 12.94, where σ_b is the location of the break-away point.

$$\sum_{i=1}^{n_p} \frac{1}{\sigma_b - p_i} = \sum_{i=1}^{n_z} \frac{1}{\sigma_b - z_i}$$

$$12.94$$

The values of p_i and z_i have negative real parts when the poles and zeros are in the left half of the s-plane, so that

$-z_i$ and $-p_i$ typically are positive numbers. The above formula can be solved easily for only simple systems. For more complex systems, an iterative method usually is needed to obtain a good estimate of the break-away point.

When the open-loop transfer function contains complex poles, the root locus will leave those poles at an angle (known as the *angle of emergence*) of 180 degrees plus the sum of the angles contributed by all the other poles at the complex pole in question, minus the sum of the angles of the zeros measured to that pole.

$$\text{angle of emergence} = 180°$$
$$+ \sum \text{angles from other poles}$$
$$- \sum \text{angles from zeros} \qquad 12.95$$

When the open-loop transfer function contains complex zeros, the root locus will converge to those zeros at the *angle of convergence*.

$$\text{angle of convergence} = 180°$$
$$- \sum \text{angles from poles}$$
$$+ \sum \text{angles from other zeros} \qquad 12.96$$

10   STATE-VARIABLE REPRESENTATION

A. Introduction

Up to this point, the methods used have been based on Laplace-transformed variables and *frequency domain methods*. A different procedure is the *time domain method*, referred to as the *state-space representation*. The fundamental aspects of this representation and the translation between time domain and frequency domain are covered in this section. The general solutions to state-variable problems will not be discussed, as they are more complex.

A simple example of an s-domain transfer function is

$$\frac{C}{R} = \frac{K}{s^2 + a_1 s + a_0} \qquad 12.97$$

This can be rewritten as

$$(s^2 + a_1 s + a_0)\frac{C}{K} = R \qquad 12.98$$

If this is a Laplace-transformed differential equation, the corresponding differential equation is

$$\frac{d^2 x_1}{dt^2} + a_1 \frac{dx_1}{dt} + a_0 x_1 = r(t) \qquad 12.99$$

$$x_1 = \frac{c(t)}{K} \qquad 12.100$$

The variable change is made to put the results into the standard state-variable notation.

Using the dot notation for time derivatives, $dx/dt = \dot{x}$ and $d^2 x/dt^2 = \ddot{x}$, so that

$$\ddot{x}_1 + a_1 \dot{x}_1 + a_0 x_1 = r(t) \qquad 12.101$$

Solving for $\ddot{x}_1$,

$$\ddot{x}_1 = r(t) - (a_1 \dot{x}_1 + a_0 x_1) \qquad 12.102$$

This result can be modeled by the use of integrators and amplifiers as indicated in block diagram form in figure 12.14. This constitutes a state-variable block diagram for the system described by equation 12.97. This system has no zeros. Therefore, it is readily transformed from the transfer function to the state-variable form.

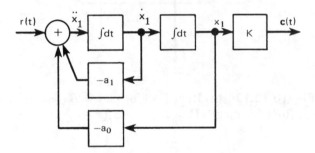

Figure 12.14   State-Variable Block Diagram

The state-variable approach is to represent a system by a set of first-order differential equations. The variables of these equations are the *state variables*. One method to generate the set of state-variable equations from a transfer function or an nth-order differential equation in one variable is given here.

step 1: Define the variable x_1 (as in the previous example) as a state variable.

step 2: Define the derivative of x_1 as a second state variable: $x_2 = \dot{x}_1$.

step 3: Continue to define new state variables up to, but not including, the highest derivative. In the previous example $\ddot{x}_1$ is the highest derivative, and $\ddot{x}_1 = \dot{x}_2$, so no further state variables can be defined.

step 4: Write the set of first-order differential equations. In the previous example, these were

$$\dot{x}_1 = x_2 \qquad 12.103$$
$$\dot{x}_2 = r(t) - a_1 x_2 - a_0 x_1 \qquad 12.104$$

Equations 12.103 and 12.104 constitute a set of first-order differential equations in two state variables. An additional equation is needed to define the output in terms of the state variables. For this example,

$$y(t) = c(t) = Kx_1 \qquad 12.105$$

The usual format for state equations and the auxiliary output equations is matrix form. The above equations are written as equations 12.106 and 12.107.

$$\begin{bmatrix} \dot{x}_1 \\ \dot{x}_2 \end{bmatrix} = \begin{bmatrix} 0 & 1 \\ -a_0 & -a_1 \end{bmatrix} \begin{bmatrix} x_1 \\ x_2 \end{bmatrix} + \begin{bmatrix} 0 \\ 1 \end{bmatrix} r(t)$$
$$12.106$$

$$y(t) = [K \quad 0] \times \begin{bmatrix} x_1 \\ x_2 \end{bmatrix} + 0 \times r(t)$$
$$12.107$$

In matrix notation, equations 12.106 and 12.107 are written as

$$\dot{\mathbf{x}} = \mathbf{A}\mathbf{x} + \mathbf{B}r(t) \qquad 12.108$$
$$y = \mathbf{C}\mathbf{x} + dr(t) \qquad 12.109$$

$\dot{\mathbf{x}}$ and $\mathbf{x}$ are column matrices, $\mathbf{A}$ is an $n \times n$ square matrix, and $\mathbf{C}$ is a row matrix.

B. Transfer Functions With Zeros

The previous example did not have any finite zeros. The output variable was a simple function of one state variable, so that it might appear that a state variable should be directly associated with the output. This is not true except for systems with no finite zeros.

Consider a transfer function of the form given by equation 12.110.

$$\frac{C}{R} = \frac{b_2 s^2 + b_1 s + b_0}{s^3 + a_2 s^2 + a_1 s} = \frac{(b_2 s^2 + b_1 s + b_0)X_1}{(s^3 + a_2 s^2 + a_1 s)X_1} \quad 12.110$$

Making variable changes,

$$R = (s^3 + a_2 s^2 + a_1 s)X_1 \qquad 12.111$$
$$r(t) = \dddot{x}_1 + a_2 \ddot{x}_1 + a_1 \dot{x}_1 \qquad 12.112$$

Therefore,

$$\dddot{x}_1 = r(t) - a_2 \ddot{x}_1 - a_1 \dot{x}_1 \qquad 12.113$$

This defines the feedback part of the state-variable block diagram indicated in figure 12.15(a).

It is then necessary to identify the feed forward portion of the diagram as indicated in figure 12.15(b). Combining figure 12.15(a) and (b) produces the state-variable block diagram as shown in figure 12.15(c).

$$C = (b_2 s^2 + b_1 s + b_0)X_1 \qquad 12.114$$

$$y(t) = c(t) = b_2 \ddot{x}_1 + b_1 \dot{x}_1 + b_0 x_1 \qquad 12.115$$

The state variables are defined as $x_2 = \dot{x}_1$ and $x_3 = \dot{x}_2$, so the matrices are

$$\begin{bmatrix} \dot{x}_1 \\ \dot{x}_2 \\ \dot{x}_3 \end{bmatrix} = \begin{bmatrix} 0 & 1 & 0 \\ 0 & 0 & 1 \\ 0 & -a_1 & -a_2 \end{bmatrix} \begin{bmatrix} x_1 \\ x_2 \\ x_3 \end{bmatrix} + \begin{bmatrix} 0 \\ 0 \\ 1 \end{bmatrix} r(t)$$
$$12.116$$

$$y(t) = [b_0 \quad b_1 \quad b_2] \begin{bmatrix} x_1 \\ x_2 \\ x_3 \end{bmatrix} \qquad 12.117$$

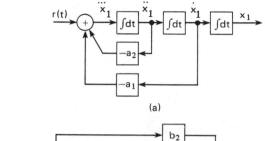
(a)

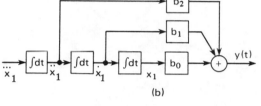
(b)

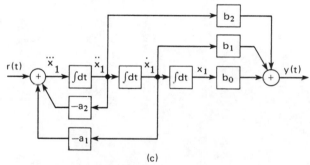
(c)

Figure 12.15 State-Variable Block Diagram Development

Example 12.5

Given the state-variable block diagram, find the equivalent open-loop frequency-domain system. Assume unity feedback.

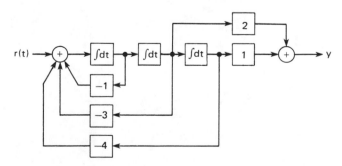

Reverse the procedure presented.

$$\frac{Y(s)}{R(s)} = \frac{C}{R} = \frac{(2s+1)X_1}{(s^3 + s^2 + 3s + 4)X_1}$$

The closed-loop equation is

$$\mathbf{G}_c = \frac{2s+1}{s^3 + s^2 + 3s + 4}$$

However, for unity feedback,

$$\mathbf{G}_c = \frac{\mathbf{G}_o}{1 + \mathbf{G}_o}$$

So,

$$\mathbf{G}_o = \frac{\mathbf{G}_c}{1 - \mathbf{G}_c}$$

$$\mathbf{G}_c = \frac{C}{R}$$

Therefore,

$$\mathbf{G}_o = \frac{C}{R - C} = \frac{(2s+1)X_1}{(s^3 + s^2 + 3s + 4)X_1 - (2s+1)X_1}$$

$$= \frac{2s+1}{s^3 + s^2 + s + 3}$$

PRACTICE PROBLEMS

Warmups

1. A separately excited DC servomotor (see figure 12.2) has $R_A = 5$ ohms (armature resistance) and $L_A = 0.1$ henrys (armature inductance). The loaded motor runs at 160 rad/sec when the applied armature voltage is 24 volts and the armature current is 1.6 amps. When the armature voltage is removed, the mechanical system has a time constant of 1.8 sec. The output impedance of the power amplifier is 3 ohms (resistive).

(a) Obtain the block diagram of the motor and load. The input to this block should be the power amplifier open-circuit voltage (i.e., include the amplifier output impedance within this block). All relationships are to be in the Laplace-transformed notation.

(b) The summing amplifier has values $R_f = R_{in} = R_\theta = 10K$ ohms. The position rheostats for reference and the output are 500 ohms each. They can be neglected when compared with the input resistances to the summing amplifier. The voltages V are 10 volts. The tachometer output is 0.05Ω, where Ω is the shaft velocity in rad/sec. The tachometer resistance is an adjustable parameter. The position measuring rheostat's swing is $-\pi < \theta < \pi$, over which the reference voltage ranges from -10 volts to $+10$ volts. The output rheostat voltage ranges from $+10$ volts to -10 volts. Thus, if the two rheostats are in the same position, they produce equal and opposite voltages. Obtain the relationships between $C(s)$ and $R(s)$ and between $E(s)$ and $R(s)$ for this system.

(c) Obtain the system block diagram in the form of figure 12.3. The power amplifier can be considered to be a voltage amplifier. Its gain is an adjustable parameter.

2. A control system has the open-loop transfer function

$$\frac{10(s + 2)}{s(s + 20)}$$

(a) Plot the amplitude asymptotes for this system's Bode diagram. (b) Find the steady-state position error to a unit step input for a closed-loop system. (Assume unity feedback.) (c) Find the steady-state velocity error to a unit ramp input for a closed-loop system.

3. A control system has unity gain and the open-loop transfer function

$$\frac{50s}{s^2 + 10s + 400}$$

(a) Construct the asymptotic Bode-amplitude plot. (b) Find the closed-loop steady-state error to a unit

step input. (c) Find the closed-loop steady-state error to a unit ramp input.

4. A particular control system has

$$\mathbf{GH}(s) = \frac{5s^2 + 16s + 25}{15s^4 + 225s^3 + 5s^2}$$

Determine the steady-state errors to unit step, ramp, and parabolic inputs.

5. Determine the stability of the following two systems using the Routh-Hurwitz method. If either system is unstable, determine the number of poles in the right half of the s-plane.

$$\mathbf{GH}(s) = \frac{40(s + 1)}{s^2(s + 10)(s + 20)} \qquad \text{(a)}$$

$$\mathbf{GH}(s) = \frac{s + 8}{s(s^2 + s + 1)} \qquad \text{(b)}$$

6. A control system has a single open-loop time constant of 0.25 sec and a variable gain. Show the root locus for this system and determine the gain necessary to produce a closed-loop time constant of 0.1 sec.

7. A feedback control system has one open-loop pole and one open-loop zero. The pole is in the left half of the s-plane. Show all possible root loci for this system.

8. A system with unity feedback has an open-loop transfer function of

$$\frac{K(s + 9)}{s(s + 3)}$$

Find the values of gain K for which the damping ratio of the closed-loop system will be $\zeta = 0.707$. Find the closed-loop system time response to a unit step for both cases.

9. A unity feedback, open-loop system has

$$\mathbf{G} = \frac{K(s + 2)}{s^2(s + 10)(s + 20)}$$

Determine the asymptotes for large values of K. Estimate the low K region by treating the two poles at the origin and the zero at $s = -2$ as the total system. From these results, sketch the remainder of the root locus. (a) Using the value of s where the high K asymptotes cross the imaginary axis, calculate the value of K for this point. (Note that the imaginary part of K indicates that this point is not actually on the root locus.)

(b) Use the Routh-Hurwitz method to determine the gain value for the crossing of the imaginary axis. (c) Determine the frequency corresponding to the gain in part (b).

10. For the system of problem 4, assume unity feedback and obtain the state-variable block diagram for the closed-loop system.

Concentrates

1. A unity feedback control system has

$$G = \frac{20}{s(s+10)}$$

The closed-loop response is considered too slow, and tachometer feedback is suggested to improve the response time. The tachometer feedback is to be applied in parallel with the unity feedback. If the tachometer has a transfer function ks, determine the value of k necessary to result in a damping ratio of 0.707.

2. The control system shown has $H = 1$, and

$$G_1 = \frac{1}{s+1}$$

$$G_2 = \frac{80}{(s+2)(s+10)}$$

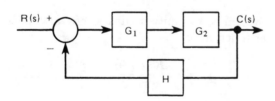

(a) Is this system stable or unstable? (b) Is the system stable if the transfer functions are changed to

$$G_1 = \frac{1}{s+1}$$

$$G_2 = \frac{80}{s+2}$$

$$H = \frac{1}{s+10}$$

How will the time output of this system differ from the time output of (a)?

3. The open-loop transfer function of a feedback system is

$$GH(s) = \frac{K(s+5)}{s(s+1)}$$

Sketch the root locus. Find the points for $K = 5$ and $K = 10$. Is this system stable?

4. A control system has an output $c(t) = 20(1 - e^{-5t})$ volts for a step error input of 10 millivolts at $t = 0$. The output is zero before $t = 0$. The feedback is proportional to the output with a value of less than or equal to one times the output. With the input $r(t) = 1$ volt, the output reaches a final value of 10 volts. (a) Determine the feedback ratio. (b) With the feedback ratio set at 0.1, determine the output as a function of time for a step input of one volt.

5. The unity feedback system shown has

$$G_p = \frac{20}{s(s+2)(s+10)}$$

$$G = G_c G_p$$

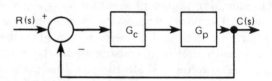

G has the following frequency description: (1) a slope of -20 dB/dec below 0.1 rad/sec, (2) a slope of -40 dB/dec between 0.1 and 1.5 rad/sec, (3) a slope of -20 dB/dec between 1.5 and 18 rad/sec, (4) a slope of -60 dB/dec for frequencies above 18 rad/sec and (5) $|G| = 1$ at 10 rad/sec. (a) Determine G. (b) Determine G_c. (c) Determine the steady-state error for $r(t) = (5 + 3t)u(t)$.

6. The open-loop frequency response data for a process is as follows:

rad/sec:	0.025	0.05	0.1	0.2	
gain (dB):	35.0	29.0	23.0	16.8	

rad/sec:	0.4	0.8	1.6	3.2	6.4
gain (dB):	10.3	2.8	-6.5	-17.2	-27.9

rad/sec:	12.8	25.6	51.2	102.4	
gain (dB):	-37.2	-44.9	-52.1	-60.3	

rad/sec:	204.8	409.6	819.2	1638	
gain (dB):	-70.4	-81.8	-93.6	-105.6	

This process is known to be minimum phase. A step response shows a dominant pole with a time constant of 1.0 second, and a high-frequency pole with a time constant around 10 milliseconds. (a) Determine the open-loop transfer function. (b) Determine the gain margin and/or phase margin. (c) For what range of feedback gains will the closed-loop system be stable?

7. An operational amplifier has a DC open-loop gain of 10^5, and open-loop poles at 10, 10^6, and 10^7 rad/sec. It is used in the non-inverting amplifier configuration as shown below, where the feedback is $\mathbf{H} = 1/k$. (a) Determine the range of k for which the amplifier will be stable. (b) Determine the frequency of oscillation when k reaches the limit.

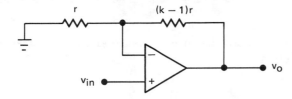

8. A particular plant can be modeled (open loop) as an integration with a time delay (or transport delay) of 1 second. (a) Determine the minimum loop gain at which the system becomes unstable. (b) Using half that loop gain, determine the open-loop gain and phase margins.

9. A control system has the open-loop transfer function

$$\mathbf{GH}(s) = \frac{10^7}{(s^2 + 10s + 100)(s + 100)}$$

This system appears to be unstable, and it is desired to compensate it with a lag compensator of the form

$$\mathbf{G}_c(s) = \frac{1}{1 + \tau s}$$

Determine the value of τ to provide 45 degrees of phase margin.

10. An asymptotic Bode plot of the open-loop characteristic of a control system has a slope of -20 dB/dec at low frequencies, a change of slope to 0 at a frequency of 2 rad/sec, another change of slope to -40 dB/dec at 8 rad/sec, goes through 0 dB at 10 rad/sec, and has one further change of slope to -60 dB/dec at a frequency of 80 rad/sec. It is known to be minimum phase, and to have complex poles with a damping ratio of 0.25. (a) Obtain the expression for the open-loop function $\mathbf{GH}(s)$. (b) Determine the frequencies at which the asymptotic Bode plot has peak (minimum or maximum) errors, and indicate the size of those errors. (c) Plot the angle portion of the Bode plot.

Timed

1. A state-variable simulation of a control system is shown. Experimental tests made on the closed-loop system show approximate critical damping at a natural frequency of 1 rad/sec. (a) Determine $C(s)/R(s)$ as a ratio of two factored polynomials. (b) The physical system has unity feedback. Obtain the transfer function of the open-loop physical system. (c) Explain which block of the simulation needs to be changed to alter the damping ratio.

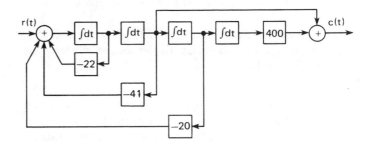

2. A unity gain feedback system has the open-loop transfer function

$$\frac{K}{(s + 3)(s + 6)(s + 7)}$$

(a) If the feedback is positive, what range of K will result in a stable system? (b) If the feedback is negative, what range of K will result in a stable system?

3. Consider the control system shown. (a) Plot the root locus as a function of K. (b) Discuss the stability. (c) Discuss the steady-state error for a step input.

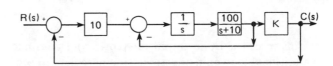

4. A control system has the closed-loop transfer function

$$\frac{K}{s^3 + 2s^2 + 4s + K}$$

For positive K, sketch the root locus and find the following: (a) any break-away points on the real axis, (b) frequency and gain at any point where instability begins, and (c) the complex frequency and gain where the damping ratio is 0.707.

5. Two transfer functions are cascaded in the forward path of a unity feedback system. These two transfer functions have the following frequency characteristics:

$\mathbf{G}_1(j\omega)$: A break frequency at 10 rad/sec where the asymptotes meet at a gain of 20 dB. Below

10 rad/sec, the slope is -20 dB/dec. Above 10 rad/sec, the slope is -40 dB/dec. The high frequency phase shift is -180 degrees.

$\mathbf{G}_2(j\omega)$: A low-frequency asymptote at -20 dB, a corner frequency at 1 rad/sec, and a high-frequency slope of -20 dB/dec. The high-frequency phase shift is -90 degrees.

(a) Determine $\mathbf{G}_1(s), \mathbf{G}_2(s)$, and $\mathbf{G}(s) = \mathbf{G}_1(s)\mathbf{G}_2(s)$. (b) Apply Routh's criterion to determine whether the system is stable.

6. A particular control system has an open-loop gain of 30 dB at a frequency of 4 rad/sec. It is known to be minimum phase and to have one pair of critically damped poles. The following open-loop phase measurements have been made, and cover the range of frequencies of interest:

ω(rad/sec)	$\frac{1}{64}$	$\frac{1}{32}$	$\frac{1}{16}$	$\frac{1}{8}$
$\angle\mathbf{GH}$(deg)	-179.2	-178.4	-176.8	-173.6

ω(rad/sec)	$\frac{1}{4}$	$\frac{1}{2}$	1	2
$\angle\mathbf{GH}$(deg)	-167.4	-156.3	-140.7	-128.0

ω(rad/sec)	4	8	16	32
$\angle\mathbf{GH}$(deg)	-126.7	-140.7	-170.9	-207.8

ω(rad/sec)	64	128	256	512
$\angle\mathbf{GH}$(deg)	-236.2	-252.7	-261.3	-265.6

(a) Construct the Bode amplitude plot. (b) Determine the additional loop gain or attenuation necessary to provide an open-loop phase margin of 45 degrees.

7. A control system has an open-loop transfer function of
$$\mathbf{GH}(s) = \frac{K}{s(s+1)(s+5)}$$
(a) Sketch the root locus. (b) Determine the value of K at the break-away point. (c) Determine the value of K that just makes the closed-loop system unstable.

8. A process has been tested and has an open-loop gain of 10 and a time constant of 1 second. Design a compensator for the feedback path so the compensated process will have a closed-loop steady-state step response error of zero, and a steady-state ramp response error of no more than 1 percent.

9. A plant has an asymptotic Bode diagram with a low-frequency slope of -20 dB/dec, which crosses the 0 dB axis at 1 radian per second. There is a change of the slope to zero at a frequency of 10 rad/sec, and another change of the slope back to -20 dB/dec at a frequency of 150 radians per second. The system is minimum phase.

A compensator has been installed and has the transfer function
$$\mathbf{G}_c = \frac{K}{s}$$

(a) Determine the value of K which will provide an open-loop phase margin of 45 degrees. (b) Using K from part (a), assuming unity feedback, determine the gain and phase of the closed-loop system at the frequency of the phase margin calculation of K.

10. An angular positioning system used with a satellite tracker uses a separately excited DC motor with its armature driven by a power amplifier. The amplifier amplifies the difference between signals proportional to the desired angle θ_r and the actual angle θ_c (these angles are in radians).

$$V_{\text{no-load}} = 5.0(\theta_r - \theta_c)$$

The amplifier has an output resistance of 2 ohms. The motor armature has a winding resistance of 3 ohms and an inductance of 0.012 henrys. The motor has a back-EMF constant of $K_v = 3$V/rad/sec (which is also its torque constant $K_T = 3$ m-N/amp).

The motor and mount have a combined moment of inertia and friction coefficients of

$$J = 0.33 \text{ meter-Newton second}^2/\text{radian}$$
$$D = 100 \text{ meter-Newton second/radian}$$

Develop a muli-path block diagram model showing the input, θ_r, and the output, θ_c, as well as additional identifiable outputs of the armature current, i, and the angular velocity, Ω.

13

ILLUMINATION AND THE NATIONAL ELECTRIC CODE

Nomenclature

CBU	coefficient of beam utilization	–
CCR	ceiling cavity ratio	–
CR	cavity ratio	–
CU	coefficient of utilization	–
E	illumination	fc
E_{rh}	reflected horizontal illumination	fc
E_{rv}	reflected vertical illumination	fc
FCR	floor cavity ratio	–
h_{cc}	ceiling cavity height	feet
Hpf	high power factor	–
h_{fc}	floor cavity height	feet
h_{rc}	room cavity height	feet
Hs	house-side lateral distance	feet
I	luminous intensity	candelas
II	initial illumination level	fc
L	length	feet
L	luminance	foot-lamberts
LC_{cc}	ceiling cavity luminance coefficient	–
LC_w	wall luminance coefficient	–
LDD	luminaire dirt depreciation factor	–
LLD	lamp lumen depreciation factor	–
LLF	light loss factor	–
MH	mounting height	feet
MMI	minimum maintained illumination level	fc
ND	part of MH	feet
Npf	normal power factor	–
R	coefficient of reflectance	–
RCR	room cavity ratio	–
RPM	room position multiplier	–
RRC	reflected radiation coefficient	–
Ss	street-side lateral distance	feet
T	coefficient of transmission	–
W	width	feet
WRRC	wall reflected radiation coefficient	–

Symbols

Φ	luminous flux	lumens
α	longitudinal angles	degrees
β	lateral angles	degrees

1 DEFINITIONS

Light is an electromagnetic phenomenon dealing with the radiation, refraction, transmission, and absorption of electromagnetic waves. Radiation patterns are similar to those of various antennas. Light radiation is usually a far-field situation, where the power density falls off as $1/\text{distance}^2$.

The language of lighting differs from that of electromagnetic radiation because only the *visible spectrum*, rather than total electromagnetic power, is involved. The human eye is sensitive to wavelengths between 380 and 780 nanometers, with maximum sensitivity at 556 nanometers. Light sources are evaluated by their power output in this range.

The unit of light power is the *lumen*, which is weighted power within the visible portion of the spectrum. The amount of light power obtained from the electromagnetic power depends on the spectral content of the electromagnetic wave. For example, a complete conversion of electromagnetic power to light power will result in 680 lumens per watt if the wavelength is exactly 556 nanometers. On the other hand, a wave with uniform power density over the total visible range from 380 to 780 nanometers will result in a light power of about 200 lumens per watt of electric power.

Luminous intensity is the basic standard quantity of light. All other quantities are derived from it. The unit of luminous intensity is the candela (cd). One candela is approximately the intensity of an ordinary wax candle as seen in the horizontal plane.

A *lumen* is a unit of total light flux. The strength of a light source is typically given in candlepower (cp).

A uniform source of one candela radiates 4π lumens of light flux. In general, light sources are quite directional. Candlepower (candelas) is used to describe the apparent strength of a source as viewed from any particular direction. Thus, it is a measure of flux-density on a solid angular basis and has units of candelas, which are equivalent to lumens per steradian. Candlepower is always used to describe a light source and its luminous intensity in some direction and is given in units of candela (cd).

Luminous flux is defined as one lumen (Lm) when a point source of one candle illuminates one square foot of a sphere having a radius of one foot which is centered on the light source. That is, the flux incident on the one-square-foot area is one lumen.

Illumination is the flux density measured in units of lumens per square foot, or footcandles. Illumination is a point quantity subject to the law of cosines. For a point source of light as shown in figure 13.1, the illumination on a point a distance D from the source, lying on a surface so that the surface normal makes an angle a with the line connecting the point on the surface with the point source is

$$E = I\frac{\cos(a)}{D^2} \qquad 13.1$$

The apparent luminous intensity of a surface, because it is either emitting light or reflecting light, is called the *luminance* or *photometric brightness*.

Luminance due to *reflected* light is typically calculated from the coefficient of reflectance, R, and the surface illumination.

$$L \text{ (footlamberts)} = E \text{ (footcandles)} \times R \qquad 13.2$$

Luminance due to transmitted light through a diffuser is typically calculated from the incident illumination and the coefficient of transmission (T).

$$L \text{ (footlamberts)} = E \text{ (footcandles)} \times T \qquad 13.3$$

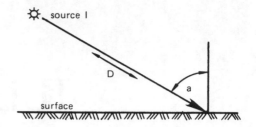

Figure 13.1   Angular Dependence of Illumination

2   LAMPS

General purpose lamps are classified according to their light-producing mechanism as *incandescent* or *vapor discharge*. Vapor discharge includes both mercury discharge and fluorescent lamps. Incandescent lamps provide light described as warm white which is considered the closest to natural light. Although they are more efficient and lower in cost, mercury vapor lamps are not widely used for interior lighting because of a less pleasing color content. Fluorescent lamps, also more efficient than incandescent lamps, are available in a wider range of colors.

A. Relative Costs

Incandescent lamps typically are much lower in cost than discharge lamps, but they must be replaced more often and have lower efficiency in terms of lumens per watt. A measure of the relative costs of each type of lamp will vary with demand and technological development, so the following information is subject to some variation on a year-to-year basis.

Taking the cost per 1000 lumen hours (multiplied by the lamp lumen depreciation factor) per watt of electrical input power and using the cost of the incandescent as the unit, approximate costs give the following relative expenses.

incandescent	1
metal halide	0.4
sodium vapor	0.1
mercury vapor	0.1
fluorescent	0.05

There is some variation depending on wattage, so these figures should be used only as a guideline.

B. Lamp Classification

Lamps are classified according to their electrical power dissipation, shape or type, size, and base. The letters used to indicate the shape of a bulb are:

A	the common shape used for domestic lighting under 200 watts
BT	bulbous tubular
F	flame shaped
G	globular or round
PAR	parabolic reflector
PS	pear-shaped
R	reflector
S	straight sides
T	tubular

The diameter of a bulb is given in eighths of inches. A lamp designated A-19, for example, indicates the common shaped bulb with a maximum diameter of 19 eighths $\left(\frac{19}{8}\right)$ or $2\frac{3}{8}$ inches.

Fluorescent tubes have the length given in inches. In this chapter, only the 48 inch and 96 inch lengths are considered. Metal halide may also be designated BU for base up, or BD for base down. High intensity discharge lamps have different initial lumen ratings depending on whether they are operated horizontally or vertically.

3 BALLAST

Ballast performs several functions. Discharge lamps, presenting negative resistance, require current limitation. To keep the power loss to a minimum, this is done with reactance. The limiting element is usually inductive, so that in *high power factor* (Hpf) *ballasts* some capacitive power factor compensation is used to achieve a power factor of 0.90 or greater. *Normal power factor* (Npf) *ballasts* can have a power factor as low as 0.45, but a value of 0.60 is typical.

Ballasts must provide the proper starting conditions for a discharge lamp. Once the lamp is started, the ballast must provide proper running conditions and, in some cases, protection from overheating. Ballasts are specifically designed for particular voltages and operating conditions.

Ballasts require some power to operate and must be considered in wiring design. Ballasts for two 40 watt fluorescent (F40T12) lamps normally require 12 watts. Normal ballast life of interior units is 12 years.

Tables in the following section give some typical ballast losses. *Auto reg ballasts* are high power factor, and *lag ballasts* are normal power factor.

4 LUMINAIRES

The terms *luminaire* and *fixture* indicate a complete lighting unit including lamps, sockets, ballasts, reflectors, diffusers, and other parts of the unit. Each type of luminaire has its own characteristics which are specified by the manufacturer.

Three important parameters are associated with a luminaire. The *coefficient of utilization* (CU) deals with the fraction of the available light reaching the surface. The *luminaire dirt depreciation factor* (LDD) depends on the environment and frequency of cleaning. The *lamp lumen depreciation factor* (LLD) indicates depreciation from aging.

A. Coefficient of Utilization (CU)

The *coefficient of utilization* is a measure of the efficiency of a luminaire. It is the ratio of the light reaching the working surface to the light being emitted from the luminaire. The coefficient of utilization takes into account the distribution of light from the luminaire, the mounting height, room conditions such as dimensions and reflectances of surfaces for interior lighting, and, for outdoor lighting, those lumens going beyond the area to be lighted.

Each type of luminaire is unique in its distribution of light. The coefficient of utilization depends on the dimensions of the space to be lighted and the pattern of luminaire placement. For interior lighting, a *room cavity ratio* (RCR) is an important parameter for determination of the coefficient of utilization. As an example, table 13.1 gives the coefficient of utilization for two T-12 fluorescent lamps when the spacing of parallel rows of luminaires does not exceed 1.3 times the mounting height. Each coefficient of utilization table is specific for a given luminaire.

B. Luminaire Dirt Depreciation Factor (LDD)

The *minimum maintained illumination level* will depend on the cleanness of a luminaire's environment and on how often the luminaire is cleaned. The minimum maintained illumination level typically occurs just before the luminaires are cleaned. Table 13.2 shows characteristic *luminaire dirt depreciation* values for interior lighting.

The types of environments in which luminaires are installed are defined according to their general cleanness.

- **Clean:** commercial offices, light assembly, or inspection
- **Medium:** industrial offices, light machining
- **Dirty:** heavy manufacturing, processing of materials

C. Lamp Lumen Depreciation Factor (LLD)

The *lamp lumen depreciation factor* is a multiplier used with initial lamp lumens to determine the lamp output

Table 13.1
Coefficients of Utilization for Two T-12 Lamps in a Fluorescent Luminaire
(maximum ratio of spacing to mounting height not to exceed 1.3)

two T-12	reflectances									
ceiling cavity	80%			50%			10%			0%
walls	50%	30%	10%	50%	30%	10%	50%	30%	10%	0%
RCR	coefficients of utilization									
1	0.88	0.84	0.81	0.79	0.77	0.74	0.69	0.68	0.66	0.64
2	0.77	0.71	0.66	0.70	0.65	0.62	0.61	0.59	0.56	0.54
3	0.68	0.61	0.56	0.61	0.56	0.52	0.54	0.51	0.48	0.46
4	0.60	0.52	0.47	0.54	0.49	0.44	0.48	0.44	0.41	0.39
5	0.52	0.45	0.39	0.48	0.42	0.37	0.43	0.38	0.35	0.33
6	0.47	0.39	0.34	0.43	0.37	0.32	0.38	0.34	0.30	0.28
7	0.42	0.34	0.29	0.38	0.32	0.28	0.34	0.30	0.26	0.24
8	0.37	0.30	0.25	0.34	0.28	0.24	0.31	0.26	0.22	0.21
9	0.33	0.26	0.21	0.31	0.25	0.21	0.31	0.23	0.19	0.18
10	0.30	0.23	0.19	0.28	0.22	0.18	0.25	0.20	0.17	0.15

Table 13.2
Luminaire Dirt Depreciation Factors (LDD)

environment	enclosed			open		
fluorescent ⟶	clean	medium	dirty	clean	medium	dirty
cleaned every year	0.88	0.83	0.77	0.94	0.90	0.84
cleaned every 2 years	0.83	0.77	0.71	0.89	0.85	0.78
cleaned every 3 years	0.80	0.74	0.66	0.87	0.80	0.74
HID						
cleaned every year	0.88	0.83	0.77	0.90	0.87	0.86
cleaned every 2 years	0.83	0.77	0.71	0.84	0.80	0.75
cleaned every 3 years	0.80	0.74	0.66	0.79	0.74	0.68

depreciation due to aging. This factor is required to determine the minimum maintained illumination level, or to design for a particular minimum level.

D. Luminaire Determination

The selection of a particular luminaire is based on loss factors, total quantity of light needed, and the quality of the light. Four of the factors affecting quality of the light are *glare*, *brightness ratios*, *diffusion*, and *color*.

- **Glare:** The greater the brightness of the light source, the greater the discomfort, interference with vision, and eye fatigue. The position of the source can produce glare. When the source is in the line of vision, glare can be a factor. Glare is reduced as the source is moved further from the line of vision.

- **Brightness ratios:** Adjacent surfaces with great contrasts in reflectance can produce glare.

- **Diffusion:** Diffusion occurs where light is coming from many directions. It is a function of the size and number of sources contributing to the illumination of a point. Perfectly diffuse light is ideal for many critical visual tasks.

- **Color:** Color has no effect on visual efficiency, but it is important in tasks involving color discrimination. Color has psychological impact, and the various colors of fluorescent lamps are named for their particular applications. (It has been found that sales in women's ready-to-wear shops improve when the most natural light is used, indicating that incandescent lamps should be the choice for such shops.)

PROFESSIONAL PUBLICATIONS ● Belmont, CA

Table 13.3
Typical Characteristics of Standard Lamps Showing Lamp Lumen Depreciation

high intensity discharge

type	lamp	ballast	rated life (hours)	initial lumens		light lumen depreciation (LLD)
				horizontal	vertical	
mercury vapor (deluxe white, auto reg ballast)						
H38JA	100	20	24,000	4,300	4,400	0.79
H39KC	175	39	24,000	8,290	8,500	0.98
H37KC	250	40	24,000	12,500	13,000	0.85
H33GL	400	59	24,000	22,100	23,000	0.86
H36GW	1,000	95	24,000	60,500	63,000	0.76
metal halide (auto reg ballast)						
MH175	175		7,500		14,000	0.73
MH250	250		7,500	19,500	20,500	0.78
MH400	400	64	15,000	30,000	32,000	0.76
MH1000	1,000	80	10,000	95,000	100,000	0.79
MH1500	1,500	120	10,000	150,000	155,000	0.92
high pressure sodium (lag ballast)						
C100S52	100	27	20,000	9,500	9,500	0.90
C150S56	150	42	20,000	16,000	16,000	0.90
C250S50	250	45	24,000	25,500	25,500	0.91
C400S51	400	80	24,000	50,000	50,000	0.90
C1000S52	1,000	100	24,000	130,000	130,000	0.92

fluorescent (cool white, two lamps per luminaire)

type	wattage per lamp		rated life (hours)	initial lumens	light lumen depreciation (LLD)
	lamp	ballast			
48″ lamps					
F40T12					
rapid start	40.0	6.0	20,000	3,150	0.84
F48T12					
slimline	38.5	4.0	9,000	3,000	0.80
high output	60.0	12.5	12,000	4,200	0.80
super high output	110.0	15.0	12,000	6,900	0.79
96″ lamps					
F96T12					
slimline	75.0	12.5	12,000	6,300	0.90
high output	100.0	22.5	12,000	9,000	0.86
super high output	215.0	15.0	12,000	15,500	0.80

Table 13.3 (continued)

Multiplying factors for lumen output for fluorescent lamp color designations are:

cool white	1.00
white and warm white	1.03
daylight	0.86
warm white deluxe and cool white deluxe	0.72
living white	0.76
supermarket white	0.75
merchandising white	0.78

filament or incandescent lamps

type	wattage	rated life (hours)	initial lumens	light lumen depreciation (LLD)
T-19	100	1,000	1,630	0.89
T-19	150	1,000	2,650	0.89
A-23	200	750	3,900	0.89
PS-25	300	750	6,300	0.89
PS-35	500	1,000	10,750	0.89
T3Q/CL	500	2,000	10,500	0.98
T3Q/CL	1,000	2,000	21,000	0.98
T3Q/CL	1,500	2,000	33,000	0.98

The choice of luminaires (and lamps) is influenced by the quality of light needed and by the level of illumination which must be maintained. Minimum levels of maintained illumination for typical usages are given in table 13.4.

The luminaire of choice depends on both the quality of the light and the minimum maintained illumination level.

5 MINIMUM MAINTAINED ILLUMINATION

The *minimum acceptable illumination level* for a specific area is one of the principal design parameters for lighting. This level must be maintained under the expected conditions of lumen depreciation, dirt factor, and coefficient of utilization for the luminaires chosen. Each of these depreciation factors is necessary in determining the number of luminaires required to meet the lighting specifications.

The *minimum maintained illumination level* (MMI) is

specified. The *initial illumination* level (II) is calculated from the coefficient of utilization (CU), the luminaire dirt depreciation factor (LDD), and the lamp lumen depreciation factor (LLD).

$$II = \frac{MMI}{CU \times LDD \times LLD} \qquad 13.4$$

Most luminaire tables assume the following reflectances.

ceiling:	80% for commercial	
	50% for industrial	
wall:	50%	
floor:	20%	

For other surfaces, the following values may be used.

light colors	
white paint	81%
ivory paint	79%
cream paint	74%
cream stone	69%

medium colors

buff paint	63%
light green paint	63%
light gray paint	58%
gray stone	56%

dark colors

tan paint	48%
light oak paint	32%
dark gray paint	26%
olive green paint	17%
dark oak paint	13%
mahogany paint	8%
natural cement	25%
red brick	13%

6 LUMEN METHOD OF ILLUMINATION DESIGN

The process for design of interior lighting based on average illumination of a working surface is interchangeably called the *lumen method* or the *zonal cavity method*.

The lumen method is used when a large area is to be provided with uniform illumination. It considers the total light flux available from the luminaires, the reflection of light from all surfaces, the coefficient of utilization, and the light loss factors, together with the total area to be illuminated. Luminaire spacing is taken into account to ensure the uniformity of illumination on the working surface.

A. Zonal Cavity Ratios

The coefficient of utilization takes into account dimensions of the space, reflectivity of the wall, ceiling, and floor surfaces, and light distribution for each specified luminaire. Luminaire spacing specified by the manufacturer has a maximum fraction of the mounting height of the luminaire so as to limit the variation of illumination to one-sixth of the average illumination.

The working surface is assumed to be 30 inches from the floor. The lumen method determines the illumination at that plane by combining the direct component from the luminaires, the component reflected from the ceiling, the component reflected from the walls, and the component reflected from the floor.

The coefficient of utilization is obtained from calculations of three zones in the room: the *room cavity*, which comprises the space between the working surface and the luminaires; the *ceiling cavity*, consisting of the space above the luminaires; and the *floor cavity* which is the space below the working surface, as illustrated in figure 13.2.

A large room has a high ratio of floor or ceiling area to wall area, so the reflectance of the walls adds little to the average illumination of the work surface. Therefore, one of the basic parameters of the lumen method is the *room cavity ratio* (RCR), the ratio of the room cavity wall area to the base (ceiling or floor) area multiplied by 2.5. A cavity ratio can be expressed as

$$\text{CR} = 2.5 \left(\frac{\text{wall area of the cavity}}{\text{area of the cavity base}} \right) \qquad 13.5$$

Table 13.4
Suggested Illumination Levels

illumination (fc)	application
3	for all emergency lighting
10	for stairways, corridors, and hallways with casual use, and for storage areas of bulk items
20	for service area stairways and corridors, elevators, and areas involving non-precision work for short periods of time
30	for stock areas and areas involving casual machining, occasional reading, and rough assembly
50	for general office background and for areas involving sorting, ordinary inspection, sustained non-precision machining, and reading over longer periods of time
100	for areas involving medium difficulty assembly, inspection, and machining
200	for areas involving precision assembly, inspection, and machining
500	for areas such as an operating room where extremely fine detail work is required

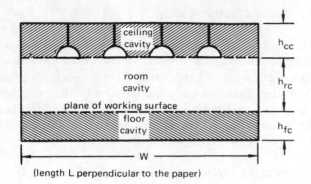

Figure 13.2 Zonal Cavities

Thus, the cavity ratios for the three zonal cavities corresponding to the room of width W and length L in figure 13.2 are:

$$\text{RCR} = 5h_{\text{rc}} \times \frac{L+W}{L \times W} \qquad 13.6$$

$$\text{CCR} = 5h_{\text{cc}} \times \frac{L+W}{L \times W} \qquad 13.7$$

$$\text{FCR} = 5h_{\text{fc}} \times \frac{L+W}{L \times W} \qquad 13.8$$

Example 13.1

A room with a ceiling height of 24 feet and a floor area of 40 feet × 60 feet has luminaires suspended 6 feet from the ceiling. Determine the cavity ratios.

Assume the floor cavity height is 30 inches = 2.5 feet = h_{fc}; the ceiling cavity height is $h_{\text{cc}} = 6$ feet. Thus, the room cavity height is $h_{\text{rc}} = 24 - 6 - 2.5 = 15.5$ feet.

$$\frac{W+L}{W \times L} = \frac{40+60}{40 \times 60} = \frac{1}{24}$$

Then,

$$\text{RCR} = 5 \times \frac{15.5}{24} = 3.23$$

$$\text{CCR} = 5 \times \frac{6}{24} = 1.25$$

$$\text{FCR} = 5 \times \frac{2.5}{24} = 0.52$$

Cavity ratios are used to determine the *effective reflectances* of the ceiling and floor cavities. These reflectances and the room cavity ratios then determine the coefficient of utilization for the source luminaires. In most cases, the effective floor reflectance is taken as 20 percent, so that only the effective ceiling cavity reflectance must be calculated.

B. Effective Ceiling Reflectance

To use a manufacturer's table for the coefficient of utilization, it is necessary to determine the effective ceiling reflectance. This is done by using the *ceiling cavity ratio* as calculated from equation 13.7, the base (ceiling) reflectance, and the reflectance of the walls within the ceiling cavity. These values are used with table 13.7.

Example 13.2

For a 24 foot high room with floor area of 40 feet × 60 feet and luminaires suspended 6 feet from the ceiling, the ceiling reflectance is 86%, and the wall reflectance is 60%. CCR = 1.25, RCR = 3.23, and FCR = 0.52. Determine the effective ceiling reflectances using table 13.5.

As none of the values given correspond to entries in table 13.5, an exact solution requires triple interpolation. Linear interpolations are used, so the order in which they are made is inconsequential.

The entries surrounding the given values are:

base reflectance	90		80	
wall reflectance	70	50	70	50
cavity ratio = 1.0	80	75	72	67
cavity ratio = 1.5	76	68	68	61

The first interpolation is taken to obtain values for a base reflectance of 86%.

$$(80-72) \times \frac{86-80}{90-80} + 72 = 76.8$$

$$(75-67) \times \frac{86-80}{90-80} + 67 = 71.8$$

$$(76-68) \times \frac{86-80}{90-80} + 68 = 72.8$$

$$(68-61) \times \frac{86-80}{90-80} + 61 = 65.2$$

base reflectance	86	
wall reflectance	70	50
CR = 1.0	76.8	71.8
CR = 1.5	72.8	65.2

The second interpolation is taken to obtain values for a wall reflectance of 60%.

$$(76.8-71.8) \times \frac{60-50}{70-50} + 71.8 = 74.3$$

$$(72.8-65.2) \times \frac{60-50}{70-50} + 65.2 = 69$$

base reflectance	86
wall reflectance	60
CR = 1.0	74.3
CR = 1.5	69.0

The final interpolation gives the effective reflectance for CR = 1.25.

$$\text{effective reflectance} = (74.3 - 69.0)\left(\frac{1.25 - 1.0}{1.5 - 1.0}\right) + 69$$
$$= 71.65$$

C. Determining the Coefficient of Utilization

When the effective ceiling reflectance is known, it is possible to determine the coefficient of utilization from a specific manufacturer's table such as the one shown in table 13.1 for two T-12 fluorescent lamps in a specific luminaire.

Tables for the coefficient of utilization are based on the assumption that the effective floor cavity reflectance is 20 percent, an assumption that is valid when the floor is a medium reflectance material. When the floor is either dark or light, adjustment to the final design is required.

The usual adjustment when the floor covering is dark is to increase the number of luminaires by 1 or 2 percent. Decrease the number of luminaires by 1 or 2 percent when the floor covering is light.

In table 13.1, the *base reflectance* used is the effective ceiling cavity reflectance, as determined in section B. When the luminaires are recessed into the ceiling, the actual ceiling reflectance is used.

Example 13.3

A room 24 feet high with a floor area of 40 feet × 60 feet having luminaires suspended 6 feet from the ceiling uses two T-12 lamps per luminaire. RCR = 3.23, and the effective ceiling reflectance is 71.65%. The wall reflectance is 60%. Determine the coefficient of utilization for the room.

Coefficients of utilization for two T-12 lamps per luminaire are given in table 13.1. In this case, interpolations are necessary for the ceiling cavity reflectance and the RCR, and an extrapolation is necessary for the wall reflectance. The interpolation of the ceiling reflectance is done first, followed by the extrapolation on the wall reflectance, and finally by the interpolation of the RCR.

Table 13.5
Effective Ceiling Reflectances

Base Refl. %	90							80							70							60							50						
Wall Refl. % (Cavity Ratio)	90	80	70	50	30	10	0	90	80	70	50	30	10	0	90	80	70	50	30	10	0	90	80	70	50	30	10	0	90	80	70	50	30	10	0
0.2	89	88	88	86	85	84	82	79	78	78	77	76	74	72	70	69	68	67	66	65	64	60	59	59	58	56	55	53	50	50	49	48	47	46	44
0.4	88	87	86	84	81	79	76	79	77	76	74	72	70	68	69	68	67	65	63	61	58	60	59	59	57	54	52	50	50	49	48	47	45	44	42
0.6	87	86	84	80	77	74	73	78	76	75	71	68	65	63	69	67	65	63	59	57	54	60	58	57	55	51	50	46	50	48	47	45	43	41	38
0.8	87	85	82	77	73	69	67	78	75	73	69	65	61	57	68	66	64	60	56	53	50	59	57	56	54	48	46	43	50	48	46	44	40	38	36
1.0	86	83	80	75	69	64	62	77	74	72	67	62	57	55	68	65	62	58	53	50	47	59	57	55	51	45	43	41	50	48	46	43	38	36	34
1.5	85	80	76	68	61	55	51	75	72	68	61	54	49	46	67	62	59	54	46	42	40	59	55	52	46	40	37	34	50	47	45	40	34	31	26
2.0	83	77	72	62	53	47	43	74	69	64	56	48	41	38	66	60	56	49	40	36	33	58	54	50	43	35	31	29	50	46	43	37	30	26	24
2.5	82	75	68	57	47	40	36	73	67	61	51	42	35	32	65	60	54	45	36	31	29	58	53	47	39	30	25	23	50	46	41	35	27	22	21
3.0	80	72	64	52	42	34	30	72	65	58	47	37	30	27	64	58	52	42	32	27	24	57	52	46	37	28	23	20	50	45	40	32	24	19	17
3.5	79	70	61	48	37	31	26	71	63	55	43	33	26	24	63	57	50	38	29	23	21	57	50	44	35	25	20	17	50	44	39	30	22	17	15
4.0	77	69	58	44	33	25	22	70	61	53	40	30	22	20	61	52	44	31	22	16	12	57	49	42	32	23	18	14	50	44	38	28	20	15	12
5.0	75	59	53	38	28	20	16	68	58	48	35	25	18	14	60	51	41	28	19	13	09	56	48	40	28	20	14	11	50	42	35	25	17	12	09
6.0	73	61	49	34	24	16	11	66	55	44	31	22	15	10	57	46	35	23	15	10	05	55	45	37	25	17	11	07	50	42	34	23	15	10	06
8.0	68	55	42	27	18	12	06	62	50	38	25	17	11	05	55	43	31	19	12	08	03	53	42	33	22	14	08	04	49	40	30	19	12	07	03
10.0	65	51	36	22	15	09	04	59	46	33	21	14	08	03	55	43	31	19	12	08	03	51	39	29	18	11	07	02	47	37	27	17	10	06	02

Base Refl. %	40							30							20							10							0						
Wall Refl. % (Cavity Ratio)	90	80	70	50	30	10	0	90	80	70	50	30	10	0	90	80	70	50	30	10	0	90	80	70	50	30	10	0	90	80	70	50	30	10	0
0.2	40	40	39	39	38	36	36	31	31	30	29	29	28	27	21	20	20	20	19	19	17	11	11	11	10	10	09	09	02	02	02	01	01	00	0
0.4	41	40	39	38	36	34	34	31	31	30	29	28	26	25	22	21	20	20	19	18	16	12	11	11	11	10	09	08	04	03	03	02	01	00	0
0.6	41	40	39	37	34	32	31	32	31	30	28	26	25	23	23	21	21	19	18	17	15	13	13	12	11	10	08	08	05	05	04	03	02	01	0
0.8	41	40	38	36	33	31	29	32	31	30	28	25	23	22	24	22	21	19	18	16	14	15	14	13	11	10	08	07	07	06	05	04	02	01	0
1.0	42	40	38	34	32	29	27	33	32	30	27	24	22	20	25	23	22	19	17	15	13	16	14	13	12	10	08	07	08	07	06	04	02	01	0
1.5	42	39	37	32	28	24	22	34	33	30	25	22	18	14	26	24	22	18	15	11	09	18	16	15	12	10	07	06	11	10	08	06	03	01	0
2.0	42	39	36	31	25	21	19	35	33	29	24	20	16	14	28	25	23	18	15	11	09	20	18	16	13	09	06	05	14	12	10	07	04	01	0
2.5	43	39	35	29	23	18	15	36	32	29	24	18	14	12	29	26	23	18	14	10	08	22	20	17	13	09	05	04	16	14	12	08	05	02	0
3.0	43	39	35	27	21	16	13	37	33	29	22	17	12	10	30	27	23	17	13	09	07	24	21	18	13	09	05	03	18	16	13	09	05	02	0
3.5	44	39	34	26	20	14	12	38	33	29	21	15	10	09	32	27	23	17	12	08	05	26	22	19	13	09	05	03	20	17	15	10	05	02	0
4.0	44	38	33	25	18	12	10	38	33	28	21	14	09	07	33	28	23	17	11	07	07	27	23	20	14	09	04	02	22	18	15	10	05	02	0
5.0	45	38	32	22	15	10	07	39	33	29	19	13	08	05	35	29	24	16	10	06	04	30	25	20	14	08	04	02	25	21	17	11	06	02	0
6.0	44	37	30	20	13	08	05	39	33	27	18	11	06	04	36	30	24	16	10	05	02	31	26	21	14	08	03	01	27	23	18	12	06	02	0
8.0	44	35	28	18	11	06	03	40	33	26	16	09	04	02	37	30	23	15	08	03	01	33	27	21	13	07	03	01	30	25	20	12	06	02	0
10.0	43	34	25	15	08	05	02	40	32	24	14	08	03	01	37	29	22	13	07	03	01	34	28	21	12	07	02	01	31	25	20	12	06	02	0

Used with permission from Lighting Handbook, by Philips Lighting Company, Somerset, New Jersey, 1984.

ceiling	from table 13.1			
cavity	80%		50%	
walls	50%	30%	50%	30%
RCR = 3	0.68	0.61	0.61	0.56
RCR = 4	0.60	0.52	0.54	0.49

interpolation 71.65%		extrapolation 71.65%	last interpolation 71.65%
50%	30%	60%	60%
0.661	0.596	0.694	RCR = 3.23
0.583	0.512	0.619	0.677 = CU

D. Determining the Number of Luminaires

The coefficient of utilization is applied to a system of luminaires to determine the illumination level at the work plane. The net flux from each luminaire is multiplied by the coefficient of utilization and by the number of luminaires to find the total luminous flux at the work plane. The total flux is then divided by the base area to yield the average illumination at the work plane.

Example 13.4

A room 24 feet high with a floor area of 40 feet × 60 feet has a coefficient of utilization of 0.677. The required illumination level is 100 footcandles. Determine the net luminous flux necessary for luminaires with two T-12 lamps per luminaire.

The total flux at the work surface is

$$100 \times 60 \times 40 = 240{,}000 \text{ lumens}$$

Total luminous flux required is

$$\text{flux required} = \frac{240{,}000}{0.677} = 355{,}505 \text{ lumens}$$

The net flux from a luminaire is determined only after taking into account a number of light loss factors such as the lamp lumen depreciation factor and luminaire dirt depreciation factors. For fluorescent and other discharge lamps, the initial lumen output is determined after 100 hours of operation, and the lamp lumen depreciation factor is determined after 70 percent of the life expectancy has elapsed. In addition to these losses, several other factors must be considered.

- Ballast performance contributes a loss of approximately 5 percent. Thus, unless better information on ballast loss is available, a multiplier of 0.95 is used.
- Voltage fluctuations cause a serious effect in incandescent lamps, with a 3 percent change in lamp lumens for each 1 percent change in line voltage. Fluorescent and other discharge lamps are less affected, having only 0.4 percent change in lumen output for each 1 percent change in line voltage.
- Lamp outage is a function of the maintenance program. Loss can be minimized by massive replacement at 70 percent of expected life.
- Ambient temperature is a factor in fluorescent lamps. Flux measurements are made at 70°F, but maximum efficiency occurs at a lower temperature. Efficiency decreases for temperatures above 70°F, so it can be worthwhile to include luminaire cooling, particularly when there are four lamps per luminaire. Some luminaires are designed to serve as cool air returns in ventilating systems, which can increase the lumens/watt up to 20 percent.

Example 13.5

A room 24 feet high with a floor area of 40 feet × 60 feet using two 40T-12 lamps per luminaire has the following environment: (1) the line voltage is known to drop 5% at the peak loading time of day, (2) burned out lamps are replaced immediately, all lamps will be replaced at the end of 70% of the expected life, and the lamps will burn 12 hours for each start, resulting in a lamp lumen depreciation of 0.84, (3) the environment is clean, and the luminaires are to be cleaned on an annual schedule. Determine the total loss factor.

As these are fluorescent lamps, the voltage fluctuation loss will be (5×0.004) or 2%, resulting in a multiplier of 0.98. The LLD factor is 0.84; from table 13.2, LDD is 0.94 (assuming an open luminaire).

Allow for ballast loss, giving an additional multiplier of 0.95. Thus, the total *light loss factor*, LLF, is

$$\text{LLF} = 0.98 \times 0.84 \times 0.94 \times 0.95 = 0.735$$

The minimum maintained illumination is the product of the number of luminaires, the initial luminous flux of the lamps, the number of lamps per luminaire, the light loss factor, and the coefficient of utilization, divided by the base area of the working surface (i.e., the floor or ceiling area).

$$\text{MMI} = \frac{(\text{no. luminaires}) \left(\frac{\text{lamps}}{\text{luminaire}} \right) (\text{initial flux}) \times \text{LLF} \times \text{CU}}{\text{area of the working surface}}$$

13.9

This relationship is rearranged to provide the design equation for determining the number of luminaires required.

$$\text{no. luminaires} = $$

$$\frac{\text{MMI} \times (\text{area of the working surface in sq. ft})}{\left(\frac{\text{lamps}}{\text{luminaire}}\right)(\text{initial flux in lumens}) \times \text{LLF} \times \text{CU}}$$

$$13.10$$

Example 13.6

A room 24 feet high with a floor area of 40 feet × 60 feet using two T-12 lamps per luminaire has a loss factor of 0.735, a coefficient of utilization of 0.674, and requires a minimum maintained illumination level of 100 foot-candles. If 48 inch fluorescent lamps with 3000 initial lumens per lamp are used, determine the number of luminaires required.

Using the previously calculated values,

$$\text{MMI} = 100 \text{ fc}$$
$$A = 2400 \text{ sq. ft}$$
$$\text{lamps/luminaires} = 2$$
$$\text{LLF} = 0.735$$
$$\text{CU} = 0.674$$
$$\text{initial lamp lumens} = 3000$$

Then,

$$\text{no. luminaires} = \frac{100 \times 2400}{2 \times 3000 \times 0.735 \times 0.674}$$
$$= 80.7 \text{ luminaires, or } 81$$

E. Placement of Luminaires

The final step in the lumen method is specifying the placement of the luminaires. The manufacturer's specifications in table 13.1 give the maximum spacing to mounting height, ensuring that the variation is no more than one-sixth of the average. It is customary to place these first and last rows of luminaires no further than one-third of the row spacing from the walls. Spacing from a wall depends on the reflectance of the wall, which usually is not calculated.

Fluorescent luminaires usually are placed in essentially continuous rows, with the long dimension of the luminaire oriented to the direction of the row. The direction of the rows depends on the general direction of vision within the room, being preferably perpendicular to the

prevailing direction of vision. This often will be along the shorter dimension of the room.

Example 13.7

A room 24 feet high with a floor area of 40 feet × 60 feet using two 40T-12 lamps per luminaire requires 81 luminaires. Assuming the luminaires are 52 inches long and are to be placed in rows across the short (40 foot) dimension of the room, specify a layout for this room.

$$\text{luminaires per row} = \frac{40 \times 12}{52}$$
$$= 9.23$$

The maximum number of luminaires per row is nine. Thus, with 81 luminaires required, there will be nine rows of nine luminaires. With nine rows, there will be eight even spaces for the interior of the layout, and two spacings of one-third the interior spacing for the distance from the walls. This results in a total dimension of $8\frac{2}{3}$ times the interior row spacing, which is required to be 60 feet (the long dimension of the room). The interior row spacing is then calculated.

$$\text{interior row spacing} = \frac{\text{room length}}{\text{no. rows} - 1 + \frac{2}{3}}$$
$$= \frac{60 \text{ ft}}{8.667} = 6 \text{ ft } 11 \text{ in}$$

Table 13.1 restricts the maximum spacing to less than 1.3 times the mounting height. The mounting height corresponds to the room cavity height (h_{rc} in figure 13.2) which is 15.5 feet, and the resulting design value of 6 feet 11 inches is well below this value.

7 POINT METHOD OF INTERIOR ILLUMINATION DESIGN

When a particular area requires a higher illumination level than that of the average illumination of the working surface, calculating the level of illumination at a point is necessary, and supplementary local sources must be specified. In such cases, the average illumination has been set by a lumen method design, so that the placement and data on the ambient lighting system are known.

The illumination level at certain points must be calculated to determine whether there is a need for supplementary lighting or for alteration of the ambient lighting system. To obtain the local or point illumination, two components of illumination, the direct and the reflected, must be calculated. The calculations for the reflected component are similar to those for the lumen method, but they exclude the direct component.

The two types of luminaire arrays considered in the calculation of the direct component are an array of point sources and an array of line sources, corresponding to fluorescent fixtures arranged in rows. In both cases, the mathematics are specific to the luminaire, so tables published by manufacturers for each luminaire are used in the calculations.

A. The Reflected Component

The reflected component is considered first because its calculation is similar to the lumen method. The reflected component is dependent on the position within the room of the point in question. For this reason, the room in figure 13.3 has been divided into a grid system which takes into account the influence of position in the room. The grid uses 11 lines in each direction with the first and last lines corresponding to the walls and the sixth line the centerline. The division is always into 11 grid lines, regardless of the length to width ratio.

The amount of reflected illumination is dependent on room position and the room cavity ratio to take into account the geometric properties. These geometric properties are then obtained from a table to provide a *room position multiplier*, RPM. Table 13.6 gives these values. Because of its symmetry, only one quarter of the grid is necessary in the tabled values. Interpolation between points often is necessary when using the table, particularly for points near the corners and along the walls. Interpolation between room cavity ratios also is required.

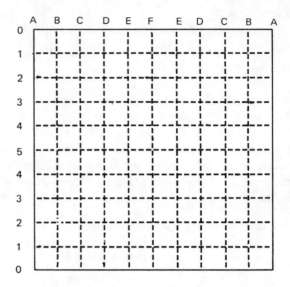

Figure 13.3 Room Position Grid

Table 13.6
Room Position Multipliers

	A	B	C	D	E	F		A	B	C	D	E	F
Room Cavity Ratio = 1							**Room Cavity Ratio = 6**						
0	.24	.42	.47	.48	.48	.48	0	.20	.23	.26	.28	.29	.30
1	.42	.74	.81	.83	.84	.84	1	.23	.26	.29	.31	.33	.36
2	.47	.81	.90	.92	.93	.93	2	.26	.29	.35	.37	.38	.40
3	.48	.83	.92	.94	.95	.95	3	.28	.31	.37	.39	.41	.43
4	.48	.84	.93	.95	.96	.97	4	.29	.33	.38	.41	.43	.45
5	.48	.84	.93	.95	.97	.97	5	.30	.36	.40	.43	.45	.47
Room Cavity Ratio = 2							**Room Cavity Ratio = 7**						
0	.24	.36	.42	.44	.46	.46	0	.18	.21	.23	.25	.26	.27
1	.36	.51	.60	.63	.66	.68	1	.21	.23	.26	.28	.29	.30
2	.42	.60	.68	.72	.78	.83	2	.23	.26	.30	.32	.33	.34
3	.44	.63	.72	.77	.82	.85	3	.25	.28	.32	.34	.35	.36
4	.46	.66	.78	.82	.85	.86	4	.26	.29	.33	.35	.37	.37
5	.46	.68	.83	.85	.86	.87	5	.27	.30	.34	.36	.37	.38
Room Cavity Ratio = 3							**Room Cavity Ratio = 8**						
0	.23	.32	.37	.40	.42	.42	0	.17	.18	.21	.22	.22	.23
1	.32	.40	.48	.51	.53	.57	1	.18	.20	.23	.25	.26	.26
2	.37	.48	.58	.61	.64	.67	2	.21	.23	.26	.27	.28	.29
3	.40	.51	.61	.65	.69	.71	3	.22	.25	.27	.29	.30	.30
4	.42	.53	.64	.69	.73	.75	4	.22	.26	.28	.30	.31	.32
5	.42	.57	.67	.71	.75	.77	5	.23	.26	.29	.30	.31	.32
Room Cavity Ratio = 4							**Room Cavity Ratio = 9**						
0	.22	.28	.32	.35	.37	.37	0	.15	.17	.18	.19	.20	.20
1	.28	.33	.40	.42	.44	.48	1	.17	.18	.20	.21	.22	.23
2	.32	.40	.48	.50	.52	.57	2	.18	.20	.23	.24	.25	.25
3	.35	.42	.50	.54	.58	.61	3	.19	.21	.24	.25	.26	.26
4	.37	.44	.52	.58	.62	.64	4	.20	.22	.25	.26	.26	.27
5	.37	.48	.57	.61	.64	.66	5	.20	.23	.25	.26	.27	.27
Room Cavity Ratio = 5							**Room Cavity Ratio = 10**						
0	.21	.25	.28	.31	.33	.33	0	.14	.16	.16	.17	.18	.18
1	.25	.29	.33	.36	.38	.42	1	.16	.17	.18	.19	.19	.20
2	.28	.33	.40	.42	.44	.48	2	.16	.18	.19	.21	.22	.22
3	.31	.36	.42	.46	.49	.52	3	.17	.19	.21	.22	.23	.23
4	.33	.38	.44	.49	.52	.54	4	.18	.19	.22	.23	.23	.24
5	.33	.42	.48	.52	.54	.56	5	.18	.20	.22	.23	.24	.25

Used with permission from Lighting Handbook, by Philips Lighting Company, Somerset, NJ, 1984.

B. Reflected Radiation Coefficient (RRC)

The method for determining the *reflected component* of illumination is the same as that for average illumination, except that the reflected radiation coefficient is substituted for the coefficient of utilization.

The *reflected radiation coefficient* depends on the wall luminance and the effective luminance of the ceiling cavity. These are taken into account by luminance coefficients that are listed by manufacturers in wall luminance coefficient and ceiling cavity luminance coefficient, LC_{cc}, tables. The reflected radiation coefficient is obtained from equation 13.11.

$$RRC = LC_w + RPM(LC_{cc} - LC_w) \qquad 13.11$$

Values for LC_w and LC_{cc} are obtained from manufacturers' listings, such as table 13.7. Parameters for ceiling and wall reflectances and the room cavity ratio are found on the lower part of the table.

C. Reflected Illumination on Vertical Surfaces

The method of determining the component of reflected illumination on vertical surfaces again uses the same formula (equation 13.6) as the average illumination, with the coefficient of utilization replaced by the *wall reflected radiation coefficient* (WRRC). This is

Table 13.7

Direct Illumination Components for Luminaire Height 6 feet Above Work Plane

	Direct Illumination Components														
β	5	15	25	35	45	55	65	75	5	15	25	35	45	55	65	75
α	**Vertical Surface Illumination Footcandles at a Point On a Plane Parallel to Luminaires**								**Vertical Surface Illumination Footcandles at a Point On a Plane Perpendicular to Luminaires**							
0-10	.9	2.6	3.6	3.9	3.3	1.9	.7	.1	.9	.8	.7	.5	.3	.1	...	...
0-20	1.8	5.0	7.0	7.7	6.6	3.8	1.5	.2	3.6	3.2	2.7	1.9	1.2	.5	.1	...
0-30	2.6	7.2	10.1	11.3	9.8	5.7	2.3	.3	7.7	7.0	5.8	4.3	2.7	1.1	.3	...
0-40	3.2	9.0	12.8	14.5	12.9	7.7	3.2	.5	12.6	11.6	9.7	7.5	4.9	2.1	.6	...
0-50	3.7	10.3	14.9	17.1	15.7	9.6	4.3	.7	17.8	16.6	14.2	11.2	7.7	3.4	1.1	.1
0-60	4.0	11.2	16.3	18.8	17.6	11.3	5.5	1.0	22.6	21.2	18.4	14.7	10.4	5.1	1.9	.2
0-70	4.1	11.6	17.0	19.8	18.9	12.7	6.8	1.4	26.2	24.7	21.8	17.8	13.1	7.2	3.2	.3
0-80	4.1	11.7	17.3	20.2	19.4	13.3	7.4	1.9	28.2	26.7	23.8	19.7	14.9	8.7	4.3	.8
0-90	4.1	11.7	17.3	20.2	19.4	13.4	7.5	2.0	28.6	27.1	24.2	20.1	15.3	9.1	4.7	1.1

α	**F.C. at a Point on Workplane**							
0-10	10.6	9.5	7.6	5.5	3.3	1.3	.3	...
0-20	20.6	18.5	14.9	10.9	6.6	2.6	.7	...
0-30	29.4	26.5	21.6	16.0	9.8	4.0	1.1	...
0-40	36.5	33.1	27.4	20.6	12.9	5.4	1.5	...
0-50	41.8	38.1	31.9	24.3	15.7	6.7	2.0	.1
0-60	45.2	41.3	34.8	26.8	17.6	7.9	2.6	.2
0-70	46.9	43.0	36.4	28.3	18.9	8.9	3.2	.3
0-80	47.4	43.6	36.9	28.8	19.4	9.3	3.5	.4
0-90	47.5	43.7	37.0	28.8	19.4	9.3	3.5	.4

Category III

2 T-12 Lamps—Any Loading
For T-10 Lamps—C.U. x 1.02

Luminance Coefficients for 20% Effective Floor Cavity Reflectance

Ceiling Cavity		**Reflectances**											
		80		50		10		80		50		10	
Walls		50	30	50	30	50	30	50	30	50	30	50	30
WDRC	RCR	**Wall Luminance Coefficients**						**Ceiling Cavity Luminance Coefficients**					
.281	1	.246	.140	.220	.126	.190	.109	.230	.209	.135	.124	.025	.023
.266	2	.232	.127	.209	.115	.182	.102	.222	.190	.130	.113	.024	.021
.245	3	.216	.115	.196	.105	.172	.095	.215	.176	.127	.105	.024	.020
.226	4	.202	.102	.183	.097	.161	.088	.209	.164	.124	.099	.023	.019
.212	5	.191	.097	.173	.090	.154	.082	.204	.156	.121	.094	.023	.018
.196	6	.178	.090	.163	.084	.145	.076	.200	.149	.118	.090	.022	.017
.182	7	.168	.083	.153	.078	.136	.071	.194	.144	.115	.087	.022	.017
.170	8	.158	.077	.145	.072	.130	.066	.190	.139	.113	.085	.021	.016
.159	9	.150	.072	.138	.068	.123	.062	.185	.135	.110	.082	.021	.016
.149	10	.141	.068	.130	.064	.116	.059	.180	.131	.107	.080	.020	.016

The direct illumination components are based on data for F40 lamps producing 3100 lumens.

Used with permission from Lighting Handbook, by Philips Lighting Company, Somerset, NJ, 1984.

determined by using the *wall direct radiation coefficient* (WDRC), as found in the left column of table 13.7, the wall luminance coefficient, and the average wall reflectance.Thus, this coefficient is produced from the same information as the reflected radiation coefficient by following the procedure indicated. The wall reflected radiation coefficient is obtained from equation 13.12.

$$WRRC =$$

$$\frac{\text{wall luminance coefficient (table 13.7)}}{\text{average wall reflectance}} - WDRC$$

$$13.12$$

• Horizontal surface

$$E_{\rm rh} = (\text{number of luminaires}) \times (\text{lamps per luminaire})$$
$$\times (\text{lumens per lamp})$$
$$\times \frac{LLF \times RRC}{\text{work plane horizontal area}} \quad 13.13$$

• Vertical surface

$$E_{\rm rv} = FC_{\rm rh} \times \frac{WRRC}{RRC} \quad 13.14$$

Example 13.8

A room with ceiling height of 24 feet and floor dimensions 40 feet × 60 feet has the following parameters: luminaires are hung 6 feet from the ceiling; there are 10 rows of two 40T-12 luminaires with spacing of 6 feet 3 inches running in the shorter direction; the LLF is

0.704; the effective ceiling reflectance is 72%, the wall reflectance is 60%, and the RCR is 3.23.

Determine the reflected horizontal component of illumination at a point which is 6 feet from two walls.

Find the room position. Taking the long dimension (60 feet) as the lettered positions, the point is located at grid line B. Along the short dimension (40 feet), 6 feet is halfway between one and two of the numbered grid lines.

Referring to table 13.6, interpolations between RCR's of 3 and 4, and between grid positions B1 and B2 must be made. For RCR = 3, the interpolation between the grid positions yields a room position multiplier RPM of 0.44 and for RCR = 4, an RPM of 0.365. Interpolating between RCR's of 3 and 4 to the actual value of 3.23, the RPM obtained is 0.423.

With ceiling reflectance of 72%, it is necessary to interpolate table 13.7 (lower) between 80% and 50%. With wall reflectance of 60%, it is necessary to extrapolate from 50% and 30%. Interpolation between RCR's of 3 and 4 is necessary.

Using equation 13.11, the reflected radiation coefficient is obtained by

$$RRC = 0.257 + 0.423(0.209 - 0.257) = 0.237$$

The reflected illumination is then found using the maintained minimum illumination formula, equation 13.9, substituting RRC for CU.

$$E_{\rm rh} = \frac{90 \times 2 \times 3150 \times 0.704 \times 0.237}{2400} = 39.4 \text{ fc}$$

	wall				ceiling			
ceil. refl.	80		50		80		50	
wall refl.	50	30	50	30	50	30	50	30
RCR = 3	0.216	0.115	0.196	0.105	0.215	0.176	0.127	0.105
RCR = 4	0.202	0.102	0.183	0.097	0.209	0.164	0.124	0.099
wall refl.	60		60		60		60	
RCR = 3	0.267		0.242		0.235		0.138	
RCR = 4	0.252		0.226		0.232		0.137	
ceil. refl.	72				72			
wall refl.	60				60			
RCR = 3	0.260				0.209			
RCR = 4	0.245				0.207			

RCR = 3.23	wall luminance coef. $LC_w = 0.257$	ceiling luminance coef. $LC_{cc} = 0.209$

D. Line Source Calculation of Direct Illumination Component

This method is particularly suited to continuous rows of fluorescent luminaires. It is based on vector integration too complex to include here. The method is based on an angular coordinate system. Imagine a vertical plane through the point where the illumination is to be determined and parallel to the rows of luminaires. Measure the angles from the point in that plane to the ends of the rows, naming these angles α_1 and α_2, as indicated in figure 13.4.

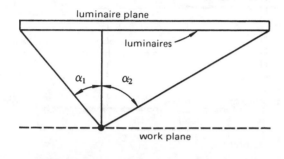

Figure 13.4 Longitudinal Angles of a Luminaire Array, Shown in the Side View

From the point where the illumination is to be determined, construct another vertical plane perpendicular to the rows of luminaires. Measure the angles from the vertical line through the point to the center line of each row of luminaires. Name these angles by giving them

the row subscript together with the variable name β as indicated in figure 13.5. The angles named α are *longitudinal angles*, and those named β are the *lateral* or *latitudinal angles*. The direct component of illumination is determined directly from a table of direct illumination components for the specific luminaire (see table 13.7).

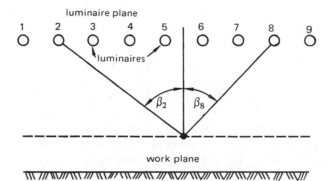

Figure 13.5 Latitudinal Angles of a Luminaire Array, Shown in the End View

The tables usually will give the same two values of longitudinal angles α for all the rows with a full set of lateral angles β. A table must be constructed to account for the contributions of each row, as demonstrated in example 13.9. Interpolation of lateral angle values usually is not necessary, but interpolation of the final sums for longitudinal angles is required if the lateral angles differ significantly from the table values.

room cavity height: 15.5 feet   point of calculation 6 feet from each of 2 walls

row	lateral distance (feet)	lateral angle β (degrees)	near longitudinal angle α_1 20°	bracketing for longitudinal angles 60° α_2	70° α_2'
1	4.125	15	18.5 fc	41.3 fc	43.0 fc
2	2.125	8	20.6	45.2	46.9
3	8.375	28	14.9	34.8	36.4
4	14.625	43	6.6	17.6	18.9
5	20.875	53	2.6	7.9	8.9
6	27.125	60	0.7	2.6	3.2
7	33.375	65	0.7	2.6	3.2
8	39.625	68	0.7	2.6	3.2
9	45.875	71	—	0.2	0.3
10	52.125	74	—	0.2	0.3
			65.3	155.0	164.3

65° interpolation 159.7

The values determined from the table are based on a room height (from the working plane to the luminaire plane) of six feet. A correction must be made for other mounting heights—by multiplying by 6 feet and dividing by the room cavity height.

$$\frac{6}{h_{\text{rc}}}$$

Example 13.9

For a room with ceiling height of 24 feet, floor dimensions 40 feet × 60 feet, luminaires hung 6 feet from the ceiling, 10 rows of two 40T-12 luminaires (table 13.7) running in the shorter direction, interior spacing of 6 feet 3 inches and spacing of 1 foot 10.5 inches from the wall for the edge rows, determine the direct illumination for the point that is 6 feet from two walls.

Latitudinal or lateral angles for the arrangement from the luminaire rows to the point are found to be 14.9 degrees, 7.8 degrees, etc., which are rounded to the nearest degrees. The longitudinal angles are found to be about 20 degrees and 65 degrees. The table contains a column for the 20 degree angle, but requires 60 degree and 70 degree columns for interpolation. In choosing values from latitudinal (lateral) columns, the value nearest the given angle is satisfactory, as errors tend to cancel each other.

Total uncorrected direct component footcandles:

$$159.7 + 65.3 = 225 \text{ fc}$$

Using the correction factor 6/15.5, the direct illumination is

$$225 \times \frac{6}{15.5} = 87 \text{ fc illumination}$$

8   STREET LIGHTING

Outdoor lighting differs from interior lighting due to the absence of reflected light from ceilings and walls. The coefficient of utilization is generally smaller for outdoor than for interior applications. Another difference is in the types of luminaires and lamps typically employed. While interior lighting is dominated by fluorescent fixtures, outdoor luminaires are usually mercury vapor and other discharge lamps, such as sodium and metal halide. They are broadly classified as *roadway* and *flood* lighting.

All the considerations developed for interior lighting are applicable to outdoor lighting. The most significant differences are in determining the coefficient of use

for average illumination calculations and the need for iso-candela or iso-footcandle curves as in figure 13.8 for point calculations.

Figure 13.6 shows the basic dimensions involved in street lighting: mounting height, street-side lateral distance, and house-side lateral distance. An additional dimension is the spacing between luminaires which is indicated in figure 13.7 for three variations of luminaire arrangements.

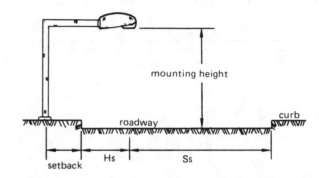

Figure 13.6   Basic Dimensions for Street Lighting

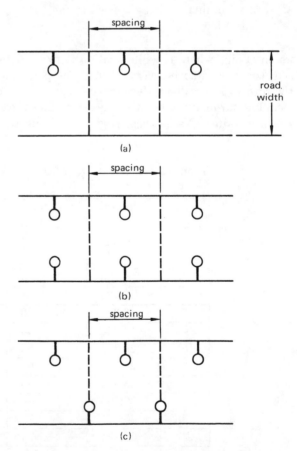

Figure 13.7   Street Luminaire Spacing and Placements:(a) One Side, (b) Opposite Sides, (c) Staggered

A. Recommended Levels For Roadway Lighting

Lighting levels depend on the density of motor traffic, parking demand, and the amount of pedestrian traffic in an area. The three area classifications are:

- *Commercial*: areas where there is high demand for off-street and on-street parking, heavy foot traffic, and periods of peak traffic as in any densely developed business area.

- *Intermediate*: areas of industrial or moderate business development, including densely developed areas such as apartment complexes, hospitals, public libraries, and recreational centers.

- *Residential*: single-family and townhouse areas or a mixture of light commercial and residential areas. Areas with less activity are also included in this classification.

The five classifications of roadway are:

- *Freeway*: all divided highways with full control of access and no crossings at grade.

- *Major and Expressway*: principal routes for through traffic, including divided arterial highways with partial control of access but having interchanges on grade at major crossroads.

- *Collector*: roadways which are used for traffic movement in local areas, principally within residential, commercial, or industrial areas.

- *Local*: roadways providing direct access to residential, commercial, or industrial areas, and carrying no through traffic.

- *Alley*: alleys.

B. Average Illumination Design

For roadside lighting, the coefficient of utilization is determined from two curves for a specified luminaire. Figure 13.8 shows the two curves with the iso-footcandles' contours.

To develop the concepts of roadway illumination design, it is necessary to know the *mounting height* (MH) and *spacing between luminaires*.

There are two coefficient of utilization curves in figure 13.8, both starting at point zero on the left edge of the graph. This point corresponds to the location of the luminaire. Points on the graph above the zero line correspond to the house side, and points below correspond to the street side. The coefficient of utilization is read on the upper scale for both curves. Note that the lateral distances on the left scale of the graph are normalized to the mounting height. For any situation, both coefficients of utilization are found and then added together to obtain the *effective coefficient of utilization*.

Example 13.10

A roadway is 60 feet wide, and the luminaires are suspended 6 feet from one side of the roadway. The luminaire is pictured in figure 13.8, and its mounting height is 30 feet. Determine the coefficient of utilization for this application.

$$\frac{Ss}{MH} = \frac{54}{30} = 1.8$$

From the lower section of figure 13.8, the Ss coefficient of utilization is 0.5.

$$\frac{Hs}{MH} = \frac{6}{30} = 0.2$$

From the upper section of figure 13.8, the Hs coefficient of utilization is 0.04. The total coefficient of utilization is

$$0.5 + 0.04 = 0.54$$

For roadway luminaires, the luminaire dirt depreciation depends on the following cases of cleaning frequency and environments.

cleaning frequency	environment		
	clean	medium	dirty
every year	0.95	0.92	0.87
every 3 years	0.86	0.77	0.67
every 6 years	0.77	0.62	0.49

Table 13.8
Recommended Illumination Levels

roadway	commercial	intermediate	residential
freeway	0.6 fc	0.6 fc	0.6 fc
major and expressway	2.0 fc	1.4 fc	1.0 fc
collector	1.2 fc	0.9 fc	0.6 fc
local	0.9 fc	0.6 fc	0.4 fc
alley	0.6 fc	0.4 fc	0.2 fc

Coefficient of Utilization and Iso Footcandle* Plot
Based on Mounting Height of 30 ft.

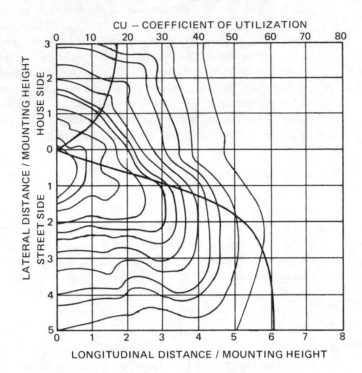

LONGITUDINAL DISTANCE / MOUNTING HEIGHT

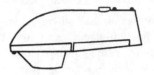

LUMINAIRE AND TEST LAMP DATA

Luminaire: sample

Refractor:

Reflector·

Socket Position:

Other Variables:

Lamp:
LCL: 5¾ inches
Lumen Rating: 25500

Adjustment Factor for Iso Ft-c values other
than 30 ft. mounting height.

MTG.HT.	20	21	22	23	24	25	26	27	28	29
FACTOR:	2.25	2.04	1.86	1.70	1.56	1.44	1.33	1.23	1.15	1.07

MTG.HT.	30	31	32	33	34	35	36	37	38	39
FACTOR:	1.00	.94	.88	.83	.78	.73	.69	.66	.62	.59

MTG.HT.	40	41	42	43	44	45	46	47	48	49
FACTOR:	.56	.54	.51	.49	.46	.44	.43	.41	.39	.37

*NOTE: Iso footcandle, candela, and Iso candela values
as printed are per 1000 lamp lumens. For
actual values multiply by $\dfrac{\text{Latest Lamp Rating}}{1000}$

At time of test this was $\dfrac{25500}{1000} = 25.5$

Initial Average Horizontal Ft-c. = $\dfrac{\text{Lamp Lumens} \times \text{C of U}}{\text{Spacing} \times \text{Width of Street}}$

Spacing = $\dfrac{\text{Lamp Lumens} \times \text{C of U}}{\text{Initial Ft-c Level} \times \text{Width of Street}}$

LIGHT FLUX DATA

Distribution by zones: (% of lamp lumens)

Downward Street Side:	62.0%]
Downward House Side:	17.3%
Upward Steet Side:	1.5%
Upward House Side:	1.0%
Total:	81.8%

Candela* Data: (per 1000 lamp lumens)
Max. Cd: 600   at 73°V,   60°H
Nader Cd: 110   at 0°V,   0°H

Distribution Classification: (ANSI D 12.2 1972)
Lateral: Type IV
Vertical: Medium
Control: Semi-cutoff

Figure 13.8   Photometric Data for a Typical Roadside Luminaire

Environmental definitions are given here.

- *Clean*: clean pavement, grass, no open loose ground, few adhesive particles in the atmosphere, average or low car and truck traffic.

- *Medium*: similar to clean except more adhesives in the atmosphere and heavier than average truck traffic.

- *Dirty*: even more adhesives in the atmosphere and heavy truck traffic.

Each luminaire serves an area indicated by the dashed vertical lines in figure 13.7. For the opposite sides layout, there are two luminaires for each area. For the staggered layout there also are two luminaires for each area. The area is (roadway width) × (spacing).

The minimum maintained illumination formula, equation 13.9, is applicable here and is repeated for convenience.

$$MMI =$$

$$\frac{(\text{no. luminaires})(\frac{\text{lamps}}{\text{luminaires}})(\text{initial flux}) \times LLF \times CU}{\text{area of the working surface}}$$

Here, the light loss factor (LLF) consists of the product of lamp lumen depreciation factor and the luminaire dirt depreciation factor. The area of the working surface is (roadway width) × (spacing).

Thus, this formula can be rearranged to obtain the spacing.

$$\text{spacing} =$$

$$\frac{(\text{no. luminaires})\left(\frac{\text{lamps}}{\text{luminaire}}\right)(\text{initial flux}) \times LLF \times CU}{MMI \times \text{roadway width}} \quad 13.15$$

Example 13.11

The roadway of example 13.10 is in a residential area. Assume an OV-25 (figure 13.8) luminaire with a high pressure sodium lamp C250S50, 250 watts. Determine the spacing to give the proper illumination level as recommended in table 13.8, assuming the street is a collector type and the environment is clean. The luminaires are cleaned on a three-year schedule.

From table 13.8, the recommended illumination level is 0.6 fc. From example 13.10, CU = 0.54. The C250S50 lamp has initial lumens of 25,500 and an LLD of 0.91. The luminaire dirt depreciation (LDD) is 0.86. Therefore, the light loss factor is

$$LLF = LLD \times LDD$$
$$= 0.91 \times 0.86 = 0.78$$

Using equation 13.15, the spacing is

$$\text{spacing} = \frac{1 \times 1 \times 25,500 \times 0.78 \times 0.54}{0.6 \times 60} = 298 \text{ ft}$$

These simple examples do not take into account a number of possible variations. The effects of other mounting heights and the uniformity of the lighting must also be considered.

C. Luminaire Classifications

Minimum mounting height and spacing are determined by the initial candlepower of the luminaire lamp(s) and by the luminaire's *light distribution*. Light distribution is classified by *lateral light distribution*, *vertical light distribution*, and *candlepower distribution*, called *control* in figure 13.8.

The vertical and lateral distribution classifications are made on the basis of an iso-candela diagram on a rectangular coordinate grid which has superimposed a series of longitudinal roadway lines and a series of transverse roadway lines, both in multiples of the mounting height.

Lateral light distributions have seven classifications: Types I through V with additional Type I-4-way and Type II-4-way for intersections. Lateral light distributions are divided into two groups: center of the roadway or side of the roadway. These groups are then subdivided according to the spread (which is half the maximum candlepower) of the iso-candela contours.

Vertical light distribution is split into three groups: short, medium, and long. These classifications depend on the vertical angle of maximum candlepower in relation to mounting height.

Candlepower distribution is divided into three categories: *cutoff* (first choice for glare control) up to 10 percent, *semi-cutoff* (second choice for glare control) from 10 percent to 30 percent, and *non-cutoff* with no limitations.

D. Spacing and Mounting Height

The location, spacing, and mounting height of luminaires involves the uniformity ratio between lowest and average illumination level and glare considerations. The minimum mounting heights are shown in table 13.9.

Table 13.9
Minimum Mounting Heights

maximum luminaire candlepower	cutoff	semi-cutoff	non-cutoff
under 5,000	20	20	20
5,000–10,000	20	25	30
10,000–15,000	25	30	35
over 15,000	30	35	40

Once a luminaire is selected, the mounting height and spacing can be determined from the above table and the spacing formula (equation 13.15). The following maximum spacing to mounting height ratios must be used to ensure that the ratio of minimum illumination to average is not less than 1/3 (1/6 for local residential streets).

vertical distribution	$\dfrac{\text{maximum spacing}}{\text{mounting height}}$
short	4.5
medium	7.5
long	12.0

In addition, the overhang of a street-side mounted luminaire is not to exceed 1/4 of the mounting height.

The staggered arrangement is preferred. Opposite spacing is used where the width of the street exceeds twice the mounting height. One-side layouts should be used only on very narrow streets. The recommended spacing and distribution types are shown in table 13.10.

At intersections, the illumination level should be the sum of the levels for the intersecting roadways.

Some typical acceptable roadway lighting configurations for two-lane roads are given in table 13.10.

A formula for the minimum mounting height is

$$\text{minimum MH (feet)} = \frac{\text{maximum candlepower}}{1000} + \text{ND}$$

$$13.16$$

ND depends on the vertical distribution: for short distribution, ND = 5 feet; for medium, ND = 10 feet; and for long distribution, ND = 15 feet.

Example 13.12

Re-examine the design of example 13.11 and modify it to account for correct mounting height and spacing.

From figure 13.8, the vertical distribution is medium for this luminaire, so from equation 13.16 with 25,500 initial lumens, the minimum MH is 35.5 feet.

From example 13.10, the roadway is 60 feet wide, and the luminaires are suspended 6 feet from one side of the roadway. With a mounting height of 36 feet, the width of the roadway is 60/36 = 1.67, and is more than 1.5 × MH, so that the luminaire placement must be staggered on opposite sides. Here assume a staggered arrangement. The coefficient of utilization for this application is then found as follows:

$$\frac{Ss}{MH} = \frac{54}{36} = 1.5$$

From figure 13.8 (lower), the Ss coefficient of utilization is 0.45.

$$\frac{Hs}{MH} = \frac{6}{36} = 0.17$$

From figure 13.8 (upper), the Hs coefficient of utilization is 0.01.

The total CU is

$$0.45 + 0.01 = 0.46$$

The recommended illumination level is 0.6 fc, and the LLF remains at 0.78. From equation 13.15,

$$\text{spacing} = \frac{2 \times 1 \times 25,500 \times 0.78 \times 0.46}{0.6 \times 60} = 508 \text{ ft}$$

The spacing/MH is 248/36 = 6.9, which is just acceptable for this luminaire to maintain the ratio of minimum illumination to average at 1/3 or more. If this ratio were higher, the mounting height would have to be increased for a non-residential application.

E. Point Calculations

It is necessary to check for illumination uniformity by making point calculations, which involve the use of the iso-footcandle contours. The points of minimum illumination can be found by overlaying the lighted zone for one luminaire with its traverse boundaries. At some

Table 13.10
Recommended Spacing and Distribution Types

roadway width	luminaire location	type of spacing	lateral distribution type
up to 1.5 × MH	side of roadway	staggered or one side	II, III, or II-4-way for intersections
beyond 1.5 × MH	side of roadway	staggered or opposite	III or IV
up to 2 × MH	center of roadway		I with I-4-way or V for intersections

point on this boundary, the point of minimum illumination will occur, and it is to be found by common sense and the luminaire pattern chosen. For instance, it will be at the opposite curb for one-sided patterns and at the middle of the street for staggered or opposite-sides patterns.

Then, it will be necessary to read the iso-footcandle contours and make the adjustments indicated in figure 13.8. The minimum illumination level will be twice the calculated value for one-side and staggered arrangements, and four times the value for opposite-sides arrangements.

9 FLOODLIGHTING

Much of what has been covered up to this point also applies to floodlighting. Some changes are necessary, particularly in the coefficient of utilization, although this is not applicable unless extensive calculations are made. The basic formula for floodlighting is

$$\text{no. floodlights} = \frac{\text{illumination level} \times \text{square feet}}{\text{beam lumens} \times (\text{lamps/floodlight}) \times \text{LLF} \times \text{CBU}}$$

13.17

Beam lumens are the total lumens contained within the beam spread. The beam spread is the angle between the two directions in which the candlepower is 10 percent of the maximum candlepower (which is typically at the center of the beam). This information is available as catalog and photometric data.

CBU is the *coefficient of beam utilization*, which is the ratio of the lumens striking the surface to the beam lumens.

Beam spreads are categorized by seven types.

- *type 1*: beam spread is 10 degrees to 18 degrees

- *type 2*: beam spread is 18 degrees to 29 degrees

- *type 3*: beam spread is 29 degrees to 46 degrees

- *type 4*: beam spread is 46 degrees to 70 degrees

- *type 5*: beam spread is 70 degrees to 100 degrees

- *type 6*: beam spread is 100 degrees to 130 degrees

- *type 7*: beam spread is 130 degrees or greater

Table 13.11
Some Acceptable Lighting Configurations for Two-Lane Roads, with LLF = 0.7

Road Type: Two-lane, major rural, 10-foot shoulders
Road Width: 24 feet (two 12-foot lanes)
Desired Illumination Level: 0.9 average horizontal footcandles
Desired Uniformity Ratio: 3:1

Light Distribution			Lamp Data Mercury		Pole Data		Arrangement	Spacing (feet)	Illumination* (footcandles)		Uniformity Ratio
Type	Vertical	Control	Designation	Light Output (lumens)	Mounting Height (feet)	Overhang (feet)			Average	Minimum	
II	S	C.O.	H39KB 175 W Clear	7,700	30	4	One Side	92	0.9	0.36	2.5:1
II	M	Semi C.O.	H37KB 250 W Clear	12,100	30	4	One Side	135	0.9	0.49	1.8:1
II	M	C.O.	H33CD 400 W Clear	21,000	30	4	One Side	100	2.13	0.71	3.0:1
II	M	C.O.	H33CD 400 W Clear	21,000	30	4	Staggered	150	1.41	0.47	3.0:1
III	L	Semi C.O.	H35NA 700 W Clear	39,000	35	4	One Side	200	1.49	0.51	2.9:1
III	L	Semi C.O.	H35NA 700 W Clear	39,000	35	4	Staggered	210	1.47	0.49	3.0:1

* Based on 70 percent maintenance factor.

Used with permission from Westinghouse Lighting Handbook, by the Westinghouse Electric Corporation, 1969, revised May, 1978.

Floodlights not having a perfectly conical beam shape are classified by the minimum and maximum angles, such as a type 5 × 6.

Design of floodlighting is best accomplished mechanically by pictorial drafting. This is perhaps more of an art than a science where the illumination is to be at a certain level, and consists of a trial and error method of laying out circular and/or elliptical areas within the beam spreads of the various floodlights.

Because the candlepower at the beam edge is only 10 percent of that at the middle, it is necessary to have some overlap of beams at the lighted surface. The overlap will account for the coefficient of beam utilization, which should lie between 0.6 and 0.9. If the coefficient of beam utilization is less than 0.6, the plan may be more expensive than necessary. If greater than 0.9, the resulting lighting pattern tends to be spotty.

10   WIRING DESIGN AND THE NATIONAL ELECTRIC CODE

Safety of persons and property is a primary concern with electrical wiring. As a dynamic and highly concentrated form of energy, electricity constitutes a hazard in the vicinity of its use. The *National Electric Code*,[1] hereafter referred to as the Code, recommends safeguards to protect against these hazards. The Code has been adopted as the basis of many state and local ordinances regulating electrical installations.

The principal concern in this section is the design of wiring systems, including conductor specifications and circuit protection. More complete treatment of Code requirements is found in other reference books.

The Code jurisdiction does not deal with the distribution system of a utility company, but begins with the *service* provided by the utility. Services can range from the common 115/230 V single-phase service typically provided to residences, to 13.8 kV three-phase service provided to large industrial users. All portions beyond the "last pole" of the utility company are covered by the Code.

The wiring design can be divided into three parts: the *service entrance*, *feeders*, and *branch circuits*.

A. Service

The *service entrance* is composed of the overhead *service drop* or the underground *lateral conductors*, *raceways*, *service disconnect*, *main overcurrent protection*,

---

[1] National Fire Protection Association, *National Electric Code* 1996, 470 Atlantic Avenue, Boston, MA 02210.

grounding electrode conductor, the *main bonding jumper*, and any protective housings. It extends from the service drop or the lateral conductors to the service disconnect. Service requirements are covered primarily in Code article 230, but other articles are pertinent to the service entrance. Further references to the Code articles are given in brackets [].

Of particular interest to electrical designers are sections [230 D] covering service entrance conductors, and [230 G] covering service equipment-overcurrent protection.

[230-42] deals with size and rating of the conductors, stating that conductors shall be of sufficient size to carry the loads in accordance with article [220], with ampacity determined from tables [310-16] through [310-19] and all applicable notes to these tables. Article [220] discusses the determination of loads but refers to [430-22], [430-24], [430-25], and [430-26] for motor loads and to article [440] for air conditioning and refrigeration. Also of interest are [440-6] and [440 D] dealing with ampacity and branch circuit conductors, respectively.

In [230 G] on overcurrent protection, section [230-90] requires that service entrance overcurrent protection be no greater than the ampacity of the conductors. Exception no. 1 refers to articles [430-52], [430-62], and [430-63] on motor circuits, and exception no. 2 permits protection at the next higher fuse or circuit breaker rating, in turn referring to sections [240-3], exception no. 1, and [240-6] dealing with the standard ratings for fuses and inverse-time circuit breakers. Fuses in grounded neutrals are prohibited in [230-90(b)], which permits circuit breakers in grounded neutrals only when all ungrounded lines are broken with the same circuit breaker.

Other applicable references are service disconnects [230-F], the grounding electrode conductor [250-53], and the main bonding jumper [250-79(c)].

B. Feeders and Branch Circuits

Feeders are used in large installations where the service is remote from parts of the load. A feeder is defined in [100] as "all conductors between the service equipment and the final branch overcurrent device." Small installations, such as residences, often have no feeders but have branch circuits radiating from the service equipment.

Feeders are covered generally in article [215] – Feeders, article [220] – Branch-Circuit and Feeder Calculations (part B), article [225] – Outside Branch Circuits and Feeders, and in article [430] – Motor Circuits and Controllers.

Branch circuits are defined as "the circuit conductors between the final overcurrent device protecting the circuit and the outlets." Branch circuits are covered in articles [210] – Branch Circuits, [220] – Branch and Feeder Calculations (part A), [225] – Outside Branch Circuits and Feeders, and [430] on motors.

C. Conductors for General Wiring

Ampacity (the current-carrying capacity of electrical conductors expressed in amperes [100]) ratings for conductors are specified for various types of insulations, uses, and conditions in [310]. Conductors that are integral parts of equipment are not included.

In general, the minimum wire size for copper conductors is AWG[2] #14. For aluminum or copper-clad aluminum, it is AWG #12 [310-5]. Important exceptions are for fixture wire [410-24] and flexible cords [400-12] which may be AWG #18. Other exceptions are given in [310-5].

Conductor types are established by the insulation used, so the insulation type, together with the metal (copper or aluminum) used, is descriptive of the conductor. Table [310-13] lists the conductor types for nominal voltages up to 600 volts. (Higher voltage conductor types are covered in tables [310-63] through [310-86] but are not considered here.) The meanings of the letters in the insulation and conductor types are:

A	asbestos
C	cotton
H	heat resistant
L	lead covered
MI	mineral insulated
R	rubber
RU	latex rubber
S	silicone rubber
T	thermoplastic
V	varnished cambric
W	moisture resistant

Table [310-13] gives the maximum operating temperature and the approved applications for each certified insulation type. Note that wet locations, including underground installations and raceways or conduits subject to moisture condensation, require a W in the insulation type.

Tables [310-16] through [310-19] give the allowable ampacities for various types. Table [310-16] is the most applicable in branch circuit design. This table and the footnotes to tables [310-16] through [310-19] are very important in the design of branch circuits and are equally applicable to feeders and services below a voltage of 600 volts nominal.

[110-3(b)] requires that "listed or labeled equipment shall be used or installed in accordance with any instructions included in the listing or labeling." This requirement incorporates the listings of Underwriters Laboratories (UL) and other laboratories into the provisions of the Code.[3] Its effect on the selection of conductors for branch circuits is significant.

A basic UL rule requires that terminations use conductors rated at 75°C. This includes fuses and circuit breakers which set the current rating of a circuit. Because the maximum circuit rating for lighting branch circuits is 50 amps [210-23], branch circuit conductors must be rated at at least 75°C.

D. Cables and Raceways

For the direct application of tables [310-16] and [310-18], a cable or raceway can contain no more than three conductors. Additional conductors require an ampacity derating according to the requirements of [note 8] following table [310-19]. According to [note 10(a)], the neutral of a normally balanced three-phase circuit is not counted as an extra conductor. [Note 10(b)] and [note 10(c)] do require that the neutral be counted as an extra conductor in cases of unbalanced loads and electric discharge lighting loads.

[300-20] requires that all phases, the neutral conductor, and the associated grounding conductor for a single circuit be grouped together in metallic enclosures and raceways, including conduits and tubing.

[300-3] permits the grouping of any number of separate circuits of nominal voltage less than 600 within the same raceway or cable.

Appendix C gives the number of conductors permitted in a conduit or tubing. [Table C1] lists the percentage of the interior that can be filled with conductor (including insulation); [table C] is for fixture wire; [table C3] shows the number of conductors for conduits or tubing of trade sizes using general purpose wiring. [Tables C2 and C3]

---

[2] Wire sizes are designated by American Wire Gage (AWG) numbers, or in thousands of circular mils of cross-sectional area (MCM). AWG numbers begin with the largest at 0000, which has a cross-sectional area of 212,000 circular mils, followed by 000, 00, 0, 1, 2, etc., with increasing numbers corresponding to smaller cross-section area and lower ampacity. Larger conductors begin at 250 MCM and increase in increments of 50,000 circular mils of cross-sectional area.

[3] *A Guide to Basic Electrical Design*, J.F. McPartland, Electrical Construction and Maintenance, July 1978, part 3, p 72 ff.

are based on identical conductors within the conduit or actual sized tubing. When sizes or types are mixed, [table C1] is used to determine the minimum size of conduit or tubing.

Associated information, which is useful in calculations, is given in chapter 9 [table 4] on trade sizes of conduit and tubing and in [tables 5, 6, and 7] listing dimensions of approved conductors.

E. Overcurrent Protection [240]

Overcurrent protection prevents overheating of conductors and equipment, and is covered in article [240]. [240-2] gives references for specific equipment. Of particular interest are items [210-20] on branch circuits, and [430 parts C, D, E, and H] and table [430-152] on motors, motor circuits, and controllers. Diagram [430-1] gives a directory through the article.

The basic code rule is that overcurrent devices must be rated no higher than the ampacity of the conductors after all derating factors have been included. Exceptions to this rule are given in [240-3]. Circuit breakers (CB) and fuses are the devices used for overcurrent protection. Standard ratings are given in [240-6], and tap rules are given in [240-21].

F. Branch Circuit Design—Electric Discharge Lighting

Article [100] defines a multi-wire branch circuit as consisting of "two or more ungrounded conductors having a potential difference between them, and an identified grounded conductor having equal potential difference between it and each ungrounded conductor of the circuit and which is connected to the neutral conductor of the system." This permits 120/240 volt single-phase and three-phase, four-wire branch circuits, such as 120/208 wye connected circuits, to be classified as multi-wire branch circuits. [210-4] requires that these circuits supply only line-to-neutral loads except when all ungrounded conductors of the circuit are opened simultaneously by the branch circuit overcurrent protection device [exception no. 2 to 210-4].

[210-6(a)] limits the maximum voltage to ground for dwelling units and motel guest room branch circuits supplying lampholders, fixtures, or standard receptacles rated to 120 volts. [210-6(b)] permits 270 volts to ground for industrial installations supplying only lighting fixtures with mogul-base screw-shell lampholders or other approved types of lampholders.

[210-6(d)] permits voltages up to 600 volts between conductors for branch circuits supplying only the ballasts of electric-discharge lamps that are permanently mounted in fixtures at least 22 feet above the ground in outdoor applications and at least 18 feet above the floor in other structures such as tunnels.

[210-10] permits the use of tapped conductors for AC circuits having two or more ungrounded conductors and a grounded neutral conductor. [210-19(c)] requires that such tapped conductors have an ampacity sufficient for the load, but that AWG #14 is the minimum tap conductor size. For circuits rated at less than 40 amps, the minimum ampacity of tap conductors is 15 amps; and for circuits rated at 40 or 50 amps, the minimum ampacity of tap conductors is 20 amperes. Such taps, as used for individual lampholders or fixtures, are restricted further to extend no more than 18 inches beyond the lampholder or fixture.

[210-20] requires that branch circuit protection be provided by overcurrent devices with ratings or settings not exceeding those specified by [240-3]. [210-21(a)] requires that lampholders connected to branch circuits having circuit ratings in excess of 20 amperes be of a heavy-duty type with ratings of 660 watts if of the admedium type and at least 750 watts if of other types.

[210-22] requires that the total load of a branch circuit not exceed the rating of the branch circuit, which is determined from the rating or setting of the overcurrent device. Further requirements restrict the load to 80 percent of the rating of the branch circuit for lighting loads. In addition, [210-22(b)] requires that circuits supplying lighting units with ballasts, transformers, or autotransformers have the load computed using the actual current rather than the power and voltage. This includes all discharge lighting systems. [Exception no. 2] to [210-22(c)] permits only one derating for lighting circuit conductor ampacity to satisfy the 80 percent figure in this article and the requirements of [note 8] to tables [310-16] through [310-19]. If the derating of that note exceeds that of this article, the severest restriction applies (i.e., if note 8 is required, a derating factor of 70 percent would prevail over the 80 percent factor of this article).

[210-23] specifies the permissible load for various rated branch circuits. In particular, 25, 30, 40, and 50 amp circuits require heavy-duty lampholders in fixed lighting

units. They are permitted only in occupancies other than dwellings. Table [210-24] summarizes the branch circuit requirements and is useful in wiring design.

[220-3] dealt with the computation of branch circuit loads. [220-3(b)] and table [220-3(b)] describe the minimum unit load for listed occupancies on the basis of volt-amperes per square foot of floor area. These unit loads are minimum values and typically will be less than those determined from a lighting design procedure as outlined earlier in this chapter. In any particular case, the actual volt-amperes per square foot must be no less than that given in table [220-3(b)].

Article [410-18] requires that all fixed metal lighting fixtures be grounded. [250-91] permits equipment grounding conductors to be wires and metal raceways including rigid conduit and EMT. Wire sizes for grounding conductors are given in table [250-95].

Example 13.13

Example 13.8 is a lighting design calling for 90 luminaires with two 48T-12 fluorescent bulbs per luminaire. Consider the branch circuit design for that example. A 120/208 volt four-wire three-phase service is available.

The room involved has 2400 square feet of floor area, and the light power, at 40 watts per bulb and 180 bulbs, is nominally 7200 watts, or 3 watts per square foot. This is not necessarily a listed occupancy of table [220-3(b)], but would qualify for commercial-industrial buildings or schools, among others.

However, the lighting power for fluorescent and other electric discharge lamps cannot be calculated from the wattage of the lamps, as there are inductive components involved in the ballasting. [210-22(b)] requires that the load be calculated from the actual current rather than from the wattage for electric discharge lighting units. Garland gives a "rule of thumb" of adding 25 percent to the wattage to account for the ballast effects. This is the same as assuming a power factor of 0.8. Thus, for 40 watt fluorescent bulbs, the volt-amperes per bulb would be $VI = 40\,\text{watts} \times 1.25 = 50\,\text{volt-amperes}$, which is to be used to determine the current load.

Then the load per bulb is $50\,\text{VA}/120\,\text{volts} = 0.417\,\text{amps}$ (using a 120 volt line-to-neutral circuit). It is sensible to divide the load into three equal parts, assigning one part to each phase. Thus, each phase would supply a current of

$$30\,\text{luminaires} \times 2\frac{\text{bulbs}}{\text{luminaire}} \times 0.417\frac{\text{amps}}{\text{bulb}} = 25\,\text{amps}$$

A circuit rated at more than 20 amps [210-20(a)] requires lampholders to be of a heavy-duty type, meaning admedium or mogul. Most fluorescent bulbs are supplied with medium bipin bases, and circuit ratings are to be restricted to 20 amp ratings for most fluorescent applications.

Here divide each phase into two branches, each having a load of 12.5 amps. [210-22(c)] requires a minimum of 80 percent derating for lighting circuits. [Note 8] to tables [310-16] through [310-19] can cause a further derating if more than six current-carrying conductors are routed through the same raceway or cable.

The luminaire layout indicates that with ten rows and with the two branch circuits per phase, it would be sensible to use two multi-conductor branch circuits with each serving five rows of 9 luminaires each. Then, each cable or raceway would contain the three phase conductors and a neutral. [Note 10(c)] to tables [310-16] through [310-19] requires that the neutral here be counted as a current carrying conductor.

[220-22] (which actually applies to feeders) requires that the portion of the neutral load supplying electric discharge lighting units cannot be reduced in ampacity from that portion of the phase currents. This in effect requires that the neutral have the same ampacity as the ungrounded conductors. Here, there is an 80 percent derating from [note 8] of tables [310-16] through [310-19]. The total derating will be 80 percent for the ampacity of the conductors, as [210-22(c)] only requires 80 percent derating when [note 8] of tables [310-16] through [310-19] has not been in force. Then, the required ampacity is

$$12.5\,\text{amps} \times 1.25 = 15.625\,\text{amps}$$

$$\frac{12.5}{0.8} = 15.625\,\text{amps}$$

As the overcurrent protection device cannot operate continuously at more than 80 percent of its rating [210-23], such devices here must be rated at

$$\frac{15.625}{0.8} = 19.53\,\text{amps}$$

This capacity is not standard, so [240-3(b)] permits the next larger size, which is 20 amps.

(Note that [210-3] restricts branch lighting circuits serving more than one load to ratings of 15, 20, 30, 40, and 50 amperes.)

The branch ampacity is over 15 amps, so the footnote (†) to table [310-16], which limits the overcurrent protection device for AWG #14 wire to 15 amps, eliminates the possibility of using 90°C conductors with their higher ampacity to achieve a smaller wire size. This a 75°C AWG #12 conductor is specified for the branch conductors. [Table C1] of [appendix C] allows up to seven AWG #12 conductors of the 90 degree temperature class in 1/2 inch electrical metal tubing.

If the luminaires are of the type that permit the circuit conductors to be run through the luminaires, tap conductors will not be needed, as the branch conductors will be within 18 inches of the fixture, allowing fixture wire of AWG #18 [240-4]. This wire should have insulation to protect it in environments up to 90°C, as [410-68] permits units to reach such temperatures. Table [402-3] gives information on fixture wire, and TFN is a suitable choice. If the maximum operating temperature is known, a lower rated fixture wire could be specified.

If the luminaires are to be supplied by individual taps from overhead branch circuits, table [210-24] requires AWG #14 conductors to be used. Again a 90°C conductor should be specified in the absence of luminaire temperature limits.

The design specifications are:

> Overcurrent protection: 20 amps (6 required)
>
> Branch conductors: AWG #12, 75°C type (type T is adequate)
>
> (a) Fixture wire: AWG #13, 90°C type (type TFN is adequate)
>
> (b) Tap conductors: AWG #14, 90°C type (type THHN is adequate)
>
> 1/2 inch EMT (electrical metal tubing) which will be used as the grounding conductor.

(a) or (b) above is to be used depending on the luminaire. Two four-wire branch circuits rated at 20 amps each are required.

PRACTICE PROBLEMS

Warmups

1. A room with a floor area of 40 feet × 60 feet and a 24 foot ceiling height has luminaires mounted flush with the ceiling. Determine the cavity ratios.

2. A room with floor area 40 feet × 60 feet and a 24 foot ceiling height has luminaires suspended 12 feet from the ceiling. Determine the cavity ratios.

3. For the room of problem 2, the ceiling reflectance is 90% and the wall reflectance is 75%. Determine the effective ceiling reflectance.

4. For the room of problem 3, use two T-12 lamps per luminaire, and determine the coefficient of utilization.

5. For the room of problem 4, the required illumination level is 100 footcandles. Determine the net luminous flux necessary for luminaires with two T-12 lamps.

6. The room of problem 5 has the following environment: (1) the line voltage is known to drop 5% at the peak loading time of day; (2) high quality material has been used in the luminaires; (3) all burned out lamps are replaced immediately; (4) all lamps will be replaced at the end of 70% of the expected life, and the lamps will burn 12 hours for each start, resulting in a lamp lumen depreciation of 0.84. The luminaire is as in table 13.1, the environment is clean, and the luminaires are to be cleaned on an annual schedule. Determine the total loss factor.

7. The room of problem 6 requires a minimum maintained illumination level of 100 footcandles. If two 4 foot T-12 lamps with 3000 initial lumens per lamp are used per luminaire, determine the number of luminaires required.

8. For the room of problem 7, assume the luminaires are 52 inches long, and are to be placed in rows across the short (40 foot) dimension of the room. Provide a layout for this room.

9. For the room of problem 8, determine the reflected horizontal component of illumination at a point that is six feet from two walls.

10. Determine the direct illumination at the point of calculation of problem 9.

Concentrates

1. A display room is 56 feet × 28 feet with a 14 foot ceiling and has display counters oriented along the length of the room. The ceiling reflectance is 90%, and the wall reflectance is 50%. Design a lighting system using warm white lamps in the 2T-12 luminaire of table 13.1. The maintained minimum illumination is to be 90 footcandles 30 inches from the floor. The environment is clean, and the cleaning schedule is annual. Make a sketch of the resulting design.

2. The floor area of an industrial plant is 96 feet × 180 feet and the ceiling height is 29 feet. The long axis of the machines is parallel to the long dimension of the building. The illumination proposal is to use 8 foot (100 inch) luminaires with four 4 foot lamps per luminaire, and to finish any rows with 4 foot (50 inch) luminaires with two 4 foot lamps per luminaire. The basic design is

    minimum maintained illumination: 100 fc
    initial lumens per lamp: 3100 Lm
    coefficient of utilization: 0.75
    maintenance factor (LLF): 0.62
    mounting height: 18 feet
    8 foot luminaire wattage (including ballast): 285 W
    4 foot luminaire wattage (including ballast): 190 W

(a) Determine the number of 8 foot and 4 foot luminaires required, (b) sketch the layout, giving spacings, and (c) determine the operating cost if energy is 1.5 cents per kilowatt hour and the factory is operated 17 hours a day for 50 weeks a year.

3. Design the lighting for a 40 foot wide roadway in a commercial area where the roadway is a major carrier, the environment is medium, and the luminaires are cleaned every three years. Use the luminaire from figure 13.8.

Timed

1. Two proposed lighting installations are being considered to maintain an illumination level of 30 footcandles in a warehouse with 20,000 square feet of floor area. Assume an electric energy cost of 75 mils and 3200 hours of lighting per year. Assume a coefficient of utilization of 78% and a maintenance factor of 72%. All costs involved in installation are prorated into the fixture costs. Replacement costs include lamps, installation, and cleaning. The two proposed systems are:

    luminaire A: incandescent, 150 W, 115 V, 2600 initial Lm, 750 hours life, fixture cost = $75, and replacement cost = $4.35.

    luminaire B: mercury vapor, 175 W, 115 V, 7700 initial Lm, 24,000 hour life, fixture = $162.50, replacement cost = $10, and ballast loss 25 W.

Compare the two systems with respect to installation cost and annual cost of operation.

2. A parking lot of dimensions 420 feet × 360 feet is to use 40-foot poles with a maximum of four luminaires per pole. The minimum maintained illumination level is to be 6.25 footcandles (average). 1000 W, 65,000 lumen lamps are to be used. Assume an LLD of 74%, an LDD of 48%, and a 25 W/lamp ballast loss. Design for a coefficient of beam utilization of 65%.

(a) Determine the layout, giving a sketch and the recommended luminaire type (i.e., beam angle), (b) specify the wiring, assuming 220 V, three-phase voltage (wire size and distribution), and (c) specify the protective breakers (voltage, current, and estimated surge current assuming proportionality to average current from 1.75 to 2).

ENGINEERING
ECONOMIC
14 ANALYSIS

Nomenclature

A	annual amount or annuity	$
B	present worth of all benefits	$
BV_j	book value at the end of the jth year	$
C	cost, or present worth of all costs	$
d	declining balance depreciation rate	decimal
D_j	depreciation in year j	$
D.R.	present worth of after-tax depreciation recovery	$
e	natural logarithm base (2.718)	–
EAA	equivalent annual amount	$
EUAC	equivalent uniform annual cost	$
f	federal income tax rate	decimal
F	future amount, or future worth	$
G	uniform gradient amount	$
i	effective rate per period (usually per year)	decimal
k	number of compounding periods per year	–
n	number of compounding periods, or life of asset	–
P	present worth, or present value	$
P_t	present worth after taxes	$
ROR	rate of return	decimal
ROI	return on investments	$
r	nominal rate per year (rate per annum)	decimal
s	state income tax rate	decimal
S_n	expected salvage value in year n	$
t	composite tax rate, or time	decimal
z	a factor equal to $(1+i)/(1-d)$	decimal
ϕ	effective rate per period	decimal

1 EQUIVALENCE

Industrial decision-makers using engineering economics are concerned with the timing of a project's cash flows as well as with the total profitability of that project. In this situation, a method is required to compare projects involving receipts and disbursements occurring at different times.

By way of illustration, consider $100 placed in a bank account that pays 5% effective annual interest at the end of each year. After the first year, the account will have grown to $105. After the second year, the account will have grown to $110.25.

Assume that you will have no need for money during the next two years and that any money received would immediately go into your 5% bank account. Then, which of the following options would be more desirable?

> **option a:** $100 now
> **option b:** $105 to be delivered in one year
> **option c:** $110.25 to be delivered in two years

In light of the previous illustration, none of the options is superior under the assumptions given. If the first option is chosen, you will immediately place $100 into a 5% account, and in two years the account will have grown to $110.25. In fact, the account will contain $110.25 at the end of two years regardless of the option chosen. Therefore, these alternatives are said to be *equivalent*.

2 CASH FLOW DIAGRAMS

Although they are not always necessary in simple problems (and they are often unwieldy in very complex problems), *cash flow diagrams* may be drawn to help visualize and simplify problems having diverse receipts and disbursements.

The conventions below are used to standardize cash flow diagrams.

- The horizontal (time) axis is marked off in equal increments, one per period, up to the duration or horizon of the project.

- All disbursements and receipts (cash flows) are assumed to take place at the end of the year in which they occur. This is known as the *year-end convention*. The exception to the year-end convention is any initial cost (purchase cost) which occurs at $t = 0$.

- Two or more transfers in the same year are placed end-to-end, and these may be combined.

- Expenses incurred before $t = 0$ are called *sunk costs*. Sunk costs are not relevant to the problem.

- Receipts are represented by arrows directed upward. Disbursements are represented by arrows directed downward. The arrow length is proportional to the magnitude of the cash flow.

Example 14.1

A mechanical device will cost $20,000 when purchased. Maintenance will cost $1000 each year. The device will generate revenues of $5000 each year for five years after which the salvage value is expected to be $7000. Draw and simplify the cash flow diagram.

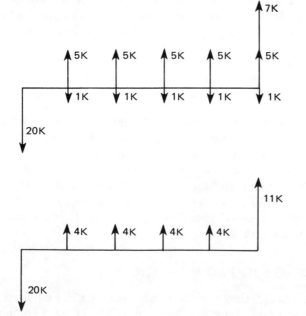

3 TYPICAL PROBLEM FORMAT

With the exception of some investment and rate of return problems, the typical problem involving engineering economics will have the following characteristics:

- An interest rate will be given.

- Two or more alternatives will be competing for funding.

- Each alternative will have its own cash flows.

- It is necessary to select the best alternative.

Example 14.2

Investment A costs $10,000 today and pays back $11,500 two years from now. Investment B costs $8000 today and pays back $4500 each year for two years. If an interest rate of 5% is used, which alternative is superior?

The solution to this example is not difficult, but it will be postponed until methods of calculating equivalence have been covered.

4 CALCULATING EQUIVALENCE

It was previously illustrated that $100 now is equivalent at 5% interest to $105 in one year. The equivalence of any present amount, P, at $t = 0$ to any future amount, F, at $t = n$ is called the *future worth* and can be calculated from equation 14.1.

$$F = P(1 + i)^n \qquad 14.1$$

The factor $(1 + i)^n$ is known as the *compound amount factor* and has been tabulated at the end of this chapter for various combinations of i and n. Rather than actually writing the formula for the compound amount factor, the convention is to use the standard functional notation $(F/P, i\%, n)$. Thus,

$$F = P(F/P, i\%, n) \qquad 14.2$$

Similarly, the equivalence of any future amount to any present amount is called the *present worth* and can be calculated from

$$P = F(1 + i)^{-n} = F(P/F, i\%, n) \qquad 14.3$$

The factor $(1+i)^{-n}$ is known as the *present worth factor*, with functional notation $(P/F, i\%, n)$. Tabulated values are also given for this factor at the end of this chapter.

Example 14.3

How much should you put into a 10% savings account in order to have $10,000 in five years?

This problem could also be stated: What is the equivalent present worth of $10,000 five years from now if money is worth 10%?

$$P = F(1+i)^{-n} = 10,000(1+0.10)^{-5}$$
$$= 6209$$

The factor 0.6209 would usually be obtained from the tables.

A cash flow which repeats regularly each year is known as an *annual amount*. When annual costs are incurred due to the functioning of a piece of equipment, they are often known as *operating and maintenance* (O&M) *costs*. The annual costs associated with operating a business in general are known as *general, selling, and administrative* (GS&A) *expenses*. Although the equivalent value for each of the n annual amounts could be calculated and then summed, it is much easier to use one of the *uniform series factors*, as illustrated in example 14.4.

Example 14.4

Maintenance costs for a machine are $250 each year. What is the present worth of these maintenance costs over a 12-year period if the interest rate is 8% ?

Notice that

$$(P/A, 8\%, 12) = (P/F, 8\%, 1) + (P/F, 8\%, 2)$$
$$+ \cdots + (P/F, 8\%, 12)$$

Then,

$$P = A(P/A, i\%, n) = -250(7.5361)$$
$$= -1884$$

A common complication involves a uniformly increasing cash flow. Such an increasing cash flow should be handled with the *uniform gradient factor*, $(P/G, i\%, n)$. The uniform gradient factor finds the present worth of a uniformly increasing cash flow which starts in year 2 (not year 1) as shown in example 14.5.

Example 14.5

Maintenance on an old machine is $100 this year but is expected to increase by $25 each year thereafter. What is the present worth of five years of maintenance? Use an interest rate of 10%.

In this problem, the cash flow must be broken down into parts. Notice that the five-year gradient factor is used even though there are only four non-zero gradient cash flows.

$$P = A(P/A, 10\%, 5) + G(P/G, 10\%, 5)$$
$$= -100(3.7908) - 25(6.8618)$$
$$= -551$$

Various combinations of the compounding and discounting factors are possible. For instance, the annual cash flow that would be equivalent to a uniform gradient may be found from

$$A = G(P/G, i\%, n)(A/P, i\%, n) \qquad 14.4$$

Formulas for all of the compounding and discounting factors are contained in table 14.1. Normally, it will not be necessary to calculate factors from the formulas. The tables at the end of this chapter are adequate for solving most problems.

5 THE MEANING OF PRESENT WORTH AND i

It is clear that $100 invested in a 5% bank account will allow you to remove $105 one year from now. If this investment is made, you will clearly receive a *return on investment* (ROI) of $5. The cash flow diagram and the present worth of the two transactions are

$$P = -100 + 105(P/F, 5\%, 1)$$
$$= -100 + 105(0.9524)$$
$$= 0$$

Table 14.1
Discount Factors for Discrete Compounding

factor name	converts	symbol	formula
single payment compound amount	P to F	$(F/P, i\%, n)$	$(1+i)^n$
present worth	F to P	$(P/F, i\%, n)$	$(1+i)^{-n}$
uniform series sinking fund	F to A	$(A/F, i\%, n)$	$\dfrac{i}{(1+i)^n - 1}$
capital recovery	P to A	$(A/P, i\%, n)$	$\dfrac{i(1+i)^n}{(1+i)^n - 1}$
compound amount	A to F	$(F/A, i\%, n)$	$\dfrac{(1+i)^n - 1}{i}$
equal series present worth	A to P	$(P/A, i\%, n)$	$\dfrac{(1+i)^n - 1}{i(1+i)^n}$
uniform gradient	G to P	$(P/G, i\%, n)$	$\dfrac{(1+i)^n - 1}{i^2(1+i)^n} - \dfrac{n}{i(1+i)^n}$

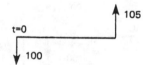

Figure 14.1　Cash Flow Diagram

Notice that the present worth is zero even though you did receive a 5% return on your investment.

However, if you are offered $120 for the use of $100 over a one-year period, the cash flow diagram and present worth (at 5%) would be

$$P = -100 + 120(P/F, 5\%, 1)$$
$$= -100 + 120(0.9524)$$
$$= 14.29$$

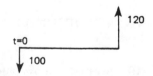

Figure 14.2　Cash Flow Diagram

Therefore, it appears that the present worth of an alternative is equal to the equivalent value at $t = 0$ of the increase in return above that which you would be able to earn in an investment offering $i\%$ per period. In the above case, $14.29 is the present worth of ($20−$5), the difference in the two ROIs.

Alternatively, the actual earned interest rate, called *rate of return*, ROR, can be defined as the rate which makes the present worth of the alternative zero.

The *present worth* is also the amount that you would have to be given to dissuade you from making an investment, since placing the initial investment amount along with the present worth into a bank account earning $i\%$ will yield the same eventual ROI. Relating this to the previous paragraphs, you could be dissuaded against investing $100 in an alternative which would return $120 in one year by a $t = 0$ payment of $14.29. Clearly, ($100 + $14.29) invested at $t = 0$ will also yield $120 in one year at 5%.

The selection of the interest rate is difficult in engineering economics problems. Usually it is taken as the average rate of return that an individual or business organization has realized in past investments. Fortunately, an interest rate is usually given. A company may not know what effective interest rate to use in an economic analysis. In such a case, the company can establish a minimum acceptable return on its investment. This *minimum attractive rate of return* (MARR) should be used as the effective interest rate i in economic analyses.

It should be obvious that alternatives with negative present worths are undesirable, and that alternatives with positive present worths are desirable because they increase the average earning power of invested capital.

6 CHOICE BETWEEN ALTERNATIVES

A variety of methods exist for selecting a superior alternative from among a group of proposals. Each method has its own merits and applications.

A. Present Worth Method

The *present worth method* has already been implied. When two or more alternatives are capable of performing the same functions, the superior alternative will have the largest present worth. This method is suitable for ranking the desirability of alternatives. The present worth method is restricted to evaluating alternatives that are mutually exclusive and which have the same lives.

Returning to example 14.2, the present worth of each alternative should be found in order to determine which alternative is superior.

Example 14.2, continued

$$P(A) = -10,000 + 11,500(P/F, 5\%, 2)$$
$$= 431$$
$$P(B) = -8000 + 4500(P/A, 5\%, 2)$$
$$= 367$$

Alternative A is superior and should be chosen.

B. Capitalized Cost Method

The present worth of a project with an infinite life is known as the *capitalized cost* or *life cycle cost*. Capitalized cost is the amount of money at $t = 0$ needed to perpetually support the project on the earned interest only. Capitalized cost is a positive number when expenses exceed income.

$$\frac{\text{capitalized}}{\text{cost}} = \frac{\text{initial}}{\text{cost}} + \frac{\text{annual costs}}{i} \qquad 14.5$$

Capitalized cost is the present worth of an infinitely-lived project. Normally, it would be difficult to work with an infinite stream of cash flows since most economics tables don't list factors for periods in excess of 100 years. However, the (A/P) discounting factor approaches the interest rate as n becomes large. Since the (P/A) and (A/P) factors are reciprocals of each other, we would expect to divide an infinite series of equal cash flows by the interest rate in order to calculate the present worth of the infinite series. This is the basis of equation 14.5.

Equation 14.5 can be used when the annual costs are equal in every year. The "annual cost" in that equation is assumed to be the same each year. If the operating and maintenance costs occur irregularly instead of annually, or if the costs vary from year to year, it will be necessary to somehow determine a cash flow of *equal annual amounts* (EAA) which is equivalent to the stream of original costs.

The equal annual amount may be calculated in the usual manner by first finding the present worth of all the actual costs, and then multiplying the present worth by the interest rate (the (A/P) factor for an infinite series). However, it is not necessary to convert the present worth to an equal annual amount, since equation 14.6 will convert the equal annual amount back to the present worth.

$$\frac{\text{capitalized}}{\text{cost}} = \frac{\text{initial}}{\text{cost}} + \frac{\text{EAA}}{i} \qquad 14.6$$

In comparing two alternatives, each of which is infinitely lived, the superior alternative will have the lowest capitalized cost.

C. Annual Cost Method

Alternatives that accomplish the same purpose but which have unequal lives must be compared by the *annual cost method*. The annual cost method assumes that each alternative will be replaced by an identical twin at the end of its useful life (infinite renewal). This method, which may also be used to rank alternatives according to their desirability, is also called the *annual return method* and *capital recovery method*.

Restrictions are that the alternatives must be mutually exclusive and infinitely renewed up to the duration of the longest-lived alternative. The calculated annual cost is known as the *equivalent uniform annual cost* (EUAC). Cost is a positive number when expenses exceed income.

Example 14.6

Which of the following alternatives is superior over a 30-year period if the interest rate is 7%?

	A	B
type	brick	wood
life	30 years	10 years
cost	$1800	$450
maintenance	$5/year	$20/year

$$\text{EUAC}(A) = 1800(A/P, 7\%, 30) + 5 = 150$$
$$\text{EUAC}(B) = 450(A/P, 7\%, 10) + 20 = 84$$

Alternative B is superior since its annual cost of operation is the lowest. It is assumed that three wood facilities, each with a life of 10 years and a cost of $450, will be built to span the 30-year period.

D. Benefit-Cost Ratio Method

The *benefit-cost ratio method* is often used in municipal project evaluations where benefits and costs accrue to different segments of the community. With this method, the present worth of all benefits (regardless of the beneficiary) is divided by the present worth of all costs. The project is considered acceptable if the ratio exceeds *one*.

When the benefit-cost ratio method is used, disbursements by the initiators or sponsors are *costs*. Disbursements by the users of the project are known as *disbenefits*. It is often difficult to determine whether a cash flow is a cost or a disbenefit (whether to place it in the numerator or denominator of the benefit-cost ratio calculation).

Regardless of where the cash flow is placed, an acceptable project will always have a benefit-cost ratio greater than one, although the actual numerical result will depend on the placement. For this reason, the benefit-cost ratio method should not be used to rank competing projects.

The benefit-cost ratio method may be used to rank alternative proposals only if an *incremental analysis* is used. First, determine that the ratio is greater than one for each alternative. Then, calculate the ratio of benefits to costs for each possible pair of alternatives by using equation 14.7. If the ratio exceeds one, alternative 2 is superior to alternative 1. Otherwise, alternative 1 is superior.

$$\text{benefit-to-cost ratio} = \frac{B_2 - B_1}{C_2 - C_1} \qquad 14.7$$

E. Rate of Return Method

Perhaps no method of analysis is less understood than the *rate of return method*, (ROR). As was stated previously, the ROR is the interest rate that would yield identical profits if all money were invested at that rate. The present worth of any such investment is zero.

The ROR is defined as the interest rate that will discount all cash flows to a total present worth equal to the initial required investment. This definition is used to determine the ROR of an alternative. The advantage of the ROR method is that no knowledge of an interest rate is required.

To find the ROR of an alternative, proceed as follows:

step 1: Set up the problem as if to calculate the present worth.

step 2: Arbitrarily select a reasonable value for i. Calculate the present worth.

step 3: Choose another value of i (not too close to the original value) and again solve for the present worth.

step 4: Interpolate or extrapolate the value of i that gives a zero present worth.

step 5: For increased accuracy, repeat steps (2) and (3) with two more values that straddle the value found in step (4).

A common, although incorrect, method of calculating the ROR involves dividing the annual receipts or returns by the initial investment. This technique ignores such items as salvage, depreciation, taxes, and the time value of money. This technique also fails when the annual returns vary.

Once a rate of return is known for an investment alternative, it is typically compared to the *minimum attractive rate of return* (MARR) specified by a company. However, ROR should not be used to rank alternatives. When two alternatives have RORs exceeding the MARR, it is not sufficient to select the alternative with the higher ROR.

An *incremental analysis*, also known as a *rate of return on added investment study*, should be performed if ROR is to be used to select between investments. In an incremental analysis, the cash flows for the investment with the lower initial cost are subtracted from the cash flows for the higher-priced alternative on a year-by-year basis. This produces, in effect, a third alternative representing the cost and benefits of the added investment. The added expense of the higher-priced investment is not warranted unless the ROR of this third alternative exceeds the MARR as well.

Example 14.7

What is the return on invested capital if $1000 is invested now with $500 being returned in year 4 and $1000 being returned in year 8?

First, set up the problem as a present worth calculation.

$$P = -1000 + 500(P/F, i\%, 4) + 1000(P/F, i\%, 8)$$

Arbitrarily select $i = 5\%$. The present worth is then found to be $88.15. Next take a higher value of i to reduce the present worth. If $i = 10\%$, the present worth is –$192. The ROR is found from simple interpolation to be approximately 6.6%.

7 TREATMENT OF SALVAGE VALUE IN REPLACEMENT STUDIES

An investigation into the retirement of an existing process or piece of equipment is known as a *replacement study*. Replacement studies are similar in most respects to other alternative comparison problems: an interest rate is given, two alternatives exist, and one of the previously mentioned methods of comparing alternatives is used to choose the superior alternative.

In replacement studies, the existing process or piece of equipment is known as the *defender*. The new process or piece of equipment being considered for purchase is known as the *challenger*.

Because most defenders still have some market value when they are retired, the problem of what to do with the salvage arises. It seems logical to use the salvage value of the defender to reduce the initial purchase cost of the challenger. This is consistent with what would actually happen if the defender were to be retired.

By convention, however, the salvage value is subtracted from the defender's present value. This does not seem logical, but it is done to keep all costs and benefits related to the defender with the defender. In this case, the salvage value is treated as an opportunity cost which would be incurred if the defender is not retired.

If the defender and the challenger have the same lives and a present worth study is used to choose the superior alternative, the placement of the salvage value will have no effect on the net difference between present worths for the challenger and defender. Although the values of the two present worths will be different depending on the placement, the difference in present worths will be the same.

If the defender and the challenger have different lives, an annual cost comparison must be made. Since the salvage value would be spread over a different number of years depending on its placement, it is important to abide by the conventions listed in this section.

There are a number of ways to handle salvage value. The best way is to think of the EUAC of the defender as the cost of keeping the defender from now until next year. In addition to the usual operating and maintenance costs, that cost would include an opportunity interest cost incurred by not selling the defender and also a drop in the salvage value if the defender is kept for one additional year. Specifically,

$$
\begin{aligned}
\text{EUAC(defender)} = \ &\text{maintenance costs} \\
&+ i\,(\text{current salvage value}) \\
&+ (\text{current salvage–next} \\
&\qquad \text{year's salvage}) \qquad 14.8
\end{aligned}
$$

It is important in retirement studies not to double count the salvage value. That is, it would be incorrect to add the salvage value to the defender and at the same time subtract it from the challenger.

8 BASIC INCOME TAX CONSIDERATIONS

Assume that an organization pays $f\%$ of its profits to the federal government as income taxes. If the organization also pays a state income tax of $s\%$, and if state taxes paid are recognized by the federal government as expenses, then the composite tax rate is

$$ t = s + f - sf \qquad 14.9 $$

The basic principles used to incorporate taxation into economic analyses are listed below.

- Initial purchase cost is unaffected by income taxes.

- Salvage value is unaffected by income taxes.

- Deductible expenses, such as operating costs, maintenance costs, and interest payments, are reduced by $t\%$ (i.e., multiplied by the quantity $(1 - t)$).

- Revenues are reduced by $t\%$ (i.e., multiplied by the quantity $(1 - t)$).

- Depreciation is multiplied by t and added to the appropriate year's cash flow, increasing that year's present worth.

Income taxes and depreciation have no bearing on municipal or governmental projects since municipalities, states, and the U.S. government pay no taxes.

Example 14.8

A corporation which pays 53% of its revenue in income taxes invests $10,000 in a project that will result in $3000 annual revenue for eight years. If the annual expenses are $700, salvage after eight years is $500, and 9% interest is used, what is the after-tax present worth? Disregard depreciation.

$$P_t = -10,000 + 3000(P/A, 9\%, 8)(1 - 0.53)$$
$$- 700(P/A, 9\%, 8)(1 - 0.53)$$
$$+ 500(P/F, 9\%, 8)$$
$$= -3766$$

It is interesting that the alternative evaluated in example 14.8 is undesirable if income taxes are considered but is desirable if income taxes are omitted.

9 DEPRECIATION

Although depreciation calculations may be considered independently, it is important to recognize that depreciation has no effect on engineering economic calculations unless income taxes are also considered.

Generally, tax regulations do not allow the cost of equipment[1] to be treated as a deductible expense in the year of purchase. Rather, portions of the cost may be allocated to each year of the item's economic life (which may be different from the actual useful life). Each year, the book value (which is initially equal to the purchase price) is reduced by the depreciation in that year. Theoretically, the book value of an item will equal the market value at any time within the economic life of that item.

Since tax regulations allow the depreciation in any year to be handled as if it were an actual operating expense, and since operating expenses are deductible from the income base prior to taxation, the after-tax profits will be increased. If D is the depreciation, the net result to the after-tax cash flow will be the addition of tD.

The present worth of all depreciation over the economic life of the item is called the *depreciation recovery*. Although originally established to do so, depreciation recovery can never fully replace an item at the end of its life.

Depreciation is often confused with amortization and depletion. While depreciation spreads the cost of a fixed asset over a number of years, *amortization* spreads the cost of an intangible asset (e.g., a patent) over some basis such as time or expected units of production.

Depletion is another artificial deductible operating expense designed to compensate mining organizations for decreasing mineral reserves. Since original and remaining quantities of minerals are seldom known accurately, the *depletion allowance* is calculated as a fixed percentage of the organization's gross income. These percentages are usually in the 10% to 20% range and apply to such mineral deposits as oil, natural gas, coal, uranium, and most metal ores.

There are four common methods of calculating depreciation. The book value of an asset depreciated with the *straight line* (SL) *method* (also known as the *fixed percentage method*) decreases linearly from the initial purchase at $t = 0$ to the estimated salvage at $t = n$. The depreciated amount is the same each year. The quantity $(C - S_n)$ in equation 14.10 is known as the *depreciation base*.

$$D_j = \frac{C - S_n}{n} \qquad 14.10$$

Double declining balance[2] (DDB) depreciation is independent of salvage value. Furthermore, the book value never stops decreasing, although the depreciation decreases in magnitude. Usually, any remaining book value is written off in the last year of the asset's estimated life. Unlike any of the other depreciation methods, DDB depends on accumulated depreciation.

$$D_j = \frac{2(C - \sum_{i=1}^{j-1} D_i)}{n} \qquad 14.11$$

In *sum-of-the-years' digits* (SOYD) depreciation, the digits from 1 to n inclusive, are summed. The total, T, can also be calculated from

$$T = \frac{1}{2} n(n + 1) \qquad 14.12$$

The depreciation can be found from

$$D_j = \frac{(C - S_n)(n - j + 1)}{T} \qquad 14.13$$

The *sinking fund method* is seldom used in industry because the initial depreciation is low. The formula for sinking fund depreciation (which increases each year) is

$$D_j = (C - S_n)(A/F, i\%, n)(F/P, i\%, j\text{-}1) \qquad 14.14$$

---

[1]The IRS tax regulations allow depreciation on almost all forms of *property* except land. The following types of property are distinguished: *real* (e.g., buildings used for business, etc.), *residential* (e.g., buildings used as rental property), and *personal* (e.g., equipment used for business). Personal property does *not* include items for personal use, despite its name. *Tangible* personal property is distinguished from *intangible* property (e.g., goodwill, copyrights, patents, trademarks, franchises, and agreements not to compete).

[2] Double declining balance depreciation is a particular form of *declining balance depreciation*, as defined by the IRS tax regulations. Declining balance depreciation also includes 125% declining balance and 150% declining balance depreciations which can be calculated by substituting 1.25 and 1.50, respectively for the 2 in equation 14.11.

The above discussion gives the impression that any form of depreciation may be chosen regardless of the nature and circumstances of the purchase. In reality, the IRS tax regulations place restrictions on the higher-rate ("accelerated") methods such as DDB and SOYD. Furthermore, the *Economic Recovery Act of 1981* substantially changed the laws relating to personal and corporate income taxes.

Property placed into service in 1981 or after must use an *accelerated cost recovery system* (ACRS or MACRS). Other methods (straight-line, declining balance, etc.) cannot be used except in special cases.

Property placed into service in 1980 or before must continue to be depreciated according to the method originally chosen (e.g., straight-line, declining balance, or sum-of-the-years' digits). ACRS and MACRS cannot be used.

Under ACRS and MACRS, the cost recovery amount in the jth year of an asset's cost recovery period is calculated by multiplying the initial cost by a factor.

$$D_j = (\text{initial cost})(\text{factor}) \qquad 14.15$$

The initial cost used is not reduced by the asset's salvage value for either the regular or alternate ACRS and MACRS calculations. The factor used depends on the asset's cost recovery period. Such factors are subject to continuing legislation changes. Current tax publications should be consulted before using an accelerated method.

Example 14.9

An asset is purchased for $9000. Its estimated economic life is 10 years, after which it will be sold for $1000. Find the depreciation in the first three years using SL, DDB, and SOYD.

SL:
$$D = \frac{(9000 - 1000)}{10}$$
$$= 800 \text{ each year}$$

DDB:
$$D_1 = \frac{2(9000)}{10}$$
$$= 1800 \text{ in year 1}$$
$$D_2 = \frac{2(9000 - 1800)}{10}$$
$$= 1440 \text{ in year 2}$$
$$D_3 = \frac{2(9000 - 3240)}{10}$$
$$= 1152 \text{ in year 3}$$

SOYD:
$$T = \frac{1}{2}(10)(11) = 55$$
$$D_1 = \left(\frac{10}{55}\right)(9000 - 1000)$$
$$= 1455 \text{ in year 1}$$
$$D_2 = \left(\frac{9}{55}\right)(8000)$$
$$= 1309 \text{ in year 2}$$
$$D_3 = \left(\frac{8}{55}\right)(8000)$$
$$= 1164 \text{ in year 3}$$

Example 14.10

For the asset described in example 14.9, calculate the book value during the first three years if SOYD depreciation is used.

The book value at the beginning of year 1 is $9000. Then,

$$BV_1 = 9000 - 1455 = 7545$$
$$BV_2 = 7545 - 1309 = 6236$$
$$BV_3 = 6236 - 1164 = 5072$$

Example 14.11

For the asset described in example 14.9, calculate the after-tax depreciation recovery with SL and SOYD depreciation methods. Use 6% interest with 48% income taxes.

SL:
$$D.R. = 0.48(800)(P/A, 6\%, 10)$$
$$= 2826$$

SOYD: The depreciation series can be thought of as a constant 1454 term with a negative 145 gradient.

$$D.R. = 0.48(1454)(P/A, 6\%, 10)$$
$$- 0.48(145)(P/G, 6\%, 10)$$
$$= 3076$$

Finding book values, depreciation, and depreciation recovery is particularly difficult with DDB depreciation, since all previous years' quantities seem to be required. It appears that the depreciation in the sixth year cannot be calculated unless the values of depreciation for the first five years are calculated first. Questions asking for depreciation or book value in the middle or at the end of an asset's economic life may be solved from the following equations:

$$d = \frac{2}{n} \qquad \text{14.16}$$

$$z = \frac{1+i}{1-d} \qquad \text{14.17}$$

$$(P/EG) = \frac{z^n - 1}{z^n(z-1)} \qquad \text{14.18}$$

Then, assuming that the remaining book value after n periods is written off in one lump sum, the present worth of the depreciation recovery is

$$D.R. = t \left[\frac{(d)(C)}{(1-d)}(P/EG) \right] \qquad \text{14.19}$$

$$D_j = (d)(C)(1-d)^{j-1} \qquad \text{14.20}$$

$$BV_j = C(1-d)^j \qquad \text{14.21}$$

Example 14.12

What is the after-tax present worth of the asset described in example 14.8 if SL, SOYD, and DDB depreciation methods are used?

The after-tax present worth, neglecting depreciation, was previously found to be -3766.

Using SL, the depreciation recovery is

$$D.R. = (0.53) \left(\frac{10,000 - 500}{8} \right)(P/A, 9\%, 8)$$

$$= 3483$$

Using SOYD, the depreciation recovery is calculated as follows:

$$T = \frac{1}{2}(8)(9) = 36$$

Depreciation base $= (10,000 - 500) = 9500$

$$D_1 = \frac{8}{36}(9500) = 2111$$

$$G = \text{gradient} = \frac{1}{36}(9500)$$

$$= 264$$

$$D.R. = (0.53)\,[2111(P/A, 9\%, 8)$$
$$- 264(P/G, 9\%, 8)]$$

$$= 3829$$

Using DDB, the depreciation recovery is calculated as follows:

$$d = \frac{2}{8} = 0.25$$

$$z = \frac{1.09}{0.75} = 1.4533$$

$$(P/EG) = \frac{(1.4533)^8 - 1}{(1.4533)^8(0.4533)} = 2.095$$

$$D.R. = 0.53 \left[\frac{(0.25)(10,000)}{0.75}(2.095) \right]$$

$$= 3701$$

The after-tax present worths, including depreciation recovery, are:

SL: $P_t = -3766 + 3483 = -283$
SOYD: $P_t = -3766 + 3829 = 63$
DDB: $P_t = -3766 + 3701 = -65$

10   ADVANCED INCOME TAX CONSIDERATIONS

There are a number of specialized techniques that are needed infrequently. These techniques are related more to the accounting profession than to the engineering profession.

A. Tax Credit

A *tax credit* (also known as an *investment tax credit* or an *investment credit*) is a one-time credit against income taxes. The investment tax credit is calculated as a fraction of the initial purchase price of certain types of equipment purchased for industrial, commercial, and manufacturing use.

$$\text{credit} = (\text{initial cost})(\text{fraction}) \qquad \text{14.22}$$

Since the investment tax credit reduces the buyer's tax liability, the credit should only be used in after-tax analyses.

B. Gain (or Loss) on the Sale of a Depreciated Asset

If an asset is sold for more (or less) than its current book value, the difference between selling price and book value is taxable income (deductible expense). The gain is taxed at capital gains rates.

11 RATE AND PERIOD CHANGES

All of the foregoing calculations were based on compounding once a year at an *effective interest rate, i.* However, some problems specify compounding more frequently than annually. In such cases, a *nominal interest rate, r,* will be given. The nominal rate does not include the effect of compounding and is not the same as the effective rate, *i.* A nominal rate may be used to calculate the effective rate by using equation 14.23 or 14.24.

$$i = \left(1 + \frac{r}{k}\right)^k - 1 \qquad 14.23$$

$$= (1 + \phi)^k - 1 \qquad 14.24$$

A problem may also specify an effective rate per period, ϕ, (e.g., per month). However, that will be a simple problem since compounding for n periods at an effective rate per period is not affected by the definition or length of the period.

The following rules may be used to determine which interest rate is given in a problem:

- Unless specifically qualified in the problem, the interest rate given is an annual rate.

- If the compounding is annually, the rate given is the effective rate. If compounding is other than annually, the rate given is the nominal rate.

- If the type of compounding is not specified, assume annual compounding.

In the case of continuous compounding, the appropriate discount factors may be calculated from the formulas in table 14.2.

Table 14.2
Discount Factors for
Continuous Compounding

(F/P)	e^{rn}
(P/F)	e^{-rn}
(A/F)	$\dfrac{e^r - 1}{e^{rn} - 1}$
(F/A)	$\dfrac{e^{rn} - 1}{e^r - 1}$
(A/P)	$\dfrac{e^r - 1}{1 - e^{-rn}}$
(P/A)	$\dfrac{1 - e^{-rn}}{e^r - 1}$

Example 14.13

A savings and loan offers $5\frac{1}{4}\%$ compounded daily. What is the annual effective rate?

method 1:

$$r = 0.0525, \ k = 365$$

$$i = \left(1 + \frac{0.0525}{365}\right)^{365} - 1 = 0.0539$$

method 2: Assume daily compounding is the same as continuous compounding.

$$i = (F/P) - 1$$
$$= e^{0.0525} - 1 = 0.0539$$

12 PROBABILISTIC PROBLEMS

Thus far, all of the cash flows included in the examples have been known exactly. If the cash flows are not known exactly but are given by some implicit or explicit probability distribution, the problem is *probabilistic.*

Probabilistic problems typically possess the following characteristics:

- There is a chance of extreme loss that must be minimized.

- There are multiple alternatives that must be chosen from. Each alternative gives a different degree of protection against the loss or failure.

- The outcome is independent of the alternative chosen. Thus, as illustrated in example 14.15 the size of the dam that is chosen for construction will not alter the rainfall in successive years. However, it will alter the effects on the down-stream watershed areas.

Probabilistic problems are typically solved using annual costs and expected values. An *expected value* is similar to an 'average value' since it is calculated as the mean of the given probability distribution. If cost 1 has a probability of occurrence of p_1, cost 2 has a probability of occurrence of p_2, and so on, the expected value is

$$E(\text{cost}) = p_1(\text{cost 1}) + p_2(\text{cost 2}) + \cdots \qquad 14.25$$

Example 14.14

Flood damage in any year is given according to the table below. What is the present worth of flood damage for a 10-year period? Use 6%.

damage	probability
0	0.75
$10,000	0.20
$20,000	0.04
$30,000	0.01

The expected value of flood damage is

$$E(\text{damage}) = (0)(0.75) + (10{,}000)(0.20)$$
$$+ (20{,}000)(0.04) + (30{,}000)(0.01)$$
$$= 3100$$

$$\text{present worth} = 3100(P/A, 6\%, 10)$$
$$= 22{,}816$$

Probabilities in probabilistic problems may be given in the problem (as in the example above) or they may have to be obtained from some named probability distribution. In either case, the probabilities are known explicitly and such problems are known as *explicit probability problems*.

Example 14.15

A dam is being considered on a river which periodically overflows and causes $600,000 damage. The damage is essentially the same each time the river causes flooding. The project horizon is 40 years. A 10% interest rate is being used.

Three different designs are available, each with different costs and storage capacities.

design alternative	cost	maximum capacity
A	500,000	1 unit
B	625,000	1.5 units
C	900,000	2.0 units

The U.S. Weather Service has provided a statistical analysis of annual rainfall in the area draining into the river.

units annual rainfall	probability
0	0.10
0.1–0.5	0.60
0.6–1.0	0.15
1.1–1.5	0.10
1.6–2.0	0.04
2.1 or more	0.01

Which design alternative would you choose assuming the dam is essentially empty at the start of each rainfall season?

The sum of the construction cost and the expected damage needs to be minimized. If alternative A is chosen, it will have a capacity of 1 unit. Its capacity will be exceeded (causing $600,000 damage) when the annual rainfall exceeds 1 unit. Therefore, the annual cost of A is

$$\text{EUAC}(A) = 500{,}000(A/P, 10\%, 40)$$
$$+ 600{,}000(0.10 + 0.04 + 0.01)$$
$$= 141{,}150$$

Similarly,

$$\text{EUAC}(B) = 625{,}000(A/P, 10\%, 40)$$
$$+ 600{,}000(0.04 + 0.01)$$
$$= 93{,}940$$

$$\text{EUAC}(C) = 900{,}000(A/P, 10\%, 40)$$
$$+ 600{,}000(0.01)$$
$$= 98{,}070$$

Alternative B should be chosen.

In other problems, a probability distribution will not be given even though some parameter (such as the life of an alternative) is not known with certainty. Such problems are known as *implicit probability problems* since they require a reasonable assumption about the probability distribution.

Implicit probability problems typically involve items whose *expected time to failure* are known. The key to such problems is in recognizing that an expected time to failure is not the same as a fixed life.

Reasonable assumptions can be made about the form of probability distributions in implicit probability problems.

One such reasonable assumption is that of a *rectangular distribution*. A rectangular distribution is one that is assumed to give an equal probability of failure in each year. Such an assumption is illustrated in example 14.16.

Example 14.16

A bridge is needed for 20 years. Failure of the bridge at any time will require a 50% reinvestment. Assume that each alternative has an annual probability of failure that is inversely proportional to its expected time to failure. Evaluate the two design alternatives below using 6% interest.

design alternative	initial cost	expected time to failure	annual costs	salvage at $t = 20$
A	15,000	9 years	1200	0
B	22,000	43 years	1000	0

For alternative A, the probability of failure in any year is 1/9. Similarly, the annual failure probability for alternative B is 1/43.

$$EUAC(A) = 15,000(A/P, 6\%, 20)$$
$$+ 15,000(0.5)\left(\frac{1}{9}\right) + 1200$$
$$= 3341$$

$$EUAC(B) = 22,000(A/P, 6\%, 20)$$
$$+ 22,000(0.5)\left(\frac{1}{43}\right) + 1000$$
$$= 3174$$

Alternative B should be chosen.

13 ESTIMATING ECONOMIC LIFE

As assets grow older, their operating and maintenance costs typically increase each year. Eventually, the cost to keep an asset in operation becomes prohibitive, and the asset is retired or replaced. However, it is not always obvious when an asset should be retired or replaced.

As the asset's maintenance is increasing each year, the amortized cost of its initial purchase is decreasing. It is the sum of these two costs that should be evaluated to determine the point at which the asset should be retired or replaced. Since an asset's initial purchase price is likely to be high, the amortized cost will be the controlling factor in those years when the maintenance costs are low. Therefore, the EUAC of the asset will decrease in the initial part of its life.

However, as the asset grows older, the change in its amortized cost decreases while maintenance increases. Eventually the sum of the two costs reaches a minimum and then starts to increase. The age of the asset at the minimum cost point is known as the *economic life* of the asset. The economic life is, generally, less than the mission and technological lifetimes of the asset.

The determination of an asset's economic life is illustrated by example 14.17.

Example 14.17

A bus in a municipal transit system has the characteristics listed below. When should the city replace its buses if money can be borrowed at 8% ?

initial cost: $120,000

year	maintenance cost	salvage value
1	35,000	60,000
2	38,000	55,000
3	43,000	45,000
4	50,000	25,000
5	65,000	15,000

If the bus is kept for one year and then sold, the annual cost will be

$$EUAC(1) = 120,000(A/P, 8\%, 1) + 35,000(A/F, 8\%, 1)$$
$$- 60,000(A/F, 8\%, 1)$$
$$= 104,600$$

If the bus is kept for two years and then sold, the annual cost will be

$$EUAC(2) = [120,000 + 35,000(P/F, 8\%, 1)](A/P, 8\%, 2)$$
$$+ (38,000 - 55,000)(A/F, 8\%, 2)$$
$$= 77,300$$

If the bus is kept for three years and then sold, the annual cost will be

$$EUAC(3) = [120,000 + 35,000(P/F, 8\%, 1)$$
$$+ 38,000(P/F, 8\%, 2)](A/P, 8\%, 3)$$
$$+ (43,000 - 45,000)(A/F, 8\%, 3)$$
$$= 71,200$$

This process is continued until EUAC begins to increase. In this example, EUAC(4) is 71,700. Therefore, the bus should be retired after three years.

14 BASIC COST ACCOUNTING

Cost accounting is the system that determines the cost of manufactured products. Cost accounting is called *job cost accounting* if costs are accumulated by part number or contract. It is called *process cost accounting* if costs are accumulated by departments or manufacturing processes.

Three types of costs (direct material, direct labor, and all indirect costs) make up the total manufacturing cost of a product.

Direct material costs are the costs of all materials that go into the product, priced at the original purchase cost.

Indirect material and labor costs are generally limited to costs incurred in the factory, excluding costs incurred in the office area. Examples of indirect materials are cleaning fluids, assembly lubricants, and temporary routing tags. Examples of indirect labor are stock-picking, inspection, expediting, and supervision labor.

Here are some important points concerning basic cost accounting:

- The sum of direct material and direct labor costs is known as the *prime cost*.

- Indirect costs may be called *indirect manufacturing expenses* (IME).

- Indirect costs may also include the overhead sector of the company (e.g., secretaries, engineers, and corporate administration). In this case, the indirect cost is usually called *burden* or *overhead*. Burden may also include the EUAC of non-regular costs which must be spread evenly over several years.

- The cost of a product is usually known in advance from previous manufacturing runs or by estimation. Any deviation from this known cost is called a *variance*. Variance may be broken down into *labor variance* and *material variance*.

- Indirect cost per item is not easily measured. The method of allocating indirect costs to a product is as follows:

step 1: Estimate the total expected indirect (and overhead) costs for the upcoming year.

step 2: Decide on some convenient vehicle for allocating the overhead to production. Usually, this vehicle is either the number of units expected to be produced or the number of direct hours expected to be worked in the upcoming year.

step 3: Estimate the quantity or size of the overhead vehicle.

step 4: Divide expected overhead costs by the expected overhead vehicle to obtain the unit overhead.

step 5: Regardless of the true size of the overhead vehicle during the upcoming year, one unit of overhead cost is allocated per product.

- Although estimates of production for the next year are always somewhat inaccurate, the cost of the product is assumed to be independent of forecasting errors. Any difference between true cost and calculated cost goes into a variance account.

- *Burden (overhead) variance* will be caused by errors in forecasting both the actual overhead for the upcoming year and the vehicle size. In the former case, the variance is called *burden budget variance*; in the latter, it is called *burden capacity variance*.

Example 14.18

A small company expects to produce 8000 items in the upcoming year. The current material cost is $4.54 each. 16 minutes of direct labor are required per unit. Workers are paid $7.50 per hour. 2133 direct labor hours are forecast for the product. Miscellaneous overhead costs are estimated at $45,000.

Find the expected direct material cost, the direct labor cost, the prime cost, the burden as a function of production and direct labor, and the total cost.

The direct material cost was given as $4.54.

The direct labor cost is $(16/60)(\$7.50) = \$2.00$.

The prime cost is $\$4.54 + \$2.00 = \$6.54$.

If the burden vehicle is production, the burden rate is $\$45,000/8000 = \$5.63$ per item, making the total cost $\$4.54 + \$2.00 + \$5.63 = \$12.17$.

If the burden vehicle is direct labor hours, the burden rate is $(45,000/2133) = \$21.10$ per hour, making the total cost $\$4.54 + \$2.00 + (16/60)(\$21.10) = \$12.17$.

Example 14.19

The actual performance of the company in example 14.18 is given by the following figures:

$$\text{actual production: } 7560$$
$$\text{actual overhead costs: } \$47,000$$

What are the burden budget variance and the burden capacity variance?

The burden capacity variance is

$$\$45,000 - 7560(\$5.63) = \$2437$$

The burden budget variance is

$$\$47,000 - \$45,000 = \$2000$$

The overall burden variance is

$$\$47,000 - 7560(\$5.63) = \$4437$$

15   BREAK-EVEN ANALYSIS

Break-even analysis is a method of determining when costs exactly equal revenue. If the manufactured quantity is less than the break-even quantity, a loss is incurred. If the manufactured quantity is greater than the break-even quantity, a profit is incurred.

Consider the following special variables:

f a fixed cost which does not vary with production

a an incremental cost which is the cost to produce one additional item. It may also be called the *marginal cost* or *differential cost*.

Q the quantity sold

p the incremental revenue

R the total revenue

C the total cost

Assuming no change in the inventory, the *break-even point* can be found from $C = R$, where

$$C = f + aQ \qquad 14.26$$

$$R = pQ \qquad 14.27$$

An alternate form of the break-even problem is to find the number of units per period for which two alternatives have the same total costs. Fixed costs are to be spread over a period longer than one year. One of the alternatives will have a lower cost if production is less than the break-even point. The other will have a lower cost for production greater than the break-even point.

The *cost per unit* problem is a variation of the break-even problem. In the typical cost per unit problem, data will be available to determine the direct labor and material costs per unit, but some method is needed to additionally allocate part of the annual overhead (burden) and initial facility purchase/construction costs.

Annual overhead is allocated to the unit cost simply by dividing the overhead by the number of units produced each year. The initial purchase/construction cost is multiplied by the appropriate (A/P) factor before similarly dividing by the production rate. The total unit cost is the sum of the direct labor, direct material, pro rata share of overhead, and pro rata share of the equivalent annual facility investment costs.

Example 14.20

Two plans are available for a company to obtain automobiles for its salesmen. How many miles must the cars be driven each year for the two plans to have the same costs? Use an interest rate of 10%. Assume the year-end convention applies to the insurance.

   Plan A   Lease the cars and pay $0.15 per mile.

   Plan B   Purchase the cars for $5000. Each car has an economic life of three years, after which it can be sold for $1,200. Gas and oil cost $0.04 per mile. Insurance is $500 per year.

Let x be the number of miles driven per year. Then, the EUAC for both alternatives is

$$\text{EUAC}(A) = 0.15x$$
$$\text{EUAC}(B) = 0.04x + 500 + 5000(A/P, 10\%, 3)$$
$$- 1200(A/F, 10\%, 3)$$
$$= 0.04x + 2148$$

Setting $\text{EUAC}(A)$ and $\text{EUAC}(B)$ equal and solving for x yields 19,527 miles per year as the break-even point.

16   HANDLING INFLATION

It is important to perform economic studies in terms of *constant value dollars*. One method of converting all cash flows to constant value dollars is to divide the flows by some annual *economic indicator* or price index. Such indicators would normally be given to you as part of a problem.

If indicators are not available, this method can still be used by assuming that inflation is relatively constant at a decimal rate e per year. Then, all cash flows can be converted to $t = 0$ dollars by dividing by $(1+e)^n$ where n is the year of the cash flow.

Example 14.21

What is the uninflated present worth of $2000 in two years if the average inflation rate is 6% and i is 10% ?

$$P = \frac{\$2000}{(1.10)^2(1.06)^2} = \$1471.07$$

An alternative is to replace i with a value corrected for inflation. This corrected value, i', is

$$i' = i + e + ie \qquad 14.28$$

This method has the advantage of simplifying the calculations. However, pre-calculated factors may not be available for the non-integer values of i'. Therefore, table 14.1 will have to be used to calculate the factors.

Example 14.22

Repeat example 14.21 using i'.

$$i' = 0.10 + 0.06 + (0.10)(0.06)$$

$$= 0.166$$

$$P = \frac{\$2000}{(1.166)^2} = \$1471.07$$

17   LEARNING CURVES

The more products that are made, the more efficient the operation becomes due to experience gained. Therefore, direct labor costs decrease. Usually, a *learning curve* is specified by the decrease in cost each time the quantity produced doubles. If there is a 20% decrease per doubling, the curve is said to be an 80% learning curve.

Consider the following special variables:

T_1 time or cost for the first item
T_n time or cost for the nth item
n total number of items produced
b learning curve constant

Then, the time to produce the nth item is given by

$$T_n = T_1(n)^{-b} \qquad 14.29$$

Table 14.3
Learning Curve Constants

learning curve	b
80%	0.322
85%	0.234
90%	0.152
95%	0.074

The total time to produce units from quantity n_1 to n_2 inclusive is

$$\int_{n_1}^{n_2} T_n \, dn \approx \frac{T_1}{(1-b)}$$
$$\left[\left(n_2 + \frac{1}{2} \right)^{1-b} - \left(n_1 - \frac{1}{2} \right)^{1-b} \right] \qquad 14.30$$

The average time per unit over the production from n_1 to n_2 is the above total time from equation 14.30 divided by the quantity produced, $(n_2 - n_1 + 1)$.

It is important to remember that learning curve reductions apply only to direct labor costs. They are not applied to indirect labor or direct material costs.

Example 14.23

A 70% learning curve is used with an item whose first production time was 1.47 hours. How long will it take to produce the 11th item? How long will it take to produce the 11th through 27th items?

First, find b.

$$\frac{T_2}{T_1} = 0.7 = (2)^{-b}$$
$$b = 0.515$$

Then,

$$T_{11} = 1.47(11)^{-0.515} = 0.428 \text{ hours}$$

The time to produce the 11th item through 27th item is approximately

$$T = \frac{1.47}{1 - 0.515} \left[(27.5)^{1-0.515} - (10.5)^{1-0.515} \right]$$
$$= 5.643 \text{ hours}$$

18   ECONOMIC ORDER QUANTITY

The *economic order quantity* (EOQ) is the order quantity which minimizes the inventory costs per unit time. Although there are many different EOQ models, the simplest is based on the following assumptions:

- Reordering is instantaneous. The time between order placement and receipt is zero.

- Shortages are not allowed.

- Demand for the inventory item is deterministic (i.e., is not a random variable).

- Demand is constant with respect to time.
- An order is placed when the on-hand quantity is zero.

The following special variables are used:

a the constant depletion rate $\left(\dfrac{\text{items}}{\text{unit time}}\right)$

h the inventory storage cost $\left(\dfrac{\$}{\text{item-unit time}}\right)$

H the total inventory storage cost between orders (\$)
K the fixed cost of placing an order (\$)
Q_0 the order quantity

If the original quantity on hand is Q_0, the stock will be depleted at

$$t* = \frac{Q_0}{a} \qquad 14.31$$

The total inventory storage cost between t_0 and t^* is

$$H = \frac{1}{2}h\frac{Q_0^2}{a} \qquad 14.32$$

The total inventory and ordering cost per unit time is

$$C_t = \frac{aK}{Q_0} + \frac{1}{2}hQ_0 \qquad 14.33$$

C_t can be minimized with respect to Q_0. The EOQ and time between orders are:

$$Q_0^* = \sqrt{2\frac{aK}{h}} \qquad 14.34$$

$$t^* = \frac{Q_0^*}{a} \qquad 14.35$$

19   CONSUMER LOANS

Many consumer loans cannot be handled by the equivalence formulas presented up to this point. Many different arrangements can be made between lender and borrower. Four of the most common consumer loan arrangements are presented below. Refer to a real estate or investment analysis book for more complex loans.

A. Simple Interest

Interest due does not compound with a *simple interest* loan. The interest due is merely proportional to the

length of time the principal is outstanding. Because of this, simple interest loans are seldom made for long periods (e.g., longer than one year).

Example 14.24

A \$12,000 simple interest loan is taken out at 16% per year. The loan matures in one year with no intermediate payments. How much will be due at the end of the year?

$$\text{amount due} = (1 + 0.16)(\$12,000)$$
$$= \$13,920$$

For loans less than one year, it is commonly assumed that a year consists of 12 months of 30 days each.

Example 14.25

\$4000 is borrowed for 75 days at 16% per annum simple interest. How much will be due at the end of 75 days?

$$\text{amount due} = \$4000 + (0.16)\left(\frac{75}{360}\right)(4000)$$
$$= \$4133$$

B. Loans With Constant Amount Paid Towards Principal

With this loan type, the payment is not the same each period. The amount paid towards the principal is constant, but the interest varies from period to period. The following special symbols are used:

BAL_j balance after the jth payment
LV total value loaned (cost minus down payment)
j payment or period number
N total number of payments to pay off the loan
PI_j jth interest payment
PP_j jth principal payment
PT_j jth total payment
ϕ effective rate per period (r/k)

The equations which govern this type of loan are:

$$BAL_j = LV - (j)(PP) \qquad 14.36$$

$$PI_j = \phi(BAL_{j-1}) \qquad 14.37$$

$$PT_j = PP + PI_j \qquad 14.38$$

C. Direct Reduction Loans

This is the typical "interest paid on unpaid balance" loan. The amount of the periodic payment is constant, but the amounts paid towards the principal and interest both vary.

The same symbols are used with this type of loan as are listed above.

$$N = -\frac{\ln\left[\dfrac{-\phi(LV)}{PT} + 1\right]}{\ln(1 + \phi)} \quad\quad 14.39$$

$$BAL_{j-1} = PT\left[\frac{1 - (1 + \phi)^{j-1-N}}{\phi}\right] \quad 14.40$$

$$PI_j = \phi(BAL_{j-1}) \quad\quad 14.41$$

$$PP_j = PT - PI_j \quad\quad 14.42$$

$$BAL_j = BAL_{j-1} - PP_j \quad\quad 14.43$$

Example 14.26

A \$45,000 loan is financed at 9.25% per annum. The monthly payment is \$385. What are the amounts paid toward interest and principal in the 14th period? What is the remaining principal balance after the 14th payment has been made?

The effective rate per month is

$$\phi = \frac{r}{k} = \frac{0.0925}{12}$$

$$= 0.007708$$

$$N = \frac{-\ln\left[\dfrac{-(0.007708)(45,000)}{385} + 1\right]}{\ln(1 + 0.007708)} = 301$$

$$BAL_{13} = 385\left[\frac{1 - (1 + 0.007708)^{14-1-301}}{0.007708}\right]$$

$$= \$44,476.39$$

$$PI_{14} = (0.007708)(\$44,476.39) = \$342.82$$

$$PP_{14} = \$385 - \$342.82 = \$42.18$$

$$BAL_{14} = \$44,476.39 - \$42.18 = \$44,434.21$$

Equation 14.39 calculates the number of payments necessary to pay off a loan. This equation can be solved with effort for the total periodic payment (PT) or the

initial value of the loan (LV). It is easier, however, to use the $(A/P, \phi, n)$ factor to find the payment and loan value.

$$PT = (LV)(A/P, \phi\%, n) \quad\quad 14.44$$

If the loan is repaid in yearly installments, then i is the effective annual rate. If the loan is paid off monthly, then i should be replaced by the effective rate per month (ϕ from equation 14.24). For monthly payments, n is the number of months in the payback period.

D. Direct Reduction Loan With Balloon Payment

This type of loan has a constant periodic payment, but the duration of the loan is insufficient to completely pay back the principal. Therefore, all remaining unpaid principal must be paid back in a lump sum when the loan matures. This large payment is known as a *balloon payment*.

Equations 14.39 through 14.43 can also be used with this type of loan. The remaining balance after the last payment is the balloon payment. This balloon payment must be repaid along with the last regular payment calculated.

20　SENSITIVITY ANALYSIS

Data analysis and forecasts in economic studies represent judgment on costs which will occur in the future. There are always uncertainties about these costs. However, these uncertainties are insufficient reason not to make the best possible estimates of the costs. Nevertheless, a decision between alternatives often can be made more confidently if it is known whether or not the conclusion is sensitive to moderate changes in data forecasts. Sensitivity analysis provides this extra dimension to an economic analysis.

The sensitivity of a decision is determined by inserting a range of estimates for critical cash flows. If radical changes can be made to a cash flow without changing the decision, the decision is said to be insensitive to uncertainties regarding that cash flow. However, if a small change in the estimate of a cash flow will alter the decision, that decision is said to be very sensitive to changes in the estimate.

An established semantic tradition distinguishes between risk analysis and uncertainty analysis. Risk analysis addresses variables which have a known or estimated probability distribution. In this regard, statistics and probability theory can be used to determine the probability of a cash flow varying between given limits. On the other hand, uncertainty analysis is concerned with situations in which there is not enough information to determine the probability or frequency distribution for the variables involved.

As a first step, sensitivity analysis should be applied one at a time to the dominant cost factors. Dominant cost factors are those which have the most significant impact on the present value of the alternative. If warranted, additional investigation can be used to determine the sensitivity to several cash flows varying simultaneously. Significant judgment is needed, however, to successfully determine the proper combinations of cash flows to vary.

It is common to plot the dependency of the present value on the cash flow being varied on a two-dimensional graph. Simple linear interpolation is used (within reason) to determine the critical value of the cash flow being varied.

PRACTICE PROBLEMS

Warmups

1. How much will be accumulated at 6% if $1000 is invested for 10 years?

2. What is the present worth of $2000 at 6% which becomes available in four years?

3. How much will it take to accumulate $2000 in 20 years at 6%?

4. What year-end annual amount over seven years at 6% is equivalent to $500 invested now?

5. $50 is invested at the end of each year for 10 years. What will be the accumulated amount at the end of 10 years at 6%?

6. How much should be deposited at 6% at the start of each year for 10 years in order to empty the fund by drawing out $200 at the end of each year for 10 years?

7. How much should be deposited at 6% at the start of each year for five years to accumulate $2000 on the date of the last deposit?

8. How much will be accumulated at 6% in 10 years if three payments of $100 are deposited every other year for four years, with the first payment occuring at $t = 0$?

9. $500 is compounded monthly at 6% annual rate. How much will be accumulated in five years?

10. What is the rate of return on an $80 investment that pays back $120 in seven years?

Concentrates

1. A new machine will cost $17,000 and will have a value of $14,000 in five years. Special tooling will cost $5000 and it will have a resale value of $2500 after five years. Maintenance will be $200 per year. What will the average cost of ownership be during the next five years if interest is 6%?

2. An old highway bridge can be strengthened at a cost of $9000, or it can be replaced for $40,000. The present salvage value of the old bridge is $13,000. It is estimated that the reinforced bridge will last for 20 years with an annual cost of $500 and will have a salvage value of $10,000 at the end of 20 years. The estimated salvage of the new bridge after 25 years is $15,000. The maintenance for the new bridge will be $100 annually. Which is the best alternative at 8% interest?

3. A firm expects to receive $32,000 each year for 15 years from the sale of a product. It will require an initial investment of $150,000. Expenses will run $7530 per year. Salvage is zero and straight-line depreciation is used. The tax rate is 48%. What is the after-tax rate of return?

4. A public works project has initial costs of $1,000,000, benefits of $1,500,000, and disbenefits of $300,000. (a) What is the benefit/cost ratio? (b) What is the excess of benefits over costs?

5. A speculator in land pays $14,000 for property that he expects to hold for 10 years. $1000 is spent in renovation and a monthly rent of $75 is collected from the tenants. Taxes are $150 per year and maintenance costs are $250. What must be the sale price in 10 years to realize a 10% rate of return? Use the year-end convention.

6. What is the effective interest rate for a payment plan of 30 equal payments of $89.30 per month when a lump sum of $2000 would have been an outright purchase?

7. An apartment complex is purchased for $500,000. What is the depreciation in each of the first three years if the salvage value is $100,000 in 25 years? Use (a) straight line, (b) sum-of-the-years' digits, and (c) double declining balance.

8. Equipment is purchased for $12,000 and is expected to be sold after 10 years for $2000. The estimated maintenance is $1000 for the first year, but is expected to increase $200 each year thereafter. Using 10%, find the present worth and the annual cost.

9. One of five grades of pipe with average lives (in years) and costs (in dollars) of $(9, 1500)$, $(14, 1600)$, $(30, 1750)$, $(52, 1900)$, and $(86, 2100)$ is to be chosen for a 20-year project. A failure of the pipe at any time during the project will result in a cost equal to 35% of the original cost. Annual costs are 4% of the initial cost, and the pipes are not recoverable. At 6%, which pipe is superior? Note: The lives are average values, not absolute replacement times.

10. A grain combine with a 20-year life can remove seven pounds of rock from its harvest per hour. Any rocks left in its output will cause $25,000 damage in subsequent processes. Several investments are available to increase its removal capacity. At 10%, what should be done?

rock rate (lb/hr)	probability of exceeding rock rate	required investment to meet rock rate
7	0.15	no cost
8	0.10	$15,000
9	0.07	$20,000
10	0.03	$30,000

Timed (1 hour allowed for each)

1. A structure which costs $10,000 has the operating costs and salvage values given.

    year 1: maintenance $2000, salvage $8000
    year 2: maintenance $3000, salvage $7000
    year 3: maintenance $4000, salvage $6000
    year 4: maintenance $5000, salvage $5000
    year 5: maintenance $6000, salvage $4000

(a) What is the economic life of the structure? (b) Assuming that the structure has been owned and operated for four years, what is the cost of owning the structure for exactly one more year? Use 20% as the interest rate.

2. A man purchases a car for $5000 for personal use, intending to drive 15,000 miles per year. It costs him $200 per year for insurance and $150 per year for maintenance. He gets 15 mpg and gasoline costs $0.60 per gallon. The resale value after five years is $1000. Because of unexpected business driving (5000 miles per year extra), his insurance is increased to $300 per year and maintenance to $200. Salvage is reduced to $500. Use 10% interest to answer the following questions. (a) The man's company offers $0.10 per mile reimbursement. Is that adequate? (b) How many miles must be driven per year at $0.10 per mile to justify the company buying a car for the man's use? The cost would be $5000, but insurance, maintenance, and salvage would be $250, $200, and $800, respectively.

APPENDIX A

Standard Cash Flow Factors

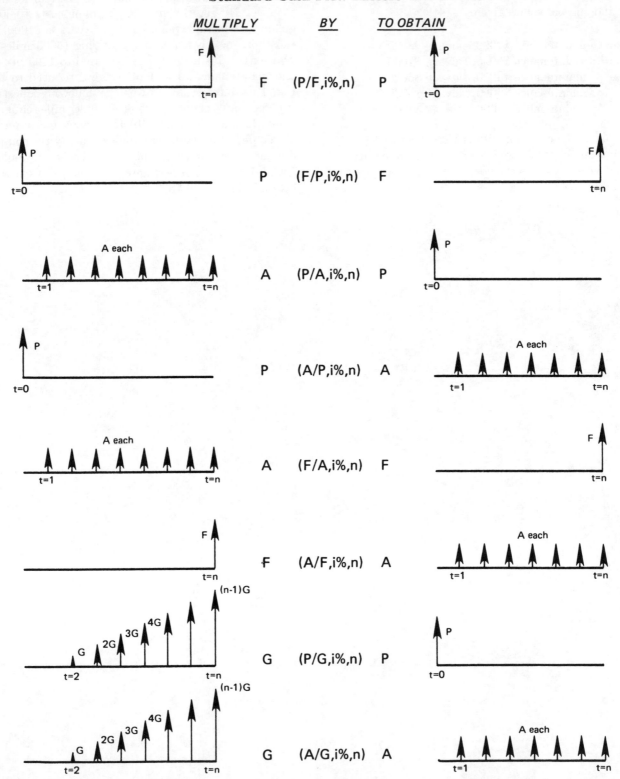

MULTIPLY	BY	TO OBTAIN
F	(P/F,i%,n)	P
P	(F/P,i%,n)	F
A	(P/A,i%,n)	P
P	(A/P,i%,n)	A
A	(F/A,i%,n)	F
F	(A/F,i%,n)	A
G	(P/G,i%,n)	P
G	(A/G,i%,n)	A

APPENDIX B

Standard Interest Tables

I = 0.50 %

N	(P/F)	(P/A)	(P/G)	(F/P)	(F/A)	(A/P)	(A/F)	(A/G)	N
1	.9950	0.9950	− 0.0000	1.0050	1.0000	1.0050	1.0000	− 0.0000	1
2	.9901	1.9851	0.9901	1.0100	2.0050	0.5038	0.4988	0.4988	2
3	.9851	2.9702	2.9604	1.0151	3.0150	0.3367	0.3317	0.9967	3
4	.9802	3.9505	5.9011	1.0202	4.0301	0.2531	0.2481	1.4938	4
5	.9754	4.9259	9.8026	1.0253	5.0503	0.2030	0.1980	1.9900	5
6	.9705	5.8964	14.6552	1.0304	6.0755	0.1696	0.1646	2.4855	6
7	.9657	6.8621	20.4493	1.0355	7.1059	0.1457	0.1407	2.9801	7
8	.9609	7.8230	27.1755	1.0407	8.1414	0.1278	0.1228	3.4738	8
9	.9561	8.7791	34.8244	1.0459	9.1821	0.1139	0.1089	3.9668	9
10	.9513	9.7304	43.3865	1.0511	10.2280	0.1028	0.0978	4.4589	10
11	.9466	10.6770	52.8526	1.0564	11.2792	0.0937	0.0887	4.9501	11
12	.9419	11.6189	63.2136	1.0617	12.3356	0.0861	0.0811	5.4406	12
13	.9372	12.5562	74.4602	1.0670	13.3972	0.0796	0.0746	5.9302	13
14	.9326	13.4887	86.5835	1.0723	14.4642	0.0741	0.0691	6.4190	14
15	.9279	14.4166	99.5743	1.0777	15.5365	0.0694	0.0644	6.9069	15
16	.9233	15.3399	113.4238	1.0831	16.6142	0.0652	0.0602	7.3940	16
17	.9187	16.2586	128.1231	1.0885	17.6973	0.0615	0.0565	7.8803	17
18	.9141	17.1728	143.6634	1.0939	18.7858	0.0582	0.0532	8.3658	18
19	.9096	18.0824	160.0360	1.0994	19.8797	0.0553	0.0503	8.8504	19
20	.9051	18.9874	177.2322	1.1049	20.9791	0.0527	0.0477	9.3342	20
21	.9006	19.8880	195.2434	1.1104	22.0840	0.0503	0.0453	9.8172	21
22	.8961	20.7841	214.0611	1.1160	23.1944	0.0481	0.0431	10.2993	22
23	.8916	21.6757	233.6768	1.1216	24.3104	0.0461	0.0411	10.7806	23
24	.8872	22.5629	254.0820	1.1272	25.4320	0.0443	0.0393	11.2611	24
25	.8828	23.4456	275.2686	1.1328	26.5591	0.0427	0.0377	11.7407	25
26	.8784	24.3240	297.2281	1.1385	27.6919	0.0411	0.0361	12.2195	26
27	.8740	25.1980	319.9523	1.1442	28.8304	0.0397	0.0347	12.6975	27
28	.8697	26.0677	343.4332	1.1499	29.9745	0.0384	0.0334	13.1747	28
29	.8653	26.9330	367.6625	1.1556	31.1244	0.0371	0.0321	13.6510	29
30	.8610	27.7941	392.6324	1.1614	32.2800	0.0360	0.0310	14.1265	30
31	.8567	28.6508	418.3348	1.1672	33.4414	0.0349	0.0299	14.6012	31
32	.8525	29.5033	444.7618	1.1730	34.6086	0.0339	0.0289	15.0750	32
33	.8482	30.3515	471.9055	1.1789	35.7817	0.0329	0.0279	15.5480	33
34	.8440	31.1955	499.7583	1.1848	36.9606	0.0321	0.0271	16.0202	34
35	.8398	32.0354	528.3123	1.1907	38.1454	0.0312	0.0262	16.4915	35
36	.8356	32.8710	557.5598	1.1967	39.3361	0.0304	0.0254	16.9621	36
37	.8315	33.7025	587.4934	1.2027	40.5328	0.0297	0.0247	17.4317	37
38	.8274	34.5299	618.1054	1.2087	41.7354	0.0290	0.0240	17.9006	38
39	.8232	35.3531	649.3883	1.2147	42.9441	0.0283	0.0233	18.3686	39
40	.8191	36.1722	681.3347	1.2208	44.1588	0.0276	0.0226	18.8359	40
41	.8151	36.9873	713.9372	1.2269	45.3796	0.0270	0.0220	19.3022	41
42	.8110	37.7983	747.1886	1.2330	46.6065	0.0265	0.0215	19.7678	42
43	.8070	38.6053	781.0815	1.2392	47.8396	0.0259	0.0209	20.2325	43
44	.8030	39.4082	815.6087	1.2454	49.0788	0.0254	0.0204	20.6964	44
45	.7990	40.2072	850.7631	1.2516	50.3242	0.0249	0.0199	21.1595	45
46	.7950	41.0022	886.5376	1.2579	51.5758	0.0244	0.0194	21.6217	46
47	.7910	41.7932	922.9252	1.2642	52.8337	0.0239	0.0189	22.0831	47
48	.7871	42.5803	959.9188	1.2705	54.0978	0.0235	0.0185	22.5437	48
49	.7832	43.3635	997.5116	1.2768	55.3683	0.0231	0.0181	23.0035	49
50	.7793	44.1428	1035.6966	1.2832	56.6452	0.0227	0.0177	23.4624	50
51	.7754	44.9182	1074.4670	1.2896	57.9284	0.0223	0.0173	23.9205	51
52	.7716	45.6897	1113.8162	1.2961	59.2180	0.0219	0.0169	24.3778	52
53	.7677	46.4575	1153.7372	1.3026	60.5141	0.0215	0.0165	24.8343	53
54	.7639	47.2214	1194.2236	1.3091	61.8167	0.0212	0.0162	25.2899	54
55	.7601	47.9814	1235.2686	1.3156	63.1258	0.0208	0.0158	25.7447	55
60	.7414	51.7256	1448.6458	1.3489	69.7700	0.0193	0.0143	28.0064	60
65	.7231	55.3775	1675.0272	1.3829	76.5821	0.0181	0.0131	30.2475	65
70	.7053	58.9394	1913.6427	1.4178	83.5661	0.0170	0.0120	32.4680	70
75	.6879	62.4136	2163.7525	1.4536	90.7265	0.0160	0.0110	34.6679	75
80	.6710	65.8023	2424.6455	1.4903	98.0677	0.0152	0.0102	36.8474	80
85	.6545	69.1075	2695.6389	1.5280	105.5943	0.0145	0.0095	39.0065	85
90	.6383	72.3313	2976.0769	1.5666	113.3109	0.0138	0.0088	41.1451	90
95	.6226	75.4757	3265.3298	1.6061	121.2224	0.0132	0.0082	43.2633	95
100	.6073	78.5426	3562.7934	1.6467	129.3337	0.0127	0.0077	45.3613	100

I = 0.75 %

N	(P/F)	(P/A)	(P/G)	(F/P)	(F/A)	(A/P)	(A/F)	(A/G)	N
1	.9926	0.9926	-0.0000	1.0075	1.0000	1.0075	1.0000	-0.0000	1
2	.9852	1.9777	0.9852	1.0151	2.0075	0.5056	0.4981	0.4981	2
3	.9778	2.9556	2.9408	1.0227	3.0226	0.3383	0.3308	0.9950	3
4	.9706	3.9261	5.8525	1.0303	4.0452	0.2547	0.2472	1.4907	4
5	.9633	4.8894	9.7058	1.0381	5.0756	0.2045	0.1970	1.9851	5
6	.9562	5.8456	14.4866	1.0459	6.1136	0.1711	0.1636	2.4782	6
7	.9490	6.7946	20.1808	1.0537	7.1595	0.1472	0.1397	2.9701	7
8	.9420	7.7366	26.7747	1.0616	8.2132	0.1293	0.1218	3.4608	8
9	.9350	8.6716	34.2544	1.0696	9.2748	0.1153	0.1078	3.9502	9
10	.9280	9.5996	42.6064	1.0776	10.3443	0.1042	0.0967	4.4384	10
11	.9211	10.5207	51.8174	1.0857	11.4219	0.0951	0.0876	4.9253	11
12	.9142	11.4349	61.8740	1.0938	12.5076	0.0875	0.0800	5.4110	12
13	.9074	12.3423	72.7632	1.1020	13.6014	0.0810	0.0735	5.8954	13
14	.9007	13.2430	84.4720	1.1103	14.7034	0.0755	0.0680	6.3786	14
15	.8940	14.1370	96.9876	1.1186	15.8137	0.0707	0.0632	6.8606	15
16	.8873	15.0243	110.2973	1.1270	16.9323	0.0666	0.0591	7.3413	16
17	.8807	15.9050	124.3887	1.1354	18.0593	0.0629	0.0554	7.8207	17
18	.8742	16.7792	139.2494	1.1440	19.1947	0.0596	0.0521	8.2989	18
19	.8676	17.6468	154.8671	1.1525	20.3387	0.0567	0.0492	8.7759	19
20	.8612	18.5080	171.2297	1.1612	21.4912	0.0540	0.0465	9.2516	20
21	.8548	19.3628	188.3253	1.1699	22.6524	0.0516	0.0441	9.7261	21
22	.8484	20.2112	206.1420	1.1787	23.8223	0.0495	0.0420	10.1994	22
23	.8421	21.0533	224.6682	1.1875	25.0010	0.0475	0.0400	10.6714	23
24	.8358	21.8891	243.8923	1.1964	26.1885	0.0457	0.0382	11.1422	24
25	.8296	22.7188	263.8029	1.2054	27.3849	0.0440	0.0365	11.6117	25
26	.8234	23.5422	284.3888	1.2144	28.5903	0.0425	0.0350	12.0800	26
27	.8173	24.3595	305.6387	1.2235	29.8047	0.0411	0.0336	12.5470	27
28	.8112	25.1707	327.5416	1.2327	31.0282	0.0397	0.0322	13.0128	28
29	.8052	25.9759	350.0867	1.2420	32.2609	0.0385	0.0310	13.4774	29
30	.7992	26.7751	373.2631	1.2513	33.5029	0.0373	0.0298	13.9407	30
31	.7932	27.5683	397.0602	1.2607	34.7542	0.0363	0.0288	14.4028	31
32	.7873	28.3557	421.4675	1.2701	36.0148	0.0353	0.0278	14.8636	32
33	.7815	29.1371	446.4746	1.2796	37.2849	0.0343	0.0268	15.3232	33
34	.7757	29.9128	472.0712	1.2892	38.5646	0.0334	0.0259	15.7816	34
35	.7699	30.6827	498.2471	1.2989	39.8538	0.0326	0.0251	16.2387	35
36	.7641	31.4468	524.9924	1.3086	41.1527	0.0318	0.0243	16.6946	36
37	.7585	32.2053	552.2969	1.3185	42.4614	0.0311	0.0236	17.1493	37
38	.7528	32.9581	580.1511	1.3283	43.7798	0.0303	0.0228	17.6027	38
39	.7472	33.7053	608.5451	1.3383	45.1082	0.0297	0.0222	18.0549	39
40	.7416	34.4469	637.4693	1.3483	46.4465	0.0290	0.0215	18.5058	40
41	.7361	35.1831	666.9144	1.3585	47.7948	0.0284	0.0209	18.9556	41
42	.7306	35.9137	696.8709	1.3686	49.1533	0.0278	0.0203	19.4040	42
43	.7252	36.6389	727.3297	1.3789	50.5219	0.0273	0.0198	19.8513	43
44	.7198	37.3587	758.2815	1.3893	51.9009	0.0268	0.0193	20.2973	44
45	.7145	38.0732	789.7173	1.3997	53.2901	0.0263	0.0188	20.7421	45
46	.7091	38.7823	821.6283	1.4102	54.6898	0.0258	0.0183	21.1856	46
47	.7039	39.4862	854.0056	1.4207	56.1000	0.0253	0.0178	21.6280	47
48	.6986	40.1848	886.8404	1.4314	57.5207	0.0249	0.0174	22.0691	48
49	.6934	40.8782	920.1243	1.4421	58.9521	0.0245	0.0170	22.5089	49
50	.6883	41.5664	953.8486	1.4530	60.3943	0.0241	0.0166	22.9476	50
51	.6831	42.2496	988.0050	1.4639	61.8472	0.0237	0.0162	23.3850	51
52	.6780	42.9276	1022.5852	1.4748	63.3111	0.0233	0.0158	23.8211	52
53	.6730	43.6006	1057.5810	1.4859	64.7859	0.0229	0.0154	24.2561	53
54	.6680	44.2686	1092.9842	1.4970	66.2718	0.0226	0.0151	24.6898	54
55	.6630	44.9316	1128.7869	1.5083	67.7688	0.0223	0.0148	25.1223	55
60	.6387	48.1734	1313.5189	1.5657	75.4241	0.0208	0.0133	27.2665	60
65	.6153	51.2963	1507.0910	1.6253	83.3709	0.0195	0.0120	29.3801	65
70	.5927	54.3046	1708.6065	1.6872	91.6201	0.0184	0.0109	31.4634	70
75	.5710	57.2027	1917.2225	1.7514	100.1833	0.0175	0.0100	33.5163	75
80	.5500	59.9944	2132.1472	1.8180	109.0725	0.0167	0.0092	35.5391	80
85	.5299	62.6838	2352.6375	1.8873	118.3001	0.0160	0.0085	37.5318	85
90	.5104	65.2746	2577.9961	1.9591	127.8790	0.0153	0.0078	39.4946	90
95	.4917	67.7704	2807.5694	2.0337	137.8225	0.0148	0.0073	41.4277	95
100	.4737	70.1746	3040.7453	2.1111	148.1445	0.0143	0.0068	43.3311	100

I = 1.00 %

N	(P/F)	(P/A)	(P/G)	(F/P)	(F/A)	(A/P)	(A/F)	(A/G)	N
1	.9901	0.9901	-0.0000	1.0100	1.0000	1.0100	1.0000	-0.0000	1
2	.9803	1.9704	0.9803	1.0201	2.0100	0.5075	0.4975	0.4975	2
3	.9706	2.9410	2.9215	1.0303	3.0301	0.3400	0.3300	0.9934	3
4	.9610	3.9020	5.8044	1.0406	4.0604	0.2563	0.2463	1.4876	4
5	.9515	4.8534	9.6103	1.0510	5.1010	0.2060	0.1960	1.9801	5
6	.9420	5.7955	14.3205	1.0615	6.1520	0.1725	0.1625	2.4710	6
7	.9327	6.7282	19.9168	1.0721	7.2135	0.1486	0.1386	2.9602	7
8	.9235	7.6517	26.3812	1.0829	8.2857	0.1307	0.1207	3.4478	8
9	.9143	8.5660	33.6959	1.0937	9.3685	0.1167	0.1067	3.9337	9
10	.9053	9.4713	41.8435	1.1046	10.4622	0.1056	0.0956	4.4179	10
11	.8963	10.3676	50.8067	1.1157	11.5668	0.0965	0.0865	4.9005	11
12	.8874	11.2551	60.5687	1.1268	12.6825	0.0888	0.0788	5.3815	12
13	.8787	12.1337	71.1126	1.1381	13.8093	0.0824	0.0724	5.8607	13
14	.8700	13.0037	82.4221	1.1495	14.9474	0.0769	0.0669	6.3384	14
15	.8613	13.8651	94.4810	1.1610	16.0969	0.0721	0.0621	6.8143	15
16	.8528	14.7179	107.2734	1.1726	17.2579	0.0679	0.0579	7.2886	16
17	.8444	15.5623	120.7834	1.1843	18.4304	0.0643	0.0543	7.7613	17
18	.8360	16.3983	134.9957	1.1961	19.6147	0.0610	0.0510	8.2323	18
19	.8277	17.2260	149.8950	1.2081	20.8109	0.0581	0.0481	8.7017	19
20	.8195	18.0456	165.4664	1.2202	22.0190	0.0554	0.0454	9.1694	20
21	.8114	18.8570	181.6950	1.2324	23.2392	0.0530	0.0430	9.6354	21
22	.8034	19.6604	198.5663	1.2447	24.4716	0.0509	0.0409	10.0998	22
23	.7954	20.4558	216.0660	1.2572	25.7163	0.0489	0.0389	10.5626	23
24	.7876	21.2434	234.1800	1.2697	26.9735	0.0471	0.0371	11.0237	24
25	.7798	22.0232	252.8945	1.2824	28.2432	0.0454	0.0354	11.4831	25
26	.7720	22.7952	272.1957	1.2953	29.5256	0.0439	0.0339	11.9409	26
27	.7644	23.5596	292.0702	1.3082	30.8209	0.0424	0.0324	12.3971	27
28	.7568	24.3164	312.5047	1.3213	32.1291	0.0411	0.0311	12.8516	28
29	.7493	25.0658	333.4863	1.3345	33.4504	0.0399	0.0299	13.3044	29
30	.7419	25.8077	355.0021	1.3478	34.7849	0.0387	0.0287	13.7557	30
31	.7346	26.5423	377.0394	1.3613	36.1327	0.0377	0.0277	14.2052	31
32	.7273	27.2696	399.5858	1.3749	37.4941	0.0367	0.0267	14.6532	32
33	.7201	27.9897	422.6291	1.3887	38.8690	0.0357	0.0257	15.0995	33
34	.7130	28.7027	446.1572	1.4026	40.2577	0.0348	0.0248	15.5441	34
35	.7059	29.4086	470.1583	1.4166	41.6603	0.0340	0.0240	15.9871	35
36	.6989	30.1075	494.6207	1.4308	43.0769	0.0332	0.0232	16.4285	36
37	.6920	30.7995	519.5329	1.4451	44.5076	0.0325	0.0225	16.8682	37
38	.6852	31.4847	544.8835	1.4595	45.9527	0.0318	0.0218	17.3063	38
39	.6784	32.1630	570.6616	1.4741	47.4123	0.0311	0.0211	17.7428	39
40	.6717	32.8347	596.8561	1.4889	48.8864	0.0305	0.0205	18.1776	40
41	.6650	33.4997	623.4562	1.5038	50.3752	0.0299	0.0199	18.6108	41
42	.6584	34.1581	650.4514	1.5188	51.8790	0.0293	0.0193	19.0424	42
43	.6519	34.8100	677.8312	1.5340	53.3978	0.0287	0.0187	19.4723	43
44	.6454	35.4555	705.5853	1.5493	54.9318	0.0282	0.0182	19.9006	44
45	.6391	36.0945	733.7037	1.5648	56.4811	0.0277	0.0177	20.3273	45
46	.6327	36.7272	762.1765	1.5805	58.0459	0.0272	0.0172	20.7524	46
47	.6265	37.3537	790.9938	1.5963	59.6263	0.0268	0.0168	21.1758	47
48	.6203	37.9740	820.1460	1.6122	61.2226	0.0263	0.0163	21.5976	48
49	.6141	38.5881	849.6237	1.6283	62.8348	0.0259	0.0159	22.0178	49
50	.6080	39.1961	879.4176	1.6446	64.4632	0.0255	0.0155	22.4363	50
51	.6020	39.7981	909.5186	1.6611	66.1078	0.0251	0.0151	22.8533	51
52	.5961	40.3942	939.9175	1.6777	67.7689	0.0248	0.0148	23.2686	52
53	.5902	40.9844	970.6057	1.6945	69.4466	0.0244	0.0144	23.6823	53
54	.5843	41.5687	1001.5743	1.7114	71.1410	0.0241	0.0141	24.0945	54
55	.5785	42.1472	1032.8148	1.7285	72.8525	0.0237	0.0137	24.5049	55
60	.5504	44.9550	1192.8061	1.8167	81.6697	0.0222	0.0122	26.5333	60
65	.5237	47.6266	1358.3903	1.9094	90.9366	0.0210	0.0110	28.5217	65
70	.4983	50.1685	1528.6474	2.0068	100.6763	0.0199	0.0099	30.4703	70
75	.4741	52.5871	1702.7340	2.1091	110.9128	0.0190	0.0090	32.3793	75
80	.4511	54.8882	1879.8771	2.2167	121.6715	0.0182	0.0082	34.2492	80
85	.4292	57.0777	2059.3701	2.3298	132.9790	0.0175	0.0075	36.0801	85
90	.4084	59.1609	2240.5675	2.4486	144.8633	0.0169	0.0069	37.8724	90
95	.3886	61.1430	2422.8811	2.5735	157.3538	0.0164	0.0064	39.6265	95
100	.3697	63.0289	2605.7758	2.7048	170.4814	0.0159	0.0059	41.3426	100

I = 1.50 %

N	(P/F)	(P/A)	(P/G)	(F/P)	(F/A)	(A/P)	(A/F)	(A/G)	N
1	.9852	0.9852	-0.0000	1.0150	1.0000	1.0150	1.0000	-0.0000	1
2	.9707	1.9559	0.9707	1.0302	2.0150	0.5113	0.4963	0.4963	2
3	.9563	2.9122	2.8833	1.0457	3.0452	0.3434	0.3284	0.9901	3
4	.9422	3.8544	5.7098	1.0614	4.0909	0.2594	0.2444	1.4814	4
5	.9283	4.7826	9.4229	1.0773	5.1523	0.2091	0.1941	1.9702	5
6	.9145	5.6972	13.9956	1.0934	6.2296	0.1755	0.1605	2.4566	6
7	.9010	6.5982	19.4018	1.1098	7.3230	0.1516	0.1366	2.9405	7
8	.8877	7.4859	25.6157	1.1265	8.4328	0.1336	0.1186	3.4219	8
9	.8746	8.3605	32.6125	1.1434	9.5593	0.1196	0.1046	3.9008	9
10	.8617	9.2222	40.3675	1.1605	10.7027	0.1084	0.0934	4.3772	10
11	.8489	10.0711	48.8568	1.1779	11.8633	0.0993	0.0843	4.8512	11
12	.8364	10.9075	58.0571	1.1956	13.0412	0.0917	0.0767	5.3227	12
13	.8240	11.7315	67.9454	1.2136	14.2368	0.0852	0.0702	5.7917	13
14	.8118	12.5434	78.4994	1.2318	15.4504	0.0797	0.0647	6.2582	14
15	.7999	13.3432	89.6974	1.2502	16.6821	0.0749	0.0599	6.7223	15
16	.7880	14.1313	101.5178	1.2690	17.9324	0.0708	0.0558	7.1839	16
17	.7764	14.9076	113.9400	1.2880	19.2014	0.0671	0.0521	7.6431	17
18	.7649	15.6726	126.9435	1.3073	20.4894	0.0638	0.0488	8.0997	18
19	.7536	16.4262	140.5084	1.3270	21.7967	0.0609	0.0459	8.5539	19
20	.7425	17.1686	154.6154	1.3469	23.1237	0.0582	0.0432	9.0057	20
21	.7315	17.9001	169.2453	1.3671	24.4705	0.0559	0.0409	9.4550	21
22	.7207	18.6208	184.3798	1.3876	25.8376	0.0537	0.0387	9.9018	22
23	.7100	19.3309	200.0006	1.4084	27.2251	0.0517	0.0367	10.3462	23
24	.6995	20.0304	216.0901	1.4295	28.6335	0.0499	0.0349	10.7881	24
25	.6892	20.7196	232.6310	1.4509	30.0630	0.0483	0.0333	11.2276	25
26	.6790	21.3986	249.6065	1.4727	31.5140	0.0467	0.0317	11.6646	26
27	.6690	22.0676	267.0002	1.4948	32.9867	0.0453	0.0303	12.0992	27
28	.6591	22.7267	284.7958	1.5172	34.4815	0.0440	0.0290	12.5313	28
29	.6494	23.3761	302.9779	1.5400	35.9987	0.0428	0.0278	12.9610	29
30	.6398	24.0158	321.5310	1.5631	37.5387	0.0416	0.0266	13.3883	30
31	.6303	24.6461	340.4402	1.5865	39.1018	0.0406	0.0256	13.8131	31
32	.6210	25.2671	359.6910	1.6103	40.6883	0.0396	0.0246	14.2355	32
33	.6118	25.8790	379.2691	1.6345	42.2986	0.0386	0.0236	14.6555	33
34	.6028	26.4817	399.1607	1.6590	43.9331	0.0378	0.0228	15.0731	34
35	.5939	27.0756	419.3521	1.6839	45.5921	0.0369	0.0219	15.4882	35
36	.5851	27.6607	439.8303	1.7091	47.2760	0.0362	0.0212	15.9009	36
37	.5764	28.2371	460.5822	1.7348	48.9851	0.0354	0.0204	16.3112	37
38	.5679	28.8051	481.5954	1.7608	50.7199	0.0347	0.0197	16.7191	38
39	.5595	29.3646	502.8576	1.7872	52.4807	0.0341	0.0191	17.1246	39
40	.5513	29.9158	524.3568	1.8140	54.2679	0.0334	0.0184	17.5277	40
41	.5431	30.4590	546.0814	1.8412	56.0819	0.0328	0.0178	17.9284	41
42	.5351	30.9941	568.0201	1.8688	57.9231	0.0323	0.0173	18.3267	42
43	.5272	31.5212	590.1617	1.8969	59.7920	0.0317	0.0167	18.7227	43
44	.5194	32.0406	612.4955	1.9253	61.6889	0.0312	0.0162	19.1162	44
45	.5117	32.5523	635.0110	1.9542	63.6142	0.0307	0.0157	19.5074	45
46	.5042	33.0565	657.6979	1.9835	65.5684	0.0303	0.0153	19.8962	46
47	.4967	33.5532	680.5462	2.0133	67.5519	0.0298	0.0148	20.2826	47
48	.4894	34.0426	703.5462	2.0435	69.5652	0.0294	0.0144	20.6667	48
49	.4821	34.5247	726.6884	2.0741	71.6087	0.0290	0.0140	21.0484	49
50	.4750	34.9997	749.9636	2.1052	73.6828	0.0286	0.0136	21.4277	50
51	.4680	35.4677	773.3629	2.1368	75.7881	0.0282	0.0132	21.8047	51
52	.4611	35.9287	796.8774	2.1689	77.9249	0.0278	0.0128	22.1794	52
53	.4543	36.3830	820.4986	2.2014	80.0938	0.0275	0.0125	22.5517	53
54	.4475	36.8305	844.2184	2.2344	82.2952	0.0272	0.0122	22.9217	54
55	.4409	37.2715	868.0285	2.2679	84.5296	0.0268	0.0118	23.2894	55
60	.4093	39.3803	988.1674	2.4432	96.2147	0.0254	0.0104	25.0930	60
65	.3799	41.3378	1109.4752	2.6320	108.8028	0.0242	0.0092	26.8393	65
70	.3527	43.1549	1231.1658	2.8355	122.3638	0.0232	0.0082	28.5290	70
75	.3274	44.8416	1352.5600	3.0546	136.9728	0.0223	0.0073	30.1631	75
80	.3039	46.4073	1473.0741	3.2907	152.7109	0.0215	0.0065	31.7423	80
85	.2821	47.8607	1592.2095	3.5450	169.6652	0.0209	0.0059	33.2676	85
90	.2619	49.2099	1709.5439	3.8189	187.9299	0.0203	0.0053	34.7399	90
95	.2431	50.4622	1824.7224	4.1141	207.6061	0.0198	0.0048	36.1602	95
100	.2256	51.6247	1937.4506	4.4320	228.8030	0.0194	0.0044	37.5295	100

I = 2.00 %

N	(P/F)	(P/A)	(P/G)	(F/P)	(F/A)	(A/P)	(A/F)	(A/G)	N
1	.9804	0.9804	-0.0000	1.0200	1.0000	1.0200	1.0000	-0.0000	1
2	.9612	1.9416	0.9612	1.0404	2.0200	0.5150	0.4950	0.4950	2
3	.9423	2.8839	2.8458	1.0612	3.0604	0.3468	0.3268	0.9868	3
4	.9238	3.8077	5.6173	1.0824	4.1216	0.2626	0.2426	1.4752	4
5	.9057	4.7135	9.2403	1.1041	5.2040	0.2122	0.1922	1.9604	5
6	.8880	5.6014	13.6801	1.1262	6.3081	0.1785	0.1585	2.4423	6
7	.8706	6.4720	18.9035	1.1487	7.4343	0.1545	0.1345	2.9208	7
8	.8535	7.3255	24.8779	1.1717	8.5830	0.1365	0.1165	3.3961	8
9	.8368	8.1622	31.5720	1.1951	9.7546	0.1225	0.1025	3.8681	9
10	.8203	8.9826	38.9551	1.2190	10.9497	0.1113	0.0913	4.3367	10
11	.8043	9.7868	46.9977	1.2434	12.1687	0.1022	0.0822	4.8021	11
12	.7885	10.5753	55.6712	1.2682	13.4121	0.0946	0.0746	5.2642	12
13	.7730	11.3484	64.9475	1.2936	14.6803	0.0881	0.0681	5.7231	13
14	.7579	12.1062	74.7999	1.3195	15.9739	0.0826	0.0626	6.1786	14
15	.7430	12.8493	85.2021	1.3459	17.2934	0.0778	0.0578	6.6309	15
16	.7284	13.5777	96.1288	1.3728	18.6393	0.0737	0.0537	7.0799	16
17	.7142	14.2919	107.5554	1.4002	20.0121	0.0700	0.0500	7.5256	17
18	.7002	14.9920	119.4581	1.4282	21.4123	0.0667	0.0467	7.9681	18
19	.6864	15.6785	131.8139	1.4568	22.8406	0.0638	0.0438	8.4073	19
20	.6730	16.3514	144.6003	1.4859	24.2974	0.0612	0.0412	8.8433	20
21	.6598	17.0112	157.7959	1.5157	25.7833	0.0588	0.0388	9.2760	21
22	.6468	17.6580	171.3795	1.5460	27.2990	0.0566	0.0366	9.7055	22
23	.6342	18.2922	185.3309	1.5769	28.8450	0.0547	0.0347	10.1317	23
24	.6217	18.9139	199.6305	1.6084	30.4219	0.0529	0.0329	10.5547	24
25	.6095	19.5235	214.2592	1.6406	32.0303	0.0512	0.0312	10.9745	25
26	.5976	20.1210	229.1987	1.6734	33.6709	0.0497	0.0297	11.3910	26
27	.5859	20.7069	244.4311	1.7069	35.3443	0.0483	0.0283	11.8043	27
28	.5744	21.2813	259.9392	1.7410	37.0512	0.0470	0.0270	12.2145	28
29	.5631	21.8444	275.7064	1.7758	38.7922	0.0458	0.0258	12.6214	29
30	.5521	22.3965	291.7164	1.8114	40.5681	0.0446	0.0246	13.0251	30
31	.5412	22.9377	307.9538	1.8476	42.3794	0.0436	0.0236	13.4257	31
32	.5306	23.4683	324.4035	1.8845	44.2270	0.0426	0.0226	13.8230	32
33	.5202	23.9886	341.0508	1.9222	46.1116	0.0417	0.0217	14.2172	33
34	.5100	24.4986	357.8817	1.9607	48.0338	0.0408	0.0208	14.6083	34
35	.5000	24.9986	374.8826	1.9999	49.9945	0.0400	0.0200	14.9961	35
36	.4902	25.4888	392.0405	2.0399	51.9944	0.0392	0.0192	15.3809	36
37	.4806	25.9695	409.3424	2.0807	54.0343	0.0385	0.0185	15.7625	37
38	.4712	26.4406	426.7764	2.1223	56.1149	0.0378	0.0178	16.1409	38
39	.4619	26.9026	444.3304	2.1647	58.2372	0.0372	0.0172	16.5163	39
40	.4529	27.3555	461.9931	2.2080	60.4020	0.0366	0.0166	16.8885	40
41	.4440	27.7995	479.7535	2.2522	62.6100	0.0360	0.0160	17.2576	41
42	.4353	28.2348	497.6010	2.2972	64.8622	0.0354	0.0154	17.6237	42
43	.4268	28.6616	515.5253	2.3432	67.1595	0.0349	0.0149	17.9866	43
44	.4184	29.0800	533.5165	2.3901	69.5027	0.0344	0.0144	18.3465	44
45	.4102	29.4902	551.5652	2.4379	71.8927	0.0339	0.0139	18.7034	45
46	.4022	29.8923	569.6621	2.4866	74.3306	0.0335	0.0135	19.0571	46
47	.3943	30.2866	587.7985	2.5363	76.8172	0.0330	0.0130	19.4079	47
48	.3865	30.6731	605.9657	2.5871	79.3535	0.0326	0.0126	19.7556	48
49	.3790	31.0521	624.1557	2.6388	81.9406	0.0322	0.0122	20.1003	49
50	.3715	31.4236	642.3606	2.6916	84.5794	0.0318	0.0118	20.4420	50
51	.3642	31.7878	660.5727	2.7454	87.2710	0.0315	0.0115	20.7807	51
52	.3571	32.1449	678.7849	2.8003	90.0164	0.0311	0.0111	21.1164	52
53	.3501	32.4950	696.9900	2.8563	92.8167	0.0308	0.0108	21.4491	53
54	.3432	32.8383	715.1815	2.9135	95.6731	0.0305	0.0105	21.7789	54
55	.3365	33.1748	733.3527	2.9717	98.5865	0.0301	0.0101	22.1057	55
60	.3048	34.7609	823.6975	3.2810	114.0515	0.0288	0.0088	23.6961	60
65	.2761	36.1975	912.7085	3.6225	131.1262	0.0276	0.0076	25.2147	65
70	.2500	37.4986	999.8343	3.9996	149.9779	0.0267	0.0067	26.6632	70
75	.2265	38.6771	1084.6393	4.4158	170.7918	0.0259	0.0059	28.0434	75
80	.2051	39.7445	1166.7868	4.8754	193.7720	0.0252	0.0052	29.3572	80
85	.1858	40.7113	1246.0241	5.3829	219.1439	0.0246	0.0046	30.6064	85
90	.1683	41.5869	1322.1701	5.9431	247.1567	0.0240	0.0040	31.7929	90
95	.1524	42.3800	1395.1033	6.5617	278.0850	0.0236	0.0036	32.9189	95
100	.1380	43.0984	1464.7527	7.2446	312.2323	0.0232	0.0032	33.9863	100

I = 3.00 %

N	(P/F)	(P/A)	(P/G)	(F/P)	(F/A)	(A/P)	(A/F)	(A/G)	N
1	.9709	0.9709	-0.0000	1.0300	1.0000	1.0300	1.0000	-0.0000	1
2	.9426	1.9135	0.9426	1.0609	2.0300	0.5226	0.4926	0.4926	2
3	.9151	2.8286	2.7729	1.0927	3.0909	0.3535	0.3235	0.9803	3
4	.8885	3.7171	5.4383	1.1255	4.1836	0.2690	0.2390	1.4631	4
5	.8626	4.5797	8.8888	1.1593	5.3091	0.2184	0.1884	1.9409	5
6	.8375	5.4172	13.0762	1.1941	6.4684	0.1846	0.1546	2.4138	6
7	.8131	6.2303	17.9547	1.2299	7.6625	0.1605	0.1305	2.8819	7
8	.7894	7.0197	23.4806	1.2668	8.8923	0.1425	0.1125	3.3450	8
9	.7664	7.7861	29.6119	1.3048	10.1591	0.1284	0.0984	3.8032	9
10	.7441	8.5302	36.3088	1.3439	11.4639	0.1172	0.0872	4.2565	10
11	.7224	9.2526	43.5330	1.3842	12.8078	0.1081	0.0781	4.7049	11
12	.7014	9.9540	51.2482	1.4258	14.1920	0.1005	0.0705	5.1485	12
13	.6810	10.6350	59.4196	1.4685	15.6178	0.0940	0.0640	5.5872	13
14	.6611	11.2961	68.0141	1.5126	17.0863	0.0885	0.0585	6.0210	14
15	.6419	11.9379	77.0002	1.5580	18.5989	0.0838	0.0538	6.4500	15
16	.6232	12.5611	86.3477	1.6047	20.1569	0.0796	0.0496	6.8742	16
17	.6050	13.1661	96.0280	1.6528	21.7616	0.0760	0.0460	7.2936	17
18	.5874	13.7535	106.0137	1.7024	23.4144	0.0727	0.0427	7.7081	18
19	.5703	14.3238	116.2788	1.7535	25.1169	0.0698	0.0398	8.1179	19
20	.5537	14.8775	126.7987	1.8061	26.8704	0.0672	0.0372	8.5229	20
21	.5375	15.4150	137.5496	1.8603	28.6765	0.0649	0.0349	8.9231	21
22	.5219	15.9369	148.5094	1.9161	30.5368	0.0627	0.0327	9.3186	22
23	.5067	16.4436	159.6566	1.9736	32.4529	0.0608	0.0308	9.7093	23
24	.4919	16.9355	170.9711	2.0328	34.4265	0.0590	0.0290	10.0954	24
25	.4776	17.4131	182.4336	2.0938	36.4593	0.0574	0.0274	10.4768	25
26	.4637	17.8768	194.0260	2.1566	38.5530	0.0559	0.0259	10.8535	26
27	.4502	18.3270	205.7309	2.2213	40.7096	0.0546	0.0246	11.2255	27
28	.4371	18.7641	217.5320	2.2879	42.9309	0.0533	0.0233	11.5930	28
29	.4243	19.1885	229.4137	2.3566	45.2189	0.0521	0.0221	11.9558	29
30	.4120	19.6004	241.3613	2.4273	47.5754	0.0510	0.0210	12.3141	30
31	.4000	20.0004	253.3609	2.5001	50.0027	0.0500	0.0200	12.6678	31
32	.3883	20.3888	265.3993	2.5751	52.5028	0.0490	0.0190	13.0169	32
33	.3770	20.7658	277.4642	2.6523	55.0778	0.0482	0.0182	13.3616	33
34	.3660	21.1318	289.5437	2.7319	57.7302	0.0473	0.0173	13.7018	34
35	.3554	21.4872	301.6267	2.8139	60.4621	0.0465	0.0165	14.0375	35
36	.3450	21.8323	313.7028	2.8983	63.2759	0.0458	0.0158	14.3688	36
37	.3350	22.1672	325.7622	2.9852	66.1742	0.0451	0.0151	14.6957	37
38	.3252	22.4925	337.7956	3.0748	69.1594	0.0445	0.0145	15.0182	38
39	.3158	22.8082	349.7942	3.1670	72.2342	0.0438	0.0138	15.3363	39
40	.3066	23.1148	361.7499	3.2620	75.4013	0.0433	0.0133	15.6502	40
41	.2976	23.4124	373.6551	3.3599	78.6633	0.0427	0.0127	15.9597	41
42	.2890	23.7014	385.5024	3.4607	82.0232	0.0422	0.0122	16.2650	42
43	.2805	23.9819	397.2852	3.5645	85.4839	0.0417	0.0117	16.5660	43
44	.2724	24.2543	408.9972	3.6715	89.0484	0.0412	0.0112	16.8629	44
45	.2644	24.5187	420.6325	3.7816	92.7199	0.0408	0.0108	17.1556	45
46	.2567	24.7754	432.1856	3.8950	96.5015	0.0404	0.0104	17.4441	46
47	.2493	25.0247	443.6515	4.0119	100.3965	0.0400	0.0100	17.7285	47
48	.2420	25.2667	455.0255	4.1323	104.4084	0.0396	0.0096	18.0089	48
49	.2350	25.5017	466.3031	4.2562	108.5406	0.0392	0.0092	18.2852	49
50	.2281	25.7298	477.4803	4.3839	112.7969	0.0389	0.0089	18.5575	50
51	.2215	25.9512	488.5535	4.5154	117.1808	0.0385	0.0085	18.8258	51
52	.2150	26.1662	499.5191	4.6509	121.6962	0.0382	0.0082	19.0902	52
53	.2088	26.3750	510.3742	4.7904	126.3471	0.0379	0.0079	19.3507	53
54	.2027	26.5777	521.1157	4.9341	131.1375	0.0376	0.0076	19.6073	54
55	.1968	26.7744	531.7411	5.0821	136.0716	0.0373	0.0073	19.8600	55
60	.1697	27.6756	583.0526	5.8916	163.0534	0.0361	0.0061	21.0674	60
65	.1464	28.4529	631.2010	6.8300	194.3328	0.0351	0.0051	22.1841	65
70	.1263	29.1234	676.0869	7.9178	230.5941	0.0343	0.0043	23.2145	70
75	.1089	29.7018	717.6978	9.1789	272.6309	0.0337	0.0037	24.1634	75
80	.0940	30.2008	756.0865	10.6409	321.3630	0.0331	0.0031	25.0353	80
85	.0811	30.6312	791.3529	12.3357	377.8570	0.0326	0.0026	25.8349	85
90	.0699	31.0024	823.6302	14.3005	443.3489	0.0323	0.0023	26.5667	90
95	.0603	31.3227	853.0742	16.5782	519.2720	0.0319	0.0019	27.2351	95
100	.0520	31.5989	879.8540	19.2186	607.2877	0.0316	0.0016	27.8444	100

I = 4.00 %

N	(P/F)	(P/A)	(P/G)	(F/P)	(F/A)	(A/P)	(A/F)	(A/G)	N
1	.9615	0.9615	-0.0000	1.0400	1.0000	1.0400	1.0000	-0.0000	1
2	.9246	1.8861	0.9246	1.0816	2.0400	0.5302	0.4902	0.4902	2
3	.8890	2.7751	2.7025	1.1249	3.1216	0.3603	0.3203	0.9739	3
4	.8548	3.6299	5.2670	1.1699	4.2465	0.2755	0.2355	1.4510	4
5	.8219	4.4518	8.5547	1.2167	5.4163	0.2246	0.1846	1.9216	5
6	.7903	5.2421	12.5062	1.2653	6.6330	0.1908	0.1508	2.3857	6
7	.7599	6.0021	17.0657	1.3159	7.8983	0.1666	0.1266	2.8433	7
8	.7307	6.7327	22.1806	1.3686	9.2142	0.1485	0.1085	3.2944	8
9	.7026	7.4353	27.8013	1.4233	10.5828	0.1345	0.0945	3.7391	9
10	.6756	8.1109	33.8814	1.4802	12.0061	0.1233	0.0833	4.1773	10
11	.6496	8.7605	40.3772	1.5395	13.4864	0.1141	0.0741	4.6090	11
12	.6246	9.3851	47.2477	1.6010	15.0258	0.1066	0.0666	5.0343	12
13	.6006	9.9856	54.4546	1.6651	16.6268	0.1001	0.0601	5.4533	13
14	.5775	10.5631	61.9618	1.7317	18.2919	0.0947	0.0547	5.8659	14
15	.5553	11.1184	69.7355	1.8009	20.0236	0.0899	0.0499	6.2721	15
16	.5339	11.6523	77.7441	1.8730	21.8245	0.0858	0.0458	6.6720	16
17	.5134	12.1657	85.9581	1.9479	23.6975	0.0822	0.0422	7.0656	17
18	.4936	12.6593	94.3498	2.0258	25.6454	0.0790	0.0390	7.4530	18
19	.4746	13.1339	102.8933	2.1068	27.6712	0.0761	0.0361	7.8342	19
20	.4564	13.5903	111.5647	2.1911	29.7781	0.0736	0.0336	8.2091	20
21	.4388	14.0292	120.3414	2.2788	31.9692	0.0713	0.0313	8.5779	21
22	.4220	14.4511	129.2024	2.3699	34.2480	0.0692	0.0292	8.9407	22
23	.4057	14.8568	138.1284	2.4647	36.6179	0.0673	0.0273	9.2973	23
24	.3901	15.2470	147.1012	2.5633	39.0826	0.0656	0.0256	9.6479	24
25	.3751	15.6221	156.1040	2.6658	41.6459	0.0640	0.0240	9.9925	25
26	.3607	15.9828	165.1212	2.7725	44.3117	0.0626	0.0226	10.3312	26
27	.3468	16.3296	174.1385	2.8834	47.0842	0.0612	0.0212	10.6640	27
28	.3335	16.6631	183.1424	2.9987	49.9676	0.0600	0.0200	10.9909	28
29	.3207	16.9837	192.1206	3.1187	52.9663	0.0589	0.0189	11.3120	29
30	.3083	17.2920	201.0618	3.2434	56.0849	0.0578	0.0178	11.6274	30
31	.2965	17.5885	209.9556	3.3731	59.3283	0.0569	0.0169	11.9371	31
32	.2851	17.8736	218.7924	3.5081	62.7015	0.0559	0.0159	12.2411	32
33	.2741	18.1476	227.5634	3.6484	66.2095	0.0551	0.0151	12.5396	33
34	.2636	18.4112	236.2607	3.7943	69.8579	0.0543	0.0143	12.8324	34
35	.2534	18.6646	244.8768	3.9461	73.6522	0.0536	0.0136	13.1198	35
36	.2437	18.9083	253.4052	4.1039	77.5983	0.0529	0.0129	13.4018	36
37	.2343	19.1426	261.8399	4.2681	81.7022	0.0522	0.0122	13.6784	37
38	.2253	19.3679	270.1754	4.4388	85.9703	0.0516	0.0116	13.9497	38
39	.2166	19.5845	278.4070	4.6164	90.4091	0.0511	0.0111	14.2157	39
40	.2083	19.7928	286.5303	4.8010	95.0255	0.0505	0.0105	14.4765	40
41	.2003	19.9931	294.5414	4.9931	99.8265	0.0500	0.0100	14.7322	41
42	.1926	20.1856	302.4370	5.1928	104.8196	0.0495	0.0095	14.9828	42
43	.1852	20.3708	310.2141	5.4005	110.0124	0.0491	0.0091	15.2284	43
44	.1780	20.5488	317.8700	5.6165	115.4129	0.0487	0.0087	15.4690	44
45	.1712	20.7200	325.4028	5.8412	121.0294	0.0483	0.0083	15.7047	45
46	.1646	20.8847	332.8104	6.0748	126.8706	0.0479	0.0079	15.9356	46
47	.1583	21.0429	340.0914	6.3178	132.9454	0.0475	0.0075	16.1618	47
48	.1522	21.1951	347.2446	6.5705	139.2632	0.0472	0.0072	16.3832	48
49	.1463	21.3415	354.2689	6.8333	145.8337	0.0469	0.0069	16.6000	49
50	.1407	21.4822	361.1638	7.1067	152.6671	0.0466	0.0066	16.8122	50
51	.1353	21.6175	367.9289	7.3910	159.7738	0.0463	0.0063	17.0200	51
52	.1301	21.7476	374.5638	7.6866	167.1647	0.0460	0.0060	17.2232	52
53	.1251	21.8727	381.0686	7.9941	174.8513	0.0457	0.0057	17.4221	53
54	.1203	21.9930	387.4436	8.3138	182.8454	0.0455	0.0055	17.6167	54
55	.1157	22.1086	393.6890	8.6464	191.1592	0.0452	0.0052	17.8070	55
60	.0951	22.6235	422.9966	10.5196	237.9907	0.0442	0.0042	18.6972	60
65	.0781	23.0467	449.2014	12.7987	294.9684	0.0434	0.0034	19.4909	65
70	.0642	23.3945	472.4789	15.5716	364.2905	0.0427	0.0027	20.1961	70
75	.0528	23.6804	493.0408	18.9453	448.6314	0.0422	0.0022	20.8206	75
80	.0434	23.9154	511.1161	23.0498	551.2450	0.0418	0.0018	21.3718	80
85	.0357	24.1085	526.9384	28.0436	676.0901	0.0415	0.0015	21.8569	85
90	.0293	24.2673	540.7369	34.1193	827.9833	0.0412	0.0012	22.2826	90
95	.0241	24.3978	552.7307	41.5114	1012.7846	0.0410	0.0010	22.6550	95
100	.0198	24.5050	563.1249	50.5049	1237.6237	0.0408	0.0008	22.9800	100

I = 5.00 %

N	(P/F)	(P/A)	(P/G)	(F/P)	(F/A)	(A/P)	(A/F)	(A/G)	N
1	.9524	0.9524	-0.0000	1.0500	1.0000	1.0500	1.0000	-0.0000	1
2	.9070	1.8594	0.9070	1.1025	2.0500	0.5378	0.4878	0.4878	2
3	.8638	2.7232	2.6347	1.1576	3.1525	0.3672	0.3172	0.9675	3
4	.8227	3.5460	5.1028	1.2155	4.3101	0.2820	0.2320	1.4391	4
5	.7835	4.3295	8.2369	1.2763	5.5256	0.2310	0.1810	1.9025	5
6	.7462	5.0757	11.9680	1.3401	6.8019	0.1970	0.1470	2.3579	6
7	.7107	5.7864	16.2321	1.4071	8.1420	0.1728	0.1228	2.8052	7
8	.6768	6.4632	20.9700	1.4775	9.5491	0.1547	0.1047	3.2445	8
9	.6446	7.1078	26.1268	1.5513	11.0266	0.1407	0.0907	3.6758	9
10	.6139	7.7217	31.6520	1.6289	12.5779	0.1295	0.0795	4.0991	10
11	.5847	8.3064	37.4988	1.7103	14.2068	0.1204	0.0704	4.5144	11
12	.5568	8.8633	43.6241	1.7959	15.9171	0.1128	0.0628	4.9219	12
13	.5303	9.3936	49.9879	1.8856	17.7130	0.1065	0.0565	5.3215	13
14	.5051	9.8986	56.5538	1.9799	19.5986	0.1010	0.0510	5.7133	14
15	.4810	10.3797	63.2880	2.0789	21.5786	0.0963	0.0463	6.0973	15
16	.4581	10.8378	70.1597	2.1829	23.6575	0.0923	0.0423	6.4736	16
17	.4363	11.2741	77.1405	2.2920	25.8404	0.0887	0.0387	6.8423	17
18	.4155	11.6896	84.2043	2.4066	28.1324	0.0855	0.0355	7.2034	18
19	.3957	12.0853	91.3275	2.5270	30.5390	0.0827	0.0327	7.5569	19
20	.3769	12.4622	98.4884	2.6533	33.0660	0.0802	0.0302	7.9030	20
21	.3589	12.8212	105.6673	2.7860	35.7193	0.0780	0.0280	8.2416	21
22	.3418	13.1630	112.8461	2.9253	38.5052	0.0760	0.0260	8.5730	22
23	.3256	13.4886	120.0087	3.0715	41.4305	0.0741	0.0241	8.8971	23
24	.3101	13.7986	127.1402	3.2251	44.5020	0.0725	0.0225	9.2140	24
25	.2953	14.0939	134.2275	3.3864	47.7271	0.0710	0.0210	9.5238	25
26	.2812	14.3752	141.2585	3.5557	51.1135	0.0696	0.0196	9.8266	26
27	.2678	14.6430	148.2226	3.7335	54.6691	0.0683	0.0183	10.1224	27
28	.2551	14.8981	155.1101	3.9201	58.4026	0.0671	0.0171	10.4114	28
29	.2429	15.1411	161.9126	4.1161	62.3227	0.0660	0.0160	10.6936	29
30	.2314	15.3725	168.6226	4.3219	66.4388	0.0651	0.0151	10.9691	30
31	.2204	15.5928	175.2333	4.5380	70.7608	0.0641	0.0141	11.2381	31
32	.2099	15.8027	181.7392	4.7649	75.2988	0.0633	0.0133	11.5005	32
33	.1999	16.0025	188.1351	5.0032	80.0638	0.0625	0.0125	11.7566	33
34	.1904	16.1929	194.4168	5.2533	85.0670	0.0618	0.0118	12.0063	34
35	.1813	16.3742	200.5807	5.5160	90.3203	0.0611	0.0111	12.2498	35
36	.1727	16.5469	206.6237	5.7918	95.8363	0.0604	0.0104	12.4872	36
37	.1644	16.7113	212.5434	6.0814	101.6281	0.0598	0.0098	12.7186	37
38	.1566	16.8679	218.3378	6.3855	107.7095	0.0593	0.0093	12.9440	38
39	.1491	17.0170	224.0054	6.7048	114.0950	0.0588	0.0088	13.1636	39
40	.1420	17.1591	229.5452	7.0400	120.7998	0.0583	0.0083	13.3775	40
41	.1353	17.2944	234.9564	7.3920	127.8398	0.0578	0.0078	13.5857	41
42	.1288	17.4232	240.2389	7.7616	135.2318	0.0574	0.0074	13.7884	42
43	.1227	17.5459	245.3925	8.1497	142.9933	0.0570	0.0070	13.9857	43
44	.1169	17.6628	250.4175	8.5572	151.1430	0.0566	0.0066	14.1777	44
45	.1113	17.7741	255.3145	8.9850	159.7002	0.0563	0.0063	14.3644	45
46	.1060	17.8801	260.0844	9.4343	168.6852	0.0559	0.0059	14.5461	46
47	.1009	17.9810	264.7281	9.9060	178.1194	0.0556	0.0056	14.7226	47
48	.0961	18.0772	269.2467	10.4013	188.0254	0.0553	0.0053	14.8943	48
49	.0916	18.1687	273.6418	10.9213	198.4267	0.0550	0.0050	15.0611	49
50	.0872	18.2559	277.9148	11.4674	209.3480	0.0548	0.0048	15.2233	50
51	.0831	18.3390	282.0673	12.0408	220.8154	0.0545	0.0045	15.3808	51
52	.0791	18.4181	286.1013	12.6428	232.8562	0.0543	0.0043	15.5337	52
53	.0753	18.4934	290.0184	13.2749	245.4990	0.0541	0.0041	15.6823	53
54	.0717	18.5651	293.8208	13.9387	258.7739	0.0539	0.0039	15.8265	54
55	.0683	18.6335	297.5104	14.6356	272.7126	0.0537	0.0037	15.9664	55
60	.0535	18.9293	314.3432	18.6792	353.5837	0.0528	0.0028	16.6062	60
65	.0419	19.1611	328.6910	23.8399	456.7980	0.0522	0.0022	17.1541	65
70	.0329	19.3427	340.8409	30.4264	588.5285	0.0517	0.0017	17.6212	70
75	.0258	19.4850	351.0721	38.8327	756.6537	0.0513	0.0013	18.0176	75
80	.0202	19.5965	359.6460	49.5614	971.2288	0.0510	0.0010	18.3526	80
85	.0158	19.6838	366.8007	63.2544	1245.0871	0.0508	0.0008	18.6346	85
90	.0124	19.7523	372.7488	80.7304	1594.6073	0.0506	0.0006	18.8712	90
95	.0097	19.8059	377.6774	103.0347	2040.6935	0.0505	0.0005	19.0689	95
100	.0076	19.8479	381.7492	131.5013	2610.0252	0.0504	0.0004	19.2337	100

I = 6.00 %

N	(P/F)	(P/A)	(P/G)	(F/P)	(F/A)	(A/P)	(A/F)	(A/G)	N
1	.9434	0.9434	-0.0000	1.0600	1.0000	1.0600	1.0000	-0.0000	1
2	.8900	1.8334	0.8900	1.1236	2.0600	0.5454	0.4854	0.4854	2
3	.8396	2.6730	2.5692	1.1910	3.1836	0.3741	0.3141	0.9612	3
4	.7921	3.4651	4.9455	1.2625	4.3746	0.2886	0.2286	1.4272	4
5	.7473	4.2124	7.9345	1.3382	5.6371	0.2374	0.1774	1.8836	5
6	.7050	4.9173	11.4594	1.4185	6.9753	0.2034	0.1434	2.3304	6
7	.6651	5.5824	15.4497	1.5036	8.3938	0.1791	0.1191	2.7676	7
8	.6274	6.2098	19.8416	1.5938	9.8975	0.1610	0.1010	3.1952	8
9	.5919	6.8017	24.5768	1.6895	11.4913	0.1470	0.0870	3.6133	9
10	.5584	7.3601	29.6023	1.7908	13.1808	0.1359	0.0759	4.0220	10
11	.5268	7.8869	34.8702	1.8983	14.9716	0.1268	0.0668	4.4213	11
12	.4970	8.3838	40.3369	2.0122	16.8699	0.1193	0.0593	4.8113	12
13	.4688	8.8527	45.9629	2.1329	18.8821	0.1130	0.0530	5.1920	13
14	.4423	9.2950	51.7128	2.2609	21.0151	0.1076	0.0476	5.5635	14
15	.4173	9.7122	57.5546	2.3966	23.2760	0.1030	0.0430	5.9260	15
16	.3936	10.1059	63.4592	2.5404	25.6725	0.0990	0.0390	6.2794	16
17	.3714	10.4773	69.4011	2.6928	28.2129	0.0954	0.0354	6.6240	17
18	.3503	10.8276	75.3569	2.8543	30.9057	0.0924	0.0324	6.9597	18
19	.3305	11.1581	81.3062	3.0256	33.7600	0.0896	0.0296	7.2867	19
20	.3118	11.4699	87.2304	3.2071	36.7856	0.0872	0.0272	7.6051	20
21	.2942	11.7641	93.1136	3.3996	39.9927	0.0850	0.0250	7.9151	21
22	.2775	12.0416	98.9412	3.6035	43.3923	0.0830	0.0230	8.2166	22
23	.2618	12.3034	104.7007	3.8197	46.9958	0.0813	0.0213	8.5099	23
24	.2470	12.5504	110.3812	4.0489	50.8156	0.0797	0.0197	8.7951	24
25	.2330	12.7834	115.9732	4.2919	54.8645	0.0782	0.0182	9.0722	25
26	.2198	13.0032	121.4684	4.5494	59.1564	0.0769	0.0169	9.3414	26
27	.2074	13.2105	126.8600	4.8223	63.7058	0.0757	0.0157	9.6029	27
28	.1956	13.4062	132.1420	5.1117	68.5281	0.0746	0.0146	9.8568	28
29	.1846	13.5907	137.3096	5.4184	73.6398	0.0736	0.0136	10.1032	29
30	.1741	13.7648	142.3588	5.7435	79.0582	0.0726	0.0126	10.3422	30
31	.1643	13.9291	147.2864	6.0881	84.8017	0.0718	0.0118	10.5740	31
32	.1550	14.0840	152.0901	6.4534	90.8898	0.0710	0.0110	10.7988	32
33	.1462	14.2302	156.7681	6.8406	97.3432	0.0703	0.0103	11.0166	33
34	.1379	14.3681	161.3192	7.2510	104.1838	0.0696	0.0096	11.2276	34
35	.1301	14.4982	165.7427	7.6861	111.4348	0.0690	0.0090	11.4319	35
36	.1227	14.6210	170.0387	8.1473	119.1209	0.0684	0.0084	11.6298	36
37	.1158	14.7368	174.2072	8.6361	127.2681	0.0679	0.0079	11.8213	37
38	.1092	14.8460	178.2490	9.1543	135.9042	0.0674	0.0074	12.0065	38
39	.1031	14.9491	182.1652	9.7035	145.0585	0.0669	0.0069	12.1857	39
40	.0972	15.0463	185.9568	10.2857	154.7620	0.0665	0.0065	12.3590	40
41	.0917	15.1380	189.6256	10.9029	165.0477	0.0661	0.0061	12.5264	41
42	.0865	15.2245	193.1732	11.5570	175.9505	0.0657	0.0057	12.6883	42
43	.0816	15.3062	196.6017	12.2505	187.5076	0.0653	0.0053	12.8446	43
44	.0770	15.3832	199.9130	12.9855	199.7580	0.0650	0.0050	12.9956	44
45	.0727	15.4558	203.1096	13.7646	212.7435	0.0647	0.0047	13.1413	45
46	.0685	15.5244	206.1938	14.5905	226.5081	0.0644	0.0044	13.2819	46
47	.0647	15.5890	209.1681	15.4659	241.0986	0.0641	0.0041	13.4177	47
48	.0610	15.6500	212.0351	16.3939	256.5645	0.0639	0.0039	13.5485	48
49	.0575	15.7076	214.7972	17.3775	272.9584	0.0637	0.0037	13.6748	49
50	.0543	15.7619	217.4574	18.4202	290.3359	0.0634	0.0034	13.7964	50
51	.0512	15.8131	220.0181	19.5254	308.7561	0.0632	0.0032	13.9137	51
52	.0483	15.8614	222.4823	20.6969	328.2814	0.0630	0.0030	14.0267	52
53	.0456	15.9070	224.8525	21.9387	348.9783	0.0629	0.0029	14.1355	53
54	.0430	15.9500	227.1316	23.2550	370.9170	0.0627	0.0027	14.2402	54
55	.0406	15.9905	229.3222	24.6503	394.1720	0.0625	0.0025	14.3411	55
60	.0303	16.1614	239.0428	32.9877	533.1282	0.0619	0.0019	14.7909	60
65	.0227	16.2891	246.9450	44.1450	719.0829	0.0614	0.0014	15.1601	65
70	.0169	16.3845	253.3271	59.0759	967.9322	0.0610	0.0010	15.4613	70
75	.0126	16.4558	258.4527	79.0569	1300.9487	0.0608	0.0008	15.7058	75
80	.0095	16.5091	262.5493	105.7960	1746.5999	0.0606	0.0006	15.9033	80
85	.0071	16.5489	265.8096	141.5789	2342.9817	0.0604	0.0004	16.0620	85
90	.0053	16.5787	268.3946	189.4645	3141.0752	0.0603	0.0003	16.1891	90
95	.0039	16.6009	270.4375	253.5463	4209.1042	0.0602	0.0002	16.2905	95
100	.0029	16.6175	272.0471	339.3021	5638.3681	0.0602	0.0002	16.3711	100

I = 7.00 %

N	(P/F)	(P/A)	(P/G)	(F/P)	(F/A)	(A/P)	(A/F)	(A/G)	N
1	.9346	0.9346	-0.0000	1.0700	1.0000	1.0700	1.0000	-0.0000	1
2	.8734	1.8080	0.8734	1.1449	2.0700	0.5531	0.4831	0.4831	2
3	.8163	2.6243	2.5060	1.2250	3.2149	0.3811	0.3111	0.9549	3
4	.7629	3.3872	4.7947	1.3108	4.4399	0.2952	0.2252	1.4155	4
5	.7130	4.1002	7.6467	1.4026	5.7507	0.2439	0.1739	1.8650	5
6	.6663	4.7665	10.9784	1.5007	7.1533	0.2098	0.1398	2.3032	6
7	.6227	5.3893	14.7149	1.6058	8.6540	0.1856	0.1156	2.7304	7
8	.5820	5.9713	18.7889	1.7182	10.2598	0.1675	0.0975	3.1465	8
9	.5439	6.5152	23.1404	1.8385	11.9780	0.1535	0.0835	3.5517	9
10	.5083	7.0236	27.7156	1.9672	13.8164	0.1424	0.0724	3.9461	10
11	.4751	7.4987	32.4665	2.1049	15.7836	0.1334	0.0634	4.3296	11
12	.4440	7.9427	37.3506	2.2522	17.8885	0.1259	0.0559	4.7025	12
13	.4150	8.3577	42.3302	2.4098	20.1406	0.1197	0.0497	5.0648	13
14	.3878	8.7455	47.3718	2.5785	22.5505	0.1143	0.0443	5.4167	14
15	.3624	9.1079	52.4461	2.7590	25.1290	0.1098	0.0398	5.7583	15
16	.3387	9.4466	57.5271	2.9522	27.8881	0.1059	0.0359	6.0897	16
17	.3166	9.7632	62.5923	3.1588	30.8402	0.1024	0.0324	6.4110	17
18	.2959	10.0591	67.6219	3.3799	33.9990	0.0994	0.0294	6.7225	18
19	.2765	10.3356	72.5991	3.6165	37.3790	0.0968	0.0268	7.0242	19
20	.2584	10.5940	77.5091	3.8697	40.9955	0.0944	0.0244	7.3163	20
21	.2415	10.8355	82.3393	4.1406	44.8652	0.0923	0.0223	7.5990	21
22	.2257	11.0612	87.0793	4.4304	49.0057	0.0904	0.0204	7.8725	22
23	.2109	11.2722	91.7201	4.7405	53.4361	0.0887	0.0187	8.1369	23
24	.1971	11.4693	96.2545	5.0724	58.1767	0.0872	0.0172	8.3923	24
25	.1842	11.6536	100.6765	5.4274	63.2490	0.0858	0.0158	8.6391	25
26	.1722	11.8258	104.9814	5.8074	68.6765	0.0846	0.0146	8.8773	26
27	.1609	11.9867	109.1656	6.2139	74.4838	0.0834	0.0134	9.1072	27
28	.1504	12.1371	113.2264	6.6488	80.6977	0.0824	0.0124	9.3289	28
29	.1406	12.2777	117.1622	7.1143	87.3465	0.0814	0.0114	9.5427	29
30	.1314	12.4090	120.9718	7.6123	94.4608	0.0806	0.0106	9.7487	30
31	.1228	12.5318	124.6550	8.1451	102.0730	0.0798	0.0098	9.9471	31
32	.1147	12.6466	128.2120	8.7153	110.2182	0.0791	0.0091	10.1381	32
33	.1072	12.7538	131.6435	9.3253	118.9334	0.0784	0.0084	10.3219	33
34	.1002	12.8540	134.9507	9.9781	128.2588	0.0778	0.0078	10.4987	34
35	.0937	12.9477	138.1353	10.6766	138.2369	0.0772	0.0072	10.6687	35
36	.0875	13.0352	141.1990	11.4239	148.9135	0.0767	0.0067	10.8321	36
37	.0818	13.1170	144.1441	12.2236	160.3374	0.0762	0.0062	10.9891	37
38	.0765	13.1935	146.9730	13.0793	172.5610	0.0758	0.0058	11.1398	38
39	.0715	13.2649	149.6883	13.9948	185.6403	0.0754	0.0054	11.2845	39
40	.0668	13.3317	152.2928	14.9745	199.6351	0.0750	0.0050	11.4233	40
41	.0624	13.3941	154.7892	16.0227	214.6096	0.0747	0.0047	11.5565	41
42	.0583	13.4524	157.1807	17.1443	230.6322	0.0743	0.0043	11.6842	42
43	.0545	13.5070	159.4702	18.3444	247.7765	0.0740	0.0040	11.8065	43
44	.0509	13.5579	161.6609	19.6285	266.1209	0.0738	0.0038	11.9237	44
45	.0476	13.6055	163.7559	21.0025	285.7493	0.0735	0.0035	12.0360	45
46	.0445	13.6500	165.7584	22.4726	306.7518	0.0733	0.0033	12.1435	46
47	.0416	13.6916	167.6714	24.0457	329.2244	0.0730	0.0030	12.2463	47
48	.0389	13.7305	169.4981	25.7289	353.2701	0.0728	0.0028	12.3447	48
49	.0363	13.7668	171.2417	27.5299	378.9990	0.0726	0.0026	12.4387	49
50	.0339	13.8007	172.9051	29.4570	406.5289	0.0725	0.0025	12.5287	50
51	.0317	13.8325	174.4915	31.5190	435.9860	0.0723	0.0023	12.6146	51
52	.0297	13.8621	176.0037	33.7253	467.5050	0.0721	0.0021	12.6967	52
53	.0277	13.8898	177.4447	36.0861	501.2303	0.0720	0.0020	12.7751	53
54	.0259	13.9157	178.8173	38.6122	537.3164	0.0719	0.0019	12.8500	54
55	.0242	13.9399	180.1243	41.3150	575.9286	0.0717	0.0017	12.9215	55
60	.0173	14.0392	185.7677	57.9464	813.5204	0.0712	0.0012	13.2321	60
65	.0123	14.1099	190.1452	81.2729	1146.7552	0.0709	0.0009	13.4760	65
70	.0088	14.1604	193.5185	113.9894	1614.1342	0.0706	0.0006	13.6662	70
75	.0063	14.1964	196.1035	159.8760	2269.6574	0.0704	0.0004	13.8136	75
80	.0045	14.2220	198.0748	224.2344	3189.0627	0.0703	0.0003	13.9273	80
85	.0032	14.2403	199.5717	314.5003	4478.5761	0.0702	0.0002	14.0146	85
90	.0023	14.2533	200.7042	441.1030	6287.1854	0.0702	0.0002	14.0812	90
95	.0016	14.2626	201.5581	618.6697	8823.8535	0.0701	0.0001	14.1319	95
100	.0012	14.2693	202.2001	867.7163	12381.6618	0.0701	0.0001	14.1703	100

PROFESSIONAL PUBLICATIONS ● Belmont, CA

I = 8.00 %

N	(P/F)	(P/A)	(P/G)	(F/P)	(F/A)	(A/P)	(A/F)	(A/G)	N
1	.9259	0.9259	-0.0000	1.0800	1.0000	1.0800	1.0000	-0.0000	1
2	.8573	1.7833	0.8573	1.1664	2.0800	0.5608	0.4808	0.4808	2
3	.7938	2.5771	2.4450	1.2597	3.2464	0.3880	0.3080	0.9487	3
4	.7350	3.3121	4.6501	1.3605	4.5061	0.3019	0.2219	1.4040	4
5	.6806	3.9927	7.3724	1.4693	5.8666	0.2505	0.1705	1.8465	5
6	.6302	4.6229	10.5233	1.5869	7.3359	0.2163	0.1363	2.2763	6
7	.5835	5.2064	14.0242	1.7138	8.9228	0.1921	0.1121	2.6937	7
8	.5403	5.7466	17.8061	1.8509	10.6366	0.1740	0.0940	3.0985	8
9	.5002	6.2469	21.8081	1.9990	12.4876	0.1601	0.0801	3.4910	9
10	.4632	6.7101	25.9768	2.1589	14.4866	0.1490	0.0690	3.8713	10
11	.4289	7.1390	30.2657	2.3316	16.6455	0.1401	0.0601	4.2395	11
12	.3971	7.5361	34.6339	2.5182	18.9771	0.1327	0.0527	4.5957	12
13	.3677	7.9038	39.0463	2.7196	21.4953	0.1265	0.0465	4.9402	13
14	.3405	8.2442	43.4723	2.9372	24.2149	0.1213	0.0413	5.2731	14
15	.3152	8.5595	47.8857	3.1722	27.1521	0.1168	0.0368	5.5945	15
16	.2919	8.8514	52.2640	3.4259	30.3243	0.1130	0.0330	5.9046	16
17	.2703	9.1216	56.5883	3.7000	33.7502	0.1096	0.0296	6.2037	17
18	.2502	9.3719	60.8426	3.9960	37.4502	0.1067	0.0267	6.4920	18
19	.2317	9.6036	65.0134	4.3157	41.4463	0.1041	0.0241	6.7697	19
20	.2145	9.8181	69.0898	4.6610	45.7620	0.1019	0.0219	7.0369	20
21	.1987	10.0168	73.0629	5.0338	50.4229	0.0998	0.0198	7.2940	21
22	.1839	10.2007	76.9257	5.4365	55.4568	0.0980	0.0180	7.5412	22
23	.1703	10.3711	80.6726	5.8715	60.8933	0.0964	0.0164	7.7786	23
24	.1577	10.5288	84.2997	6.3412	66.7648	0.0950	0.0150	8.0066	24
25	.1460	10.6748	87.8041	6.8485	73.1059	0.0937	0.0137	8.2254	25
26	.1352	10.8100	91.1842	7.3964	79.9544	0.0925	0.0125	8.4352	26
27	.1252	10.9352	94.4390	7.9881	87.3508	0.0914	0.0114	8.6363	27
28	.1159	11.0511	97.5687	8.6271	95.3388	0.0905	0.0105	8.8289	28
29	.1073	11.1584	100.5738	9.3173	103.9659	0.0896	0.0096	9.0133	29
30	.0994	11.2578	103.4558	10.0627	113.2832	0.0888	0.0088	9.1897	30
31	.0920	11.3498	106.2163	10.8677	123.3459	0.0881	0.0081	9.3584	31
32	.0852	11.4350	108.8575	11.7371	134.2135	0.0875	0.0075	9.5197	32
33	.0789	11.5139	111.3819	12.6760	145.9506	0.0869	0.0069	9.6737	33
34	.0730	11.5869	113.7924	13.6901	158.6267	0.0863	0.0063	9.8208	34
35	.0676	11.6546	116.0920	14.7853	172.3168	0.0858	0.0058	9.9611	35
36	.0626	11.7172	118.2839	15.9682	187.1021	0.0853	0.0053	10.0949	36
37	.0580	11.7752	120.3713	17.2456	203.0703	0.0849	0.0049	10.2225	37
38	.0537	11.8289	122.3579	18.6253	220.3159	0.0845	0.0045	10.3440	38
39	.0497	11.8786	124.2470	20.1153	238.9412	0.0842	0.0042	10.4597	39
40	.0460	11.9246	126.0422	21.7245	259.0565	0.0839	0.0039	10.5699	40
41	.0426	11.9672	127.7470	23.4625	280.7810	0.0836	0.0036	10.6747	41
42	.0395	12.0067	129.3651	25.3395	304.2435	0.0833	0.0033	10.7744	42
43	.0365	12.0432	130.8998	27.3666	329.5830	0.0830	0.0030	10.8692	43
44	.0338	12.0771	132.3547	29.5560	356.9496	0.0828	0.0028	10.9592	44
45	.0313	12.1084	133.7331	31.9204	386.5056	0.0826	0.0026	11.0447	45
46	.0290	12.1374	135.0384	34.4741	418.4261	0.0824	0.0024	11.1258	46
47	.0269	12.1643	136.2739	37.2320	452.9002	0.0822	0.0022	11.2028	47
48	.0249	12.1891	137.4428	40.2106	490.1322	0.0820	0.0020	11.2758	48
49	.0230	12.2122	138.5480	43.4274	530.3427	0.0819	0.0019	11.3451	49
50	.0213	12.2335	139.5928	46.9016	573.7702	0.0817	0.0017	11.4107	50
51	.0197	12.2532	140.5799	50.6537	620.6718	0.0816	0.0016	11.4729	51
52	.0183	12.2715	141.5121	54.7060	671.3255	0.0815	0.0015	11.5318	52
53	.0169	12.2884	142.3923	59.0825	726.0316	0.0814	0.0014	11.5875	53
54	.0157	12.3041	143.2229	63.8091	785.1141	0.0813	0.0013	11.6403	54
55	.0145	12.3186	144.0065	68.9139	848.9232	0.0812	0.0012	11.6902	55
60	.0099	12.3766	147.3000	101.2571	1253.2133	0.0808	0.0008	11.9015	60
65	.0067	12.4160	149.7387	148.7798	1847.2481	0.0805	0.0005	12.0602	65
70	.0046	12.4428	151.5326	218.6064	2720.0801	0.0804	0.0004	12.1783	70
75	.0031	12.4611	152.8448	321.2045	4002.5566	0.0802	0.0002	12.2658	75
80	.0021	12.4735	153.8001	471.9548	5886.9354	0.0802	0.0002	12.3301	80
85	.0014	12.4820	154.4925	693.4565	8655.7061	0.0801	0.0001	12.3772	85
90	.0010	12.4877	154.9925	1018.9151	12723.9386	0.0801	0.0001	12.4116	90
95	.0007	12.4917	155.3524	1497.1205	18701.5069	0.0801	0.0001	12.4365	95
100	.0005	12.4943	155.6107	2199.7613	27484.5157	0.0800	0.0000	12.4545	100

PROFESSIONAL PUBLICATIONS ● Belmont, CA

I = 9.00 %

N	(P/F)	(P/A)	(P/G)	(F/P)	(F/A)	(A/P)	(A/F)	(A/G)	N
1	.9174	0.9174	-0.0000	1.0900	1.0000	1.0900	1.0000	-0.0000	1
2	.8417	1.7591	0.8417	1.1881	2.0900	0.5685	0.4785	0.4785	2
3	.7722	2.5313	2.3860	1.2950	3.2781	0.3951	0.3051	0.9426	3
4	.7084	3.2397	4.5113	1.4116	4.5731	0.3087	0.2187	1.3925	4
5	.6499	3.8897	7.1110	1.5386	5.9847	0.2571	0.1671	1.8282	5
6	.5963	4.4859	10.0924	1.6771	7.5233	0.2229	0.1329	2.2498	6
7	.5470	5.0330	13.3746	1.8280	9.2004	0.1987	0.1087	2.6574	7
8	.5019	5.5348	16.8877	1.9926	11.0285	0.1807	0.0907	3.0512	8
9	.4604	5.9952	20.5711	2.1719	13.0210	0.1668	0.0768	3.4312	9
10	.4224	6.4177	24.3728	2.3674	15.1929	0.1558	0.0658	3.7978	10
11	.3875	6.8052	28.2481	2.5804	17.5603	0.1469	0.0569	4.1510	11
12	.3555	7.1607	32.1590	2.8127	20.1407	0.1397	0.0497	4.4910	12
13	.3262	7.4869	36.0731	3.0658	22.9534	0.1336	0.0436	4.8182	13
14	.2992	7.7862	39.9633	3.3417	26.0192	0.1284	0.0384	5.1326	14
15	.2745	8.0607	43.8069	3.6425	29.3609	0.1241	0.0341	5.4346	15
16	.2519	8.3126	47.5849	3.9703	33.0034	0.1203	0.0303	5.7245	16
17	.2311	8.5436	51.2821	4.3276	36.9737	0.1170	0.0270	6.0024	17
18	.2120	8.7556	54.8860	4.7171	41.3013	0.1142	0.0242	6.2687	18
19	.1945	8.9501	58.3868	5.1417	46.0185	0.1117	0.0217	6.5236	19
20	.1784	9.1285	61.7770	5.6044	51.1601	0.1095	0.0195	6.7674	20
21	.1637	9.2922	65.0509	6.1088	56.7645	0.1076	0.0176	7.0006	21
22	.1502	9.4424	68.2048	6.6586	62.8733	0.1059	0.0159	7.2232	22
23	.1378	9.5802	71.2359	7.2579	69.5319	0.1044	0.0144	7.4357	23
24	.1264	9.7066	74.1433	7.9111	76.7898	0.1030	0.0130	7.6384	24
25	.1160	9.8226	76.9265	8.6231	84.7009	0.1018	0.0118	7.8316	25
26	.1064	9.9290	79.5863	9.3992	93.3240	0.1007	0.0107	8.0156	26
27	.0976	10.0266	82.1241	10.2451	102.7231	0.0997	0.0097	8.1906	27
28	.0895	10.1161	84.5419	11.1671	112.9682	0.0989	0.0089	8.3571	28
29	.0822	10.1983	86.8422	12.1722	124.1354	0.0981	0.0081	8.5154	29
30	.0754	10.2737	89.0280	13.2677	136.3075	0.0973	0.0073	8.6657	30
31	.0691	10.3428	91.1024	14.4618	149.5752	0.0967	0.0067	8.8083	31
32	.0634	10.4062	93.0690	15.7633	164.0370	0.0961	0.0061	8.9436	32
33	.0582	10.4644	94.9314	17.1820	179.8003	0.0956	0.0056	9.0718	33
34	.0534	10.5178	96.6935	18.7284	196.9823	0.0951	0.0051	9.1933	34
35	.0490	10.5668	98.3590	20.4140	215.7108	0.0946	0.0046	9.3083	35
36	.0449	10.6118	99.9319	22.2512	236.1247	0.0942	0.0042	9.4171	36
37	.0412	10.6530	101.4162	24.2538	258.3759	0.0939	0.0039	9.5200	37
38	.0378	10.6908	102.8158	26.4367	282.6298	0.0935	0.0035	9.6172	38
39	.0347	10.7255	104.1345	28.8160	309.0665	0.0932	0.0032	9.7090	39
40	.0318	10.7574	105.3762	31.4094	337.8824	0.0930	0.0030	9.7957	40
41	.0292	10.7866	106.5445	34.2363	369.2919	0.0927	0.0027	9.8775	41
42	.0268	10.8134	107.6432	37.3175	403.5281	0.0925	0.0025	9.9546	42
43	.0246	10.8380	108.6758	40.6761	440.8457	0.0923	0.0023	10.0273	43
44	.0226	10.8605	109.6456	44.3370	481.5218	0.0921	0.0021	10.0958	44
45	.0207	10.8812	110.5561	48.3273	525.8587	0.0919	0.0019	10.1603	45
46	.0190	10.9002	111.4103	52.6767	574.1860	0.0917	0.0017	10.2210	46
47	.0174	10.9176	112.2115	57.4176	626.8628	0.0916	0.0016	10.2780	47
48	.0160	10.9336	112.9625	62.5852	684.2804	0.0915	0.0015	10.3317	48
49	.0147	10.9482	113.6661	68.2179	746.8656	0.0913	0.0013	10.3821	49
50	.0134	10.9617	114.3251	74.3575	815.0836	0.0912	0.0012	10.4295	50
51	.0123	10.9740	114.9420	81.0497	889.4411	0.0911	0.0011	10.4740	51
52	.0113	10.9853	115.5193	88.3442	970.4908	0.0910	0.0010	10.5158	52
53	.0104	10.9957	116.0593	96.2951	1058.8349	0.0909	0.0009	10.5549	53
54	.0095	11.0053	116.5642	104.9617	1155.1301	0.0909	0.0009	10.5917	54
55	.0087	11.0140	117.0362	114.4083	1260.0918	0.0908	0.0008	10.6261	55
60	.0057	11.0480	118.9683	176.0313	1944.7921	0.0905	0.0005	10.7683	60
65	.0037	11.0701	120.3344	270.8460	2998.2885	0.0903	0.0003	10.8702	65
70	.0024	11.0844	121.2942	416.7301	4619.2232	0.0902	0.0002	10.9427	70
75	.0016	11.0938	121.9646	641.1909	7113.2321	0.0901	0.0001	10.9940	75
80	.0010	11.0998	122.4306	986.5517	10950.5741	0.0901	0.0001	11.0299	80
85	.0007	11.1038	122.7533	1517.9320	16854.8003	0.0901	0.0001	11.0551	85
90	.0004	11.1064	122.9758	2335.5266	25939.1842	0.0900	0.0000	11.0726	90
95	.0003	11.1080	123.1287	3593.4971	39916.6350	0.0900	0.0000	11.0847	95
100	.0002	11.1091	123.2335	5529.0408	61422.6755	0.0900	0.0000	11.0930	100

I = 10.00 %

N	(P/F)	(P/A)	(P/G)	(F/P)	(F/A)	(A/P)	(A/F)	(A/G)	N
1	.9091	0.9091	−0.0000	1.1000	1.0000	1.1000	1.0000	−0.0000	1
2	.8264	1.7355	0.8264	1.2100	2.1000	0.5762	0.4762	0.4762	2
3	.7513	2.4869	2.3291	1.3310	3.3100	0.4021	0.3021	0.9366	3
4	.6830	3.1699	4.3781	1.4641	4.6410	0.3155	0.2155	1.3812	4
5	.6209	3.7908	6.8618	1.6105	6.1051	0.2638	0.1638	1.8101	5
6	.5645	4.3553	9.6842	1.7716	7.7156	0.2296	0.1296	2.2236	6
7	.5132	4.8684	12.7631	1.9487	9.4872	0.2054	0.1054	2.6216	7
8	.4665	5.3349	16.0287	2.1436	11.4359	0.1874	0.0874	3.0045	8
9	.4241	5.7590	19.4215	2.3579	13.5795	0.1736	0.0736	3.3724	9
10	.3855	6.1446	22.8913	2.5937	15.9374	0.1627	0.0627	3.7255	10
11	.3505	6.4951	26.3963	2.8531	18.5312	0.1540	0.0540	4.0641	11
12	.3186	6.8137	29.9012	3.1384	21.3843	0.1468	0.0468	4.3884	12
13	.2897	7.1034	33.3772	3.4523	24.5227	0.1408	0.0408	4.6988	13
14	.2633	7.3667	36.8005	3.7975	27.9750	0.1357	0.0357	4.9955	14
15	.2394	7.6061	40.1520	4.1772	31.7725	0.1315	0.0315	5.2789	15
16	.2176	7.8237	43.4164	4.5950	35.9497	0.1278	0.0278	5.5493	16
17	.1978	8.0216	46.5819	5.0545	40.5447	0.1247	0.0247	5.8071	17
18	.1799	8.2014	49.6395	5.5599	45.5992	0.1219	0.0219	6.0526	18
19	.1635	8.3649	52.5827	6.1159	51.1591	0.1195	0.0195	6.2861	19
20	.1486	8.5136	55.4069	6.7275	57.2750	0.1175	0.0175	6.5081	20
21	.1351	8.6487	58.1095	7.4002	64.0025	0.1156	0.0156	6.7189	21
22	.1228	8.7715	60.6893	8.1403	71.4027	0.1140	0.0140	6.9189	22
23	.1117	8.8832	63.1462	8.9543	79.5430	0.1126	0.0126	7.1085	23
24	.1015	8.9847	65.4813	9.8497	88.4973	0.1113	0.0113	7.2881	24
25	.0923	9.0770	67.6964	10.8347	98.3471	0.1102	0.0102	7.4580	25
26	.0839	9.1609	69.7940	11.9182	109.1818	0.1092	0.0092	7.6186	26
27	.0763	9.2372	71.7773	13.1100	121.0999	0.1083	0.0083	7.7704	27
28	.0693	9.3066	73.6495	14.4210	134.2099	0.1075	0.0075	7.9137	28
29	.0630	9.3696	75.4146	15.8631	148.6309	0.1067	0.0067	8.0489	29
30	.0573	9.4269	77.0766	17.4494	164.4940	0.1061	0.0061	8.1762	30
31	.0521	9.4790	78.6395	19.1943	181.9434	0.1055	0.0055	8.2962	31
32	.0474	9.5264	80.1078	21.1138	201.1378	0.1050	0.0050	8.4091	32
33	.0431	9.5694	81.4856	23.2252	222.2515	0.1045	0.0045	8.5152	33
34	.0391	9.6086	82.7773	25.5477	245.4767	0.1041	0.0041	8.6149	34
35	.0356	9.6442	83.9872	28.1024	271.0244	0.1037	0.0037	8.7086	35
36	.0323	9.6765	85.1194	30.9127	299.1268	0.1033	0.0033	8.7965	36
37	.0294	9.7059	86.1781	34.0039	330.0395	0.1030	0.0030	8.8789	37
38	.0267	9.7327	87.1673	37.4043	364.0434	0.1027	0.0027	8.9562	38
39	.0243	9.7570	88.0908	41.1448	401.4478	0.1025	0.0025	9.0285	39
40	.0221	9.7791	88.9525	45.2593	442.5926	0.1023	0.0023	9.0962	40
41	.0201	9.7991	89.7852	49.7852	487.8518	0.1020	0.0020	9.1596	41
42	.0183	9.8174	90.5047	54.7637	537.6370	0.1019	0.0019	9.2188	42
43	.0166	9.8340	91.2019	60.2401	592.4007	0.1017	0.0017	9.2741	43
44	.0151	9.8491	91.8508	66.2641	652.6408	0.1015	0.0015	9.3258	44
45	.0137	9.8628	92.4544	72.8905	718.9048	0.1014	0.0014	9.3740	45
46	.0125	9.8753	93.0157	80.1795	791.7953	0.1013	0.0013	9.4190	46
47	.0113	9.8866	93.5372	88.1975	871.9749	0.1011	0.0011	9.4610	47
48	.0103	9.8969	94.0217	97.0172	960.1723	0.1010	0.0010	9.5001	48
49	.0094	9.9063	94.4715	106.7190	1057.1896	0.1009	0.0009	9.5365	49
50	.0085	9.9148	94.8889	117.3909	1163.9085	0.1009	0.0009	9.5704	50
51	.0077	9.9226	95.2761	129.1299	1281.2994	0.1008	0.0008	9.6020	51
52	.0070	9.9296	95.6351	142.0429	1410.4293	0.1007	0.0007	9.6313	52
53	.0064	9.9360	95.9679	156.2472	1552.4723	0.1006	0.0006	9.6586	53
54	.0058	9.9418	96.2763	171.8719	1708.7195	0.1006	0.0006	9.6840	54
55	.0053	9.9471	96.5619	189.0591	1880.5914	0.1005	0.0005	9.7075	55
60	.0033	9.9672	97.7010	304.4816	3034.8164	0.1003	0.0003	9.8023	60
65	.0020	9.9796	98.4705	490.3707	4893.7073	0.1002	0.0002	9.8672	65
70	.0013	9.9873	98.9870	789.7470	7887.4696	0.1001	0.0001	9.9113	70
75	.0008	9.9921	99.3317	1271.8954	12708.9537	0.1001	0.0001	9.9410	75
80	.0005	9.9951	99.5606	2048.4002	20474.0021	0.1000	0.0000	9.9609	80
85	.0003	9.9970	99.7120	3298.9690	32979.6903	0.1000	0.0000	9.9742	85
90	.0002	9.9981	99.8118	5313.0226	53120.2261	0.1000	0.0000	9.9831	90
95	.0001	9.9988	99.8773	8556.6760	85556.7605	0.1000	0.0000	9.9889	95
100	.0001	9.9993	99.9202	13780.6123	137796.1234	0.1000	0.0000	9.9927	100

I = 12.00 %

N	(P/F)	(P/A)	(P/G)	(F/P)	(F/A)	(A/P)	(A/F)	(A/G)	N
1	.8929	0.8929	-0.0000	1.1200	1.0000	1.1200	1.0000	-0.0000	1
2	.7972	1.6901	0.7972	1.2544	2.1200	0.5917	0.4717	0.4717	2
3	.7118	2.4018	2.2208	1.4049	3.3744	0.4163	0.2963	0.9246	3
4	.6355	3.0373	4.1273	1.5735	4.7793	0.3292	0.2092	1.3589	4
5	.5674	3.6048	6.3970	1.7623	6.3528	0.2774	0.1574	1.7746	5
6	.5066	4.1114	8.9302	1.9738	8.1152	0.2432	0.1232	2.1720	6
7	.4523	4.5638	11.6443	2.2107	10.0890	0.2191	0.0991	2.5515	7
8	.4039	4.9676	14.4714	2.4760	12.2997	0.2013	0.0813	2.9131	8
9	.3606	5.3282	17.3563	2.7731	14.7757	0.1877	0.0677	3.2574	9
10	.3220	5.6502	20.2541	3.1058	17.5487	0.1770	0.0570	3.5847	10
11	.2875	5.9377	23.1288	3.4785	20.6546	0.1684	0.0484	3.8953	11
12	.2567	6.1944	25.9523	3.8960	24.1331	0.1614	0.0414	4.1897	12
13	.2292	6.4235	28.7024	4.3635	28.0291	0.1557	0.0357	4.4683	13
14	.2046	6.6282	31.3624	4.8871	32.3926	0.1509	0.0309	4.7317	14
15	.1827	6.8109	33.9202	5.4736	37.2797	0.1468	0.0268	4.9803	15
16	.1631	6.9740	36.3670	6.1304	42.7533	0.1434	0.0234	5.2147	16
17	.1456	7.1196	38.6973	6.8660	48.8837	0.1405	0.0205	5.4353	17
18	.1300	7.2497	40.9080	7.6900	55.7497	0.1379	0.0179	5.6427	18
19	.1161	7.3658	42.9979	8.6128	63.4397	0.1358	0.0158	5.8375	19
20	.1037	7.4694	44.9676	9.6463	72.0524	0.1339	0.0139	6.0202	20
21	.0926	7.5620	46.8188	10.8038	81.6987	0.1322	0.0122	6.1913	21
22	.0826	7.6446	48.5543	12.1003	92.5026	0.1308	0.0108	6.3514	22
23	.0738	7.7184	50.1776	13.5523	104.6029	0.1296	0.0096	6.5010	23
24	.0659	7.7843	51.6929	15.1786	118.1552	0.1285	0.0085	6.6406	24
25	.0588	7.8431	53.1046	17.0001	133.3339	0.1275	0.0075	6.7708	25
26	.0525	7.8957	54.4177	19.0401	150.3339	0.1267	0.0067	6.8921	26
27	.0469	7.9426	55.6369	21.3249	169.3740	0.1259	0.0059	7.0049	27
28	.0419	7.9844	56.7674	23.8839	190.6989	0.1252	0.0052	7.1098	28
29	.0374	8.0218	57.8141	26.7499	214.5828	0.1247	0.0047	7.2071	29
30	.0334	8.0552	58.7821	29.9599	241.3327	0.1241	0.0041	7.2974	30
31	.0298	8.0850	59.6761	33.5551	271.2926	0.1237	0.0037	7.3811	31
32	.0266	8.1116	60.5010	37.5817	304.8477	0.1233	0.0033	7.4586	32
33	.0238	8.1354	61.2612	42.0915	342.4294	0.1229	0.0029	7.5302	33
34	.0212	8.1566	61.9612	47.1425	384.5210	0.1226	0.0026	7.5965	34
35	.0189	8.1755	62.6052	52.7996	431.6635	0.1223	0.0023	7 6577	35
36	.0169	8.1924	63.1970	59.1356	484.4631	0.1221	0.0021	7.7141	36
37	.0151	8.2075	63.7406	66.2318	543.5987	0.1218	0.0018	7.7661	37
38	.0135	8.2210	64.2394	74.1797	609.8305	0.1216	0.0016	7.8141	38
39	.0120	8.2330	64.6967	83.0812	684.0102	0.1215	0.0015	7.8582	39
40	.0107	8.2438	65.1159	93.0510	767.0914	0.1213	0.0013	7.8988	40
41	.0096	8.2534	65.4997	104.2171	860.1424	0.1212	0.0012	7.9361	41
42	.0086	8.2619	65.8509	116.7231	964.3595	0.1210	0.0010	7.9704	42
43	.0076	8.2696	66.1722	130.7299	1081.0826	0.1209	0.0009	8.0019	43
44	.0068	8.2764	66.4659	146.4175	1211.8125	0.1208	0.0008	8.0308	44
45	.0061	8.2825	66.7342	163.9876	1358.2300	0.1207	0.0007	8.0572	45
46	.0054	8.2880	66.9792	183.6661	1522.2176	0.1207	0.0007	8.0815	46
47	.0049	8.2928	67.2028	205.7061	1705.8838	0.1206	0.0006	8.1037	47
48	.0043	8.2972	67.4068	230.3908	1911.5898	0.1205	0.0005	8.1241	48
49	.0039	8.3010	67.5929	258.0377	2141.9806	0.1205	0.0005	8.1427	49
50	.0035	8.3045	67.7624	289.0022	2400.0182	0.1204	0.0004	8.1597	50
51	.0031	8.3076	67.9169	323.6825	2689.0204	0.1204	0.0004	8.1753	51
52	.0028	8.3103	68.0576	362.5243	3012.7029	0.1203	0.0003	8.1895	52
53	.0025	8.3128	68.1856	406.0273	3375.2272	0.1203	0.0003	8.2025	53
54	.0022	8.3150	68.3022	454.7505	3781.2545	0.1203	0.0003	8.2143	54
55	.0020	8.3170	68.4082	509.3206	4236.0050	0.1202	0.0002	8.2251	55
60	.0011	8.3240	68.8100	897.5969	7471.6411	0.1201	0.0001	8.2664	60
65	.0006	8.3281	69.0581	1581.8725	13173.9374	0.1201	0.0001	8.2922	65
70	.0004	8.3303	69.2103	2787.7998	23223.3319	0.1200	0.0000	8.3082	70
75	.0002	8.3316	69.3031	4913.0558	40933.7987	0.1200	0.0000	8.3181	75
80	.0001	8.3324	69.3594	8658.4831	72145.6925	0.1200	0.0000	8.3241	80
85	.0001	8.3328	69.3935	15259.2057	127151.7140	0.1200	0.0000	8.3278	85
90	.0000	8.3330	69.4140	26891.9342	224091.1185	0.1200	0.0000	8.3300	90
95	.0000	8.3332	69.4263	47392.7766	394931.4719	0.1200	0.0000	8.3313	95
100	.0000	8.3332	69.4336	83522.2657	696010.5477	0.1200	0.0000	8.3321	100

I = 15.00 %

N	(P/F)	(P/A)	(P/G)	(F/P)	(F/A)	(A/P)	(A/F)	(A/G)	N
1	.8696	0.8696	-0.0000	1.1500	1.0000	1.1500	1.0000	-0.0000	1
2	.7561	1.6257	0.7561	1.3225	2.1500	0.6151	0.4651	0.4651	2
3	.6575	2.2832	2.0712	1.5209	3.4725	0.4380	0.2880	0.9071	3
4	.5718	2.8550	3.7864	1.7490	4.9934	0.3503	0.2003	1.3263	4
5	.4972	3.3522	5.7751	2.0114	6.7424	0.2983	0.1483	1.7228	5
6	.4323	3.7845	7.9368	2.3131	8.7537	0.2642	0.1142	2.0972	6
7	.3759	4.1604	10.1924	2.6600	11.0668	0.2404	0.0904	2.4498	7
8	.3269	4.4873	12.4807	3.0590	13.7268	0.2229	0.0729	2.7813	8
9	.2843	4.7716	14.7548	3.5179	16.7858	0.2096	0.0596	3.0922	9
10	.2472	5.0188	16.9795	4.0456	20.3037	0.1993	0.0493	3.3832	10
11	.2149	5.2337	19.1289	4.6524	24.3493	0.1911	0.0411	3.6549	11
12	.1869	5.4206	21.1849	5.3503	29.0017	0.1845	0.0345	3.9082	12
13	.1625	5.5831	23.1352	6.1528	34.3519	0.1791	0.0291	4.1438	13
14	.1413	5.7245	24.9725	7.0757	40.5047	0.1747	0.0247	4.3624	14
15	.1229	5.8474	26.6930	8.1371	47.5804	0.1710	0.0210	4.5650	15
16	.1069	5.9542	28.2960	9.3576	55.7175	0.1679	0.0179	4.7522	16
17	.0929	6.0472	29.7828	10.7613	65.0751	0.1654	0.0154	4.9251	17
18	.0808	6.1280	31.1565	12.3755	75.8364	0.1632	0.0132	5.0843	18
19	.0703	6.1982	32.4213	14.2318	88.2118	0.1613	0.0113	5.2307	19
20	.0611	6.2593	33.5822	16.3665	102.4436	0.1598	0.0098	5.3651	20
21	.0531	6.3125	34.6448	18.8215	118.8101	0.1584	0.0084	5.4883	21
22	.0462	6.3587	35.6150	21.6447	137.6316	0.1573	0.0073	5.6010	22
23	.0402	6.3988	36.4988	24.8915	159.2764	0.1563	0.0063	5.7040	23
24	.0349	6.4338	37.3023	28.6252	184.1678	0.1554	0.0054	5.7979	24
25	.0304	6.4641	38.0314	32.9190	212.7930	0.1547	0.0047	5.8834	25
26	.0264	6.4906	38.6918	37.8568	245.7120	0.1541	0.0041	5.9612	26
27	.0230	6.5135	39.2890	43.5353	283.5688	0.1535	0.0035	6.0319	27
28	.0200	6.5335	39.8283	50.0656	327.1041	0.1531	0.0031	6.0960	28
29	.0174	6.5509	40.3146	57.5755	377.1697	0.1527	0.0027	6.1541	29
30	.0151	6.5660	40.7526	66.2118	434.7451	0.1523	0.0023	6.2066	30
31	.0131	6.5791	41.1466	76.1435	500.9569	0.1520	0.0020	6.2541	31
32	.0114	6.5905	41.5006	87.5651	577.1005	0.1517	0.0017	6.2970	32
33	.0099	6.6005	41.8184	100.6998	664.6655	0.1515	0.0015	6.3357	33
34	.0086	6.6091	42.1033	115.8048	765.3654	0.1513	0.0013	6.3705	34
35	.0075	6.6166	42.3586	133.1755	881.1702	0.1511	0.0011	6.4019	35
36	.0065	6.6231	42.5872	153.1519	1014.3457	0.1510	0.0010	6.4301	36
37	.0057	6.6288	42.7916	176.1246	1167.4975	0.1509	0.0009	6.4554	37
38	.0049	6.6338	42.9743	202.5433	1343.6222	0.1507	0.0007	6.4781	38
39	.0043	6.6380	43.1374	232.9248	1546.1655	0.1506	0.0006	6.4985	39
40	.0037	6.6418	43.2830	267.8635	1779.0903	0.1506	0.0006	6.5168	40
41	.0032	6.6450	43.4128	308.0431	2046.9539	0.1505	0.0005	6.5331	41
42	.0028	6.6478	43.5286	354.2495	2354.9969	0.1504	0.0004	6.5478	42
43	.0025	6.6503	43.6317	407.3870	2709.2465	0.1504	0.0004	6.5609	43
44	.0021	6.6524	43.7235	468.4950	3116.6334	0.1503	0.0003	6.5725	44
45	.0019	6.6543	43.8051	538.7693	3585.1285	0.1503	0.0003	6.5830	45
46	.0016	6.6559	43.8778	619.5847	4123.8977	0.1502	0.0002	6.5923	46
47	.0014	6.6573	43.9423	712.5224	4743.4824	0.1502	0.0002	6.6006	47
48	.0012	6.6585	43.9997	819.4007	5456.6047	0.1502	0.0002	6.6080	48
49	.0011	6.6596	44.0506	942.3108	6275.4055	0.1502	0.0002	6.6146	49
50	.0009	6.6605	44.0958	1083.6574	7217.7163	0.1501	0.0001	6.6205	50
51	.0008	6.6613	44.1360	1246.2061	8301.3737	0.1501	0.0001	6.6257	51
52	.0007	6.6620	44.1715	1433.1370	9547.5798	0.1501	0.0001	6.6304	52
53	.0006	6.6626	44.2031	1648.1075	10980.7167	0.1501	0.0001	6.6345	53
54	.0005	6.6631	44.2311	1895.3236	12628.8243	0.1501	0.0001	6.6382	54
55	.0005	6.6636	44.2558	2179.6222	14524.1479	0.1501	0.0001	6.6414	55
60	.0002	6.6651	44.3431	4383.9987	29219.9916	0.1500	0.0000	6.6530	60
65	.0001	6.6659	44.3903	8817.7874	58778.5826	0.1500	0.0000	6.6593	65
70	.0001	6.6663	44.4156	17735.7200	118231.4669	0.1500	0.0000	6.6627	70
75	.0000	6.6665	44.4292	35672.8680	237812.4532	0.1500	0.0000	6.6646	75
80	.0000	6.6666	44.4364	71750.8794	478332.5293	0.1500	0.0000	6.6656	80
85	.0000	6.6666	44.4402	144316.6470	962104.3133	0.1500	0.0000	6.6661	85
90	.0000	6.6666	44.4422	290272.3252	1935142.1680	0.1500	0.0000	6.6664	90
95	.0000	6.6667	44.4433	583841.3276	3892268.8509	0.1500	0.0000	6.6665	95
100	.0000	6.6667	44.4438	1174313.4507	7828749.6713	0.1500	0.0000	6.6666	100

I = 20.00 %

N	(P/F)	(P/A)	(P/G)	(F/P)	(F/A)	(A/P)	(A/F)	(A/G)	N
1	.8333	0.8333	-0.0000	1.2000	1.0000	1.2000	1.0000	-0.0000	1
2	.6944	1.5278	0.6944	1.4400	2.2000	0.6545	0.4545	0.4545	2
3	.5787	2.1065	1.8519	1.7280	3.6400	0.4747	0.2747	0.8791	3
4	.4823	2.5887	3.2986	2.0736	5.3680	0.3863	0.1863	1.2742	4
5	.4019	2.9906	4.9061	2.4883	7.4416	0.3344	0.1344	1.6405	5
6	.3349	3.3255	6.5806	2.9860	9.9299	0.3007	0.1007	1.9788	6
7	.2791	3.6046	8.2551	3.5832	12.9159	0.2774	0.0774	2.2902	7
8	.2326	3.8372	9.8831	4.2998	16.4991	0.2606	0.0606	2.5756	8
9	.1938	4.0310	11.4335	5.1598	20.7989	0.2481	0.0481	2.8364	9
10	.1615	4.1925	12.8871	6.1917	25.9587	0.2385	0.0385	3.0739	10
11	.1346	4.3271	14.2330	7.4301	32.1504	0.2311	0.0311	3.2893	11
12	.1122	4.4392	15.4667	8.9161	39.5805	0.2253	0.0253	3.4841	12
13	.0935	4.5327	16.5883	10.6993	48.4966	0.2206	0.0206	3.6597	13
14	.0779	4.6106	17.6008	12.8392	59.1959	0.2169	0.0169	3.8175	14
15	.0649	4.6755	18.5095	15.4070	72.0351	0.2139	0.0139	3.9588	15
16	.0541	4.7296	19.3208	18.4884	87.4421	0.2114	0.0114	4.0851	16
17	.0451	4.7746	20.0419	22.1861	105.9306	0.2094	0.0094	4.1976	17
18	.0376	4.8122	20.6805	26.6233	128.1167	0.2078	0.0078	4.2975	18
19	.0313	4.8435	21.2439	31.9480	154.7400	0.2065	0.0065	4.3861	19
20	.0261	4.8696	21.7395	38.3376	186.6880	0.2054	0.0054	4.4643	20
21	.0217	4.8913	22.1742	46.0051	225.0256	0.2044	0.0044	4.5334	21
22	.0181	4.9094	22.5546	55.2061	271.0307	0.2037	0.0037	4.5941	22
23	.0151	4.9245	22.8867	66.2474	326.2369	0.2031	0.0031	4.6475	23
24	.0126	4.9371	23.1760	79.4968	392.4842	0.2025	0.0025	4.6943	24
25	.0105	4.9476	23.4276	95.3962	471.9811	0.2021	0.0021	4.7352	25
26	.0087	4.9563	23.6460	114.4755	567.3773	0.2018	0.0018	4.7709	26
27	.0073	4.9636	23.8353	137.3706	681.8528	0.2015	0.0015	4.8020	27
28	.0061	4.9697	23.9991	164.8447	819.2233	0.2012	0.0012	4.8291	28
29	.0051	4.9747	24.1406	197.8136	984.0680	0.2010	0.0010	4.8527	29
30	.0042	4.9789	24.2628	237.3763	1181.8816	0.2008	0.0008	4.8731	30
31	.0035	4.9824	24.3681	284.8516	1419.2579	0.2007	0.0007	4.8908	31
32	.0029	4.9854	24.4588	341.8219	1704.1095	0.2006	0.0006	4.9061	32
33	.0024	4.9878	24.5368	410.1863	2045.9314	0.2005	0.0005	4.9194	33
34	.0020	4.9898	24.6038	492.2235	2456.1176	0.2004	0.0004	4.9308	34
35	.0017	4.9915	24.6614	590.6682	2948.3411	0.2003	0.0003	4.9406	35
36	.0014	4.9929	24.7108	708.8019	3539.0094	0.2003	0.0003	4.9491	36
37	.0012	4.9941	24.7531	850.5622	4247.8112	0.2002	0.0002	4.9564	37
38	.0010	4.9951	24.7894	1020.6747	5098.3735	0.2002	0.0002	4.9627	38
39	.0008	4.9959	24.8204	1224.8096	6119.0482	0.2002	0.0002	4.9681	39
40	.0007	4.9966	24.8469	1469.7716	7343.8578	0.2001	0.0001	4.9728	40
41	.0006	4.9972	24.8696	1763.7259	8813.6294	0.2001	0.0001	4.9767	41
42	.0005	4.9976	24.8890	2116.4711	10577.3553	0.2001	0.0001	4.9801	42
43	.0004	4.9980	24.9055	2539.7653	12693.8263	0.2001	0.0001	4.9831	43
44	.0003	4.9984	24.9196	3047.7183	15233.5916	0.2001	0.0001	4.9856	44
45	.0003	4.9986	24.9316	3657.2620	18281.3099	0.2001	0.0001	4.9877	45
46	.0002	4.9989	24.9419	4388.7144	21938.5719	0.2000	0.0000	4.9895	46
47	.0002	4.9991	24.9506	5266.4573	26327.2863	0.2000	0.0000	4.9911	47
48	.0002	4.9992	24.9581	6319.7487	31593.7436	0.2000	0.0000	4.9924	48
49	.0001	4.9993	24.9644	7583.6985	37913.4923	0.2000	0.0000	4.9935	49
50	.0001	4.9995	24.9698	9100.4382	45497.1908	0.2000	0.0000	4.9945	50
51	.0001	4.9995	24.9744	10920.5258	54597.6289	0.2000	0.0000	4.9953	51
52	.0001	4.9996	24.9783	13104.6309	65518.1547	0.2000	0.0000	4.9960	52
53	.0001	4.9997	24.9816	15725.5571	78622.7856	0.2000	0.0000	4.9966	53
54	.0001	4.9997	24.9844	18870.6685	94348.3427	0.2000	0.0000	4.9971	54
55	.0000	4.9998	24.9868	22644.8023	113219.0113	0.2000	0.0000	4.9976	55
60	.0000	4.9999	24.9942	56347.5144	281732.5718	0.2000	0.0000	4.9989	60
65	.0000	5.0000	24.9975	140210.6469	701048.2346	0.2000	0.0000	4.9995	65
70	.0000	5.0000	24.9989	348888.9569	1744439.7847	0.2000	0.0000	4.9998	70
75	.0000	5.0000	24.9995	868147.3693	4340731.8466	0.2000	0.0000	4.9999	75

I = 25.00 %

N	(P/F)	(P/A)	(P/G)	(F/P)	(F/A)	(A/P)	(A/F)	(A/G)	N
1	.8000	0.8000	0.0	1.2500	1.0000	1.2500	1.0000	0.0	1
2	.6400	1.4400	0.6400	1.5625	2.2500	0.6944	0.4444	0.4444	2
3	.5120	1.9520	1.6640	1.9531	3.8125	0.5123	0.2623	0.8525	3
4	.4096	2.3616	2.8928	2.4414	5.7656	0.4234	0.1734	1.2249	4
5	.3277	2.6893	4.2035	3.0518	8.2070	0.3718	0.1218	1.5631	5
6	.2621	2.9514	5.5142	3.8147	11.2588	0.3388	0.0888	1.8683	6
7	.2097	3.1611	6.7725	4.7684	15.0735	0.3163	0.0663	2.1424	7
8	.1678	3.3289	7.9469	5.9605	19.8419	0.3004	0.0504	2.3872	8
9	.1342	3.4631	9.0207	7.4506	25.8023	0.2888	0.0388	2.6048	9
10	.1074	3.5705	9.9870	9.3132	33.2529	0.2801	0.0301	2.7971	10
11	.0859	3.6564	10.8460	11.6415	42.5661	0.2735	0.0235	2.9663	11
12	.0687	3.7251	11.6020	14.5519	54.2077	0.2684	0.0184	3.1145	12
13	.0550	3.7801	12.2617	18.1899	68.7596	0.2645	0.0145	3.2437	13
14	.0440	3.8241	12.8334	22.7374	86.9495	0.2615	0.0115	3.3559	14
15	.0352	3.8593	13.3260	28.4217	109.6868	0.2591	0.0091	3.4530	15
16	.0281	3.8874	13.7482	35.5271	138.1085	0.2572	0.0072	3.5366	16
17	.0225	3.9099	14.1085	44.4089	173.6357	0.2558	0.0058	3.6084	17
18	.0180	3.9279	14.4147	55.5112	218.0446	0.2546	0.0046	3.6698	18
19	.0144	3.9424	14.6741	69.3889	273.5558	0.2537	0.0037	3.7222	19
20	.0115	3.9539	14.8932	86.7362	342.9447	0.2529	0.0029	3.7667	20
21	.0092	3.9631	15.0777	108.4202	429.6809	0.2523	0.0023	3.8045	21
22	.0074	3.9705	15.2326	135.5253	538.1011	0.2519	0.0019	3.8365	22
23	.0059	3.9764	15,3625	169.4066	673.6264	0.2515	0.0015	3.8634	23
24	.0047	3.9811	15.4711	211.7582	843.0329	0.2512	0.0012	3.8861	24
25	.0038	3.9849	15.5618	264.6978	1054.7912	0.2509	0.0009	3.9052	25
26	.0030	3.9879	15.6373	330.8722	1319.4890	0.2508	0.0008	3.9212	26
27	.0024	3.9903	15.7002	413.5903	1650.3612	0.2506	0.0006	3.9346	27
28	.0019	3.9923	15.7524	516.9879	2063.9515	0.2505	0.0005	3.9457	28
29	.0015	3.9938	15.7957	646.2349	2580.9394	0.2504	0.0004	3.9551	29
30	.0012	3.9950	15.8316	807.7936	3227.1743	0.2503	0.0003	3.9628	30
31	.0010	3.9960	15.8614	1009.7420	4034.9678	0.2502	0.0002	3.9693	31
32	.0008	3.9968	15.8859	1262.1774	5044.7098	0.2502	0.0002	3.9746	32
33	.0006	3.9975	15.9062	1577.7218	6306.8872	0.2502	0.0002	3.9791	33
34	.0005	3.9980	15.9229	1972.1523	7884.6091	0.2501	0.0001	3.9828	34
35	.0004	3.9984	15.9367	2465.1903	9856.7613	0.2501	0.0001	3.9858	35
36	.0003	3.9987	15.9481	3081.4879	12321.9516	0.2501	0.0001	3.9883	36
37	.0003	3.9990	15.9574	3851.8599	15403.4396	0.2501	0.0001	3.9904	37
38	.0002	3.9992	15.9651	4814.8249	19255.2994	0.2501	0.0001	3.9921	38
39	.0002	3.9993	15.9714	6018.5311	24070.1243	0.2500	0.0000	3.9935	39
40	.0001	3.9995	15.9766	7523.1638	30088.6554	0.2500	0.0000	3.9947	40
41	.0001	3.9996	15.9809	9403.9548	37611.8192	0.2500	0.0000	3.9956	41
42	.0001	3.9997	15.9843	11754.9435	47015.7740	0.2500	0.0000	3.9964	42
43	.0001	3.9997	15.9872	14693.6794	58770.7175	0.2500	0.0000	3.9971	43
44	.0001	3.9998	15.9895	18367.0992	73464.3969	0.2500	0.0000	3.9976	44
45	.0000	3.9998	15.9915	22958.8740	91831.4962	0.2500	0.0000	3.9980	45
46	.0000	3.9999	15.9930	28698.5925	114790.3702	0.2500	0.0000	3.9984	46
47	.0000	3.9999	15.9943	35873.2407	143488.9627	0.2500	0.0000	3.9987	47
48	.0000	3.9999	15.9954	44841.5509	179362.2034	0.2500	0.0000	3.9989	48
49	.0000	3.9999	15.9962	56051.9386	224203.7543	0.2500	0.0000	3.9991	49
50	.0000	3.9999	15.9969	70064.9232	280255.6929	0.2500	0.0000	3.9993	50
51	.0000	4.0000	15.9975	87581.1540	350320.6161	0.2500	0.0000	3.9994	51
52	.0000	4.0000	15.9980	109476.4425	437901.7701	0.2500	0.0000	3.9995	52
53	.0000	4.0000	15.9983	136845.5532	547378.2126	0.2500	0.0000	3.9996	53
54	.0000	4.0000	15.9986	171056.9414	684223.7658	0.2500	0.0000	3.9997	54
55	.0000	4.0000	15.9989	213821.1768	855280.7072	0.2500	0.0000	3.9997	55
60	.0000	4.0000	15.9996	652530.4468	2610117.7872	0.2500	0.0000	3.9999	60

15 SYSTEMS OF UNITS

1 INTRODUCTION

The purpose of this chapter is to eliminate some of the confusion regarding the many units available for each engineering variable. In particular, an effort has been made to clarify the use of the so-called English systems, which for years have used the *pound* unit both for force and mass—a practice that has resulted in confusion for even those familiar with it.[1]

It is expected that most engineering problems will be stated and solved in either the English gravitational or SI system, both of which are consistent systems. Therefore, a discussion of these two systems occupies the majority of this chapter. However, other systems are briefly mentioned in case they are encountered through work in other areas (e.g., physics, chemistry, electricity, etc.).

2 COMMON UNITS OF MASS

The choice of a mass unit is the major factor in determining which system of units will be used in solving a problem. It is obvious that one will not easily end up with a force in pounds if the rest of the problem is stated in meters and kilograms. Actually, the choice of a mass unit determines more than whether a conversion factor will be necessary to convert from one system to another (e.g., between the SI and English systems). An inappropriate choice of a mass unit may actually require a conversion factor *within* the system of units.

The common units of mass are the gram, pound, kilogram, and slug.[2] There is nothing mysterious about

these units. All represent different quantities of matter, as Fig. 15.1 illustrates. In particular, note that the pound and slug do not represent the same quantity of matter.[3]

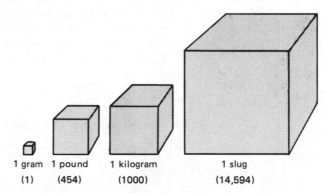

| 1 gram | 1 pound | 1 kilogram | 1 slug |
| (1) | (454) | (1000) | (14,594) |

Figure 15.1 Common Units of Mass

3 MASS AND WEIGHT

The SI system uses *kilograms* for mass and *newtons* for weight (force). The units are different, and there is no confusion between the variables. However, for years, the term *pound* has been used for both mass and weight. This usage has obscured the distinction between the two: mass is a constant property of an object; weight varies with the gravitational field. Even the conventional use of the abbreviations *lbm* and *lbf* (to distinguish between pounds-mass and pounds-force) has not helped eliminate the confusion.

It is true that an object with a mass of one pound will have an earthly weight of one pound, but this is true only on the earth. The weight of the same object will be much less on the moon. Therefore, care must be taken when working with mass and force in the same problem.

---

[1] This chapter does not attempt to recommend, legitimize, defend, or ridicule the English system. The fact that the English system is still in widespread use, particularly in the United States, is sufficient to require a familiarity with it for the foreseeable future.

[2] Normally, one does not distinguish between a unit and a multiple of that unit, as is done here with the gram and the kilogram. However, these two units actually are bases for different consistent systems.

[3] A slug is equal to 32.1740 pounds-mass.

The relationship that converts mass to weight is familiar to every engineering student:

$$W = mg \qquad 15.1$$

Equation 15.1 illustrates that an object's weight will depend on the local acceleration of gravity as well as the object's mass. The mass will be constant, but gravity will depend on location. Mass and weight are not the same.

4 ACCELERATION OF GRAVITY

Gravitational acceleration on the earth's surface is usually taken as $32.2\,\text{ft/sec}^2$ or $9.81\,\text{m/s}^2$. These values are rounded from the more exact standard values of $32.1740\,\text{ft/sec}^2$ and $9.8066\,\text{m/s}^2$. However, the need for greater accuracy must be evaluated on a problem-by-problem basis. Usually, three significant digits are adequate, since gravitational acceleration is not constant anyway, but is affected by location (primarily latitude and altitude) and major geographical features.

The term *standard gravity*, g_0, is derived from the acceleration at essentially any point at sea level and approximately 45 degrees N latitude. If additional accuracy is needed, the gravitational acceleration can be calculated from Eq. 15.2. This equation neglects the effects of large land and water masses. (ϕ is the latitude in degrees.)

$$g_{\text{surface}} = g'[1 + (5.305 \times 10^{-3})\sin^2 \phi$$
$$- (5.9 \times 10^{-6})\sin^2 2\phi] \qquad 15.2$$
$$g' = 32.0881\,\text{ft/sec}^2$$
$$= 9.78045\,\text{m/s}^2$$

If the effects of the earth's rotation are neglected, the gravitational acceleration at an altitude, h, above the earth's surface is given by Eq. 15.3. R_e is the earth's radius.

$$g_h = g_{\text{surface}} \left(\frac{R_e}{R_e + h} \right)^2 \qquad 15.3$$
$$R_e = 3960\,\text{miles}$$
$$= 6.37 \times 10^6\,\text{m}$$

5 CONSISTENT SYSTEMS OF UNITS

A set of units used in a calculation is said to be *consistent* if no conversion factors are needed.[4] For example, a moment is calculated as the product of a force and a lever arm length.

$$M = F \times d \qquad 15.4$$

---

[4] The terms *homogeneous* and *coherent* are also used to describe a consistent set of units.

A calculation using Eq. 15.4 would be consistent if M was in newton-meters, F was in newtons, and d was in meters. The calculation would be inconsistent if M was in ft-kips, F was in kips, and d was in inches (because a conversion factor of $1/12$ would be required).

The concept of a consistent calculation can be extended to a system of units. A *consistent system of units* is one in which Newton's second law of motion can be written without conversion factors. Newton's second law simply states that the force required to accelerate an object is proportional to the acceleration of the object. The constant of proportionality is the object's mass.

$$F = ma \qquad 15.5$$

Notice that Eq. 15.5 is $F = ma$, not $F = Wa/g$ or $F = ma/g_c$. Equation 15.5 is consistent: it requires no conversion factors. This means that, in a consistent system where conversion factors are not used, once the units of m and a have been selected, the units of F are fixed. This has the effect of establishing units of work and energy, power, fluid properties, etc.

It should be mentioned that the decision to work with a consistent set of units is desirable but unnecessary, depending often on tradition and environment. Problems in fluid flow and thermodynamics are routinely solved in the United States with inconsistent units. This causes no more of a problem than working with inches and feet when calculating a moment. It is necessary only to use the proper conversion factors.

6 THE ENGLISH ENGINEERING SYSTEM

Through common and widespread use, pounds-mass (lbm) and pounds-force (lbf) have become the standard units for mass and force in the *English Engineering System*.

There are subjects in the United States where the practice of using pounds for mass is firmly entrenched. For example, most thermodynamics, fluid flow, and heat transfer problems have traditionally been solved using the units of lbm/ft^3 for density, BTU/lbm for enthalpy, and BTU/lbm-$^\circ$F for specific heat. Unfortunately, some equations contain both lbm-related and lbf-related variables, as does the steady flow conservation of energy equation, which combines enthalpy in BTU/lbm with pressure in lbf/ft^2.

The units of pounds-mass and pounds-force are as different as the units of gallons and feet, and they cannot be canceled. A mass conversion factor, g_c, is needed to make the equations containing lbf and lbm dimensionally consistent. This factor is known as the *gravitational constant* and has a value of $32.1740\,\text{lbm-ft/lbf-sec}^2$. The numerical value is the same as the standard acceleration of gravity, but g_c is not the local gravitational acceleration, g.[5] g_c is a conversion constant, just as 12.0 is the conversion factor between feet and inches.

The English Engineering System is an inconsistent system, as defined according to Newton's second law. $F = ma$ cannot be written if lbf, lbm, and ft/sec² are the units used. The g_c term must be included.

$$F \text{ in lbf} = \frac{(m \text{ in lbm})(a \text{ in ft/sec}^2)}{g_c \text{ in } \frac{\text{lbm-ft}}{\text{lbf-sec}^2}} \qquad 15.6$$

It is important to note in Eq. 15.6 that g_c does more than "fix the units." Since g_c has a numerical value of 32.174, it actually changes the calculation numerically. A force of 1.0 pound will not accelerate a 1.0 pound mass at the rate of 1.0 ft/sec².

In the English Engineering System, work and energy are typically measured in ft-lbf (mechanical systems) or in British thermal units, BTU (thermal and fluid systems). One BTU is equal to 778.26 ft-lbf.

Example 15.1

Calculate the weight, in lbf, of a 1.0 lbm object in a gravitational field of 27.5 ft/sec².

From Eq. 15.6,

$$F = \frac{ma}{g_c} = \frac{(1 \text{ lbm})(27.5 \text{ ft/sec}^2)}{32.2 \frac{\text{lbm-ft}}{\text{lbf-sec}^2}} = 0.854 \text{ lbf}$$

7 OTHER FORMULAS AFFECTED BY INCONSISTENCY

It is not a significant burden to include g_c in a calculation, but it may be difficult to remember when g_c should be used. Knowing when to include the gravitational constant can be learned through repeated exposure to the formulas in which it is needed, but it is safer to carry the units along in every calculation.

The following is a representative (but not exhaustive) listing of formulas that require the g_c term. In all cases, it is assumed that the standard English Engineering System units will be used.

- Kinetic Energy

$$E = \frac{mv^2}{2g_c} \text{ (in ft-lbf)} \qquad 15.7$$

- Potential Energy

$$E = \frac{mgz}{g_c} \text{ (in ft-lbf)} \qquad 15.8$$

- Pressure at a Depth

$$p = \frac{\rho g h}{g_c} \text{ (in lbf/ft}^2) \qquad 15.9$$

Example 15.2

A rocket with a mass of 4000 lbm travels at 27,000 ft/sec. What is its kinetic energy in ft-lbf?

From Eq. 15.7,

$$E_k = \frac{mv^2}{2g_c} = \frac{(4000 \text{ lbm}) \left(27,000 \frac{\text{ft}}{\text{sec}^2}\right)^2}{(2) \left(32.2 \frac{\text{lbm-ft}}{\text{lbf-sec}^2}\right)}$$

$$= 4.53 \times 10^{10} \text{ ft-lbf}$$

8 WEIGHT AND WEIGHT DENSITY

Weight is a force exerted on an object due to its placement in a gravitational field. If a consistent set of units is used, Eq. 15.1 can be used to calculate the weight of a mass. In the English Engineering System, however, Eq. 15.10 must be used.

$$W = \frac{mg}{g_c} \qquad 15.10$$

Both sides of Eq. 15.10 can be divided by the volume of an object to derive the *weight density*, γ, of the object. Equation 15.11 illustrates that the weight density (in lbf/ft³) can also be calculated by multiplying the mass density (in lbm/ft³) by g/g_c. Since g and g_c usually have the same numerical values, the only effect of Eq. 15.12 is to change the units of density.

$$\frac{W}{V} = \frac{m}{V} \left(\frac{g}{g_c}\right) \qquad 15.11$$

$$\gamma = \frac{W}{V} = \left(\frac{m}{V}\right)\left(\frac{g}{g_c}\right) = \frac{\rho g}{g_c} \qquad 15.12$$

It is important to recognize that weight does not occupy volume. Only mass has volume. The concept of weight density has evolved to simplify certain calculations, particularly fluid calculations. For example, pressure at a depth is calculated from Eq. 15.13. (Compare this to Eq. 15.9.)

$$p = \gamma h \qquad 15.13$$

9 THE ENGLISH GRAVITATIONAL SYSTEM

Not all English systems are inconsistent. Pounds can still be used as the unit of force as long as pounds are

---

[5] It is acceptable (and recommended) that g_c be rounded to the same number of significant digits as g. Therefore, a value of 32.2 for g_c would typically be used.

not used as the unit of mass. Such is the case with the consistent *English Gravitational System*.

If acceleration is given in ft/sec^2, the units of mass for a consistent system of units can be determined from Newton's second law.

$$\text{units of } m = \frac{\text{units of } F}{\text{units of } a}$$

$$= \frac{\text{lbf}}{\text{ft/sec}^2} = \frac{\text{lbf-sec}^2}{\text{ft}} \qquad 15.14$$

The combination of units in Eq. 15.14 is known as a *slug*. Slugs and pounds-mass are not the same, as Fig. 15.1 illustrates. However, both are units for the same quantity: mass. Equation 15.15 will convert between slugs and pounds-mass.

$$\text{number of slugs} = \frac{\text{number of lbm}}{g_c} \qquad 15.15$$

It is important to recognize that the number of slugs is not derived by dividing the number of pounds-mass by the local gravity. g_c is used regardless of the local gravity. The conversion between feet and inches is not dependent on local gravity; neither is the conversion between slugs and pounds-mass.

Since the English Gravitational System is consistent, Eq. 15.16 can be used to calculate weight. Notice that the local gravitational acceleration is used.

$$W \text{ in lbf} = (m \text{ in slugs}) \left(g \text{ in } \frac{\text{ft}}{\text{sec}^2} \right) \qquad 15.16$$

10 THE ABSOLUTE ENGLISH SYSTEM

The obscure *Absolute English System* takes the approach that mass must have units of pounds-mass (lbm) and the units of force can be derived from Newton's second law.

$$\text{units of } F = (\text{units of } m)(\text{units of } a)$$

$$= (\text{lbm}) \left(\frac{\text{ft}}{\text{sec}}^2 \right)$$

$$= \frac{\text{lbm-ft}}{\text{sec}^2} \qquad 15.17$$

The units for F cannot be simplified any more than they are in Eq. 15.17. This particular combination of units is known as a *poundal*.[6] A poundal is not the same as a pound.

Poundals have not seen widespread use in the United States. The English Gravitational System (using slugs for mass) has greatly eclipsed the Absolute English System in popularity. Both are consistent systems, and

there seems little need for poundals in modern engineering.

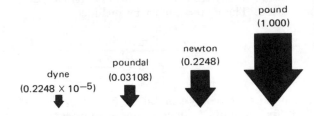

Figure 15.2   Common Force Units

11 METRIC SYSTEMS OF UNITS

Strictly speaking, a *metric system* is any system of units that is based on meters or parts of meters. This broad definition includes *mks systems* (based on meters and kilograms) as well as *cgs systems* (based on centimeters and grams).

Metric systems avoid the pounds-mass versus pounds-force ambiguity in two ways. First, the concept of weight is not used at all. All quantities of matter are specified as mass. Second, force and mass units do not share a common name.

The term *metric system* is not explicit enough to define which units are to be used for any given variable. For example, within the cgs system there is variation in how certain electrical and magnetic quantities are represented (resulting in the ESU and EMU systems). Also, within the mks system, it is common engineering practice today to use kilocalories as the unit of thermal energy, while the SI system requires the use of joules. Thus, there is a lack of uniformity even within the metricated engineering community.[7]

The "metric" parts of this book are based on the SI system, which is the most developed and codified of the so-called metric systems.[8] It is expected that there will be occasional variances with local engineering custom, but it is difficult to anticipate such variances within a book that must be consistent.[9]

---

[6]A poundal is equal to 0.03108 pounds-force.

[7]In the "field test" of the metric system conducted over the past 200 years, other conventions are to use kilograms-force (kgf) instead of newtons and kgf/cm^2 for pressure (instead of pascals).
[8]The SI System is an outgrowth of the *General Conference of Weights and Measures*, an international treaty organization which established the *Systeme International d'Unites (International System of Units)* in 1960. The United States subscribed to this treaty in 1975.
[9]Conversion to pure SI units is essentially complete in Australia, Canada, New Zealand, and South Africa. The use of non-standard metric units is more common among European civil engineers, who have had little need to deal with the inertial properties of mass. However, even the American Society of Civil Engineers declared its support for the SI system in 1985.

12 THE cgs SYSTEM

The *cgs system* is used widely by chemists and physicists. It is named for the three primary units used to construct its derived variables: the centimeter, the gram, and the second.

When Newton's second law is written in the cgs system, the following combination of units results.

$$\text{units of force} = (m \text{ in g})(a \text{ in cm/s}^2)$$
$$= \frac{\text{g·cm}}{\text{s}^2} \qquad 15.18$$

This combination of units for force is known as a *dyne*. Energy variables in the cgs system have units of dyne·cm or, equivalently, gm·cm^2/sec^2. This combination is known as an *erg*. There is no uniformly accepted unit of power in the cgs system, although calories per second is frequently used.

The fundamental volume unit in the cgs system is the cubic centimeter (cc). Since this is the same volume as one thousandth of a liter, units of milliliters (ml) are also used.

13 THE SI (mks) SYSTEM

The *SI system* is an *mks system* (so named because it uses the meter, kilogram, and second as base units). All other units are derived from the base units, which are completely listed in Table 15.1. The SI system is fully consistent, and there is only one recognized unit for each physical quantity (variable).

Three types of units are used: base units, supplementary units, and derived units. The *base units* (Table 15.1) are dependent only on accepted standards or reproducible phenomena. The *supplementary units* (Table 15.2) have not yet been classified as being base units or derived units. The *derived units* (Tables 15.3 and 15.4) are made up of combinations of base and supplementary units.

Table 15.1
SI Base Units

quantity	name	symbol
length	meter	m
mass	kilogram	kg
time	second	s
electric current	ampere	A
temperature	kelvin	K
amount of substance	mole	mol
luminous intensity	candela	cd

In addition, there is a set of non-SI units that may be used. This concession is primarily due to the significance and widespread acceptance of these units. Use of the non-SI units listed in Table 15.5 will usually create an inconsistent expression requiring conversion factors.

Table 15.2
SI Supplementary Units

quantity	name	symbol
plane angle	radian	rad
solid angle	steradian	sr

Table 15.3
Some SI Derived Units With Special Names

quantity	name	symbol	expressed in terms of other units
frequency	hertz	Hz	
force	newton	N	
pressure, stress	pascal	Pa	N/m^2
energy, work, quantity of heat	joule	J	N·m
power, radiant flux	watt	W	J/s
quantity of electricity, electric charge	coulomb	C	
electric potential, potential difference, electromotive force	volt	V	W/A
electric capacitance	farad	F	C/V
electric resistance	ohm	Ω	V/A
electric conductance	siemen	S	A/V
magnetic flux	weber	Wb	V·s
magnetic flux density	tesla	T	Wb/m^2
inductance	henry	H	Wb/A
luminous flux	lumen	lm	
illuminance	lux	lx	lm/m^2

The units of force can be derived from Newton's second law.

$$\text{units of force} = (m \text{ in kg})(a \text{ in m/s}^2)$$
$$= \frac{\text{kg·m}}{\text{s}^2} \qquad 15.19$$

This combination of units for force is known as a *newton*.

Energy variables in the SI system have units of N·m, or equivalently, kg·m^2/s^2. Both of these combinations are known as a *joule*. The units of power are joules per second, equivalent to a *watt*.

Table 15.4
Some SI Derived Units

quantity	description	expressed in terms of other units
area	square meter	m^2
volume	cubic meter	m^3
speed—linear	meter per second	m/s
—angular	radian per second	rad/s
acceleration—linear	meter per second squared	m/s^2
—angular	radian per second squared	rad/s^2
density, mass density	kilogram per cubic meter	kg/m^3
concentration (of amount of substance)	mole per cubic meter	mol/m^3
specific volume	cubic meter per kilogram	m^3/kg
luminance	candela per square meter	cd/m^2
absolute viscosity	pascal second	$Pa{\cdot}s$
kinematic viscosity	square meters per second	m^2/s
moment of force	newton meter	$N{\cdot}m$
surface tension	newton per meter	N/m
heat flux density, irradiance	watt per square meter	W/m^2
heat capacity, entropy	joule per kelvin	J/K
specific heat capacity, specific entropy	joule per kilogram kelvin	$J/(kg{\cdot}K)$
specific energy	joule per kilogram	J/kg
thermal conductivity	watt per meter kelvin	$W/(m{\cdot}K)$
energy density	joule per cubic meter	J/m^3
electric field strength	volt per meter	V/m
electric charge density	coulomb per cubic meter	C/m^3
surface density of charge, flux density	coulomb per square meter	C/m^2
permittivity	farad per meter	F/m
current density	ampere per square meter	A/m^2
magnetic field strength	ampere per meter	A/m
permeability	henry per meter	H/m
molar energy	joule per mole	J/mol
molar entropy, molar heat capacity	joule per mole kelvin	$J/(mol{\cdot}K)$
radiant intensity	watt per steradian	W/sr

Table 15.5
Acceptable Non-SI Units

quantity	unit name	symbol name	relationship to SI units
area	hectare	ha	1 ha = 10 000 m^2
energy	kilowatt-hour	kWh	1 kWh = 3.6 MJ
mass	metric ton[a]	t	1 t = 1000 kg
plane angle	degree (of arc)	°	1° = 0.017 453 rad
speed of rotation	revolution per minute	r/min	1 r/min = $2\pi/60$ rad/s
temperature interval	degree Celsius	°C	1°C = 1 K
time	minute	min	1 min = 60 s
	hour	h	1 h = 3600 s
	day (mean solar)	d	1 d = 86 400 s
	year (calendar)	a	1 a = 31 536 000 s
velocity	kilometer per hour	km/h	1 km/h = 0.278 m/s
volume	liter[b]	l	1 l = 0.001 m^3

[a] The International name for metric ton is *tonne*. The metric ton is equal to the *megagram* (Mg).

[b] The international symbol for liter is the lowercase l, which can be easily confused with the numeral 1. Several English-speaking countries have adopted the script l and uppercase L as a symbol for liter in order to avoid any misinterpretation.

Example 15.3

A 10-kg block hangs from a cable. What is the tension in the cable? (Standard gravity equals 9.81 m/s^2.)

$$F = mg = (10\,\text{kg})\left(9.81\,\frac{\text{m}}{\text{s}^2}\right)$$

$$= 98.1\,\frac{\text{kg}{\cdot}\text{m}}{\text{s}^2}$$

$$= 98.1\,\text{N}$$

Example 15.4

A 10-kg block is raised vertically 3 meters. What is the change in potential energy?

$$\Delta E_p = mg\Delta h = (10\,\text{kg})\left(9.81\,\frac{\text{m}}{\text{s}^2}\right)(3\,\text{m})$$

$$= 294\,\frac{\text{kg}{\cdot}\text{m}^2}{\text{s}^2} = 294\,\text{N}{\cdot}\text{m}$$

$$= 294\,\text{J}$$

14   RULES FOR USING THE SI SYSTEM

In addition to having standardized units, the SI system also has rigid syntax rules for writing the units and combinations of units. Each unit is abbreviated with a specific symbol. The following rules for writing and combining these symbols should be adhered to.

- The expressions for derived units in symbolic form are obtained by using the mathematical signs of multiplication and division. For example, units of velocity are m/s. Units of torque are N·m (not N-m or Nm).

- Scaling of most units is done in multiples of 1000.

Table 15.6
SI Prefixes

prefix	symbol	value
exa	E	10^{18}
peta	P	10^{15}
tera	T	10^{12}
giga	G	10^{9}
mega	M	10^{6}
kilo	k	10^{3}
hecto	h	10^{2}
deca	da	10^{1}
deci	d	10^{-1}
centi	c	10^{-2}
milli	m	10^{-3}
micro	μ	10^{-6}
nano	n	10^{-9}
pico	p	10^{-12}
femto	f	10^{-15}
atto	a	10^{-18}

- The symbols are always printed in roman type, regardless of the type used in the rest of the text. The only exception to this is in the use of the symbol for liter, where the use of the lowercase el (l) may be confused with the numeral one (1). In this case, "liter" should be written out in full, or the script ℓ or L used. There is no problem with such symbols as cl (centiliter) or ml (milliliter).

- Symbols are not pluralized: 1 kg, 45 kg (not 45 kgs).

- A period after a symbol is not used, except when the symbol occurs at the end of a sentence.

- When symbols consist of letters, there is always a full space between the quantity and the symbols: 45 kg (not 45kg). However, when the first character of a symbol is not a letter, no space is left: 32°C (not 32° C or 32 °C); or 42° 12′ 45″ (not 42 ° 12 ′ 45 ″).

- All symbols are written in lowercase, except when the unit is derived from a proper name: m for meter; s for second; A for ampere, Wb for weber, N for newton, W for watt.

- Prefixes are printed without spacing between the prefix and the unit symbol (e.g., km is the symbol for kilometer).

- In text, symbols should be used when associated with a number. However, when no number is involved, the unit should be spelled out. Examples: The area of the carpet is 16 m², not 16 square meters. Carpet is sold by the square meter, not by the m².

- A practice in some countries is to use a comma as a decimal marker, while the practice in North America, the United Kingdom, and some other countries is to use a period (or dot) as the decimal marker. Furthermore, in some countries that use the decimal comma, a dot is frequently used to divide long numbers into groups of three. Because of these differing practices, spaces must be used instead of commas to separate long lines of digits into easily readable blocks of three digits with respect to the decimal marker: 32 453.246 072 5. A space (half-space preferred) is optional with a four-digit number: 1 234 or 1 234.

- Where a decimal fraction of a unit is used, a zero should always be placed before the decimal marker: 0.45 kg (not .45 kg). This practice draws attention to the decimal marker and helps avoid errors of scale.

- Some confusion may arise with the word "tonne" (1 000 kg). When this word occurs in French text of Canadian origin, the meaning may be a ton of 2 000 pounds.

15 PRIMARY DIMENSIONS

Regardless of the system of units chosen, each variable representing a physical quantity will have the same *primary dimensions*. For example, velocity may be expressed in miles per hour (mph) or meters per second (m/s), but both units have dimensions of length per unit time. Length and time are two of the primary dimensions, as neither can be broken down into more basic dimensions. The concept of primary dimensions is useful when converting little-used variables between different systems of units, as well as in correlating experimental results (i.e., dimensional analysis).

There are three different sets of primary dimensions in use.[10] In the $ML\theta T$ system, the primary dimensions are mass (M), length (L), time (θ), and temperature (T). Notice that all symbols are uppercase. In order to avoid confusion between time and temperature, the Greek letter theta is used for time.[11]

---

[10] One of these, the $FL\theta T$ system, is not discussed here but appears in Table 15.7.

[11] This is the most common usage. There is a lack of consistency in the engineering world about the symbols for the primary dimensions in dimensional analysis. Some writers use t for time instead of θ. Some use H for heat instead of Q. And, in the worst mix-up of all, some have reversed the use of T and θ.

PROFESSIONAL PUBLICATIONS, INC. • Belmont, CA

Table 15.7
Dimensions of Common Variables

variable	dimensional system		
	$ML\theta T$	$FL\theta T$	$FML\theta TQ$
mass (m)	M	$F\theta^2/L$	M
force (F)	ML/θ^2	F	F
length (L)	L	L	L
time (θ)	θ	θ	θ
temperature (T)	T	T	T
work (W)	ML^2/θ^2	FL	FL
heat (Q)	ML^2/θ^2	FL	Q
L/θ^2			
acceleration (a)	L/θ^2	L/θ^2	
frequency (N)	$1/\theta$	$1/\theta$	$1/\theta$
area (A)	L^2	L^2	L^2
coefficient of thermal expansion (β)	$1/T$	$1/T$	$1/T$
density (ρ)	M/L^3	$F\theta^2/L^4$	M/L^3
dimensional constant (g_c)	1.0	1.0	$ML/\theta^2 F$
specific heat at constant pressure (c_p); at constant volume (c_v)	$L^2/\theta^2 T$	$L^2/\theta^2 T$	Q/MT
heat transfer coefficient (h); overall (U)	$M/\theta^3 T$	$F/\theta LT$	$Q/\theta L^2 T$
power (P)	ML^2/θ^3	FL/θ	FL/θ
heat flow rate ($\dot{Q}$)	ML^2/θ^3	FL/θ	Q/θ
kinematic viscosity (ν)	L^2/θ	L^2/θ	L^2/θ
mass flow rate ($\dot{m}$)	M/θ	$F\theta/L$	M/θ
mechanical equivalent of heat (J)			FL/Q
pressure (p)	$M/L\theta^2$	F/L^2	F/L^2
surface tension (σ)	M/θ^2	F/L	F/L
angular velocity (ω)	$1/\theta$	$1/\theta$	$1/\theta$
volumetric flow rate ($\dot{m}/\rho = \dot{V}$)	L^3/θ	L^3/θ	L^3/θ
conductivity (k)	$ML/\theta^3 T$	$F/\theta T$	$Q/L\theta T$
thermal diffusivity (α)	L^2/θ	L^2/θ	L^2/θ
velocity (v)	L/θ	L/θ	L/θ
viscosity, absolute (μ)	$M/L\theta$	$F\theta/L^2$	$F\theta/L^2$
volume (V)	L^3	L^3	L^3

All other physical quantities can be derived from these primary dimensions.[12] For example, work in the SI system has units of N·m. Since a newton is a kg·m/s², the primary dimensions of work are ML^2/θ^2. The primary dimensions for many important engineering variables are shown in Table 15.7. If it is more convenient to stay with traditional English units, it may be more desirable to work in the $FML\theta TQ$ system (sometimes called the *engineering dimensional system*). This system adds the primary dimensions of force (F) and heat (Q). Thus, work (ft-lbf in the English system) has the primary dimensions of FL. (Compare this to the primary dimensions for work in the $ML\theta T$ system.) Thermodynamic variables are similarly simplified.

Dimensional analysis will be more conveniently carried out when one of the four-dimension systems ($ML\theta T$ or $FL\theta T$) is used. Whether the $ML\theta T$, $FL\theta T$, or $FML\theta TQ$ system is used depends on what is being derived and who will be using it, and whether or not a consistent set of variables is desired. Conversion constants such as g_c and J will almost certainly be required if the $ML\theta T$ system is used to generate variables for use in the English systems. It is also much more convenient to use the $FML\theta TQ$ system when working in the fields of thermodynamics, fluid flow, heat transfer, etc.

16 DIMENSIONLESS GROUPS

A *dimensionless group* is derived as a ratio of two forces or other quantities. Considerable use of dimensionless groups is made in certain subjects, notably fluid mechanics and heat transfer. For example, the Reynolds number, Mach number, and Froude number are used to distinguish between distinctly different flow regimes in pipe flow, compressible flow, and open channel flow, respectively.

Table 15.8 contains information about the most common dimensionless groups used in fluid mechanics and heat transfer.

---

[12]A *primary dimension* is the same as a *base unit* in the SI system. The SI system adds several other base units, as shown in Table 15.1, to deal with variables that are difficult to derive in terms of the four primary base units.

Table 15.8
Common Dimensionless Groups

name	symbol	formula	interpretation
Biot number	Bi	$\dfrac{hL}{k_s}$	$\dfrac{\text{surface conductance}}{\text{internal conduction of solid}}$
Cauchy number	Ca	$\dfrac{v^2}{B_s/\rho} = \dfrac{v^2}{a^2}$	$\dfrac{\text{inertia force}}{\text{compressive force}} = \text{Mach number}^2$
Eckert number	Ec	$\dfrac{v^2}{2c_p\Delta T}$	$\dfrac{\text{temperature rise due to energy conversion}}{\text{temperature difference}}$
Euler number	Eu	$\dfrac{\Delta p}{\rho V^2}$	$\dfrac{\text{pressure force}}{\text{inertia force}}$
Fourier number	Fo	$\dfrac{kt}{\rho c_p L^2} = \dfrac{\alpha t}{L^2}$	$\dfrac{\text{rate of conduction of heat}}{\text{rate of storage of energy}}$
Froude number[a]	Fr	$\dfrac{v^2}{gL}$	$\dfrac{\text{inertia force}}{\text{gravity force}}$
Graetz number[a]	Gz	$\dfrac{D}{L} \cdot \dfrac{v\rho c_p D}{k}$	$\dfrac{\dfrac{(\text{Re})(\text{Pr})}{L/D}}{\text{heat transfer by convection in entrance region}}$ over $\text{heat transfer by conduction}$
Grashof number[a]	Gr	$\dfrac{g\beta\Delta T L^3}{\nu^2}$	$\dfrac{\text{buoyancy force}}{\text{viscous force}}$
Knudsen number	Kn	$\dfrac{\lambda}{L}$	$\dfrac{\text{mean free path of molecules}}{\text{characteristic length of object}}$
Lewis number[a]	Le	$\dfrac{\alpha}{D_c}$	$\dfrac{\text{thermal diffusivity}}{\text{molecular diffusivity}}$
Mach number	M	$\dfrac{v}{a}$	$\dfrac{\text{macroscopic velocity}}{\text{speed of sound}}$
Nusselt number	Nu	$\dfrac{hL}{k}$	$\dfrac{\text{temperature gradient at wall}}{\text{overall temperature difference}}$
Péclet number	Pé	$\dfrac{v\rho c_p D}{k}$	$\dfrac{(\text{Re})(\text{Pr})}{\dfrac{\text{heat transfer by convection}}{\text{heat transfer by conduction}}}$
Prandtl number	Pr	$\dfrac{\mu c_p}{k} = \dfrac{\nu}{\alpha}$	$\dfrac{\text{diffusion of momentum}}{\text{diffusion of heat}}$
Reynolds number	Re	$\dfrac{\rho v L}{\mu} = \dfrac{v L}{\nu}$	$\dfrac{\text{inertia force}}{\text{viscous force}}$
Schmidt number	Sc	$\dfrac{\mu}{\rho D_c} = \dfrac{\nu}{D_c}$	$\dfrac{\text{diffusion of momentum}}{\text{diffusion of mass}}$
Sherwood number[a]	Sh	$\dfrac{k_D L}{D_c}$	$\dfrac{\text{mass diffusivity}}{\text{molecular diffusivity}}$
Stanton number	St	$\dfrac{h}{v\rho c_p} = \dfrac{h}{c_p G}$	$\dfrac{\text{heat transfer at wall}}{\text{energy transported by stream}}$
Stokes number	Sk	$\dfrac{\Delta p L}{\mu v}$	$\dfrac{\text{pressure force}}{\text{viscous force}}$
Strouhal number[a]	Sl	$\dfrac{L}{t_v}$	$\dfrac{\text{frequency of vibration}}{\text{characteristic frequency}}$
Weber number	We	$\dfrac{\rho v^2 L}{\sigma}$	$\dfrac{\text{inertia force}}{\text{surface tension force}}$

[a]Multiple definitions exist.

PROFESSIONAL PUBLICATIONS, INC. ● Belmont, CA

PRACTICE PROBLEMS

For the problems that follow, use these values:

$$g_{earth} = 32.2 \, ft/sec^2; \; 9.81 \, m/s^2$$

$$g_{moon} = 5.47 \, ft/sec^2; \; 1.67 \, m/s^2$$

English Systems

1. A 10-lbm object is acted upon by a 4-lbf force. What is the acceleration in ft/min^2?

2. If mud is 70% sand (density of $140 \, lbm/ft^3$) and 30% water by weight, what is the density of mud in $slugs/ft^3$?

3. How much does a 10-lbm object weigh on the moon?

4. (a) How much does a 10-slug object weigh on the moon? (b) How much does it weigh on the earth?

5. What force is acting if a 40-slug object accelerates at $8 \, ft/sec^2$?

SI System

6. A 7-kg mass is suspended by a rope. What is the tension in the rope?

7. What is the acceleration of a body that increases in velocity from 20 m/s to 40 m/s in 3 s?

8. What force accelerates a 3-kg mass at the rate of $4 \, m/s^2$?

9. What is the decrease in height of a 4-kg body whose kinetic energy increases from 20 J to 100 J as a result of moving downward along an inclined plane?

10. A 10-kg body freefalls from rest for 10 m. What is the increase in kinetic energy?

11. A force of 25 N causes an acceleration of $2 \, m/s^2$ when acting on a 7-kg mass. What is the frictional force also acting?

12. A stone is dropped over a cliff and falls for 6 s before reaching the bottom. How high above the bottom is the cliff?

13. A 10-kg object starts from a standstill and slides without friction down a gentle incline of 1:5 for a distance of 10 m. What is (a) the decrease in potential energy, (b) the final kinetic energy, and (c) the final velocity?

14. A vehicle is lifted by a hydraulic jack with a hydraulic cylinder diameter of 350 mm. The lifting force is provided by a 150-N force applied to a 35-mm plunger.

What is (a) the pressure in the liquid, and (b) the mass that can be lifted?

15. A copper laboratory tank has a diameter (at 20°C) of 0.48 m. What will be the new diameter if the tank is filled with 80°C alcohol? (The coefficient of linear expansion of copper is $1.6 \times 10^{-5} \, 1/°C$.)

16. A rigid container contains $0.15 \, m^3$ of 20°C air at a pressure of 1500 kPa. What will be the pressure if the temperature drops to 3°C?

17. A 1-kg iron container holds 5 kg of water in thermal equilibrium at 15°C. Two liters of 80°C water are subsequently added to the container. What is the final temperature of the container and the water? (Assume the specific heat constant for iron is 500 J/kg·°C.)

SI Applications

18. A steel ($E = 30 \times 10^6$ psi) rod 10 m long and 5 cm in diameter is subjected to a compressive load of $F = 1000$ N. Find the deformation of the rod.

19. Find the heat flux through a composite wall composed of 10 cm of brick ($k = 0.69 \, W/m·K$), 3 cm of insulation ($k = 0.043 \, W/m·K$), and 6 cm of oak ($k = 0.21 \, W/m·K$), if the outside temperature is 15°C ($h_{outside \, air} = 34 \, W/m^2·K$) and the inside temperature is 25°C? ($h_{inside \, air} = 9.37 \, W/m^2·K$).

20. Steam (1000 kg/s) at 400°C and 400 kPa enters a turbine and leaves at 30°C. (a) What is the quality of the exiting steam? (b) Assuming a turbine efficiency of 85%, what power can be obtained?

At 400°C and 400 kPa (superheated steam),

$$h = 3273.4 \, kJ/kg$$

$$s = 7.8992 \, kJ/kg·K$$

At 30°C,

$$h_f = 125.8 \, kJ/kg$$

$$h_{fg} = 2430.4 \, kJ/kg$$

$$h_g = 2556.2 \, kJ/kg$$

$$s_f = 0.4367 \, kJ/kg·K$$

$$s_{fg} = 8.0174 \, kJ/kg·K$$

$$h_g = 8.4541 \, kJ/kg·K$$

21. What is the total force on the bottom of a swimming pool if the pool is 20 m by 15 m and it is filled to a depth of 4 m? (At 20°C, $\rho_{water} \approx 1000 \, kg/m^3$ and $p_{atm} = 1.0 \times 10^5$ Pa.)

22. What is the specific volume (in m^3/kg) of a 37.80°C steam mixture with a quality of 95%? ($v_f = 0.01613$ ft^3/lbm, $v_{fg} = 350.3\,ft^3/lbm$.)

23. (a) Find the speed of sound in air at 0°C. (b) What percent faster is it in air at 25°C? ($R_{air} = 287.0304$ J/kg·K; $k_{air} = 1.4$.)

24. How much heat must be added to a piece of pine wood ($\rho_{pine} = 625\,kg/m^3$, $c_{pine} = 2.5\,kJ/kg$·°C) 2″ × 4″ × 5 ft to bring it from 25°C to 205°C? Assume dry wood.

25. A force of 1 N is required to turn a 6-cm-diameter doorknob. What is the torque required?

26. A ship is modeled at a scale of 1:50. What ship speed (km/hr) is represented by a model speed of 30 m/s?

27. Find the area moment of inertia about the x-axis for the object shown.

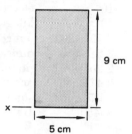

28. A 100-kg box is dragged 70 m down the sidewalk at constant velocity. What work is done if the coefficient of sliding friction between the box and the pavement is $\mu = 0.3$?

29. A train with mass of 8×10^3 kg is traveling at 100 km/hr. (a) What is its kinetic energy? (b) If the train runs out of fuel while traveling up a grade, how high will it climb before stopping, assuming frictionless rails?

30. A 73-kg man stands on a ladder. If the distance from his heel to the ball of his foot, where the load is carried, is 20 cm, what is the bending moment resisted by his ankles?

APPENDIX A
Conversion Factors

multiply	by	to obtain
acres	0.4047	hectares
	43,560.0	square feet
	1.5625×10^{-3}	square miles
ampere hours	3600.0	coulombs
angstrom units	3.937×10^{-9}	inches
	1×10^{-4}	microns
astronomical units	1.496×10^{8}	kilometers
atmospheres	76.0	centimeters of mercury
atomic mass unit	9.316×10^{8}	electron-volts
	1.492×10^{-10}	joules
	1.66×10^{-27}	kilograms
BeV (also GeV)	10^{9}	electron-volts
BTU	3.93×10^{-4}	horsepower-hours
	778.3	foot-pounds
	2.93×10^{-4}	kilowatt hours
	1.0×10^{-5}	therms
BTU/hr	0.2161	foot-pounds/sec
	3.929×10^{-4}	horsepower
	0.293	watts
bushels	2150.4	cubic inches
calories, gram (mean)	3.9683×10^{-3}	BTU (mean)
centares	1.0	square meters
centimeters	1×10^{-5}	kilometers
	1×10^{-2}	meters
	10.0	millimeters
	3.281×10^{-2}	feet
	0.3937	inches
chains	792.0	inches
coulombs	1.036×10^{-5}	faradays
cubic centimeters	0.06102	cubic inches
	2.113×10^{-3}	pints (U.S. liquid)
cubic feet	0.02832	cubic meters
	7.4805	gallons
cubic feet/min	62.43	pounds H_2O/min
cubic feet/sec	448.831	gallons/min
	0.64632	millions of gallons per day
cubits	18.0	inches
days	86,400.0	seconds
degrees (angle)	1.745×10^{-2}	radians
degrees/sec	0.1667	revolutions/min
dynes	1×10^{-5}	newtons
electron volts	1.074×10^{-9}	atomic mass units
	10^{-9}	BeV (also GeV)
	1.602×10^{-19}	joules
	1.783×10^{-36}	kilograms
	10^{-6}	MeV
faradays/sec	96,500	amperes (absolute)
fathoms	6.0	feet
feet	30.48	centimeters
	0.3048	meters
	1.645×10^{-4}	miles (nautical)
	1.894×10^{-4}	miles (statute)

multiply	by	to obtain
feet/min	0.5080	centimeters/sec
feet/sec	0.592	knots
	0.6818	miles/hour
foot-pounds	1.285×10^{-3}	BTU
	5.051×10^{-7}	horsepower-hr
	3.766×10^{-7}	kilowatt-hours
foot-pound/sec	4.6272	BTU/hr
	1.818×10^{-3}	horsepower
	1.356×10^{-3}	kilowatts
furlongs	660.0	feet
	0.125	miles (U.S.)
gallons	0.1337	cubic feet
	3.785	liters
gallons H_2O	8.3453	pounds H_2O
gallons/min	8.0208	cubic feet/hr
	0.002228	cubic feet/sec
GeV (also BeV)	10^{9}	electron-volts
grams	10^{-3}	kilograms
	3.527×10^{-2}	ounces (avoirdupois)
	3.215×10^{-2}	ounces (troy)
	2.205×10^{-3}	pounds
hectares	2.471	acres
	1.076×10^{5}	square feet
horsepower	2545.0	BTU/hr
	42.44	BTU/min
	550	foot-pounds/sec
	0.7457	kilowatts
	745.7	watts
horsepower-hour	2545.0	BTU
	1.976×10^{6}	foot-pounds
	0.7457	kilowatt hours
hours	4.167×10^{-2}	days
	5.952×10^{-3}	weeks
inches	2.540	centimeters
	1.578×10^{-5}	miles
inches, H_2O	5.199	lbf/ft^2
	0.0361	psi
	0.0735	inches, mercury
inches, mercury	70.7	lbf/ft^2
	0.491	psi
	13.60	inches, H_2O
joules	6.705×10^{9}	atomic mass units
	9.480×10^{-4}	BTU
	1×10^{7}	ergs
	6.242×10^{18}	electron-volts
	1.113×10^{-17}	kilograms
kilograms	6.025×10^{26}	atomic mass units
	5.610×10^{35}	electron-volts
	8.987×10^{16}	joules
	2.205	pounds
kilometers	3281.0	feet
	1000.0	meters
	0.6214	miles
kilometers/hr	0.5396	knots

APPENDIX A   (continued)
Conversion Factors

multiply	by	to obtain	multiply	by	to obtain
kilowatts	3412.9	BTU/hr	pints (liquid)	473.2	cubic centimeters
	737.6	foot-pound/sec		28.87	cubic inches
	1.341	horsepower		0.125	gallons
kilowatt-hours	3413.0	BTU		0.5	quarts (liquid)
knots	6076.0	feet/hr	poise	0.002089	pound-sec/ft^2
	1.0	nautical miles/hr	pounds	0.4536	kilograms
	1.151	statute miles/hr		16.0	ounces
light years	5.9×10^{12}	miles		14.5833	ounces (troy)
links (surveyor)	7.92	inches		1.21528	pounds (troy)
liters	1000.0	cubic centimeters	pounds/ft^2	0.006944	pounds/in^2
	61.02	cubic inches	pounds/in^2	2.308	feet, H_2O
	0.2642	gallons (U.S. liquid)		27.7	inches, H_2O
	1000.0	milliliters		2.037	inches, mercury
	2.113	pints		144	pounds/ft^2
MeV	10^6	electron-volts	quarts (dry)	67.20	cubic inches
meters	100.0	centimeters	quarts (liquid)	57.75	cubic inches
	3.281	feet		0.25	gallons
	1×10^{-3}	kilometers		0.9463	liters
	5.396×10^{-4}	miles (nautical)	radians	57.30	degrees
	6.214×10^{-4}	miles (statute)		3438.0	minutes
	1000.0	millimeters	revolutions	360.0	degrees
microns	1×10^{-6}	meters	revolutions/min	6.0	degrees/sec
miles (nautical)	6076.27	feet	rods	16.5	feet
	1.853	kilometers		5.029	meters
	1.1516	miles (statute)	rods (surveyor)	5.5	yards
miles (statute)	5280.0	feet	seconds	1.667×10^{-2}	minutes
	1.609	kilometers	square meters/sec	10^6	centistokes
	0.8684	miles (nautical)		10.76	square feet/sec
miles/hr	88.0	feet/min		10^4	stokes
milligrams/liter	1.0	parts/million	slugs	32.174	pounds
milliliters	1×10^{-3}	liters	stokes	0.001076	square feet/sec
millimeters	3.937×10^{-2}	inches	tons (long)	1016.0	kilograms
newtons	1×10^5	dynes		2240.0	pounds
ohms				1.120	tons (short)
(international)	1.0005	ohms (absolute)	tons (short)	907.1848	kilograms
ounces	28.349527	grams		2000.0	pounds
	6.25×10^{-2}	pounds		0.89287	tons (long)
ounces (troy)	1.09714	ounces (avoirdupois)	volts (absolute)	3.336×10^{-3}	statvolts
parsecs	3.06×10^{13}	kilometers	watts	3.4129	BTU/hr
	1.9×10^{13}	miles		1.341×10^{-3}	horsepower
pascal-sec	1×10^{-3}	centipoise	yards	0.9144	meters
	0.1	poise		4.934×10^{-4}	miles (nautical)
	47.88	pound force-sec/ft^2		5.682×10^{-4}	miles (statute)
	1.488	pound mass/ft-sec			
	47.88	slug/ft-sec			

APPENDIX B
Common SI Unit Conversion Factors

multiply	by	to obtain
AREA		
circular mil	506.7	square micrometer
square feet	0.0929	square meter
square kilometer	0.3861	square mile
square meter	10.764	square feet
	1.196	square yard
square micrometer	0.001974	circular mil
square mile	2.590	square kilometer
square yard	0.8361	square meter
ENERGY		
BTU (international)	1.0551	kilojoule
erg	0.1	microjoule
foot-pound	1.3558	joule
horsepower-hour	2.6485	megajoule
joule	0.7376	foot-pound
	0.10197	meter-kilogram force
kilogram-calorie (international)	4.1868	kilojoule
kilojoule	0.9478	BTU
	0.2388	kilogram-calorie
kilowatt hour	3.6	megajoule
megajoule	0.3725	horsepower-hour
	0.2278	kilowatt-hour
	0.009478	therm
meter-kilogram force	9.8067	joule
microjoule	10.0	erg
therm	105.506	megajoule
FORCE		
dyne	10.0	micronewton
kilogram force	9.8067	newton
kip	4448.2	newton
micronewton	0.1	dyne
newton	0.10197	kilogram force
	0.0002248	kip
	3.597	ounce force
	0.2248	pound force
ounce force	0.2780	newton
pound force	4.4482	newton
HEAT		
BTU/ft^2-hr	3.1546	watt/m^2
BTU/ft^2-°F	5.6783	watt/m^2·°C
BTU/ft^3	0.0373	megajoule/m^3
BTU/ft^3-°F	0.06707	megajoule/m^3·°C
BTU/hr	0.2931	watt
BTU/lbm	2326	joule/kg
BTU/lbm-°F	4186.8	joule/kg·°C
BTU-inch/ft^2-hr-°F	0.1442	watt/meter·°C
joule/kg	0.000430	BTU/lbm
joule/kg·°C	0.0002388	BTU/lbm-°F
megajoule/m^3	26.839	BTU/ft^3
megajoule/m^3·°C	14.911	BTU/ft^3-°F

PROFESSIONAL PUBLICATIONS, INC. ● Belmont, CA

APPENDIX B (continued)
Common SI Unit Conversion Factors

watt	3.4121	BTU/hr
watt/m·°C	6.933	BTU-inch/ft^2-hr-°F
watt/m^2	0.3170	BTU/ft^2-hr
watt/m^2·°C	0.1761	BTU/ft^2-hr

LENGTH

Angstrom	0.1	nanometer
foot	0.3048	meter
inch	25.4	millimeter
kilometer	0.6214	mile
	0.540	mile (international nautical)
meter	3.2808	foot
	1.0936	yard
micrometer	1.0	micron
micron	1.0	micrometer
mil	0.0254	millimeter
mile	1.6093	kilometer
mile (international nautical)	1.852	kilometer
millimeter	0.0394	inch
	39.370	mil
nanometer	10.0	Angstrom
yard	0.9144	meter

MASS (weight)

grain	64.799	milligram
gram	0.0353	ounce (avoirdupois)
	0.03215	ounce (troy)
kilogram	2.2046	pounds-mass
	0.068522	slug
	0.0009842	ton (long—2240 lbm)
	0.001102	ton (short—2000 lbm)
milligram	0.0154	grain
ounce (avoirdupois)	28.350	gram
ounce (troy)	31.1035	gram
pounds-mass	0.4536	kilogram
slug	14.5939	kilogram
ton (long—2240 lbm)	1016.047	kilogram
ton (short—2000 lbm)	907.185	kilogram

PRESSURE

bar	100.0	kilopascal
inch, H$_2$O (20°C)	0.2486	kilopascal
inch, Hg (20°C)	3.3741	kilopascal
kilogram force/cm^2	98.067	kilopascal
kilopascal	0.01	bar
	4.0219	inch, H$_2$O (20°C)
	0.2964	inch, Hg (20°C)
	0.0102	kilogram force/cm^2
	7.528	millimeter Hg (20°C)
	0.1450	psi
	0.009869	standard atmosphere (760 torr)
	7.5006	torr
millimeter Hg (20°C)	0.13284	kilopascal
psi	6.8948	kilopascal
standard atmosphere (760 torr)	101.325	kilopascal
torr	0.13332	kilopascal

PROFESSIONAL PUBLICATIONS, INC. ● Belmont, CA

APPENDIX B   (continued)
Common SI Unit Conversion Factors

POWER

BTU (international)/hr	0.2931	watt
foot-pound/sec	1.3558	watt
horsepower	0.7457	kilowatt
kilowatt	1.341	horsepower
	0.2843	tons of refrigeration
meter-kilogram force/second	9.8067	watt
tons of refrigeration	3.517	kilowatt
watt	3.4122	BTU (international)/hr
	0.7376	foot-pound/sec
	0.10197	meter-kilogram force/second

TEMPERATURE

Celsius	$\left(\frac{9}{5}{}^{\circ}C + 32\right)$	Fahrenheit
Fahrenheit	$\frac{5}{9}({}^{\circ}F - 32)$	Celsius
Kelvin	$\frac{9}{5}$	Rankine
Rankine	$\frac{5}{9}$	Kelvin

TORQUE

gram force centimeter	0.098067	millinewton·meter
kilogram force meter	9.8067	newton-meter
millinewton	10.197	gram force centimeter
newton-meter	0.10197	kilogram force meter
	0.7376	foot pound
	8.8495	inch pound
foot-pound	1.3558	newton-meter
inch-pound	0.1130	newton-meter

VELOCITY

feet/sec	0.3048	meters/second
kilometers/hr	0.6214	miles/hr
meters/second	3.2808	feet/sec
	2.2369	miles/hr
miles/hr	1.60934	kilometers/hr
	0.44704	meters/second

VISCOSITY

centipoise	1.0	millipascal-seconds
centistoke	1.0	micrometer^2/second
millipascal-second	1.0	centipoise
micrometer^2/second	1.0	centistoke

APPENDIX B (continued)
Common SI Unit Conversion Factors

VOLUME (capacity)

cubic centimeter	0.06102	cubic inch
cubic foot	28.3168	liter
cubic inch	16.3871	cubic centimeter
cubic meter	1.308	cubic yard
cubic yard	0.7646	cubic meter
gallon (U.S.)	3.785	liter
liter	0.2642	gallon (U.S.)
	2.113	pint (U.S. fluid)
	1.0567	quart (U.S. fluid)
	0.03531	cubic foot
milliliter	0.0338	ounce (U.S. fluid)
ounce (U.S. fluid)	29.574	milliliter
pint (U.S. fluid)	0.4732	liter
quart (U.S. fluid)	0.9464	liter

VOLUME FLOW (gas–air)

cubicmeter/sec	2119.	standard cubic foot/min
liter/sec	2.119	standard cubic foot/min
microliter/sec	0.000127	standard cubic foot/hr
milliliter/sec	0.002119	standard cubic foot/min
	0.127133	standard cubic foot/hr
standard cubic foot/min	0.0004719	cubic meter/sec
	0.4719	liter/sec
	471.947	milliliter/sec
standard cubic foot/hr	7866.0	microliter/sec
	7.8658	milliliter/sec

VOLUME FLOW (liquid)

gallon/hr (U.S.)	0.001052	liter/sec
gallon/min (U.S.)	0.06309	liter/sec
liter/sec	951.02	gallon/hr (U.S.)
	15.850	gallon/min (U.S.)

PROFESSIONAL PUBLICATIONS, INC. ● Belmont, CA

16 MANAGEMENT THEORIES

1 INTRODUCTION

Effective management techniques are based on behavioral science studies. Behavioral science is an outgrowth of the *human relations theories* of the 1930's, in which the happiness of employees was the goal (i.e., a happy employee is a productive employee...). Current behavioral science theories emphasize minimizing tensions that inhibit productivity.

There is no evidence yet that employees want a social aspect to their jobs. Nor is there evidence that employees desire job enlargement or autonomy. Behavioral science makes these assumptions anyway.[1]

2 BEHAVIORAL SCIENCE KEY WORDS

Cognitive system: method we use to interpret our environment.

Collaboration: influence through a mutual agreement, relationship, respect, or understanding, without a formal or contractual authority relationship.

Equilibrium: maintaining the status quo of a group or an individual.

Job enrichment: letting employees have more control over their activities and working conditions.

Manipulation: influencing others by recognizing and building upon their needs.

MBO: management by objectives—setting job responsibilities and standards for each group and employee.

Normative judgment: judging others according to our own values.

Paternalism: corporate subsidy—showering employees with benefits and expecting submission in return.

Personal map: a person's expectations of his environment.

Selective perception: seeing what we want to see—a form of defense mechanism, since things first must be perceived to be ignored.

Superordinate goals: goals which are outside of the individual, such as corporate goals.

3 HISTORY OF BEHAVIORAL SCIENCE STUDIES AND THEORIES

A. Hawthorne Experiments

From 1927 to 1932, the Hawthorne (Chicago) Works of the Western Electric Company experimented with working conditions in an attempt to determine what factors affected output.[2] Six average employees were chosen to assemble and inspect phone relays.

Many factors were investigated in this exhaustive test. After weeks of observation without making any changes (to establish a baseline), Western Electric varied the number and length of breaks, the length of the work day, the length of the work week, and the illumination level of the lighting. Group incentive plans were tried, and in several tests, the company even provided food during breaks and lunch periods.

The employees reacted in the ways they thought they should. Output (as measured in relay production) increased after every change was implemented. In effect, the employees reacted to the attention they received, regardless of the working conditions.

---

[1] In an exhaustive literature survey up through 1955, researchers found no conclusive relationships between satisfaction and productivity. There was, however, a relationship between lack of satisfaction and absenteeism and turnover.

[2] Experiments were conducted by Elton Mayo from the Harvard Business School.

Western Electric concluded that there was no relationship between illumination and other conditions to productivity. The increase in productivity during the testing procedure was attributed to the sense of value each employee felt in being part of an important test.

The employees also became a social group, even after hours. Leadership and common purpose developed, and even though the employees were watched more than ever, they felt no supervision anxiety since they were, in effect, free to react in any way they wanted.

One employee summed up the test when she said, "It was fun."

B. Bank Wiring Observation Room Experiments

In an attempt to devise an experiment which would not suffer from the problems associated with the Hawthorne studies, Western Electric conducted experiments in 1931 and 1932 on the effects of wage incentives.

The group of nine wiremen, three soldermen, and two inspectors was interdependent. This was supposed to prevent any individual from slacking off. However, wage incentives failed to improve productivity. In fact, fast employees slowed down to protect their slower friends. Illicit activities, such as job trading and helping, also occurred.

The group was reacting to the notion of a *proper day's work*. When the day's work (or what the group considered to be a day's work) was assured, the whole group slacked off. The group also varied what it reported as having been accomplished and claimed more unavoidable delays than actually occurred. The output was essentially constant.

Western Electric concluded that social groups form as protection against arbitrary management decisions, even when such decisions have never been experienced. The effort to form the social groups, to protect slow workers, and to develop the notion of a proper day's work is not conscious. It develops automatically when the company fails to communicate to the contrary.

C. Need Hierarchy Theory

During World War II, Dr. Abraham Maslow's *need hierarchy theory* was implemented into leadership training for the U.S. Air Force. This theory claims that certain needs become dominant when lesser needs are satisfied. Although some needs can be sublimated and others overlap, the need hierarchy theory generally requires the lower-level needs to be satisfied before the higher-level needs are realized. (The ego and self-fulfillment needs rarely are satisfied.)

Table 16.1
The Need Hierarchy

(In order of lower to higher needs)

1. Physiological needs: air, food, water.

2. Safety needs: protection against danger, threat, deprivation, arbitrary management decisions. Need for security in a dependent relationship.

3. Social needs: belonging, association, acceptance, giving and receiving of love and friendship.

4. Ego needs: self-respect and confidence, achievement, self-image. Group image and reputation, status, recognition, appreciation.

5. Self-fulfillment needs: realizing self-potential, self-development, creativity.

The need hierarchy theory explains why money is a poor motivator of an affluent individual. The theory does not explain how management should apply the need hierarchy to improve productivity.

D. Theory of Influence

In 1948, the Human Relations Program (under the direction of Donald C. Pelz) at Detroit Edison studied the effectiveness of its supervisors. The most effective supervisors were those who helped their employees benefit. Supervisors who were close to their employees (and sided with them in disputes) were effective only if they were influential enough to help the employees. The study results were formulated into the *theory of influence*.

- Employees think well of supervisors who help them reach their goals and meet their needs.

- An influential supervisor will be able to help employees.

- An influential supervisor who is also a disciplinarian will breed dissatisfaction.

- A supervisor with no influence will not be able to affect worker satisfaction in any way.

The implication of the theory of influence is that whether or not a supervisor is effective depends on his influence. Training of supervisors is useless unless they have the power to implement what they have learned. Also, increases in supervisor influence are necessary to increase employee satisfaction.

PROFESSIONAL PUBLICATIONS ● Belmont, CA

E. Herzberg Motivation Studies

Frederick W. Herzberg interviewed 200 technical personnel in 11 firms during the late 1950's. Herzberg was especially interested in exceptional occurrences resulting in increases in job satisfaction and performance. From those interviews, Herzberg formulated his *motivation-maintenance theory*.

According to this theory, there are satisfiers and dissatisfiers which influence employee behavior. The *dissatisfiers* (also called *maintenance/motivation factors*) do not motivate employees; they can only dissatisfy them. However, the dissatisfiers must be eliminated before the satisfiers work. Dissatisfiers include company policy, administration, supervision, salary, working conditions (environment), and interpersonal relations.

Satisfiers (also known as *motivators*) determine job satisfaction. Common satisfiers are achievement, recognition, the type of work itself, responsibility, and advancement.

An interesting conclusion based on the motivation-maintenance theory is that fringe benefits and company paternalism do not motivate employees since they are related to dissatisfiers only.

F. Theory X and Theory Y

During the 1950's, Douglas McGregor (Sloan School of Industrial Management at MIT) introduced the concept that management had two ways of thinking about its employees. One way of thinking, which was largely pessimistic, was theory X. The other theory, theory Y, was largely optimistic.

Theory X is based on the assumption that the average employee inherently dislikes and avoids work. Therefore, employees must be coerced into working by threats of punishment. Rewards are not sufficient. The average employee wants to be directed, avoids responsibility, and seeks the security of an employer-employee relationship.

This assumption is supported by much evidence. Employees exist in a continuum of wants, needs, and desires. Many of the need satisfiers (salary, fringe benefits, etc.) are effective only off the job. Therefore, work is considered a punishment or a price paid for off-the-job satisfaction.

Theory X is pessimistic about the effectiveness of employers to satisfy or motivate their employees. By satisfying the physiological and safety (lower level) needs, employers have shifted the emphasis to higher level needs which they cannot satisfy. Employees, unable to derive satisfaction from their work, behave according to theory X.

Theory Y, on the other hand, assumes that the expenditure of effort is natural and is not inherently disliked. It assumes that the average employee can learn to accept and enjoy responsibility. Creativity is widely distributed among employees, and the potentials of average employees are only partially realized.

Theory Y places the blame for worker laziness, indifference, and lack of cooperation in the lap of management, since the integration of individual and organization needs is required. This theory is not fully validated, nor is its full use ever likely to be implemented.[3]

4 JOB ENRICHMENT

In an effort to make their employees happier, companies have tried to enrich the jobs performed by employees. Enrichment is a subjective result felt by employees when their jobs are made more flexible or are enlarged. Adding flexibility to a job allows an employee to move from one task to another, rather than doing the same thing continually. Horizontal job enlargement adds new production activities to a job. Vertical job enlargement adds planning, inspection, and other nonproduction tasks to the job.

There are advantages to keeping a job small in scope. Learning time is low, employee mental effort is reduced, and the pay rate can be lower for untrained labor. Supervision is reduced. Such simple jobs, however, also result in high turnover, absenteeism, and lower pride in job (and subsequent low quality rates).

Job enlargement generally results in better quality products, reduces inspection and material handling, and counteracts the disadvantages previously mentioned. However, training time is greater, tooling costs are higher, and inventory records are more complex.

5 QUALITY IMPROVEMENT PROGRAMS

A. Zero Defects Program

Employees have been conditioned to believe that they are not perfect and that errors are natural. However, we demand zero defects from some professions (e.g., doctors, lawyers, engineers). The philosophy of a zero defect program is to expect zero defects from everybody.

---

[3] Theory Y is not synonymous with soft management. Rather than emphasize tough management (as does theory X), theory Y depends on commitment of employees to achieve mutual goals.

Zero defects programs develop a constant, conscious desire to do the job right the first time. This is accomplished by giving employees constant awareness that their jobs are important, that the product is important, and that management thinks their efforts are important.

Zero defects programs try to correct the faults of other types of employee programs.[4] Programs are based on what the employee has for his own: pride and desire. The programs present the challenge of perfection and explain the importance of that perfection. Management sets an example by expecting zero defects of itself. Standards of performance are set and are related to each employee. Employees are checked against these performance requirements periodically, and recognition is given when goals are met.

B. Quality Circles/Team Programs

Quality circle programs are voluntary or required programs in which employees within a department actively participate in measuring and improving quality and performance. It involves periodic meetings on a weekly or a monthly basis. Workers are encouraged to participate in volunteering ideas for improvement.

---

[4] Motivational programs are not honest, according to the zero-defects theory, since management tries to convince employees to do what management wants. Wage incentive programs encourage employee dishonesty and errors by emphasizing quantity, not quality. Theory X management, with its implied punitive action if goals are not achieved, never has been effective.

17 ENGINEERING LICENSING

Purpose of Registration

As an engineer, you may have to obtain your professional engineering license through procedures that have been established by the state in which you work. These procedures are designed to protect the public by preventing unqualified individuals from legally practicing as engineers.

There are many reasons for wanting to become a professional engineer. Among them are the following:

- You may wish to become an independent consultant. By law, consulting engineers must be registered.

- Your company may require a professional engineering license as a requisite for employment or advancement.

- Your state may require registration as a professional engineer if you want to use the title *engineer*.

The Registration Procedure

The registration procedure is similar in most states. You will probably take two eight-hour written examinations. The first examination is the *Fundamentals of Engineering* examination, also known as the *Intern Engineer* exam and the *Engineer-In-Training* exam. The initials F.E., I.E., and E-I-T are also used. The second examination is the *Professional Engineering* (P.E.) exam, which differs from the E-I-T exam in format and content.

If you have advanced degrees, other licenses, or significant experience in engineering, you may be allowed to skip the E-I-T examination. However, actual details of registration, experience requirements, minimum education levels, fees, and examination schedules vary from state to state. You should contact your state's Board of Registration for Professional Engineers.

Reciprocity Among States

All states use the NCEES P.E. examination.[1] If you take and pass the P.E. examination in one state, your license will probably be honored by other states that have used the same NCEES examination. It will not be necessary to retake the P.E. examination.

The simultaneous administration of identical examinations in multiple states has led to the term *Uniform Examination*. However, each state is free to choose its own minimum passing score or to add special questions to the NCEES examination. Therefore, this Uniform Examination does not automatically ensure reciprocity among states.

Of course, you may apply for and receive a professional engineering license from another state. However, a license from one state will not permit you to practice engineering in another state. You must have a professional engineering license from each state in which you work.

Applying for the Examination

Each state charges different fees, requires different qualifications, and uses different forms. Therefore, it will be necessary for you to request an application and an information packet from the state in which you plan to take the exam. It generally is sufficient to phone for this information. Telephone numbers for all of the U.S. state boards of registration are given in the accompanying list.

---

[1] The National Council of Examiners for Engineering and Surveying (NCEES) in Clemson, South Carolina produces, distributes, and grades the national examinations. It does not distribute applications to take the P.E. examination.

Phone Numbers of State Boards of Registration

Wisconsin	(608) 266-5511
Wyoming	(307) 777-6155

Alabama	(334) 242-5568
Alaska	(907) 465-2540
Arizona	(602) 255-4053
Arkansas	(501) 324-9085
California	(916) 263-2222
Colorado	(303) 894-7788
Connecticut	(860) 566-3290
Delaware	(302) 577-6500
District of Columbia	(202) 727-7454
Florida	(904) 488-9912
Georgia	(404) 656-3926
Guam	(671) 646-3115
Hawaii	(808) 586-3000
Idaho	(208) 334-3860
Illinois	(217) 782-8556
Indiana	(317) 232-2980
Iowa	(515) 281-5602
Kansas	(913) 296-3053
Kentucky	(502) 573-2680
Louisiana	(504) 295-8522
Maine	(207) 287-3236
Maryland	(410) 333-6322
Massachusetts	(617) 727-9957
Michigan	(517) 241-9253
Minnesota	(612) 296-2388
Mississippi	(601) 359-6160
Missouri	(573) 751-0047
Montana	(406) 444-4285
Nebraska	(402) 471-2407
Nevada	(702) 688-1231
New Hampshire	(603) 271-2219
New Jersey	(201) 504-6460
New Mexico	(505) 827-7561
New York	(518) 474-3846
North Carolina	(919) 881-2293
North Dakota	(701) 258-0786
Ohio	(614) 466-3650
Oklahoma	(405) 521-2874
Oregon	(503) 378-4180
Pennsylvania	(717) 783-7049
Puerto Rico	(787) 722-0058
Rhode Island	(401) 277-2565
South Carolina	(803) 896-4422
South Dakota	(605) 394-2510
Tennessee	(615) 741-3221
Texas	(512) 440-7723
Utah	(801) 530-6551
Vermont	(802) 828-2875
Virginia	(804) 367-8512
Virgin Islands	(809) 774-3130
Washington	(360) 753-6966
West Virginia	(304) 558-3554

Examination Dates

The NCEES examinations are administered on the same weekend in all states. Each state determines independently whether to offer the examination on Thursday, Friday, or Saturday of the examination period. The upcoming examination dates are given in the accompanying listing.

National P.E. Examination Dates

year	Spring dates	Fall dates
1998	April 24–25	October 30–31
1999	April 23–24	October 29–30
2000	April 14–15	October 27–28

Examination Format

The NCEES Professional Engineering examination in Electrical Engineering consists of two four-hour sessions separated by a one-hour lunch period. Both the morning and the afternoon sessions contain twelve problems. Each multiple-choice problem is made up of ten individual multiple-choice questions, each with four answer options. Most states do not have required problems.

Each examinee is given an exam booklet that contains problems for civil, mechanical, electrical, and chemical engineers. Some states, such as California, will allow you to work problems only from the electrical part of the booklet. Other states will allow you to work problems from the entire booklet. Read the examination instructions on this point carefully.

NCEES has released the following analysis of subjects on a typical electrical P.E. exam. However, there is no guarantee that this structure will be adhered to in every case.

Technical Specialty Areas (TSA) in the Discipline of Electrical Engineering

major work behavior	number of items
fundamental design of generation systems	1
final design and applications of generation systems	1

fundamental design of transmission and distribution systems	1
final design and applications of transmission and distribution systems	2
final design and applications of rotating machines	1
final design and applications of instrumentation	1
final design and applications of lightning protection and grounding	1
design control systems	2
design electronic devices	2
applications of electronic devices	1
design of instrumentation	1
applications of instrumentation	1
design of digital systems	2
design of computer systems	2
applications of digital systems	1
design of communication systems	2
applications of communication systems	1
design of biomedical systems	1

Since the examination structure is not rigid, it is not possible to give the exact number of problems that will appear in each subject area. There is no guarantee that any single subject will appear. Many problems require a variety of approaches, including design, analysis, applications, and operations. Economic aspects may appear in some problems.

The examination is open book. With rare exceptions, all forms of solution aids are allowed in the examination, including nomographs, specialty slide rules, and pre-programmed and programmable calculators. Since their use says little about the depth of your knowledge, such aids should be used only to check your work.

Most states do not limit the number and types of books you can bring into the exam.[2] Loose-leaf papers (including Post-it™ notes) and writing tablets are usually forbidden, although you may be able to bring in loose reference pages in a three-ring binder. References used in the afternoon session need not be the same as for the morning session.

Any battery-powered, silent calculator may be used. Most states have no restrictions on programmable or pre-programmed calculators. To ensure exam security,

---

[2] Check with your state to see if review books can be brought into the examination. Most states do not have any restrictions. Some states ban only collections of solved problems, such as Schaum's Outline Series. A few states prohibit all review books.

however, printers and calculators with alphanumeric memories and other word processing functions may be prohibited at the state's discretion. Printers cannot be used.

You will not be permitted to share books, calculators, or any other items with other examinees.

You will receive the results of your examination by mail. Allow 12–16 weeks for notification. Your score may or may not be revealed to you, depending on your state's procedure.

Preparing for the Exam

You should develop an examination strategy early in the preparation process. This strategy will depend on your background. One of the following two general strategies is typically used:

- Examinees who have recently completed academic studies found a broad approach to be successful. Their strategy was to review the fundamentals in a broad range of undergraduate electrical engineering subjects. The examination includes enough fundamental problems to give merit to this strategy.

- Working engineers who have been away from classroom work for a long time found it better to concentrate on the subjects in which they had extensive professional experience. By studying the list of examination subjects, they were able to choose those subjects that would give them a high probability of finding enough problems that they could solve.

Do not make the mistake of studying only a few subjects in hopes of finding enough problems to work. The more subjects you are familiar with, the better your chances of passing the examination will be. More important than strategy are fast recall and stamina. You must be able to recall solution procedures, formulas, and important data quickly, and this sharpness must be maintained for eight hours.

In order to develop this recall and stamina, you should work the sample problems at the end of each chapter and compare your answers to the solutions in the accompanying *Solutions Manual*. This will enable you to become familar with problem types and solution methods. You will not have time in the exam to derive solution methods; you must know them instinctively.

It is imperative that you develop and adhere to a review outline and schedule. If you are not taking a classroom review course where the order of preparation is determined by the lectures, you should use the accompanying *Outline of Subjects for Self-Study* to schedule your preparation.

Outline of Subjects for Self-Study
(Subjects do not have to be studied in the order listed.)

subject	chapter	recommended number of weeks	date to be started	date to be completed	complete
mathematics	1	1	_____	_____	☐
linear circuit analysis	2	1	_____	_____	☐
waveforms, power, and measurements	3	1	_____	_____	☐
time and frequency response	4	1	_____	_____	☐
power systems	5	1	_____	_____	☐
transmission lines	6	1	_____	_____	☐
rotating machines	7	1	_____	_____	☐
fundamental semiconductor circuits	8	1	_____	_____	☐
amplifier applications	9	1	_____	_____	☐
waveshaping, logic, and data conversion	10	1	_____	_____	☐
digital logic	11	1	_____	_____	☐
control systems	12	1	_____	_____	☐
illumination and the national electric code	13	1	_____	_____	☐
economic analysis	14	1	_____	_____	☐
systems of units	15	0	_____	_____	☐
management theories	16	0	_____	_____	☐
engineering licensing	17	0	_____	_____	☐
postscripts	18	0	_____	_____	☐

It is unnecessary to take a large quantity of books to the examination. This book, a dictionary, and three to five other references of your choice should be sufficient. The examination is very fast-paced. You will not have time to use books with which you are not thoroughly familiar.

To minimize time spent searching for often-used formulas and data, you should prepare a one-page summary of all the important formulas and information in each subject area. You can then use these summaries during the examination instead of searching for the correct page in your book.

Items to Get for the Examination

- Obtain ten sheets of each of the following types of graph paper: 10 squares to the inch grid, semi-log (3 cycles × 10 squares to the inch grid), and full-log (3 cycles × 3 cycles).

- Obtain a long, flexible, clear plastic ruler marked in tenths of an inch or in millimeters.

- From Professional Publications, Inc., the following special study aids will prove valuable:

  · *Expanded Interest Tables for Economic Analysis Problems*

  · *Engineering Law, Design Liability, and Professional Ethics*

  · *Electrical Engineering Sample Examination*

What to Do Before the Exam

The engineers who have taken the P.E. exam in previous years have made the suggestions listed below. These suggestions will make your examination experience as comfortable and successful as possible.

- Keep a copy of your examination application. Send the original application by certified mail and request a receipt of delivery. Tape your delivery receipt in the space indicated on the first page of this book.

- Visit the exam site the day before your examination. This is especially important if you are not familiar with the area. Find the examination room, the parking area, and the rest rooms.

- Plan on arriving 30–60 minutes before the examination starts. This will assure you a convenient parking place and adequate time for site, room, and seating changes.

- If you live a considerable distance from the examination site, consider getting a hotel room in which to spend the night before.

- Take the day before the examination to relax. Don't cram the last night. Rather, get a good night's sleep.

- Be prepared to find that the examination room is not ready at the designated time. Take an interesting novel or magazine to read in the interim and at lunch.

- If you make arrangements for babysitters or transportation, allow for a delayed completion.

- Prepare your examination kit the day before. Here is a checklist of items to take with you to the examination. (Some states may not permit food or beverages in the exam room.)

  [ ] copy of your application

  [ ] proof of delivery receipt

  [ ] letter admitting you to the exam

  [ ] photographic identification

  [ ] this book

  [ ] *Phillips Lighting Handbook*

  [ ] *Elements of Power System Analysis* (W. D. Stevenson, Jr.)

  [ ] *National Electric Code* (National Fire Protection Association)

  [ ] other books of your choice, including:

    [ ] CRC math tables

    [ ] digital logic book

    [ ] linear control systems analysis book

    [ ] engineering circuit analysis book

    [ ] scientific dictionary

    [ ] standard English dictionary

  [ ] review course notes in a binder

  [ ] calculator and a spare

  [ ] spare calculator batteries or battery pack

  [ ] battery charger and 20 foot extension cord

  [ ] chair cushions (a large, thick bath mat works well)

  [ ] earplugs

  [ ] desk expander—if you are taking the exam in theater chairs with tiny, fold-up writing surfaces, you should take a long, wide board to place across the armrests

  [ ] cardboard box cut to fit your references

  [ ] twist-to-advance pencils with extra leads

[ ] number 2 pencils

[ ] large eraser

[ ] snacks

[ ] beverage in a thermos

[ ] light lunch

[ ] collection of graph paper (use only if permitted in exam)

[ ] transparent and masking tapes

[ ] sunglasses

[ ] extra prescription glasses, if you wear them

[ ] aspirin

[ ] travel pack of Kleenex and other personal needs

[ ] $2.00 in change, $10.00 in cash

[ ] light, comfortable sweater

[ ] comfortable shoes or slippers for the exam room

[ ] raincoat, boots, gloves, hat, and umbrella

[ ] local street maps

[ ] note to the parking patrol for your windshield

[ ] pad of three-hole punched scratch paper

[ ] straightedge, ruler, compass, protractor, and French curves

[ ] battery-powered desk lamp

[ ] watch (turn off all audible alarms and beeps)

[ ] extra car keys

What to Do During the Exam

Previous examinees have reported that the following strategies and techniques have helped them considerably.

● Read through all of the problems before starting your first solution. In order to save you from rereading and reevaluating each problem later in the day, you should classify each problem at the beginning of the four hour session. The following categories are suggested:

　· problems you can do easily

　· problems you can do with effort

　· problems for which you can get partial credit

　· problems you cannot do

● Do all of the problems in order of increasing difficulty. All problems on the examination are worth ten points. There is nothing to be gained by attempting the difficult or long problems if easier or shorter problems are available.

● Follow these guidelines when solving a problem:

　· Do not rewrite the problem statement.

　· Do not unnecessarily redraw any figures.

　· Use pencil only.

　· Be neat. (Print all text. Use a straightedge or template where possible.)

　· Draw a box around each answer.

　· Label each answer with a symbol.

　· Give the units.

　· List your sources whenever you use obscure solution methods or data.

　· Write on one side of the page only.

　· Use one page per problem, no matter how short the solution is.

　· Go through all calculations a second time and check for mathematical errors, or solve the problem by an alternate method.

● Remember the details of any problem that you think is impossible to solve with the information given. Your ability to point out an error may later give you the margin needed to pass.

INDEX

PROFESSIONAL PUBLICATIONS ● Belmont, CA

More Resources To Help You Pass

Studying for the exam requires a significant amount of time and effort on your part. Make every minute count toward your success with these additional study materials in the Electrical Engineering Review Series:

Solutions Manual for the Electrical Engineering Reference Manual

176 pages, 8½ × 11, paperback

Don't forget that there is a companion **Solutions Manual** that provides step-by-step solutions to the practice problems given at the end of each chapter in this reference manual. This important study aid will provide immediate feedback on your progress. Without the **Solutions Manual**, you may never know if your methods are correct.

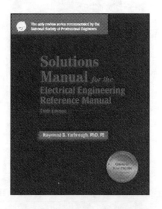

Quick Reference for the Electrical Engineering PE Exam

70 pages, 8½ × 11, paperback

Because speed is important during the exam, you will welcome the advantage provided by **Quick Reference for the Electrical Engineering PE Exam**. This handy resource gives you quick access to equations, methods, and data needed during the exam. Each exam subject is summarized using standard symbols, variable names, and nomenclature.

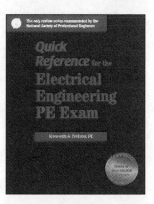

Electrical Engineering Sample Examination

48 pages, 8½ × 11, paperback

Reinforce your knowledge and increase your speed in solving the types of questions that may appear on the PE exam by taking the **Electrical Engineering Sample Examination**. This eight-hour exercise consists of 24 representative problems, 12 in essay form and 12 multiple choice, just as you'll find on the actual exam. Detailed solutions are provided so you can determine if you approached the problem correctly.

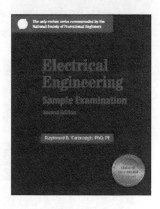

To order, contact
Professional Publications, Inc.
1250 Fifth Avenue • Belmont, CA 94002
800-426-1178 • fax 650-592-4519
www.ppi2pass.com

Resources to Help You on the PE Exam...and Beyond

Engineering Unit Conversions

160 pages, 6 × 9, hardcover

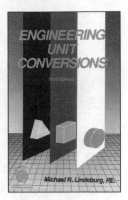

If you have ever struggled with converting grams to slugs, centistokes to square feet per second, or pounds per million gallon (lbm/MG) to milligrams per liter (mg/L), you will immediately appreciate the time-saving value of this book. With more than 4500 conversions, this is the most complete reference of its kind. Covering traditional English, conventional metric, and SI units in the fields of civil, mechanical, electrical, and chemical engineering, this book puts virtually every engineering conversion at your fingertips.

Engineering Economic Analysis

233 pages, 6 × 9, paperback

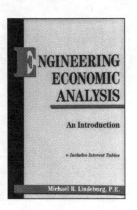

Written specifically with PE examination questions in mind, this handy book provides a capsule review of all the key principles and mathematical models needed to solve investment and cash-flow problems. As an added bonus, *Engineering Economic Analysis* contains expanded interest tables, starting at 0.25% and going up to 25.00% in quarter-percent increments—with factors for up to 100 years. No other book offers this level of detail. Practice problems with solutions show you the most efficient way to attack each type of question.

Getting Started as a Consulting Engineer

92 pages, 6 × 9, paperback

Have you ever considered being your own boss as a consulting engineer? The ability to work as a consultant is one of the main reasons that engineers obtain a PE license. If you are interested in consulting—and most engineers do consider this at some point in their careers—you will find all the information you need to get you started in this handy, no-nonsense guide. First, you'll see how consultants typically operate: how they organize their time and how they make their money. Next, you'll learn each of the necessary steps in setting up your own business, from getting appropriate business licenses and becoming incorporated to developing a referral base and negotiating contracts. Reading this short book before you launch your consulting career is essential if you want to avoid the pitfalls and surprises that can slow you down.

To order, contact
Professional Publications, Inc.
1250 Fifth Avenue • Belmont, CA 94002
800-426-1178 • fax 650-592-4519
www.ppi2pass.com